Partial Differential Equations

Partial Differential Equations
Analytical and Numerical Methods
SECOND EDITION

Mark S. Gockenbach
Michigan Technological University
Houghton, Michigan

Society for Industrial and Applied Mathematics
Philadelphia

Copyright © 2011 by the Society for Industrial and Applied Mathematics

10 9 8 7 6 5 4 3 2 1

All rights reserved. Printed in the United States of America. No part of this book may be reproduced, stored, or transmitted in any manner without the written permission of the publisher. For information, write to the Society for Industrial and Applied Mathematics, 3600 Market Street, 6th Floor, Philadelphia, PA 19104-2688 USA.

Maple is a trademark of Waterloo Maple, Inc.

Mathematica is a registered trademark of Wolfram Research, Inc.

MATLAB is a registered trademark of The MathWorks, Inc. For MATLAB product information, please contact The MathWorks, Inc., 3 Apple Hill Drive, Natick, MA 01760-2098 USA, 508-647-7000, Fax: 508-647-7001, *info@mathworks.com*, *www.mathworks.com*.

Library of Congress Cataloging-in-Publication Data
Gockenbach, Mark S.
  Partial differential equations : analytical and numerical methods / Mark S. Gockenbach. -- 2nd ed.
    p. cm.
  Includes bibliographical references and index.
  ISBN 978-0-898719-35-2
  1. Differential equations, Partial. I. Title.
  QA377.G63 2010
  515'.353--dc22
                              2010034976

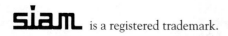 is a registered trademark.

*Dedicated to Joan,
in appreciation for her love,
support, and perseverance.*

■ ■ ■

Contents

Preface			xv
1	**Classification of Differential Equations**		**1**
2	**Models in One Dimension**		**9**
	2.1	Heat flow in a bar; Fourier's law	9
		2.1.1 Boundary and initial conditions for the heat equation	13
		2.1.2 Steady-state heat flow	14
		2.1.3 Diffusion	16
	2.2	The hanging bar	20
		2.2.1 Boundary conditions for the hanging bar	22
	2.3	The wave equation for a vibrating string	26
	2.4	Advection; kinematic waves	28
		2.4.1 Initial/boundary conditions for the advection equation	29
		2.4.2 The advection-diffusion equation	31
		2.4.3 Conservation laws	32
		2.4.4 Burgers's equation	33
	2.5	Suggestions for further reading	34
3	**Essential Linear Algebra**		**35**
	3.1	Linear systems as linear operator equations	35
	3.2	Existence and uniqueness of solutions to $\mathbf{Ax} = \mathbf{b}$	42
		3.2.1 Existence	42
		3.2.2 Uniqueness	45
		3.2.3 The Fredholm alternative	48
	3.3	Basis and dimension	52
	3.4	Orthogonal bases and projections	57
		3.4.1 The L^2 inner product	60
		3.4.2 The projection theorem	62
	3.5	Eigenvalues and eigenvectors of a symmetric matrix	68
		3.5.1 The transpose of a matrix and the dot product	70
		3.5.2 Special properties of symmetric matrices	72
		3.5.3 The spectral method for solving $\mathbf{Ax} = \mathbf{b}$	73
	3.6	Preview of methods for solving ODEs and PDEs	76
	3.7	Suggestions for further reading	77

4 Essential Ordinary Differential Equations — 79

- 4.1 Background — 79
  - 4.1.1 Converting a higher-order equation to a first-order system — 79
  - 4.1.2 The general solution of a homogeneous linear second-order ODE — 81
  - 4.1.3 The Wronskian test — 83
- 4.2 Solutions to some simple ODEs — 85
  - 4.2.1 The general solution of a second-order homogeneous ODE with constant coefficients — 85
  - 4.2.2 Variation of parameters — 87
  - 4.2.3 A special inhomogeneous second-order linear ODE — 89
  - 4.2.4 First-order linear ODEs — 91
  - 4.2.5 Euler equations — 93
- 4.3 Linear systems with constant coefficients — 97
  - 4.3.1 Homogeneous systems — 98
  - 4.3.2 Inhomogeneous systems and variation of parameters — 102
  - 4.3.3 Duhamel's principle — 104
- 4.4 Numerical methods for initial value problems — 111
  - 4.4.1 Euler's method — 111
  - 4.4.2 Improving on Euler's method: Runge–Kutta methods — 113
  - 4.4.3 Numerical methods for systems of ODEs — 116
  - 4.4.4 Automatic step control and Runge–Kutta–Fehlberg methods — 118
- 4.5 Stiff systems of ODEs — 122
  - 4.5.1 A simple example of a stiff system — 124
  - 4.5.2 The backward Euler method — 126
- 4.6 Suggestions for further reading — 130

5 Boundary Value Problems in Statics — 131

- 5.1 The analogy between BVPs and linear algebraic systems — 131
  - 5.1.1 A note about direct integration — 137
- 5.2 Introduction to the spectral method; eigenfunctions — 140
  - 5.2.1 Eigenpairs of $-\frac{d^2}{dx^2}$ under Dirichlet conditions — 140
  - 5.2.2 Representing functions in terms of eigenfunctions — 142
  - 5.2.3 Eigenfunctions under other boundary conditions; other Fourier series — 144
- 5.3 Solving the BVP using Fourier series — 149
  - 5.3.1 A special case — 149
  - 5.3.2 The general case — 151
  - 5.3.3 Other boundary conditions — 154
  - 5.3.4 Inhomogeneous boundary conditions — 157
  - 5.3.5 Summary — 159
- 5.4 Finite element methods for BVPs — 165
  - 5.4.1 The principle of virtual work and the weak form of a BVP — 166

		5.4.2	The equivalence of the strong and weak forms of the BVP 170
	5.5	\multicolumn{2}{l}{The Galerkin method . . . . 174}
	5.6	\multicolumn{2}{l}{Piecewise polynomials and the finite element method . . . . 181}
		5.6.1	Examples using piecewise linear finite elements 185
		5.6.2	Inhomogeneous Dirichlet conditions 189
	5.7	\multicolumn{2}{l}{Suggestions for further reading . . . . 194}

6 Heat Flow and Diffusion — 195

- 6.1 Fourier series methods for the heat equation 195
  - 6.1.1 The homogeneous heat equation 198
  - 6.1.2 Nondimensionalization 201
  - 6.1.3 The inhomogeneous heat equation 203
  - 6.1.4 Inhomogeneous boundary conditions 204
  - 6.1.5 Steady-state heat flow and diffusion 206
  - 6.1.6 Separation of variables 207
- 6.2 Pure Neumann conditions and the Fourier cosine series 211
  - 6.2.1 One end insulated; mixed boundary conditions 211
  - 6.2.2 Both ends insulated; Neumann boundary conditions . . . 213
  - 6.2.3 Pure Neumann conditions in a steady-state BVP 218
- 6.3 Periodic boundary conditions and the full Fourier series 225
  - 6.3.1 Eigenpairs of $-\frac{d^2}{dx^2}$ under periodic boundary conditions 227
  - 6.3.2 Solving the BVP using the full Fourier series 229
  - 6.3.3 Solving the IBVP using the full Fourier series 232
- 6.4 Finite element methods for the heat equation 237
  - 6.4.1 The method of lines for the heat equation 240
- 6.5 Finite elements and Neumann conditions 245
  - 6.5.1 The weak form of a BVP with Neumann conditions . . . 245
  - 6.5.2 Equivalence of the strong and weak forms of a BVP with Neumann conditions 246
  - 6.5.3 Piecewise linear finite elements with Neumann conditions 248
  - 6.5.4 Inhomogeneous Neumann conditions 252
  - 6.5.5 The finite element method for an IBVP with Neumann conditions 253
- 6.6 Suggestions for further reading 257

7 Waves — 259

- 7.1 The homogeneous wave equation without boundaries 259
- 7.2 Fourier series methods for the wave equation 265
  - 7.2.1 Fourier series solutions of the homogeneous wave equation 267
  - 7.2.2 Fourier series solutions of the inhomogeneous wave equation 270
  - 7.2.3 Other boundary conditions 275
- 7.3 Finite element methods for the wave equation 279

		7.3.1	The wave equation with Dirichlet conditions 281

        7.3.1   The wave equation with Dirichlet conditions 281
        7.3.2   The wave equation under other boundary conditions . . . 285
  7.4 Resonance 290
        7.4.1   The wave equation with a periodic boundary condition . . 291
        7.4.2   The wave equation with a localized source 293
  7.5 Finite difference methods for the wave equation 297
        7.5.1   Finite difference approximation of derivatives 298
        7.5.2   The wave equation 299
        7.5.3   Neumann boundary conditions 303
  7.6 Comparison of the heat and wave equations 305
  7.7 Suggestions for further reading 308

8 First-Order PDEs and the Method of Characteristics   **309**

  8.1 The simplest PDE and the method of characteristics 309
        8.1.1   Changing variables 312
        8.1.2   An inhomogeneous PDE 315
  8.2 First-order quasi-linear PDEs 319
        8.2.1   Linear equations 320
        8.2.2   Noncharacteristic initial curves 321
        8.2.3   Semilinear equations 323
        8.2.4   Quasi-linear equations 324
  8.3 Burgers's equation 327
  8.4 Suggestions for further reading 334

9 Green's Functions   **335**

  9.1 Green's functions for BVPs in ODEs: Special cases 336
        9.1.1   The Green's function and the inverse of a differential
                operator 341
        9.1.2   Symmetry of the Green's function; reciprocity 341
  9.2 Green's functions for BVPs in ODEs: The symmetric case 344
        9.2.1   Derivation of the Green's function 345
        9.2.2   Properties of the Green's function; inhomogeneous
                boundary conditions 352
  9.3 Green's functions for BVPs in ODEs: The general case 357
  9.4 Introduction to Green's functions for IVPs 362
        9.4.1   The Green's function for first-order linear ODEs 362
        9.4.2   The Green's function for higher-order ODEs 363
        9.4.3   Interpretation of the causal Green's function 365
  9.5 Green's functions for the heat equation 368
        9.5.1   The Gaussian kernel 369
        9.5.2   The Green's function on a bounded interval 372
        9.5.3   Properties of the Green's function 375
        9.5.4   Green's functions under other boundary conditions 375
  9.6 Green's functions for the wave equation 376
        9.6.1   The Green's function on the real line 377
        9.6.2   The Green's function on a bounded interval 378
  9.7 Suggestions for further reading 384

10	**Sturm–Liouville Eigenvalue Problems**		**385**
	10.1	Introduction	385
		10.1.1 How Sturm–Liouville problems arise	386
		10.1.2 Boundary conditions for the Sturm–Liouville problem	388
	10.2	Properties of the Sturm–Liouville operator	389
		10.2.1 Symmetry	389
		10.2.2 Existence of eigenvalues and eigenfunctions	390
	10.3	Numerical methods for Sturm–Liouville problems	396
		10.3.1 The weak form	396
	10.4	Examples of Sturm–Liouville problems	401
		10.4.1 A guitar string with variable density	401
		10.4.2 Heat flow with a variable thermal conductivity	404
	10.5	Robin boundary conditions	409
		10.5.1 Eigenvalues under Robin conditions	410
		10.5.2 The nonphysical case	415
	10.6	Finite element methods for Robin boundary conditions	418
		10.6.1 A BVP with a Robin condition	419
		10.6.2 A Sturm–Liouville problem with a Robin condition	422
	10.7	The theory of Sturm–Liouville problems: An outline	425
		10.7.1 Facts about the eigenvalues	426
		10.7.2 Facts about the eigenfunctions	431
	10.8	Suggestions for further reading	436
11	**Problems in Multiple Spatial Dimensions**		**437**
	11.1	Physical models in two or three spatial dimensions	437
		11.1.1 The divergence theorem	437
		11.1.2 The heat equation for a three-dimensional domain	440
		11.1.3 Boundary conditions for the three-dimensional heat equation	442
		11.1.4 The heat equation in a bar	442
		11.1.5 The heat equation in two dimensions	443
		11.1.6 The wave equation for a three-dimensional domain	443
		11.1.7 The wave equation in two dimensions	444
		11.1.8 Equilibrium problems and Laplace's equation	444
		11.1.9 Advection and other first-order PDEs	445
		11.1.10 Green's identities and the symmetry of the Laplacian	447
	11.2	Fourier series on a rectangular domain	450
		11.2.1 Dirichlet boundary conditions	450
		11.2.2 Solving a boundary value problem	455
		11.2.3 Time-dependent problems	457
		11.2.4 Other boundary conditions for the rectangle	458
		11.2.5 Neumann boundary conditions	459
		11.2.6 Dirichlet and Neumann problems for Laplace's equation	462
		11.2.7 Fourier series methods for a rectangular box in three dimensions	465
	11.3	Fourier series on a disk	469

		11.3.1	The Laplacian in polar coordinates 470
		11.3.2	Separation of variables in polar coordinates 472
		11.3.3	Bessel's equation 473
		11.3.4	Properties of the Bessel functions 476
		11.3.5	The eigenfunctions of the negative Laplacian on the disk 478
		11.3.6	Solving PDEs on a disk 482
	11.4	Finite elements in two dimensions 486	
		11.4.1	The weak form of a BVP in multiple dimensions 487
		11.4.2	Galerkin's method 487
		11.4.3	Piecewise linear finite elements in two dimensions . . . 488
		11.4.4	Finite elements and Neumann conditions 495
		11.4.5	Inhomogeneous boundary conditions 498
	11.5	The free-space Green's function for the Laplacian 500	
		11.5.1	The free-space Green's function in two dimensions . . . 500
		11.5.2	The free-space Green's function in three dimensions . . . 505
	11.6	The Green's function for the Laplacian on a bounded domain 508	
		11.6.1	Reciprocity 510
		11.6.2	The Green's function for a disk 511
		11.6.3	Inhomogeneous boundary conditions 514
		11.6.4	The Poisson integral formula 515
	11.7	Green's function for the wave equation 518	
		11.7.1	The free-space Green's function 518
		11.7.2	The wave equation in two-dimensional space 521
		11.7.3	Huygen's principle 523
		11.7.4	The Green's function for the wave equation on a bounded domain 523
	11.8	Green's functions for the heat equation 527	
		11.8.1	The free-space Green's function 527
		11.8.2	The Green's function on a bounded domain 528
	11.9	Suggestions for further reading 531	
12	**More about Fourier Series**		**533**
	12.1	The complex Fourier series 534	
		12.1.1	Complex inner products 535
		12.1.2	Orthogonality of the complex exponentials 536
		12.1.3	Representing functions with complex Fourier series . . . 537
		12.1.4	The complex Fourier series of a real-valued function . . . 537
	12.2	Fourier series and the FFT 540	
		12.2.1	Using the trapezoidal rule to estimate Fourier coefficients 541
		12.2.2	The discrete Fourier transform 543
		12.2.3	A note about using packaged FFT routines 547
		12.2.4	Fast transforms and other boundary conditions; the discrete sine transform 548
		12.2.5	Computing the DST using the FFT 549
	12.3	Relationship of sine and cosine series to the full Fourier series 552	

12.4	Pointwise convergence of Fourier series	556
	12.4.1 Modes of convergence for sequences of functions	556
	12.4.2 Pointwise convergence of the complex Fourier series . . .	558
12.5	Uniform convergence of Fourier series	571
	12.5.1 Rate of decay of Fourier coefficients	571
	12.5.2 Uniform convergence	574
	12.5.3 A note about Gibbs's phenomenon	577
12.6	Mean-square convergence of Fourier series	579
	12.6.1 The space $L^2(-\ell, \ell)$	579
	12.6.2 Mean-square convergence of Fourier series	582
	12.6.3 Cauchy sequences and completeness	584
12.7	A note about general eigenvalue problems	588
12.8	Suggestions for further reading	592

13 More about Finite Element Methods — 593

13.1	Implementation of finite element methods	593
	13.1.1 Describing a triangulation	594
	13.1.2 Computing the stiffness matrix	596
	13.1.3 Computing the load vector	598
	13.1.4 Quadrature	598
13.2	Solving sparse linear systems	603
	13.2.1 Gaussian elimination for dense systems	603
	13.2.2 Direct solution of banded systems	605
	13.2.3 Direct solution of general sparse systems	607
	13.2.4 Iterative solution of sparse linear systems	608
	13.2.5 The conjugate gradient algorithm	611
	13.2.6 Convergence of the CG algorithm	614
	13.2.7 Preconditioned CG	614
13.3	An outline of the convergence theory for finite element methods	617
	13.3.1 The Sobolev space $H_0^1(\Omega)$	617
	13.3.2 Best approximation in the energy norm	619
	13.3.3 Approximation by piecewise polynomials	620
	13.3.4 Elliptic regularity and L^2 estimates	620
13.4	Finite element methods for eigenvalue problems	623
13.5	Suggestions for further reading	628

Appendix A Proof of Theorem 3.47 — 629

Appendix B Shifting the Data in Two Dimensions — 633

B.1	Inhomogeneous Dirichlet conditions on a rectangle	633
B.2	Inhomogeneous Neumann conditions on a rectangle	636

Bibliography — 641

Index — 647

Preface

This introductory text on partial differential equations (PDEs) has several features that are not found in other texts at this level, including:

- equal emphasis on classical and modern techniques,
- the explicit use of the language and results of linear algebra,
- examples and exercises analyzing realistic experiments (with correct physical parameters and units),
- a recognition that mathematical software forms a part of the arsenal of both students and professional mathematicians.

In this preface, I will discuss these features and also describe the expanded coverage offered by this second edition: much more material on Green's functions, new chapters on the method of characteristics and on Sturm–Liouville eigenvalue problems, and a section on the finite difference method for the wave equation.

Classical and modern techniques

Undergraduate courses on PDEs tend to focus on Fourier series methods and separation of variables. These techniques are still useful after two centuries because they offer a great deal of insight into those problems to which they apply. However, the subject of PDEs has driven much of the research in both pure and applied mathematics in the last century, and students ought to be exposed to some more modern techniques as well.

The limitation of the Fourier series technique is its restricted applicability: it can be used only for equations with constant coefficients and only on certain simple geometries. To complement the classical topic of Fourier series, I present the finite element method, a modern, powerful, and flexible approach to solving PDEs. Although many introductory texts include some discussion of finite elements (or finite differences, a competing computational methodology), the modern approach tends to receive less attention and a subordinate place in the exposition. In this text, I have put equal weight on Fourier series and finite elements.

Linear algebra

Both linear and nonlinear differential equations occur as models of physical phenomena of great importance in science and engineering. However, most introductory texts focus on linear equations, and mine is no exception. There are several reasons why this should be so. The study of PDEs is difficult, and it makes sense to begin with the simpler linear equations before moving on to the more difficult nonlinear equations. Moreover, linear equations are much better understood. Finally, much of what is known about nonlinear differential equations depends on the analysis of linear differential equations, so this material is prerequisite for moving on to nonlinear equations.

Because we focus on linear equations, linear algebra is extremely useful. Indeed, no discussion of Fourier series or finite element methods can be complete unless it puts the results in the proper linear algebraic framework. For example, both methods produce the best approximate solution from certain finite-dimensional subspaces, and the projection theorem is therefore central to both techniques. Symmetry is another key feature exploited by both methods.

While many texts de-emphasize the linear algebraic nature of the concepts and solution techniques, I have chosen to make it explicit. This decision, I believe, leads to a more cohesive course and a better preparation for future study. However, it presents certain challenges. Linear algebra does not seem to receive the attention it deserves in many engineering and science programs, and so many students will take a course based on this text without the "prerequisites." Therefore, I present a fairly complete overview of the necessary material in Chapter 3, *Essential Linear Algebra*.

Both faculty previewing this text and students taking a course from it will soon realize that there is too much material in Chapter 3 to cover thoroughly in the couple of weeks it can reasonably occupy in a semester course. From experience I know that conscientious students dislike moving so quickly through material that they cannot master it. However, one of the keys to using this text is to avoid getting bogged down in Chapter 3. Students should try to get from it the "big picture" and the following two essential ideas:

- How to compute a best approximation to a vector from a subspace, with and without an orthogonal basis (Section 3.4).

- How to solve a matrix-vector equation when the matrix is symmetric and its eigenvalues and eigenvectors are known (Section 3.5).

Having at least begun to grasp these ideas, students should move on to Chapter 4 (or Chapter 5—see below) even if some details are not clear. The concepts from linear algebra will become much clearer as they are used throughout the remainder of the text.

I have taught this course several times using this approach, and, although students often find it frustrating at the beginning, the results seem to be good.

Realistic problems

The subject of PDEs is easier to grasp if one keeps in mind certain standard physical experiments modeled by the equations under consideration. I have used these models to introduce the equations and to aid in understanding their solutions. The models also show, of course, that the subject of PDEs is worth studying!

To make the applications as meaningful as possible, I have included many examples and exercises posed in terms of meaningful experiments with realistic physical parameters.

Software

There exists powerful mathematical software that can be used to illuminate the material presented in this book. Computer software is useful for at least three reasons.

- It removes the need to do tedious computations that are necessary to compute solutions. Just as a calculator eliminates the need to use a table and interpolation to compute a logarithm, a computer algebra system can eliminate the need to perform integration by parts several times in order to evaluate an integral. With the more mechanical obstacles removed, there is more time to focus on concepts.

- Problems that simply cannot be solved (in a reasonable time) by hand can often be done with the assistance of a computer. This allows for more interesting assignments.

- Graphical capabilities allow students to visualize the results of their computations, improving understanding and interpretation.

I expect students to use a software package such as MATLAB®, *Mathematica*®, or *Maple*™ to reproduce the examples from the text and to solve the exercises.

I prefer not to introduce a particular software package in the text itself, for at least two reasons. The explanation of the features and usage of the software can detract from the mathematics. Also, if the book is based on a particular software package, then it can be difficult to use with a different package. For these reason, my text does not mention any software packages except in a few footnotes. However, since the use of software is, in my opinion, essential for a modern course, I have written tutorials for MATLAB, *Mathematica*, and *Maple* that explain the various capabilities of these programs that are relevant to this book. These tutorials can be found on the book's web site, http://www.siam.org/books/ot122.

Outline

The core material in this text is found in Chapters 5–7, which present Fourier series and finite element techniques for the three most important differential equations of mathematical physics: Laplace's equation, the heat equation, and the wave equation. In addition to Fourier series and finite elements, Section 7.5 introduces finite difference methods in the context of the wave equation, where such methods have some significant advantages. Chapters 5–7 are restricted to problems in a single spatial dimension.

Several introductory chapters set the stage for this core. Chapter 1 briefly defines the basic terminology and notation that will be used in the text. Chapter 2 then derives the differential equations that will be discussed in the text, in the process explaining the meaning of various physical parameters that appear in the equations and introducing the associated boundary conditions and initial conditions. Chapter 2 is restricted to one spatial dimension (models in multiple spatial dimensions are covered in Chapter 11).

Chapter 3, which has already been discussed above, presents the concepts and techniques from linear algebra that will be used in subsequent chapters. I want to reiterate that perhaps the most important key to using this text effectively is to move through Chapter 3 expeditiously. The rudimentary understanding that students obtain in going through Chapter 3 will grow as the concepts are used in the rest of the book.

Chapter 4 presents the background material on ordinary differential equations (ODEs) that is needed in later chapters. This chapter is much easier to understand than the previous one, because much of the material is a review for many students. Only the last two sections, on numerical methods and stiff systems, are likely to be new. I prefer to cover the material in Chapter 4 on a "just-in-time" basis, moving from Chapter 3 to Chapter 5 and referring back to Chapter 4 as needed.

Following the core material in Chapters 5–7, there are several chapters that are largely independent. Chapter 8 presents the method of characteristics for first-order PDEs, focusing on linear, semilinear, and quasi-linear equations. There is a section about the inviscid Burgers's equation, which is the only significant discussion of nonlinear PDEs in the text.

Chapter 9 presents Green's functions for both stationary and time-dependent equations, still in one spatial dimension, while Chapter 10 discusses Sturm–Liouville eigenvalue problems. Chapter 10 is independent of Chapter 9, with the exception of Section 10.7, which outlines the theory of Sturm–Liouville problems and refers to results from Chapter 9.

Chapter 11 extends the models and techniques developed in the first part of the book to two and three spatial dimensions. Included are Fourier series for rectangular and circular domains in two dimensions, finite element methods in two dimensions, and Green's functions in two and three spatial dimensions.

The last two chapters provide a more in-depth treatment of Fourier series (Chapter 12) and finite elements (Chapter 13). In addition to the standard theory of Fourier series, Chapter 12 shows how to use the fast Fourier transform to efficiently compute Fourier series solutions of the PDEs, explains the relationships among the various types of Fourier series, and discusses the extent to which the Fourier series method can be extended to complicated geometries and equations with nonconstant coefficients. Sections 12.4–12.6 present a careful mathematical treatment of the convergence of Fourier series, and they have a different flavor from the remainder of the book. In particular, they are less suited for an audience of science and engineering students and have been included as a reference for the curious student.

Chapter 13 gives some advice on implementing finite element computations, discusses the solution of the resulting sparse linear systems, and briefly outlines the convergence theory for finite element methods. It also shows how to use finite elements to solve general eigenvalue problems. The tutorials on the book's web page include programs implementing two-dimensional finite element methods, as described in Section 13.1, in each of the supported software packages (MATLAB, *Mathematica*, and *Maple*). The sections on sparse systems and the convergence theory are both little more than outlines, pointing students toward more advanced concepts. Both of these topics, of course, could easily justify a dedicated semester-long course, and I had no intention of going into detail. I hope that the material on implementation of finite elements (in Section 13.1) will encourage some students to experiment with two-dimensional calculations, which are already too tedious to carry out by hand. This sort of information seems to be lacking from most books accessible to students at this level.

Possible course outlines

As mentioned above, the core of the textbook is Chapters 5–7, and it is assumed that instructors will cover most of this material. A reasonable schedule, which assumes that material from Chapter 4 (Essential Ordinary Differential Equations) is incorporated as needed into lectures from later chapters, is the following.

Chapter	Lectures
1	1
2	3
3	5–6
5	6–7
6	7–8
7	5–6
Total	27–31

With a well-prepared class, it may not be necessary to spend so much time on Chapter 3. Alternatively, it may be possible to give Chapter 2 as a reading assignment, particularly for the intended audience of science and engineering students, who will typically be comfortable with the physical modeling represented in the differential equations.

In a 14–15 week course with 42–45 lectures, the above schedule leaves three or more weeks to address other topics (plus several class days for exams). The focus on Fourier series and finite element methods can be maintained by covering the first few sections of Chapter 11 and material from Chapters 12 and 13. On the other hand, the introduction of PDEs can be broadened by including material on characteristics (Chapter 8) and Green's functions (Chapter 9 and the second half of Chapter 11). Finally, Chapter 10 gives more information about eigenfunction expansions (generalized Fourier series) in the context of Sturm–Liouville problems, and also introduces new material on the finite element method (including its use for eigenvalue problems and the treatment of different boundary conditions).

Chapters 8–13 can be studied in any order (with the one exception, noted above, that Section 10.7 depends on Chapter 9). Since it is not assumed that these chapters will be read in order, there is some repetition in the exposition and in a few exercises.

Resources on the Web

This textbook has an accompanying web page:

> http://www.siam.org/books/ot122

There the reader will find tutorials for MATLAB, *Mathematica*, and *Maple*. Solutions for selected odd-numbered exercises are also available.

I will maintain a list of errata on the web page. Readers are encouraged to send me notice of errors by email (msgocken@mtu.edu).

Graphics and software

The graphs in this book were generated with MATLAB. For MATLAB product information, please contact

> The MathWorks, Inc.
> 3 Apple Hill Drive
> Natick, MA 01760-2098, USA
> Tel: 508-647-7000
> Fax: 508-647-7001
> E-mail: info@mathworks.com

As mentioned above, the use of *Mathematica* and *Maple* is also supported by the tutorials found on the book's web page. For *Mathematica* product information, contact

> Wolfram Research, Inc.
> 100 Trade Center Drive
> Champaign, IL 61820-7237, USA
> Tel: 800-965-3726
> Fax: 217-398-0747
> E-mail: info@wolfram.com

For *Maple* product information, contact

> Waterloo Maple Inc.
> 615 Kumpf Drive
> Waterloo, Ontario
> Canada, N2V 1K8
> Tel: 1-800-267-6583
> Fax: 519-747-5284
> E-mail: info@maplesoft.com

Acknowledgments

The various physical parameters used in the examples and exercises were derived (sometimes by interpolation) from tables in the *CRC Handbook of Chemistry and Physics* [44].

This book began when I was visiting Rice University in 1998–1999 and taught a course using the lecture notes of Professor William W. Symes. To satisfy my personal predilections, I rewrote the notes significantly, and for the convenience of myself and my students, I typeset them in the form of a book, which was the first version of this text. Although the final result bears, in some ways, little resemblance to Symes's original notes, I am indebted to him for the idea of recasting the undergraduate PDE course in more modern terms. His example was the inspiration for this project, and I benefited from his advice throughout the writing of the first edition.

Mark S. Gockenbach
msgocken@mtu.edu

Chapter 1

Classification of Differential Equations

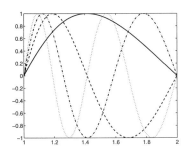

Loosely speaking, a differential equation is an equation specifying a relation between the derivatives of a function or between one or more derivatives and the function itself. We will call the function appearing in such an equation the *unknown function*. We use this terminology because the typical task involving a differential equation, and the focus of this book, is to *solve* the differential equation, that is, to find a function whose derivatives are related as specified by the differential equation. In carrying out this task, everything else about the relation, other than the unknown function, is regarded as known. Any function satisfying the differential equation is called a *solution*. In other words, a solution of a differential equation is a function that, when substituted for the unknown function, causes the equation to be satisfied.

Differential equations fall into several natural and widely used categories.

1. **Ordinary versus partial:**

    (a) If the unknown function has a single independent variable, say t, then the equation is an *ordinary differential equation* (ODE). In this case, only "ordinary" derivatives are involved. Examples of ODEs are

    $$\frac{du}{dt} = 3u \qquad (1.1)$$

    and

    $$a\frac{d^2u}{dt^2} + b\frac{du}{dt} + cu = f(t). \qquad (1.2)$$

    In the second example, a, b, and c are regarded as known constants, and $f(t)$ as a known function of t. In both equations, the unknown is $u = u(t)$.

    (b) If the unknown function has two or more independent variables, the equation is called a *partial differential equation* (PDE). Examples include

    $$\frac{\partial^2 u}{\partial x^2} + \frac{\partial^2 u}{\partial y^2} = 0 \qquad (1.3)$$

and
$$\frac{\partial^2 u}{\partial t^2} - c^2 \frac{\partial^2 u}{\partial x^2} = 0. \qquad (1.4)$$

In (1.3), the unknown is $u = u(x,y)$, while in (1.4), it is $u = u(x,t)$. In (1.4), c is a known constant.

2. **Order:** The *order* of a differential equation is the order of the highest derivative appearing in the equation. Most differential equations arising in science and engineering are first or second order. Example (1.1) above is first order, while examples (1.2), (1.3), and (1.4) are second order.

3. **Linear versus nonlinear:**

   (a) As the examples suggest, a differential equation has, on each side of the equal sign, algebraic expressions involving the unknown function and its derivatives, and possibly other functions and constants regarded as known. A differential equation is *linear* if those terms involving the unknown function contain only products of a *single factor* of the unknown function or one of its derivatives with other (known) constants or functions of the independent variables.[1] In linear differential equations, the unknown function and its derivatives do *not* appear raised to a power other than 1, or as the argument of a nonlinear function (like sin, exp, log, etc.).

   For example, the general linear second-order ODE has the form
   $$a_2(t)\frac{d^2 u}{dt^2} + a_1(t)\frac{du}{dt} + a_0(t)u = f(t).$$

   Here we require that $a_2(t) \neq 0$ in order that the equation truly be of second order. Example (1.2) is a special case of this general equation. In a linear differential equation, the unknown or its derivatives can be multiplied by constants or by functions of the independent variable, but not by functions of the unknown.

   As another example, the general linear second-order PDE in two independent variables is
   $$a_{11}(x,y)\frac{\partial^2 u}{\partial x^2} + a_{12}(x,y)\frac{\partial^2 u}{\partial x \partial y} + a_{22}(x,y)\frac{\partial^2 u}{\partial y^2}$$
   $$+ a_1(x,y)\frac{\partial u}{\partial x} + a_2(x,y)\frac{\partial u}{\partial y} + a_0(x,y)u = f(x,y).$$

   Examples (1.3) and (1.4) are of this form (although the independent variables are called x and t in (1.4), not x and y). Not all of a_{11}, a_{12}, and a_{22} can be zero in order for this equation to be second order.

   A linear differential equation is *homogeneous* if the zero function is a solution. For example, $u(t) \equiv 0$ satisfies (1.1), $u(x,y) \equiv 0$ satisfies (1.3), and $u(x,t) \equiv 0$ satisfies (1.4), so these are examples of homogeneous linear differential equations. A homogeneous linear equation has another important property:

---

[1] In Section 3.1, we will give a more precise definition of linearity.

Chapter 1. Classification of Differential Equations

whenever u and v are solutions of the equation, so is $\alpha u + \beta v$ for all real numbers α and β. For example, suppose $u(t)$ and $v(t)$ are solutions of (1.1) (that is, $du/dt = 3u$ and $dv/dt = 3v$), and $w = \alpha u + \beta v$. Then

$$\frac{dw}{dt} = \frac{d}{dt}[\alpha u + \beta v] = \alpha \frac{du}{dt} + \beta \frac{dv}{dt} = \alpha 3u + \beta 3v = 3(\alpha u + \beta v) = 3w.$$

Thus w is also a solution of (1.1).

A linear differential equation which is not homogeneous is called *inhomogeneous*. For example,

$$\frac{du}{dt} = 3u + \sin(2\pi t) \tag{1.5}$$

is linear inhomogeneous, since $u(t) \equiv 0$ does not satisfy this equation. Example (1.2) might be homogeneous or not, depending on whether $f(t) \equiv 0$ or not.

It is always possible to group all terms involving the unknown function in a linear differential equation on the left-hand side and all terms not involving the unknown function on the right-hand side. For example, (1.5) is equivalent to

$$\frac{du}{dt} - 3u = \sin(2\pi t). \tag{1.6}$$

"Equivalent to" in this context means that (1.5) and (1.6) have exactly the same solutions: if $u(t)$ solves one of these two equations, it solves the other as well. When the equation is written this way, with all of the terms involving the unknown on the left and those not involving the unknown on the right, homogeneity is easy to recognize: the equation is homogeneous if and only if the right-hand side is identically zero.

(b) Differential equations which are not linear are termed *nonlinear*. For example,

$$\frac{d^2u}{dt^2} + u^2 = 0 \tag{1.7}$$

is nonlinear. This is clear, as the unknown function u appears raised to the second power. Another way to see that (1.7) is nonlinear is as follows: if the equation were linear, it would be linear homogeneous, since the zero function is a solution. However, suppose that $u(t)$ and $v(t)$ are nonzero solutions, so that

$$\frac{d^2u}{dt^2} + u^2 = 0, \qquad \frac{d^2v}{dt^2} + v^2 = 0,$$

and $w = u + v$. Then

$$\begin{aligned}\frac{d^2w}{dt^2} + w^2 &= \frac{d^2}{dt^2}[u+v] + (u+v)^2 \\ &= \frac{d^2u}{dt^2} + \frac{d^2v}{dt^2} + u^2 + 2uv + v^2 \\ &= \left(\frac{d^2u}{dt^2} + u^2\right) + \left(\frac{d^2v}{dt^2} + v^2\right) + 2uv \\ &= 2uv.\end{aligned}$$

Thus w does not satisfy the equation, so the equation must be nonlinear.

The terms "homogeneous" and "inhomogeneous" are used only for linear differential equations.

Both linear and nonlinear differential equations occur as models of physical phenomena of great importance in science and engineering. However, linear differential equations are much better understood, and most of what is known about nonlinear differential equations depends on the analysis of linear differential equations. This book will focus almost exclusively on the solution of linear differential equations.

4. **Constant versus nonconstant coefficients:** Linear differential equations like

$$m\frac{d^2x}{dt^2} + c\frac{dx}{dt} + kx = f(t)$$

have *constant coefficients* if the *coefficients* m, c, and k are constants rather than (nonconstant) functions of the independent variable. The *right-hand side* $f(t)$ may depend on the independent variable; it is only the quantities which multiply the unknown function and its derivatives which must be constant, in order for the equation to have constant coefficients. Some techniques that are effective for constant-coefficient problems are very difficult to apply to problems with nonconstant coefficients.

As a convention, we will explicitly write the independent variable when a coefficient is a function rather than a constant. The only function in a differential equation for which we will omit the independent variable is the unknown function. Thus

$$-\frac{d}{dx}\left[k(x)\frac{du}{dx}\right] = f(x)$$

is an ODE with a nonconstant coefficient, namely $k(x)$.

We will only apply the phrase "constant coefficients" to linear differential equations.

5. **Scalar equation versus system of equations:** A single differential equation in one unknown function will be referred to as a scalar equation (all of the examples we have seen to this point have been scalar equations). A system of differential equations consists of several equations for one or more unknown functions. Here is a system of three first-order linear, constant-coefficient ODEs for three unknown functions $x_1(t)$, $x_2(t)$, and $x_3(t)$:

$$\frac{dx_1}{dt} = 2x_1 - x_2 + x_3,$$
$$\frac{dx_2}{dt} = x_1 + x_2,$$
$$\frac{dx_3}{dt} = -x_1 + x_2 - x_3.$$

It is important to realize that, since a differential equation is an equation involving functions, a solution will satisfy the equation for all values of the independent variable(s),

Chapter 1. Classification of Differential Equations

or at least all values in some restricted domain. The equation is to be interpreted as an *identity*, such as the familiar trigonometric identity

$$\cos^2(t) + \sin^2(t) = 1. \tag{1.8}$$

To say that (1.8) is an identity is to say that it is satisfied for all values of the independent variable t. Similarly, $u(t) = e^{3t}$ is a solution of (1.1) because this function u satisfies

$$\frac{du}{dt}(t) = 3u(t)$$

for *all* values of t.

The careful reader will note the close analogy between the uses of many terms introduced in this section ("unknown function," "solution," "linear" versus "nonlinear") and the uses of similar terms in discussing algebraic equations. The analogy between linear differential equations and linear algebraic systems is an important theme in this book.

Exercises

1. Classify each of the following differential equations according to the categories described in this chapter (ODE or PDE, linear or nonlinear, etc.):

    (a)
    $$\frac{dx}{dt} + tx = 0;$$

    (b)
    $$\frac{\partial u}{\partial t} - \frac{\partial^2 u}{\partial x^2} = (1+t)\sin(x);$$

    (c)
    $$\frac{\partial w}{\partial t} + w\frac{\partial w}{\partial x} = 0.$$

2. Repeat Exercise 1 for the following equations:

    (a)
    $$\frac{\partial v}{\partial t} + 3\frac{\partial v}{\partial x} = x + 3t;$$

    (b)
    $$\frac{dx}{dt} + \frac{1}{t}x = 0;$$

    (c)
    $$\frac{\partial u}{\partial t} - \frac{\partial}{\partial x}\left[2u\frac{\partial u}{\partial x}\right] = 0.$$

3. Repeat Exercise 1 for the following equations:

    (a)
    $$\frac{d^2\theta}{dt^2} + \sin(\theta) = 0;$$

(b)
$$\frac{\partial u}{\partial t} - 2\frac{\partial u}{\partial x} = 1+x;$$

(c)
$$\rho(x,y)\frac{\partial^2 u}{\partial t^2} - T\left\{\frac{\partial^2 u}{\partial x^2} + \frac{\partial^2 u}{\partial y^2}\right\} = f(x,t).$$

4. Repeat Exercise 1 for the following equations:

(a)
$$\frac{d^2 x}{dt^2} - 2\frac{dx}{dt} + tx = 0;$$

(b)
$$\rho(x)c(x)\frac{\partial v}{\partial t} - \frac{\partial}{\partial x}\left(\kappa(x)\frac{\partial v}{\partial x}\right) = 0;$$

(c)
$$\frac{d^2 x}{dt^2} + 2\frac{dx}{dt} + 3x = \sin(\pi t).$$

5. Determine whether each of the functions below is a solution of the corresponding differential equation in Exercise 1:

   (a) $x(t) = e^{\frac{1}{2}t}$;
   (b) $u(x,t) = t\sin(x)$;
   (c) $w(x,t) = t(1-x)$.

6. Determine whether each of the functions below is a solution of the corresponding differential equation in Exercise 2:

   (a) $v(x,t) = tx$;
   (b) $x(t) = \frac{1}{t}$;
   (c) $u(x,t) = x^2 t$.

7. Find a function $f(t)$ so that $u(t) = t\sin(t)$ is a solution of the ODE
$$\frac{du}{dt} - \frac{1}{t}u = f(t), \qquad t > 0.$$
Is there only one such function f? Why or why not?

8. Find all solutions to the simple ODE
$$\frac{du}{dt} = f(t),$$
where f is a given continuous function.

Chapter 1. Classification of Differential Equations

9. Suppose u is a nonzero solution of a homogeneous linear differential equation. What is another nonzero solution?

10. Suppose u is a solution of

$$a(t)\frac{d^2u}{dt^2} + b(t)\frac{du}{dt} + c(t)u = f(t) \tag{1.9}$$

and v is a (nonzero) solution of

$$a(t)\frac{d^2u}{dt^2} + b(t)\frac{du}{dt} + c(t)u = 0.$$

Explain how to produce infinitely many different solutions of (1.9).

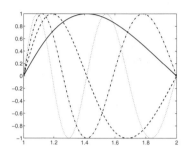

Chapter 2
Models in One Dimension

In this chapter, we present several different and interesting physical processes that can be modeled by ODEs or PDEs. For now we restrict ourselves to phenomena that can be described (at least approximately) as occurring in a single spatial dimension: heat flow or mechanical vibration in a long, thin bar, vibration of a string, diffusion of chemicals in a pipe, and so forth. In Chapters 5–7, we will learn methods for solving the resulting differential equations.

In Chapter 11, we consider similar experiments occurring in multiple spatial dimensions.

2.1 Heat flow in a bar; Fourier's law

We begin by considering the distribution of heat energy (or, equivalently, of temperature) in a long, thin bar. We assume that the cross sections of the bar are uniform, and that the temperature varies only in the longitudinal direction. In particular, we assume that the bar is perfectly insulated, except possibly at the ends, so that no heat escapes through the side. By making several simplifying assumptions, we derive a linear PDE whose solution is the temperature in the bar as a function of spatial position and time.

We show, when we treat multiple space dimensions in Chapter 11, that if the initial temperature is constant in each cross section, and if any heat source depends only on the longitudinal coordinate, then all subsequent temperature distributions depend only on the longitudinal coordinate. Therefore, in this regard, there is no modeling error in adopting a one-dimensional model. (There is modeling error associated with some of the other assumptions we make.)

We will now derive the model, which is usually called the heat equation. We begin by defining a coordinate system (see Figure 2.1). The variable x denotes position in the bar in the longitudinal direction, and we assume that one end of the bar is at $x = 0$. The length of the bar will be ℓ, so the other end is at $x = \ell$. We denote by A the area of a cross section of the bar.

The temperature of the bar is determined by the amount of heat energy in the bar and its material properties. Moreover, the relationship between heat energy and temperature is nonlinear, so accurately modeling this relationship results in a nonlinear differential

Figure 2.1. *The bar and its coordinate system.*

equation. To simplify matters, we can assume that we are interested only in modeling temperatures near a reference temperature and that an approximately linear relationship exists between the heat energy in the bar and its temperature near the reference temperature.

The relationship between heat energy and temperature is described by the *specific heat* of the material. (Specific heat is also called *heat capacity* or *specific heat capacity*.) If the specific heat of a material at T Kelvin is c Joules per gram per Kelvin, it means that c J of energy are required to raise the temperature of 1 g of the material by 1 K. Assuming a linear relationship means that we assume the heat energy required is proportional to the temperature difference; that is, $2c$ J are required to raise the temperature of 1 g by 2 K, $3c$ J are required to raise it by 3 K, and so forth. Such an assumption is valid only if the temperatures remain near T.

We now consider the small part of the bar between x and $x + \Delta x$. We will denote by E_0 the total heat energy in this part of the bar when its temperature is T_0. Then, the total heat energy in this part of the bar, when its temperature is given by $u(x,t)$, is

$$E_0 + \int_x^{x+\Delta x} A\rho c (u(s,t) - T_0) \, ds, \tag{2.1}$$

where ρ is the *density* (mass per unit volume) of the bar, c is the specific heat, and u is given in Kelvin. The values of ρ and c are characteristics of the material of which the bar is made, and, in general, can depend on x (if the bar is heterogeneous). The integral is required in (2.1) because the temperature is not assumed constant in the bar, and it represents the amount of energy required to change the temperature in that part of the bar from T_0 to $u(x,t)$ (see Exercise 2.1.3).

We can simplify (2.2) as follows:

$$E_0 + \int_x^{x+\Delta x} A\rho c (u(s,t) - T_0) \, ds = E_0 - \int_x^{x+\Delta x} A\rho c T_0 \, ds + \int_x^{x+\Delta x} A\rho c u(s,t) \, ds$$

$$= E_0 - A\rho c T_0 \Delta x + \int_x^{x+\Delta x} A\rho c u(s,t) \, ds$$

$$= \tilde{E}_0 + \int_x^{x+\Delta x} A\rho c u(s,t) \, ds.$$

We thus have the following expression for the total heat between x and $x + \Delta x$:

$$\tilde{E}_0 + \int_x^{x+\Delta x} A\rho c u(s,t) \, ds. \tag{2.2}$$

The fact that we do not know $\tilde{E}_0$ causes no difficulty, because we are going to model the *change* in the energy (and hence temperature).

Specifically, we examine the rate of change, with respect to time, of the heat energy in the part of the bar between x and $x + \Delta x$. The total heat in $[x, x + \Delta x]$ is given by (2.2),

2.1. Heat flow in a bar; Fourier's law

so its rate of change is

$$\frac{d}{dt}\int_x^{x+\Delta x} A\rho c u(s,t)\,ds$$

(the constant $\tilde{E}_0$ differentiates to zero). To work with this expression, we need the following result, which allows us to move the derivative past the integral sign.

Theorem 2.1. *Let $F:(a,b)\times[c,d]\to\mathbf{R}$ and $\partial F/\partial x$ be continuous, and define $\phi:(a,b)\to\mathbf{R}$ by*

$$\phi(x)=\int_c^d F(x,y)\,dy.$$

Then ϕ is continuously differentiable, and

$$\frac{d\phi}{dx}(x)=\int_c^d \frac{\partial F}{\partial x}(x,y)\,dy.$$

That is,

$$\frac{d}{dx}\int_c^d F(x,y)\,dy=\int_c^d \frac{\partial F}{\partial x}(x,y)\,dy.$$

By Theorem 2.1, we have

$$\frac{d}{dt}\int_x^{x+\Delta x} A\rho c u(s,t)\,ds = \int_x^{x+\Delta x} A\rho c \frac{\partial u}{\partial t}(s,t)\,ds. \tag{2.3}$$

If we assume that there are no internal sources or sinks of heat, the heat contained in $[x,x+\Delta x]$ can change only because heat flows through the cross sections at x and $x+\Delta x$. The rate at which heat flows through the cross section at x is called the *heat flux* and is denoted $q(x,t)$. It has units of energy per unit area per unit time, and it is a signed quantity: if heat energy is flowing in the positive x-direction, then q is positive.

The net heat entering $[x,x+\Delta x]$ at time t, through the two cross sections, is

$$A(q(x,t)-q(x+\Delta x,t))=-\int_x^{x+\Delta x} A\frac{\partial q}{\partial x}(s,t)\,ds; \tag{2.4}$$

the last equation follows from the fundamental theorem of calculus. (See Figure 2.2.) Equating (2.3) and (2.4) yields

$$\int_x^{x+\Delta x} A\rho c \frac{\partial u}{\partial t}(s,t)\,ds = -\int_x^{x+\Delta x} A\frac{\partial q}{\partial x}(s,t)\,ds$$

or

$$\int_x^{x+\Delta x}\left\{A\rho c\frac{\partial u}{\partial t}(s,t)+A\frac{\partial q}{\partial x}(s,t)\right\}ds=0.$$

Since this holds for all x and Δx, it follows that the integrand must be zero:

$$A\rho c\frac{\partial u}{\partial t}(x,t)+A\frac{\partial q}{\partial x}(x,t)=0,\qquad 0<x<\ell. \tag{2.5}$$

Figure 2.2. *The heat flux into and out of a part of the bar.*

(The key point here is that the integral above equals zero over *every* small interval. This is possible only if the integrand is itself zero. If the integrand were positive, say, at some point in $[0,\ell]$, and were continuous, then the integral over a small interval around that point would be positive.)

Naturally, heat will flow from regions of higher temperature to regions of lower temperature; note that the sign of $\partial u/\partial x$ indicates whether temperature is increasing or decreasing as x increases, and hence indicates the direction of heat flow. We now make a simplifying assumption, which is called *Fourier's law of heat conduction*: the heat flux q is proportional to the temperature gradient, $\partial u/\partial x$. We call the magnitude of the constant of proportionality the *thermal conductivity* κ and obtain the equation

$$q(x,t) = -\kappa \frac{\partial u}{\partial x}(x,t) \tag{2.6}$$

(the units of κ can be determined from the units of the heat flux and those of the temperature gradient; see Exercise 2.1.1). The negative sign is necessary so that heat flows from hot regions to cold regions. Substituting Fourier's law into the differential equation (2.5), we can eliminate q and find a PDE for u:

$$\rho c \frac{\partial u}{\partial t} - \kappa \frac{\partial^2 u}{\partial x^2} = 0, \ 0 < x < \ell, \ \text{for all } t. \tag{2.7}$$

(We have canceled a common factor of A.) Here we have assumed that κ is constant, which would be true if the bar were homogeneous. It is possible that κ depends on x (in which case ρ and c probably do as well); then we obtain

$$\rho(x)c(x)\frac{\partial u}{\partial t} - \frac{\partial}{\partial x}\left(\kappa(x)\frac{\partial u}{\partial x}\right) = 0, \ 0 < x < \ell, \ \text{for all } t.$$

We call (2.7) the *heat equation*.

If the bar contains internal sources (or sinks) of heat (such as chemical reactions that produce heat), we collect all such sources into a single *source function* $f(x,t)$ (in units of heat energy per unit time per unit volume). Then the total heat added to $[x, x+\Delta x]$ during the time interval $[t, t+\Delta t]$ is

$$\int_t^{t+\Delta t} \int_x^{x+\Delta x} Af(s,\tau)\,ds\,d\tau.$$

The time rate of change of this contribution to the total heat energy (at time t) is

$$\int_x^{x+\Delta x} Af(s,t)\,ds. \tag{2.8}$$

2.1. Heat flow in a bar; Fourier's law

The rate of change of total heat in $[x, x + \Delta x]$ is now given by the sum of (2.4) and (2.8), so we obtain, by the above reasoning, the inhomogeneous[2] heat equation

$$\rho c \frac{\partial u}{\partial t} - \kappa \frac{\partial^2 u}{\partial x^2} = f(x,t), \; 0 < x < \ell, \; \text{for all } t. \tag{2.9}$$

2.1.1 Boundary and initial conditions for the heat equation

The heat equation by itself is not a complete model of heat flow. We must know how heat flows through the ends of the bar, and we must know the temperature distribution in the bar at some initial time.

Two possible boundary conditions for heat flow in a bar correspond to perfect insulation and perfect thermal contact. If the ends of the bar are perfectly insulated, so that the heat flux across the ends is zero, then we have

$$-\kappa \frac{\partial u}{\partial x}(0,t) = 0, \; -\kappa \frac{\partial u}{\partial x}(\ell,t) = 0 \; \text{for all } t$$

(no heat flows into the left end or out of the right end). On the other hand, if the ends of the bar are kept fixed at temperature zero[3] (through perfect thermal contact with an ice bath, for instance), we obtain

$$u(0,t) = u(\ell,t) = 0 \; \text{for all } t.$$

Either of these boundary conditions can be inhomogeneous (that is, have a nonzero right-hand side), and we could, of course, have mixed conditions (one end insulated, the other held at fixed temperature).

A boundary condition that specifies the value of the solution is called a *Dirichlet condition*, while a condition that specifies the value of the derivative is called a *Neumann condition*. A problem with a Dirichlet condition at one end and a Neumann condition at the other is said to have *mixed boundary conditions*. As noted above, we will also use the terms homogeneous and inhomogeneous to refer to the boundary conditions. The condition $u(\ell) = 10$ is called an inhomogeneous Dirichlet condition, for example.

To completely determine the temperature as a function of time and space, we must know the temperature distribution at some initial time t_0. This is an *initial condition*:

$$u(x,t_0) = \psi(x), \; 0 < x < \ell.$$

Putting together the PDE, the boundary conditions, and the initial condition, we obtain an *initial-boundary value problem* (IBVP) for the heat equation. For example, with both

---

[2] The term homogeneous is used in two completely different ways in this section and throughout the book. A material can be (physically) homogeneous, which implies that the coefficients in the differential equations will be constants. On the other hand, a linear differential equation can be (mathematically) homogeneous, which means that the right-hand side is zero. These two uses of the word homogeneous are unrelated and potentially confusing, but the usage is standard and so the reader must understand from the context the sense in which the word is used.

[3] Since changing u by an additive constant does not affect the differential equation, we can use Celsius as the temperature scale rather than Kelvin. (See Exercise 2.1.6.)

ends of the bar held at fixed temperatures, we have the IBVP

$$\rho c \frac{\partial u}{\partial t} - \kappa \frac{\partial^2 u}{\partial x^2} = f(x,t),\ 0 < x < \ell,\ t > t_0,$$
$$u(x,t_0) = \psi(x),\ 0 < x < \ell, \qquad (2.10)$$
$$u(0,t) = 0,\ t > t_0,$$
$$u(\ell,t) = 0,\ t > t_0.$$

2.1.2 Steady-state heat flow

An important special case modeled by the heat equation is *steady-state heat flow*—a situation in which the temperature is constant with respect to time (although not necessarily with respect to space). In this case, the temperature function u can be thought of as a function of x alone, $u = u(x)$, the partial derivative with respect to t is zero, and the differential equation becomes

$$-\kappa \frac{d^2 u}{dx^2} = f(x),\ 0 < x < \ell.$$

In the steady-state case, any source term f must be independent of time (otherwise, the equation could not possibly be satisfied by a function u that is independent of time). Boundary conditions have the same meaning as in the time-dependent case, although in the case of inhomogeneous boundary conditions, the boundary data must be constant. On the other hand, it obviously does not make sense to impose an initial condition on a steady-state temperature distribution.

Collecting these observations, we see that a steady-state heat flow problem takes the form of a *boundary value problem* (BVP). For example, if the temperature is fixed at the two endpoints of the bar, we have the following Dirichlet problem:

$$-\kappa \frac{d^2 u}{dx^2} = f(x),\ 0 < x < \ell,$$
$$u(0) = 0,$$
$$u(\ell) = 0.$$

We remark that, when only one spatial dimension is taken into account, a steady-state (or *equilibrium*) problem results in an ODE rather than a PDE. Moreover, these problems, at least in their simplest form, can be solved directly by two integrations. Nevertheless, we will (in Chapter 5) devote a significant amount of effort toward developing methods for solving these ODEs, since the techniques can be generalized to multiple spatial dimensions, and also form the foundation for techniques for solving time-dependent problems.

Example 2.2. The thermal conductivity of an aluminum alloy is $1.5\,\text{W}/(\text{cm}\,\text{K})$. We will calculate the steady-state temperature of an aluminum bar of length 1 m (insulated along the sides) with its left end fixed at 20 degrees Celsius and its right end fixed at 30 degrees. If we write $u = u(x)$ for the steady-state temperature, then u satisfies the BVP

$$-1.5 \frac{d^2 u}{dx^2} = 0,\ 0 < x < 100,$$
$$u(0) = 20,$$
$$u(100) = 30.$$

2.1. Heat flow in a bar; Fourier's law

(We changed the units of the length of the bar to centimeters, so as to be consistent with the units of the thermal conductivity.) The differential equation implies that d^2u/dx^2 is zero, so, integrating once, du/dx is constant, say

$$\frac{du}{dx}(x) = C_1, \; 0 < x < 100.$$

Integrating a second time yields

$$u(x) = C_1 x + C_2, \; 0 < x < \ell.$$

The boundary condition $u(0) = 20$ implies that $C_2 = 20$, and the second boundary condition then yields

$$100 C_1 + 20 = 30 \Rightarrow C_1 = \frac{1}{10}.$$

Thus

$$u(x) = 20 + \frac{x}{10}. \quad \blacksquare$$

The reader should notice that the thermal conductivity of the bar does not affect the solution in the previous example. This is because the bar is homogeneous (that is, the thermal conductivity of the bar is constant throughout). This should be contrasted with the next example.

Example 2.3. A metal bar of length 100 cm is manufactured so that its thermal conductivity is given by the formula

$$\kappa(x) = 1.4 + 0.0025x, \; 0 < x < 100$$

(the units of $\kappa(x)$ are W/(cm K)). We will find the steady-state temperature of the bar under the same conditions as in the previous example. That is, we will solve the BVP

$$-\frac{d}{dx}\left((1.4 + 0.0025x)\frac{du}{dx}\right) = 0, \; 0 < x < 100,$$
$$u(0) = 20,$$
$$u(100) = 30.$$

Integrating both sides of the differential equation once yields

$$(1.4 + 0.0025x)\frac{du}{dx}(x) = C, \; 0 < x < 100,$$

where C is a constant. Equivalently, we have

$$\frac{du}{dx}(x) = \frac{C}{1.4 + 0.0025x}, \; 0 < x < 100.$$

Integrating a second time yields

$$u(x) = u(0) + \int_0^x \frac{C}{1.4 + 0.0025s} \, ds = 20 + \int_0^x \frac{400C}{560 + s} \, ds$$
$$= 20 + 400C \ln\left(1 + \frac{x}{560}\right).$$

Applying the boundary condition $u(100) = 30$, we obtain

$$C = \frac{10}{400\ln\left(\frac{33}{28}\right)}.$$

With this value for C, we have

$$u(x) = 20 + \frac{10\ln\left(1 + \frac{x}{560}\right)}{\ln\left(\frac{33}{28}\right)}. \quad \blacksquare$$

The steady-state temperatures from the two previous examples are shown in Figure 2.3, which shows that the heterogeneity of the second bar makes a small but discernible difference in the temperature distribution. In both cases, heat energy is flowing from the right end of the bar to the left, and the heterogeneous bar has a higher thermal conductivity at the right end. This leads to a higher temperature in the middle of the bar.

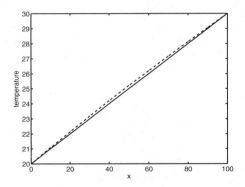

Figure 2.3. *The steady-state temperature from Example 2.2 (solid line) and Example 2.3 (dashed curve).*

2.1.3 Diffusion

One of the facts that makes mathematics so useful is that different physical phenomena can lead to essentially the same mathematical model. In this section, we introduce another experiment that is modeled by the heat equation. Of course, the meaning of the quantities appearing in the equation will be different than in the case of heat flow.

We suppose that a certain chemical is in solution in a straight pipe having a uniform cross section. We assume that the pipe has length ℓ, and we establish a coordinate system as in the preceding section, with one end of the pipe at $x = 0$ and the other at $x = \ell$. Assuming the concentration varies only in the x-direction, we can define $u(x,t)$ to be the concentration, in units of mass per volume, of the chemical at time t in the cross section located at x. Then the total mass of the chemical in the part of the bar between x and $x + \Delta x$ (at time t) is

$$\int_x^{x+\Delta x} Au(s,t)\,ds, \tag{2.11}$$

2.1. Heat flow in a bar; Fourier's law

where A is the cross-sectional area of the pipe. The chemical will tend to *diffuse* from regions of high concentration to regions of low concentration (just as heat energy flows from hot regions to cooler regions). We will assume[4] that the rate of diffusion is proportional to the *concentration gradient*

$$\frac{\partial u}{\partial x}(x,t);$$

the meaning of this assumption is that there exists a constant[5] $D > 0$ such that, at time t, the chemical moves across the cross section at x at a rate of

$$-D\frac{\partial u}{\partial x}(x,t).$$

The units of this mass flux are mass per area per time (for example, g/cm^{2}s). Since the units of $\partial u/\partial x$ are mass/length^4, the *diffusion coefficient* D must have unit of area per time (for example, cm^2/s).

With this assumption, the equation modeling diffusion is derived exactly as was the heat equation, with a similar result. The total amount of the chemical contained in the part of the pipe between x and $x + \Delta x$ is given by (2.11), so the rate at which this total mass is changing is

$$\frac{\partial}{\partial t}\left[\int_x^{x+\Delta x} Au(s,t)\,ds\right] = \int_x^{x+\Delta x} A\frac{\partial u}{\partial t}(s,t)\,ds.$$

This same quantity can be computed from the fluxes at the two cross sections at x and $x + \Delta x$. At the left, mass is entering at a rate of

$$-AD\frac{\partial u}{\partial x}(x,t),$$

while at the right it enters at a rate of

$$AD\frac{\partial u}{\partial x}(x+\Delta x,t).$$

We therefore have

$$\int_x^{x+\Delta x} A\frac{\partial u}{\partial t}(s,t)\,ds = -AD\frac{\partial u}{\partial x}(x,t) + AD\frac{\partial u}{\partial x}(x+\Delta x,t) = \int_x^{x+\Delta x} AD\frac{\partial^2 u}{\partial x^2}(s,t)\,ds.$$

The result is the *diffusion equation*

$$\frac{\partial u}{\partial t} = D\frac{\partial^2 u}{\partial x^2},\ 0 < x < \ell,\ t > t_0.$$

If the chemical is added to the interior of the pipe, this can be accounted for by a function $f(x,t)$, where mass is added to the part of the pipe between x and $x + \Delta x$ at a rate of

$$\int_x^{x+\Delta x} Af(s,t)\,ds$$

---

[4]This assumption, which is analogous to Fourier's law of heat conduction, is called Fick's law in the diffusion setting.

[5]This constant varies with temperature and pressure; see the *CRC Handbook of Chemistry and Physics* [44, page 6–179].

(mass per unit time—f has units of mass per volume per time). We then obtain the inhomogeneous diffusion equation

$$\frac{\partial u}{\partial t} - D\frac{\partial^2 u}{\partial x^2} = f(x,t),\ 0 < x < \ell,\ t > t_0.$$

Just as in the case of heat flow, we can consider steady-state diffusion. The result is the ODE

$$-D\frac{d^2 u}{dx^2} = f(x),\ 0 < x < \ell,$$

with appropriate boundary conditions. Boundary conditions for the diffusion equation are explored in the exercises.

Exercises

1. Determine the units of the thermal conductivity κ from (2.6).

2. In the *CRC Handbook of Chemistry and Physics* [44], there is a table labeled "Heat Capacity of Selected Solids" which "gives the molar heat capacity at constant pressure of representative metals... as a function of temperature in the range 200 to 600 K" (see [44, page 12–190]). For example, the entry for iron is as follows:

Temp. (K)	200	250	300	350	400	500	600
c (J/mole·K)	21.59	23.74	25.15	26.28	27.39	29.70	32.05

   (The units given in this table can be converted to the more typical units J/(g K) using the *molar mass* of the material.) As this table indicates, the specific heat of a material depends on its temperature. How would the heat equation change if we did not ignore the dependence of the specific heat on temperature?

3. Verify that the integral in (2.1) has units of energy.

4. Suppose u represents the temperature distribution in a homogeneous bar, as discussed in this section, and assume that both ends of the bar are perfectly insulated.

   (a) What is the IBVP modeling this situation?

   (b) Show (mathematically) that the total heat energy in the bar is constant with respect to time. (Of course, this is obvious from a physical point of view. The fact that the mathematical model implies that the total heat energy is constant is one confirmation that the model is not completely divorced from reality.)

5. Suppose we have a means of "pumping" heat energy into a bar through one of the ends. If we add r Joules per second through the end at $x = \ell$, what would the corresponding boundary condition be?

6. In our derivation of the heat equation, we assumed that temperature was measured on the Kelvin scale. Explain what changes must be made (in the PDE, initial condition, or boundary conditions) to use degrees Celsius instead.

2.1. Heat flow in a bar; Fourier's law

7. The thermal conductivity of iron is 0.802 W/(cm K). Consider an iron bar of length 1 m and radius 1 cm, with the lateral boundary completely insulated, and assume that the temperature of one end of the bar is held fixed at 20 degrees Celsius, while the temperature of the other end is held fixed at 30 degrees. Assume that no heat energy is added to or removed from the interior of the bar.

    (a) What is the (steady-state) temperature distribution in the bar?

    (b) At what rate is heat energy flowing through the bar?

8. (a) Show that the function

    $$u(x,t) = e^{-\kappa \theta^2 t/(\rho c)} \sin(\theta x)$$

    is a solution to the homogeneous heat equation

    $$\rho c \frac{\partial u}{\partial t} - \kappa \frac{\partial^2 u}{\partial x^2} = 0, \ 0 < x < \ell, \ -\infty < x < \infty.$$

    (b) What values of θ will cause u to also satisfy homogeneous Dirichlet conditions at $x = 0$ and $x = \ell$?

9. In this exercise, we consider a boundary condition for a bar that may be more realistic than a simple Dirichlet condition. Assume that, as usual, the side of a bar is completely insulated and that the ends are placed in a bath maintained at constant temperature. Assume that the heat flows out of or into the ends in accordance with *Newton's law of cooling*: the heat flux is proportional to the difference in temperature between the end of the bar and the surrounding medium. What are the resulting boundary conditions?

10. Verify that Theorem 2.1 holds for $(a,b) \times (c,d) = (0,1) \times [0,1]$ and F defined by $F(x,y) = \cos(xy)$.

11. Suppose a chemical is diffusing in a pipe, and both ends of the pipe are sealed. What are the appropriate boundary conditions for the diffusion equation? What initial conditions are required? Write down a complete IBVP for the diffusion equation under these conditions.

12. Suppose that a chemical contained in a pipe of length ℓ has an initial concentration distribution of $u(x,0) = \psi(x)$. At time zero, the ends of the pipe are sealed, and no mass is added to or removed from the interior of the pipe.

    (a) Write down the IBVP describing the diffusion of the chemical.

    (b) Show *mathematically* that the total mass of the chemical in the pipe is constant. (Derive this fact from the equations rather than from common sense.)

    (c) Describe the ultimate steady-state concentration.

    (d) Give a formula for the steady-state concentration in terms of ψ.

13. Suppose a pipe of length ℓ and radius r joins two large reservoirs, each containing a (well-mixed) solution of the same chemical. Let the concentration in one reservoir be u_0 and in the other be u_ℓ, and assume that $u(0,t) = u_0$ and $u(\ell,t) = u_\ell$.

(a) Find the steady-state rate at which the chemical diffuses through the pipe (in units of mass per time).

(b) How does this rate vary with the length ℓ and the radius r?

14. Consider the previous exercise, in which the chemical is carbon monoxide (CO) and the solution is CO in air. Suppose that $u_0 = 0.01\,\text{g/cm}^3$ and $u_\ell = 0.015\,\text{g/cm}^3$, and that the diffusion coefficient of CO in air is $0.208\,\text{cm}^2/\text{s}$. If the bar is 1 m long and its radius is 2 cm, find the steady-state rate at which CO diffuses through the pipe (in units of mass per time).

2.2 The hanging bar

Suppose that a bar, with uniform cross-sectional area A and length ℓ, hangs vertically and stretches due to a force (perhaps gravity) acting upon it. We assume that the deformation occurs only in the vertical direction; this assumption is reasonable only if the bar is long and thin. Normal materials tend to contract horizontally when they are stretched vertically, but both this contraction and the coupling between horizontal and vertical motion are small compared to the elongation when the bar is thin (see Lin and Segel [45, Chapter 12]).

With the assumption of purely vertical deformation, we can describe the movement of the bar in terms of a *displacement function* $u(x,t)$. Specifically, suppose that the top of the bar is fixed at $x = 0$, and let down be the positive x-direction. Let the cross section of the bar originally at x move to $x + u(x,t)$ at time t (see Figure 2.4). We will derive a PDE describing the dynamics of the bar by applying Newton's second law of motion.

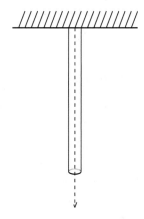

Figure 2.4. *The hanging bar and its coordinate system.*

We assume that the bar is *elastic*, which means that the internal forces in the bar depend on the local relative change in length. The deformed length of the part of the bar originally between x and $x + \Delta x$ (at time t) is

$$(x + \Delta x + u(x + \Delta x, t)) - (x + u(x,t)) = \Delta x + u(x + \Delta x, t) - u(x,t).$$

2.2. The hanging bar

Since the original length is Δx, the change in length of this part is

$$u(x+\Delta x,t)-u(x,t),$$

and the relative change in length is

$$\frac{u(x+\Delta x,t)-u(x,t)}{\Delta x} \doteq \frac{\partial u}{\partial x}(x,t).$$

This explains the definition of the *strain*, the local relative change in length, as the dimensionless quantity

$$\frac{\partial u}{\partial x}(x,t).$$

As we noted above, to say that the bar is elastic is to say that the internal restoring force of the deformed bar depends only on the strain. We now make the further assumption that the deformations are small, and therefore that the internal forces are, to good approximation, proportional to the strain. This is equivalent to assuming that the bar is linearly elastic, or *Hookean*.

Under the assumption that the bar is Hookean, we can write an expression for the total internal force acting on P, the part of the bar between x and $x+\Delta x$. We denote by $k(x)$ the *stiffness* of the bar at x, that is, the constant of proportionality in Hooke's law, with units of force per unit area. (In the engineering literature, k is called the *Young's modulus* or the *modulus of elasticity*. Values for various materials can be found in reference books, such as [44].) Then the total internal force is

$$Ak(x+\Delta x)\frac{\partial u}{\partial x}(x+\Delta x,t)-Ak(x)\frac{\partial u}{\partial x}(x,t). \tag{2.12}$$

The first term in this expression is the force exerted by the part of the bar below $x+\Delta x$ on P, and the second term is the force exerted by the part of the bar above x on P. The signs are correct; if the strains are positive, then the bar has been stretched, and the internal restoring force is pulling down (the positive direction) at $x+\Delta x$ and up at x.

Now, (2.12) is equal (by the fundamental theorem of calculus) to

$$\int_x^{x+\Delta x} A\frac{\partial}{\partial x}\left(k(s)\frac{\partial u}{\partial x}(s,t)\right)ds.$$

We now assume that all external forces are lumped into a *body force* given by a force density f (which has units of force per unit volume). Then the total external force on P (at time t) is

$$\int_x^{x+\Delta x} f(s,t)A\,ds,$$

and the sum of the forces acting on part P is

$$\int_x^{x+\Delta x} A\frac{\partial}{\partial x}\left(k(s)\frac{\partial u}{\partial x}(s,t)\right)ds + \int_x^{x+\Delta x} Af(s,t)\,ds.$$

Newton's second law states that the total force acting on P must equal the mass of P times its acceleration. This law takes the form

$$\int_x^{x+\Delta x} A\frac{\partial}{\partial x}\left(k(s)\frac{\partial u}{\partial x}(s,t)\right)ds + \int_x^{x+\Delta x} Af(s,t)\,ds = \int_x^{x+\Delta x} A\rho(s)\frac{\partial^2 u}{\partial t^2}(s,t)\,ds,$$

where $\rho(x)$ is the density of the bar at x (in units of mass per volume). We can rewrite this as

$$\int_x^{x+\Delta x} \left[\rho(s)\frac{\partial^2 u}{\partial t^2}(s,t) - \frac{\partial}{\partial x}\left(k(s)\frac{\partial u}{\partial x}(s,t)\right) - f(s,t)\right] ds = 0$$

(note how the factor of A cancels). This integral must be zero for every $x \in [0, \ell)$ and every $\Delta x > 0$. It follows (by the reasoning introduced on pages 11–12) that the integrand must be identically zero; this gives the equation

$$\rho(x)\frac{\partial^2 u}{\partial t^2} - \frac{\partial}{\partial x}\left(k(x)\frac{\partial u}{\partial x}\right) - f(x,t) = 0$$

or

$$\rho(x)\frac{\partial^2 u}{\partial t^2} - \frac{\partial}{\partial x}\left(k(x)\frac{\partial u}{\partial x}\right) = f(x,t). \tag{2.13}$$

The PDE (2.2) is called the *wave equation*.

If the bar is in equilibrium, then the displacement does not depend on t, and we can write $u = u(x)$. In this case, the acceleration $\partial^2 u/\partial t^2$ is zero, and the forcing function f must also be independent of time. We then obtain the following ODE for the equilibrium displacement of the bar:

$$-\frac{d}{dx}\left(k(x)\frac{du}{dx}\right) = f(x). \tag{2.14}$$

This is the same equation that governs steady-state heat flow! Just as in the case of steady-state heat flow, the resulting BVPs can be solved with two integrations (see Examples 2.2 and 2.3).

If the bar is homogeneous, so that ρ and k are constants, these last two differential equations can be written as

$$\rho\frac{\partial^2 u}{\partial t^2} - k\frac{\partial^2 u}{\partial x^2} = f(x,t)$$

and

$$-k\frac{d^2 u}{dx^2} = f(x),$$

respectively.

2.2.1 Boundary conditions for the hanging bar

Equation (2.14) by itself does not determine a unique displacement; we need boundary conditions, as well as initial conditions if the problem is time dependent. The statement of the problem explicitly gives us one boundary condition: $u(0) = 0$ (the top end of the bar cannot move). Moreover, we can deduce a second boundary condition from force balance at the other end of the bar. If the bottom of the bar is unsupported, then there is no contact

2.2. The hanging bar

force applied at $x = \ell$. On the other hand, the analysis that led to (2.12) shows that the part of the bar above $x = \ell$ (which is all of the bar) exerts an internal force of

$$-Ak(\ell)\frac{du}{dx}(\ell)$$

on the surface at $x = \ell$. Since there is nothing to balance this force, we must have

$$-Ak(\ell)\frac{du}{dx}(\ell) = 0,$$

or simply

$$\frac{du}{dx}(\ell) = 0.$$

Since the wave equation involves the *second* time derivative of u, we need two initial conditions to uniquely determine the motion of the bar: the initial displacement and the initial velocity. We thus arrive at the following IBVP for the wave equation:

$$\begin{aligned}
\rho(x)\frac{\partial^2 u}{\partial t^2} - \frac{\partial}{\partial x}\left(k(x)\frac{\partial u}{\partial x}\right) &= f(x,t),\ 0 < x < \ell,\ t > t_0, \\
u(x,t_0) &= \psi(x),\ 0 < x < \ell, \\
\frac{\partial u}{\partial t}(x,t_0) &= \gamma(x),\ 0 < x < \ell, \\
u(0,t) &= 0,\ t > t_0, \\
\frac{\partial u}{\partial x}(\ell,t) &= 0,\ t > t_0.
\end{aligned} \qquad (2.15)$$

The corresponding steady-state BVP (expressing mechanical equilibrium) is

$$\begin{aligned}
-\frac{d}{dx}\left(k(x)\frac{du}{dx}\right) &= f(x),\ 0 < x < \ell, \\
u(0) &= 0, \\
\frac{du}{dx}(\ell) &= 0.
\end{aligned} \qquad (2.16)$$

There are several other sets of boundary conditions that might be of interest in connection with the differential equations (2.2) or (2.14). For example, if both ends of the bar are fixed (not allowed to move), we have the boundary conditions

$$u(0) = 0,\ u(\ell) = 0$$

(recall that u is the displacement, so the condition $u(\ell) = 0$ indicates that the cross section at the end of the bar corresponding to $x = \ell$ does not move from its original position). If both ends of the bar are free, the corresponding boundary conditions are

$$\frac{du}{dx}(0) = 0,\ \frac{du}{dx}(\ell) = 0.$$

Any of the above boundary conditions can be inhomogeneous. For example, we could fix one end of the bar at $x = 0$ and stretch the other to $x = \ell + \Delta\ell$. This experiment corresponds to the boundary conditions $u(0) = 0$, $u(\ell) = \Delta\ell$. As another example, if one end of the bar (say $x = 0$) is fixed and a force F is applied to the other end ($x = \ell$), then the applied force determines the value of $du/dx(\ell)$. Indeed, as indicated above, the restoring force of the bar on the $x = \ell$ cross section is

$$-Ak(\ell)\frac{du}{dx}(\ell),$$

and this must balance the applied force F:

$$-Ak(\ell)\frac{du}{dx}(\ell) + F = 0.$$

This leads to the boundary condition

$$k(\ell)\frac{du}{dx}(\ell) = \frac{F}{A},$$

and the quantity F/A has units of pressure (force per unit area). For mathematical purposes, it is simplest to write an inhomogeneous boundary condition of this type as

$$\frac{du}{dx}(\ell) = C,$$

but for solving a practical problem, it is essential to recognize that

$$C = \frac{F}{Ak(\ell)}.$$

Exercises

1. Consider the following experiment: A bar is hanging with the top end fixed at $x = 0$, and the bar is stretched by a pressure (force per unit area) p applied uniformly to the free (bottom) end. What are the boundary conditions describing this situation?

2. Suppose that a homogeneous bar (that is, a bar with constant stiffness k) of length ℓ has its top end fixed at $x = 0$, and the bar is stretched to a length $\ell + \Delta\ell$ by a pressure p applied to the bottom end. Take down to be the positive x-direction.

   (a) Explain why p and $\Delta\ell$ have the same sign.
   (b) Explain why p and $\Delta\ell$ cannot both be chosen arbitrarily (even subject to the requirement that they have the same sign). Give both physical and mathematical reasons.
   (c) Suppose p is specified. Find $\Delta\ell$ (in terms of p, k, and ℓ).
   (d) Suppose $\Delta\ell$ is specified. Find p (in terms of $\Delta\ell$, k, and ℓ).

3. A certain type of stainless steel has a stiffness of 195 GPa. (A *Pascal* (Pa) is the standard unit of pressure, or force per unit area. The Pascal is a derived unit: one Pascal equals one Newton per square meter. The *Newton* is the standard unit of force: one Newton equals one kilogram meter per second squared. Finally, GPa is short for gigaPascal, or 10^9 Pascals.)

2.2. The hanging bar

(a) Explain in words (including units) what a stiffness of 195 GPa means.

(b) Suppose a pressure of 1 GPa is applied to the end of a homogeneous, circular cylindrical bar of this stainless steel, and that the other end is fixed. If the original length of the bar is 1 m and its radius is 1 cm, what will its length be, in the equilibrium state, after the pressure has been applied?

(c) Verify the result of 3(b) by formulating and solving the BVP representing this experiment.

4. Consider a circular cylindrical bar, of length 1 m and radius 1 cm, made from an aluminum alloy with stiffness 70 GPa. If the top end of the bar ($x = 0$) is fixed, what total force must be applied to the other end ($x = 1$) to stretch the bar to a length of 1.01 m?

5. Write the wave equation for the bar of Exercise 3, given that the density of the stainless steel is 7.9 g/cm^3. (Warning: Use consistent units!) What must the units of the forcing function f be? Verify that the two terms on the left side of the differential equation have the same units as f.

6. Suppose that a 1 m bar of the stainless steel described in Exercise 3, with density 7.9 g/cm^3, is supported at the bottom but free at the top. Let the cross-sectional area of the bar be 0.1 m^2. A weight of 1000 kg is placed on top of the bar, exerting pressure on it via gravity (the gravitational constant is 9.8 m/s^2).

The purpose of this problem is to compute and compare the effects on the bar of the mass on the top and the weight of the bar itself.

(a) Write down three BVPs:
   i. First, take into account the weight of the bar (which means that gravity induces a body force), but ignore the mass on the top (so the top end of the bar is free—the boundary condition is a homogeneous Neumann condition).
   ii. Next, take into account the mass on the top, but ignore the effect of the weight of the bar (so there is no body force).
   iii. Last, take both effects into account.

(b) Explain why the third BVP can be solved by solving the first two and adding the results.

(c) Solve the first two BVPs by direct integration. Compare the two displacements. Which is more significant, the weight of the bar or the mass on top?

(d) How would the situation change if the cross-sectional area of the bar were changed to 0.2 m^2?

7. (a) Show that the function
$$u(x,t) = \cos(c\theta t)\sin(\theta x)$$
is a solution to the homogeneous wave equation
$$\frac{\partial^2 u}{\partial t^2} - c^2 \frac{\partial^2 u}{\partial x^2} = 0, \ 0 < x < \ell, \text{ for all } t.$$

(b) What values of θ will cause u to also satisfy homogeneous Dirichlet conditions at $x = 0$ and $x = \ell$?

2.3 The wave equation for a vibrating string

We now present an argument that the wave equation (2.2) also describes the small transverse vibrations of an elastic string (such as a guitar string). In the course of the following derivation, we make several a priori unjustified assumptions which are significant enough that the end result ought to be viewed with some skepticism. However, a careful analysis leads to the same model (see the article by Antman [2]).

For simplicity, we will assume that the string in question is homogeneous, so that any material properties are constant throughout the string. We suppose that the string is stretched to length ℓ and that its two endpoints are not allowed to move. We further suppose that the string vibrates in the xy-plane, occupying the interval $[0,\ell]$ on the x-axis when at rest, and that the point at $(x,0)$ in the reference configuration moves to $(x,u(x,t))$ at time t. We are thus postulating that the motion of the string is entirely in the transverse direction (this is one of the severe assumptions that we mentioned in the previous paragraph). Granted this assumption, we now derive the differential equation satisfied by the displacement u.

Since, by assumption, a string does not resist bending, the internal restoring force of the string under tension is tangent to the string itself at every point. We will denote the *magnitude* of this restoring force by $T(x,t)$. In Figure 2.5, we display a part of the deformed string, corresponding to the part of the string between x and $x + \Delta x$ in the reference configuration, together with the internal forces at the ends of this part, and their magnitudes. In the absence of any external forces, the sum of these internal forces must balance the mass times acceleration of this part of the string. To write down these equations, we must decompose the internal force into its horizontal and vertical components.

Figure 2.5. *A part of the deformed string, with the forces acting on its ends.*

We write $\mathbf{n} = \mathbf{n}(x,t)$ for the force at the left endpoint and θ for the angle this force vector makes with the horizontal. We then have

$$n_1^2 + n_2^2 = T(x,t)^2$$

with

$$n_1 = -T(x,t)\cos(\theta), \; n_2 = -T(x,t)\sin(\theta).$$

Assuming that $|\partial u/\partial x| \ll 1$ at every point, and noting that

$$\tan(\theta) = \frac{\partial u}{\partial x}(x,t),$$

2.3. The wave equation for a vibrating string

a reasonable approximation is

$$\cos(\theta) \doteq 1, \; \sin(\theta) \doteq \tan(\theta) = \frac{\partial u}{\partial x}(x,t).$$

We then obtain

$$\mathbf{n}(x,t) \doteq \left(-T(x,t), -T(x,t)\frac{\partial u}{\partial x}(x,t) \right).$$

Similarly, the force at the right end of the part of the string under consideration is

$$\mathbf{n}(x+\Delta x, t) \doteq \left(T(x+\Delta x,t), T(x+\Delta x,t)\frac{\partial u}{\partial x}(x+\Delta x,t) \right).$$

We then see that

$$T(x+\Delta x, t) - T(x,t)$$

is (an approximation to) the horizontal component of the total force on the part of the string, and

$$T(x+\Delta x,t)\frac{\partial u}{\partial x}(x+\Delta x,t) - T(x,t)\frac{\partial u}{\partial x}(x,t)$$

is (an approximation to) the vertical component.

By assumption, the horizontal component of the acceleration of the string is zero, and so Newton's second law yields

$$T(x+\Delta x, t) - T(x,t) = 0;$$

thus the tension is constant throughout the string at each point in time. We will assume that the length of the string never changes much from its length in the reference configuration (which is ℓ), and therefore that it makes sense to assume that T is independent of t as well, that is, that T is constant. Applying Newton's second law to the vertical component yields

$$\int_x^{x+\Delta x} \rho \frac{\partial^2 u}{\partial t^2}(s,t) ds = T\frac{\partial u}{\partial x}(x+\Delta x,t) - T\frac{\partial u}{\partial x}(x,t)$$
$$= \int_x^{x+\Delta x} T\frac{\partial^2 u}{\partial x^2}(s,t) ds,$$

where ρ is the density of the string (in units of mass per length). Since this holds for all x and Δx sufficiently small, we obtain the differential equation

$$\rho \frac{\partial^2 u}{\partial t^2} = T\frac{\partial^2 u}{\partial x^2}, \; 0 < x < \ell, \; t > 0. \tag{2.17}$$

We recognize this as the homogeneous wave equation. It is usual to write this in the form

$$\frac{\partial^2 u}{\partial t^2} - c^2 \frac{\partial^2 u}{\partial x^2} = 0, \; 0 < x < \ell, \; t > 0,$$

where $c^2 = T/\rho$. The significance of the parameter c will become clear in Chapter 7.

In the case that an external body force is applied to the string (in the vertical direction), the equation becomes

$$\frac{\partial^2 u}{\partial t^2} - c^2 \frac{\partial^2 u}{\partial x^2} = f(x,t), \ 0 < x < \ell, \ t > 0. \tag{2.18}$$

Exercise 1 asks the reader to determine the units of f.

The natural boundary conditions for the vibrating string, as suggested above, are homogeneous Dirichlet conditions:

$$u(0,t) = u(\ell,t) = 0, \ t > 0.$$

One can also imagine that one or both ends of the string are allowed to move freely in the vertical direction (perhaps an end of the string slides along a frictionless pole). In this case, the appropriate boundary condition is a homogeneous Neumann condition (see Exercise 5 below).

Exercises

1. What units must $f(x,t)$ have in (2.18)?

2. What are the units of the tension T in the derivation of the wave equation for the string?

3. What are the units of the parameter c in (2.17)?

4. Suppose the only external force applied to the string is the force due to gravity. What form does (2.18) take in this case? (Let g be the acceleration due to gravity, and take g to be constant.)

5. Explain why a homogeneous Neumann condition models an end of the string that is allowed to move freely in the vertical direction.

6. Suppose that an elastic string is fixed at both ends, as in this section, and it sags under the influence of an external force $f(x)$ (f is constant with respect to time). What differential equation and side conditions does the equilibrium displacement of the string satisfy? Assume that f is given in units of force per length.

7. Let $f : \mathbf{R} \to \mathbf{R}$ be smooth (at least twice continuously differentiable), and define $u(x,t) = f(x-ct)$, $v(x,t) = f(x+ct)$. Prove that both u and v are solutions of the homogeneous wave equation.

2.4 Advection; kinematic waves

We return to the situation described in Section 2.1.3, in which a chemical solution is found in a straight pipe of uniform cross section. There we assumed that the chemical was diffusing in the solution, with the fluid itself stationary. Now, however, we study the case in which the solution is flowing through the pipe. As before, we write $u(x,t)$ for the concentration of the chemical at the spatial point x and time t.

2.4. Advection; kinematic waves

We begin with the simplest case in which the concentration at x changes only because the solution is flowing (that is, there is no diffusion); moreover, we assume that the velocity of the solution is constant. As usual, we consider a small section of the pipe occupying the interval $[x, x + \Delta x]$; in that part of the pipe, the total mass of the chemical is

$$\int_x^{x+\Delta x} Au(s,t)\,dt,$$

where A is the cross-sectional area of the pipe. Therefore, the rate of change of the mass in the given interval is

$$\frac{\partial}{\partial t}\left[\int_x^{x+\Delta x} Au(s,t)\,dt\right] = \int_x^{x+\Delta x} A\frac{\partial u}{\partial t}(s,t)\,dt. \tag{2.19}$$

On the other hand, the same rate of the change can be computed by considering the rates at which mass enters and leaves the section. If the velocity of the solution is c, then the rate at which mass enters the interval $[x, x + \Delta x]$ through the cross section at x is

$$cAu(x,t),$$

while the rate at which mass leaves through the cross section at $x + \Delta x$ is

$$cAu(x+\Delta x,t).$$

Therefore, the rate of change of mass in $[x, x + \Delta x]$ is

$$Au(x,t)c - Au(x+\Delta x,t)c = -\int_x^{x+\Delta x} cA\frac{\partial u}{\partial x}(s,t)\,dt. \tag{2.20}$$

Equating (2.19) and (2.20) yields

$$\int_x^{x+\Delta x} A\frac{\partial u}{\partial t}(s,t)\,dt = -\int_x^{x+\Delta x} cA\frac{\partial u}{\partial x}(s,t)\,dt$$

$$\Rightarrow \int_x^{x+\Delta x} \left\{A\frac{\partial u}{\partial t}(s,t) + cA\frac{\partial u}{\partial x}(s,t)\right\}dt = 0.$$

By the usual reasoning, this implies that u satisfies the PDE

$$\frac{\partial u}{\partial t} + c\frac{\partial u}{\partial x} = 0. \tag{2.21}$$

Equation (2.21) is called the one-dimensional *advection equation* (or sometimes the *convection equation*).

2.4.1 Initial/boundary conditions for the advection equation

We assume that the pipe described above occupies the interval $[0, \ell]$. We need an initial condition:

$$u(x,0) = u_0(x),\ 0 < x < \ell.$$

Assuming the solution is flowing from left to right, we need to know the concentration of the solution entering the pipe at $x = 0$:

$$u(0,t) = \phi(t), \ t > 0.$$

It should be clear from physical considerations that we do not need a boundary condition at $x = \ell$; indeed, such a boundary condition would be superfluous, since the concentration $u(\ell,t)$ is determined by the other conditions.

A complete IBVP for the advection equation is

$$\begin{aligned}\frac{\partial u}{\partial t} + c\frac{\partial u}{\partial x} &= 0, \ 0 < x < \ell, \ t > 0, \\ u(x,0) &= u_0(x), \ 0 < x < \ell, \\ u(0,t) &= \phi(t), \ t > 0.\end{aligned} \quad (2.22)$$

It is straightforward to describe the solution to this IBVP on physical grounds, at least in simple cases. Since the chemical does not diffuse, but merely moves with the solution, the concentration distribution defined by $u_0(x)$ simply moves to the right at velocity c. Figure 2.6 shows an example, in which the initial concentration is nonzero only between $x = 1$ and $x = 2$. Also, the boundary condition is $u(0,t) = 0, t > 0$, so that no more chemical enters the pipe.

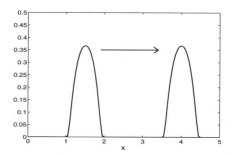

Figure 2.6. *The solution to the advection equation when the initial concentration is nonzero only between $x = 1$ and $x = 2$, and no chemical enters the pipe for $t > 0$.*

If the boundary value $\phi(t)$ is nonzero, then the IBVP is not quite as simple. However, we can pose a simpler problem by pretending that the pipe is infinite in extent:

$$\begin{aligned}\frac{\partial u}{\partial t} + c\frac{\partial u}{\partial x} &= 0, \ -\infty < x < \infty, \ t > 0, \\ u(x,0) &= u_0(x), \ -\infty < x < \infty.\end{aligned} \quad (2.23)$$

In this problem, the initial concentration is given on the entire real line, and this initial distribution simply moves to the right (assuming $c > 0$) at speed c. We can work out the formula for the solution on first principles: The cross section of solution that is found at the point x at time t was ct units to the left at time $t = 0$. Therefore, $u(x,t) = u(x - ct, 0) =$

2.4. Advection; kinematic waves

$u_0(x - ct)$. It is easy to verify that this is the desired solution:

$$u(x,t) = u_0(x-ct) \Rightarrow \frac{\partial u}{\partial x}(x,t) = \frac{du_0}{dx}(x-ct), \ \frac{\partial u}{\partial t}(x,t) = \frac{du_0}{dx}(x-ct)(-c)$$

$$\Rightarrow \frac{\partial u}{\partial t}(x,t) + c\frac{\partial u}{\partial x}(x,t) = -c\frac{du_0}{dx}(x-ct) + c\frac{du_0}{dx}(x-ct) = 0,$$

$$u(x,t) = u_0(x-ct) \Rightarrow u(x,0) = u_0(x).$$

The IBVP (2.22) is nearly as easy to solve, but the solution must be expressed in terms of both u_0 and ϕ; for certain (x,t), the value of $u(x,t)$ depends on u_0, and for others it depends on ϕ (see Exercise 2.4.1).

The solution of (2.22) or (2.23) can be viewed as a kind of wave, called a *kinematic wave*. Kinematic waves are caused by the transport of mass or some other quantity and should be distinguished from dynamic waves. A *dynamic wave* is the result of the oscillation of particles about an equilibrium, such as in the hanging bar (Section 2.2) or the vibrating string (Section 2.3). Interestingly enough, a dynamic wave can take the form of a *traveling wave* (described by a function of the form $u(x,t) = f(x \pm ct)$), which is mathematically the same as a kinematic wave. For instance, when one end of a rope is fastened to a wall and the other end is quickly "flicked" up and down, a traveling wave results (see Example 7.5 in Section 7.2). However, the mechanism underlying a traveling (dynamic) wave is quite different from that of a kinematic wave.

Since solutions of the advection equation can be regarded as waves which move in one direction (right if $c > 0$, left if $c < 0$), (2.21) is also called the *one-way wave equation*.

2.4.2 The advection-diffusion equation

Realistically, in the experiment described above (a solution flowing through a pipe), the chemical would be subject to diffusion as well as advection. Derivation of the appropriate equation would take into account two ways that the chemical enters the interval $[x, x + \Delta x]$ through the cross section at x: The flow would carry chemical into the interval at the rate $cu(x,t)$, and diffusion would impose the rate $-D\frac{\partial u}{\partial x}(x,t)$, where D is the diffusion coefficient described in Section 2.1.3. Thus the total rate at which the chemical enters $[x, x + \Delta x]$ at x is

$$cu(x,t) - D\frac{\partial u}{\partial x}(x,t).$$

Similarly, at $x + \Delta x$, the rate is

$$-cu(x+\Delta x,t) + D\frac{\partial u}{\partial x}(x+\Delta x,t).$$

Following the derivation at the beginning of the section, we end up with the PDE

$$\frac{\partial u}{\partial t} + c\frac{\partial u}{\partial x} - D\frac{\partial^2 u}{\partial x^2} = 0, \qquad (2.24)$$

which is called the *advection-diffusion equation* (or *convection-diffusion equation*).

Initial and boundary conditions for the advection-diffusion equation are the same as for the diffusion (heat) equation. A possible IBVP (with Dirichlet boundary conditions) is

$$\frac{\partial u}{\partial t} + c\frac{\partial u}{\partial x} - D\frac{\partial^2 u}{\partial x^2} = 0,\ 0 < x < \ell,\ t > t_0,$$
$$u(x,t_0) = \psi(x),\ 0 < x < \ell,$$
$$u(0,t) = 0,\ t > t_0,$$
$$u(\ell,t) = 0,\ t > t_0.$$

Neumann or mixed boundary conditions are also possible.

2.4.3 Conservation laws

The advection equation is a simple example of a *conservation law*; it was derived from the fact that the mass of the chemical is conserved, so the only way for the total mass in $[x, x + \Delta x]$ to change is for mass to enter or leave through the endpoints x and $x + \Delta x$. There are many similar situations where a quantity—mass, momentum, energy—is conserved, and the corresponding PDE can be derived in the same way.

Let $u(x,t)$ be the (linear) density of the quantity under consideration, so that the total quantity in the interval $[\overline{x}, \overline{x} + \Delta x]$ at time t is

$$\int_{\overline{x}}^{\overline{x}+\Delta x} u(s,t)\,ds.$$

Suppose further that the flux of u, that is, the rate at which the quantity moves past the point x at time t, is $f(u(x,t))$. (In the simple advection equation above, the flux was simply $cu(x,t)$; that is, the function f would be $f(u) = cu$ in that example.) The rate of change in $[\overline{x}, \overline{x} + \Delta x]$ can be written two ways:

$$\frac{\partial}{\partial t}\int_{\overline{x}}^{\overline{x}+\Delta x} u(x,t)\,dx = \int_{\overline{x}}^{\overline{x}+\Delta x} \frac{\partial u}{\partial t}(x,t)\,dx$$

and

$$f(\overline{x},t) - f(\overline{x}+\Delta x,t) = -\int_{\overline{x}}^{\overline{x}+\Delta x} \frac{\partial}{\partial x}(f(u(x,t)))\,dx.$$

The two expressions for the rate of change must be equal, so we obtain

$$\int_{\overline{x}}^{\overline{x}+\Delta x} \frac{\partial u}{\partial t}(x,t)\,dx = -\int_{\overline{x}}^{\overline{x}+\Delta x} \frac{\partial}{\partial x}(f(u(x,t)))\,dx$$
$$\Rightarrow \int_{\overline{x}}^{\overline{x}+\Delta x} \left\{\frac{\partial u}{\partial t}(x,t) + \frac{\partial}{\partial x}(f(u(x,t)))\right\} = 0$$
$$\Rightarrow \frac{\partial u}{\partial t}(x,t) + \frac{\partial}{\partial x}(f(u(x,t))) = 0.$$

Thus a one-dimensional conservation law takes the form

$$\frac{\partial u}{\partial t} + \frac{\partial}{\partial x}(f(u)) = 0. \tag{2.25}$$

2.4. Advection; kinematic waves

Except for the simplest flux laws (such as $f(u) = cu$), (2.25) is a nonlinear PDE.

The simplest auxiliary conditions that define a unique solution are initial conditions on the entire real line, leading to the IVP

$$\frac{\partial u}{\partial t} + \frac{\partial}{\partial x}(f(u)) = 0, \quad -\infty < x < \infty, \, t > 0,$$

$$u(x,0) = u_0(x), \quad -\infty < x < \infty.$$

We can also pose an IBVP on a restricted interval by adding a boundary condition at one endpoint:

$$\frac{\partial u}{\partial t} + \frac{\partial}{\partial x}(f(u)) = 0, \, 0 < x < \ell,$$

$$u(x,0) = u_0(x), \, 0 < x < \ell,$$

$$u(0,t) = \phi(t), \, t > 0.$$

2.4.4 Burgers's equation

We mention here two other PDEs that are frequently used as model problems in the study of nonlinear equations. They are *Burgers's equation*

$$\frac{\partial u}{\partial t} + u\frac{\partial u}{\partial x} = \nu \frac{\partial^2 u}{\partial x^2} \qquad (2.26)$$

and the *inviscid Burgers's equation*

$$\frac{\partial u}{\partial t} + u\frac{\partial u}{\partial x} = 0. \qquad (2.27)$$

Burgers's equation was originally proposed as a model of some aspects of turbulence, and it can also be derived as governing certain disturbances in one-dimensional fluid flow (for this latter derivation, the reader can consult Section 5.5 of Logan [46]). However, any derivation is quite involved, and the equation (or the related inviscid version) is often now used simply to illustrate certain mathematical aspects of nonlinear PDEs. We will discuss the behavior of solutions to the inviscid Burgers's equation in Section 8.3.

Exercises

1. Determine the solution of the IBVP (2.22).

2. Find the solution $u = u(x,t)$, $0 \le x \le 10$, $t > 0$, of the following IBVP:

$$\frac{\partial u}{\partial t} + 2\frac{\partial u}{\partial x} = 0, \, 0 < x < 10, \, t > 0,$$

$$u(x,0) = 1, \, 0 < x < \ell,$$

$$u(0,t) = e^{-t}, \, t > 0.$$

3. Suppose u solves the IVP

$$\frac{\partial u}{\partial t} + c\frac{\partial u}{\partial x} = 0, \quad -\infty < x < \infty, \; t > 0,$$
$$u(x,0) = \phi(x), \quad -\infty < x < \infty,$$

where $c > 0$. Suppose further that $\phi(x) = 0$ for all $x \leq a$. For what values of (x,t) is $u(x,t) = 0$ guaranteed to hold?

4. Suppose $\nu > 0$ and that $\phi = \phi(x,t)$ satisfies the homogeneous diffusion equation

$$\frac{\partial \phi}{\partial t} - \nu \frac{\partial^2 \phi}{\partial x^2} = 0.$$

Prove that u defined by

$$u(x,t) = -\frac{2\nu}{\phi(x,t)}\frac{\partial \phi}{\partial x}(x,t)$$

is a solution of Burgers's equation (2.26). (The transformation from ϕ to u is called the *Hopf–Cole* transformation.)

2.5 Suggestions for further reading

If the reader wishes to learn more about the use of mathematical models in the sciences, an excellent place to start is the text by Lin and Segel [45], which combines modeling with the analysis of the models by a number of different analytical techniques. Lin and Segel cover the models that form the basis for this book, as well as many others. A more advanced text, which focuses almost entirely on the derivation of differential equations from the basic principles of continuum mechanics, is Gurtin [28].

This book focuses on standard PDE models that are used to illustrate the subject. For a presentation of a number of less standard PDEs arising in a variety of applications, see Mattheij, Rienstra, and ten Thije Boonkkamp [49].

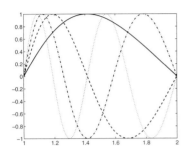

Chapter 3
Essential Linear Algebra

The solution techniques presented in this book can be described by analogy to techniques for solving

$$\mathbf{A}\mathbf{x} = \mathbf{b},$$

where $\mathbf{A}$ is an $n \times n$ matrix ($\mathbf{A} \in \mathbf{R}^{n \times n}$) and $\mathbf{x}$ and $\mathbf{b}$ are n-vectors ($\mathbf{x}, \mathbf{b} \in \mathbf{R}^n$). Recall that such a matrix-vector equation represents the following system of n linear equations in the n unknowns $x_1, x_2, \ldots, x_n$:

$$\begin{aligned}
a_{11}x_1 + a_{12}x_2 + \cdots + a_{1n}x_n &= b_1, \\
a_{21}x_1 + a_{22}x_2 + \cdots + a_{2n}x_n &= b_2, \\
&\vdots \\
a_{n1}x_1 + a_{n2}x_2 + \cdots + a_{nn}x_n &= b_n.
\end{aligned}$$

Before we discuss methods for solving differential equations, we review the fundamental facts about systems of linear (algebraic) equations.

3.1 Linear systems as linear operator equations

To fully appreciate the point of view taken in this book, it is necessary to understand the equation $\mathbf{A}\mathbf{x} = \mathbf{b}$ not just as a system of linear equations, but as a finite-dimensional linear operator equation. In other words, we must view the matrix $\mathbf{A}$ as defining an *operator* (or *mapping*, or simply *function*) from $\mathbf{R}^n$ to $\mathbf{R}^n$ via matrix multiplication: $\mathbf{A}$ maps $\mathbf{x} \in \mathbf{R}^n$ to $\mathbf{y} = \mathbf{A}\mathbf{x} \in \mathbf{R}^n$. (More generally, if $\mathbf{A}$ is not square, say $\mathbf{A} \in \mathbf{R}^{m \times n}$, then $\mathbf{A}$ defines a mapping from $\mathbf{R}^n$ to $\mathbf{R}^m$, since $\mathbf{A}\mathbf{x} \in \mathbf{R}^m$ for each $\mathbf{x} \in \mathbf{R}^n$.)

The following language is useful in discussing operator equations.

Definition 3.1. *Let X and Y be sets. A function (operator, mapping) f from X to Y is a rule for associating with each $x \in X$ a unique $y \in Y$, denoted $y = f(x)$. The set X is called the* domain *of f, and the* range *of f is the set*

$$\mathcal{R}(f) = \{f(x) \in Y : x \in X\}.$$

We write $f : X \to Y$ ("f maps X into Y") to indicate that f is a function from X to Y.

The reader should recognize the difference between the range of a function $f : X \to Y$ and the set Y (which is sometimes called the *codomain* of f). The set Y merely identifies the *type* of the output values $f(x)$; for example, if $Y = \mathbf{R}$, then every $f(x)$ is a real number. On the other hand, the range of f is the set of elements of Y that are actually attained by f. As a simple example, consider $f : \mathbf{R} \to \mathbf{R}$ defined by $f(x) = x^2$. The codomain of f is $\mathbf{R}$, but the range of f consists of the set of nonnegative numbers:

$$\mathcal{R}(f) = [0, \infty).$$

In many cases, it is quite difficult to determine the range of a function. The codomain, on the other hand, must be specified as part of the definition of the function.

A set is just a collection of objects (*elements*); most useful sets have operations defined on their elements. The most important sets used in this book are *vector spaces*.

Definition 3.2. *A vector space V is a set on which two operations are defined*, addition *(if $\mathbf{u}, \mathbf{v} \in V$, then $\mathbf{u} + \mathbf{v} \in V$) and* scalar multiplication *(if $\mathbf{u} \in V$ and α is a scalar, then $\alpha \mathbf{u} \in V$). The elements of the vector space are called* vectors. *(In this book, the scalars are usually real numbers, and we assume this unless otherwise stated. Occasionally we use the set of complex numbers as the scalars. Vectors will always be denoted by lower case boldface letters.)*

The two operations must satisfy the following algebraic properties:

1. $\mathbf{u} + \mathbf{v} = \mathbf{v} + \mathbf{u}$ *for all $\mathbf{u}, \mathbf{v} \in V$.*

2. $(\mathbf{u} + \mathbf{v}) + \mathbf{w} = \mathbf{u} + (\mathbf{v} + \mathbf{w})$ *for all $\mathbf{u}, \mathbf{v}, \mathbf{w} \in V$.*

3. *There is a* zero *vector $\mathbf{0}$ in V with the property that $\mathbf{u} + \mathbf{0} = \mathbf{u}$ for all $\mathbf{u} \in V$.*

4. *For each $\mathbf{u} \in V$, there is a vector $-\mathbf{u} \in V$ such that $\mathbf{u} + (-\mathbf{u}) = \mathbf{0}$.*

5. $\alpha(\mathbf{u} + \mathbf{v}) = \alpha \mathbf{u} + \alpha \mathbf{v}$ *for all $\mathbf{u}, \mathbf{v} \in V$ and for all scalars α.*

6. $(\alpha + \beta)\mathbf{u} = \alpha \mathbf{u} + \beta \mathbf{u}$ *for all $\mathbf{u} \in V$ and for all scalars α, β.*

7. $\alpha(\beta \mathbf{u}) = (\alpha \beta) \mathbf{u}$ *for all $\mathbf{u} \in V$ and for all scalars α, β.*

8. $1\mathbf{u} = \mathbf{u}$ *for all $\mathbf{u} \in V$.*

For every vector space considered in this book, the verification of these vector space properties is straightforward and will be taken for granted.

Example 3.3. The most common example of a vector space is (real) *Euclidean n-space*:

$$\mathbf{R}^n = \{(u_1, u_2, \ldots, u_n) : u_i \in \mathbf{R}, i = 1, 2, \ldots, n\}.$$

Vectors in $\mathbf{R}^n$ are usually written in column form,

$$\mathbf{u} = \begin{bmatrix} u_1 \\ u_2 \\ \vdots \\ u_n \end{bmatrix},$$

3.1. Linear systems as linear operator equations

as it is convenient at times to think of $\mathbf{u} \in \mathbf{R}^n$ as an $n \times 1$ matrix. Addition and scalar multiplication are defined componentwise:

$$\mathbf{u} + \mathbf{v} = (u_1, u_2, \ldots, u_n) + (v_1, v_2, \ldots, v_n) = (u_1 + v_1, u_2 + v_2, \ldots, u_n + v_n),$$
$$\alpha \mathbf{u} = \alpha(u_1, u_2, \ldots, u_n) = (\alpha u_1, \alpha u_2, \ldots, \alpha u_n). \quad \blacksquare$$

Example 3.4. Apart from Euclidean n-space, the most common vector spaces are function spaces—vector spaces in which the vectors are functions. Functions (with common domains) can be added together and multiplied by scalars, and the algebraic properties of a vector space are easily verified. Therefore, when defining a *function space*, one must only check that any desired properties of the functions are preserved by addition and scalar multiplication. Here are some important examples.

1. $C[a,b]$ is defined to be the set of all continuous, real-valued functions defined on the interval $[a,b]$. The sum of two continuous functions is also continuous, as is any scalar multiple of a continuous function. Therefore, $C[a,b]$ is a vector space.

2. $C^1[a,b]$ is defined to be the set of all real-valued, continuously differentiable functions defined on the interval $[a,b]$. (A function is continuously differentiable if its derivative exists and is continuous.) The sum of two continuously differentiable functions is also continuously differentiable, and the same is true for a scalar multiple of a continuously differentiable function. Therefore, $C^1[a,b]$ is a vector space.

3. For any positive integer k, $C^k[a,b]$ is the space of real-valued functions defined on $[a,b]$ that have k continuous derivatives. $\quad \blacksquare$

Many vector spaces that are encountered in practice are *subspaces* of other vector spaces.

Definition 3.5. *Let V be a vector space, and suppose W is a subset of V with the following properties:*

1. *The zero vector belongs to W.*

2. *Every linear combination of vectors in W is also in W. That is, if* $\mathbf{x}, \mathbf{y} \in W$ *and* $\alpha, \beta \in \mathbf{R}$, *then*
$$\alpha \mathbf{x} + \beta \mathbf{y} \in W.$$

Then we call W a subspace *of V.*

A subspace of a vector space is a vector space in its own right, as the reader can verify by checking that all the properties of a vector space are satisfied for a subspace.

Example 3.6. We define
$$C_D^2[a,b] = \{u \in C^2[a,b] : u(a) = u(b) = 0\}.$$

The set $C_D^2[a,b]$ is a subset of $C^2[a,b]$, and this subset contains the zero function (hopefully this is obvious to the reader). Also, if $u, v \in C_D^2[a,b]$, $\alpha, \beta \in \mathbf{R}$, and $w = \alpha u + \beta v$, then

- $w \in C^2[a,b]$ (since $C^2[a,b]$ is a vector space); and
- $w(a) = \alpha u(a) + \beta v(a) = \alpha \cdot 0 + \beta \cdot 0 = 0$, and similarly $w(b) = 0$.

Therefore, $w \in C_D^2[a,b]$, which shows that $C_D^2[a,b]$ is a subspace of $C^2[a,b]$. ∎

Example 3.7. We define

$$C_N^2[a,b] = \left\{ u \in C^2[a,b] : \frac{du}{dx}(a) = \frac{du}{dx}(b) = 0 \right\}.$$

The set $C_N^2[a,b]$ is also a subset of $C^2[a,b]$, and it can be shown to be a subspace. Clearly the zero function belongs to $C_N^2[a,b]$. If $u,v \in C_N^2[a,b]$, $\alpha, \beta \in \mathbf{R}$, and $w = \alpha u + \beta v$, then

$$\frac{dw}{dx}(a) = \alpha \frac{du}{dx}(a) + \beta \frac{dw}{dx}(a) = \alpha \cdot 0 + \beta \cdot 0 = 0.$$

Similarly,
$$\frac{dw}{dx}(b) = 0,$$

and so $w \in C_N^2[a,b]$. This shows that $C_N^2[a,b]$ is a subspace of $C^2[a,b]$. ∎

The previous two examples will be used throughout this book. The letters "D" and "N" stand for Dirichlet and Neumann, respectively (see Section 2.1, for example).

The following provides an important *nonexample* of a subspace.

Example 3.8. We define

$$W = \left\{ u \in C^2[a,b] : u(a) = \gamma,\ u(b) = \delta \right\},$$

where γ and δ are *nonzero* real numbers. Then, although W is a subset of $C^2[a,b]$, it is not a subspace. For example, the zero function does not belong to W, since it does not satisfy the boundary conditions. Also, if $u, v \in W$ and $\alpha, \beta \in \mathbf{R}$, then, with $w = \alpha u + \beta v$, we have

$$w(a) = \alpha u(a) + \beta v(a) = \alpha \gamma + \beta \gamma = (\alpha + \beta)\gamma.$$

Thus $w(a)$ does not equal γ, except in the special case that $\alpha + \beta = 1$. Similarly, $w(b)$ does not satisfy the boundary condition at the right endpoint. ∎

The concept of a vector space allows us to define linearity, which describes many simple processes and is indispensable in modeling and analysis.

Definition 3.9. *Suppose X and Y are vector spaces, and $f : X \to Y$ is an operator with domain X and range Y. Then f is* linear *if and only if*

$$f(\alpha \mathbf{x} + \beta \mathbf{z}) = \alpha f(\mathbf{x}) + \beta f(\mathbf{z}) \text{ for all } \alpha, \beta \in \mathbf{R},\ \mathbf{x}, \mathbf{z} \in X. \tag{3.1}$$

This condition can be expressed as the following two conditions, which together are equivalent to (3.1):

1. $f(\alpha \mathbf{x}) = \alpha f(\mathbf{x})$ *for all* $\mathbf{x} \in X$ *and all* $\alpha \in \mathbf{R}$;
2. $f(\mathbf{x} + \mathbf{z}) = f(\mathbf{x}) + f(\mathbf{z})$ *for all* $\mathbf{x}, \mathbf{z} \in X$.

3.1. Linear systems as linear operator equations

A linear operator is thus a particularly simple kind of operator; its simplicity can be appreciated by comparing the property of linearity with common nonlinear operators; for example, a linear operator f must satisfy $f(x+y) = f(x) + f(y)$, but

$$\sqrt{x+y} \neq \sqrt{x} + \sqrt{y}, \ \sin(x+y) \neq \sin(x) + \sin(y), \ \text{etc.}$$

Example 3.10. The operator defined by a matrix $\mathbf{A} \in \mathbf{R}^{m \times n}$ via matrix-vector multiplication,

$$f(\mathbf{x}) = \mathbf{A}\mathbf{x},$$

is linear; the reader should verify this if necessary (see Exercise 3.1.13). Moreover, it can be shown that every linear operator mapping $\mathbf{R}^n$ into $\mathbf{R}^m$ can be represented by a matrix $\mathbf{A} \in \mathbf{R}^{m \times n}$ in this way (see Exercise 3.1.14). This explains why the study of (finite-dimensional) linear algebra is largely the study of matrices. In this book, matrices will be denoted by upper case boldface letters. ∎

Example 3.11. To show that the sine function is not linear, we observe that

$$\sin\left(2\frac{\pi}{2}\right) = \sin(\pi) = 0,$$

while

$$2\sin\left(\frac{\pi}{2}\right) = 2 \cdot 1 = 2. \quad \blacksquare$$

Example 3.12. Differentiation defines an operator

$$\frac{d}{dx} : C^1[a,b] \to C[a,b],$$

and this operator is well known to be linear. For example,

$$\frac{d}{dx}\left[2\sin(x) - 3e^x\right] = 2\cos(x) - 3e^x,$$

since

$$\frac{d}{dx}[\sin(x)] = \cos(x), \ \frac{d}{dx}\left[e^x\right] = e^x.$$

In general, the kth derivative operator defines a linear operator mapping $C^k[a,b]$ into $C[a,b]$. This is why linearity is so important in the study of differential equations. ∎

Since a matrix $\mathbf{A} \in \mathbf{R}^{n \times n}$ defines a linear operator from $\mathbf{R}^n$ to $\mathbf{R}^n$, the linear system $\mathbf{A}\mathbf{x} = \mathbf{b}$ can be regarded as a *linear operator equation*. From this point of view, the questions posed by the system are, Is there a vector $\mathbf{x} \in \mathbf{R}^n$ whose image under $\mathbf{A}$ is the given vector $\mathbf{b}$? If so, is there only one such vector $\mathbf{x}$? In the next section, we will explore these questions.

The point of view of a linear operator equation is also useful in discussing differential equations. For example, consider the steady-state heat flow problem

$$-\kappa \frac{\partial^2 u}{\partial x^2} = f(x),\ 0 < x < \ell,$$
$$u(0) = 0,$$
$$u(\ell) = 0$$

(3.2)

from Section 2.1. We define a differential operator $L_D : C_D^2[0,\ell] \to C[0,\ell]$ by

$$L_D u = -\kappa \frac{d^2 u}{dx^2}.$$

Then the BVP (3.2) is equivalent to the operator equation

$$L_D u = f$$

(the reader should notice how the Dirichlet boundary conditions are enforced by the definition of the domain of L_D). This and similar examples will be discussed throughout this chapter and in detail in Section 5.1.

Exercises

1. In elementary algebra and calculus courses, it is often said that $f : \mathbf{R} \to \mathbf{R}$ is linear if and only if it has the form $f(x) = ax + b$, where a and b are constants. Does this agree with Definition 3.9? If not, what is the form of a linear function $f : \mathbf{R} \to \mathbf{R}$?

2. Show explicitly that $f : [0,\infty) \to \mathbf{R}$ defined by $f(x) = \sqrt{x}$ is not linear.

3. Let $f : \mathbf{R}^2 \to \mathbf{R}^2$ be defined by

$$f(\mathbf{x}) = \begin{bmatrix} x_1^2 + x_2^2 \\ x_2 - x_1^2 \end{bmatrix}.$$

   Prove that f is not linear.

4. Define $f : \mathbf{R}^2 \to \mathbf{R}$ by $f(\mathbf{x}) = x_1 x_2$. Is f linear or nonlinear?

5. For each of the following sets of functions, determine whether or not it is a vector space. (Define addition and scalar multiplication in the obvious way.) If it is not, state what property of a vector space fails to hold.

   (a) $\{f \in C[0,1] : f(0) = 0\}$.
   (b) $\{f \in C[0,1] : f(0) = 1\}$.
   (c) $\{f \in C[0,1] : \int_0^1 f(x)\,dx = 0\}$.
   (d) $\mathcal{P}_n$, the set of all polynomials of degree n or less.
   (e) The set of all polynomials of degree exactly n.

6. Prove that the differential operator $L : C^1[a,b] \to C[a,b]$ defined by
$$Lu = u\frac{du}{dx}$$
is not a linear operator.

7. Prove that the differential operator $L : C^1[a,b] \to C[a,b]$ defined by
$$Lu = \frac{du}{dx} + u^3$$
is not a linear operator.

8. Prove that the differential operator $M : C^2[a,b] \to C[a,b]$ defined by
$$Mu = \frac{d^2u}{dx^2} - 2\frac{du}{dx} + 3u$$
is a linear operator.

9. Define $K : C^2[a,b] \to C[a,b]$ by
$$Ku = x^2\frac{d^2u}{dx^2} - 2x\frac{du}{dx} + 3u.$$
Is K linear or nonlinear?

10. Let X and Y be vector spaces, and let $f : X \to Y$ be linear. Prove that $f(0) = 0$. (Notice that both X and Y contain a zero vector; in the equation $f(0) = 0$, the 0 on the left is the zero vector in X, while the 0 on the right is the zero vector in Y.)

11. Let ρ, c, and κ be constants, and define $L : C^2(\mathbf{R}^2) \to C(\mathbf{R}^2)$ by
$$Lu = \rho c \frac{\partial u}{\partial t} - \kappa \frac{\partial^2 u}{\partial x^2}.$$
Prove that L is a linear operator.

12. Define $K : C^2(\mathbf{R}^2) \to C(\mathbf{R}^2)$ by
$$Ku = -\frac{\partial}{\partial x}\left(u\frac{\partial u}{\partial x}\right) - \frac{\partial}{\partial y}\left(u\frac{\partial u}{\partial y}\right).$$
Is K linear or nonlinear?

13. (a) Let $\mathbf{A} \in \mathbf{R}^{2\times 2}$. Prove that
$$\mathbf{A}(\alpha\mathbf{x} + \beta\mathbf{y}) = \alpha\mathbf{A}\mathbf{x} + \beta\mathbf{A}\mathbf{y} \text{ for all } \alpha, \beta \in \mathbf{R},\ \mathbf{x},\mathbf{y} \in \mathbf{R}^2.$$
(Write
$$\mathbf{A} = \begin{bmatrix} a_{11} & a_{12} \\ a_{21} & a_{22} \end{bmatrix},\ \mathbf{x} = \begin{bmatrix} x_1 \\ x_2 \end{bmatrix},\ \mathbf{y} = \begin{bmatrix} y_1 \\ y_2 \end{bmatrix}$$
and write out $\mathbf{A}(\alpha\mathbf{x} + \beta\mathbf{y})$ explicitly.)

(b) Now repeat part 13(a) for $\mathbf{A} \in \mathbf{R}^{n \times n}$. The proof is not difficult when one uses summation notation. For example, if the (i,j)-entry of $\mathbf{A}$ is denoted a_{ij}, then

$$(\mathbf{Ax})_i = \sum_{j=1}^{n} a_{ij} x_j.$$

Write $(\mathbf{A}(\alpha \mathbf{x} + \beta \mathbf{y}))_i$ and $(\alpha \mathbf{Ax} + \beta \mathbf{Ay})_i$ in summation notation, and show that the first can be rewritten as the second using the elementary properties of arithmetic.

14. (a) Suppose $f : \mathbf{R}^2 \to \mathbf{R}^2$ is linear. Prove that there is a matrix $\mathbf{A} \in \mathbf{R}^{2 \times 2}$ such that f is given by $f(\mathbf{x}) = \mathbf{Ax}$. (Hint: Each $\mathbf{x} \in \mathbf{R}^2$ can be written as $\mathbf{x} = x_1 \mathbf{e}_1 + x_2 \mathbf{e}_2$, where

$$\mathbf{e}_1 = \begin{bmatrix} 1 \\ 0 \end{bmatrix}, \mathbf{e}_2 = \begin{bmatrix} 0 \\ 1 \end{bmatrix}.$$

Since f is linear, we have $f(\mathbf{x}) = x_1 f(\mathbf{e}_1) + x_2 f(\mathbf{e}_2)$. The desired matrix $\mathbf{A}$ can be expressed in terms of the vectors $f(\mathbf{e}_1), f(\mathbf{e}_2)$.)

(b) Now show that if $f : \mathbf{R}^n \to \mathbf{R}^m$ is linear, then there exists a matrix $\mathbf{A} \in \mathbf{R}^{m \times n}$ such that $f(\mathbf{x}) = \mathbf{Ax}$ for all $\mathbf{x} \in \mathbf{R}^n$.

3.2 Existence and uniqueness of solutions to Ax = b

We will now discuss the linear system $\mathbf{Ax} = \mathbf{b}$, where $\mathbf{A} \in \mathbf{R}^{n \times n}$ and $\mathbf{b} \in \mathbf{R}^n$, as a linear operator equation. We consider three fundamental questions.

1. Does a solution to the equation exist?
2. If a solution exists, is it unique?
3. If a unique solution exists, how can we compute it?

It turns out that the first two questions are intimately linked; the purpose of this section is to shed some light on these two questions and the connection between them. We will also briefly discuss how this point of view can be carried over to the case of a linear differential equation. The third question will be deferred to later in this chapter.

3.2.1 Existence

The existence of a solution to $\mathbf{Ax} = \mathbf{b}$ is equivalent to the condition that $\mathbf{b}$ lie in $\mathcal{R}(\mathbf{A})$, the range of $\mathbf{A}$. This begs the question, What sort of a set is $\mathcal{R}(\mathbf{A})$?

If $\mathbf{y}, \mathbf{w} \in \mathcal{R}(\mathbf{A})$, say $\mathbf{y} = \mathbf{Ax}, \mathbf{w} = \mathbf{Az}$, then

$$\alpha \mathbf{y} + \beta \mathbf{w} = \alpha \mathbf{Ax} + \beta \mathbf{Az}$$
$$= \mathbf{A}(\alpha \mathbf{x} + \beta \mathbf{z}).$$

This shows that $\alpha \mathbf{y} + \beta \mathbf{w} \in \mathcal{R}(\mathbf{A})$. Moreover, the zero vector lies in $\mathcal{R}(\mathbf{A})$, since $\mathbf{A0} = \mathbf{0}$. It follows that $\mathcal{R}(\mathbf{A})$ is a subspace of $\mathbf{R}^n$ (possibly the entire space $\mathbf{R}^n$—every vector space

3.2. Existence and uniqueness of solutions to $\mathbf{Ax} = \mathbf{b}$

is a subspace of itself). Every linear operator has this property; if $\mathbf{f}: X \to Y$ is a linear operator, then $\mathcal{R}(\mathbf{f})$ is a subspace of the vector space Y. (The same need not be true for a nonlinear operator.)

The geometry of subspaces of $\mathbf{R}^n$ is particularly simple: the proper subspaces of $\mathbf{R}^n$ (i.e., those that are not the entire space) are lower-dimensional spaces: lines in $\mathbf{R}^2$, lines and planes in $\mathbf{R}^3$, and so forth (we cannot visualize these objects in dimensions greater than three, but we can understand them by analogy). Since every subspace must contain the zero vector, not every line in $\mathbf{R}^2$, for example, is a subspace, but only those passing through the origin are.

With this understanding of the geometry of $\mathbf{R}^n$, we obtain the following conclusions.

- If $\mathcal{R}(\mathbf{A}) = \mathbf{R}^n$, then $\mathbf{Ax} = \mathbf{b}$ has a solution for each $\mathbf{b} \in \mathbf{R}^n$. (This is a tautology.)
- If $\mathcal{R}(\mathbf{A}) \neq \mathbf{R}^n$, then $\mathbf{Ax} = \mathbf{b}$ fails to have a solution for almost every $\mathbf{b} \in \mathbf{R}^n$. This is because a lower-dimensional subspace comprises very little of $\mathbf{R}^n$ (think of a line contained in the plane or in three-dimensional space).

Example 3.13. We consider the equation $\mathbf{Ax} = \mathbf{b}$, where $\mathbf{A} \in \mathbf{R}^{2 \times 2}$ is given by

$$\mathbf{A} = \begin{bmatrix} 1 & 2 \\ 2 & 4 \end{bmatrix}.$$

For any $\mathbf{x} \in \mathbf{R}^2$, we have

$$\mathbf{Ax} = \begin{bmatrix} 1 & 2 \\ 2 & 4 \end{bmatrix} \begin{bmatrix} x_1 \\ x_2 \end{bmatrix} = \begin{bmatrix} x_1 + 2x_2 \\ 2x_1 + 4x_2 \end{bmatrix} = (x_1 + 2x_2) \begin{bmatrix} 1 \\ 2 \end{bmatrix}.$$

This calculation shows that every vector $\mathbf{b}$ in the range of $\mathbf{A}$ is a multiple of the vector $(1, 2)$. Therefore, the subspace $\mathcal{R}(\mathbf{A})$ is a line in the plane $\mathbf{R}^2$ (see Figure 3.1). Since this line is a very small part of $\mathbf{R}^2$, the system $\mathbf{Ax} = \mathbf{b}$ fails to have a solution for almost every $\mathbf{b} \in \mathbf{R}^2$. ∎

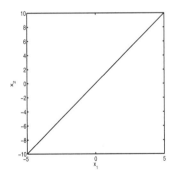

Figure 3.1. *The range of the matrix* $\mathbf{A}$ *in Example* 3.13.

As mentioned at the beginning of this chapter, there is a close analogy between linear (algebraic) systems and linear differential equations. The reader should think carefully about the similarities between the following example and the previous one.

Example 3.14. We define the linear differential operator $L_N : C_N^2[0,\ell] \to C[0,\ell]$ by

$$L_N u = -\frac{d^2 u}{dx^2},$$

where

$$C_N^2[0,\ell] = \left\{ u \in C^2[0,\ell] : \frac{du}{dx}(0) = \frac{du}{dx}(\ell) = 0 \right\}$$

(as defined in the previous section). If $f \in \mathcal{R}(L_N)$, then there exists $u \in C_N^2[0,\ell]$ such that $L_N u = f$. It follows that

$$\int_0^\ell f(x)\,dx = -\int_0^\ell \frac{d^2 u}{dx^2}(x)\,dx = -\left[\frac{du}{dx}(x)\right]_0^\ell = -\frac{du}{dx}(\ell) + \frac{du}{dx}(0) = 0.$$

This shows that $f \in C[0,\ell]$ cannot belong to the range of L_N unless it satisfies the special condition

$$\int_0^\ell f(x)\,dx = 0. \tag{3.3}$$

In fact, $\mathcal{R}(L_N)$ is the set of all such f, as the reader is asked to show in Exercise 3.1.12.

Because the space $C[0,\ell]$ is infinite dimensional, we cannot visualize this situation (as we could in the previous example). However, the reader should appreciate that most functions $f \in C[0,\ell]$ do not satisfy condition (3.3). (For example, the reader is invited to write down a quadratic polynomial at random and call it $f(x)$. Chances are that this quadratic will not satisfy (3.3).) Therefore, the range of L_N is only a small part of $C[0,\ell]$, and for most choices of $f \in C[0,\ell]$, there is no solution to $L_N u = f$. ■

Example 3.15. We now define $L_D : C_D^2[0,\ell] \to C[0,\ell]$ by

$$L_D u = -\frac{d^2 u}{dx^2},$$

where

$$C_D^2[0,\ell] = \left\{ u \in C^2[0,\ell] : u(0) = u(\ell) = 0 \right\}.$$

The reader should recall from the previous section that the BVP

$$-\frac{d^2 u}{dx^2} = f(x),\ 0 < x < \ell,$$
$$u(0) = 0,$$
$$u(\ell) = 0$$

can be written as the linear operator equation $L_D u = f$. We will show that $\mathcal{R}(A)$ is all of $C[0,\ell]$ by showing that we can solve $L_D u = f$ for every $f \in C[0,\ell]$. The idea is to integrate twice and use the boundary conditions to determine the constants of integration. We have

$$\frac{d^2 u}{dx^2}(x) = -f(x),\ 0 < x < \ell,$$

3.2. Existence and uniqueness of solutions to $\mathbf{Ax}=\mathbf{b}$

so
$$\frac{du}{dx}(x) = -\int_0^x f(s)\,ds + C_1, \; 0 < x < \ell.$$

We then integrate again to obtain
$$u(x) = -\int_0^x \int_0^z f(s)\,ds\,dz + C_1 x + C_2, \; 0 < x < \ell.$$

The reader should notice the use of the dummy variables of integration s and z.

The first boundary condition, $u(0) = 0$, implies that $C_2 = 0$. We then have
$$u(\ell) = -\int_0^\ell \int_0^z f(s)\,ds\,dz + C_1 \ell;$$

since $u(\ell) = 0$, we obtain
$$C_1 = \frac{1}{\ell} \int_0^\ell \int_0^z f(s)\,ds\,dz$$

and so
$$u(x) = -\int_0^x \int_0^z f(s)\,ds\,dz + \frac{x}{\ell} \int_0^\ell \int_0^z f(s)\,ds\,dz, \; 0 < x < \ell. \tag{3.4}$$

The reader can verify directly (by differentiating twice) that this formula defines a solution of the BVP represented by $L_D u = f$. This shows that we can solve $L_D u = f$ for any $f \in C[0,\ell]$, and so $\mathcal{R}(L_D) = C[0,\ell]$. ∎

3.2.2 Uniqueness

The linearity of a matrix operator $\mathbf{A}$ implies that nonuniqueness of solutions to $\mathbf{Ax} = \mathbf{b}$, if it occurs, has a special structure. Suppose $\mathbf{x}$ and $\mathbf{z}$ in $\mathbf{R}^n$ are both solutions to $\mathbf{Ax} = \mathbf{b}$ (i.e., $\mathbf{Ax} = \mathbf{b}$ and $\mathbf{Az} = \mathbf{b}$ both hold). Then
$$\mathbf{Ax} = \mathbf{Az} \Rightarrow \mathbf{Ax} - \mathbf{Az} = \mathbf{0} \Rightarrow \mathbf{A}(\mathbf{x}-\mathbf{z}) = \mathbf{0},$$

where the last step follows from the linearity of $\mathbf{A}$. If $\mathbf{x} \neq \mathbf{z}$, then $\mathbf{w} = \mathbf{x} - \mathbf{z}$ is a nonzero vector satisfying $\mathbf{Aw} = \mathbf{0}$.

On the other hand, suppose $\mathbf{x}$ is a solution to $\mathbf{Ax} = \mathbf{b}$ and $\mathbf{w}$ is a nonzero vector satisfying $\mathbf{Aw} = \mathbf{0}$. Then
$$A(\mathbf{x}+\mathbf{w}) = \mathbf{Ax} + \mathbf{Aw} = \mathbf{b} + \mathbf{0} = \mathbf{b},$$

and in this case there cannot be a unique solution to $\mathbf{Ax} = \mathbf{b}$.

Because of the above observations, we define the *null space* of $\mathbf{A}$ to be
$$\mathcal{N}(\mathbf{A}) = \{\mathbf{x} \in \mathbf{R}^n \,:\, \mathbf{Ax} = \mathbf{0}\}.$$

Since $\mathbf{A0} = \mathbf{0}$ always holds for a linear operator $\mathbf{A}$, we always have $\mathbf{0} \in \mathcal{N}(\mathbf{A})$. Moreover, if $\mathbf{x}, \mathbf{z} \in \mathcal{N}(\mathbf{A})$ and $\alpha, \beta \in \mathbf{R}$, then
$$\mathbf{Ax} = \mathbf{0}, \; \mathbf{Az} = \mathbf{0}$$

and so
$$\mathbf{A}(\alpha \mathbf{x} + \beta \mathbf{z}) = \alpha \mathbf{A}\mathbf{x} + \beta \mathbf{A}\mathbf{z} = \alpha \cdot \mathbf{0} + \beta \cdot \mathbf{0} = \mathbf{0}.$$

Therefore, $\alpha \mathbf{x} + \beta \mathbf{z} \in \mathcal{N}(\mathbf{A})$. This shows that $\mathcal{N}(\mathbf{A})$ is a subspace of $\mathbf{R}^n$. If $\mathbf{0}$ is the only vector in $\mathcal{N}(\mathbf{A})$, we say that $\mathcal{N}(\mathbf{A})$ is *trivial*.

Our observations above lead to the following conclusion: If $\mathbf{A}\mathbf{x} = \mathbf{b}$ has a solution, it is unique if and only if $\mathcal{N}(\mathbf{A})$ is trivial. Furthermore, nothing in the above discussion depends on $\mathbf{A}$'s being a matrix operator; the same arguments can be made for any linear operator, such as a differential operator.

If $\mathcal{N}(\mathbf{A})$ is nontrivial and $\mathbf{A}\mathbf{x} = \mathbf{b}$ has a solution, then the equation has in fact infinitely many solutions. To see this, suppose $\mathbf{x} \in \mathbf{R}^n$ satisfies $\mathbf{A}\mathbf{x} = \mathbf{b}$ and $\mathbf{w} \in \mathcal{N}(\mathbf{A})$, $\mathbf{w} \neq \mathbf{0}$. Then, for each $\alpha \in \mathbf{R}$, we have

$$A(\mathbf{x} + \alpha \mathbf{w}) = \mathbf{A}\mathbf{x} + \alpha \mathbf{A}\mathbf{w} = \mathbf{A}\mathbf{x} + \alpha \mathbf{0} = \mathbf{A}\mathbf{x} = \mathbf{b}.$$

Since $\mathbf{x} + \alpha \mathbf{w}$ is different for each different choice of the real number α, this shows that the equation has infinitely many solutions. Moreover, it easily follows that the set of all solutions to $\mathbf{A}\mathbf{x} = \mathbf{b}$ is, in this case,

$$\mathbf{x} + \mathcal{N}(\mathbf{A}) = \{\mathbf{x} + \mathbf{w} \, : \, \mathbf{w} \in \mathcal{N}(\mathbf{A})\}.$$

Once again, the same properties hold for any linear operator equation.

Example 3.16. Let $\mathbf{A} \in \mathbf{R}^{4 \times 4}$ be defined by

$$\mathbf{A} = \begin{bmatrix} 1 & 3 & -1 & 2 \\ 0 & 1 & 4 & 2 \\ 2 & 7 & 2 & 6 \\ 1 & 4 & 3 & 4 \end{bmatrix}.$$

Consider the equation $\mathbf{A}\mathbf{x} = \mathbf{b}$, where $\mathbf{b} \in \mathbf{R}^4$ is arbitrary. Using the standard elimination algorithm,[6] the system $\mathbf{A}\mathbf{x} = \mathbf{b}$ can be shown to be equivalent to the system

$$\begin{aligned} x_1 \quad - \quad 13x_3 \quad - \quad 4x_4 &= b_1 - 3b_2, \\ x_2 \quad + \quad 4x_3 \quad + \quad 2x_4 &= b_2, \\ 0 &= b_3 - b_2 - 2b_1, \\ 0 &= b_4 - b_2 - b_1. \end{aligned}$$

We see that the system is inconsistent unless the conditions

$$\begin{aligned} b_3 - b_2 - 2b_1 &= 0, \\ b_4 - b_2 - b_1 &= 0 \end{aligned} \tag{3.5}$$

hold. If these conditions are satisfied by $\mathbf{b}$, then

$$\begin{aligned} x_1 &= b_1 - 3b_2 + 13x_3 + 4x_4, \\ x_2 &= b_2 - 4x_3 - 2x_4, \end{aligned}$$

---

[6]We assume that the reader is familiar with Gaussian elimination, the standard row reduction algorithm for solving $\mathbf{A}\mathbf{x} = \mathbf{b}$. For a review, see any introductory text on linear algebra, such as the text by Lay [41].

3.2. Existence and uniqueness of solutions to $\mathbf{Ax} = \mathbf{b}$

where x_3 and x_4 can take on any value. Setting $x_3 = s$ and $x_4 = t$, every vector of the form

$$\mathbf{x} = \begin{bmatrix} b_1 - 3b_2 \\ b_2 \\ 0 \\ 0 \end{bmatrix} + s \begin{bmatrix} 13 \\ 4 \\ 1 \\ 0 \end{bmatrix} + t \begin{bmatrix} 4 \\ -2 \\ 0 \\ 1 \end{bmatrix}$$

is a solution of the system. We have that

$$\mathbf{x} = \begin{bmatrix} b_1 - 3b_2 \\ b_2 \\ 0 \\ 0 \end{bmatrix}$$

is one solution of $\mathbf{Ax} = \mathbf{b}$, and

$$\mathcal{N}(\mathbf{A}) = \left\{ s \begin{bmatrix} 13 \\ 4 \\ 1 \\ 0 \end{bmatrix} + t \begin{bmatrix} 4 \\ -2 \\ 0 \\ 1 \end{bmatrix} : s, t \in \mathbf{R} \right\}. \quad \blacksquare$$

Example 3.17. We compute the null space of the operator L_N defined in Example 3.14. If $L_N u = 0$, then u satisfies the BVP

$$-\frac{d^2 u}{dx^2} = 0, \ 0 < x < \ell,$$

$$\frac{du}{dx}(0) = 0,$$

$$\frac{du}{dx}(\ell) = 0.$$

The differential equation implies that $u(x) = C_1 x + C_2$ for some constants C_1 and C_2. Each of the two boundary conditions leads to the conclusion that $C_1 = 0$; however, the constant C_2 can have any value, and both the differential equation and the two boundary conditions will be satisfied. This shows that every constant function $u(x) = C_2$ satisfies $L_N u = 0$. That is, the null space of L_N is the space of constant functions. $\blacksquare$

Since the null space of L_N is not trivial, we know from the above discussion that the operator equation $L_N u = f$ cannot have a unique solution—if there is one solution, there must in fact be infinitely many. However, since the null space is the set of constant functions, all solutions to $L_N u = f$ differ by a constant.

Example 3.18. The null space of the operator L_D defined in Example 3.15 is trivial. To see this, the reader should note that the differential equation

$$-\frac{d^2 u}{dx^2} = 0$$

again implies that $u(x) = C_1 x + C_2$, but now the boundary conditions ($u(0) = 0$, $u(\ell) = 0$) force $C_1 = C_2 = 0$. Therefore, the only solution of $L_D u = 0$ is the zero function. $\blacksquare$

Since the null space of L_D is trivial, we see that $L_D u = f$ has at most one solution for any right-hand side f.

3.2.3 The Fredholm alternative

One of the fundamental results of linear algebra is that, in a certain sense, existence and uniqueness are equivalent for a square system. To be precise, if $\mathbf{A} \in \mathbf{R}^{n \times n}$, and $\mathbf{Ax} = \mathbf{b}$ has a solution for each $\mathbf{b} \in \mathbf{R}^n$ (i.e., if $\mathcal{R}(\mathbf{A}) = \mathbf{R}^n$), then the solution is unique for each $\mathbf{b} \in \mathbf{R}^n$ (i.e., $\mathcal{N}(\mathbf{A})$ is trivial). On the other hand, if the solution to $\mathbf{Ax} = \mathbf{b}$, whenever it exists, is unique (i.e., if $\mathcal{N}(\mathbf{A})$ is trivial), then $\mathbf{Ax} = \mathbf{b}$ has a solution for each $\mathbf{b} \in \mathbf{R}^n$ (i.e., $\mathcal{R}(\mathbf{A}) = \mathbf{R}^n$).

Moreover, in the case that $\mathcal{N}(\mathbf{A})$ is not trivial, we can give a condition that $\mathbf{b} \in \mathbf{R}^n$ must satisfy in order for $\mathbf{Ax} = \mathbf{b}$ to have a solution. We collect these facts in the following theorem.

Theorem 3.19 (The Fredholm alternative). *Suppose* $\mathbf{A} \in \mathbf{R}^{n \times n}$. *Then exactly one of the following is true:*

1. *The null space of* $\mathbf{A}$ *is trivial, and for each* $\mathbf{b} \in \mathbf{R}^n$ *there exists a unique solution* $\mathbf{x} \in \mathbf{R}^n$ *to* $\mathbf{Ax} = \mathbf{b}$.

2. *The null space of* $\mathbf{A}$ *is nontrivial, and the equation* $\mathbf{Ax} = \mathbf{b}$ *has a solution if and only if* $\mathbf{b}$ *satisfies the following condition:*

$$\mathbf{w} \in \mathbf{R}^n, \ \mathbf{A}^T \mathbf{w} = 0 \Rightarrow \mathbf{w} \cdot \mathbf{b} = 0, \tag{3.6}$$

or, equivalently,

$$\mathbf{w} \in \mathcal{N}(\mathbf{A}^T) \Rightarrow \mathbf{w} \cdot \mathbf{b} = 0. \tag{3.7}$$

If this condition is satisfied, then the equation $\mathbf{Ax} = \mathbf{b}$ *has infinitely many solutions.*

In the statement of the Fredholm alternative, we used the *transpose* of $\mathbf{A}$, $\mathbf{A}^T$, which is the matrix whose rows are the columns of $\mathbf{A}$ and vice versa. We also used the dot product, defined for two vectors in $\mathbf{R}^n$:

$$\mathbf{w} \cdot \mathbf{b} = \sum_{i=1}^{n} w_i b_i.$$

We will discuss the dot product and its significance in Section 3.4.

We will have more to say later about condition (3.6). For now it is sufficient to understand that, if $\mathcal{N}(\mathbf{A})$ is not trivial, then $\mathbf{Ax} = \mathbf{b}$ has a solution only if the right-hand side vector $\mathbf{b}$ satisfies a certain *compatibility condition*.

Example 3.20. In Example 3.16, $\mathcal{N}(A)$ is nontrivial, and we saw directly that $\mathbf{Ax} = \mathbf{b}$ has a solution if and only if $\mathbf{b}$ satisfies conditions (3.5). These conditions can be written as

$$\mathbf{w}_1 \cdot \mathbf{b} = 0,$$
$$\mathbf{w}_2 \cdot \mathbf{b} = 0,$$

where

$$\mathbf{w}_1 = \begin{bmatrix} -2 \\ -1 \\ 1 \\ 0 \end{bmatrix}, \ \mathbf{w}_2 = \begin{bmatrix} -1 \\ -1 \\ 0 \\ 1 \end{bmatrix}.$$

3.2. Existence and uniqueness of solutions to $\mathbf{Ax} = \mathbf{b}$

It is straightforward to show that

$$\mathcal{N}(\mathbf{A}^T) = \{\alpha \mathbf{w}_1 + \beta \mathbf{w}_2 \,:\, \alpha, \beta \in \mathbf{R}\}. \qquad \blacksquare \tag{3.8}$$

Example 3.21. In Examples 3.14 and 3.17, we showed that $\mathcal{N}(L_N)$ is nontrivial and that $L_N u = f$ has a solution if and only if the compatibility condition (3.3) is satisfied. Although the Fredholm alternative, as stated in Theorem 3.19, does not apply to this situation, an analogous statement can in fact be made. We will see how close the analogy is when we define a "dot product" for functions (see Sections 3.4.1 and 6.2.3). $\blacksquare$

If $\mathbf{A} \in \mathbf{R}^{n \times n}$, and $\mathcal{N}(\mathbf{A})$ is trivial, then $\mathbf{A}$ is called *nonsingular* (or *invertible*), and there exists a matrix $\mathbf{A}^{-1} \in \mathbf{R}^{n \times n}$ (the *inverse* of $\mathbf{A}$), with the property that

$$\mathbf{A}\mathbf{A}^{-1} = \mathbf{A}^{-1}\mathbf{A} = \mathbf{I} = \begin{bmatrix} 1 & 0 & \cdots & 0 \\ 0 & 1 & \cdots & 0 \\ \vdots & \vdots & \ddots & \vdots \\ 0 & 0 & \cdots & 1 \end{bmatrix}.$$

The matrix $\mathbf{I}$ is the $n \times n$ *identity matrix*; it has the property that $\mathbf{Ix} = \mathbf{x}$ for all $\mathbf{x} \in \mathbf{R}^n$.

If we know $\mathbf{A}^{-1}$, and we wish to solve $\mathbf{Ax} = \mathbf{b}$, we need only compute a matrix-vector product, as the following calculation shows:

$$\mathbf{Ax} = \mathbf{b}$$
$$\Rightarrow \mathbf{A}^{-1}\mathbf{Ax} = \mathbf{A}^{-1}\mathbf{b}$$
$$\Rightarrow \mathbf{Ix} = \mathbf{A}^{-1}\mathbf{b}$$
$$\Rightarrow \mathbf{x} = \mathbf{A}^{-1}\mathbf{b}.$$

Thus the solution of $\mathbf{Ax} = \mathbf{b}$ is $\mathbf{x} = \mathbf{A}^{-1}\mathbf{b}$.

Since $\mathbf{A}^{-1}$ defines a linear operator, we see that the solution $\mathbf{x}$ to $\mathbf{Ax} = \mathbf{b}$ depends linearly on the right-hand side $\mathbf{b}$. In fact, we can demonstrate this for any linear operator equation. We assume that X and Y are vector spaces and $\mathbf{f} : X \to Y$ is a linear operator with a trivial null space. Then, for each $\mathbf{y} \in \mathcal{R}(\mathbf{f})$, there is a unique solution $\mathbf{x} \in X$ to $\mathbf{f}(\mathbf{x}) = \mathbf{y}$. Let us define the *solution operator* $\mathbf{S} : \mathcal{R}(\mathbf{f}) \to X$ by the condition that $\mathbf{x} = \mathbf{Sy}$ is the solution to $\mathbf{f}(\mathbf{x}) = \mathbf{y}$. (If $\mathcal{R}(\mathbf{f}) = Y$, so that $\mathbf{f}$ is invertible, then $\mathbf{S}$ is nothing more than $\mathbf{f}^{-1}$.) We wish to show that $\mathbf{S}$ is linear.

We will show that if $\mathbf{y}, \mathbf{z} \in \mathcal{R}(\mathbf{f})$ and $\alpha, \beta \in \mathbf{R}$, then

$$\mathbf{S}(\alpha \mathbf{y} + \beta \mathbf{z}) = \alpha \mathbf{Sy} + \beta \mathbf{Sz}.$$

We define $\mathbf{x} = \mathbf{Sy}$ and $\mathbf{w} = \mathbf{Sz}$; then

$$\mathbf{f}(\mathbf{x}) = \mathbf{y}, \ \mathbf{f}(\mathbf{w}) = \mathbf{z}.$$

By the linearity of $\mathbf{f}$, we have

$$\mathbf{f}(\alpha \mathbf{x} + \beta \mathbf{w}) = \alpha \mathbf{f}(\mathbf{x}) + \beta \mathbf{f}(\mathbf{z}) = \alpha \mathbf{y} + \beta \mathbf{z}.$$

This last equation is equivalent to

$$\mathbf{S}(\alpha \mathbf{y} + \beta \mathbf{z}) = \alpha \mathbf{x} + \beta \mathbf{w} = \alpha \mathbf{Sy} + \beta \mathbf{Sz}.$$

Thus, the linearity of **f** implies that the solution operator is also necessarily linear. To put it another way, the solution to a linear operator equation depends linearly on the right-hand side of the equation. In the context of differential equations, this property is usually called the *principle of superposition*.

Example 3.22. The BVP

$$-\kappa \frac{d^2u}{dx^2} = f(x), \ 0 < x < \ell,$$
$$u(0) = 0,$$
$$u(\ell) = 0$$

can be written as the linear operator equation $L_D u = f$, as explained in Examples 3.15 and 3.18, and we showed in those examples that there is a unique solution u for each $f \in C[0,\ell]$. Since integration is a linear operation, formula (3.4) shows that u depends linearly on f. ∎

For many linear differential equations, we cannot find an explicit formula for the solution that makes this linear dependence obvious. Even in those cases, we know from the above argument that the solution depends linearly on the right-hand side.

Exercises

1. Let
$$\mathbf{A} = \begin{bmatrix} 1 & 2 \\ -1 & -2 \end{bmatrix}.$$
   Graph $\mathcal{R}(\mathbf{A})$ in the plane.

2. (a) Fill in the missing steps in Example 3.16.

   (b) Let **A** be the matrix in Example 3.16. Compute the solution set of the equation $\mathbf{A}^T \mathbf{w} = \mathbf{0}$, and show that the result is (3.8).

3. For each of the following matrices **A**, determine if $\mathbf{Ax} = \mathbf{b}$ has a unique solution for each **b**, that is, determine if **A** is nonsingular. For each matrix **A** which is singular, find a vector **b** such that $\mathbf{Ax} = \mathbf{b}$ has a solution and a vector **c** such that $\mathbf{Ax} = \mathbf{c}$ does not have a solution.

   (a) $\mathbf{A} = \begin{bmatrix} 1 & -1 \\ -2 & 4 \end{bmatrix}.$

   (b) $\mathbf{A} = \begin{bmatrix} 1 & 2 \\ 3 & 4 \end{bmatrix}.$

   (c) $\mathbf{A} = \begin{bmatrix} 1 & 1 \\ 1 & 1 \end{bmatrix}.$

4. For each of the following matrices **A**, determine if $\mathbf{Ax} = \mathbf{b}$ has a unique solution for each **b**, that is, determine if **A** is nonsingular. For each matrix **A** which is singular, find a vector **b** such that $\mathbf{Ax} = \mathbf{b}$ has a solution and a vector **c** such that $\mathbf{Ax} = \mathbf{c}$ does not have a solution.

3.2. Existence and uniqueness of solutions to $\mathbf{Ax} = \mathbf{b}$

(a) $\mathbf{A} = \begin{bmatrix} 1 & -1 & 2 \\ 0 & 2 & 2 \\ 1 & 1 & 4 \end{bmatrix}$.

(b) $\mathbf{A} = \begin{bmatrix} 1 & -1 & 2 \\ 0 & 2 & 2 \\ 1 & 1 & 3 \end{bmatrix}$.

(c) $\mathbf{A} = \begin{bmatrix} 1 & 2 & -1 \\ 2 & 4 & -2 \\ -11 & -2 & 1 \end{bmatrix}$.

5. Suppose $\mathbf{A} \in \mathbf{R}^{n \times n}$ and $\mathbf{b} \in \mathbf{R}^n$, $\mathbf{b} \neq 0$. Is the solution set of the equation $\mathbf{Ax} = \mathbf{b}$,

$$\{\mathbf{x} \in \mathbf{R}^n : \mathbf{Ax} = \mathbf{b}\},$$

a subspace of $\mathbf{R}^n$? Why or why not?

6. Let $D : C^1[a,b] \to C[a,b]$ be the derivative operator. What is the null space of D?

7. Let $L : C^2[a,b] \to C[a,b]$ be the second derivative operator. What is the null space of L?

8. As shown in Chapter 2, differential equations involving the second derivative operator are sometimes paired with mixed boundary conditions. Define the set

$$C_m^2[a,b] = \left\{ u \in C^2[a,b] : u(a) = \frac{du}{dx}(b) = 0 \right\}$$

and define $L_m : C_m^2[a,b] \to C[a,b]$ by

$$L_m u = -\frac{d^2 u}{dx^2}.$$

The reader should note that $C_m^2[a,b]$ is a subspace of $C^2[a,b]$.

(a) Determine the null space of L_m.

(b) Explain how to solve $L_m u = f$ for u, when f is a given function in $C[a,b]$. (Hint: Integrate twice.) Is this differential equation solvable for every $f \in C[a,b]$?

9. Repeat Exercise 8, but with

$$C_{\tilde{m}}^2[a,b] = \left\{ u \in C^2[a,b] : \frac{du}{dx}(a) = u(b) = 0 \right\}$$

and $L_{\tilde{m}} : C_{\tilde{m}}^2[a,b] \to C[a,b]$ defined by

$$L_{\tilde{m}} u = -\frac{d^2 u}{dx^2}.$$

10. Consider the following system of *nonlinear* equations:

$$x_1^2 + x_2^2 = 1,$$
$$x_2 - x_1^2 = 0.$$

This system can be written as $f(\mathbf{x}) = \mathbf{b}$, where $f : \mathbf{R}^2 \to \mathbf{R}^2$ is defined by

$$f(\mathbf{x}) = \begin{bmatrix} x_1^2 + x_2^2 \\ x_2 - x_1^2 \end{bmatrix}$$

and

$$\mathbf{b} = \begin{bmatrix} 1 \\ 0 \end{bmatrix}.$$

(a) Show that $f(\mathbf{x}) = \mathbf{b}$ has exactly two solutions. (Hint: A graph is useful.)

(b) Show that the only solution to $f(\mathbf{x}) = \mathbf{0}$ is $\mathbf{x} = \mathbf{0}$. (Yet, as part (a) shows, the solution to $f(\mathbf{x}) = \mathbf{b}$ is not unique.)

This example illustrates that the properties of linear systems do not necessarily carry over to nonlinear systems.

11. Let $D : C^1[a,b] \to C[a,b]$ be the differentiation operator:

$$Df = \frac{df}{dx}.$$

(a) Show that the range of D is all of $C[a,b]$.

(b) Part (a) is equivalent to saying that, for every $f \in C[a,b]$, the (differential) equation $Du = f$ has a solution. Is this solution unique? Why or why not?

12. Let L_N be defined as in Example 3.14, and suppose $f \in C[0,\ell]$ satisfies

$$\int_0^\ell f(x)\,dx = 0.$$

Show that $f \in \mathcal{R}(L_N)$.

3.3 Basis and dimension

The matrix-vector product $\mathbf{Ax}$ is equivalent to a linear combination of the columns of the matrix $\mathbf{A}$. If $\mathbf{A}$ has columns $\mathbf{v}_1, \mathbf{v}_2, \ldots, \mathbf{v}_n$, then

$$\mathbf{Ax} = x_1 \mathbf{v}_1 + x_2 \mathbf{v}_2 + \cdots + x_n \mathbf{v}_n.$$

The reader should write out a specific example if this is not clear (see Exercise 3.3.1). The quantities $x_1, x_2, \ldots, x_n$ are scalars and $\mathbf{v}_1, \mathbf{v}_2, \ldots, \mathbf{v}_n$ are vectors; an expression such as $x_1 \mathbf{v}_1 + x_2 \mathbf{v}_2 + \cdots + x_n \mathbf{v}_n$ is called a *linear combination* of the vectors $\mathbf{v}_1, \mathbf{v}_2, \ldots, \mathbf{v}_n$ because the vectors are combined using the linear operations of addition and scalar multiplication.

3.3. Basis and dimension

When $\mathbf{A} \in \mathbf{R}^{n \times n}$ is nonsingular, each $\mathbf{b} \in \mathbf{R}^n$ can be written in a unique way as a linear combination of the columns of $\mathbf{A}$ (that is, the equation $\mathbf{A}\mathbf{x} = \mathbf{b}$ has a unique solution). The following definition is related.

Definition 3.23. *Let V be a vector space, and suppose $\mathbf{v}_1, \mathbf{v}_2, \ldots, \mathbf{v}_n$ are vectors in V with the property that each $\mathbf{v} \in V$ can be written in a unique way as a linear combination of $\{\mathbf{v}_1, \mathbf{v}_2, \ldots, \mathbf{v}_n\}$. Then $\{\mathbf{v}_1, \mathbf{v}_2, \ldots, \mathbf{v}_n\}$ is called a* basis *of V. Moreover, we say that n is the* dimension *of V.*

A vector space can have many different bases, but it can be shown that each contains the same number of vectors, so the concept of dimension is well defined.

We now present several examples of bases.

Example 3.24. The standard basis for $\mathbf{R}^n$ is $\{\mathbf{e}_1, \mathbf{e}_2, \ldots, \mathbf{e}_n\}$, where every entry of $\mathbf{e}_j$ is zero except the jth, which is one. Then we obviously have, for any $\mathbf{x} \in \mathbf{R}^n$,

$$\mathbf{x} = x_1 \mathbf{e}_1 + x_2 \mathbf{e}_2 + \cdots + x_n \mathbf{e}_n,$$

and it is not hard to see that this representation is unique. For example, for $\mathbf{x} \in \mathbf{R}^3$,

$$\mathbf{x} = x_1 \begin{bmatrix} 1 \\ 0 \\ 0 \end{bmatrix} + x_2 \begin{bmatrix} 0 \\ 1 \\ 0 \end{bmatrix} + x_3 \begin{bmatrix} 0 \\ 0 \\ 1 \end{bmatrix}. \quad \blacksquare$$

Example 3.25. An alternate basis for $\mathbf{R}^3$ is $\{\mathbf{v}_1, \mathbf{v}_2, \mathbf{v}_3\}$, where

$$\mathbf{v}_1 = \begin{bmatrix} \frac{1}{\sqrt{3}} \\ \frac{1}{\sqrt{3}} \\ \frac{1}{\sqrt{3}} \end{bmatrix}, \; \mathbf{v}_2 = \begin{bmatrix} \frac{1}{\sqrt{2}} \\ 0 \\ -\frac{1}{\sqrt{2}} \end{bmatrix}, \; \mathbf{v}_3 = \begin{bmatrix} \frac{1}{\sqrt{6}} \\ -\frac{2}{\sqrt{6}} \\ \frac{1}{\sqrt{6}} \end{bmatrix}.$$

It may not be obvious to the reader why one would want to use this basis instead of the much simpler basis $\{\mathbf{e}_1, \mathbf{e}_2, \mathbf{e}_3\}$. However, it is easy to check that

$$\mathbf{v}_1 \cdot \mathbf{v}_2 = 0, \; \mathbf{v}_1 \cdot \mathbf{v}_3 = 0, \; \mathbf{v}_2 \cdot \mathbf{v}_3 = 0,$$

and this property makes the basis $\{\mathbf{v}_1, \mathbf{v}_2, \mathbf{v}_3\}$ almost as easy to use as $\{\mathbf{e}_1, \mathbf{e}_2, \mathbf{e}_3\}$. We explore this topic in the next section. $\blacksquare$

Example 3.26. The set $\mathcal{P}_n$ is the vector space of all polynomials of degree n or less (see Exercise 3.1.5). The standard basis is $\{1, x, x^2, \ldots, x^n\}$. To see that this is indeed a basis, we first note that every polynomial $p \in \mathcal{P}_n$ can be written as a linear combination of $1, x, x^2, \ldots, x^n$:

$$p(x) = c_0 \cdot 1 + c_1 x + c_2 x^2 + \cdots + c_n x^n$$

(this is just the definition of a polynomial of degree n). Showing that this representation is unique is a little subtle. If we also had

$$p(x) = d_0 \cdot 1 + d_1 x + d_2 x^2 + \cdots + d_n x^n,$$

then, subtraction would yield

$$(c_0 - d_0) + (c_1 - d_1)x + \cdots + (c_n - d_n)x^n = 0$$

for every x. However, a nonzero polynomial of degree n can have at most n roots, so it must be the case that $(c_0 - d_0) + (c_1 - d_1)x + \cdots + (c_n - d_n)x^n$ is the zero polynomial. That is, $c_0 = d_0, c_1 = d_1, \ldots, c_n = d_n$ must hold. ∎

Example 3.27. An alternate basis for $\mathcal{P}_2$ is

$$\left\{1, x - \frac{1}{2}, x^2 - x + \frac{1}{6}\right\}$$

(the advantage of this basis will be discussed in Example 3.39 in the next section). To show that this is indeed a basis, we must show that, given any $p(x) = c_0 + c_1 x + c_2 x^2$, there is a unique choice of the scalars a_0, a_1, a_2 such that

$$a_0 \cdot 1 + a_1 \left(x - \frac{1}{2}\right) + a_2 \left(x^2 - x + \frac{1}{6}\right) = c_0 + c_1 x + c_2 x^2.$$

This equation is equivalent to the three linear equations

$$a_0 - \frac{1}{2}a_1 + \frac{1}{6}a_2 = c_0,$$

$$a_1 - a_2 = c_1,$$

$$a_2 = c_2.$$

The reader can easily verify that this system has a unique solution, regardless of the values of c_0, c_1, c_2. ∎

Example 3.28. Yet another basis for $\mathcal{P}_2$ is $\{L_1, L_2, L_3\}$, where

$$L_1(x) = 2\left(x - \frac{1}{2}\right)(x - 1),$$

$$L_2(x) = -4x(x - 1),$$

$$L_3(x) = 2x\left(x - \frac{1}{2}\right).$$

If we write $x_1 = 0$, $x_2 = 1/2$, and $x_3 = 1$, then the property

$$L_i(x_j) = \begin{cases} 1, & i = j, \\ 0, & i \neq j \end{cases} \tag{3.9}$$

holds. From this property, the properties of a basis can be verified (see Exercise 3.3.7). ∎

There are two essential properties of a basis $\{\mathbf{v}_1, \mathbf{v}_2, \ldots, \mathbf{v}_n\}$ of a vector space V. First, every vector in V can be represented as a linear combination of the basis vectors. Second, this representation is unique. The following two definitions provide concise ways to express these two properties.

3.3. Basis and dimension

Definition 3.29. *Let V be a vector space, and suppose $\{v_1, v_2, \ldots, v_n\}$ is a collection of vectors in V. The* span *of $\{v_1, v_2, \ldots, v_n\}$ is the set of all linear combinations of these vectors:*

$$\text{span}\{v_1, v_2, \ldots, v_n\} = \{\alpha_1 v_1 + \alpha_2 v_2 + \cdots + \alpha_n v_n : \alpha_1, \alpha_2, \ldots, \alpha_n \in \mathbf{R}\}.$$

Thus, one of the properties of a basis $\{v_1, v_2, \ldots, v_n\}$ of a vector space V is that

$$V = \text{span}\{v_1, v_2, \ldots, v_n\}.$$

The reader should also be aware that, for *any* vectors $v_1, v_2, \ldots, v_n$ in a vector space V, $\text{span}\{v_1, v_2, \ldots, v_n\}$ is a subspace of V (possibly the entire space V, as in the case of a basis).

Definition 3.30. *A set of vectors $\{v_1, v_2, \ldots, v_n\}$ is called* linearly independent *if the only scalars $c_1, c_2, \ldots, c_n$ satisfying*

$$c_1 v_1 + c_2 v_2 + \cdots + c_n v_n = 0$$

are $c_1 = c_2 = \cdots = c_n = 0$.

It can be shown that the uniqueness part of the definition of a basis is equivalent to the linear independence of the basis vectors. Therefore, a basis for a vector space is a linearly independent spanning set.

A third quality of a basis is the number of vectors in it—the dimension of the vector space. It can be shown that any two of these properties imply the third. That is, if V has dimension n, then any two of the following statements about $\{v_1, v_2, \ldots, v_k\}$ imply the third:

- $k = n$;
- $\{v_1, v_2, \ldots, v_k\}$ is linearly independent;
- $\{v_1, v_2, \ldots, v_k\}$ spans V.

Thus if $\{v_1, v_2, \ldots, v_k\}$ is known to satisfy two of the above properties, then it is a basis for V.

Before leaving the topic of basis, we wish to remind the reader of the fact indicated in the opening paragraphs of this section, which is so fundamental that we express it formally as a theorem.

Theorem 3.31. *Let $\mathbf{A}$ be an $n \times n$ matrix. Then $\mathbf{A}$ is nonsingular if and only if the columns of $\mathbf{A}$ form a basis for $\mathbf{R}^n$.*

Thus, when A is nonsingular, its columns form a basis for $\mathbf{R}^n$, and *solving $\mathbf{A}\mathbf{x} = \mathbf{b}$ is equivalent to finding the weights that express $\mathbf{b}$ as a linear combination of this basis*. This fact answers the following important questions: Suppose we have a basis $v_1, v_2, \ldots, v_n$ for $\mathbf{R}^n$ and a vector $\mathbf{b} \in \mathbf{R}^n$. Then, of course, $\mathbf{b}$ is a linear combination of the basis vectors. How do we find the weights in this linear combination? How expensive is it to do (that is, how much work is required)?

To find the scalars $x_1, x_2, \ldots, x_n$ in the equation

$$x_1 \mathbf{v}_1 + x_2 \mathbf{v}_2 + \cdots + x_n \mathbf{v}_n = \mathbf{b},$$

we define[7] $\mathbf{A} = [\mathbf{v}_1 | \mathbf{v}_2 | \cdots | \mathbf{v}_n]$ and solve $\mathbf{Ax} = \mathbf{b}$ via Gaussian elimination. The expense of computing $\mathbf{x}$ can be measured by counting the number of arithmetic operations—the number of additions, subtractions, multiplications, and divisions—required. The total number of operations required to solve $\mathbf{Ax} = \mathbf{b}$ is a polynomial in n, and it is convenient to report just the leading term in the polynomial, which can be shown to be $(2/3)n^3$ (the lower-order terms are not very significant when n is large). We usually express this saying that the operation count is

$$O\left(\frac{2}{3}n^3\right)$$

("on the *order* of $(2/3)n^3$").

In the next section, we discuss a certain special type of basis for which it is much easier to express a vector in terms of the basis.

Exercises

1. (a) Let
    $$\mathbf{A} = \begin{bmatrix} 3 & -1 & 2 \\ -1 & 3 & 4 \\ 2 & 0 & -3 \end{bmatrix}, \mathbf{x} = \begin{bmatrix} 2 \\ -2 \\ 3 \end{bmatrix}.$$

    Compute both $\mathbf{Ax}$ and

    $$2\begin{bmatrix} 3 \\ -1 \\ 2 \end{bmatrix} - 2\begin{bmatrix} -1 \\ 3 \\ 0 \end{bmatrix} + 3\begin{bmatrix} 2 \\ 4 \\ -3 \end{bmatrix},$$

    and verify that they are equal.

    (b) Let $\mathbf{A} \in \mathbf{R}^{n \times n}$ and $\mathbf{x} \in \mathbf{R}^n$, and suppose the columns of $\mathbf{A}$ are the vectors $\mathbf{v}_1, \mathbf{v}_2, \ldots, \mathbf{v}_n \in \mathbf{R}^n$, so that the (i, j)-entry of $\mathbf{A}$ is $(\mathbf{v}_j)_i$. Compute both $(\mathbf{Ax})_i$ and $(x_1 \mathbf{v}_1 + x_2 \mathbf{v}_2 + \cdots + x_n \mathbf{v}_n)_i$, and verify that they are equal.

2. Is
    $$\left\{ \begin{bmatrix} 1 \\ 0 \\ 1 \end{bmatrix}, \begin{bmatrix} 1 \\ 1 \\ 1 \end{bmatrix}, \begin{bmatrix} 2 \\ 2 \\ 4 \end{bmatrix} \right\}$$

    a basis for $\mathbf{R}^3$? (Hint: As explained in the last paragraphs of this section, the three given vectors form a basis for $\mathbf{R}^n$ if and only if $\mathbf{Ax} = \mathbf{b}$ has a unique solution for every $\mathbf{b} \in \mathbf{R}^n$, where $\mathbf{A}$ is the 3×3 matrix whose columns are the three given vectors.)

3. Is
    $$\left\{ \begin{bmatrix} 1 \\ 0 \\ 1 \end{bmatrix}, \begin{bmatrix} 1 \\ 1 \\ 1 \end{bmatrix}, \begin{bmatrix} 1 \\ 2 \\ 1 \end{bmatrix} \right\}$$

    a basis for $\mathbf{R}^3$? (See the hint for the previous exercise.)

---

[7] This notation means that A is the matrix whose columns are the vectors $\mathbf{v}_1, \mathbf{v}_2, \ldots, \mathbf{v}_n$.

3.4. Orthogonal bases and projections

4. Show that $\{x^2+1, x+1, x^2-x+1\}$ is a basis for $\mathcal{P}_2$, the space of polynomials of degree 2 or less. (Hint: Verify directly that the definition holds.)

5. Prove that $\{1-x+2x^2, 1-2x^2, 1-3x+7x^2\}$ is linearly dependent and hence is not a basis for $\mathcal{P}_2$.

6. Prove that $\{1+x, 1-x+2x^2, x-x^3\}$ does not span $\mathcal{P}_3$ by finding a polynomial $p \in \mathcal{P}_3$ such that $p \notin \text{span}\{1+x, 1-x+2x^2, x-x^3\}$.

7. Show that $\{L_1, L_2, L_3\}$, defined in Example 3.28, is a basis for $\mathcal{P}_2$. (Hint: Use (3.9) to show that
$$p(x) = p(x_1)L_1(x) + p(x_2)L_2(x) + p(x_3)L_3(x)$$
holds for every $p \in \mathcal{P}_2$.)

8. Let V be the space of all continuous, complex-valued functions defined on the real line:
$$V = \{f : \mathbf{R} \to \mathbf{C} : f \text{ is continuous}\}.$$
Define W to be the subspace of V spanned by e^{ix} and e^{-ix}, where $i = \sqrt{-1}$. Show that $\{\cos(x), \sin(x)\}$ is another basis for W. (Hint: Use Euler's formula: $e^{i\theta} = \cos(\theta) + i\sin(\theta)$.)

9. Let $L : C^2[a,b] \to C[a,b]$ be the second derivative operator. Find a basis for the null space of L.

10. Let $L_N : C_N^2[a,b] \to C[a,b]$ be the second derivative operator. Find a basis for the null space of L_N.

3.4 Orthogonal bases and projections

At the end of the last section, we discussed the question of expressing a vector in terms of a given basis. This question is important for the following reason, which we can only describe in general terms at the moment: Many problems that are posed in vector spaces admit a special basis, in terms of which the problem is easy to solve. That is, for many problems, there exists a special basis with the property that if all vectors are expressed in terms of that basis, then a very simple calculation will produce the final solution. For this reason, it is important to be able to take a vector (perhaps expressed in terms of a standard basis) and express it in terms of a different basis. In the latter part of this section, we will study one type of problem for which it is advantageous to use a special basis, and we will discuss another such problem in the next section.

It is quite easy to express a vector in terms of a basis if that basis is *orthogonal*. We wish to describe the concept of an orthogonal basis and show some important examples. Before we can do so, we must introduce the idea of an inner product, which is a generalization of the Euclidean dot product.

The dot product plays a special role in the geometry of $\mathbf{R}^2$ and $\mathbf{R}^3$. The reason for this is the fact that two vectors $\mathbf{x}, \mathbf{y}$ in $\mathbf{R}^2$ or $\mathbf{R}^3$ are perpendicular if and only if
$$\mathbf{x} \cdot \mathbf{y} = 0.$$

Indeed, one can show that
$$\mathbf{x} \cdot \mathbf{y} = \|\mathbf{x}\| \|\mathbf{y}\| \cos(\theta),$$
where θ is the angle between the two vectors (see Figure 3.2).

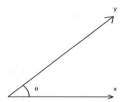

Figure 3.2. *The angle between two vectors.*

From elementary Euclidean geometry, we know that, if $\mathbf{x}$ and $\mathbf{y}$ are perpendicular, then
$$\|\mathbf{x} + \mathbf{y}\|^2 = \|\mathbf{x}\|^2 + \|\mathbf{y}\|^2$$
(the Pythagorean theorem). Using the dot product, we can give a purely algebraic proof of the Pythagorean theorem. By definition, $\|\mathbf{u}\| = \sqrt{\mathbf{u} \cdot \mathbf{u}}$, so
$$\begin{aligned}\|\mathbf{x} + \mathbf{y}\|^2 = (\mathbf{x} + \mathbf{y}) \cdot (\mathbf{x} + \mathbf{y}) &= \mathbf{x} \cdot \mathbf{x} + \mathbf{x} \cdot \mathbf{y} + \mathbf{y} \cdot \mathbf{x} + \mathbf{y} \cdot \mathbf{y} \\ &= \mathbf{x} \cdot \mathbf{x} + 2\mathbf{x} \cdot \mathbf{y} + \mathbf{y} \cdot \mathbf{y} \\ &= \|\mathbf{x}\|^2 + 2\mathbf{x} \cdot \mathbf{y} + \|\mathbf{y}\|^2.\end{aligned}$$

This calculation shows that $\|\mathbf{x} + \mathbf{y}\|^2 = \|\mathbf{x}\|^2 + \|\mathbf{y}\|^2$ holds if and only if $\mathbf{x} \cdot \mathbf{y} = 0$.

Seen this way, the Pythagorean theorem is an algebraic property that can be deduced in $\mathbf{R}^n$, $n > 3$, even though in those spaces we cannot visualize vectors or what it means for vectors to be perpendicular. We prefer to use the word *orthogonal* instead of perpendicular: Vectors $\mathbf{x}$ and $\mathbf{y}$ in $\mathbf{R}^n$ are orthogonal if $\mathbf{x} \cdot \mathbf{y} = 0$.

In the course of solving differential equations, we deal with function spaces in addition to Euclidean spaces, and our methods are heavily dependent on the existence of an *inner product*—the analogue of the dot product in more general vector spaces. Here is the definition.

Definition 3.32. *Let V be a real vector space. A (real) inner product on V is a function, usually denoted $(\cdot, \cdot)$ or $(\cdot, \cdot)_V$, taking two vectors from V and producing a real number. This function must satisfy the following three properties:*

1. $(\mathbf{u}, \mathbf{v}) = (\mathbf{v}, \mathbf{u})$ *for all vectors* $\mathbf{u}$ *and* $\mathbf{v}$;

2. $(\alpha \mathbf{u} + \beta \mathbf{v}, \mathbf{w}) = \alpha(\mathbf{u}, \mathbf{w}) + \beta(\mathbf{v}, \mathbf{w})$ *and* $(\mathbf{w}, \alpha \mathbf{u} + \beta \mathbf{v}) = \alpha(\mathbf{w}, \mathbf{u}) + \beta(\mathbf{w}, \mathbf{v})$ *for all vectors* $\mathbf{u}$, $\mathbf{v}$, *and* $\mathbf{w}$, *and all real numbers* α *and* β;

3. $(\mathbf{u}, \mathbf{u}) \geq 0$ *for all vectors* $\mathbf{u}$, *and* $(\mathbf{u}, \mathbf{u}) = 0$ *if and only if* $\mathbf{u}$ *is the zero vector.*

It should be easy to check that these properties hold for the ordinary dot product on Euclidean n-space.

3.4. Orthogonal bases and projections

Given an *inner product space* (a vector space with an inner product), we define orthogonality just as in Euclidean space: two vectors are orthogonal if and only if their inner product is zero. It can then be shown that the Pythagorean theorem holds (see Exercise 3.4.3).

An *orthogonal basis* for an inner product space V is a basis $\{v_1, v_2, \ldots, v_n\}$ with the property that
$$i \neq j \Rightarrow (v_i, v_j) = 0$$
(that is, every vector in the basis is orthogonal to every other vector in the basis). We now demonstrate the first special property of an orthogonal basis. Suppose $\{v_1, v_2, \ldots, v_n\}$ is an orthogonal basis for an inner product space V and x is any vector in V. Then there exist scalars $\alpha_1, \alpha_2, \ldots, \alpha_n$ such that
$$x = \alpha_1 v_1 + \alpha_2 v_2 + \cdots + \alpha_n v_n. \tag{3.10}$$

To deduce the value of α_i, we take the inner product of both sides of (3.10) with v_i:
$$\begin{aligned}(v_i, x) &= (v_i, \alpha_1 v_1 + \alpha_2 v_2 + \cdots + \alpha_n v_n) \\ &= \alpha_1 (v_i, v_1) + \alpha_2 (v_i, v_2) + \cdots + \alpha_n (v_i, v_n) \\ &= \alpha_i (v_i, v_i).\end{aligned}$$

The last step follows from the fact that every inner product (v_i, v_j) vanishes except (v_i, v_i). We then obtain
$$\alpha_i = \frac{(v_i, x)}{(v_i, v_i)}, \quad i = 1, 2, \ldots, n,$$
and so
$$x = \frac{(v_1, x)}{(v_1, v_1)} v_1 + \frac{(v_2, x)}{(v_2, v_2)} v_2 + \cdots + \frac{(v_n, x)}{(v_n, v_n)} v_n. \tag{3.11}$$

This formula shows that it is easy to express a vector in terms of an orthogonal basis. Assuming that we compute $(v_1, v_1), (v_2, v_2), \ldots, (v_n, v_n)$ once and for all, it requires just n inner products to find the weights in the linear combination. In the case of Euclidean n-vectors, a dot product requires $2n - 1$ arithmetic operations (n multiplications and $n - 1$ additions), so the total cost is just $O(2n^2)$. If n is large, this is *much* less costly than the $O(2n^3/3)$ operations required for a nonorthogonal basis. We also remark that if the basis is *orthonormal*—each basis vector is normalized to have length one—then (3.11) simplifies to
$$x = (v_1, x) v_1 + (v_2, x) v_2 + \cdots + (v_n, x) v_n. \tag{3.12}$$

Example 3.33. The basis $\{v_1, v_2, v_3\}$ for $\mathbf{R}^3$, where

$$v_1 = \begin{bmatrix} \frac{1}{\sqrt{3}} \\ \frac{1}{\sqrt{3}} \\ \frac{1}{\sqrt{3}} \end{bmatrix}, \quad v_2 = \begin{bmatrix} \frac{1}{\sqrt{2}} \\ 0 \\ -\frac{1}{\sqrt{2}} \end{bmatrix}, \quad v_3 = \begin{bmatrix} \frac{1}{\sqrt{6}} \\ -\frac{2}{\sqrt{6}} \\ \frac{1}{\sqrt{6}} \end{bmatrix},$$

is orthonormal, as can be verified directly. If

$$x = \begin{bmatrix} 1 \\ 2 \\ 3 \end{bmatrix},$$

then
$$\mathbf{x} = (\mathbf{v}_1 \cdot \mathbf{x})\mathbf{v}_1 + (\mathbf{v}_2 \cdot \mathbf{x})\mathbf{v}_2 + (\mathbf{v}_3 \cdot \mathbf{x})\mathbf{v}_3 = \frac{6}{\sqrt{3}}\mathbf{v}_1 - \frac{2}{\sqrt{2}}\mathbf{v}_2 + 0\mathbf{v}_3. \qquad \blacksquare$$

3.4.1 The L^2 inner product

We have seen that functions can be regarded as vectors, at least in a formal sense: functions can be added together and multiplied by scalars. (See Example 3.4 in Section 3.1.) We will now show more directly that functions are not so different from Euclidean vectors. In the process, we show that a suitable inner product can be defined for functions.

Suppose we have a function $g \in C[a,b]$—a continuous function defined on the interval $[a,b]$. By sampling g on a grid, we can produce a vector that approximates the function g. Let $x_i = a + i\Delta x$, $\Delta x = (b-a)/N$, and define a vector $G \in \mathbf{R}^N$ by

$$G_i = g(x_i), \ i = 0, 1, \ldots, N-1.$$

Then G can be regarded as an approximation to g (see Figure 3.3). Given another function $f(x)$ and the corresponding vector $F \in \mathbf{R}^N$, we have

$$F \cdot G = \sum_{i=0}^{N-1} F_i G_i = \sum_{i=0}^{N-1} f(x_i) g(x_i).$$

Refining the discretization (increasing N) leads to a sampled function that obviously represents the original function more accurately. Therefore, we ask, What happens to $F \cdot G$ as $N \to \infty$? The dot product

$$F \cdot G = \sum_{i=0}^{N-1} f(x_i) g(x_i)$$

does not converge to any value as $N \to \infty$, but a simple modification induces convergence. We replace the ordinary dot product by the following scaled dot product, for which we introduce a new notation:

$$(F, G) = \sum_{i=0}^{N-1} F_i G_i \Delta x.$$

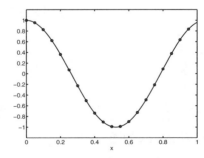

Figure 3.3. *Approximating a function $g(x)$ by a vector $G \in \mathbf{R}^N$; the components of G are the values of g at the indicated points.*

3.4. Orthogonal bases and projections

Then, when F and G are sampled functions as above, we have

$$(F,G) = \sum_{i=0}^{N-1} f(x_i)g(x_i)\Delta x \to \int_a^b f(x)g(x)dx \text{ as } N \to \infty.$$

Based on this observation, we argue that a natural inner product $(\cdot,\cdot)$ on $C[a,b]$ is

$$(f,g) = \int_a^b f(x)g(x)dx. \tag{3.13}$$

Just as the dot product defines a norm on Euclidean n-space ($\|x\| = \sqrt{x \cdot x}$), so the inner product (3.13) defines a norm for functions:

$$\|f\| = \sqrt{(f,f)} = \sqrt{\int_0^\ell |f(x)|^2 dx}. \tag{3.14}$$

For completeness, we give the definition of norm. Norms measure the size or magnitude of vectors, and the definition is intended to describe abstractly the properties that any reasonable notion of size ought to have.

Definition 3.34. *Let V be a vector space. A norm on V is a real-valued function with domain V, usually denoted by $\|\cdot\|$ or $\|\cdot\|_V$, and satisfying the following properties:*

1. $\|\mathbf{v}\| \geq 0$ *for all* $\mathbf{v} \in V$ *and* $\|\mathbf{v}\| = 0$ *if and only if* $\mathbf{v} = \mathbf{0}$;
2. $\|\alpha \mathbf{v}\| = |\alpha| \|\mathbf{v}\|$ *for all scalars α and all* $\mathbf{v} \in V$;
3. $\|\mathbf{u} + \mathbf{v}\| \leq \|\mathbf{u}\| + \|\mathbf{v}\|$ *for all* $\mathbf{u}, \mathbf{v} \in V$.

The last property is called the triangle inequality.

For Euclidean vectors in the plane, the triangle inequality expresses the fact that one side of a triangle cannot be longer than the sum of the other two sides.

The inner product defined by (3.13) is called the L^2 inner product.[8] Two functions in $C[a,b]$ are said to be orthogonal if $(f,g) = 0$. This condition does not have a direct geometric meaning, as the analogous condition does for Euclidean vectors in $\mathbf{R}^2$ or $\mathbf{R}^3$, but, as we argued above, orthogonality is still important algebraically.

When we measure norm in the L^2 sense, we say that functions f and g are close (for example, that g is a good approximation to f) if

$$\sqrt{\int_a^b (f(x) - g(x))^2 dx}$$

is small. This does not mean that $(f(x) - g(x))^2$ is small for every $x \in [a,b]$ ($(f(x) - g(x))^2$ can be large in places, as long as this difference is large only over very small intervals), but rather it implies that $(f(x) - g(x))^2$ is small on the average over the interval $[a,b]$. For this reason, we often use the term "mean-square" in referring to the L^2 norm (for example, we might say "g is close to f in the mean-square sense").

---

[8] The "L" refers to the French mathematician Lebesgue, and the "2" to the exponent in the formula for the L^2 norm of a function. The symbol L^2 is read "L-two."

Example 3.35. If $f : [0,1] \to \mathbf{R}$ is defined by $f(x) = x(1-x)$, then

$$\|f\| = \sqrt{\int_0^1 x^2(1-x)^2\,dx} = \sqrt{\frac{1}{30}} = \frac{1}{\sqrt{30}} \doteq 0.1826.$$

With $g : [0,1] \to \mathbf{R}$ defined by

$$g(x) = \frac{8}{\pi^3} \sin(\pi x),$$

we have

$$\|f - g\| = \sqrt{\int_0^1 \left(x(1-x) - \frac{8}{\pi^3}\sin(\pi x)\right)^2 dx} = \sqrt{\frac{(\pi^6 - 960)}{30\pi^6}} \doteq 0.006940.$$

These two functions differ by less than 4% in the mean-square sense (cf. Figure 3.4). ∎

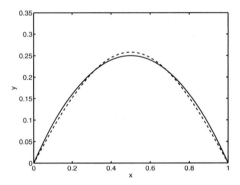

Figure 3.4. *The functions of Example 3.35:* $y = x(1-x)$ *(solid curve) and* $y = (8/\pi^3)\sin(\pi x)$ *(dashed curve).*

3.4.2 The projection theorem

The projection theorem is about approximating a vector v in a vector space V by a vector from a subspace W. More specifically, the projection theorem answers the question, Is there a vector $w \in W$ closest to v (the *best approximation* to v from W), and if so, how can it be computed? Since this theorem is so important, and its proof is so informative, we will formally state and prove the theorem.

Theorem 3.36. *Let V be a vector space with inner product $(\cdot,\cdot)$, let W be a finite-dimensional subspace of V, and let $\mathbf{v} \in V$.*

1. *There is a unique $\mathbf{u} \in W$ such that*

$$\|\mathbf{v} - \mathbf{u}\| = \min_{\mathbf{w} \in W} \|\mathbf{v} - \mathbf{w}\|.$$

That is, there is a unique best approximation to $\mathbf{v}$ from W. We also call $\mathbf{u}$ the projection of $\mathbf{v}$ onto W, and write $\mathbf{u} = \mathrm{proj}_W \mathbf{v}$.

3.4. Orthogonal bases and projections

2. *A vector $\mathbf{u} \in W$ is the best approximation to $\mathbf{v}$ from W if and only if*

$$(\mathbf{v} - \mathbf{u}, \mathbf{z}) = 0 \text{ for all } \mathbf{z} \in W. \tag{3.15}$$

3. *If $\{\mathbf{w}_1, \mathbf{w}_2, \ldots, \mathbf{w}_n\}$ is a basis for W, then*

$$\text{proj}_W \mathbf{v} = \sum_{i=1}^{n} x_i \mathbf{w}_i, \tag{3.16}$$

where

$$\mathbf{G}\mathbf{x} = \mathbf{b}, \ G_{ij} = (\mathbf{w}_j, \mathbf{w}_i), \ b_i = (\mathbf{w}_i, \mathbf{v}). \tag{3.17}$$

The equations represented by $\mathbf{G}\mathbf{x} = \mathbf{b}$ are called the normal equations, *and the matrix $\mathbf{G}$ is called the* Gram *matrix.*

4. *If $\{\mathbf{w}_1, \mathbf{w}_2, \ldots, \mathbf{w}_n\}$ is an orthogonal basis for W, then the best approximation to $\mathbf{v}$ from W is*

$$\text{proj}_W \mathbf{v} = \sum_{i=1}^{n} \frac{(\mathbf{w}_i, \mathbf{v})}{(\mathbf{w}_i, \mathbf{w}_i)} \mathbf{w}_i. \tag{3.18}$$

If the basis is orthonormal, this simplifies to

$$\text{proj}_W \mathbf{v} = \sum_{i=1}^{n} (\mathbf{w}_i, \mathbf{v}) \mathbf{w}_i. \tag{3.19}$$

Proof. We will prove the second conclusion first. Suppose that $\mathbf{u} \in W$, and $\mathbf{z}$ is any other vector in W. Then, since W is closed under addition and scalar multiplication, we have that $\mathbf{u} + t\mathbf{z} \in W$ for all real numbers t. On the other hand, every other vector $\mathbf{w}$ in W can be written as $\mathbf{u} + t\mathbf{z}$ for some $\mathbf{z} \in W$ and some $t \in \mathbf{R}$ (just take $\mathbf{z} = \mathbf{w} - \mathbf{u}$ and $t = 1$). Therefore, $\mathbf{u} \in W$ is closest to $\mathbf{v}$ if and only if

$$\|\mathbf{v} - \mathbf{u}\| \leq \|\mathbf{v} - (\mathbf{u} + t\mathbf{z})\| \text{ for all } \mathbf{z} \in W, \ t \in \mathbf{R}. \tag{3.20}$$

Since $\|\mathbf{x}\|^2 = (\mathbf{x}, \mathbf{x})$, this last inequality is equivalent to

$$\begin{aligned}(\mathbf{v} - \mathbf{u}, \mathbf{v} - \mathbf{u}) &\leq (\mathbf{v} - (\mathbf{u} + t\mathbf{z}), \mathbf{v} - (\mathbf{u} + t\mathbf{z})) \\ &= ((\mathbf{v} - \mathbf{u}) - t\mathbf{z}, (\mathbf{v} - \mathbf{u}) - t\mathbf{z}) \\ &= (\mathbf{v} - \mathbf{u}, \mathbf{v} - \mathbf{u}) - 2t(\mathbf{v} - \mathbf{u}, \mathbf{z}) + t^2(\mathbf{z}, \mathbf{z})\end{aligned}$$

or to

$$t^2(\mathbf{z}, \mathbf{z}) - 2t(\mathbf{v} - \mathbf{u}, \mathbf{z}) \geq 0 \text{ for all } \mathbf{z} \in W, \ t \in \mathbf{R}.$$

If we regard $\mathbf{z}$ as fixed, then

$$t^2(\mathbf{z}, \mathbf{z}) + 2t(\mathbf{v} - \mathbf{u}, \mathbf{z})$$

is a simple quadratic in t, and the inequality holds if and only if $(\mathbf{v}-\mathbf{u},\mathbf{z}) = 0$. It follows that the inequality holds for all $\mathbf{z}$ and all t if and only if (3.15) holds. In addition, provided $\mathbf{z} \neq 0$, (3.20) holds as an equation only when $t = 0$ (since $(\mathbf{z},\mathbf{z}) > 0$ for $\mathbf{z} \neq 0$). That is, if $\mathbf{w} \in W$ and $\mathbf{w} \neq \mathbf{u}$, then

$$\|\mathbf{v}-\mathbf{u}\| < \|\mathbf{v}-\mathbf{w}\|.$$

Thus, if the best approximation problem has a solution, it is unique.

We now prove the remaining conclusions to the theorem. Since W is a finite-dimensional subspace, it has a basis $\{\mathbf{w}_1, \mathbf{w}_2, \ldots, \mathbf{w}_n\}$. A vector $\mathbf{u} \in W$ solves the best approximation problem if and only if (3.15) holds; moreover, it is straightforward to show that (3.15) is equivalent to

$$(\mathbf{v}-\mathbf{u},\mathbf{w}_i) = 0 \text{ for } i = 1,2,\ldots,n \tag{3.21}$$

(see Exercise 3.4.5). Any vector $\mathbf{u} \in W$ can be written as

$$\mathbf{u} = \sum_{j=1}^{n} x_j \mathbf{w}_j. \tag{3.22}$$

Thus, $\mathbf{u} \in W$ is a solution if and only if (3.22) holds and

$$\left(\mathbf{v} - \sum_{j=1}^{n} x_j \mathbf{w}_j, \mathbf{w}_i\right) = 0, \ i = 1,2,\ldots,n, \tag{3.23}$$

which simplifies to

$$\sum_{j=1}^{n} (\mathbf{w}_j, \mathbf{w}_i) x_j = (\mathbf{w}_i, \mathbf{v}), \ i = 1,2,\ldots,n. \tag{3.24}$$

If we define $\mathbf{G} \in \mathbf{R}^{n \times n}$ by $G_{ij} = (\mathbf{w}_j, \mathbf{w}_i)$ and $\mathbf{b} \in \mathbf{R}^n$ by $b_i = (\mathbf{w}_i, \mathbf{v})$, then (3.24) is equivalent to $\mathbf{Gx} = \mathbf{b}$. It can be shown that $\mathbf{G}$ is nonsingular (see Exercise 3.4.6), so the unique best approximation to $\mathbf{v}$ from W is given by (3.22), where $\mathbf{x}$ solves $\mathbf{Gx} = \mathbf{b}$.

If the basis is orthogonal, then $(\mathbf{w}_j, \mathbf{w}_i) = 0$ unless $j = i$. In this case, $\mathbf{G}$ is the diagonal matrix with diagonal entries $(\mathbf{w}_1, \mathbf{w}_1), (\mathbf{w}_2, \mathbf{w}_2), \ldots, (\mathbf{w}_n, \mathbf{w}_n)$, and $\mathbf{Gx} = \mathbf{b}$ is equivalent to the n simple equations

$$(\mathbf{w}_i, \mathbf{w}_i) x_i = (\mathbf{v}, \mathbf{w}_i), \ i = 1,2,\ldots,n,$$

that is, to

$$x_i = \frac{(\mathbf{v}, \mathbf{w}_i)}{(\mathbf{w}_i, \mathbf{w}_i)}, \ i = 1,2,\ldots,n.$$

This completes the proof. □

We now present two examples of best approximation problems that commonly occur in scientific and computational practice.

3.4. Orthogonal bases and projections

Example 3.37. We assume that data points

$$(x_1, y_1) = (0.10, 1.7805),$$
$$(x_2, y_2) = (0.30, 2.2285),$$
$$(x_3, y_3) = (0.40, 2.3941),$$
$$(x_4, y_4) = (0.75, 3.2226),$$
$$(x_5, y_5) = (0.90, 3.5697)$$

have been collected in the laboratory, and there is a theoretical reason to believe that $y_i = ax_i + b$ ought to hold for some choices of $a, b \in \mathbf{R}$. Of course, due to measurement error, this relationship is unlikely to hold exactly for any choice of a and b, but we would like to find a and b that come as close as possible to satisfying it. If we define

$$\mathbf{y} = \begin{bmatrix} 1.7805 \\ 2.2285 \\ 2.3941 \\ 3.2226 \\ 3.5697 \end{bmatrix}, \quad \mathbf{x} = \begin{bmatrix} 0.10 \\ 0.30 \\ 0.40 \\ 0.75 \\ 0.90 \end{bmatrix}, \quad \mathbf{e} = \begin{bmatrix} 1 \\ 1 \\ 1 \\ 1 \\ 1 \end{bmatrix},$$

then one way to pose this problem is to choose a and b so that $a\mathbf{x} + b\mathbf{e}$ is as close as possible to $\mathbf{y}$ in the Euclidean norm. That is, we find the best approximation to $\mathbf{y}$ from $W = \text{span}\{\mathbf{x}, \mathbf{e}\}$.

The Gram matrix is

$$\mathbf{G} = \begin{bmatrix} \mathbf{x} \cdot \mathbf{x} & \mathbf{e} \cdot \mathbf{x} \\ \mathbf{x} \cdot \mathbf{e} & \mathbf{e} \cdot \mathbf{e} \end{bmatrix} = \begin{bmatrix} 1.6325 & 2.45 \\ 2.45 & 5 \end{bmatrix},$$

and the right-hand side of the normal equations is

$$\mathbf{b} = \begin{bmatrix} \mathbf{x} \cdot \mathbf{y} \\ \mathbf{e} \cdot \mathbf{y} \end{bmatrix} \doteq \begin{bmatrix} 7.4370 \\ 13.196 \end{bmatrix}.$$

Solving the normal equations yields

$$\begin{bmatrix} a \\ b \end{bmatrix} \doteq \begin{bmatrix} 2.2411 \\ 1.5409 \end{bmatrix}.$$

The resulting linear model is displayed, together with the original data, in Figure 3.5. ∎

Example 3.38. One advantage of working with polynomials instead of transcendental functions like e^x is that polynomials are very easy to evaluate—only the basic arithmetic operations are required. For this reason, it is often desirable to approximate a more complicated function by a polynomial. Considering $f(x) = e^x$ as a function in $C[0, 1]$, we find the best quadratic approximation, in the mean-square sense, to f. A basis for the space $\mathcal{P}_2$ of polynomials of degree 2 or less is $\{1, x, x^2\}$. It is easy to verify that the normal equations for the problem of finding the best approximation to f from $\mathcal{P}_2$ are (in matrix-vector form)

$$\mathbf{G}\alpha = \mathbf{b},$$

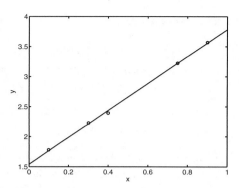

Figure 3.5. *The data from Example 3.37 and the approximate linear relationship.*

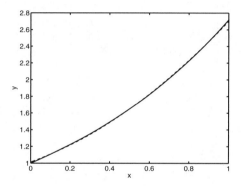

Figure 3.6. *The function $f(x) = e^x$ (solid curve) and the best quadratic approximation $y = p_2(x)$ (dashed curve) (see Example 3.38).*

where

$$\mathbf{G} = \begin{bmatrix} 1 & 1/2 & 1/3 \\ 1/2 & 1/3 & 1/4 \\ 1/3 & 1/4 & 1/5 \end{bmatrix}, \mathbf{b} = \begin{bmatrix} e-1 \\ 1 \\ e-2 \end{bmatrix}.$$

Since

$$\mathbf{G}^{-1}\mathbf{b} \doteq \begin{bmatrix} 1.013 \\ 0.8511 \\ 0.8392 \end{bmatrix},$$

the best quadratic approximation is

$$e^x \doteq p_2(x) = 1.013 + 0.8511x + 0.8392x^2.$$

The exponential function and the quadratic approximation are graphed in Figure 3.6. ∎

3.4. Orthogonal bases and projections

Example 3.39. An orthonormal basis for $\mathcal{P}_2$ (on the interval $[0,1]$) is $\{q_1, q_2, q_3\}$, where

$$q_1(x) = 1, \; q_2(x) = 2\sqrt{3}\left(x - \frac{1}{2}\right), \; q_3(x) = 6\sqrt{5}\left(x^2 - x + \frac{1}{6}\right).$$

The best approximation p_2 to e^x (which was calculated in the previous exercise) can be computed by the formula

$$p_2(x) = (q_1, f)q_1(x) + (q_2, f)q_2(x) + (q_3, f)q_3(x).$$

A direct calculation shows that the same result is obtained (see Exercise 3.4.10). ∎

The concept of best approximation is central in this book, since the two main solution techniques we discuss, Fourier series and finite elements, both produce a best approximation to the true solution of a BVP.

Exercises

1. (a) Show that the basis $\{v_1, v_2, v_3\}$ from Example 3.33 is an orthonormal basis for $\mathbf{R}^3$.

   (b) Express the vector
   $$\mathbf{x} = \begin{bmatrix} 1 \\ 2 \\ 1 \end{bmatrix}$$
   as a linear combination of v_1, v_2, v_3.

2. Is the basis $\{L_1, L_2, L_3\}$ for $\mathcal{P}_2$ (on the interval $[0, 1]$) given in Example 3.28 an orthogonal basis?

3. Let V be an inner product space. Prove that $\mathbf{x}, \mathbf{y} \in V$ satisfy
   $$\|\mathbf{x} + \mathbf{y}\|^2 = \|\mathbf{x}\|^2 + \|\mathbf{y}\|^2$$
   if and only if $(\mathbf{x}, \mathbf{y}) = 0$.

4. Use the results of this section to show that any orthonormal set containing n vectors in $\mathbf{R}^n$ is a basis for $\mathbf{R}^n$. (Hint: Since the dimension of $\mathbf{R}^n$ is n, it suffices to show either that the orthogonal set spans $\mathbf{R}^n$ or that it is linearly independent. Showing linear independence is probably easier.)

5. Let W be a subspace of an inner product space V and let $\{\mathbf{w}_1, \mathbf{w}_2, \ldots, \mathbf{w}_n\}$ be a basis for W. Show that, for $\mathbf{y} \in V$,
   $$(\mathbf{y}, \mathbf{z}) = 0 \text{ for all } \mathbf{z} \in W$$

holds if and only if
$$(\mathbf{y}, \mathbf{w}_i) = 0, \ i = 1, 2, \ldots, n,$$
holds.

6. Let $\{\mathbf{w}_1, \mathbf{w}_2, \ldots, \mathbf{w}_n\}$ be a linearly independent set in an inner product space V, and define $\mathbf{G} \in \mathbf{R}^{n \times n}$ by
$$G_{ij} = (\mathbf{w}_j, \mathbf{w}_i), \ i, j = 1, 2, \ldots, n.$$
Prove that $\mathbf{G}$ is invertible. (Hint: By the Fredholm alternative, it suffices to show that the only solution to $\mathbf{G}\mathbf{x} = 0$ is $\mathbf{x} = 0$. Assume that $\mathbf{x} \in \mathbf{R}^n$ satisfies $\mathbf{G}\mathbf{x} = 0$. This implies that $\mathbf{x} \cdot \mathbf{G}\mathbf{x} = 0$. Show that $\mathbf{x} \cdot \mathbf{G}\mathbf{x} = (\mathbf{u}, \mathbf{u})$, where $\mathbf{u} \in V$ is given by
$$\mathbf{u} = x_1 \mathbf{w}_1 + x_2 \mathbf{w}_2 + \cdots + x_n \mathbf{w}_n,$$
and show that $\mathbf{u}$ cannot be the zero vector unless $\mathbf{x} = \mathbf{0}$.)

7. Consider the following data:

x	1.000	1.250	1.400	1.500	1.900	2.100	2.250	2.600	2.900
y	2.009	2.491	2.836	2.971	3.909	4.193	4.506	5.253	5.803

Find the relationship $y = mx + c$ that best fits these data. Plot the data and the best fit line.

8. Consider the following data:

x	1.000	1.250	1.400	1.500	1.900	2.100	2.250	2.600	2.900
y	1.248	1.637	1.982	2.224	3.477	4.230	4.848	6.513	8.233

Find the relationship $y = c_2 x^2 + c_1 x + c_0$ that best fits these data. If possible, plot the data and the best fit parabola.

9. Using the orthonormal basis $\{q_1, q_2, q_3\}$ for $\mathcal{P}_2$ given in Example 3.39, find the quadratic polynomial $p(x)$ that best approximates $g(x) = \sin(\pi x)$, in the mean-square sense, over the interval $[0, 1]$. Produce a graph of g and the quadratic approximation.

10. (a) Verify that $\{q_1, q_2, q_3\}$, defined in Example 3.39, is an orthonormal basis for $\mathcal{P}_2$ on the interval $[0, 1]$.

    (b) Using this orthonormal basis, find the best approximation to $f(x) = e^x$ (on the interval $[0, 1]$) in the mean-square sense, and verify that the same result is obtained as in Example 3.38.

3.5 Eigenvalues and eigenvectors of a symmetric matrix

The eigenvectors of a matrix operator are special vectors for which the action of the matrix is particularly simple. In this section, we discuss the properties of eigenvalues and eigenvectors of symmetric matrices. Understanding this topic will set the stage for the Fourier series method for solving differential equations, which takes advantage of the simple action of a *differential* operator on its eigen*functions*.

3.5. Eigenvalues and eigenvectors of a symmetric matrix

Definition 3.40. *Let $\mathbf{A} \in \mathbf{R}^{n \times n}$. We say that the scalar λ is an* eigenvalue *of $\mathbf{A}$ if there exists a nonzero vector $\mathbf{x}$ such that*

$$\mathbf{A}\mathbf{x} = \lambda \mathbf{x}.$$

As we discuss below, the scalar λ may be complex even if $\mathbf{A}$ has only real entries; if λ is complex and $\mathbf{A}$ is real, then $\mathbf{x}$ must have complex entries.

As we are about to demonstrate, eigenvalues are naturally expressed as the roots of a polynomial, with the coefficients of the polynomial computed from the entries in the matrix. A polynomial with real coefficients can have complex roots. For this reason, it is natural to allow complex eigenvalues and eigenvectors in the above definition. However, we will show below that a *symmetric* matrix can have only real eigenvalues, which will simplify matters. Moreover, the eigenvalues and eigenvectors of a symmetric matrix have other useful properties, which we also explore below.

The following calculation is essential for understanding eigenvalues and eigenvectors:

$$\mathbf{A}\mathbf{x} = \lambda \mathbf{x}, \; \mathbf{x} \neq 0$$
$$\Leftrightarrow \lambda \mathbf{x} - \mathbf{A}\mathbf{x} = 0, \; \mathbf{x} \neq 0$$
$$\Leftrightarrow (\lambda \mathbf{I} - \mathbf{A})\mathbf{x} = 0, \; \mathbf{x} \neq 0.$$

This last condition is possible if and only if $\lambda \mathbf{I} - \mathbf{A}$ is a singular matrix; that is, if and only if

$$\det(\lambda \mathbf{I} - \mathbf{A}) = 0,$$

where $\det(\mathbf{B})$ is the determinant[9] of the square matrix $\mathbf{B}$. In principle, then, we can find the eigenvalues of $\mathbf{A}$ by solving the equation $\det(\lambda \mathbf{I} - \mathbf{A}) = 0$.

It is not hard to show that $p_{\mathbf{A}}(\lambda) = \det(\lambda \mathbf{I} - \mathbf{A})$ is a polynomial of degree n (the *characteristic polynomial* of $\mathbf{A}$), and so $\mathbf{A}$ has n eigenvalues (counted according to multiplicity as roots of $p_{\mathbf{A}}(\lambda)$). Any or all of the eigenvalues can be complex, even if $\mathbf{A}$ has real entries, since a polynomial with real coefficients can have complex roots.

Example 3.41. Let

$$\mathbf{A} = \begin{bmatrix} 1 & 1 \\ 1 & 2 \end{bmatrix}.$$

Then

$$p_{\mathbf{A}}(\lambda) = \det(\lambda \mathbf{I} - \mathbf{A}) = \begin{vmatrix} \lambda - 1 & -1 \\ -1 & \lambda - 2 \end{vmatrix} = (\lambda - 1)(\lambda - 2) - 1 = \lambda^2 - 3\lambda + 1.$$

Therefore, the eigenvalues are

$$\frac{3 \pm \sqrt{5}}{2}. \quad \blacksquare$$

---

[9] We assume that the reader is familiar with the elementary properties of determinants (such as the computation of determinants of small matrices). The most important of these is that a matrix is singular if and only if its determinant is zero. Any introductory text on linear algebra, such as [41], can be consulted for details.

Example 3.42. Let
$$\mathbf{A} = \begin{bmatrix} 1 & 1 \\ -1 & 1 \end{bmatrix}.$$

Then
$$p_\mathbf{A}(\lambda) = \det(\lambda \mathbf{I} - \mathbf{A}) = \begin{vmatrix} \lambda - 1 & -1 \\ 1 & \lambda - 1 \end{vmatrix} = (\lambda - 1)^2 + 1 = \lambda^2 - 2\lambda + 2.$$

Therefore, the eigenvalues are $1 \pm i$, where $i = \sqrt{-1}$. ∎

Example 3.43. Let
$$\mathbf{A} = \begin{bmatrix} 1 & 1 & 1 \\ 0 & 1 & 1 \\ 0 & 0 & 1 \end{bmatrix}.$$

Then
$$p_\mathbf{A}(\lambda) = \det(\lambda \mathbf{I} - \mathbf{A}) = \begin{vmatrix} \lambda - 1 & -1 & -1 \\ 0 & \lambda - 1 & -1 \\ 0 & 0 & \lambda - 1 \end{vmatrix} = (\lambda - 1)^3.$$

Therefore, the only eigenvalue is 1, which is an eigenvalue of multiplicity 3. In this example, it is instructive to compute the eigenvectors. We must solve $(\lambda \mathbf{I} - \mathbf{A})x = 0$ for $\lambda = 1$, that is, $(\mathbf{I} - \mathbf{A})x = 0$. We have

$$\mathbf{I} - \mathbf{A} = \begin{bmatrix} 1 & 0 & 0 \\ 0 & 1 & 0 \\ 0 & 0 & 1 \end{bmatrix} - \begin{bmatrix} 1 & 1 & 1 \\ 0 & 1 & 1 \\ 0 & 0 & 1 \end{bmatrix} = \begin{bmatrix} 0 & -1 & -1 \\ 0 & 0 & -1 \\ 0 & 0 & 0 \end{bmatrix},$$

and a straightforward calculation shows that the solution space (the *eigenspace*) is

$$\{(\alpha, 0, 0) : \alpha \in \mathbf{R}\} = \mathrm{span}\{(1, 0, 0)\}.$$

Thus, in spite of the fact that the eigenvalue has multiplicity 3, there is only one linearly independent eigenvector corresponding to the eigenvalue. ∎

The fact that the matrix in the previous example has only one eigenvector for an eigenvalue of multiplicity 3 is significant. It means that there is not a basis of $\mathbf{R}^3$ consisting of eigenvectors of $\mathbf{A}$.

3.5.1 The transpose of a matrix and the dot product

The theory of eigenvalues and eigenvectors is greatly simplified when $\mathbf{A} \in \mathbf{R}^{n \times n}$ is symmetric, that is, when $\mathbf{A}^T = A$. In order to appreciate this, we need to understand the relationship of the transpose of a matrix to the Euclidean inner product (or dot product).

3.5. Eigenvalues and eigenvectors of a symmetric matrix

We consider a linear operator defined by $\mathbf{A} \in \mathbf{R}^{m \times n}$ and perform the following calculation:

$$(\mathbf{Au}) \cdot \mathbf{v} = \sum_{i=1}^{m} (\mathbf{Au})_i v_i = \sum_{i=1}^{m} \left(\sum_{j=1}^{n} A_{ij} u_j \right) v_i = \sum_{i=1}^{m} \sum_{j=1}^{n} A_{ij} u_j v_i$$

$$= \sum_{j=1}^{n} \sum_{i=1}^{m} A_{ij} v_i u_j$$

$$= \sum_{j=1}^{n} \left(\sum_{i=1}^{m} A_{ij} v_i \right) u_j$$

$$= \sum_{j=1}^{n} (\mathbf{A}^T \mathbf{v})_j u_j$$

$$= \mathbf{u} \cdot \left(\mathbf{A}^T \mathbf{v} \right).$$

(The above manipulations are just applications of the elementary properties of arithmetic—basically that numbers can be added in any order, and multiplication distributes over addition.) Thus the transpose $\mathbf{A}^T$ of $\mathbf{A} \in \mathbf{R}^{m \times n}$ satisfies the following fundamental property:

$$(\mathbf{Au}) \cdot \mathbf{v} = \mathbf{u} \cdot \left(\mathbf{A}^T \mathbf{v} \right) \text{ for all } \mathbf{u} \in \mathbf{R}^n, \mathbf{v} \in \mathbf{R}^m. \tag{3.25}$$

For a symmetric matrix $\mathbf{A} \in \mathbf{R}^{n \times n}$, this simplifies to

$$(\mathbf{Au}) \cdot \mathbf{v} = \mathbf{u} \cdot (\mathbf{Av}) \text{ for all } \mathbf{u}, \mathbf{v} \in \mathbf{R}^n.$$

When we allow for complex scalars and vectors with complex entries, we must modify the dot product. If $\mathbf{x}, \mathbf{y} \in \mathbf{C}^n$, then we define

$$\mathbf{x} \cdot \mathbf{y} = \sum_{i=1}^{n} x_i \overline{y_i}.$$

The second vector in a dot product must thus be conjugated; this is necessary so that $\mathbf{x} \cdot \mathbf{x}$ will be real, allowing the norm to be defined. We will use the same notation for the dot product whether the vectors are in $\mathbf{R}^n$ or $\mathbf{C}^n$. The complex dot product has the following properties, which form the definition of an inner product on a complex vector space:

1. $\mathbf{x} \cdot \mathbf{x} \geq 0$ for all $\mathbf{x} \in \mathbf{C}^n$, and $\mathbf{x} \cdot \mathbf{x} = 0$ if and only if $\mathbf{x} = \mathbf{0}$.

2. $\mathbf{x} \cdot \mathbf{y} = \overline{\mathbf{y} \cdot \mathbf{x}}$ for all $\mathbf{x}, \mathbf{y} \in \mathbf{C}^n$.

3. $(\alpha \mathbf{x} + \beta \mathbf{y}) \cdot \mathbf{z} = \alpha \mathbf{x} \cdot \mathbf{z} + \beta \mathbf{y} \cdot \mathbf{z}$ for all $\mathbf{x}, \mathbf{y}, \mathbf{z} \in \mathbf{C}^n$, $\alpha, \beta \in \mathbf{C}$. Together with the second property, this implies that

    $$\mathbf{z} \cdot (\alpha \mathbf{x} + \beta \mathbf{y}) = \overline{\alpha} \mathbf{z} \cdot \mathbf{x} + \overline{\beta} \mathbf{z} \cdot \mathbf{y}$$

    for all $\mathbf{x}, \mathbf{y}, \mathbf{z} \in \mathbf{C}^n$, $\alpha, \beta \in \mathbf{C}$.

Complex vector spaces and inner products are discussed in more detail in Section 12.1.

3.5.2 Special properties of symmetric matrices

We can now derive the special properties of the eigenvalues and eigenvectors of a symmetric matrix.

Theorem 3.44. *If $A \in \mathbf{R}^{n \times n}$ is symmetric, then every eigenvalue of A is real. Moreover, each eigenvalue corresponds to a real eigenvector.*

Proof. Suppose $A\mathbf{x} = \lambda \mathbf{x}$, $\mathbf{x} \neq \mathbf{0}$, where for the moment we do not exclude the possibility that λ and $\mathbf{x}$ might be complex. Then

$$(A\mathbf{x}) \cdot \mathbf{x} = (\lambda \mathbf{x}) \cdot \mathbf{x} = \lambda(\mathbf{x} \cdot \mathbf{x})$$

and

$$\mathbf{x} \cdot (A\mathbf{x}) = \mathbf{x} \cdot (\lambda \mathbf{x}) = \overline{\lambda}(\mathbf{x} \cdot \mathbf{x}).$$

But $(A\mathbf{x}) \cdot \mathbf{x} = \mathbf{x} \cdot (A\mathbf{x})$ when A is symmetric, so

$$\lambda(\mathbf{x} \cdot \mathbf{x}) = \overline{\lambda}(\mathbf{x} \cdot \mathbf{x}).$$

Since $\mathbf{x} \cdot \mathbf{x} \neq 0$, this yields $\lambda = \overline{\lambda}$, which implies that λ is real. Let $\mathbf{x} = \mathbf{u} + i\mathbf{v}$, where $\mathbf{u}, \mathbf{v} \in \mathbf{R}^n$. Then

$$A\mathbf{x} = \lambda \mathbf{x} \Leftrightarrow A(\mathbf{u} + i\mathbf{v}) = \lambda(\mathbf{u} + i\mathbf{v})$$
$$\Leftrightarrow A\mathbf{u} + iA\mathbf{v} = \lambda \mathbf{u} + i\lambda \mathbf{v}$$
$$\Leftrightarrow \begin{cases} A\mathbf{u} = \lambda \mathbf{u}, \\ A\mathbf{v} = \lambda \mathbf{v}. \end{cases}$$

Since $\mathbf{x} \neq \mathbf{0}$, we must have either $\mathbf{u} \neq \mathbf{0}$ or $\mathbf{v} \neq \mathbf{0}$ (or both), so one of $\mathbf{u}$, $\mathbf{v}$ (or both) must be a real eigenvector of A corresponding to λ. $\square$

From this point on, we will only discuss eigenvalues and eigenvectors for real symmetric matrices. According to the last theorem, then, we will not need to use complex numbers or vectors.

Theorem 3.45. *Let $A \in \mathbf{R}^{n \times n}$ be symmetric, and let $\mathbf{x}_1$, $\mathbf{x}_2$ be eigenvectors of A corresponding to distinct eigenvalues λ_1, λ_2. Then $\mathbf{x}_1$ and $\mathbf{x}_2$ are orthogonal.*

Proof. We have

$$(A\mathbf{x}_1) \cdot \mathbf{x}_2 = \mathbf{x}_1 \cdot (A\mathbf{x}_2).$$

But

$$(A\mathbf{x}_1) \cdot \mathbf{x}_2 = (\lambda_1 \mathbf{x}_1) \cdot \mathbf{x}_2 = \lambda_1(\mathbf{x}_1 \cdot \mathbf{x}_2)$$

and

$$\mathbf{x}_1 \cdot (A\mathbf{x}_2) = \mathbf{x}_1 \cdot (\lambda_2 \mathbf{x}_2) = \lambda_2(\mathbf{x}_1 \cdot \mathbf{x}_2).$$

Therefore

$$\lambda_1(\mathbf{x}_1 \cdot \mathbf{x}_2) = \lambda_2(\mathbf{x}_1 \cdot \mathbf{x}_2),$$

and since $\lambda_1 \neq \lambda_2$, this implies that $(\mathbf{x}_1, \mathbf{x}_2) = 0$. $\square$

3.5. Eigenvalues and eigenvectors of a symmetric matrix

Example 3.46. Consider
$$\mathbf{A} = \begin{bmatrix} 1/3 & 1/3 & 1/3 \\ 1/3 & 1/3 & 1/3 \\ 1/3 & 1/3 & 1/3 \end{bmatrix}.$$

A straightforward calculation shows that
$$\det(\lambda \mathbf{I} - \mathbf{A}) = \lambda^3 - \lambda^2 = \lambda^2(\lambda - 1),$$

so the eigenvalues of $\mathbf{A}$ are $0, 0, 1$. Another straightforward calculation shows that there are two linearly independent eigenvectors corresponding to $\lambda = 0$, namely,

$$\mathbf{x}_1 = \begin{bmatrix} -1 \\ 1 \\ 0 \end{bmatrix}, \mathbf{x}_2 = \begin{bmatrix} 0 \\ -1 \\ 1 \end{bmatrix},$$

and a single (independent) eigenvector for $\lambda = 1$,

$$\mathbf{x}_3 = \begin{bmatrix} 1 \\ 1 \\ 1 \end{bmatrix}.$$

By inspection, the eigenvectors for $\lambda = 0$ are orthogonal to the eigenvector for $\lambda = 1$. ∎

Here is another special property of symmetric matrices. Example 3.43 shows that this result is not true for nonsymmetric matrices.

Theorem 3.47. *Let $\mathbf{A} \in \mathbf{R}^{n \times n}$ be symmetric, and suppose $\mathbf{A}$ has an eigenvalue μ of (algebraic) multiplicity k (meaning that μ is a root of multiplicity k of the characteristic polynomial of $\mathbf{A}$). Then $\mathbf{A}$ has k linearly independent eigenvectors corresponding to μ.*

The proof of this theorem is rather involved and does not generalize to differential operators (as do the proofs given above). We therefore relegate it to Appendix A.

If μ is an eigenvalue of multiplicity k, as in the previous theorem, then we can choose the k linearly independent eigenvectors corresponding to μ to be orthonormal.[10] We thus obtain the following corollary.

Corollary 3.48 (the spectral theorem for symmetric matrices). *Let $\mathbf{A} \in \mathbf{R}^{n \times n}$ be symmetric. Then there is an orthonormal basis $\{\mathbf{u}_1, \mathbf{u}_2, \ldots, \mathbf{u}_n\}$ of $\mathbf{R}^n$ consisting of eigenvectors of $\mathbf{A}$.*

3.5.3 The spectral method for solving $\mathbf{Ax} = \mathbf{b}$

When $\mathbf{A} \in \mathbf{R}^{n \times n}$ is symmetric and the eigenvalues and eigenvectors of $\mathbf{A}$ are known, there is a simple method[11] for solving $\mathbf{Ax} = \mathbf{b}$.

---

[10] It is always possible to replace any linearly independent set with an orthonormal set spanning the same subspace. The technique for doing this is called the Gram–Schmidt procedure; it is explained in elementary linear algebra texts such as [41].

[11] This method is not normally taught in elementary linear algebra courses or books, because it is more difficult to find the eigenvalues and eigenvectors of $\mathbf{A}$ than to just solve $\mathbf{Ax} = \mathbf{b}$ by other means.

Let $\mathbf{A}$ be symmetric with eigenvalues $\lambda_1, \lambda_2, \ldots, \lambda_n$ and orthonormal eigenvectors $\mathbf{u}_1, \mathbf{u}_2, \ldots, \mathbf{u}_n$. For any $\mathbf{b} \in \mathbf{R}^n$, we can write

$$\mathbf{b} = (\mathbf{u}_1 \cdot \mathbf{b})\mathbf{u}_1 + (\mathbf{u}_2 \cdot \mathbf{b})\mathbf{u}_2 + \cdots + (\mathbf{u}_n \cdot \mathbf{b})\mathbf{u}_n.$$

We can also write

$$\mathbf{x} = \alpha_1 \mathbf{u}_1 + \alpha_2 \mathbf{u}_2 + \cdots + \alpha_n \mathbf{u}_n.$$

(Of course, $\alpha_i = \mathbf{u}_i \cdot \mathbf{x}$, but, as we are thinking of $\mathbf{x}$ as the unknown, we cannot compute α_i from this formula.) We then have

$$\begin{aligned}\mathbf{Ax} = \mathbf{b} &\Rightarrow \mathbf{A}(\alpha_1 \mathbf{u}_1 + \cdots + \alpha_n \mathbf{u}_n) = (\mathbf{u}_1 \cdot \mathbf{b})\mathbf{u}_1 + \cdots + (\mathbf{u}_n \cdot \mathbf{b})\mathbf{u}_n \\ &\Rightarrow \alpha_1 \mathbf{A}\mathbf{u}_1 + \cdots + \alpha_n \mathbf{A}\mathbf{u}_n = (\mathbf{u}_1 \cdot \mathbf{b})\mathbf{u}_1 + \cdots + (\mathbf{u}_n \cdot \mathbf{b})\mathbf{u}_n \\ &\Rightarrow \alpha_1 \lambda_1 \mathbf{u}_1 + \cdots + \alpha_n \lambda_n \mathbf{u}_n = (\mathbf{u}_1 \cdot \mathbf{b})\mathbf{u}_1 + \cdots + (\mathbf{u}_n \cdot \mathbf{b})\mathbf{u}_n.\end{aligned}$$

Since $\mathbf{b}$ can have only one expansion in terms of the basis $\{\mathbf{u}_1, \ldots, \mathbf{u}_n\}$, we must have

$$\alpha_i \lambda_i = (\mathbf{u}_i \cdot \mathbf{b}), \ i = 1, 2, \ldots, n,$$

that is,

$$\alpha_i = \frac{(\mathbf{u}_i \cdot \mathbf{b})}{\lambda_i}, \ i = 1, 2, \ldots, n.$$

Thus we obtain the solution

$$\mathbf{x} = \frac{(\mathbf{u}_1 \cdot \mathbf{b})}{\lambda_1}\mathbf{u}_1 + \frac{(\mathbf{u}_2 \cdot \mathbf{b})}{\lambda_2}\mathbf{u}_2 + \cdots + \frac{(\mathbf{u}_n \cdot \mathbf{b})}{\lambda_n}\mathbf{u}_n. \qquad (3.26)$$

We see that all of the eigenvalues of $\mathbf{A}$ must be nonzero in order to apply this method, which is only sensible: if 0 is an eigenvalue of $\mathbf{A}$, then $\mathbf{A}$ is singular and $\mathbf{Ax} = \mathbf{b}$ either has no solution or has infinitely many solutions.

If we already have the eigenvalues and eigenvectors of $\mathbf{A}$, then this method for solving $\mathbf{Ax} = \mathbf{b}$ is simpler and less expensive than the usual method of Gaussian elimination.

Example 3.49. Let

$$\mathbf{A} = \begin{bmatrix} 11 & -4 & -1 \\ -4 & 14 & -4 \\ -1 & -4 & 11 \end{bmatrix}, \ \mathbf{b} = \begin{bmatrix} 1 \\ 2 \\ 1 \end{bmatrix}.$$

A direct calculation shows that the eigenvalues of $\mathbf{A}$ are

$$\lambda_1 = 6, \ \lambda_2 = 12, \ \lambda_3 = 18,$$

and the corresponding (orthonormal) eigenvectors are

$$\mathbf{u}_1 = \begin{bmatrix} \frac{1}{\sqrt{3}} \\ \frac{1}{\sqrt{3}} \\ \frac{1}{\sqrt{3}} \end{bmatrix}, \ \mathbf{u}_2 = \begin{bmatrix} \frac{1}{\sqrt{2}} \\ 0 \\ -\frac{1}{\sqrt{2}} \end{bmatrix}, \ \mathbf{u}_3 = \begin{bmatrix} \frac{1}{\sqrt{6}} \\ -\frac{2}{\sqrt{6}} \\ \frac{1}{\sqrt{6}} \end{bmatrix}.$$

3.5. Eigenvalues and eigenvectors of a symmetric matrix

(According to Theorem 3.45, the eigenvectors are automatically orthogonal in this case; however, we had to ensure that each eigenvector was normalized.) The solution to $\mathbf{Ax} = \mathbf{b}$ is

$$\mathbf{x} = \frac{\mathbf{u}_1 \cdot \mathbf{b}}{\lambda_1}\mathbf{u}_1 + \frac{\mathbf{u}_2 \cdot \mathbf{b}}{\lambda_2}\mathbf{u}_2 + \frac{\mathbf{u}_3 \cdot \mathbf{b}}{\lambda_3}\mathbf{u}_3 = \frac{4}{6\sqrt{3}}\mathbf{u}_1 + 0\mathbf{u}_2 - \frac{2}{18\sqrt{6}}\mathbf{u}_3 \doteq \begin{bmatrix} 0.20370 \\ 0.25926 \\ 0.20370 \end{bmatrix}. \quad \blacksquare$$

Exercises

1. Let
$$A = \begin{bmatrix} 164 & -48 \\ -48 & 136 \end{bmatrix}, \mathbf{b} = \begin{bmatrix} 116 \\ 88 \end{bmatrix}.$$

   Compute, by hand, the eigenvalues and eigenvectors of $\mathbf{A}$, and use them to solve $\mathbf{Ax} = \mathbf{b}$ for $\mathbf{x}$ (use the "spectral method").

2. Repeat Exercise 1 for
$$\mathbf{A} = \begin{bmatrix} 3 & -1 \\ -1 & 3 \end{bmatrix}, \mathbf{b} = \begin{bmatrix} 11 \\ -9 \end{bmatrix}.$$

3. Repeat Exercise 1 for
$$\mathbf{A} = \begin{bmatrix} 1 & 1 & 0 \\ 1 & 0 & 1 \\ 0 & 1 & 1 \end{bmatrix}, \mathbf{b} = \begin{bmatrix} -1 \\ 1 \\ 3 \end{bmatrix}.$$

4. Repeat Exercise 1 for
$$\mathbf{A} = \begin{bmatrix} 7 & -2 & 1 \\ -2 & 10 & -2 \\ 1 & -2 & 7 \end{bmatrix}, \mathbf{b} = \begin{bmatrix} 6 \\ 12 \\ 18 \end{bmatrix}.$$

5. Let $\mathbf{A} \in \mathbf{R}^{n \times n}$ be symmetric, and suppose the eigenvalues and (orthonormal) eigenvectors of $\mathbf{A}$ are already known. How many arithmetic operations are required to solve $\mathbf{Ax} = \mathbf{b}$ using the spectral method?

6. A symmetric matrix $\mathbf{A} \in \mathbf{R}^{n \times n}$ is called *positive definite* if
$$\mathbf{x} \neq \mathbf{0} \Rightarrow (\mathbf{Ax}) \cdot \mathbf{x} > 0.$$

   Use the spectral theorem to show that $\mathbf{A}$ is positive definite if and only if all of the eigenvalues of $\mathbf{A}$ are positive.

7. Let $\mathbf{L}$ be the $n \times n$ matrix defined by the condition that
$$L_{ij} = \begin{cases} \frac{2}{h^2}, & i = j, \\ -\frac{1}{h^2}, & |i - j| = 1, \\ 0 & \text{otherwise}, \end{cases}$$

where $h = 1/(n+1)$. For example, with $n = 5$,

$$\mathbf{L} = \begin{bmatrix} 72 & -36 & 0 & 0 & 0 \\ -36 & 72 & -36 & 0 & 0 \\ 0 & -36 & 72 & -36 & 0 \\ 0 & 0 & -36 & 72 & -36 \\ 0 & 0 & 0 & -36 & 72 \end{bmatrix}.$$

(a) For each $j = 1, 2, \ldots, n$, define the discrete sine wave $\mathbf{s}^{(j)}$ of frequency j by

$$\mathbf{s}_k^{(j)} = \sin(jk\pi h).$$

Show that $\mathbf{s}^{(j)}$ is an eigenvector of $\mathbf{L}$, and find the corresponding eigenvalue λ_j. (Hint: Compute $\mathbf{L}\mathbf{s}^{(j)}$ and apply the addition formula for the sine function.)

(b) What is the relationship between the frequency j and the magnitude of λ_j?

(c) The discrete sine waves are orthogonal (since they are the eigenvectors of a symmetric matrix corresponding to distinct eigenvalues) and thus form an orthogonal basis for $\mathbf{R}^n$. Moreover, it can be shown that every $\mathbf{s}^{(j)}$ has the same norm:

$$\left\| \mathbf{s}^{(j)} \right\| = \frac{1}{\sqrt{2h}}, \; j = 1, 2, \ldots, n.$$

Therefore, $\{\sqrt{2h}\mathbf{s}^{(1)}, \sqrt{2h}\mathbf{s}^{(2)}, \ldots, \sqrt{2h}\mathbf{s}^{(n)}\}$ is an orthonormal basis for $\mathbf{R}^n$. We will call a vector $\mathbf{x} \in \mathbf{R}^n$ smooth or rough depending on whether its components in the discrete sine wave basis are heavily weighted toward the low or the high frequencies, respectively. Show that the solution $\mathbf{x}$ of $\mathbf{L}\mathbf{x} = \mathbf{b}$ is smoother than $\mathbf{b}$.

3.6 Preview of methods for solving ODEs and PDEs

The close formal relation between the concepts of linear differential equation and linear algebraic system suggests that ideas about linear operators might play a role in solving the former, as is the case for the latter. In order to make this parallel explicit, we will apply the machinery of linear algebra in the context of differential equations as follows: view solutions as vectors in a vector space, identify the linear operators which define linear differential equations, and understand how facts such as the Fredholm alternative appear in the context of differential equations.

In the chapters to follow we will accomplish all of this. In the process, we will develop the following three general classes of methods for solving linear ODEs and PDEs; each class is closely analogous to a method for solving linear algebraic systems.

1. **The method of Fourier series:** Differential operators, like d^2/dx^2, can have eigenvalues and eigenfunctions; for example,

$$\frac{d^2}{dx^2}[\sin(\omega x)] = -\omega^2 \sin(\omega x).$$

The method of Fourier series is a spectral method, using the eigenfunctions of the differential operator.

2. **The method of Green's functions:** A Green's function for a differential equation is the solution to a special form of the equation (just as $\mathbf{A}^{-1}$ is the solution to $\mathbf{AB} = I$) that allows one to immediately write down the solution to the equation (just as we could write down $\mathbf{x} = \mathbf{A}^{-1}\mathbf{b}$).

3. **The method of finite elements:** This is a direct numerical method that can be used when the first two methods fail (or are intractable). It can be compared with Gaussian elimination, the standard direct numerical method for solving $A\mathbf{x} = \mathbf{b}$. Like Gaussian elimination, finite element methods do not produce a formula for the solution; however, again like Gaussian elimination, they are broadly applicable.

In Chapters 5–7, we introduce Fourier series and finite element methods in one spatial dimension, extending these methods to multiple dimensions in Chapter 11. Chapters 10 and 12 examine spectral methods and Fourier series in more depth, while Chapter 13 gives more details about finite element methods. Chapter 9 introduces the method of Green's functions.

An approach that has no analogue in linear algebra is the method of characteristics, which transforms a first-order PDE into a family of ODEs. This technique is explained in Chapter 8.

3.7 Suggestions for further reading

A good introductory text, which assumes no prior knowledge of linear algebra and is written at a very accessible level, is Lay [41]. An alternative is Strang [60]; this book also assumes no background in linear algebra, but it is written at a somewhat more demanding level. It is noteworthy for its many insights into the applications of linear algebra and for its conversational tone. More advanced references include Meyer [51] and the author's text [24].

Anyone seriously interested in applied mathematics must become familiar with the computational aspects of linear algebra. The text by Strang mentioned above includes material on the numerical aspects of the subject. There are also more specialized references. A good introductory text is Hager [31], while more advanced treatments include Demmel [17] and Trefethen and Bau [65]. An encyclopedic reference is Golub and Van Loan [26].

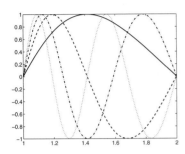

Chapter 4
Essential Ordinary Differential Equations

In an ordinary differential equation (ODE), there is a single independent variable. Commonly ODEs model changes over time, so the independent variable is t (time). Our interest in ODEs derives from the following fact: both the Fourier series method and the finite element method reduce time-dependent PDEs into systems of ODEs. In the case of the Fourier series method, the system is completely decoupled, so the "system" is really just a sequence of scalar ODEs. In Section 4.2, we learn how to solve the scalar ODEs that arise in the Fourier series method.

The finite element method, on the other hand, results in coupled systems of ODEs. In Section 4.3, we discuss the solution of linear, coupled systems of first-order ODEs. Although we present an explicit solution technique in that section, the emphasis is really on the properties of the solutions, as the systems that arise in practice are destined to be solved by numerical rather than analytical means. In Sections 4.4 and 4.5, we introduce some simple numerical methods that are adequate for our purposes.

4.1 Background

4.1.1 Converting a higher-order equation to a first-order system

We begin our discussion of ODEs with a simple observation: It is always possible to convert a single ODE of order two or more to a system of first-order ODEs. We illustrate this on the following second-order equation:

$$a\frac{d^2u}{dt^2} + b\frac{du}{dt} + cu = f(t). \tag{4.1}$$

We define

$$x_1(t) = u(t), \; x_2(t) = \frac{du}{dt}(t).$$

Then we have

$$\frac{dx_1}{dt} = x_2, \; \frac{dx_2}{dt} = \frac{d^2u}{dt^2} = -\frac{c}{a}u - \frac{b}{a}\frac{du}{dt} + \frac{1}{a}f(t) = -\frac{c}{a}x_1 - \frac{b}{a}x_2 + \frac{1}{a}f(t).$$

We can write this as a first-order system using matrix-vector notation:

$$\frac{d\mathbf{x}}{dt} = \mathbf{A}\mathbf{x} + \mathbf{f}(t), \ \mathbf{A} = \begin{bmatrix} 0 & 1 \\ -\frac{c}{a} & -\frac{b}{a} \end{bmatrix}, \ \mathbf{f}(t) = \begin{bmatrix} 0 \\ \frac{1}{a}f(t) \end{bmatrix}. \quad (4.2)$$

In (4.2), $\mathbf{x}$ is the vector-valued function

$$\mathbf{x}(t) = \begin{bmatrix} x_1(t) \\ x_2(t) \end{bmatrix}.$$

The same technique will convert any scalar equation to a first-order system. If the unknown is u and the equation is of order m, we define

$$x_1 = u, \ x_2 = \frac{du}{dt}, \ldots, x_m = \frac{d^{m-1}u}{dt^{m-1}}.$$

The first $m-1$ equations will be

$$\frac{dx_1}{dt} = x_2, \ \frac{dx_2}{dt} = x_3, \ldots, \frac{dx_{m-1}}{dt} = x_m,$$

and the last equation will be the original ODE (expressed in the new variables). If the original scalar equation is linear, then the resulting system will also be linear, and it can be written in a matrix-vector form if desired.

An mth-order ODE typically has infinitely many solutions; in fact, an mth-order linear ODE has an m-dimensional subspace of solutions. To narrow down a unique solution, m auxiliary conditions are required. When the independent variable is time, the auxiliary conditions are usually *initial conditions*:

$$u(t_0) = \alpha_0, \ \frac{du}{dt}(t_0) = \alpha_1, \ldots, \frac{d^{m-1}u}{dt^{m-1}}(t_0) = \alpha_{m-1}.$$

These initial conditions translate immediately into initial conditions for the new unknowns $x_1, x_2, \ldots, x_m$:

$$x_1(t_0) = \alpha_0, \ x_2(t_0) = \alpha_1, \ldots, x_m(t_0) = \alpha_{m-1}.$$

We can apply a similar technique to convert an mth-order system of ODEs to a first-order system. For example, suppose $\mathbf{x}(t)$ is a vector-valued function, say $\mathbf{x}(t) \in \mathbf{R}^n$. Consider a second-order system

$$\frac{d^2\mathbf{x}}{dt^2} = \mathbf{f}\left(t, \mathbf{x}, \frac{d\mathbf{x}}{dt}\right). \quad (4.3)$$

If we define

$$\mathbf{y}(t) = \mathbf{x}(t), \ \mathbf{z}(t) = \frac{d\mathbf{x}}{dt}(t),$$

then we obtain the system

$$\frac{d\mathbf{y}}{dt} = \mathbf{z},$$
$$\frac{d\mathbf{z}}{dt} = \mathbf{f}(t, \mathbf{y}, \mathbf{z}).$$

4.1. Background

The original system (4.3) consists of n second-order ODEs in n unknowns (the components of $\mathbf{x}(t)$). We have rewritten this original system as $2n$ first-order ODEs in $2n$ unknowns (the n components of $\mathbf{y}(t)$ and the n components of $\mathbf{z}(t)$).

The fact that any ODE can be written as a first-order system has the following benefit: Any theory or algorithm developed for first-order systems of ODEs is automatically applicable to *any* ODE. This leads to a considerable simplification in the study of this subject.

4.1.2 The general solution of a homogeneous linear second-order ODE

A homogeneous linear second-order ODE has the form

$$a(t)\frac{d^2u}{dt^2} + b(t)\frac{du}{dt} + c(t)u = 0, \tag{4.4}$$

where a, b, c are continuous functions and we will assume that $a(t) \neq 0$ in the interval of interest. A basic result in the theory of differential equations is that there is a unique solution to any *initial value problem* (IVP) consisting of (4.4) together with two *initial conditions*

$$u(t_0) = v_1, \quad \frac{du}{dt}(t_0) = v_2. \tag{4.5}$$

Conversely, given any solution x of (4.4), if we define $v_1 = u(t_0)$, $v_2 = du/dt(t_0)$, then x is the solution of the IVP (4.4)–(4.5). Therefore, if we can find a formula representing the solution of the IVP (4.4)–(4.5), then it represents all solutions to (4.4), and we will call this formula the general solution of (4.4).

We will now show that we can find the general solution of (4.4) by finding two *linearly independent* solutions u_1, u_2 of the ODE (below we explain what linear independence means for functions). We choose linearly independent vectors $\mathbf{v}, \mathbf{w}$ in $\mathbf{R}^2$ and solve the two IVPs

$$\begin{aligned} a(t)\frac{d^2u}{dt^2} + b(t)\frac{du}{dt} + c(t)u &= 0, \\ u(t_0) &= v_1, \\ \frac{du}{dt}(t_0) &= v_2 \end{aligned} \tag{4.6}$$

and

$$\begin{aligned} a(t)\frac{d^2u}{dt^2} + b(t)\frac{du}{dt} + c(t)u &= 0, \\ u(t_0) &= w_1, \\ \frac{du}{dt}(t_0) &= w_2 \end{aligned} \tag{4.7}$$

to get solutions u_1, u_2, respectively. We first show that $\{u_1, u_2\}$ is linearly independent, which means that the only scalars c_1 and c_2 such that $c_1 u_1 + c_2 u_2 = 0$ are $c_1 = c_2 = 0$. (The reader should notice that $c_1 u_1 + c_2 u_2$ is a function, and therefore the equation $c_1 u_1 + c_2 u_2 = 0$

means that $c_1 u_1 + c_2 u_2$ is the zero function.) If $c_1 u_1 + c_2 u_2 = 0$, then, by differentiating both sides, we obtain
$$c_1 \frac{du_1}{dt} + c_2 \frac{du_2}{dt} = 0.$$
Evaluating these two equations at $t = t_0$, we obtain
$$c_1 u_1(t_0) + c_2 u_2(t_0) = 0,$$
$$c_1 \frac{du_1}{dt}(t_0) + c_2 \frac{du_2}{dt}(t_0) = 0,$$
or, taking into account the initial conditions satisfied by u_1, u_2,
$$c_1 v_1 + c_2 w_1 = 0,$$
$$c_1 v_2 + c_2 w_2 = 0.$$
This in turn is equivalent to the equation $\mathbf{Ac} = 0$, where $\mathbf{A}$ is the matrix whose columns are the vectors $\mathbf{v}, \mathbf{w}$. Since $\{\mathbf{v}, \mathbf{w}\}$ is linearly independent by assumption, $\mathbf{A}$ is nonsingular, and therefore $\mathbf{Ac} = 0$ implies that $\mathbf{c} = 0$. This shows that $\{u_1, u_2\}$ is linearly independent.

Now suppose that z_1, z_2 are any initial values and we wish to solve
$$a(t)\frac{d^2 u}{dt^2} + b(t)\frac{du}{dt} + c(t)x = 0,$$
$$u(t_0) = z_1, \qquad (4.8)$$
$$\frac{du}{dt}(t_0) = z_2.$$

Since the ODE is linear, $u = c_1 u_1 + c_2 u_2$ is a solution for any scalars c_1, c_2. We choose $\mathbf{c} = (c_1, c_2)$ to solve $\mathbf{Ac} = \mathbf{z}$, where, as above, $\mathbf{A} = [\mathbf{v}|\mathbf{w}]$ (this equation is equivalent to the system $c_1 v_1 + c_2 w_1 = z_1$, $c_1 v_2 + c_2 w_2 = z_2$). Notice that $\mathbf{Ac} = \mathbf{z}$ has a unique solution because $\mathbf{A}$ is nonsingular. We then have
$$u(t_0) = c_1 u_1(t_0) + c_2 u_2(t_0) = c_1 v_1 + c_2 w_1 = z_1$$
and
$$\frac{du}{dt}(t_0) = c_1 \frac{du_1}{dt}(t_0) + c_2 \frac{du_2}{dt}(t_0) = c_1 v_2 + c_2 w_2 = z_2,$$
and therefore x solves the IVP (4.8). This shows that every solution of (4.4) can be expressed in the form $c_1 u_1 + c_2 u_2$, and therefore that $c_1 u_1 + c_2 u_2$ is the general solution of (4.4).

In the above discussion, we chose two linearly independent solutions of (4.4) in a special fashion, by solving (4.6) and (4.7). Let us suppose that we find solutions u_1, u_2 of (4.4) by some other fashion. We wish to show that $\{u_1, u_2\}$ is linearly independent if and only if $\{\mathbf{v}, \mathbf{w}\}$ is linearly independent, where
$$\mathbf{v} = \begin{bmatrix} u_1(t_0) \\ \frac{du_1}{dt}(t_0) \end{bmatrix}, \quad \mathbf{w} = \begin{bmatrix} u_2(t_0) \\ \frac{du_2}{dt}(t_0) \end{bmatrix}.$$

We have already seen that if $\{\mathbf{v}, \mathbf{w}\}$ is linearly independent, then $\{u_1, u_2\}$ is linearly independent. Therefore, let us suppose that $\{\mathbf{v}, \mathbf{w}\}$ is linearly dependent. Then there exist scalars c_1, c_2, not both zero, such that $c_1 \mathbf{v} + c_2 \mathbf{w} = 0$. We define
$$u = c_1 u_1 + c_2 u_2$$

4.1. Background

and notice that x is a solution of the linear ODE (4.4). Also,

$$u(t_0) = c_1 u_1(t_0) + c_2 u_2(t_0) = c_1 v_1 + c_2 w_1 = 0,$$
$$\frac{du}{dt}(t_0) = c_1 \frac{du_1}{dt}(t_0) + c_2 \frac{du_2}{dt}(t_0) = c_1 v_2 + c_2 w_2 = 0.$$

Therefore, u solves the IVP

$$a(t)\frac{d^2 u}{dt^2} + b(t)\frac{du}{dt} + c(t)u = 0,$$
$$u(t_0) = 0,$$
$$\frac{du}{dt}(t_0) = 0.$$

But, by the theoretical result mentioned above, this IVP has a unique solution, which is obviously the zero function. Therefore, $c_1 u_1 + c_2 u_2 = 0$ and c_1, c_2 are not both zero. This shows that $\{u_1, u_2\}$ is linearly dependent, and we have completed the proof that $\{u_1, u_2\}$ is linearly independent if and only if $\{\mathbf{v}, \mathbf{w}\}$ is linearly independent.

4.1.3 The Wronskian test

The result derived in the previous paragraph is called the *Wronskian* test and is usually expressed as follows. If u_1, u_2 are two solutions to the linear ODE (4.4), then $\{u_1, u_2\}$ is linearly independent if and only if the matrix

$$\begin{bmatrix} u_1(t_0) & u_2(t_0) \\ \frac{du_1}{dt}(t_0) & \frac{du_2}{dt}(t_0) \end{bmatrix}$$

is nonsingular for any t_0. Using determinants, we can say that $\{u_1, u_2\}$ is linearly independent if and only if

$$W(t_0) = \begin{vmatrix} u_1(t_0) & u_2(t_0) \\ \frac{du_1}{dt}(t_0) & \frac{du_2}{dt}(t_0) \end{vmatrix} \neq 0$$

for any t_0. If $W(t_0) \neq 0$, then the general solution of (4.4) is $u = c_1 u_1 + c_2 u_2$.

We will present examples of the Wronskian test and of general solutions to some simple ODEs in the next section.

Exercises

1. Write the ODE
$$2\frac{d^2 u}{du^2} + u = 0$$
as a system of first-order ODEs.

2. Write the ODE
$$\frac{d^2 u}{du^2} + c\sin(u) = 0$$
as a system of first-order ODEs.

3. Write the ODE
$$\frac{d^4 u}{dt^4} - 2\frac{d^2 u}{dt^2} + u = \sin(t)$$
as a system of first-order ODEs.

4. Write the ODE
$$\frac{d^2 u}{dt^2} - 2u\frac{du}{dt} + t^2 u = 0$$
as a system of first-order ODEs.

5. Write the following system of second-order ODEs as a system of first-order ODEs:
$$m_1 \frac{d^2 p}{dt^2} = f_1\left(p, q, \frac{dp}{dt}, \frac{dq}{dt}\right),$$
$$m_2 \frac{d^2 q}{dt^2} = f_2\left(p, q, \frac{dp}{dt}, \frac{dq}{dt}\right).$$

Here m_1 and m_2 are constants, and f_1 and f_2 are real-valued functions of four variables.

6. Let m_1, m_2, k_1, k_2, k_3 be constants. Write
$$m_1 \frac{d^2 u_1}{dt^2} = -(k_1 + k_2) u_1 + k_2 u_2,$$
$$m_2 \frac{d^2 u_2}{dt^2} = k_2 u_1 - (k_2 + k_3) u_2$$
as a system of first-order ODEs.

7. Let $u_1(t) = e^{-t}$, $u_2(t) = e^{-2t}$.

    (a) Verify that u_1, u_2 are solutions to the ODE
    $$\frac{d^2 u}{dt^2} + 3\frac{du}{dt} + 2u = 0.$$

    (b) Use the Wronskian test to prove that $\{u_1, u_2\}$ is linearly independent.

8. Let $u_1(t) = t^2$, $u_2(t) = t^{-3}$.

    (a) Verify that u_1, u_2 are solutions to the ODE
    $$t^2 \frac{d^2 u}{dt^2} + 2t\frac{du}{dt} - 6u = 0.$$

    (b) Use the Wronskian test to prove that $\{u_1, u_2\}$ is linearly independent.

4.2. Solutions to some simple ODEs

9. The Wronskian test, which is an "if and only if" statement, applies to solutions of a linear ODE. If applied to arbitrary functions u_1, u_2, the "if" part still holds, but not the "only if" part.

   (a) Let u_1, u_2 be two functions defined on an interval, let t_0 be a point in that interval, and suppose
   $$\begin{vmatrix} u_1(t_0) & u_2(t_0) \\ \frac{du_1}{dt}(t_0) & \frac{du_2}{dt}(t_0) \end{vmatrix} \neq 0.$$
   Prove that $\{u_1, u_2\}$ is linearly independent.

   (b) Let $u_1(t) = \cos(t)$, $u_2(t) = \cos(2t)$. Show that $\{u_1, u_2\}$ is linearly independent, yet
   $$\begin{vmatrix} u_1(0) & u_2(0) \\ \frac{du_1}{dt}(0) & \frac{du_2}{dt}(0) \end{vmatrix} = 0.$$

4.2 Solutions to some simple ODEs

In this section, we show how to solve some simple first- and second-order ODEs that will arise later in the text.

4.2.1 The general solution of a second-order homogeneous ODE with constant coefficients

In the ensuing chapters, we will often encounter the second-order linear homogeneous ODE with constant coefficients,

$$a\frac{d^2u}{dt^2} + b\frac{du}{dt} + cu = 0, \quad (4.9)$$

so we will present a simple method for computing its general solution. The method is based on the following idea: When faced with a differential equation, one can sometimes guess the general form of the solution and, by substituting this general form into the equation, determine the specific form.

In this case, we assume that the solution of (4.9) is of the form $u(t) = e^{rt}$. Substituting into (4.9) yields

$$ar^2 e^{rt} + bre^{rt} + ce^{rt} = 0;$$

since the exponential is never zero, this equation holds if and only if

$$ar^2 + br + c = 0.$$

This quadratic is called the *characteristic polynomial* of the ODE (4.9), and its roots are called the *characteristic roots* of the ODE.

The characteristic roots are given by the quadratic formula

$$r_1 = \frac{-b - \sqrt{b^2 - 4ac}}{2a}, \quad r_2 = \frac{-b + \sqrt{b^2 - 4ac}}{2a}.$$

We distinguish three cases as follows.

1. **The characteristic roots are real and unequal (i.e., $b^2 - 4ac > 0$).** In this case, we have two solutions $u_1(t) = e^{r_1 t}$, $u_2(t) = e^{r_2 t}$. We can use the Wronskian test to verify that $\{u_1, u_2\}$ is linearly independent. We have

$$\frac{du_1}{dt}(t) = r_1 e^{r_1 t}, \quad \frac{du_2}{dt}(t) = r_2 e^{r_2 t},$$

and therefore

$$W(0) = \begin{vmatrix} u_1(0) & u_2(0) \\ \frac{du_1}{dt}(0) & \frac{du_2}{dt}(0) \end{vmatrix} = \begin{vmatrix} 1 & 1 \\ r_1 & r_2 \end{vmatrix} = r_2 - r_1 \neq 0$$

(since $r_2 \neq r_1$). Therefore, $\{u_1, u_2\}$ is linearly independent, and the general solution of (4.9) is

$$u(t) = c_1 e^{r_1 t} + c_2 e^{r_2 t}. \tag{4.10}$$

2. **The characteristic roots are complex (i.e., $b^2 - 4ac < 0$).** In this case, the characteristic roots are also unequal; in fact, they form a complex conjugate pair. The analysis of the previous case applies, and we could write the general solution in the form (4.10) in this case as well. However, with r_1, r_2 complex, $e^{r_1 t}, e^{r_2 t}$ are complex-valued functions, and this is undesirable.

With $r_1 = \mu - \lambda i$, $r_2 = \mu + \lambda i$, we have (by Euler's formula)

$$e^{r_1 t} = e^{\mu t}(\cos(\lambda t) - i\sin(\lambda t)),$$
$$e^{r_2 t} = e^{\mu t}(\cos(\lambda t) + i\sin(\lambda t)),$$

and so

$$\frac{1}{2}\left(e^{r_1 t} + e^{r_2 t}\right) = e^{\mu t} \cos(\lambda t),$$
$$\frac{i}{2}\left(e^{r_1 t} - e^{r_2 t}\right) = e^{\mu t} \sin(\lambda t).$$

This shows (because the equation is linear) that

$$u_1(t) = e^{\mu t} \cos(\lambda t), \; u_2(t) = e^{\mu t} \sin(\lambda t)$$

are also solutions of (4.9) in this case. We will write the general solution in the form

$$u(t) = c_1 e^{\mu t} \cos(\lambda t) + c_2 e^{\mu t} \sin(\lambda t).$$

As in the previous case, we can show that $\{u_1, u_2\}$ is linearly independent (see Exercise 4.1.2).

3. **The characteristic polynomial has a single (repeated) real root (i.e., $b^2 - 4ac = 0$).** In this case the root is $r = -b/(2a)$. We cannot write the general solution with the single solution e^{rt}; by an inspired guess, we try te^{rt} as the second solution. Indeed, with $u(t) = te^{rt}$, we have

$$\frac{du}{dt}(t) = (1+rt)e^{rt}, \; \frac{d^2 u}{dt^2}(t) = (2+rt)re^{rt},$$

4.2. Solutions to some simple ODEs

and so

$$a\frac{d^2u}{dt^2}(t) + b\frac{du}{dt}(t) + cu(t)$$
$$= a(2+rt)re^{rt} + b(1+rt)e^{rt} + cte^{rt}$$
$$= \left(ar^2 + br + c\right)te^{rt} + (2ar+b)e^{rt}$$
$$= 0 \text{ (since } ar^2 + br + c = 0 \text{ and } 2ar + b = 2a(-b/(2a)) + b = 0).$$

This shows that te^{rt} is also a solution of (4.9), and we write the general solution in this case as

$$u(t) = c_1 e^{rt} + c_2 t e^{rt},$$

where $r = -b/(2a)$. Once again, it is not hard to show that this is indeed the general solution of (4.9) in this case (see Exercise 4.1.3).

Example 4.1. The characteristic polynomial of

$$\frac{d^2u}{dt^2} - \frac{du}{dt} + u = 0$$

is $r^2 - r + 1$, which has roots

$$r_1 = \frac{1}{2} + \frac{\sqrt{3}}{2}i, \; r_2 = \frac{1}{2} - \frac{\sqrt{3}}{2}i.$$

Therefore this example falls in case 4.2.1 above, and the general solution of the ODE is

$$u(t) = \left(c_1 \cos\left(\frac{\sqrt{3}t}{2}\right) + c_2 \sin\left(\frac{\sqrt{3}t}{2}\right)\right)e^{t/2}. \quad \blacksquare$$

Example 4.2. The ODE

$$\frac{d^2u}{dt^2} - 2\frac{du}{dt} + u = 0$$

has characteristic polynomial $r^2 - 2r + 1$. The characteristic roots are

$$r_1 = r_2 = 1,$$

so case 4.2.1 above applies. The general solution is

$$u(t) = c_1 e^t + c_2 t e^t. \quad \blacksquare$$

4.2.2 Variation of parameters

We have shown above how to solve any linear homogeneous second-order ODE with constant coefficients; that is, we can find the general solution

$$u(t) = c_1 u_1(t) + c + 2u_2(t)$$

of (4.9). We now consider the inhomogeneous version of the same equation:

$$a\frac{d^2u}{dt^2} + b\frac{du}{dt} + cu = f(t). \tag{4.11}$$

This equation does not have a unique solution; if u_p is one solution of (4.11), then, for any scalars c_1, c_2, $u = u_p + c_1 u_1 + c_2 u_2$ is another solution. To see this, let $L : C^2(\mathbf{R}) \to C(\mathbf{R})$ be defined by

$$Lu = a\frac{d^2u}{dt^2} + b\frac{du}{dt} + cu.$$

By assumption, we have $L(u_p) = f$ and $L(c_1 u_1 + c_2 u_2) = 0$. By linearity, then, we have

$$L(u_p + c_1 u_1 + c_2 u_2) = L(u_p) + L(c_1 u_1 + c_2 u_2) = f + 0 = f.$$

Moreover, it is not difficult to prove that $u = u_p + c_1 u_1 + c_2 u_2$ is the general solution to (4.11) (see Exercise 4.2.14).

Therefore, to find the general solution of (4.11), it suffices to find the general solution of the homogeneous equation (4.9) (which we already know how to do) and to find any one solution u_p of (4.11). Such a solution u_p is often called a *particular solution*.

There is a trick for finding a particular solution of (4.11), called the method of *variation of parameters*. We look for a solution of the form

$$u(t) = c_1(t) u_1(t) + c_2(t) u_2(t)$$

(notice that this solution looks like the general solution of the homogeneous ODE, except that the parameters c_1, c_2 are allowed to vary with t—hence the name). We substitute this formula for u into (4.11), yielding one equation for the two unknowns c_1, c_2. We are free to impose a second equation, which (for reasons that will become clear) we take to be

$$\frac{dc_1}{dt} u_1 + \frac{dc_2}{dt} u_2 = 0. \tag{4.12}$$

We then have

$$u = c_1 u_1 + c_2 u_2,$$
$$\frac{du}{dt} = c_1 \frac{du_1}{dt} + c_2 \frac{du_2}{dt} + \frac{dc_1}{dt} u_1 + \frac{dc_2}{dt} u_2$$
$$= c_1 \frac{du_1}{dt} + c_2 \frac{du_2}{dt},$$
$$\frac{d^2u}{dt^2} = c_1 \frac{d^2u_1}{dt^2} + c_2 \frac{d^2u_2}{dt^2} + \frac{dc_1}{dt} \frac{du_1}{dt} + \frac{dc_2}{dt} \frac{du_2}{dt}$$

(notice how we used (4.12) to simplify the first derivative of u). Substituting u into (4.11) and simplifying, we obtain

$$\frac{dc_1}{dt}\frac{du_1}{dt} + \frac{dc_2}{dt}\frac{du_2}{dt} = a^{-1} f(t) \tag{4.13}$$

4.2. Solutions to some simple ODEs

(see Exercise 4.2.15). Equations (4.12) and (4.13) can be written in a matrix-vector form as

$$\begin{bmatrix} u_1 & u_2 \\ \frac{du_1}{dt} & \frac{du_2}{dt} \end{bmatrix} \begin{bmatrix} \frac{dc_1}{dt} \\ \frac{dc_2}{dt} \end{bmatrix} \begin{bmatrix} 0 \\ a^{-1}f(t) \end{bmatrix}. \quad (4.14)$$

Since $\{u_1, u_2\}$ are assumed to be linearly independent, the Wronskian test implies that the matrix appearing in (4.14) is nonsingular, and therefore we can solve uniquely for dc_1/dt, dc_2/dt. We can then integrate to find c_1, c_2.

4.2.3 A special inhomogeneous second-order linear ODE

We now present an important example that we can solve using the method of variation of parameters. In Chapter 7, we will encounter the following inhomogeneous IVP:

$$\begin{aligned} \frac{d^2u}{dt^2} + \theta^2 u &= f(t), \\ u(t_0) &= \alpha, \\ \frac{du}{dt}(t_0) &= \beta. \end{aligned} \quad (4.15)$$

The coefficient θ^2 is a positive real number. In this section, we will derive a surprisingly simple formula for the solution.

By the principle of superposition, we can solve the two IVPs

$$\begin{aligned} \frac{d^2u}{dt^2} + \theta^2 u &= 0, \\ u(t_0) &= \alpha, \\ \frac{du}{dt}(t_0) &= \beta \end{aligned} \quad (4.16)$$

and

$$\begin{aligned} \frac{d^2u}{dt^2} + \theta^2 u &= f(t), \\ u(t_0) &= 0, \\ \frac{du}{dt}(t_0) &= 0 \end{aligned} \quad (4.17)$$

and add the solutions to get the solution to (4.15). It is straightforward to apply the techniques of Section 4.2.1 to derive the solution of (4.16):

$$u_1(t) = \alpha \cos(\theta(t - t_0)) + \frac{\beta}{\theta} \sin(\theta(t - t_0)) \quad (4.18)$$

(see Exercise 4.2.12).

To solve (4.17), we use the method of variation of parameters. We look for a solution of the form $u(t) = c_1(t)\cos(\theta t) + c_2(t)\sin(\theta t)$, where c_1, c_2 are determined by

$$\cos(\theta t)\frac{dc_1}{dt} + \sin(\theta t)\frac{dc_2}{dt} = 0,$$
$$-\theta\sin(\theta t)\frac{dc_1}{dt} + \theta\cos(\theta t)\frac{dc_2}{dt} = f(t).$$

The reader can verify that the solution is

$$\frac{dc_1}{dt}(t) = -\theta^{-1}f(t)\sin(\theta t),$$
$$\frac{dc_2}{dt}(t) = \theta^{-1}f(t)\cos(\theta t),$$

and hence

$$c_1(t) = -\theta^{-1}\int_{t_0}^{t}\sin(\theta s)f(s)ds,$$
$$c_2(t) = \theta^{-1}\int_{t_0}^{t}\cos(\theta s)f(s)ds.$$

By choosing t_0 for the lower limit of integration, we have $c_1(t_0) = c_2(t_0) = 0$, and therefore $u(t_0) = 0$. Moreover, since

$$\frac{du}{dt} = c_1\frac{du_1}{dt} + c_2\frac{du_2}{dt}$$

by (4.12), we also have $du/dt(t_0) = 0$. Therefore, the solution u defined by this choice of c_1, c_2 satisfies the initial conditions of the IVP (4.17).

We can rewrite the solution u as follows:

$$u(t) = c_1(t)\cos(\theta t) + c_2(t)\sin(\theta t)$$
$$= \frac{1}{\theta}\left\{-\cos(\theta t)\int_{t_0}^{t}\sin(\theta s)f(s)ds + \sin(\theta t)\int_{t_0}^{t}\cos(\theta s)f(s)ds\right\}$$
$$= \frac{1}{\theta}\int_{t_0}^{t}(\sin(\theta t)\cos(\theta s) - \cos(\theta t)\sin(\theta s))f(s)ds.$$

Using the trigonometric identity

$$\sin\alpha\cos\beta - \cos\alpha\sin\beta = \sin(\alpha-\beta),$$

and renaming the solution as u_2, we obtain

$$u_2(t) = \frac{1}{\theta}\int_{t_0}^{t}\sin(\theta(t-s))f(s)ds, \qquad (4.19)$$

the unique solution of the IVP (4.17).

We have now shown that the solution to (4.15) is

$$u(t) = \alpha\cos(\theta(t-t_0)) + \frac{\beta}{\theta}\sin(\theta(t-t_0)) + \frac{1}{\theta}\int_{t_0}^{t}\sin(\theta(t-s))f(s)ds. \qquad (4.20)$$

4.2. Solutions to some simple ODEs

Example 4.3. The solution to

$$\frac{d^2u}{dt^2} + u = \cos(t),$$
$$u(0) = 0,$$
$$\frac{du}{dt}(0) = 0$$

is

$$u(t) = \int_0^t \sin(t-s)\cos(s)\,ds = \frac{t}{2}\sin(t)$$

(the integral can be computed using the subtraction formula for sine and can be simplified using other trigonometric identities). ∎

4.2.4 First-order linear ODEs

Another type of ODE for which we can find the general solution is the first-order linear ODE

$$\frac{du}{dt} - a(t)u = f(t). \tag{4.21}$$

We first treat the simpler case in which a is a constant.

The solution is based on the idea of an *integrating factor*: we multiply the equation by a special function that allows us to solve by direct integration. We note that

$$\frac{d}{dt}\left[e^{-at}u(t)\right] = e^{-at}\frac{du}{dt}(t) - ae^{-at}u(t).$$

Therefore, we have

$$\frac{du}{dt} - au = f(t) \Rightarrow e^{-at}\frac{du}{dt} - ae^{-at}u = e^{-at}f(t)$$
$$\Rightarrow \frac{d}{dt}\left[e^{-at}u(t)\right] = e^{-at}f(t)$$
$$\Rightarrow e^{-at}u(t) = e^{-at_0}u(t_0) + \int_{t_0}^t e^{-as}f(s)\,ds$$
$$\Rightarrow u(t) = e^{a(t-t_0)}u(t_0) + \int_{t_0}^t e^{a(t-s)}f(s)\,ds.$$

This immediately gives us the solution to the IVP

$$\frac{du}{dt} - au = f(t),$$
$$u(t_0) = u_0. \tag{4.22}$$

Example 4.4. The IVP

$$\frac{du}{dt} + u = \sin(t),$$
$$u(0) = 2$$

has the solution

$$u(t) = 2e^{-t} + \int_0^t e^{-(t-s)}\sin(s)\,ds$$
$$= 2e^{-t} + \frac{1}{2}\left(\sin(t) - \cos(t) + e^{-t}\right)$$
$$= \frac{5}{2}e^{-t} + \frac{1}{2}(\sin(t) - \cos(t))$$

(the integral can be computed by two applications of integration by parts). ∎

If a is a nonconstant function of t, we can use a more complicated integrating factor. We choose A to be any antiderivative of a, such as

$$A(t) = \int_{t_0}^t a(\tau)\,d\tau.$$

The integrating factor is then $e^{-A(t)}$, since

$$e^{-A(t)}\left(\frac{du}{dt} - a(t)u\right) = e^{-A(t)}\frac{du}{dt} - a(t)e^{-A(t)}u = e^{-A(t)}\frac{du}{dt} - A'(t)e^{-A(t)}u$$
$$= \frac{d}{dt}\left[e^{-A(t)}u\right].$$

We can then solve the ODE as follows:

$$\frac{du}{dt} - a(t)u = f(t) \Rightarrow e^{-A(t)}\frac{du}{dt} - a(t)e^{-A(t)}u = e^{-A(t)}f(t)$$
$$\Rightarrow \frac{d}{dt}\left[e^{-A(t)}u\right] = e^{-A(t)}f(t)$$
$$\Rightarrow e^{-A(t)}u(t) = e^{-A(t_0)}u(t_0) + \int_{t_0}^t e^{-A(s)}f(s)\,ds$$
$$\Rightarrow u(t) = e^{A(t)}e^{-A(t_0)}u(t_0) + e^{A(t)}\int_{t_0}^t e^{-A(s)}f(s)\,ds$$
$$\Rightarrow u(t) = e^{A(t)-A(t_0)}u(t_0) + \int_{t_0}^t e^{A(t)-A(s)}f(s)\,ds.$$

Example 4.5. Consider the IVP

$$\frac{du}{dt} - \frac{1}{t}u = t^2,$$
$$u(1) = 1.$$

An antiderivative of $a(t) = 1/t$ is $A(t) = \ln(t)$, and we have $e^{-A(t)} = e^{-\ln(t)} = 1/t$, $e^{A(t)} = e^{\ln(t)} = t$, and $e^{A(t)-A(s)} = e^{\ln(t)-\ln(s)} = t/s$. Therefore, the solution of the IVP is

$$u(t) = e^{A(t)-A(1)}u_0 + \int_1^t e^{A(t)-A(s)}s^2\,ds = t\cdot 1 + \int_1^t \frac{t}{s}s^2\,ds = \frac{1}{2}t + \frac{1}{2}t^3.\quad\blacksquare$$

It should be appreciated that, although we have an explicit formula for the solution of (4.21), it will be impossible in many cases to explicitly evaluate the integrals involved in terms of elementary functions.

4.2.5 Euler equations

We showed above (Section 4.2.1) how to solve second-order linear ODEs with constant coefficients. There are relatively few second-order, nonconstant-coefficient equations for which we can obtain explicit formulas. However, it is straightforward to solve equations of the form

$$t^2 \frac{d^2u}{dt^2} + at\frac{du}{dt} + bu = 0, \ t > 0. \tag{4.23}$$

Such an equation is called an *Euler equation*, and it has solutions of the form $u(t) = t^k$. Substituting $u(t) = t^k$ into (4.23) yields

$$k(k-1)t^k + akt^k + bt^k = 0, \ t > 0,$$

or

$$k^2 + (a-1)k + b = 0. \tag{4.24}$$

This is the characteristic equation for (4.23). Noting that $t^k = e^{k\ln(t)}$, if the roots of (4.24) are complex, say $k = \alpha \pm \beta i$, then we have

$$t^{\alpha \pm \beta i} = e^{(\alpha \pm \beta i)\ln(t)},$$

which simplifies to

$$t^\alpha (\cos(\beta \ln(t)) \pm i \sin(\beta \ln(t))).$$

The usual reasoning then shows that

$$u_1(t) = t^\alpha \cos(\beta \ln(t)), \ u_2(t) = t^\alpha \sin(\beta \ln(t))$$

are linearly independent solutions of (4.23). If (4.24) has a single real root k, then $u_2(t) = t^k \ln(t)$ is a second solution, independent of $u_1(t) = t^k$. Several examples are given in the exercises.

Exercises

1. Let S be the solution set of (4.9).

    (a) Show that S is a subspace of $C^2(\mathbf{R})$, the set of twice-continuously differentiable function defined on $\mathbf{R}$.

    (b) Use the results of this section to show that S is two-dimensional for any values of a, b, and c, provided only that $a \neq 0$.

2. Suppose (4.9) has characteristic roots $\mu \pm \lambda i$, where $\mu, \lambda \in \mathbf{R}$ and $\lambda \neq 0$. Use the Wronskian test to show that $\{e^{\mu t}\cos(\lambda t), e^{\mu t}\sin(\lambda t)\}$ is linearly independent, and therefore that the general solution of (4.9) is

$$c_1 e^{\mu t}\cos(\lambda t) + c_2 e^{\mu t}\sin(\lambda t)$$

in this case.

3. Suppose (4.9) has the single characteristic root $r = -b/(2a)$. Use the Wronskian test to show that $\{e^{rt}, te^{rt}\}$ is linearly independent, and therefore that the general solution of (4.9) is $c_1 e^{rt} + c_2 t e^{rt}$ in this case.

4. For each of the following IVPs, find the general solution of the ODE and use it to solve the IVP:

   (a)
   $$\frac{d^2 u}{dt^2} + 2 \frac{du}{dt} + u = 0,$$
   $$u(0) = 1,$$
   $$\frac{du}{dt}(0) = 0;$$

   (b)
   $$3\frac{d^2 u}{dt^2} + 2 \frac{du}{dt} + u = 0,$$
   $$u(0) = -1,$$
   $$\frac{du}{dt}(0) = 3;$$

   (c)
   $$2\frac{d^2 u}{dt^2} + 2 \frac{du}{dt} + u = 0,$$
   $$u(0) = 0,$$
   $$\frac{du}{dt}(0) = 1;$$

   (d)
   $$\frac{d^2 u}{dt^2} + 3 \frac{du}{dt} + u = 0,$$
   $$u(0) = 1,$$
   $$\frac{du}{dt}(0) = 2.$$

5. The following differential equations are accompanied by *boundary conditions*—auxiliary conditions that refer to the boundary of a spatial domain rather than to an initial time. By using the general solution of the ODE, determine whether a nonzero solution to the BVP exists, and if so, whether the solution is unique.

   (a)
   $$\frac{d^2 u}{dx^2} - 2u = 0,$$
   $$u(0) = 0,$$
   $$u(1) = 0;$$

4.2. Solutions to some simple ODEs

(b)
$$\frac{d^2u}{dx^2} + 2u = 0,$$
$$u(0) = 0,$$
$$u(1) = 0;$$

(c)
$$\frac{d^2u}{dx^2} + \pi^2 u = 0,$$
$$u(0) = 0,$$
$$u(1) = 0.$$

6. Determine the values of $\lambda \in \mathbf{R}$ such that the BVP
$$\frac{d^2u}{dx^2} + \lambda u = 0,$$
$$u(0) = 0,$$
$$u(1) = 0$$
has a nonzero solution.

7. Prove directly (that is, by substituting u into the differential equation and initial condition) that
$$u(t) = e^{a(t-t_0)} u_0 + \int_{t_0}^{t} e^{a(t-s)} f(s) \, ds$$
solves
$$\frac{du}{dt} - au = f(t),$$
$$u(t_0) = u_0.$$

8. Solve the following IVPs:

(a)
$$\frac{du}{dt} = 2u - 0.1,$$
$$u(0) = 1.0;$$

(b)
$$\frac{du}{dt} = -2u + 0.1t,$$
$$u(0) = 0;$$

(c)
$$\frac{du}{dt} + u = t,$$
$$u(0) = 0.$$

9. Find the solution to the IVP
$$\frac{d^2u}{dt^2} + 4u = 1,$$
$$u(0) = 0,$$
$$\frac{du}{dt}(0) = 0.$$

10. Find the solution to the IVP
$$\frac{d^2u}{dt^2} + 9u = t,$$
$$u(0) = 0,$$
$$\frac{du}{dt}(0) = 0.$$

11. Find the solution to the IVP
$$\frac{d^2u}{dt^2} + u = e^{-t},$$
$$u(0) = 1,$$
$$\frac{du}{dt}(0) = 0.$$

12. Use the techniques of Section 4.2.1 to derive the solution to (4.16), and verify that you obtain (4.18). (Hint: One way to do this is to write the general solution of the ODE as $u(t) = c_1 \cos(\theta t) + c_2 \sin(\theta t)$ and then solve for c_1 and c_2. If you do this, you will then have to apply trigonometric identities to put the solution in the form given in (4.18). It is simpler to recognize at the beginning that the general solution could just as well be written as $u(t) = c_1 \cos(\theta(t - t_0)) + c_2 \sin(\theta(t - t_0))$.)

13. Solve each of the following IVPs using the method of integrating factors:
    (a) $\frac{dx}{dt} = tx$, $x(0) = 1$;
    (b) $\frac{dx}{dt} + \frac{3}{t}x = \frac{\sin(t)}{t^3}$, $x(1) = 1$;
    (c) $\frac{dy}{dt} + 2ty = t$, $y(0) = 1$.

14. Let $\{u_1, u_2\}$ be a linearly independent set of solutions of (4.9), and let u_p be any solution of (4.11). Prove that $u = u_p + c_1 u_1 + c_2 u_2$ is the general solution of (4.11)

by proving that every IVP of the form

$$a\frac{d^2u}{dt^2} + b\frac{du}{dt} + cu = f(t),$$

$$u(t_0) = v_1,$$

$$\frac{du}{dt}(t_0) = v_2$$

has a solution of the form $u = u_p + c_1u_1 + c_2u_2$ for some scalars c_1, c_2.

15. In the presentation of the method of variation of parameters in Section 4.2.2, prove that (4.13) is the result of substituting $u(t) = c_1(t)u_1(t) + c_2(t)u_2(t)$ into the ODE (4.11) under the assumption (4.12).

16. Find the general solution of each of the following Euler equations:

    (a) $t^2\frac{d^2u}{dt^2} + 3t\frac{du}{dt} + u = 0$;

    (b) $t^2\frac{d^2u}{dt^2} + t\frac{du}{dt} + u = 0$;

    (c) $t^2\frac{d^2u}{dt^2} + 6t\frac{du}{dt} + 4u = 0$.

17. Show that the change of variables $t = e^s$ transforms the Euler equation (4.23) into a constant-coefficient ODE.

4.3 Linear systems with constant coefficients

We now consider a first-order linear system of ODEs with constant coefficients. Such a system can be written as

$$\begin{aligned}\frac{dx_1}{dt} &= a_{11}x_1 + a_{12}x_2 + \cdots + a_{1n}x_n + f_1(t), \\ \frac{dx_2}{dt} &= a_{21}x_1 + a_{22}x_2 + \cdots + a_{2n}x_n + f_2(t), \\ &\vdots \\ \frac{dx_n}{dt} &= a_{n1}x_1 + a_{n2}x_2 + \cdots + a_{nn}x_n + f_n(t).\end{aligned} \qquad (4.25)$$

As with a linear algebraic system, there is a great advantage in using matrix-vector notation. System (4.25) can be written as

$$\frac{d\mathbf{x}}{dt} = \mathbf{A}\mathbf{x} + \mathbf{f}(t), \qquad (4.26)$$

where

$$\mathbf{A} = \begin{bmatrix} a_{11} & a_{12} & \cdots & a_{1n} \\ a_{21} & a_{22} & \cdots & a_{2n} \\ \vdots & \vdots & \ddots & \vdots \\ a_{n1} & a_{n2} & \cdots & a_{nn} \end{bmatrix}, \ \mathbf{x}(t) = \begin{bmatrix} x_1(t) \\ x_2(t) \\ \vdots \\ x_n(t) \end{bmatrix}, \ \mathbf{f}(t) = \begin{bmatrix} f_1(t) \\ f_2(t) \\ \vdots \\ f_n(t) \end{bmatrix}.$$

Although it is possible to develop a solution technique that is applicable for any matrix $\mathbf{A} \in \mathbf{R}^{n \times n}$, it is sufficient for our purposes to discuss the case in which there is a basis for $\mathbf{R}^n$ consisting of eigenvectors of $\mathbf{A}$, notably the case in which $\mathbf{A}$ is symmetric. In this case, we can develop a *spectral method*, much like the spectral method for solving $\mathbf{A}\mathbf{x} = \mathbf{b}$ in the case that $\mathbf{A}$ is symmetric (see Section 3.5.3).

As we develop an explicit solution for (4.26), the reader should concentrate on the qualitative properties (of the solution) that are revealed. These properties turn out to be more important (at least in this book) than the formula for the solution.

4.3.1 Homogeneous systems

We begin with the homogeneous version of (4.26),

$$\frac{d\mathbf{x}}{dt} = \mathbf{A}\mathbf{x}. \tag{4.27}$$

We will look for solutions to (4.26) of the special form $\mathbf{x}(t) = \alpha(t)\mathbf{u}$, where $\mathbf{u} \neq \mathbf{0}$ is a constant vector. We have

$$\mathbf{x}(t) = \alpha(t)\mathbf{u}, \ \frac{d\mathbf{x}}{dt} = A\mathbf{x} \Rightarrow \frac{d\alpha}{dt}\mathbf{u} = \alpha \mathbf{A}\mathbf{u}.$$

This last equation states that two vectors are equal, which means that $\mathbf{A}\mathbf{u}$ must be a multiple of $\mathbf{u}$ (otherwise, the two vectors would point in different directions, which is not possible if they are equal). Therefore, $\mathbf{u}$ must be an eigenvector of $\mathbf{A}$:

$$\mathbf{A}\mathbf{u} = \lambda \mathbf{u}.$$

We then obtain

$$\frac{d\alpha}{dt} = \lambda \alpha \Rightarrow \alpha(t) = Ce^{\lambda(t-t_0)},$$

where C is a constant. Therefore, if $\lambda, \mathbf{u}$ is any eigenvalue-eigenvector pair, then

$$\mathbf{x}(t) = e^{\lambda(t-t_0)}\mathbf{u}$$

is a solution of (4.27).

Example 4.6. Let

$$\mathbf{A} = \frac{1}{6}\begin{bmatrix} -5 & 4 & 1 \\ 4 & -8 & 4 \\ 1 & 4 & -5 \end{bmatrix}.$$

Then the eigenvalues of $\mathbf{A}$ are $\lambda_1 = 0$, $\lambda_2 = -1$, and $\lambda_3 = -2$, and the corresponding eigenvectors are

$$\mathbf{u}_1 = \begin{bmatrix} \frac{1}{\sqrt{3}} \\ \frac{1}{\sqrt{3}} \\ \frac{1}{\sqrt{3}} \end{bmatrix}, \ \mathbf{u}_2 = \begin{bmatrix} \frac{1}{\sqrt{2}} \\ 0 \\ -\frac{1}{\sqrt{2}} \end{bmatrix}, \ \mathbf{u}_3 = \begin{bmatrix} \frac{1}{\sqrt{6}} \\ -\frac{2}{\sqrt{6}} \\ \frac{1}{\sqrt{6}} \end{bmatrix}.$$

4.3. Linear systems with constant coefficients

We therefore know three independent solutions of

$$\frac{d\mathbf{x}}{dt} = \mathbf{A}\mathbf{x},$$

namely,

$$\mathbf{x}_1(t) = \mathbf{u}_1,$$
$$\mathbf{x}_2(t) = e^{-(t-t_0)}\mathbf{u}_2,$$
$$\mathbf{x}_3(t) = e^{-2(t-t_0)}\mathbf{u}_3. \quad \blacksquare$$

If, as in the previous example, there is a basis for $\mathbf{R}^n$ consisting of eigenvectors $\{\mathbf{u}_1, \mathbf{u}_2, \ldots, \mathbf{u}_n\}$ of $\mathbf{A}$, with corresponding eigenvalues $\lambda_1, \lambda_2, \ldots, \lambda_n$, then we can write the general solution of (4.27). Indeed, it suffices to show that we can solve

$$\frac{d\mathbf{x}}{dt} = \mathbf{A}\mathbf{x}, \quad (4.28)$$
$$\mathbf{x}(t_0) = \mathbf{x}_0.$$

Since $\{\mathbf{u}_1, \mathbf{u}_2, \ldots, \mathbf{u}_n\}$ is a basis for $\mathbf{R}^n$, there exists a vector $\mathbf{c} \in \mathbf{R}^n$ such that

$$\mathbf{x}_0 = c_1\mathbf{u}_1 + c_2\mathbf{u}_2 + \cdots + c_n\mathbf{u}_n.$$

The solution to (4.28) is then

$$\mathbf{x}(t) = c_1 e^{\lambda_1(t-t_0)}\mathbf{u}_1 + c_2 e^{\lambda_2(t-t_0)}\mathbf{u}_2 + \cdots + c_n e^{\lambda_n(t-t_0)}\mathbf{u}_n. \quad (4.29)$$

This is verified by computing $d\mathbf{x}/dt$:

$$\frac{d\mathbf{x}}{dt}(t) = c_1\lambda_1 e^{\lambda_1(t-t_0)}\mathbf{u}_1 + c_2\lambda_2 e^{\lambda_2(t-t_0)}\mathbf{u}_2 + \cdots + c_n\lambda_n e^{\lambda_n(t-t_0)}\mathbf{u}_n$$
$$= c_1 e^{\lambda_1(t-t_0)}\mathbf{A}\mathbf{u}_1 + c_2 e^{\lambda_2(t-t_0)}\mathbf{A}\mathbf{u}_2 + \cdots + c_n e^{\lambda_n(t-t_0)}\mathbf{A}\mathbf{u}_n$$
$$= \mathbf{A}\left(c_1 e^{\lambda_1(t-t_0)}\mathbf{u}_1 + c_2 e^{\lambda_2(t-t_0)}\mathbf{u}_2 + \cdots + c_n e^{\lambda_n(t-t_0)}\mathbf{u}_n\right)$$
$$= \mathbf{A}\mathbf{x}(t).$$

We also have

$$\mathbf{x}(t_0) = c_1\mathbf{u}_1 + c_2\mathbf{u}_2 + \cdots + c_n\mathbf{u}_n = \mathbf{x}_0$$

by construction.

Example 4.7. Let $\mathbf{A}$ be the matrix in Example 4.6, and let

$$\mathbf{x}_0 = \begin{bmatrix} 0 \\ 1 \\ -1 \end{bmatrix}.$$

Since the basis of eigenvectors $\{\mathbf{u}_1, \mathbf{u}_2, \mathbf{u}_3\}$ of $\mathbf{A}$ is orthonormal, as is easily checked, we have

$$\mathbf{x}_0 = (\mathbf{u}_1 \cdot \mathbf{x}_0)\mathbf{u}_1 + (\mathbf{u}_2 \cdot \mathbf{x}_0)\mathbf{u}_2 + (\mathbf{u}_3 \cdot \mathbf{x}_0)\mathbf{u}_3$$
$$= 0\mathbf{u}_1 + \frac{1}{\sqrt{2}}\mathbf{u}_2 - \frac{3}{\sqrt{6}}\mathbf{u}_3.$$

We can then immediately write down the solution to

$$\frac{d\mathbf{x}}{dt} = \mathbf{A}\mathbf{x},$$
$$\mathbf{x}(0) = \mathbf{x}_0.$$

It is

$$\mathbf{x}(t) = \frac{1}{\sqrt{2}}e^{-t}\mathbf{u}_2 - \frac{3}{\sqrt{6}}e^{-2t}\mathbf{u}_3$$
$$= \begin{bmatrix} \frac{1}{2}e^{-t} - \frac{1}{2}e^{-2t} \\ e^{-2t} \\ -\frac{1}{2}e^{-t} - \frac{1}{2}e^{-2t} \end{bmatrix}. \quad \blacksquare$$

If, as in the previous example, $\mathbf{A}$ is symmetric, then the eigenvectors

$$\{\mathbf{u}_1, \mathbf{u}_2, \ldots, \mathbf{u}_n\}$$

can be chosen to be orthonormal, in which case it is particularly simple to compute the coefficients $c_1, c_2, \ldots, c_n$ expressing $\mathbf{x}_0$ in terms of the eigenvectors. We obtain the following simple formula for the solution of (4.28):

$$\mathbf{x}(t) = \sum_{i=1}^{n} (\mathbf{x}_0 \cdot \mathbf{u}_i) e^{\lambda_i(t-t_0)} \mathbf{u}_i.$$

We have derived the solution to (4.28) in the case that $\mathbf{A}$ has n linearly independent eigenvectors. We remark that formula (4.29) might involve complex numbers if $\mathbf{A}$ is not symmetric. However, we have a complete and satisfactory solution in the case that $\mathbf{A}$ is symmetric—there are n linearly independent eigenvectors, the eigenvalues are guaranteed to be real, and the basis of eigenvectors can be chosen to be orthonormal.

We now discuss the significance of solution (4.29). The interpretation is very simple: The solution to (4.28) is the sum of n components, one in the direction of each eigenvector, and component i is

- exponentially increasing (if $\lambda_i > 0$);
- exponentially decreasing (if $\lambda_i < 0$); or
- constant (if $\lambda_i = 0$).

Moreover, the rate of increase or decrease of each component is governed by the magnitude of the corresponding eigenvalue.

4.3. Linear systems with constant coefficients

We can therefore draw the following conclusions:

1. If all of the eigenvalues of $\mathbf{A}$ are negative, then every solution $\mathbf{x}$ of (4.28) satisfies
$$\|\mathbf{x}(t)\| \to 0 \text{ as } t \to \infty.$$

2. If all of the eigenvalues of $\mathbf{A}$ are positive, then every solution $\mathbf{x}$ of (4.28) (except the zero solution) satisfies
$$\|\mathbf{x}(t)\| \to \infty \text{ as } t \to \infty.$$

3. If any of the eigenvalues of $\mathbf{A}$ is positive, then, for most initial values $\mathbf{x}_0$, the solution $\mathbf{x}$ of (4.28) satisfies
$$\|\mathbf{x}(t)\| \to \infty \text{ as } t \to \infty.$$

The only statement which requires justification is the third. Suppose an eigenvalue λ_j of $\mathbf{A}$ is positive, and suppose $\mathbf{A}$ has k linearly independent eigenvectors corresponding to λ_j. Then, unless $\mathbf{x}_0$ lies in the $(n-k)$-dimensional subspace of $\mathbf{R}^n$ spanned by the other $n-k$ eigenvectors, the solution $\mathbf{x}$ of (4.28) will contain a factor of $e^{\lambda_j(t-t_0)}$, guaranteeing that
$$\|\mathbf{x}(t)\| \to \infty \text{ as } t \to \infty.$$

Even if $k = 1$, an $(n-k)$-dimensional subspace of $\mathbf{R}^n$ is a very small part of $\mathbf{R}^n$ (comparable to a plane or a line in $\mathbf{R}^3$). Therefore, most initial vectors do not lie in this subspace.

Example 4.8. Let
$$\mathbf{A} = \begin{bmatrix} -\frac{1}{2} & 1 & \frac{1}{2} \\ 1 & -1 & 1 \\ \frac{1}{2} & 1 & -\frac{1}{2} \end{bmatrix}.$$

Then the eigenvalues of $\mathbf{A}$ are $\lambda_1 = 1$, $\lambda_2 = -1$, and $\lambda_3 = -2$, and the corresponding eigenvectors are

$$\mathbf{u}_1 = \begin{bmatrix} \frac{1}{\sqrt{3}} \\ \frac{1}{\sqrt{3}} \\ \frac{1}{\sqrt{3}} \end{bmatrix}, \quad \mathbf{u}_2 = \begin{bmatrix} \frac{1}{\sqrt{2}} \\ 0 \\ -\frac{1}{\sqrt{2}} \end{bmatrix}, \quad \mathbf{u}_3 = \begin{bmatrix} \frac{1}{\sqrt{6}} \\ -\frac{2}{\sqrt{6}} \\ \frac{1}{\sqrt{6}} \end{bmatrix}.$$

The solution of (4.28) is
$$\mathbf{x}(t) = c_1 e^{t-t_0} \mathbf{u}_1 + c_2 e^{-(t-t_0)} \mathbf{u}_2 + c_3 e^{-2(t-t_0)} \mathbf{u}_3,$$

where
$$\mathbf{x}_0 = c_1 \mathbf{u}_1 + c_2 \mathbf{u}_2 + c_3 \mathbf{u}_3.$$

The only initial values that lead to a solution $\mathbf{x}$ that does *not* grow without bound are
$$\mathbf{x}_0 \in S = \text{span}\{\mathbf{u}_2, \mathbf{u}_3\}.$$

But S is a plane, which is a very small part of Euclidean 3-space. Thus almost every initial value leads to a solution that grows exponentially. ■

Another conclusion that we can draw is somewhat more subtle than those given above, but it will be important in Chapter 6.

4. If **A** has eigenvalues of very different magnitudes, then solutions of (4.28) have components whose magnitudes change at very different rates. Such solutions can be difficult to compute efficiently using numerical methods.

We will discuss this point in more detail in Section 4.5.

4.3.2 Inhomogeneous systems and variation of parameters

We can now explain how to solve the inhomogeneous system

$$\frac{d\mathbf{x}}{dt} = \mathbf{A}\mathbf{x} + \mathbf{f}(t), \tag{4.30}$$

again considering only the case in which there is a basis of $\mathbf{R}^n$ consisting of eigenvectors of **A**. The method is a spectral method, and the reader may wish to review Section 3.5.3.

If $\{\mathbf{u}_1, \mathbf{u}_2, \ldots, \mathbf{u}_n\}$ is a basis for $\mathbf{R}^n$, then every vector in $\mathbf{R}^n$ can be written uniquely as a linear combination of these vectors. In particular, for each t, we can write $\mathbf{f}(t)$ as a linear combination of $\mathbf{u}_1, \mathbf{u}_2, \ldots, \mathbf{u}_n$:

$$\mathbf{f}(t) = c_1(t)\mathbf{u}_1 + c_2(t)\mathbf{u}_2 + \cdots + c_n(t)\mathbf{u}_n.$$

Since the vector $\mathbf{f}(t)$ depends on t, so do the weights $c_1(t), c_2(t), \ldots, c_n(t)$. These weights can be computed explicitly from **f**, which is considered to be known.

We can also write the solution of (4.30) in terms of the basis vectors:

$$\mathbf{x}(t) = a_1(t)\mathbf{u}_1 + a_2(t)\mathbf{u}_2 + \cdots + a_n(t)\mathbf{u}_n. \tag{4.31}$$

Since $\mathbf{x}(t)$ is unknown, so are the weights $a_1(t), a_2(t), \ldots, a_n(t)$. However, when these basis vectors are eigenvectors of **A**, it is easy to solve for the unknown weights. Indeed, substituting (4.31) in place of **x** yields

$$\frac{d\mathbf{x}}{dt} - \mathbf{A}\mathbf{x} = \frac{d}{dt}\left[\sum_{i=1}^{n} a_i \mathbf{u}_i\right] - \mathbf{A}\left(\sum_{i=1}^{n} a_i \mathbf{u}_i\right)$$

$$= \sum_{i=1}^{n} \frac{da_i}{dt}\mathbf{u}_i - \sum_{i=1}^{n} a_i \mathbf{A}\mathbf{u}_i$$

$$= \sum_{i=1}^{n} \frac{da_i}{dt}\mathbf{u}_i - \sum_{i=1}^{n} \lambda_i a_i \mathbf{u}_i$$

$$= \sum_{i=1}^{n} \left\{\frac{da_i}{dt} - \lambda_i a_i\right\} \mathbf{u}_i.$$

The ODE

$$\frac{d\mathbf{x}}{dt} - \mathbf{A}\mathbf{x} = \mathbf{f}(t)$$

4.3. Linear systems with constant coefficients

then implies that

$$\sum_{i=1}^{n}\left\{\frac{da_i}{dt}-\lambda_i a_i\right\}\mathbf{u}_i = \sum_{i=1}^{n} c_i(t)\mathbf{u}_i,$$

which can only hold if

$$\frac{da_i}{dt}-\lambda_i a_i = c_i(t), \; i=1,2,\ldots,n. \tag{4.32}$$

Thus, by representing the solution in terms of the eigenvectors, we reduced the system of ODEs to n scalar ODEs that can be solved independently.[12] We can also recognize this approach as another version of the method of variation of parameters (see Section 4.2.2), in that we have represented a solution of the inhomogeneous equation as a linear combination of the solutions of the homogeneous equation.

The computation of the functions $c_1(t), c_2(t), \ldots, c_n(t)$ is simplest when $\mathbf{A}$ is symmetric, so that the eigenvectors can be chosen to be orthonormal. In that case,

$$c_i(t) = \mathbf{f}(t)\cdot\mathbf{u}_i, \; i=1,2,\ldots,n.$$

We will concentrate on this case henceforth.

The initial condition $\mathbf{x}(t_0) = \mathbf{x}_0$ provides initial values for the unknowns $a_1(t), a_2(t), \ldots, a_n(t)$. Indeed, we can write

$$\mathbf{x}_0 = b_1\mathbf{u}_1 + b_2\mathbf{u}_2 + \cdots + b_n\mathbf{u}_n,$$

so $\mathbf{x}(t_0) = \mathbf{x}_0$ implies that

$$a_1(t_0)\mathbf{u}_1 + a_2(t_0)\mathbf{u}_2 + \cdots + a_n(t_0)\mathbf{u}_n = b_1\mathbf{u}_1 + b_2\mathbf{u}_2 + \cdots + b_n\mathbf{u}_n,$$

or

$$a_i(t_0) = b_i, \; i=1,2,\ldots,n.$$

When the basis is orthonormal, the coefficients $b_1, b_2, \ldots, b_n$ can be computed in the usual way:

$$b_i = \mathbf{x}_0 \cdot \mathbf{u}_i, \; i=1,2,\ldots,n.$$

Example 4.9. Consider the IVP

$$\frac{d\mathbf{x}}{dt} = \mathbf{A}\mathbf{x} + \mathbf{f}(t),$$
$$\mathbf{x}(0) = \mathbf{0},$$

where

$$\mathbf{A} = \begin{bmatrix} 1 & 1 \\ 1 & 1 \end{bmatrix}, \; \mathbf{f}(t) = \begin{bmatrix} \cos(t) \\ \sin(t) \end{bmatrix}.$$

---

[12] This is just what happened in the spectral method for solving $\mathbf{A}\mathbf{x} = \mathbf{b}$—the system of n simultaneous equations was reduced to n independent equations.

The matrix **A** is symmetric, and an orthonormal basis for $\mathbf{R}^2$ consists of the vectors

$$\mathbf{u}_1 = \begin{bmatrix} \frac{1}{\sqrt{2}} \\ -\frac{1}{\sqrt{2}} \end{bmatrix}, \mathbf{u}_2 = \begin{bmatrix} \frac{1}{\sqrt{2}} \\ \frac{1}{\sqrt{2}} \end{bmatrix}.$$

The corresponding eigenvalues are $\lambda_1 = 0$, $\lambda_2 = 2$.

We now express $\mathbf{f}(t)$ in terms of the eigenvectors:

$$\mathbf{f}(t) = c_1(t)\mathbf{u}_1 + c_2(t)\mathbf{u}_2,$$

$$c_1(t) = \mathbf{f}(t) \cdot \mathbf{u}_1 = \frac{1}{\sqrt{2}}(\cos(t) - \sin(t)),$$

$$c_2(t) = \mathbf{f}(t) \cdot \mathbf{u}_2 = \frac{1}{\sqrt{2}}(\cos(t) + \sin(t)).$$

The solution is

$$\mathbf{x}(t) = a_1(t)\mathbf{u}_1 + a_2(t)\mathbf{u}_2,$$

where

$$\frac{da_1}{dt} = \frac{1}{\sqrt{2}}(\cos(t) - \sin(t)),\ a_1(0) = 0,$$

$$\frac{da_2}{dt} = 2a_2 + \frac{1}{\sqrt{2}}(\cos(t) + \sin(t)),\ a_2(0) = 0.$$

We obtain

$$a_1(t) = \frac{1}{\sqrt{2}} \int_0^t \{\cos(s) - \sin(s)\}\, ds$$

$$= \frac{1}{\sqrt{2}}(\sin(t) + \cos(t) - 1)$$

and

$$a_2(t) = \frac{1}{\sqrt{2}} \int_0^t e^{2(t-s)} \{\cos(s) + \sin(s)\}\, ds$$

$$= \frac{1}{5\sqrt{2}}\left(3e^{2t} - 3\cos(t) - \sin(t)\right)$$

(in computing a_2, we used the results of Section 4.2.4). Finally,

$$\mathbf{x}(t) = a_1(t)\mathbf{u}_1 + a_2(t)\mathbf{u}_2 = \begin{bmatrix} \frac{1}{10}(2\cos(t) + 4\sin(t) + 3e^{2t} - 5) \\ \frac{1}{10}(-8\cos(t) - 6\sin(t) + 3e^{2t} + 5) \end{bmatrix}. \ \blacksquare$$

4.3.3 Duhamel's principle

Duhamel's principle states that if we can solve an IVP with a homogeneous differential equation, then we can use the solution to construct the solution of the inhomogeneous version of the problem. We begin with two illustrative examples.

4.3. Linear systems with constant coefficients

First example of Duhamel's principle

We begin by presenting a different method for solving the IVP

$$\frac{d\mathbf{x}}{dt} - \mathbf{A}\mathbf{x} = \mathbf{f}(t), \qquad (4.33)$$
$$\mathbf{x}(0) = \mathbf{0},$$

where $\mathbf{A} \in \mathbf{R}^{n \times n}$ is given. We continue to assume that there is a basis $\{\mathbf{u}_1, \ldots, \mathbf{u}_n\}$ for $\mathbf{R}^n$ consisting of eigenvectors of $\mathbf{A}$, where the corresponding eigenvalues are $\lambda_1, \ldots, \lambda_n$. The functions

$$\mathbf{x}_i(t) = e^{\lambda_i t} \mathbf{u}_i, \ i = 1, 2, \ldots, n,$$

form a linearly independent set of solutions of the ODE $d\mathbf{x}/dt - \mathbf{A}\mathbf{x} = 0$. We define the matrix $\mathbf{X}(t) \in \mathbf{R}^{n \times n}$ by

$$\mathbf{X}(t) = [\mathbf{x}_1(t) | \mathbf{x}_2(t) | \cdots | \mathbf{x}_n(t)].$$

Then, for any constant vector $\mathbf{a}$, the function $\mathbf{x}(t) = \mathbf{X}(t)\mathbf{a}$ is a solution of the homogeneous ODE $d\mathbf{x}/dt - \mathbf{A}\mathbf{x} = 0$ (the reader should notice that $\mathbf{X}(t)\mathbf{a}$ is just a linear combination of the columns of $\mathbf{X}(t)$, and hence of the solutions $\mathbf{x}_1, \ldots, \mathbf{x}_n$).

We can find the solution of

$$\frac{d\mathbf{x}}{dt} - \mathbf{A}\mathbf{x} = 0, \qquad (4.34)$$
$$\mathbf{x}(0) = \mathbf{c}$$

in the form $\mathbf{x}(t) = \mathbf{X}(t)\mathbf{a}$ by choosing the appropriate vector $\mathbf{a}$. We need $\mathbf{x}(0) = \mathbf{c}$, which is equivalent to $\mathbf{X}(0)\mathbf{a} = \mathbf{c}$, or $\mathbf{a} = \mathbf{X}(0)^{-1}\mathbf{c}$. Thus the solution to (4.34) is $\mathbf{x}(t) = \mathbf{X}(t)\mathbf{X}(0)^{-1}\mathbf{c}$. We can simplify this formula by defining the matrix $\mathbf{U}(t) = \mathbf{X}(t)\mathbf{X}(0)^{-1}$, in which case the solution to (4.34) is $\mathbf{x}(t) = \mathbf{U}(t)\mathbf{c}$.

We now have a formula for the solution of (4.34), given any initial vector $\mathbf{c}$. Duhamel's principle states the following: If the solution of (4.34), with $\mathbf{c} = \mathbf{f}(s)$, is $\mathbf{x} = \mathbf{x}(t; s)$, then the solution of (4.33) is

$$\mathbf{u}(t) = \int_0^t \mathbf{x}(t - s; s) \, ds. \qquad (4.35)$$

Since $\mathbf{x}(t; s) = \mathbf{U}(t)\mathbf{f}(s)$, formula (4.35) can be written explicitly as

$$\mathbf{u}(t) = \int_0^t \mathbf{U}(t - s)\mathbf{f}(s) \, ds. \qquad (4.36)$$

(Exercise 4.3.16 asks the reader to derive (4.36) by applying the method of variation of parameters to the solution of (4.34).)

We can verify the validity of Duhamel's principle by showing directly that (4.35) satisfies (4.33).[13] With **u** defined by (4.35), we have

$$\frac{d\mathbf{u}}{dt}(t) = \mathbf{x}(t-t;t) + \int_0^t \frac{d\mathbf{x}}{dt}(t;s)\,ds = \mathbf{x}(0;t) + \int_0^t \frac{d\mathbf{x}}{dt}(t;s)\,ds = \mathbf{f}(t) + \int_0^t \frac{d\mathbf{x}}{dt}(t;s)\,ds.$$

Since

$$\frac{d\mathbf{x}}{dt}(t;s) = \mathbf{A}\mathbf{x}(t;s),$$

we obtain

$$\frac{d\mathbf{u}}{dt}(t) = \mathbf{f}(t) + \int_0^t \mathbf{A}\mathbf{x}(t;s)\,ds = \mathbf{f}(t) + \mathbf{A}\int_0^t \mathbf{x}(t;s)\,ds = \mathbf{f}(t) + \mathbf{A}\mathbf{u}(t).$$

Thus, **u** satisfies the ODE (4.32). Since **u** also satisfies $\mathbf{u}(0) = \int_0^0 \mathbf{x}(t;s)\,ds = 0$, we have verified that **u** is the solution of (4.33).

Second example of Duhamel's principle

We now turn to a different IVP, the one considered in Section 4.2.3. We know that the solution to

$$\frac{d^2u}{dt^2} + \theta^2 u = 0,$$
$$u(0) = 0, \qquad (4.37)$$
$$\frac{du}{dt}(0) = c$$

is

$$u(t) = \theta^{-1}\sin(\theta t)c. \qquad (4.38)$$

We also showed in Section 4.2.3 that the solution of

$$\frac{d^2u}{dt^2} + \theta^2 u = f(t),$$
$$u(0) = 0, \qquad (4.39)$$
$$\frac{du}{dt}(0) = 0$$

---

[13]The calculation uses the following differentiation rule:

$$\frac{d}{dt}\left(\int_a^t F(t,s)\,ds\right) = F(t,t) + \int_a^t \frac{\partial F}{\partial t}(t,s)\,ds. \qquad (*)$$

This is derived from the chain rule and the fundamental theorem of calculus. If we write $h(x,y) = \int_a^x F(y,s)\,ds$, then the chain rule implies $\frac{d}{dt}(h(x,y)) = \frac{\partial h}{\partial x}(x,y)\frac{dx}{dt} + \frac{\partial h}{\partial y}(x,y)\frac{dy}{dt}$. By the fundamental theorem of calculus, Theorem 2.1, and the fact that here $x = t$, $y = t$ (and therefore $dx/dt = dy/dt = 1$), we obtain (*).

4.3. Linear systems with constant coefficients

is

$$u(t) = \int_0^t \theta^{-1} \sin(\theta(t-s)) f(s) \, ds. \tag{4.40}$$

Comparing the solutions (4.38) and (4.40), we see that (4.40) is equivalent to

$$u(t) = \int_0^t v(t-s;s) \, ds, \tag{4.41}$$

where $v(t;s)$ is the solution of (4.37) with $c = f(s)$. This suggests (via a bit of an intuitive leap) the correct form of Duhamel's principle for higher-order ODEs: If the solution of

$$\begin{aligned}
\frac{d^{(k)}u}{dt} + a_{k-1}\frac{d^{(k-1)}u}{dt} + \cdots + a_1\frac{du}{dt} + a_0 u &= 0, \\
u(0) &= 0, \\
\frac{du}{dt}(0) &= 0, \\
&\vdots \\
\frac{d^{(k-2)}u}{dt}(0) &= 0, \\
\frac{d^{(k-1)}u}{dt}(0) &= c,
\end{aligned} \tag{4.42}$$

with $c = f(s)$, is $v(t;s)$, then the solution to

$$\begin{aligned}
\frac{d^{(k)}u}{dt} + a_{k-1}\frac{d^{(k-1)}u}{dt} + \cdots + a_1\frac{du}{dt} + a_0 u &= f(t), \\
u(0) &= 0, \\
\frac{du}{dt}(0) &= 0, \\
&\vdots \\
\frac{d^{(k-2)}u}{dt}(0) &= 0, \\
\frac{d^{(k-1)}u}{dt}(0) &= 0
\end{aligned} \tag{4.43}$$

is

$$u(t) = \int_0^t v(t-s;s) \, ds. \tag{4.44}$$

Exercise 4.3.12 asks the reader to show that this is true.

Duhamel's principle, as described above, depends on the fact that the differential equation is linear and has constant coefficients. If the differential equation has nonconstant coefficients, then we must modify the procedure slightly. We assume that the solution of

$$\frac{d^{(k)}u}{dt} + a_{k-1}(t)\frac{d^{(k-1)}u}{dt} + \cdots + a_1(t)\frac{du}{dt} + a_0(t)u = 0,$$
$$u(s) = 0,$$
$$\frac{du}{dt}(s) = 0,$$
$$\vdots$$
$$\frac{d^{(k-2)}u}{dt}(s) = 0,$$
$$\frac{d^{(k-1)}u}{dt}(s) = f(s)$$
(4.45)

is $v(t;s)$ (notice that the initial time is taken to be $t = s$ rather than $t = 0$). Then the solution of

$$\frac{d^{(k)}u}{dt} + a_{k-1}(t)\frac{d^{(k-1)}u}{dt} + \cdots + a_1(t)\frac{du}{dt} + a_0(t)u = f(t),$$
$$u(0) = 0,$$
$$\frac{du}{dt}(0) = 0,$$
$$\vdots$$
$$\frac{d^{(k-2)}u}{dt}(0) = 0,$$
$$\frac{d^{(k-1)}u}{dt}(0) = 0$$
(4.46)

is

$$u(t) = \int_0^t v(t;s)\,ds. \qquad (4.47)$$

To see why this modification is necessary, let us consider the second-order case for definiteness. With constant coefficients, if $u(t) = v(t;s)$ solves the differential equation subject to $u(0) = 0$, $du/dt(0) = f(s)$, then $u(t) = v(t-s;s)$ still solves the differential equation, but now with $u(s) = 0$, $du/dt(s) = f(s)$. However, this is not true if the equation has nonconstant coefficients—$u(t) = v(t-s;s)$ cannot be guaranteed to solve the ODE. For this reason, in the case of nonconstant coefficients, we define $v(t;s)$ with initial conditions specified at $t = s$.

Exercises

1. Consider the IVP (4.28), where **A** is the matrix in Example 4.6. Find the solution for

$$\mathbf{x}_0 = \begin{bmatrix} 0 \\ 0 \\ 1 \end{bmatrix}.$$

4.3. Linear systems with constant coefficients

2. Consider the IVP (4.28), where $\mathbf{A}$ is the matrix in Example 4.6.
    (a) Explain why, no matter what the value of $\mathbf{x}_0$, the solution $\mathbf{x}(t)$ converges to a constant vector as $t \to \infty$.
    (b) Find all values of $\mathbf{x}_0$ such that the solution $\mathbf{x}$ is *equal to* a constant vector for all values of t.

3. Find the general solution to
$$\frac{d\mathbf{x}}{dt} = \mathbf{A}\mathbf{x},$$
where
$$\mathbf{A} = \begin{bmatrix} 0 & 1 \\ 1 & 0 \end{bmatrix}.$$

4. Find the general solution to
$$\frac{d\mathbf{x}}{dt} = \mathbf{A}\mathbf{x},$$
where
$$\mathbf{A} = -\frac{1}{100} \begin{bmatrix} 36 & 48 \\ 48 & 64 \end{bmatrix}.$$

5. Let $\mathbf{A}$ be the matrix in Exercise 3. Find all values of $\mathbf{x}_0$ such that the solution to (4.28) decays exponentially to zero.

6. Let $\mathbf{A}$ be the matrix in Exercise 4. Find all values of $\mathbf{x}_0$ such that the solution to (4.28) decays exponentially to zero.

7. Let $\mathbf{A}$ be the matrix of Example 4.6. Solve the IVP
$$\frac{d\mathbf{x}}{dt} = \mathbf{A}\mathbf{x} + \mathbf{f}(t),$$
$$\mathbf{x}(0) = \mathbf{0},$$
where
$$\mathbf{f}(t) = \begin{bmatrix} 1 \\ 0 \\ \sin(t) \end{bmatrix}.$$

8. Let
$$\mathbf{A} = \frac{1}{6} \begin{bmatrix} -4 & 2 & -4 \\ 2 & -4 & -4 \\ -4 & -4 & 2 \end{bmatrix}.$$
Solve the IVP
$$\frac{d\mathbf{x}}{dt} = \mathbf{A}\mathbf{x} + \mathbf{f}(t),$$
$$\mathbf{x}(0) = \mathbf{x}_0,$$
where
$$\mathbf{f}(t) = \begin{bmatrix} \cos(t) \\ 1 \\ \sin(t) \end{bmatrix}, \quad \mathbf{x}_0 = \begin{bmatrix} 1 \\ 1 \\ 1 \end{bmatrix}.$$

9. The following system has been proposed as a model of the population dynamics of two species of animal that compete for the same resource:

$$\frac{dx}{dt} = ax - by, \ x(0) = x_0,$$

$$\frac{dy}{dt} = -cx + dy, \ y(0) = y_0.$$

Here a, b, c, d are positive constants, $x(t)$ is the population of the first species at time t, and $y(t)$ is the corresponding population of the second species (x and y are measured in some convenient units, say thousands or millions of animals). The equations are easy to understand: either species increases (exponentially) if the other is not present, but, since the two species compete for the same resource, the presence of one species contributes negatively to the growth rate of the other.

(a) Solve the IVP with $b = c = 2$, $a = d = 1$, $x(0) = 2$, and $y(0) = 1$, and explain (in words) what happens to the populations of the two species over the long term.

(b) With the values of a, b, c, d given in part (a), is there an initial condition which will lead to a different (qualitative) outcome?

10. Let $\mathbf{U} = \mathbf{U}(t)$ as defined in Section 4.3.3. Prove that $\mathbf{U}(t) = \mathbf{U}(t-s)\mathbf{U}(s)$ for all $s \in (0,t)$. (Hint: Let $\mathbf{c} \in \mathbf{R}^n$ be arbitrary, and define $\mathbf{u}(t) = \mathbf{U}(t)\mathbf{c}$, $\mathbf{v}(t) = \mathbf{U}(t-s)\mathbf{U}(s)\mathbf{c}$. Prove that $\mathbf{u} = \mathbf{v}$ by showing that they solve the same IVP, where the initial time is $t = s$. Explain why this proves the desired result.)

11. Using only the fact that $v(t;s)$ satisfies (4.37) with $c = f(s)$, prove that (4.41) satisfies (4.39).

12. Prove directly that (4.44) solves (4.43). Where do you need to use the fact that the ODE has constant coefficients?

13. Prove directly that (4.47) solves (4.46).

14. Apply Duhamel's principle to derive the solution of the IVP

$$\frac{du}{dt} - a(t)u = f(t), \ u(0) = 0.$$

15. Use Duhamel's principle to find a formula for the solution of

$$\frac{d^2u}{dt^2} + 3\frac{du}{dt} + 2u = f(t),$$

$$u(0) = 0,$$

$$\frac{du}{dt}(0) = 0.$$

16. Let $\mathbf{U}(t) = \mathbf{X}(t)\mathbf{X}(0)^{-1}$ be the matrix defined above in Section 4.3.3, so that $\mathbf{x}(t) = \mathbf{U}(t)\mathbf{c}$ solves (4.34). Find the solution of (4.33) in the form $\mathbf{x}(t) = \mathbf{U}(t)\mathbf{c}(t)$ by substituting into (4.32) and solving for $\mathbf{c}(t)$. Show that the result simplifies to (4.36). (Hint: You will need the result of Exercise 10.)

4.4 Numerical methods for initial value problems

So far in this chapter, we have discussed simple classes of ODEs, for which it is possible to produce an explicit formula for the solution. However, most differential equations cannot be solved in this sense. Moreover, sometimes the only formula that can be found is difficult to evaluate, involving integrals that cannot be computed in terms of elementary functions, eigenvalues and eigenvectors that cannot be found exactly, and so forth.

In cases like these, it may be that the only way to investigate the solution is to approximate it using a *numerical method*, which is simply an algorithm producing an approximate solution. We emphasize that the use of a numerical method always implies that the computed solution will be in error. It is essential to know something about the magnitude of this error; otherwise, one cannot use the solution with any confidence.

Most numerical methods for IVPs in ODEs are designed for first-order scalar equations. These can be applied, almost without change, to first-order systems, and hence to higher-order ODEs (after they have been converted to first-order systems). Therefore, we begin by discussing the general first-order scalar IVP, which is of the form

$$\frac{du}{dt} = f(t,u), \; u(t_0) = u_0. \tag{4.48}$$

We will discuss *time-stepping* methods, which seek to find approximations to $u(t_1)$, $u(t_2),\ldots,u(t_n)$, where $t_0 < t_1 < t_2 < \cdots < t_n$ define the *grid*. The quantities $t_1 - t_0, t_2 - t_1, \ldots, t_n - t_{n-1}$ are called the *time steps*.

The basic idea of time-stepping methods is based on the fundamental theorem of calculus, which implies that if

$$\frac{du}{dt}(t) = f(t,u(t)),$$

then

$$u(t_{i+1}) = u(t_i) + \int_{t_i}^{t_{i+1}} f(s,u(s))ds. \tag{4.49}$$

Now, we cannot (in general) hope to evaluate the integral in (4.49) analytically; however, any numerical method for approximating the value of a definite integral (such methods are often referred to as *quadrature* rules) can be adapted to form the basis of a numerical method of IVPs. By the way, (4.49) shows why the process of solving an IVP numerically is often referred to as *integrating* the ODE.

4.4.1 Euler's method

The simplest method of integrating an ODE is based on the *left-endpoint rule*:

$$\int_a^b f(x)dx \doteq f(a)(b-a).$$

Applying this to (4.49), we obtain

$$u(t_{i+1}) \doteq u(t_i) + f(t_i,u(t_i))(t_{i+1} - t_i). \tag{4.50}$$

Of course, this is not a computable formula, because, except possibly on the first step ($i = 0$), we do not know $u(t_i)$ exactly. Instead, we have an approximation, $u_i \doteq u(t_i)$. Formula (4.50) suggests how to obtain an approximation u_{i+1} to $u(t_{i+1})$:

$$u_{i+1} = u_i + f(t_i, u_i)(t_{i+1} - t_i). \tag{4.51}$$

The reader should notice that (except for $i = 0$) there are two sources of error in the estimate u_{i+1}. First of all, there is the error inherent in using formula (4.50) to advance the integration by one time step. Second, there is the accumulated error due to the fact that we do not know $u(t_i)$ in (4.50), but rather only an approximation to it.

The method (4.51) is referred to as *Euler's method*.

Example 4.10. Consider the IVP

$$\frac{du}{dt} = \frac{u}{1+t^2}, \quad u(0) = 1. \tag{4.52}$$

The exact solution is

$$u(t) = e^{\tan^{-1} t},$$

and we can use this formula to determine the errors in the computed approximation to $u(t)$. Applying Euler's method with the regular grid $t_i = i \Delta t$, $\Delta t = 10/n$, we have

$$u_{i+1} = u_i + \Delta t \frac{u_i}{1+t_i^2} = u_i \frac{1 + \Delta t + t_i^2}{1 + t_i^2} = u_i \frac{1 + \Delta t + i^2 \Delta t^2}{1 + i^2 \Delta t^2}.$$

For example, with $u_0 = 1$ and $n = 100$, we obtain

$$u_1 \doteq 1.1000, \ u_2 \doteq 1.2089, \ u_3 \doteq 1.3252, \ldots.$$

In Figure 4.1, we graph the exact solution and the approximations computed using Euler's method for $n = 10, 20, 40$. As we should expect, the approximation produced by Euler's

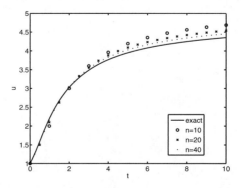

Figure 4.1. *Euler's method applied to* (4.52).

4.4. Numerical methods for initial value problems

Table 4.1. *Global error in Euler's method for* (4.52).

n	Error in $u(10)$	Error in $\frac{u(10)}{\Delta t}$
10	$-3.3384 \cdot 10^{-1}$	$-3.3384 \cdot 10^{-1}$
20	$-1.8507 \cdot 10^{-1}$	$-3.7014 \cdot 10^{-1}$
40	$-1.0054 \cdot 10^{-1}$	$-4.0218 \cdot 10^{-1}$
80	$-5.2790 \cdot 10^{-2}$	$-4.2232 \cdot 10^{-1}$
160	$-2.7111 \cdot 10^{-2}$	$-4.3378 \cdot 10^{-1}$
320	$-1.3748 \cdot 10^{-2}$	$-4.3992 \cdot 10^{-1}$
640	$-6.9235 \cdot 10^{-3}$	$-4.4310 \cdot 10^{-1}$

method gets better as Δt (the time step) decreases. In fact, Table 4.1, where we collect the errors in the approximations to $u(10)$, suggests that the *global error* (which comprises the total error after a number of steps) is $O(\Delta t)$. The symbol $O(\Delta t)$ ("big-oh of Δt") denotes a quantity that is proportional to or smaller than Δt as $\Delta t \to 0$. ■

Proving that the *order* of Euler's method is really $O(\Delta t)$ is beyond the scope of this book, but we can easily sketch the essential ideas of the proof. It can be proved that the left-endpoint rule, applied to an interval of integration of length Δt, has an error that is $O(\Delta t^2)$. In integrating an IVP, we apply the left-endpoint rule repeatedly, in fact, $n = O(1/\Delta t)$ times. Adding up $1/\Delta t$ errors of order Δt^2 gives a total error of $O(\Delta t)$. (Proving that this heuristic reasoning is correct takes some rather involved but elementary analysis that can be found in most numerical analysis textbooks.)

4.4.2 Improving on Euler's method: Runge–Kutta methods

If the results suggested in the previous section are correct, Euler's method can be used to approximate the solution to an IVP to any desired accuracy by simply making Δt small enough. Although this is true (with certain restrictions), we might hope for a more efficient method—one that would not require as many time steps to achieve a given accuracy.

To improve Euler's method, we choose a numerical integration technique that is more accurate than the left-endpoint rule. The simplest is the *midpoint rule*:

$$\int_a^b f(x)\,dx \doteq (b-a)f\left(\frac{a+b}{2}\right).$$

If $\Delta t = b - a$ (and the integrand f is sufficiently smooth), then the error in the midpoint rule is $O(\Delta t^3)$. Following the reasoning in the previous section, we expect that the corresponding method for integrating an IVP would have $O(\Delta t^2)$ global error.

It is not immediately clear how to use the midpoint rule for quadrature to integrate an IVP. The obvious equation is

$$u(t_i + \Delta t) \doteq u(t_i) + \Delta t f\left(t_i + \frac{\Delta t}{2}, u\left(t_i + \frac{\Delta t}{2}\right)\right).$$

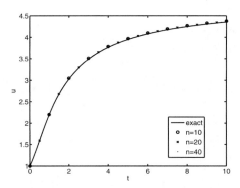

Figure 4.2. *Improved Euler method applied to* (4.52).

Table 4.2. *Global error in the improved Euler method for* (4.52).

n	Error in $u(10)$	$\frac{\text{Error in } u(10)}{\Delta t^2}$
10	$-2.1629 \cdot 10^{-2}$	$-2.1629 \cdot 10^{-2}$
20	$-1.2535 \cdot 10^{-2}$	$-5.0140 \cdot 10^{-2}$
40	$-4.7603 \cdot 10^{-3}$	$-7.6165 \cdot 10^{-2}$
80	$-1.4746 \cdot 10^{-3}$	$-9.4376 \cdot 10^{-2}$
160	$-4.1061 \cdot 10^{-4}$	$-10.512 \cdot 10^{-2}$
320	$-1.0835 \cdot 10^{-4}$	$-11.095 \cdot 10^{-2}$
640	$-2.7828 \cdot 10^{-5}$	$-11.398 \cdot 10^{-2}$

However, after reaching time t_i in the integration, we have an approximation for $u(t_i)$, but none for $u(t_i + \Delta t/2)$. To use the midpoint rule requires that we first generate an approximation for $u(t_i + \Delta t/2)$; the simplest way to do this is with an Euler step:

$$u\left(t_i + \frac{\Delta t}{2}\right) \doteq u(t_i) + \frac{\Delta t}{2} f(t_i, u(t_i)).$$

Putting this approximation together with the midpoint rule, and using the approximation $u_i \doteq u(t_i)$, we obtain the *improved Euler method*:

$$u_{i+1} = u_i + \Delta t k_2, \; k_2 = f\left(t_i + \frac{\Delta t}{2}, u_i + \frac{\Delta t}{2} k_1\right), \; k_1 = f(t_i, u_i).$$

Figure 4.2 shows the results of the improved Euler method, applied to (4.52) with $n = 10$, $n = 20$, and $n = 40$. The improvement in accuracy can be easily seen by comparison to Figure 4.1. We can see the $O(\Delta t^2)$ convergence in Table 4.2—when Δt is divided by 2 (i.e., n is doubled), the error is divided by approximately 4.

4.4. Numerical methods for initial value problems

The improved Euler method is a simple example of a *Runge–Kutta* (RK) method. An (explicit) RK method takes the following form:

$$u_{i+1} = u_i + \Delta t \sum_{j=1}^{m} \alpha_j k_j,$$

$$k_1 = f(t_i, u_i),$$
$$k_2 = f(t_i + \gamma_2 \Delta t, u_i + \beta_{21} \Delta t k_1),$$
$$k_3 = f(t_i + \gamma_3 \Delta t, u_i + \beta_{31} \Delta t k_1 + \beta_{32} \Delta t k_2), \quad (4.53)$$
$$\vdots$$
$$k_m = f\left(t_i + \gamma_m \Delta t, u_i + \sum_{l=1}^{m-1} \beta_{ml} \Delta t k_l\right)$$

with certain restrictions on the values of the parameters α_j, β_{kl}, and γ_j (such as $\alpha_1 + \cdots + \alpha_m = 1$). Although (4.53) looks complicated, it is not hard to understand the idea. A general form for a quadrature rule is

$$\int_a^b f(x)\,dx \doteq \sum_{j=1}^{m} w_j f(x_j),$$

where $w_1, w_2, \ldots, w_m$ are the quadrature weights and $x_1, x_2, \ldots, x_m \in [a,b]$ are the quadrature nodes. In the formula for u_{i+1} in (4.53), the weights are

$$\Delta t \alpha_1, \Delta t \alpha_2, \ldots, \Delta t \alpha_m,$$

and the values $k_1, k_2, \ldots, k_m$ are estimates of $f(t, u(t))$ at m nodes in the interval $[t_i, t_{i+1}]$. We will not use the general formula (4.53), but the reader should appreciate the following point: there are many RK methods, obtained by choosing various values for the parameters in (4.53). This fact is used in designing algorithms that attempt to automatically control the error. We discuss this further in Section 4.4.4.

The most popular RK method is analogous to *Simpson's rule* for quadrature:

$$\int_a^b f(x)\,dx \doteq \frac{b-a}{6}\left(f(a) + 4f\left(\frac{a+b}{2}\right) + f(b)\right).$$

Simpson's rule has an error of $O(\Delta t^5)$ ($\Delta t = b - a$), and the following related method for integrating an IVP has a global error of $O(\Delta t^4)$:

$$u_{i+1} = u_i + \frac{\Delta t}{6}(k_1 + 2k_2 + 2k_3 + k_4),$$
$$k_1 = f(t_i, u_i),$$
$$k_2 = f\left(t_i + \frac{\Delta t}{2}, u_i + \frac{\Delta t}{2}k_1\right), \quad (4.54)$$
$$k_3 = f\left(t_i + \frac{\Delta t}{2}, u_i + \frac{\Delta t}{2}k_2\right),$$
$$k_4 = f(t_{i+1}, u_i + \Delta t k_3).$$

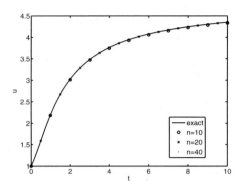

Figure 4.3. *Fourth-order Runge–Kutta method (RK4) applied to (4.52).*

Table 4.3. *Global error in the fourth-order Runge–Kutta method (RK4) for (4.52).*

n	Error in $u(10)$	$\dfrac{\text{Error in } u(10)}{\Delta t^4}$
10	$1.9941 \cdot 10^{-2}$	$1.9941 \cdot 10^{-2}$
20	$1.0614 \cdot 10^{-3}$	$1.6982 \cdot 10^{-2}$
40	$6.5959 \cdot 10^{-5}$	$1.6885 \cdot 10^{-2}$
80	$4.0108 \cdot 10^{-6}$	$1.6428 \cdot 10^{-2}$
160	$2.4500 \cdot 10^{-7}$	$1.6057 \cdot 10^{-2}$
320	$1.5100 \cdot 10^{-8}$	$1.5833 \cdot 10^{-2}$
640	$9.3654 \cdot 10^{-10}$	$1.5712 \cdot 10^{-2}$

We call this the RK4 method. Figure 4.3 and Table 4.3 demonstrate the accuracy of RK4; dividing Δt by 2 decreases the error by (approximately) a factor of 16. This is typical for $O(\Delta t^4)$ convergence.

We must note here that the improvement in efficiency of these higher-order methods is not as dramatic as it might appear at first glance. For instance, comparing Tables 4.1 and 4.3, we notice that just 10 steps of RK4 gives a smaller error than 160 steps of Euler's method. However, the improvement in efficiency is not a factor of 16, since 160 steps of Euler's method use 160 evaluations of $f(t,u)$, while 10 steps of RK4 use 40 evaluations of $f(t,u)$. In problems of realistic complexity, the evaluation of $f(t,u)$ is the most expensive part of the calculation (often f is defined by a computer simulation rather than a simple algebraic formula), and so a more reasonable comparison of these results would conclude that RK4 is four times as efficient as Euler's method for this particular example and level of error. Higher-order methods tend to use fewer steps than lower-order methods, which is enough to make the higher-order methods more efficient even though they require more work per step.

4.4.3 Numerical methods for systems of ODEs

The numerical methods presented above can be applied directly to a system of ODEs. Indeed, when the system is written in vector form, the notation is virtually identical. We now present an example.

4.4. Numerical methods for initial value problems

Example 4.11. Consider two species of animal that share a habitat, and suppose one species is prey to the other. Let $x_1(t)$ be the population of the predator species at time t, and let $x_2(t)$ be the population of the prey at the same time. The *Lotka–Volterra predator-prey* model for these two populations is

$$\frac{dx_1}{dt} = e_2 e_1 x_1 x_2 - q x_1,$$
$$\frac{dx_2}{dt} = r x_2 - e_1 x_1 x_2,$$

where e_1, e_2, q, r are positive constants. The parameters have the following interpretations:

- e_1 describes the attack rate of the predators (the rate at which the prey are killed by the predators is $e_1 x_1 x_2$);
- e_2 describes the growth rate of the predator population based on the number of prey killed (the efficiency of predators at converting predators to prey);
- q is the rate at which the predators die;
- r is the intrinsic growth rate of the prey population.

The equations describe the rate of population growth (or decline) of each species. The rate of change of the predator population is the difference between the rate of growth due to feeding on the prey and the rate of death. The rate of change of the prey population is the difference between the natural growth rate (which would govern in the absence of predators) and the death rate due to predation.

Define $\mathbf{f} : \mathbf{R} \times \mathbf{R}^2 \to \mathbf{R}^2$ by

$$\mathbf{f}(t, \mathbf{x}) = \begin{bmatrix} e_2 e_1 x_1 x_2 - q x_1 \\ r x_2 - e_1 x_1 x_2 \end{bmatrix}$$

($\mathbf{f}$ is actually independent of t; however, we include t as an independent variable so that this example fits into the general form discussed above). Then, given initial values $x_{1,0}$ and $x_{2,0}$ of the predator and prey populations, respectively, the IVP of interest is

$$\frac{d\mathbf{x}}{dt} = \mathbf{f}(t, \mathbf{x}), \qquad (4.55)$$
$$\mathbf{x}(0) = \mathbf{x}_0,$$

where

$$\mathbf{x}_0 = \begin{bmatrix} x_{1,0} \\ x_{2,0} \end{bmatrix}.$$

Suppose

$$e_1 = 0.01,\ e_2 = 0.2,\ r = 0.2,\ q = 0.4,\ x_{1,0} = 40,\ x_{2,0} = 500.$$

Using the RK4 method described above, we estimated the solution of (4.55) on the time interval $[0, 50]$. The implementation is exactly as described in (4.54), except that now the various quantities $(\mathbf{k}_i, \mathbf{u}_i, \mathbf{f}(t_i, \mathbf{u}_i))$ are vectors. The time step used was $\Delta t = 0.05$ (for a total

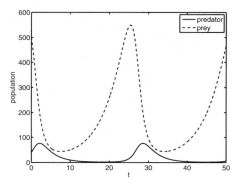

Figure 4.4. *The variation of the two populations with time (Example 4.11).*

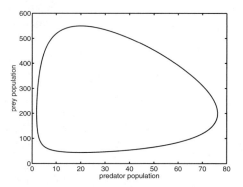

Figure 4.5. *The variation of the two populations (Example 4.11).*

of 1000 steps). In Figure 4.4, we plot the two populations versus time; this graph suggests that both populations vary periodically.

Another way to visualize the results is to graph x_2 versus x_1; such a graph is meaningful because, for an *autonomous* ODE (one in which t does not appear explicitly), the curve $(x_1(t), x_2(t))$ is determined entirely by the initial value $\mathbf{x}_0$ (see Exercise 4.4.5 for a precise formulation of this property). In Figure 4.5, we graph x_2 versus x_1. This curve is traversed in the clockwise direction (as can be determined by comparing with Figure 4.4). For example, when the prey population is large and the predator population is small (the upper left part of the curve), the predator population will begin to grow. As it does, the prey population begins to decrease (since more prey are eaten). Eventually, the prey population gets small enough that it cannot support a large number of predators (far right part of the curve), and the predator population decreases rapidly. When the predator population gets small, the prey population begins to grow, and the whole cycle begins again. ∎

4.4.4 Automatic step control and Runge–Kutta–Fehlberg methods

As the above examples show, it is possible to design numerical methods which give a predictable improvement in the error as the time step is decreased. However, these methods

4.4. Numerical methods for initial value problems

do not allow the user to choose the step size in order to attain a desired level of accuracy. For example, we know that decreasing Δt from 0.1 to 0.05 will decrease the error in an $O(\Delta t^4)$ method by approximately a factor of 16, but this does not tell us that either step size will lead to a global error less than 10^{-3}.

It would be desirable to have a method that, given a desired level of accuracy, could choose the step size in an algorithm so as to attain that accuracy. While no method guaranteed to do this is known, there is a heuristic technique that is usually successful in practice. The basic idea is quite simple. When an algorithm wishes to integrate from t_n to t_{n+1}, it uses two different methods, one of order Δt^k and another of order Δt^{k+1}. It then regards the more accurate estimate (resulting from the $O(\Delta t^{k+1})$ method) as the exact value, and compares it with the estimate from the lower-order method. If the difference is sufficiently small, then it is assumed that the step size is sufficiently small, and the step is accepted. (Moreover, if the difference is too small, then it is assumed that the step size is smaller than it needs to be, and it may be increased on the next step.) On the other hand, if the difference is too large, then it is assumed that the step is inaccurate. The step is rejected, and Δt, the step size, is reduced. This is called *automatic step control*, and it leads to an approximate solution computed on an irregular grid, since Δt can vary from one step to the next. This technique is not guaranteed to produce a global error below the desired level, since the method only controls the *local error* (the error resulting from a single step of the method). However, the relationship between the local and global errors is understood in principle, and this understanding leads to methods that usually achieve the desired accuracy.

A popular class of methods for automatic step control consists of the *Runge–Kutta–Fehlberg* (RKF) methods, which use two RK methods together to control the local error as described in the previous paragraph. The general form of RK methods, given in (4.53), shows that there are many possible RK formulas; indeed, for a given order Δt^k, there are many different RK methods that can be derived. This fact is used in the RKF methodology to choose pairs of formulas that evaluate f, the function defining the ODE, as few times as possible. For example, it can be proved that every RK method of order Δt^5 requires at least six evaluations of f. It is possible to choose six points (i.e., the values $t_i, t_i + \gamma_2 h, \ldots, t_i + \gamma_6 h$ in (4.53)) so that five of the points can be used in an $O(\Delta t^4)$ formula, and the six points together define an $O(\Delta t^5)$ formula. (One such pair of formulas is the basis of the popular RKF45 method.) This allows a very efficient implementation of automatic step control.

Exercises

1. The purpose of this exercise is to estimate the value of $u(0.5)$, where $u(t)$ is the solution of the IVP

$$\frac{du}{dt} = u + e^t, \quad u(0) = 0. \quad (4.56)$$

The solution is $u(t) = te^t$, and so $u(0.5) = e^{1/2}/2 \doteq 0.82436$.

(a) Estimate $u(0.5)$ by taking 4 steps of Euler's method.

(b) Estimate $u(0.5)$ by taking 2 steps of the improved Euler's method.

(c) Estimate $u(0.5)$ by taking 1 step of the classical fourth-order RK method.

Which estimate is more accurate? How many times did each evaluate the function $f(t,u) = u + e^t$?

2. Reproduce the results of Example 4.11.

3. The system of ODEs[14]

$$\frac{d^2x}{dt^2} = 2\frac{dy}{dt} + x - \frac{\mu_2(x+\mu_1)}{r_1(x,y)^3} - \frac{\mu_1(x-\mu_2)}{r_2(x,y)^3},$$
$$\frac{d^2y}{dt^2} = -2\frac{dx}{dt} + y - \frac{\mu_2 y}{r_1(x,y)^3} - \frac{\mu_1 y}{r_2(x,y)^3}, \quad (4.57)$$

where

$$r_1(x,y) = \sqrt{(x+\mu_1)^2 + y^2},$$
$$r_2(x,y) = \sqrt{(x-\mu_2)^2 + y^2},$$

models the orbit of a satellite about two heavenly bodies, which we will assume to be the earth and the moon.

In these equations, $(x(t), y(t))$ are the coordinates of the satellite at time t. The origin of the coordinate system is the center of mass of the earth-moon system, and the x-axis is the line through the centers of the earth and the moon. The center of the moon is at the point $(1 - \mu_1, 0)$ and the center of the earth is at $(-\mu_1, 0)$, where $\mu_1 = 1/82.45$ is the ratio of mass of the moon to the mass of the earth. The unit of length is the distance from the center of the earth to the center of the moon. We write $\mu_2 = 1 - \mu_1$.

If the satellite satisfies the following initial conditions, then its orbit is known to be periodic with period $T = 6.19216933$:

$$x(0) = 1.2, \quad \frac{dx}{dt}(0) = 0,$$
$$y(0) = 0, \quad \frac{dy}{dt}(0) = -1.04935751.$$

(a) Convert the system (4.57) of second-order ODEs to a first-order system.

(b) Use a program that implements an adaptive (automatic step control) method[15] (such as RKF45) to follow the orbit for one period. Plot the orbit in the plane, and make sure that the tolerance used by the adaptive algorithm is small enough that the orbit really appears in the plot to be periodic. Explain the variation in time steps with reference to the motion of the satellite.

(c) Assume that, in order to obtain comparable accuracy with a fixed step size, it is necessary to use the minimum step chosen by the adaptive algorithm. How many steps would be required by an algorithm with fixed step size? How many steps were used by the adaptive algorithm?[16]

---

[14]Adapted from [21, pages 153–154].
[15]Both MATLAB and *Mathematica* provide such routines, **ode45** and **NDSolve**, respectively.
[16]It is easy to determine the number of steps used by the **ode45** routine from MATLAB. It is possible but tricky to determine the number of steps used by the **NDSolve** routine of *Mathematica*.

4.4. Numerical methods for initial value problems

4. Let

$$\mathbf{A} = \frac{1}{5}\begin{bmatrix} 11 & 48 \\ 48 & 39 \end{bmatrix}.$$

The problem is to produce a graph, on the interval $[0,20]$, of the two components of the solution to

$$\frac{d\mathbf{u}}{dt} = \mathbf{A}\mathbf{u}, \qquad (4.58)$$
$$\mathbf{u}(0) = \mathbf{u}_0,$$

where

$$\mathbf{u}_0 = \begin{bmatrix} 4 \\ -3 \end{bmatrix}.$$

(a) Use the RK4 method with a step size of $\Delta t = 0.1$ to compute the solution. Graph both components on the interval $[0,20]$.

(b) Use the techniques of the previous section to compute the exact solution. Graph both components on the interval $[0,20]$.

(c) Explain the difference.

5. (a) Suppose $t_0 \in [a,b]$ and $\mathbf{f} : \mathbf{R}^n \to \mathbf{R}^n$, $\mathbf{u} : [a,b] \to \mathbf{R}^n$ satisfy

$$\frac{d\mathbf{u}}{dt} = \mathbf{f}(\mathbf{u}),$$
$$\mathbf{u}(t_0) = \mathbf{u}_0.$$

Show that $\mathbf{v} : [a + (t_1 - t_0), b + (t_1 - t_0)] \to \mathbf{R}^n$ defined by

$$\mathbf{v}(t) = \mathbf{u}(t - (t_1 - t_0))$$

satisfies

$$\frac{d\mathbf{v}}{dt} = \mathbf{f}(\mathbf{v}),$$
$$\mathbf{v}(t_1) = \mathbf{u}_0.$$

Moreover, show that the curves

$$\{\mathbf{u}(t) : t \in [a,b]\}$$

and

$$\{\mathbf{v}(t) : t \in [a + (t_1 - t_0), b + (t_1 - t_0)]\}$$

are the same.

(b) Show, by producing an explicit example (a scalar IVP will do), that the above property does not hold for a differential equation that depends explicitly on t (that is, for a *nonautonomous* ODE).

6. Consider the IVP
$$\frac{dx}{dt} = \cos(t)x,$$
$$x(0) = 1.$$

Use both Euler's method and the improved Euler method to estimate $x(10)$, using step sizes $1, 1/2, 1/4, 1/8$, and $1/16$. Verify that the error in Euler's method is $O(\Delta t)$, while the error in the improved Euler method is $O(\Delta t^2)$.

7. Consider the IVP
$$\frac{dx}{dt} = 1 + |x - 1|,$$
$$x(0) = 0.$$

(a) Use the RK4 method with step sizes $1/4, 1/8, 1/16, 1/32, 1/64$, and $1/128$ to estimate $x(1/2)$ and verify that the error is $O(\Delta t^4)$.

(b) Use the RK4 method with step sizes $1/4, 1/8, 1/16, 1/32, 1/64$, and $1/128$ to estimate $x(2)$. Is the error $O(\Delta t^4)$ in this case as well? If not, speculate as to why it is not.

8. Let
$$\mathbf{A} = \begin{bmatrix} -2 & 1 & 0 & 1 & -3 \\ 1 & 0 & 0 & -1 & 2 \\ 0 & 0 & -1 & 1 & -1 \\ 1 & -1 & 1 & 0 & 0 \\ -3 & 2 & -1 & 0 & -2 \end{bmatrix}.$$

The matrix $\mathbf{A}$ has an eigenvalue $\lambda \doteq -0.187$; let $\mathbf{x}$ be a corresponding eigenvector. The problem is to produce a graph on the interval $[0, 20]$ of the five components of the solution to
$$\frac{d\mathbf{u}}{dt} = \mathbf{A}\mathbf{u},$$
$$\mathbf{u}(0) = \mathbf{x}.$$

(a) Solve the problem using the RK4 method (in floating point arithmetic) and graph the results.

(b) What is the exact solution?

(c) Why is there a discrepancy between the computed solution and the exact solution?

(d) Can this discrepancy be eliminated by decreasing the step size in the RK4 method?

4.5 Stiff systems of ODEs

The numerical methods described in the last section, while adequate for many IVPs, can be unnecessarily inefficient for some problems. We begin with a comparison of two IVPs, and the performance of a state-of-the-art automatic step-control algorithm on them.

4.5. Stiff systems of ODEs

Example 4.12. We consider the following two IVPs:

$$\frac{d\mathbf{x}}{dt} = \mathbf{A}_1 \mathbf{x}, \tag{4.59}$$
$$\mathbf{x}(0) = \mathbf{x}_0;$$

$$\frac{d\mathbf{x}}{dt} = \mathbf{A}_2 \mathbf{x}, \tag{4.60}$$
$$\mathbf{x}(0) = \mathbf{x}_0.$$

For both IVPs, the initial value is the same:

$$\mathbf{x}_0 = \begin{bmatrix} 1 \\ 1 \\ 1 \end{bmatrix}.$$

We will describe the matrices $\mathbf{A}_1$ and $\mathbf{A}_2$ by their spectral decompositions. The two matrices are symmetric, with the same eigenvectors but different eigenvalues. The eigenvectors are

$$\mathbf{u}_1 = \begin{bmatrix} \frac{1}{\sqrt{3}} \\ \frac{1}{\sqrt{3}} \\ \frac{1}{\sqrt{3}} \end{bmatrix}, \mathbf{u}_2 = \begin{bmatrix} \frac{1}{\sqrt{2}} \\ 0 \\ -\frac{1}{\sqrt{2}} \end{bmatrix}, \mathbf{u}_3 = \begin{bmatrix} \frac{1}{\sqrt{6}} \\ -\frac{2}{\sqrt{6}} \\ \frac{1}{\sqrt{6}} \end{bmatrix},$$

while the eigenvalues of $\mathbf{A}_1$ are $\lambda_1 = -1, \lambda_2 = -2, \lambda_3 = -3$ and the eigenvalues of $\mathbf{A}_2$ are $\lambda_1 = -1, \lambda_2 = -10, \lambda_3 = -100$.

Since the initial value $\mathbf{x}_0$ is an eigenvector of both matrices, corresponding to the eigenvalue -1 in both cases, the two IVPs have exactly the same solution:

$$\mathbf{x}(t) = e^{-t} \mathbf{x}_0.$$

We solved both systems using the MATLAB command **ode45**, which implements a state-of-the-art automatic step-control algorithm based on a fourth-fifth-order scheme.[17] The results are graphed in Figure 4.6, where, as expected, we see that the two computed solutions are the same (or at least very similar).

However, examining the results of the computations performed by **ode45** reveals a surprise: the algorithm used only 52 steps for IVP (4.59) but 1124 steps for (4.60)! ∎

A detailed explanation of these results is beyond the scope of this book, but we briefly describe the reason for this behavior. These comments are illustrated by an example below. Explicit time-stepping methods for ODEs, such as those that form the basis of **ode45**, are stable only if the time step is sufficiently small (where "sufficiently small" is problem dependent). This means that the expected $O(\Delta t^k)$ behavior of the global error is not observed unless Δt is small enough. For larger values of Δt, the method is unstable and produces computed solutions that "blow up" as the integration proceeds. Moreover, the upper bound on allowable time steps gets smaller as the eigenvalues of the matrix $\mathbf{A}$ get large and negative.[18]

---
[17]The routine **ode45** was implemented by Shampine and Reichelt. See [56] for details.
[18]If the system of ODEs under consideration is nonlinear, say $\mathbf{x} = \mathbf{f}(\mathbf{x}, t)$, then it is the eigenvalues of the Jacobian matrix $\partial \mathbf{f}/\partial \mathbf{x}$ that determine the behavior.

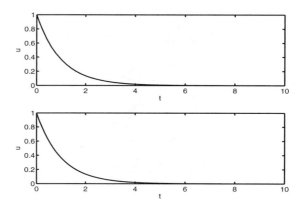

Figure 4.6. *The computed solutions to IVP* (4.59) *(top) and IVP* (4.60) *(bottom). (Each graph shows all three components of the solution; however, the exact solution, which is the same for the two IVPs, satisfies* $x_1(t) = x_2(t) = x_3(t)$.*)*

If **A** has large negative eigenvalues, then the system being modeled has some transient behavior that quickly dies out. It is the presence of the transient behavior that requires a small time step—initially, the solution is changing over a very small time scale, and the time step must be small to model this behavior accurately. If, at the same time, there are small negative eigenvalues, then the system also models components that die out more slowly. Once the transients have died out, one ought to be able to increase the time step to model the more slowly varying components of the solution. However, this is the weakness of explicit methods like Euler's method, RK4, and others: the transient behavior inherent in the *system* haunts the numerical method even when the transient components of the *solution* have died out. A time step that ought to be small enough to accurately follow the slowly varying components produces instability in the numerical method and ruins the computed solution.

A system with negative eigenvalues of widely different magnitudes (that is, a system that models components that decay at greatly different rates) is sometimes called *stiff*.[19] Stiff problems require special numerical methods; we give examples in Section 4.5.2. First, however, we discuss a simple example in which the ideas presented above can be easily understood.

4.5.1 A simple example of a stiff system

The IVP

$$\begin{aligned} \frac{du_1}{dt} &= -10^7 u_1, \ u_1(0) = 1, \\ \frac{du_2}{dt} &= -u_2, \ u_2(0) = 1 \end{aligned} \quad (4.61)$$

---

[19] Stiffness is a surprisingly subtle concept. The definition we follow here does not capture all of this subtlety; moreover, there is not a single, well-accepted definition of a stiff system. See [40, Section 6.2] for a discussion of the various definitions of stiffness.

4.5. Stiff systems of ODEs

has the solution

$$\mathbf{u}(t) = \begin{bmatrix} u_1(0)e^{-10^7 t} \\ u_2(0)e^{-t} \end{bmatrix},$$

and the system qualifies as stiff according to the description given above. We will apply Euler's method, since it is simple enough that we can completely understand its behavior. However, similar results would be obtained with RK4 or another explicit method.

To understand the behavior of Euler's method on (4.61), we write it out explicitly. We have

$$\mathbf{u}^{(n+1)} = \mathbf{u}^{(n)} + \Delta t \mathbf{f}(t, \mathbf{u}^{(n)})$$

with

$$\mathbf{f}(t, \mathbf{u}) = \begin{bmatrix} -10^7 u_1 \\ -u_2 \end{bmatrix};$$

that is,

$$u_1^{(n+1)} = u_1^{(n)} - 10^7 \Delta t u_1^{(n)} = \left(1 - 10^7 \Delta t\right) u_1^{(n)},$$
$$u_2^{(n+1)} = u_2^{(n)} - \Delta t u_2^{(n)} = (1 - \Delta t) u_2^{(n)}.$$

It follows that

$$\begin{aligned}
u_1^{(n)} &= \left(1 - 10^7 \Delta t\right) u_1^{(n-1)} \\
&= \left(1 - 10^7 \Delta t\right)^2 u_1^{(n-2)} \\
&= \left(1 - 10^7 \Delta t\right)^3 u_1^{(n-3)} \\
&= \cdots = \left(1 - 10^7 \Delta t\right)^n u_1^{(0)} \\
&= \left(1 - 10^7 \Delta t\right)^n
\end{aligned}$$

and, similarly,

$$u_2^{(n)} = (1 - \Delta t)^n.$$

Clearly, the computation depends critically on

$$|1 - 10^7 \Delta t|, \ |1 - \Delta t|.$$

If one of these quantities is larger than 1, the corresponding component will grow exponentially as the iteration progresses. At the very least, for *stability* (that is, to avoid spurious exponential growth), we need

$$|1 - 10^7 \Delta t| \leq 1, \ |1 - \Delta t| \leq 1.$$

The first inequality determines the restriction on Δt, and a little algebra shows that we must have

$$0 < \Delta t \leq 2 \cdot 10^{-7}.$$

Thus a very small time step is required for stability, and this restriction on Δt is imposed by the transient behavior in the system.

As we show below, it is possible to avoid the need for overly small time steps in a stiff system; however, we must use *implicit* methods. Euler's method and the RK methods discussed in Section 4.4 are explicit, meaning that the value $u^{(n+1)}$ is defined by a formula involving only known quantities. In particular, only $u^{(n)}$ appears in the formula; in some explicit methods (*multistep methods*), $u^{(n-1)}$, $u^{(n-2)}$,... may appear in the formula, but $u^{(n+1)}$ itself does not.

On the other hand, an implicit method defines $u^{(n+1)}$ by a formula that involves $u^{(n+1)}$ itself (as well as $u^{(n)}$ and possibly earlier computed values of u). This means that, at each step of the iteration, an algebraic equation (possibly nonlinear) must be solved to find the value of $u^{(n+1)}$. In spite of this additional computational expense, implicit methods are useful because of their improved stability properties.

4.5.2 The backward Euler method

The simplest implicit method is the *backward Euler method*. Recall that Euler's method for the IVP

$$\frac{du}{dt} = f(t,u), \; u(t_0) = u_0$$

was derived from the equivalent formulation

$$u(t_{n+1}) = u(t_n) + \int_{t_n}^{t_{n+1}} f(t,u(t))\,dt$$

by applying the left-endpoint rule for quadrature:

$$\int_a^b h(x)\,dx \doteq h(a)(b-a).$$

The result is

$$u_{n+1} = u_n + \Delta t f(t_n, u_n)$$

(see Section 4.4.1). To obtain the backward Euler method, we apply the right-hand rule,

$$\int_a^b h(x)\,dx \doteq h(b)(b-a)$$

instead. The result is

$$u_{n+1} = u_n + \Delta t f(t_{n+1}, u_{n+1}). \tag{4.62}$$

This method is indeed implicit. It is necessary to solve (4.62) for u_{n+1}. In the case of a nonlinear ODE (or a system of nonlinear ODEs), this may be difficult (requiring a numerical root-finding algorithm such as Newton's method). However, it is not difficult to implement

4.5. Stiff systems of ODEs

the backward Euler method for a linear system. We illustrate this for the linear, constant-coefficient system

$$\frac{d\mathbf{x}}{dt} = \mathbf{A}\mathbf{x} + \mathbf{f}(t), \qquad (4.63)$$
$$\mathbf{x}(0) = \mathbf{x}_0.$$

The backward Euler method takes the form

$$\mathbf{x}_{n+1} = \mathbf{x}_n + \Delta t(\mathbf{A}\mathbf{x}_{n+1} + \mathbf{f}(t_{n+1}))$$
$$\Rightarrow \mathbf{x}_{n+1} - \Delta t \mathbf{A}\mathbf{x}_{n+1} = \mathbf{x}_n + \Delta t \mathbf{f}(t_{n+1})$$
$$\Rightarrow (\mathbf{I} - \Delta t \mathbf{A})\mathbf{x}_{n+1} = \mathbf{x}_n + \Delta t \mathbf{f}(t_{n+1})$$
$$\Rightarrow \mathbf{x}_{n+1} = (\mathbf{I} - \Delta t \mathbf{A})^{-1}(\mathbf{x}_n + \Delta t \mathbf{f}(t_{n+1})).$$

Example 4.13. Applying the backward Euler method to the simple example (4.61) gives some insight into the difference between the implicit and explicit methods. We obtain the following iteration for the first component:

$$u_1^{(n+1)} = u_1^{(n)} - 10^7 \Delta t u_1^{(n+1)}$$
$$\Rightarrow \left(1 + 10^7 \Delta t\right) u_1^{(n+1)} = u_1^{(n)}$$
$$\Rightarrow u_1^{(n+1)} = \left(1 + 10^7 \Delta t\right)^{-1} u_1^{(n)}$$
$$\Rightarrow u_1^{(n)} = \left(1 + 10^7 \Delta t\right)^{-n}.$$

Similarly, for the second component, we obtain

$$u_2^{(n+1)} = (1 + \Delta t)^{-n}.$$

For *any* positive value of Δt, we have

$$\left|\left(1 + 10^7 \Delta t\right)^{-1}\right| < 1, \quad \left|(1 + \Delta t)^{-1}\right| < 1,$$

and so no instability can arise. The step size Δt will still have to be small enough for the solution to be accurate, but we do not need to take Δt excessively small to avoid spurious exponential growth. ∎

Example 4.14. We will apply both Euler's method and the backward Euler method to the IVP

$$\frac{d\mathbf{x}}{dt} = \mathbf{A}_2 \mathbf{x}, \qquad (4.64)$$
$$\mathbf{x}(0) = \mathbf{x}_0,$$

where $\mathbf{A}_2$ is the matrix from Example 4.12 and

$$\mathbf{x}_0 = \begin{bmatrix} 1 \\ -1 \\ 2 \end{bmatrix}.$$

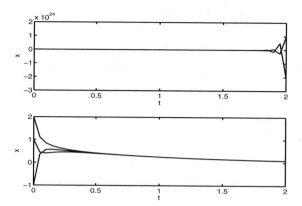

Figure 4.7. *The computed solutions to IVP (4.64) using Euler's method (top) and the backward Euler method (bottom). (Each graph shows all three components of the solution; however, in the top graph, the first and third components are indistinguishable on this scale.)*

We integrate over the interval [0, 2] with a time step of $\Delta t = 0.05$ (40 steps) in both algorithms. The results are shown in Figure 4.7. Euler's method "blows up" with this time step, while the backward Euler method produces a reasonable approximation of the true solution. ∎

Remark. There are higher-order methods designed for use with stiff ODEs, notably the trapezoidal method. It is also possible to develop automatic step-control methods for stiff systems.[20]

Exercises

1. Let
$$\mathbf{A} = \begin{bmatrix} 0 & -1 \\ -1 & 0 \end{bmatrix}, \mathbf{x}_0 = \begin{bmatrix} 1 \\ 0 \end{bmatrix},$$
and consider the IVP
$$\frac{d\mathbf{x}}{dt} = \mathbf{A}\mathbf{x},$$
$$\mathbf{x}(0) = \mathbf{x}_0.$$

   (a) Find the exact solution $\mathbf{x}(t)$.

   (b) Estimate $\mathbf{x}(1)$ using 10 steps of Euler's method and compute the norm of the error.

   (c) Estimate $\mathbf{x}(1)$ using 10 steps of the backward Euler method and compute the norm of the error.

---
[20]MATLAB includes such routines.

4.5. Stiff systems of ODEs

2. Repeat Exercise 1 for
$$\mathbf{A} = \begin{bmatrix} 1 & -1 \\ -1 & 1 \end{bmatrix}, \mathbf{x}_0 = \begin{bmatrix} 1 \\ 0 \end{bmatrix}.$$

3. Let
$$\mathbf{A} = \frac{1}{6} \begin{bmatrix} -113 & -80 & -107 \\ -80 & -140 & -80 \\ -107 & -80 & -113 \end{bmatrix}.$$

   Find the largest value of Δt such that Euler's method is stable. (Use numerical experimentation.)

4. Let $\mathbf{f} : \mathbf{R} \times \mathbf{R}^2 \to \mathbf{R}^2$ be defined by
$$\mathbf{f}(t, \mathbf{x}) = \begin{bmatrix} x_2 - 2x_1^2 + 2t^2 e^{-2t} - te^{-t} \\ -100x_2^2 + 100e^{-2t} - e^{-t} \end{bmatrix}.$$

   The nonlinear IVP
$$\frac{d\mathbf{x}}{dt} = \mathbf{f}(t, \mathbf{x}),$$
$$\mathbf{x}(0) = \begin{bmatrix} 0 \\ 1 \end{bmatrix}$$

   has the solution
$$\mathbf{x}(t) = \begin{bmatrix} te^{-t} \\ e^{-t} \end{bmatrix}.$$

   (a) Show by example that Euler's method is unstable unless Δt is chosen small enough. Determine (by trial and error) how small Δt must be in order that Euler's method is stable.

   (b) Apply the backward Euler method. Is it stable for all positive values of Δt?

5. Let
$$\mathbf{A} = \begin{bmatrix} -50.5 & 49.5 \\ 49.5 & -50.5 \end{bmatrix}.$$

   (a) Find the exact solution to
$$\frac{d\mathbf{x}}{dt} = \mathbf{A}\mathbf{x}, \ \mathbf{x}(0) = \begin{bmatrix} 1 \\ 2 \end{bmatrix}.$$

   (b) Determine experimentally (that is, by trial and error) how small Δt must be in order that Euler's method behaves in a stable manner on this IVP.

   (c) Euler's method takes the form
$$\mathbf{x}_{i+1} = \mathbf{x}_i + \Delta t \mathbf{A} \mathbf{x}_i, \ i = 0, 1, 2, \ldots.$$

Let the eigenpairs of $\mathbf{A}$ be $\lambda_1, \mathbf{u}_1$ and $\lambda_2, \mathbf{u}_2$. Show that Euler's method is equivalent to

$$y_1^{(i+1)} = (1 + \Delta t \lambda_1) y_1^{(i)},$$
$$y_2^{(i+1)} = (1 + \Delta t \lambda_2) y_2^{(i)},$$

where

$$y_1^{(i)} = \mathbf{x}_1 \cdot \mathbf{u}_1, \quad y_2^{(i)} = \mathbf{x}_i \cdot \mathbf{u}_2.$$

Find explicit formulas for $y_1^{(i)}$, $y_2^{(i)}$, and derive an upper bound on Δt that guarantees stability of Euler's method. Compare with the experimental result.

(d) Repeat with the backward Euler method in place of Euler's method.

6. Let $\mathbf{A} \in \mathbf{R}^{n \times n}$ be symmetric with *negative* eigenvalues $\lambda_1, \lambda_2, \ldots, \lambda_n$.

  (a) Consider Euler's method and the backward Euler method applied to the homogeneous linear system
  $$\frac{d\mathbf{x}}{dt} = \mathbf{A}\mathbf{x}.$$
  Show that Euler's method produces a sequence $\mathbf{x}_0, \mathbf{x}_1, \mathbf{x}_2, \ldots$ where
  $$\mathbf{x}_i = (\mathbf{I} + \Delta t \mathbf{A})^i \mathbf{x}_0,$$
  while the backward Euler method produces a sequence $\mathbf{x}_0, \mathbf{x}_1, \mathbf{x}_2, \ldots$ where
  $$\mathbf{x}_i = (\mathbf{I} - \Delta t \mathbf{A})^{-i} \mathbf{x}_0.$$

  (b) What are the eigenvalues of $\mathbf{I} + \Delta t \mathbf{A}$?

  (c) What condition must Δt satisfy in order that Euler's method be stable? (Hint: See the previous exercise.)

  (d) What are the eigenvalues of $(\mathbf{I} - \Delta t \mathbf{A})^{-1}$?

  (e) Show that the backward Euler method is stable for all positive values of Δt.

4.6 Suggestions for further reading

There are many introductory textbooks on ODEs, such as the texts by Zill [71] and by Goldberg and Potter [25]. An introductory text with a particularly strong emphasis on applied mathematics is Boyce and DiPrima [7]. A more advanced book that focuses on the theory of ODEs (and the necessary linear algebra) is Hirsch and Smale [34]. The reader is referred to Hirsch and Smale for a complete description of the theory of linear, constant-coefficient systems. (We covered only a special case, albeit an important one, in Section 4.3.)

For more information on numerical methods for ODEs, the reader is referred to Shampine [57] or Lambert [40]. Most books on numerical analysis, such as Atkinson [3] and Kincaid and Cheney [38], also cover numerical methods for ODEs, as well as the background material on quadrature and interpolation.

The text by Mattheij and Molenaar [48] gives an integrated presentation of the theory and numerical analysis of ODEs, as well as their use for modeling physical systems.

Chapter 5
Boundary Value Problems in Statics

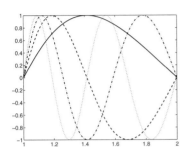

Our first examples of partial differential equations (PDEs) will arise in the study of static (equilibrium) phenomena in mechanics and heat flow. To make our introduction to the subject as simple as possible, we begin with one-dimensional examples; that is, all of the variation is assumed to occur in one spatial direction. Since the phenomena are static, time is not involved, and the single spatial variable is the only independent variable. Therefore, the "PDEs" are actually ODEs! Nevertheless, the techniques we develop for the one-dimensional problems generalize to real PDEs.

5.1 The analogy between BVPs and linear algebraic systems

As mentioned in Chapter 3, the solution methods presented in this book bear strong resemblance, at least in spirit, to methods useful for solving a linear system of the form $\mathbf{Ax} = \mathbf{b}$. In the exercises and examples of Chapter 3, we showed some of the similarities between linear systems and linear BVPs. We now review these similarities and explain the analogy further.

We will use the equilibrium displacement u of a sagging string as our first example. In Section 2.3, we showed that u satisfies the BVP

$$-T\frac{d^2u}{dx^2} = f(x),\ 0 < x < \ell,$$
$$u(0) = 0,$$
$$u(\ell) = 0,$$
(5.1)

where T is the tension in the string and f is an external force density (in units of force per length).

If $\mathbf{A} \in \mathbf{R}^{m \times n}$, then $\mathbf{A}$ defines an operator mapping $\mathbf{R}^n$ into $\mathbf{R}^m$, and given $\mathbf{b} \in \mathbf{R}^m$, we can ask the following questions:

- Is there an $\mathbf{x} \in \mathbf{R}^n$ which is mapped to $\mathbf{b}$ by $\mathbf{A}$?

- If so, is **x** unique?

- How can such an **x** be computed?

In much the same way,
$$-T\frac{d^2}{dx^2}$$
defines an operator—usually referred to as a *differential operator*. For convenience, we will refer to this operator as L. It maps one function to another, and the differential equation

$$-T\frac{d^2u}{dx^2} = f(x),\ 0 < x < \ell, \tag{5.2}$$

can be interpreted as the following question, Given a function f defined on the interval $[0, \ell]$, does there exist a function u which is mapped to f by L? That is, does there exist a function u satisfying $Lu = f$?

The analogy between $\mathbf{A}\mathbf{x} = \mathbf{b}$ and $Lu = f$ is strengthened by the fact that the differential operator L is linear, just as is the operator defined by the matrix $\mathbf{A}$. Indeed, if u and v are two functions and α and β are scalars, then

$$\begin{aligned}L(\alpha u + \beta v) &= -T\frac{d^2}{dx^2}[\alpha u + \beta v]\\ &= -\alpha T\frac{d^2u}{dx^2} - \beta T\frac{d^2v}{dx^2}\\ &= \alpha Lu + \beta Lv.\end{aligned}$$

Underlying this calculation is the fact that a space of functions can be viewed as a vector space—with functions as the vectors. The operator L is naturally defined on the space $C^2[0, \ell]$ introduced in Section 3.1. Thus L takes as input a twice-continuously differentiable function u and produces as output a continuous function Lu:

$$L : C^2[0, 1] \to C[0, 1].$$

It is possible to take the analogy of $Lu = f$ with $\mathbf{A}\mathbf{x} = \mathbf{b}$ even further. If $\mathbf{A} \in \mathbf{R}^{m \times n}$, then $\mathbf{A}\mathbf{x} = \mathbf{b}$ represents m equations in n unknowns, each equation having the form

$$a_{i1}x_1 + a_{i2}x_2 + \cdots + a_{in}x_n = b_i.$$

In the same way, $Lu = f$ represents a collection of equations, namely, the conditions that the functions

$$-T\frac{d^2u}{dx^2}$$

and f must be equal at each $x \in [0, \ell]$. Here, though, we see a major difference: $Lu = f$ represents an infinite number of equations, since there is an infinite number of points $x \in [0, \ell]$. Also, there are infinitely many unknowns, namely, the values $u(x), x \in [0, \ell]$.

The role of the boundary conditions is not difficult to explain in this context. Just as a matrix $\mathbf{A} \in \mathbf{R}^{m \times n}$ can have a nontrivial null space, in which case the equation $\mathbf{A}\mathbf{x} = \mathbf{b}$

5.1. The analogy between BVPs and linear algebraic systems

cannot have a unique solution, so a differential operator can have a nontrivial null space. It is easy to verify that the null space of L consists of all first-degree polynomial functions:

$$\mathcal{N}(L) = \{u \,:\, u(x) = ax + b, \; a, b \in \mathbf{R}\}.$$

Therefore, $Lu = f$ cannot have a unique solution, since if u is one solution and $w(x) = ax + b$, then $u + w$ is another solution. On the other hand, $Lu = f$ does have solutions for every $f \in C[0, \ell]$; for example,

$$u(x) = -\frac{1}{T} \int_0^x \int_0^s f(z) \, dz \, ds$$

solves $Lu = f$. This is equivalent to saying that the range of L is all of $C[0, \ell]$.

Example 5.1. Let $f \in C[0, 1]$ be defined by $f(x) = x$. Then, for every $a, b \in \mathbf{R}$,

$$u(x) = -\frac{1}{6T} x^3 + ax + b$$

is a solution of $Lu = f$. That is, the general solution of $Lu = f$ consists of $u(x) = -x^3/(6T)$ plus any function from the null space of L. (See page 46.) ∎

The fact that the null space of L is two dimensional suggests that the operator equation $Lu = f$ is closely analogous to a linear algebraic equation $\mathbf{A}\mathbf{x} = \mathbf{b}$, where $\mathbf{A}$ is an $(n-2) \times n$ matrix and $\mathcal{R}(A) = \mathbf{R}^{n-2}$ (that is, the columns of $\mathbf{A}$ span $\mathbf{R}^{n-2}$). Such a matrix $\mathbf{A}$ has the property that $\mathbf{A}\mathbf{x} = \mathbf{b}$ has a solution for every $\mathbf{b} \in \mathbf{R}^{n-2}$, but the solution cannot be unique. Indeed, in this case, $\mathcal{N}(A)$ is two dimensional, and therefore so is the solution set of $\mathbf{A}\mathbf{x} = \mathbf{b}$ for every $\mathbf{b} \in \mathbf{R}^n$. In order to obtain a unique solution, we must restrict the allowable vectors x by adding two equations. This could be accomplished by adding two rows (properly chosen) to the matrix $\mathbf{A}$.

In the case of the BVP (5.1) and the corresponding operator L, we "add" the boundary conditions to the operator by regarding

$$u(0) = u(\ell) = 0$$

as restricting the domain of the operator L. We define

$$L_D : C_D^2[0, \ell] \to C[0, \ell],$$

where

$$C_D^2[0, \ell] = \{u \in C^2[0, \ell] \,:\, u(0) = u(\ell) = 0\},$$

by

$$L_D u = Lu = -T \frac{d^2 u}{dx^2}.$$

This new operator L_D has a trivial null space. Indeed, if $u \in C^2[0, \ell]$ satisfies

$$-T \frac{d^2 u}{dx^2} = 0,$$

then $u(x) = ax + b$ for some $a, b \in \mathbf{R}$. If u also satisfies $u(0) = u(\ell) = 0$, then $a = b = 0$, and u is the zero function. The operator L_D, like L, has the property that $L_D u = f$ has a solution for each $f \in C[0, \ell]$ (that is, the range of L_D is $C[0, \ell]$), and this solution is unique because the null space of L_D is trivial.

We next show that L_D is a *symmetric* operator, and, as such, it has many of the properties of a symmetric matrix. Before we can demonstrate this, of course, we must explain what symmetry means for a differential operator. In Section 3.5, we saw that a matrix $\mathbf{A} \in \mathbf{R}^{n \times n}$ is symmetric if and only if

$$(\mathbf{A}\mathbf{x}) \cdot \mathbf{y} = \mathbf{x} \cdot (\mathbf{A}\mathbf{y}) \text{ for all } \mathbf{x}, \mathbf{y} \in \mathbf{R}^n.$$

We can make the analogous definition for a differential operator, using the L^2 inner product in place of the dot product.

Definition 5.2. *Let S be a subspace of $C^k[a, b]$, and let $K : S \to C[a, b]$ be a linear operator. We say that K is symmetric if*

$$(Ku, v) = (u, Kv) \text{ for all } u, v \in S, \tag{5.3}$$

where $(\cdot, \cdot)$ represents the L^2 inner product on the interval $[a, b]$. That is, K is symmetric if, whenever $u, v \in S$, and $w = Ku$, $z = Kv$, then

$$\int_a^b w(x) v(x) \, dx = \int_a^b u(x) z(x) \, dx.$$

In working with differential operators and symmetry, integration by parts is an essential technique, as the following examples show.

Example 5.3. Let L_D be the operator defined above. Then, if $u, v \in C_D^2[0, \ell]$, we have

$$\begin{aligned}
(L_D u, v) &= -T \int_0^\ell \frac{d^2 u}{dx^2}(x) v(x) \, dx \\
&= \left[-T \frac{du}{dx}(x) v(x) \right]_0^\ell + T \int_0^\ell \frac{du}{dx}(x) \frac{dv}{dx}(x) \, dx \\
&= -T \frac{du}{dx}(\ell) v(\ell) + T \frac{du}{dx}(0) v(0) + T \int_0^\ell \frac{du}{dx}(x) \frac{dv}{dx}(x) \, dx \\
&= T \int_0^\ell \frac{du}{dx}(x) \frac{dv}{dx}(x) \, dx \text{ (since } v(0) = v(\ell) = 0\text{)} \\
&= \left[T u(x) \frac{dv}{dx}(x) \right]_0^\ell - T \int_0^\ell u(x) \frac{d^2 v}{dx^2}(x) \, dx \\
&= T u(\ell) \frac{dv}{dx}(\ell) - T u(0) \frac{du}{dx}(0) - T \int_0^\ell u(x) \frac{d^2 v}{dx^2}(x) \, dx \\
&= -T \int_0^\ell u(x) \frac{d^2 v}{dx^2}(x) \, dx \text{ (since } u(0) = u(\ell) = 0\text{)} \\
&= (u, L_D v).
\end{aligned}$$

Therefore, L_D is a symmetric operator. ∎

5.1. The analogy between BVPs and linear algebraic systems

Example 5.4. Define
$$C_D^1[0,\ell] = \{u \in C^1[0,\ell] : u(0) = u(\ell) = 0\}$$
and $M_D : C_D^1[0,\ell] \to C[0,\ell]$ by
$$M_D u = \frac{du}{dx}.$$
Then
$$\begin{aligned}(M_D u, v) &= \int_0^\ell \frac{du}{dx}(x)v(x)\,dx = u(x)v(x)\Big|_0^\ell - \int_0^\ell u(x)\frac{dv}{dx}(x)\,dx \\ &= u(\ell)v(\ell) - u(0)v(0) - \int_0^\ell u(x)\frac{dv}{dx}(x)\,dx \\ &= -\int_0^\ell u(x)\frac{dv}{dx}(x)\,dx \text{ (since } u(0) = u(\ell) = 0) \\ &= -(u, M_D v).\end{aligned}$$

Therefore, M_D is not symmetric. ∎

A symmetric matrix $\mathbf{A} \in \mathbf{R}^{n \times n}$ has the following properties (see Section 3.5):

- All eigenvalues of $\mathbf{A}$ are real.
- Eigenvectors of $\mathbf{A}$ corresponding to distinct eigenvalues are orthogonal.
- There exists a basis of $\mathbf{R}^n$ consisting of eigenvectors of $\mathbf{A}$.

Analogous properties exist for a symmetric differential operator. In fact, the first two properties can be proved exactly as they were for symmetric matrices.[21]

In the following discussion, S is a subspace of $C^k[a,b]$ and $K : S \to C[a,b]$ is a symmetric linear operator. A scalar λ is an *eigenvalue* of K if there exists a nonzero function u such that
$$Ku = \lambda u.$$

Just as in the case of matrices, we cannot assume a priori that λ is real, or that the *eigenfunction* u is real valued. When working with complex-valued functions, the L^2 inner product on $[a,b]$ is defined by
$$(f,g) = \int_a^b f(x)\overline{g(x)}\,dx. \tag{5.4}$$

The properties
$$(\alpha f, g) = \alpha(f,g), \ (f, \alpha g) = \overline{\alpha}(f,g)$$
hold, just as for the complex dot product.

---
[21] The third property is more difficult. We discuss an analogous property for symmetric differential operators in the subsequent sections and delay the proof of this property until Chapter 12.

We immediately show that there is no need to allow complex numbers when working with symmetric operators. Indeed, suppose u is a nonzero function and λ is a scalar satisfying

$$Ku = \lambda u.$$

Then, since $(Ku,u) = (u,Ku)$ holds for complex-valued functions as well as real-valued functions (see Exercise 4), we have

$$(Ku,u) = (u,Ku) \Rightarrow (\lambda u, u) = (u, \lambda u) \Rightarrow \lambda(u,u) = \overline{\lambda}(u,u) \Rightarrow \lambda = \overline{\lambda} \Rightarrow \lambda \in \mathbf{R}.$$

It is easy to show that there must be a real-valued eigenfunction corresponding to λ (the proof is the same as for matrices; see Theorem 3.44). Therefore, we do not need to consider complex numbers any longer when we are dealing with symmetric operators.

We now assume that $\lambda_1, \lambda_2 \in \mathbf{R}$ are distinct eigenvalues of the operator K, with corresponding eigenfunctions $u_1, u_2 \in S$. We then have

$$\lambda_1(u_1, u_2) = (\lambda_1 u_1, u_2) = (Ku_1, u_2) = (u_1, Ku_2) = (u_1, \lambda_2 u_2) = \lambda_2(u_1, u_2).$$

Since $\lambda_1 \neq \lambda_2$, this is only possible if $(u_1, u_2) = 0$, that is, if u_1 and u_2 are orthogonal. Therefore, a symmetric differential operator has orthogonal eigenfunctions.

The symmetric differential operator L_D has a special property not shared by every symmetric operator. Suppose λ is an eigenvalue of L_D and u is a corresponding eigenfunction, normalized so that $(u,u) = 1$. Then

$$\lambda = \lambda(u,u) = (\lambda u, u) = (L_D u, u)$$

$$= -T \int_0^\ell \frac{d^2 u}{dx^2}(x) u(x) dx$$

$$= T \int_0^\ell \frac{du}{dx}(x) \frac{du}{dx}(x) dx \text{ (integration by parts)} \qquad (5.5)$$

$$= T \int_0^\ell \left(\frac{du}{dx}(x)\right)^2 dx$$

$$> 0.$$

Therefore, every eigenvalue of L_D is positive. (This explains why we prefer to work with the *negative* second derivative operator—the eigenvalues are then positive. Actually, in the computation above, it is only obvious that

$$\int_0^\ell \left(\frac{du}{dx}(x)\right)^2 dx \geq 0,$$

and therefore that $\lambda \geq 0$. How can we conclude that, in fact, the inequality must be strict? See Exercise 5.1.5.)

Summary

We have now seen that the differential operator

$$-T\frac{d^2}{dx^2},$$

5.1. The analogy between BVPs and linear algebraic systems

subject to Dirichlet conditions, is symmetric, which means that any eigenvalues must be real and the eigenfunctions will be orthogonal. We have also seen directly that any eigenvalues must be positive. Assuming we can find the eigenvalues and eigenfunctions of L_D, and that there are "enough" eigenfunctions to represent u and f, we can contemplate a spectral method for solving

$$-T\frac{d^2u}{dx^2} = f(x), \ 0 < x < \ell,$$
$$u(0) = 0, \qquad\qquad\qquad (5.6)$$
$$u(\ell) = 0.$$

We develop this method in the next two sections.

5.1.1 A note about direct integration

The BVP (5.6) is quite elementary; indeed, we already showed in Chapter 2 how to solve this and similar problems by integrating twice (see Examples 2.2 and 2.3). There are two reasons why we spend this chapter developing other methods for solving (5.6). First of all, the methods we develop, Fourier series and finite elements, form the basis of methods for solving time-dependent PDEs in one spatial variable. Such PDEs do not admit direct solution by integration. Second, both the Fourier series method and the finite element method generalize in a fairly straightforward way to problems in multiple spatial dimensions, whereas the direct integration method does not.

Exercises

1. Let M_D be the operator defined in Example 5.4.

    (a) Show that if $f \in C[0,\ell]$ is such that $M_D u = f$ has a solution, then that solution is unique.

    (b) Show that $M_D u = f$ has a solution only if $f \in C[0,\ell]$ satisfies a certain constraint. What is that constraint?

2. Define $C_l^1[0,\ell] = \{u \in C^1[0,\ell] : u(0) = 0\}$ and $M_I : C_l^1[0,\ell] \to C[0,\ell]$ by

$$M_I u = \frac{du}{dx}.$$

    (a) Show that $M_I u = f$ has a unique solution for each $f \in C[0,\ell]$. (Hint: Use the fundamental theorem of calculus.) This is equivalent to showing that $\mathcal{N}(M_I)$ is trivial and $\mathcal{R}(M_I) = C[0,\ell]$.

    (b) Show that M_I is not symmetric.

3. Consider again the differential operator equation $Lu = f$ discussed in Section 5.1. Suppose an equation $L_S u = f$ is produced by restricting the domain of L to a subspace S of $C^2[0,\ell]$ (that is, by defining $L_S u = Lu$ for all $u \in S$).

    (a) Show that $L_S u = f$ has at most one solution for each $f \in C[0,\ell]$ provided $\mathcal{N}(L) \cap S = \{0\}$.

(b) Show that $L_S u = f$ has a unique solution for each $f \in C[0,\ell]$ for either of the following choices of S:
   i. $S = \left\{ u \in C^2[0,\ell] : \frac{du}{dx}\left(\frac{\ell}{2}\right) = 0,\ \int_0^\ell u(x)\,dx = 0 \right\}$.
   ii. $S = \left\{ u \in C^2[0,\ell] : u(0) = \frac{du}{dx}(0) = 0 \right\}$.

4. Suppose S is a subspace of $C^k[a,b]$ and $K : S \to C[a,b]$ is a symmetric linear operator. Suppose u, v are complex-valued functions, with $u = f + ig$, $v = w + iz$, where $f, g, w, z \in S$. Show that
$$(Ku, v) = (u, Kv)$$
holds, where $(\cdot, \cdot)$ is defined by (5.4).

5. Explain why, in the last step of the calculation (5.5), the integral
$$\int_0^\ell \left(\frac{du}{dx}(x)\right)^2 dx$$
must be positive. (Hint: If du/dx is zero, then u must be constant. What constant functions belong to $C_D^2[0,\ell]$?)

6. Define
$$C_m^2[0,\ell] = \left\{ u \in C^2[0,\ell] : u(0) = \frac{du}{dx}(\ell) = 0 \right\}$$
and $L_m : C_m^2[0,\ell] \to C[0,\ell]$ by
$$L_m u = -\frac{d^2 u}{dx^2}.$$
   (a) Show that the null space of L_m is trivial.
   (b) Show that the range of L_m is all of $C[0,\ell]$.
   (c) Show that L_m is symmetric.
   (d) Show that all of the eigenvalues of L_m are positive.

7. Repeat Exercise 6 with $C_{\tilde{m}}^2[0,\ell]$ and $L_{\tilde{m}} : C_{\tilde{m}}^2[0,\ell] \to C[0,\ell]$ defined by
$$C_{\tilde{m}}^2[0,\ell] = \left\{ u \in C^2[0,\ell] : \frac{du}{dx}(0) = u(\ell) = 0 \right\},$$
$$L_{\tilde{m}} u = -\frac{d^2 u}{dx^2}.$$

8. Repeat Exercise 6 with $K : C_D^2[0,\ell] \to C[0,\ell]$ defined by
$$Ku = -a\frac{d^2 u}{dx^2} + bu,$$
where a and b are positive constants.

9. Consider the differential operator $M : C_D^2[0,1] \to C[0,1]$ defined by
$$Mu = -\frac{d^2 u}{dx^2} + \frac{du}{dx} + 5u.$$

5.1. The analogy between BVPs and linear algebraic systems

Show that M is not symmetric by producing two functions $u, v \in C_D^2[0,1]$ such that
$$(Mu, v) \neq (u, Mv).$$

10. Define $B : C^4[0, \ell] \to C[0, \ell]$ by
$$Bu = \frac{d^4 u}{dx^4}.$$

    (a) Determine the null space of B.

    (b) Find a set of boundary conditions such that, if S is the subspace of $C^4[0, \ell]$ consisting of functions satisfying these boundary conditions, then B, restricted to S, is symmetric and has a trivial null space.

11. In this exercise, we consider a new boundary condition, more complicated than Dirichlet or Neumann conditions, that provides a more realistic model of certain physical phenomena. This boundary condition is called a *Robin* condition; we will introduce it in the context of steady-state heat flow in a one-dimensional bar.

    Suppose the ends of a bar are uninsulated and the heat flux through each end is proportional to the difference between the temperature at the end of the bar and the surrounding temperature (assumed constant). If we let $\alpha > 0$ be the constant of proportionality and T_0, T_ℓ be the temperatures surrounding the ends of the bar at $x = 0$ and $x = \ell$, respectively, then the resulting boundary conditions are
    $$\kappa \frac{du}{dx}(0) = -\alpha(T_0 - u(0)),$$
    $$\kappa \frac{du}{dx}(\ell) = \alpha(T_\ell - u(\ell)).$$

    These can be rewritten as
    $$-\kappa \frac{du}{dx}(0) + \alpha u(0) = \alpha T_0,$$
    $$\kappa \frac{du}{dx}(\ell) + \alpha u(\ell) = \alpha T_\ell.$$

    Define
    $$C_R^2[0, \ell] = \left\{ u \in C^2[0, \ell] \, : \, -\kappa \frac{du}{dx}(0) + \alpha u(0) = 0, \, \kappa \frac{du}{dx}(\ell) + \alpha u(\ell) = 0 \right\},$$
    and let $L_R : C_R^2[0, \ell] \to C[0, \ell]$ be defined by
    $$L_R u = -\kappa \frac{d^2 u}{dx^2}.$$

    (As usual, we define the operator in terms of the *homogeneous* version of the boundary conditions.)

    (a) Prove that L_R is symmetric.

    (b) Find the null space of L_R.

5.2 Introduction to the spectral method; eigenfunctions

In this section and the next, we develop a "spectral" method for the BVP

$$-T\frac{d^2u}{dx^2} = f(x), \ 0 < x < \ell,$$
$$u(0) = 0,$$
$$u(\ell) = 0.$$
(5.7)

This BVP can be written simply as $L_D u = f$, where L_D is the symmetric linear differential operator defined in the last section.

The spectral method for a symmetric linear system $\mathbf{Ax} = \mathbf{b}$ is based on the fact that a symmetric matrix $\mathbf{A} \in \mathbf{R}^{n \times n}$ has n orthonormal eigenvectors, and therefore any vector in $\mathbf{R}^n$ (including the right-hand side $\mathbf{b}$ and the solution $\mathbf{x}$) can be written, in a simple way, as a linear combination of those eigenvectors. (The reader may wish to review Section 3.5 at this time.) We have seen, at the end of the previous section, that any eigenvalues of L_D must be real and positive and that eigenfunctions of L_D corresponding to distinct eigenvalues must be orthogonal. We now find the eigenvalue-eigenfunction pairs (eigenpairs for short) of L_D and explore the following question: Can we represent the right-hand side $f(x)$ and the solution $u(x)$ in terms of the eigenfunctions?

5.2.1 Eigenpairs of $-\frac{d^2}{dx^2}$ under Dirichlet conditions

We will show that the operator L_D has an infinite collection of eigenpairs. To simplify the calculations, we may as well assume that $T = 1$, since if λ, u is an eigenpair of

$$-\frac{d^2}{dx^2}$$

(subject to Dirichlet conditions), then $T\lambda, u$ is an eigenpair of

$$-T\frac{d^2}{dx^2}$$

under the same boundary conditions.

To find the eigenpairs, we need to solve

$$-\frac{d^2u}{dx^2} = \lambda u,$$
$$u(0) = 0,$$
$$u(\ell) = 0.$$
(5.8)

"Solving" this problem means identifying those special values of λ such that the BVP (5.8) has a *nonzero* solution u, and then determining u. (For any value of λ, the zero function is a solution of (5.8), just as $\mathbf{x} = \mathbf{0}$ is a solution of $\mathbf{Ax} = \lambda \mathbf{x}$ for any λ. In both cases, the eigenvalues are the scalars which lead to a nonzero solution.) Since we know, from the

5.2. Introduction to the spectral method; eigenfunctions

previous section, that all such λ are positive, we write $\lambda = \theta^2$. We can use the results of Section 4.2 to write the general solution of the ODE

$$\frac{d^2u}{dx^2} + \theta^2 u = 0, \tag{5.9}$$

and then try to satisfy the boundary conditions.

The characteristic roots of (5.9) are $\pm \theta i$, and the general solution is

$$u(x) = c_1 \cos(\theta x) + c_2 \sin(\theta x).$$

Therefore,
$$u(0) = c_1,$$

and the first boundary condition implies $c_1 = 0$. We then have $u(x) = c_2 \sin(\theta x)$, and the second boundary condition becomes

$$c_2 \sin(\theta \ell) = 0.$$

This equation only holds in two cases.

1. The value of c_2 is zero. But then $u(x)$ is the zero function, which cannot be an eigenfunction.

2. The value of $\lambda = \theta^2$ is such that $\sin(\theta \ell) = 0$. This holds in the cases

$$\theta \ell = \pm n\pi, \, n = 1, 2, \ldots.$$

Solving for λ, we have

$$\lambda = \frac{n^2 \pi^2}{\ell^2}, \, n = 1, 2, \ldots$$

(we discard the case that $\lambda = 0$, for then u is the zero function). It turns out that the value of c_2 is immaterial; these values of λ, and no others, produce eigenfunctions with the correct boundary conditions.

We thus see that the operator L_D has infinitely many eigenpairs:

$$\lambda_n = \frac{n^2 \pi^2}{\ell^2}, \, \psi_n(x) = \sin\left(\frac{n\pi x}{\ell}\right), \, n = 1, 2, 3, \ldots. \tag{5.10}$$

We already know that, since the eigenvalues of L_D are distinct, the eigenfunctions listed above must be orthogonal. This can also be verified directly using the trigonometric identity

$$\sin \alpha \sin \beta = \frac{1}{2}(\cos(\alpha - \beta) - \cos(\alpha + \beta))$$

(see Exercise 5.2.1). Also, for each n,

$$(\psi_n, \psi_n) = \int_0^\ell \sin^2\left(\frac{n\pi x}{\ell}\right) dx = \frac{\ell}{2},$$

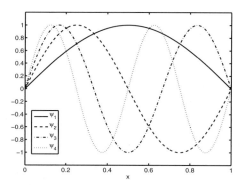

Figure 5.1. *The first four eigenfunctions* $\psi_1, \psi_2, \psi_3, \psi_4$ ($\ell = 1$).

that is,
$$\|\psi_n\| = \sqrt{\frac{\ell}{2}}, \; n = 1, 2, \ldots.$$

Thus the eigenfunctions are orthogonal, and they can be normalized by multiplying each by $\sqrt{2/\ell}$.

The first four eigenfunctions $\psi_1, \psi_2, \psi_3, \psi_4$ are graphed in Figure 5.1.

5.2.2 Representing functions in terms of eigenfunctions

We have shown that the operator $-d^2/dx^2$, subject to Dirichlet boundary conditions, has an infinite number of eigenpairs, and that the eigenfunctions are orthogonal. We now must answer the crucial question, Is it possible to represent both the solution and the right-hand side of (5.7) in terms of the eigenfunctions?

Let us take any function $f \in C[0, \ell]$. We know, from Section 3.5, how to find the best approximation to f from a finite-dimensional subspace. We define
$$F_N = \text{span}\left\{\sin\left(\frac{\pi x}{\ell}\right), \sin\left(\frac{2\pi x}{\ell}\right), \ldots, \sin\left(\frac{N\pi x}{\ell}\right)\right\}.$$

The best approximation to f from F_N is
$$f_N(x) = \frac{(\psi_1, f)}{(\psi_1, \psi_1)}\psi_1(x) + \frac{(\psi_2, f)}{(\psi_2, \psi_2)}\psi_2(x) + \cdots + \frac{(\psi_N, f)}{(\psi_N, \psi_N)}\psi_N(x) = \sum_{n=1}^{N} c_n \sin\left(\frac{n\pi x}{\ell}\right).$$

Since $(\psi_n, \psi_n) = \ell/2$ for all n, we have
$$c_n = \frac{(\psi_n, f)}{(\psi_n, \psi_n)} = \frac{2}{\ell} \int_0^\ell f(x) \sin\left(\frac{n\pi x}{\ell}\right) dx, \; n = 1, 2, \ldots, N.$$

We now show several examples of this approximation.[22] In the following examples, $\ell = 1$.

---

[22] Many specific integrals must be computed to compute these approximations. It is beyond the scope of this book to discuss methods of integration; moreover, modern technology often allows us to avoid this mechanical step. Computer algebra software, such as *Mathematica*, MATLAB, *Maple*, etc., as well as the more powerful handheld calculators, will compute many integrals symbolically.

5.2. Introduction to the spectral method; eigenfunctions

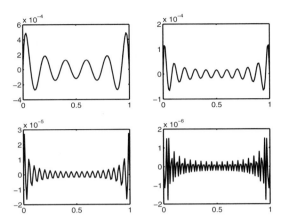

Figure 5.2. *Error in approximating $f(x) = x(1-x)$ by $f_N(x)$; see Example 5.5. The value of N is 10 (top left), 20 (top right), 40 (bottom left), and 80 (bottom right).*

Example 5.5. Let
$$f(x) = x(1-x).$$
Then
$$c_n = 2\int_0^1 x(1-x)\sin(n\pi x)\,dx = \frac{4\left(1+(-1)^{n+1}\right)}{n^3\pi^3}.$$

Figure 5.2 shows the errors in approximating $f(x)$ by $f_N(x)$ for $N = 10, 20, 40, 80$. (The approximations are so good that, if we were to plot both f and f_N on the same graph, the two curves would be virtually indistinguishable. Therefore, we plot the difference $f(x) - f_N(x)$ instead.) In this example, the approximation error is small and gets increasingly smaller as N increases. It is easy to believe that
$$f_N(x) \to f(x) \text{ as } N \to \infty. \qquad \blacksquare$$

Example 5.6. Let
$$f(x) = 1 - x.$$
Then
$$c_n = 2\int_0^1 (1-x)\sin(n\pi x)\,dx = \frac{2}{n\pi}.$$

Figure 5.3 shows $f(x)$ together with the approximations $f_N(x)$ for $N = 10, 20, 40, 80$.

The fact that the approximation is worst near $x = 0$ is due to the fact that f does not satisfy the Dirichlet condition there. Indeed, this example shows that f_N need not converge to f at every $x \in [0, \ell]$; here $f(0) = 1$, but $f_N(0) = 0$ for every N. However, it seems clear from the graphs that f_N approximates f in some meaningful sense, and that this approximation gets better as N increases.

The oscillation near $x = 0$ is known as *Gibbs's phenomenon*. $\qquad \blacksquare$

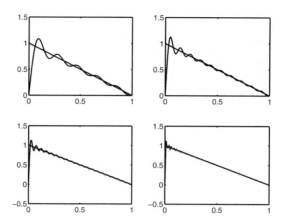

Figure 5.3. *Approximating $f(x) = 1 - x$ by $f_N(x)$; see Example 5.6. The value of N is 10 (top left), 20 (top right), 40 (bottom left), and 80 (bottom right).*

In Chapter 12, we justify the following fact: f_N converges to f as $N \to \infty$, in the sense that $\|f - f_N\| \to 0$ (that is, the L^2 norm of the error $f - f_N$ goes to zero). We write

$$f(x) = \sum_{n=1}^{\infty} c_n \sin\left(\frac{n\pi x}{\ell}\right)$$

to indicate this fact. As we saw in Example 5.6, this does not necessarily imply that $f_N(x) \to f(x)$ for all $x \in [0, \ell]$. It does imply, however, that f_N gets arbitrarily close to f, over the entire interval, in an average sense. This type of convergence (in the L^2 norm) is sometimes referred to as *mean-square* convergence.

The resulting representation,

$$f(x) = \sum_{n=1}^{\infty} c_n \sin\left(\frac{n\pi x}{\ell}\right), \quad c_n = \frac{2}{\ell} \int_0^{\ell} f(x) \sin\left(\frac{n\pi x}{\ell}\right) dx, \tag{5.11}$$

is referred to as the *Fourier sine series* of f, and the scalars $c_1, c_2, \ldots$ are called the *Fourier sine coefficients* of f.

In the next section, we will show how to use the Fourier sine series to solve the BVP (5.7). First, however, we discuss other boundary conditions and the resulting Fourier series.

5.2.3 Eigenfunctions under other boundary conditions; other Fourier series

We now consider a BVP with mixed boundary conditions,

$$\begin{aligned} -T\frac{d^2u}{dx^2} &= f(x), \ 0 < x < \ell, \\ u(0) &= 0, \\ \frac{du}{dx}(\ell) &= 0, \end{aligned} \tag{5.12}$$

5.2. Introduction to the spectral method; eigenfunctions

which models, for example, a hanging bar with the bottom end free. In order to consider a spectral method, we define the subspace

$$C_m^2[0,\ell] = \left\{ u \in C^2[0,\ell] \; : \; u(0) = \frac{du}{dx}(\ell) = 0 \right\}$$

and the operator $L_m : C_m^2[0,\ell] \to C[0,\ell]$, where

$$L_m u = -\frac{d^2 u}{dx^2}.$$

The operator L_m is symmetric with positive eigenvalues, it has a trivial null space, and the range of L_m is all of $C[0,\ell]$ (see Exercise 5.1.6). Therefore, $L_m u = f$ has a unique solution for every $f \in C[0,\ell]$, and we will show, in the next section, how to compute the solution by the spectral method. First, however, we must find the eigenvalues and eigenfunctions of L_m.

Since L_m has only positive eigenvalues, we define $\lambda = \theta^2$, where $\theta > 0$, and solve

$$\frac{d^2 u}{dx^2} + \theta^2 u = 0, \; 0 < x < \ell,$$

$$u(0) = 0,$$

$$\frac{du}{dx}(\ell) = 0.$$

The general solution of the differential equation is

$$u(x) = c_1 \cos(\theta x) + c_2 \sin(\theta x).$$

The first boundary condition, $u(0) = 0$, implies that $c_1 = 0$, and hence any nonzero eigenfunction must be a multiple of $u(x) = \sin(\theta x)$. The second boundary condition,

$$\frac{du}{dx}(\ell) = 0,$$

yields the condition that $\theta \cos(\theta \ell) = 0$. Since $\theta > 0$ by assumption, this is possible only if $\cos(\theta \ell) = 0$ or

$$\theta \ell = \frac{\pi}{2}, \frac{3\pi}{2}, \ldots, \frac{(2n-1)\pi}{2}, \ldots.$$

Solving for λ, we obtain eigenvalues $\lambda_1, \lambda_2, \ldots$, with

$$\lambda_n = \frac{(2n-1)^2 \pi^2}{4 \ell^2}, \; n = 1, 2, 3, \ldots.$$

The corresponding eigenfunctions are

$$\phi_n(x) = \sin\left(\frac{(2n-1)\pi x}{2\ell}\right), \; n = 1, 2, 3, \ldots.$$

The eigenfunctions are orthogonal, as is guaranteed since L_m is symmetric, and as can be verified directly by computing (ϕ_n, ϕ_m). The best approximation to $f \in C[0,\ell]$, using $\phi_1, \phi_2, \ldots, \phi_N$, is

$$f(x) \doteq \sum_{n=1}^{N} \frac{(\phi_n, f)}{(\phi_n, \phi_n)} \phi_n. \qquad (5.13)$$

Since
$$(\phi_n, \phi_n) = \int_0^\ell \sin^2\left(\frac{(2n-1)\pi x}{2\ell}\right) dx = \frac{\ell}{2},$$

(5.13) becomes
$$f(x) \doteq \sum_{n=1}^{N} b_n \sin\left(\frac{(2n-1)\pi x}{2\ell}\right),$$

where
$$b_n = \frac{2}{\ell} \int_0^\ell f(x) \sin\left(\frac{(2n-1)\pi x}{2\ell}\right) dx, \quad n = 1, 2, 3, \ldots.$$

The resulting representation,
$$f(x) = \sum_{n=1}^{\infty} b_n \sin\left(\frac{(2n-1)\pi x}{2\ell}\right),$$

is referred to as the *Fourier quarter-wave sine series*. It is valid in the mean-square sense.

Example 5.7. Let $f(x) = x$ and $\ell = 1$. Then
$$2\int_0^1 x \sin\left(\frac{(2n-1)\pi x}{2}\right) dx = \frac{8(-1)^{n+1}}{(2n-1)^2 \pi^2}, \quad n = 1, 2, 3, \ldots,$$

and the best approximation to f using the first N eigenfunctions is
$$f_N(x) = \sum_{n=1}^{N} b_n \sin\left(\frac{(2n-1)\pi x}{2}\right).$$

The error in approximating f on $[0,1]$ by f_N, for $N = 10, 20, 40, 80$, is graphed in Figure 5.4. ∎

In the exercises, the reader is asked to derive other Fourier series, including the appropriate series for the boundary conditions
$$\frac{du}{dx}(0) = 0, \ u(\ell) = 0.$$

In Chapter 6, we will present Fourier series for the Neumann conditions
$$\frac{du}{dx}(0) = 0, \ \frac{du}{dx}(\ell) = 0$$

and the periodic boundary conditions
$$u(-\ell) = u(\ell), \ \frac{du}{dx}(-\ell) = \frac{du}{dx}(\ell)$$

(for the interval $[-\ell, \ell]$). Generally speaking, given a symmetric differential operator, a sequence of orthogonal eigenfunctions exists, and it can be used to represent functions in a Fourier series. However, it may not be easy to find the eigenfunctions in some cases (see Exercise 5.2.4).

5.2. Introduction to the spectral method; eigenfunctions

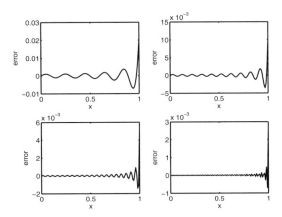

Figure 5.4. *Approximating $f(x) = x$ by a quarter-wave sine series (see Example 5.7). The value of N is 10 (top left), 20 (top right), 40 (bottom left), and 80 (bottom right).*

Exercises

1. Use the trigonometric identity

$$\sin \alpha \sin \beta = \frac{1}{2}(\cos(\alpha - \beta) - \cos(\alpha + \beta))$$

   to verify that, if $n \neq m$, then $(\psi_n, \psi_m) = 0$, where $\psi_1, \psi_2, \psi_3, \ldots$ are given in (5.10).

2. Define K as in Exercise 5.1.8. We know (from Exercise 5.1.8) that K has only real, positive eigenvalues. Find all of the eigenvalues and eigenvectors of K.

3. Repeat Exercise 2 for the differential operator $L_{\tilde{m}}$ defined in Exercise 5.1.7. The resulting eigenfunctions are the quarter-wave cosine functions.

4. **(Hard)** Define L_R as in Exercise 5.1.11.

   (a) Show that L_R has only positive eigenvalues.

   (b) Show that L_R has an infinite sequence of positive eigenvalues. Note: The equation that determines the positive eigenvalues cannot be solved analytically, but a simple graphical analysis can be used to show that they exist and to estimate their values.

   (c) For $\alpha = \kappa = 1$, find the first two eigenpairs by finding accurate estimates of the two smallest eigenvalues.

5. **(Hard)** Consider the differential operator $M : C_m^2[0, 1] \to C[0, 1]$ defined by

$$Mu = -\frac{d^2u}{dx^2} + \frac{du}{dx} + 5u$$

   (recall that $C_m^2[0, 1] = \{u \in C^2[0, 1] : u(0) = \frac{du}{dx}(1) = 0\}$). Analyze the eigenpairs of M as follows.

(a) Write down the characteristic polynomial of the ODE

$$\frac{d^2u}{dx^2} - \frac{du}{dx} + (\lambda - 5)u = 0. \tag{5.14}$$

Using the quadratic formula, find the characteristic roots.

(b) There are three cases to consider, depending on whether the discriminant in the quadratic formula is negative, zero, or positive (that is, depending on whether the characteristic roots are complex conjugate, real and repeated, or real and distinct). In each case, write down the general solution of (5.14).

(c) Show that in the case of real roots (either repeated or distinct), there is no nonzero solution.

(d) Show that there is an infinite sequence of values of λ, leading to complex conjugate roots, each yielding a nonzero solution of (5.14). Note: The equation that determines the eigenvalues cannot be solved analytically. However, a simple graphical analysis is sufficient to show the existence of the sequence of eigenvalues.

(e) Find the first two eigenpairs (that is, those corresponding to the two smallest eigenvalues), using some numerical method to compute the first two eigenvalues accurately.

(f) Show that the first two eigenfunctions are *not* orthogonal.

6. Define

$$C_b^2[0,\ell] = \left\{ u \in C^2[0,\ell] : u(\ell) = a_{11}u(0) + a_{12}\frac{du}{dx}(0), \right.$$
$$\left. \frac{du}{dx}(\ell) = a_{21}u(0) + a_{22}\frac{du}{dx}(0) \right\},$$

and define $L_b : C_b^2[0,\ell] \to C[0,\ell]$ by

$$L_b u = -\frac{d^2u}{dx^2}.$$

Find necessary and sufficient conditions on the coefficients $a_{11}, a_{12}, a_{21}, a_{22}$ for the operator L_b to be symmetric.

7. Compute the Fourier sine series, on the interval $[0,1]$, for each of the following functions:[23]

(a) $g(x) = x$;
(b) $h(x) = \frac{1}{2} - |x - \frac{1}{2}|$;
(c) $m(x) = x - x^3$;
(d) $k(x) = 7x - 10x^3 + 3x^5$.

---

[23] The use of *Mathematica* or some other program to compute the necessary integrals is recommended.

5.3. Solving the BVP using Fourier series

Graph the error that occurs when approximating each function by 10 terms of the Fourier sine series.

8. Explain why a Fourier sine series, if it converges to a function on $[0,\ell]$, defines an odd function of $x \in \mathbf{R}$. (A function $f: \mathbf{R} \to \mathbf{R}$ is *odd* if $f(-x) = -f(x)$ for all $x \in \mathbf{R}$.)

9. What is the Fourier sine series of $f(x) = \sin(3\pi x)$ on the interval $[0,1]$?

10. Repeat Exercise 7 using the quarter-wave sine series (see Section 5.2.3).

11. Repeat Exercise 7 using the quarter-wave cosine series (see Exercise 3 above).

5.3 Solving the BVP using Fourier series

We now return to the problem of solving the BVP

$$-T\frac{d^2u}{dx^2} = f(x),\ 0 < x < \ell,$$
$$u(0) = 0,$$
$$u(\ell) = 0,$$
(5.15)

which can be expressed simply as $L_D u = f$.

5.3.1 A special case

We begin with a special case that is easy to solve. Suppose f is of the form

$$f(x) = \sum_{n=1}^{N} c_n \sin\left(\frac{n\pi x}{\ell}\right),$$
(5.16)

where the coefficients $c_1, c_2, \ldots, c_N$ are known. We then look for the solution in the form

$$u(x) = \sum_{n=1}^{N} b_n \sin\left(\frac{n\pi x}{\ell}\right).$$

By construction, $u \in C_D^2[0,\ell]$, and so u satisfies the boundary conditions (regardless of the values of $b_1, b_2, \ldots, b_N$). We therefore choose the coefficients $b_1, b_2, \ldots, b_N$ so that the differential equation is satisfied.

Since u is expressed in terms of the eigenfunctions, the expression for $L_D u$ is very simple:

$$(L_D u)(x) = -T\frac{d^2u}{dx^2}(x) = \sum_{n=1}^{N} \frac{Tn^2\pi^2}{\ell^2} b_n \sin\left(\frac{n\pi x}{\ell}\right).$$

The equation $L_D u = f$ then becomes

$$\sum_{n=1}^{N} \frac{Tn^2\pi^2}{\ell^2} b_n \sin\left(\frac{n\pi x}{\ell}\right) = \sum_{n=1}^{N} c_n \sin\left(\frac{n\pi x}{\ell}\right).$$

It is easy to find the coefficients $b_1, b_2, \ldots, b_N$ satisfying this equation; we need

$$\frac{Tn^2\pi^2}{\ell^2} b_n = c_n, \; n = 1, 2, \ldots, N,$$

or

$$b_n = \frac{\ell^2 c_n}{Tn^2\pi^2}, \; n = 1, 2, \ldots, N.$$

We then have the following solution:

$$u(x) = \sum_{n=1}^{N} \frac{\ell^2 c_n}{Tn^2\pi^2} \sin\left(\frac{n\pi x}{\ell}\right).$$

Example 5.8. Suppose $\ell = 1$, $T = 1$, and

$$f(x) = \sin(\pi x) - 2\sin(2\pi x) + 5\sin(3\pi x).$$

The solution to (5.15), with this right-hand side f, is

$$u(x) = b_1 \sin(\pi x) + b_2 \sin(2\pi x) + b_3 \sin(3\pi x),$$

where

$$b_1 = \frac{1}{\pi^2}, \; b_2 = -\frac{2}{4\pi^2}, \; b_3 = \frac{5}{9\pi^2}.$$

In Figure 5.5, we display both the right-hand side f and the solution u. ∎

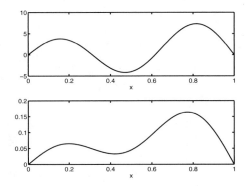

Figure 5.5. *The right-hand side f (top) and solution u (bottom) from Example 5.8.*

When f is of the special form (5.16), the solution technique described above is nothing more than the spectral method explained for matrix-vector equations in Section 3.5. Writing

$$\lambda_n = \frac{Tn^2\pi^2}{\ell^2}, \; \psi_n(x) = \sin\left(\frac{n\pi x}{\ell}\right),$$

5.3. Solving the BVP using Fourier series

we simply express the right-hand side f in terms of the (orthogonal) eigenfunctions,

$$f = \frac{(\psi_1, f)}{(\psi_1, \psi_1)}\psi_1 + \frac{(\psi_2, f)}{(\psi_2, \psi_2)}\psi_2 + \cdots + \frac{(\psi_N, f)}{(\psi_N, \psi_N)}\psi_N,$$

and then divide the coefficients by the eigenvalues to get the solution

$$u = \frac{(\psi_1, f)}{\lambda_1(\psi_1, \psi_1)}\psi_1 + \frac{(\psi_2, f)}{\lambda_2(\psi_2, \psi_2)}\psi_2 + \cdots + \frac{(\psi_N, f)}{\lambda_N(\psi_N, \psi_N)}\psi_N.$$

5.3.2 The general case

Now suppose that f is not of the special form (5.16). We learned in Section 5.2 that we can approximate f by

$$f_N(x) = \sum_{n=1}^{N} c_n \sin\left(\frac{n\pi x}{\ell}\right),$$

where

$$c_n = \frac{2}{\ell} \int_0^\ell f(x) \sin\left(\frac{n\pi x}{\ell}\right) dx, \ n = 1, 2, 3, \ldots.$$

This approximation gets better and better (at least in the mean-square sense) as $N \to \infty$.

We know how to solve $L_D u = f_N$; let us call the solution u_N:

$$u_N(x) = \sum_{n=1}^{N} \frac{\ell^2 c_n}{T n^2 \pi^2} \sin\left(\frac{n\pi x}{\ell}\right).$$

It is reasonable to believe that, since f_N gets closer and closer to f as $N \to \infty$, the function u_N will get closer and closer to the true solution u as $N \to \infty$.

Example 5.9. We consider the BVP

$$\begin{aligned} -\frac{d^2 u}{dx^2} &= x, \ 0 < x < 1, \\ u(0) &= 0, \\ u(1) &= 0. \end{aligned} \quad (5.17)$$

The exact solution is easily computed (by integration) to be

$$u(x) = \frac{1}{6}x\left(1 - x^2\right) \quad (5.18)$$

(see Exercise 5.3.5). We begin by computing the Fourier sine coefficients of $f(x) = x$. We have

$$c_n = 2\int_0^1 x \sin(n\pi x)\, dx = \frac{2(-1)^{n+1}}{n\pi}, \ n = 1, 2, \ldots.$$

The solution to $L_D u = f_N$, with

$$f_N(x) = \sum_{n=1}^{N} c_n \sin(n\pi x),$$

is then
$$u_N(x) = \sum_{n=1}^{N} \frac{c_n}{n^2\pi^2} \sin(n\pi x) = \sum_{n=1}^{N} \frac{2(-1)^{n+1}}{n^3\pi^3} \sin(n\pi x).$$

We can now compare u_N with the exact solution u. In Figure 5.6, we show the graphs of u and u_{10}; on this scale, the two curves are indistinguishable. Figure 5.7 shows the approximation error $u(x) - u_{10}(x)$.

The fact that u_N approximates u so closely is no accident. If we compute the Fourier sine coefficients $a_1, a_2, a_3, \ldots$ of u directly, we find that

$$a_n = 2\int_0^1 \frac{1}{6}x\left(1-x^2\right)\sin(n\pi x)\,dx = \frac{2(-1)^{n+1}}{n^3\pi^3}, \ n=1,2,3,\ldots.$$

These are precisely the coefficients of u_N! Therefore, u_N is the best approximation to u using the first N eigenfunctions. ∎

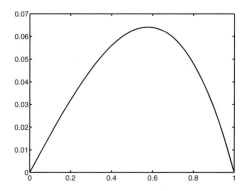

Figure 5.6. *The exact solution u of (5.17) and the approximation u_{10}.*

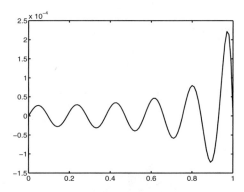

Figure 5.7. *The error in approximating the solution u of (5.17) with u_{10}.*

5.3. Solving the BVP using Fourier series

The last example is typical: By solving $L_D u = f_N$, where f_N is the best approximation to f using the first N eigenfunctions, we obtain the best approximation u_N to the solution u of $L_D u = f$. We can demonstrate this directly. We assume that u satisfies $u(0) = u(\ell) = 0$, and we write $a_1, a_2, a_3, \ldots$ for the Fourier (sine) coefficients of u:

$$a_n = \frac{2}{\ell} \int_0^\ell u(x) \sin\left(\frac{n\pi x}{\ell}\right) dx, \; n = 1, 2, 3, \ldots.$$

We can then compute the Fourier sine coefficients of

$$-T \frac{d^2 u}{dx^2}$$

by using integration by parts:

$$\frac{2}{\ell} \int_0^\ell -T \frac{d^2 u}{dx^2}(x) \sin\left(\frac{n\pi x}{\ell}\right) dx$$

$$= \frac{-2T}{\ell} \left\{ \left[\frac{du}{dx}(x) \sin\left(\frac{n\pi x}{\ell}\right)\right]_{x=0}^{\ell} - \frac{n\pi}{\ell} \int_0^\ell \frac{du}{dx}(x) \cos\left(\frac{n\pi x}{\ell}\right) dx \right\}$$

$$= \frac{-2T}{\ell} \left\{ \frac{du}{dx}(\ell) \sin(n\pi) - \frac{du}{dx}(0) \sin(0) - \frac{n\pi}{\ell} \int_0^\ell \frac{du}{dx}(x) \cos\left(\frac{n\pi x}{\ell}\right) dx \right\}$$

$$= \frac{2Tn\pi}{\ell^2} \int_0^\ell \frac{du}{dx}(x) \cos\left(\frac{n\pi x}{\ell}\right) dx \quad (\text{since } \sin(0) = \sin(n\pi) = 0) \qquad (5.19)$$

$$= \frac{2Tn\pi}{\ell^2} \left\{ \left[u(x) \cos\left(\frac{n\pi x}{\ell}\right)\right]_{x=0}^{\ell} + \frac{n\pi}{\ell} \int_0^\ell u(x) \sin\left(\frac{n\pi x}{\ell}\right) dx \right\}$$

$$= \frac{Tn^2 \pi^2}{\ell^2} \frac{2}{\ell} \int_0^\ell u(x) \sin\left(\frac{n\pi x}{\ell}\right) dx \quad (\text{since } u(0) = u(\ell) = 0)$$

$$= \frac{Tn^2 \pi^2}{\ell^2} a_n.$$

We thus see that the Fourier sine series of

$$-T \frac{d^2 u}{dx^2}$$

is[24]

$$\sum_{n=1}^\infty \frac{Tn^2 \pi^2}{\ell^2} a_n \sin\left(\frac{n\pi x}{\ell}\right).$$

---

[24] In other words,

$$-T \frac{\partial^2 u}{\partial x^2}(x,t) = -T \frac{\partial^2}{\partial x^2} \left(\sum_{n=1}^\infty a_n \sin\left(\frac{n\pi x}{\ell}\right) \right) = \sum_{n=1}^\infty a_n \left(-T \frac{\partial^2}{\partial x^2} \left(\sin\left(\frac{n\pi x}{\ell}\right) \right) \right).$$

A priori, it is not obvious that we can differentiate the infinite series term by term, but (5.19) shows that this is valid in this case. But caution is required; see Exercise 5.3.6.

Since $-T d^2 u/dx^2 = f$ by assumption, the Fourier sine series of the two functions must be the same, and so we obtain

$$\frac{T n^2 \pi^2}{\ell^2} a_n = c_n, \; n = 1, 2, 3, \ldots,$$

or

$$a_n = \frac{\ell^2 c_n}{T n^2 \pi^2}, \; n = 1, 2, 3, \ldots.$$

This is exactly what we obtained by the reasoning presented earlier.

Example 5.10. Consider an elastic string that, when stretched by a tension of 10 N, has a length 50 cm. Suppose that the density of the (stretched) string is $\rho = 0.2$ g/cm. If the string is fixed horizontally and sags under gravity, what shape does it assume?

We let $u(x)$, $0 < x < 50$, be the vertical displacement (in cm) of the string. To use consistent units, we convert 10 Newtons to 10^6 dynes (g cm/s^2). Then u satisfies the BVP

$$-T \frac{d^2 u}{dx^2} = -\rho g, \; 0 < x < \ell,$$
$$u(0) = 0,$$
$$u(\ell) = 0,$$

or

$$-10^6 \frac{d^2 u}{dx^2} = -196, \; 0 < x < 50,$$
$$u(0) = 0,$$
$$u(50) = 0$$

(using 980 cm/s^2 for the gravitational constant). To solve this, we first compute the Fourier sine coefficients of the right-hand side:

$$c_n = \frac{2}{50} \int_0^{50} -196 \sin\left(\frac{n\pi x}{50}\right) dx = \frac{392((-1)^n - 1)}{n\pi}, \; n = 1, 2, 3, \ldots.$$

We can then find the Fourier sine coefficients of the solution u:

$$a_n = \frac{50^2 c_n}{10^6 n^2 \pi^2} = \frac{49((-1)^n - 1)}{50 n^3 \pi^3}, \; n = 1, 2, 3, \ldots.$$

We plot the approximate solution u_{20} in Figure 5.8. The maximum deflection of the string is quite small, less than half of a millimeter. This is to be expected, since the tension in the string is much more than the total gravitational force acting on it. ∎

5.3.3 Other boundary conditions

We can apply the Fourier series method to BVPs with boundary conditions other than Dirichlet conditions, provided the differential operator is symmetric under the boundary

5.3. Solving the BVP using Fourier series

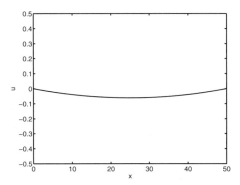

Figure 5.8. *The shape of the sagging string in Example 5.10.*

conditions, and provided we know the eigenvalues and eigenfunctions. For example, we consider the BVP

$$-T\frac{d^2u}{dx^2} = f(x), \ 0 < x < \ell,$$
$$u(0) = 0, \qquad (5.20)$$
$$\frac{du}{dx}(\ell) = 0.$$

Written as a differential operator equation, this takes the form $L_m u = f$, where L_m is defined as in Section 5.2.3, but including the factor of T:

$$L_m : C_m^2[0,\ell] \to C[0,\ell],$$
$$L_m u = -T\frac{d^2u}{dx^2}.$$

We saw, in that section, that the eigenvalues and eigenfunctions of L_m are

$$\lambda_n = \frac{T(2n-1)^2\pi^2}{4\ell^2}, \ \phi_n(x) = \sin\left(\frac{(2n-1)\pi x}{2\ell}\right), \ n = 1,2,3,\ldots.$$

Therefore, we write

$$f(x) = \sum_{n=1}^{\infty} c_n \sin\left(\frac{(2n-1)\pi x}{2\ell}\right),$$

where

$$c_n = \frac{2}{\ell}\int_0^\ell f(x)\sin\left(\frac{(2n-1)\pi x}{2\ell}\right) dx, \ n = 1,2,3,\ldots.$$

We represent the solution as

$$u(x) = \sum_{n=1}^{\infty} a_n \sin\left(\frac{(2n-1)\pi x}{2\ell}\right),$$

where the coefficients $a_1, a_2, a_3, \ldots$ are to be determined. The Fourier series of $L_m u = -T d^2 u/dx^2$ is then

$$-T \frac{d^2 u}{dx^2}(x) = \sum_{n=1}^{\infty} \frac{T(2n-1)^2 \pi^2}{4\ell^2} a_n \sin\left(\frac{(2n-1)\pi x}{2\ell}\right).$$

The fact that these are the correct Fourier coefficients can be verified by a calculation exactly like (5.19) (see Exercise 5.3.7) and is also justified in Section 5.3.5 (see (5.25)).

The BVP (5.20) can now be written as

$$\sum_{n=1}^{\infty} \frac{T(2n-1)^2 \pi^2}{4\ell^2} a_n \sin\left(\frac{(2n-1)\pi x}{2\ell}\right) = \sum_{n=1}^{\infty} c_n \sin\left(\frac{(2n-1)\pi x}{2\ell}\right).$$

From this equation, we see that

$$\frac{T(2n-1)^2 \pi^2}{4\ell^2} a_n = c_n, \quad n = 1, 2, 3, \ldots,$$

or

$$a_n = \frac{4\ell^2 c_n}{T(2n-1)^2 \pi^2}, \quad n = 1, 2, 3, \ldots.$$

Thus the solution to (5.20) is

$$u(x) = \sum_{n=1}^{\infty} \frac{4\ell^2 c_n}{T(2n-1)^2 \pi^2} \sin\left(\frac{(2n-1)\pi x}{2\ell}\right).$$

Example 5.11. Consider a circular cylindrical bar, of length 1 m and radius 1 cm, made from an aluminum alloy with stiffness 70 GPa and density 2.64 g/cm^3. Suppose the top end of the bar ($x = 0$) is fixed and the only force acting on the bar is the force due to gravity. We will find the displacement u of the bar and determine the change in length of the bar (which is just $u(1)$).

The BVP describing u is

$$\begin{aligned} -k \frac{d^2 u}{dx^2} &= \rho g, \quad 0 < x < 1, \\ u(0) &= 0, \\ \frac{du}{dx}(1) &= 0, \end{aligned} \quad (5.21)$$

where $k = 7 \cdot 10^{10}$, $\rho = 2640 \, \text{kg/m}^3$, and $g = 9.8 \, \text{m/s}^2$. The Fourier quarter-wave sine coefficients of the constant function ρg are

$$c_n = 2 \int_0^1 \rho g \sin\left(\frac{(2n-1)\pi x}{2}\right) dx = \frac{4\rho g}{(2n-1)\pi}, \quad n = 1, 2, 3, \ldots,$$

and the eigenvalues of the differential operator are

$$\frac{k(2n-1)^2 \pi^2}{4}, \quad n = 1, 2, 3, \ldots.$$

5.3. Solving the BVP using Fourier series

Therefore, the solution is

$$u(x) = \sum_{n=1}^{\infty} \frac{16\rho g}{k(2n-1)^3 \pi^3} \sin\left(\frac{(2n-1)\pi x}{2}\right)$$

$$\doteq \left(5.9136 \cdot 10^{-6}\right) \sum_{n=1}^{\infty} \frac{1}{(2n-1)^3 \pi^3} \sin\left(\frac{(2n-1)\pi x}{2}\right).$$

We then have

$$u(1) \doteq \left(5.9136 \cdot 10^{-6}\right) \sum_{n=1}^{\infty} \frac{1}{(2n-1)^3 \pi^3} \sin\left(\frac{(2n-1)\pi}{2}\right)$$

$$= \left(5.9136 \cdot 10^{-6}\right) \sum_{n=1}^{\infty} \frac{(-1)^{n+1}}{(2n-1)^3 \pi^3}$$

$$\doteq 1.85 \cdot 10^{-7}.$$

Thus the bar stretches by only about $1.85 \cdot 10^{-7}$ m, or less than two ten-thousandths of a millimeter. ∎

The Fourier series method is a fairly complicated technique for solving such a simple problem as (5.11). As we mentioned in Section 5.1.1, we could simply integrate the differential equation twice and use the boundary conditions to determine the constants of integration. The Fourier series method will prove its worth when we apply it to PDEs (either time-dependent problems (Chapters 6 and 7) or problems with multiple spatial dimensions (Chapter 11)). We introduce the Fourier series method in this simple context so that the reader can understand its essence before dealing with more complicated applications.

5.3.4 Inhomogeneous boundary conditions

We now consider the following BVP with inhomogeneous boundary conditions:

$$\begin{aligned}
-T\frac{d^2 u}{dx^2} &= f(x),\ 0 < x < \ell, \\
u(0) &= a, \\
u(\ell) &= b
\end{aligned} \quad (5.22)$$

(T constant). To apply the Fourier series method to this problem, we make a transformation of the dependent variable u to obtain a BVP with homogeneous boundary conditions. We notice that the linear function

$$p(x) = a + \frac{b-a}{\ell}x$$

satisfies the boundary conditions ($p(0) = a$, $p(\ell) = b$). Defining $v(x) = u(x) - p(x)$, we have

$$-T\frac{d^2 v}{dx^2} = -T\frac{d^2 u}{dx^2} + T\frac{d^2 p}{dx^2} = f(x) + 0 = f(x),$$

since the second derivative of a linear function is zero. Also,

$$v(0) = u(0) - p(0) = a - a = 0, \ v(\ell) = u(\ell) - p(\ell) = b - b = 0.$$

Thus $v(x)$ solves (5.7), so we know how to compute $v(x)$. We then obtain the desired solution from $u(x) = v(x) + p(x)$.

This technique, of transforming the problem to one with homogeneous boundary conditions, is referred to as the method of *shifting the data*.

Example 5.12. Consider

$$-\frac{d^2u}{dx^2} = x, \ 0 < x < 1,$$
$$u(0) = 1,$$
$$u(1) = 0.$$
(5.23)

Let $p(x) = 1 - x$, the linear function satisfying the boundary conditions, and define $v(x) = u(x) - p(x)$. Then $v(x)$ solves (5.17), and so

$$v(x) = \sum_{n=1}^{\infty} \frac{2(-1)^{n+1}}{n^3 \pi^3} \sin(n\pi x).$$

We then have

$$u(x) = 1 - x + \sum_{n=1}^{\infty} \frac{2(-1)^{n+1}}{n^3 \pi^3} \sin(n\pi x). \quad \blacksquare$$

Example 5.13. Consider again the bar of Example 5.11, and suppose that now a mass of 1000 kg is hung from the bottom end of the bar. Assume that the mass is evenly distributed over the end of the bar, resulting in a pressure of

$$p = \frac{(1000 \, \text{kg})(9.8 \, \text{m/s}^2)}{\pi \left(10^{-2}\right)^2 \text{m}^2} = \frac{9.8 \cdot 10^7}{\pi} \, \text{Pa}$$

on the end of the bar. The BVP satisfied by the displacement u is now

$$-k\frac{d^2u}{dx^2} = \rho g, \ 0 < x < 1,$$
$$u(0) = 0,$$
$$\frac{du}{dx}(1) = \frac{p}{k},$$
(5.24)

where $k = 7 \cdot 10^{10}$, $\rho = 2640 \, \text{kg/m}^3$, and $g = 9.8 \, \text{m/s}^2$. To solve this problem using the Fourier series method, we shift the data by finding a linear function satisfying the boundary conditions. The function $q(x) = (p/k)x$ satisfies

$$q(0) = 0, \ \frac{dq}{dx}(1) = \frac{p}{k},$$

5.3. Solving the BVP using Fourier series

so we define $v = u - q$. It then turns out that, since $d^2q/dx^2 = 0$, v satisfies (5.21), so

$$v(x) \doteq (5.9136 \cdot 10^{-6}) \sum_{n=1}^{\infty} \frac{1}{(2n-1)^3 \pi^3} \sin\left(\frac{(2n-1)\pi x}{2}\right)$$

and

$$u(x) \doteq \frac{p}{k}x + (5.9136 \cdot 10^{-6}) \sum_{n=1}^{\infty} \frac{1}{(2n-1)^3 \pi^3} \sin\left(\frac{(2n-1)\pi x}{2}\right).$$

We then obtain

$$u(1) \doteq \frac{p}{k} + 1.8 \cdot 10^{-7} \doteq 1.4 \cdot 10^{-3}.$$

This is 1.4 millimeters. The displacement due to gravity ($1.85 \cdot 10^{-4}$ millimeters) is insignificant compared to the displacement due to the mass hanging on the end of the bar. ∎

5.3.5 Summary

We can now summarize the method of Fourier series for solving a BVP. We assume that the BVP has been written in the form $Ku = f$, where, as usual, the boundary conditions form part of the definition of the domain of the differential operator K. For the Fourier series method to be applicable, it must be the case that

- K is symmetric;
- K has a sequence of eigenvalues $\lambda_1, \lambda_2, \lambda_3, \ldots$, with corresponding eigenfunctions $\psi_1, \psi_2, \psi_3, \ldots$ (which are necessarily orthogonal);
- any function g in $C[0, \ell]$ can be written in a Fourier series

$$g(x) = \sum_{n=1}^{\infty} d_n \psi_n(x),$$

where

$$d_n = \frac{(\psi_n, g)}{(\psi_n, \psi_n)}.$$

It then follows that the function Ku is given by the series

$$(Ku)(x) = \sum_{n=1}^{\infty} \lambda_n a_n \psi_n(x),$$

where $a_1, a_2, a_3, \ldots$ are the Fourier coefficients of u. This follows from the symmetry of K and the definition of the Fourier coefficients $b_1, b_2, b_3, \ldots$ of Ku:

$$b_n = \frac{(Ku, \psi_n)}{(\psi_n, \psi_n)} = \frac{(u, K\psi_n)}{(\psi_n, \psi_n)} = \frac{(u, \lambda_n \psi_n)}{(\psi_n, \psi_n)} = \lambda_n \frac{(u, \psi_n)}{(\psi_n, \psi_n)} = \lambda_n a_n. \quad (5.25)$$

The BVP $Ku = f$ then takes the form

$$\sum_{n=1}^{\infty} \lambda_n a_n \psi_n(x) = \sum_{n=1}^{\infty} c_n \psi_n(x),$$

whence we obtain

$$a_n = \frac{c_n}{\lambda_n}, \ n = 1, 2, 3, \ldots,$$

and

$$u(x) = \sum_{n=1}^{\infty} \frac{c_n}{\lambda_n} \psi_n(x).$$

Here is a summary of the Fourier series method for solving BVPs.

0. Pose the BVP as a differential operator equation.

1. Verify that the operator is symmetric, and find the eigenvalue/eigenfunction pairs.

2. Express the unknown function u as a series in terms of the eigenfunctions. The coefficients are the unknowns of the problem.

3. Express the right-hand side f of the differential equation as a series in terms of the eigenfunctions. Use the formula for the projection of f onto the eigenfunctions to compute the coefficients.

4. Express the left-hand side of the differential equation in a series in terms of the eigenfunctions. This is done by simply multiplying the Fourier coefficients of u by the corresponding eigenvalues.

5. Equate the coefficients in the series for the left- and right-hand sides of the differential equation, and solve for the unknowns.

As mentioned above, the Fourier series method is analogous to what we called (in Section 3.5) the spectral method for solving $Ax = b$. This method is only applicable if the matrix A is symmetric, so that the eigenvectors are orthogonal. In the same way, the method described above fails at the first step if the eigenfunctions are not orthogonal, that is, if the differential operator is not symmetric.

An even more pertinent observation is that one rarely uses the spectral method to solve $Ax = b$, for the simple reason that computing the eigenpairs of A is more costly, in most cases, than solving $Ax = b$ directly by other means. It is only in special cases that one knows the eigenpairs of a matrix. In the same way, *it is only for very special differential operators that the eigenvalues and eigenfunctions can be found.* Therefore, the method of Fourier series, which works very well when it can be used, is applicable to only a small set of problems. On most problems, such as the BVP

$$-\frac{d}{dx}\left(k(x)\frac{du}{dx}\right) = f(x), \ 0 < x < \ell,$$
$$u(0) = 0,$$
$$u(\ell) = 0$$

5.3. Solving the BVP using Fourier series

when $k(x)$ is nonconstant, it is more work to find the eigenfunctions than to just solve the problem using a different method. For this reason, we discuss a more broadly applicable method, the finite element method, beginning in Section 5.4.

Example 5.14 (homogeneous Dirichlet conditions). We will solve the following BVP, applying the procedure given above:

$$-3\frac{d^2 u}{dx^2} = x^3, \ 0 < x < 2,$$
$$u(0) = 0,$$
$$u(2) = 0.$$
(5.26)

1. We have already seen that the eigenfunctions of the negative second derivative operator, on the interval $[0,2]$ and subject to Dirichlet conditions, are

$$\sin\left(\frac{n\pi x}{2}\right), \ n = 1, 2, 3, \ldots.$$

The effect of multiplying the operator by 3 is to multiply the eigenvalues by 3, so the eigenvalues are

$$\lambda_n = \frac{3n^2\pi^2}{4}, \ n = 1, 2, 3, \ldots.$$

2. We write the unknown solution u as

$$u(x) = \sum_{n=1}^{\infty} a_n \sin\left(\frac{n\pi x}{2}\right).$$

The problem now reduces to computing $a_1, a_2, a_3, \ldots$.

3. The Fourier sine coefficients of $f(x) = x^3$ on the interval $[0,2]$ are

$$c_n = \int_0^2 x^3 \sin\left(\frac{n\pi x}{2}\right) dx = \frac{16\left(6n\pi(-1)^n - n^3\pi^3(-1)^n\right)}{n^4\pi^4}, \ n = 1, 2, 3, \ldots,$$

and so

$$x^3 = \sum_{n=1}^{\infty} \frac{16\left(6n\pi(-1)^n - n^3\pi^3(-1)^n\right)}{n^4\pi^4} \sin\left(\frac{n\pi x}{2}\right).$$

4. The Fourier sine coefficients of $-3 d^2 u / dx^2$ are just

$$\lambda_n a_n = \frac{3n^2\pi^2}{4} a_n.$$

Thus

$$-3\frac{d^2 u}{dx^2} = \sum_{n=1}^{\infty} \frac{3n^2\pi^2}{4} a_n \sin\left(\frac{n\pi x}{2}\right).$$

5. Finally, the differential equation can be written as

$$\sum_{n=1}^{\infty} \frac{3n^2\pi^2}{4} a_n \sin\left(\frac{n\pi x}{2}\right) = \sum_{n=1}^{\infty} \frac{16\left(6n\pi(-1)^n - n^3\pi^3(-1)^n\right)}{n^4\pi^4} \sin\left(\frac{n\pi x}{2}\right),$$

so we have

$$\frac{3n^2\pi^2}{4} a_n = \frac{16\left(6n\pi(-1)^n - n^3\pi^3(-1)^n\right)}{n^4\pi^4}, \quad n = 1, 2, 3, \ldots.$$

This yields

$$a_n = \frac{64\left(6n\pi(-1)^n - n^3\pi^3(-1)^n\right)}{3n^6\pi^6}, \quad n = 1, 2, 3, \ldots,$$

and so

$$u(x) = \sum_{n=1}^{\infty} \frac{64\left(6n\pi(-1)^n - n^3\pi^3(-1)^n\right)}{3n^6\pi^6} \sin\left(\frac{n\pi x}{2}\right)$$

is the solution to (5.26). ∎

Example 5.15 (inhomogeneous Dirichlet conditions). Next we solve a BVP with inhomogeneous boundary conditions:

$$-\frac{d^2u}{dx^2} = 1 - x, \ 0 < x < \frac{1}{2},$$
$$u(0) = 3,$$
$$u\left(\frac{1}{2}\right) = -1.$$
(5.27)

We must first shift the data to transform the problem into one with homogeneous boundary conditions. The function $p(x) = 3 - 8x$ satisfies the boundary conditions, so we define $v = u - p$ and solve the BVP

$$-\frac{d^2v}{dx^2} = 1 - x, \ 0 < x < \frac{1}{2},$$
$$v(0) = 0,$$
$$v\left(\frac{1}{2}\right) = 0.$$

1. The eigenpairs are now $\lambda_n = 4n^2\pi^2$, $\psi_n(x) = \sin(2n\pi x)$, $n = 1, 2, 3, \ldots$.

2. We write the solution v in the form

$$v(x) = \sum_{n=1}^{\infty} a_n \sin(2n\pi x).$$

5.3. Solving the BVP using Fourier series

3. The Fourier sine coefficients of $f(x) = 1-x$ are

$$4\int_0^{1/2} (1-x)\sin(2n\pi x)\,dx = \frac{2-(-1)^n}{n\pi}, \quad n=1,2,3,\ldots,$$

and so

$$1-x = \sum_{n=1}^{\infty} \frac{2-(-1)^n}{n\pi} \sin(2n\pi x).$$

4. The Fourier sine coefficients of $-d^2v/dx^2$ are

$$\lambda_n a_n = 4n^2\pi^2 a_n, \quad n=1,2,3,\ldots,$$

and so

$$-\frac{d^2v}{dx^2}(x) = \sum_{n=1}^{\infty} 4n^2\pi^2 a_n \sin(2n\pi x).$$

5. We thus obtain

$$\sum_{n=1}^{\infty} 4n^2\pi^2 a_n \sin(2n\pi x) = \sum_{n=1}^{\infty} \frac{2-(-1)^n}{n\pi} \sin(2n\pi x),$$

which implies that

$$4n^2\pi^2 a_n = \frac{2-(-1)^n}{n\pi}, \quad n=1,2,3,\ldots,$$

or

$$a_n = \frac{2-(-1)^n}{4n^3\pi^3}, \quad n=1,2,3,\ldots.$$

The solution v is then

$$v(x) = \sum_{n=1}^{\infty} \frac{2-(-1)^n}{4n^3\pi^3} \sin(2n\pi x).$$

This yields

$$u(x) = 3 - 8x + \sum_{n=1}^{\infty} \frac{2-(-1)^n}{4n^3\pi^3} \sin(2n\pi x). \quad \blacksquare$$

Exercises

In Exercises 1–4, solve the BVPs using the method of Fourier series, shifting the data if necessary. If possible, produce a graph of the computed solution by plotting a partial Fourier series with enough terms to give a qualitatively correct graph. (The number of terms can be determined by trial and error; if the plot no longer changes, qualitatively, when more terms are included, it can be assumed that the partial series contains enough terms.)

1. (a) $-\frac{d^2u}{dx^2} = 1$, $u(0) = u(1) = 0$;
   (b) $-\frac{d^2u}{dx^2} = 1$, $u(0) = 0$, $u(1) = 1$.

2. (a) $-\frac{d^2u}{dx^2} = e^x$, $u(0) = u(1) = 0$;
   (b) $-\frac{d^2u}{dx^2} = e^x$, $u(0) = u(1) = 1$.

3. The results of Exercises 5.2.2 and 5.2.3 will be useful for the following problems:
   (a) $-\frac{d^2u}{dx^2} = x$, $u(0) = 0$, $\frac{du}{dx}(1) = 0$;
   (b) $-\frac{d^2u}{dx^2} = x$, $\frac{du}{dx}(0) = 0$, $u(1) = 1$;
   (c) $-\frac{d^2u}{dx^2} + 2u = 1$, $u(0) = u(1) = 0$;
   (d) $-\frac{d^2u}{dx^2} + u = 0$, $u(0) = 0$, $u(1) = 1$.

4. The results of Exercises 5.2.2 and 5.2.3 will be useful for the following problems:
   (a) $-\frac{d^2u}{dx^2} = x^2$, $u(0) = u(1) = 0$;
   (b) $-\frac{d^2u}{dx^2} + 2u = \frac{1}{2} - x$, $u(0) = u(2) = 0$;
   (c) $-\frac{d^2u}{dx^2} = x + \sin(\pi x)$, $u(0) = \frac{du}{dx}(1) = 0$;
   (d) $-\frac{d^2u}{dx^2} = f(x)$, $\frac{du}{dx}(0) = u(1) = 0$, where
   $$f(x) = \begin{cases} 1, & 0 < x < \frac{1}{2}, \\ 0, & \frac{1}{2} < x < 1. \end{cases}$$

5. Solve (5.17) to get (5.18).

6. Suppose u satisfies
   $$u(0) = 0, \quad \frac{du}{dx}(\ell) = 0,$$
   and the Fourier sine series of u is
   $$u(x) = \sum_{n=1}^{\infty} b_n \sin\left(\frac{n\pi x}{\ell}\right).$$

Show that the Fourier sine coefficients of $-d^2u/dx^2$ cannot be represented in terms of $b_1, b_2, b_3, \ldots$. (In particular, we cannot obtain the Fourier sine series of $-d^2u/dx^2$ just by differentiating the Fourier sine series of u term by term.) This shows that it is not possible to use the "wrong" eigenfunctions (that is, eigenfunctions corresponding to different boundary conditions) to solve a BVP.

7. Suppose u satisfies
$$u(0) = 0, \frac{du}{dx}(\ell) = 0,$$
and has Fourier quarter-wave sine series
$$u(x) = \sum_{n=1}^{\infty} a_n \sin\left(\frac{(2n-1)\pi x}{2\ell}\right).$$
Show that the Fourier series of $-T d^2u/dx^2$ is
$$-T\frac{d^2u}{dx^2}(x) = \sum_{n=1}^{\infty} \frac{T(2n-1)^2\pi^2}{4\ell^2} a_n \sin\left(\frac{(2n-1)\pi x}{2\ell}\right).$$
(Hint: The computation of the Fourier coefficients of $-Td^2u/dx^2$ is similar to (5.19)).

8. Consider an aluminum bar of length 1 m and radius 1 cm. Suppose that the side of the bar is perfectly insulated, the ends of the bar are placed in ice baths, and heat energy is added throughout the interior of the bar at a constant rate of $0.001 \,\text{W/cm}^3$. The thermal conductivity of the aluminum alloy is $1.5 \,\text{W/(cm K)}$. Find and graph the steady-state temperature of the bar. Use the Fourier series method.

9. Repeat the previous exercise, assuming that the right end of the bar is perfectly insulated and the left is placed in an ice bath.

10. Consider the string of Example 5.10. Suppose that the right end of the string is free to move vertically (along a frictionless pole, for example), and a pressure of 2200 dynes per centimeter, in the upward direction, is applied along the string (recall that a dyne is a unit of force—one dyne equals one gram-centimeter per square second). What is the equilibrium displacement of the string? How does the answer change if the force due to gravity is taken into account?

5.4 Finite element methods for BVPs

As we mentioned near the end of the last section, the primary utility of the Fourier series method is for problems with constant coefficients, in which case the eigenpairs can be found explicitly. For problems in two or three dimensions, in order for the eigenfunctions to be explicitly computable, it is also necessary that the domain on which the equation is to be solved be geometrically simple. We discuss this further in Chapter 11.

In Chapter 10, we discuss spectral theory for problems with nonconstant coefficients, and we show that the Fourier series method applies in principle—the eigenfunctions exist, they are orthogonal, and so forth. However, without explicit formulas for the eigenfunctions, it is not easy to apply the Fourier series method to compute solutions to BVPs.

These remarks apply, for example, to the BVP

$$-\frac{d}{dx}\left(k(x)\frac{du}{dx}\right) = f(x),\ 0 < x < \ell,$$
$$u(0) = 0,$$
$$u(\ell) = 0$$
(5.28)

(where k is a positive function). The operator $K : C_D^2[0,\ell] \to C[0,\ell]$ defined by

$$Ku = -\frac{d}{dx}\left(k(x)\frac{du}{dx}\right) \quad (5.29)$$

is symmetric (see Exercise 5.4.1), and there exists an orthogonal sequence of eigenfunctions, with corresponding positive eigenvalues. However, when the coefficient $k(x)$ is not a constant, there is no simple way to find these eigenpairs. We can approximate them by appropriate numerical methods, but this requires more work than solving the original BVP by similar numerical methods.

Because of the limitations of the Fourier series approach, we now introduce the *finite element method*, one of the most powerful methods for approximating solutions to PDEs. The finite element method can handle both variable coefficients and, in multiple spatial dimensions, irregular geometries. We will still restrict ourselves to symmetric operators, although it is possible to apply the finite element method to nonsymmetric problems.

The finite element method is based on three ideas.

1. The BVP is rewritten in its *weak* or *variational form*, which expresses the problem as infinitely many scalar equations. In this form, the boundary conditions are implicit in the definition of the underlying vector space.

2. The *Galerkin method* is applied to "solve the equation on a finite-dimensional subspace." This results in an ordinary linear system (matrix-vector equation) that must be solved.

3. A basis of *piecewise polynomials* is chosen for the finite-dimensional subspace so that the matrix of the linear system is *sparse* (that is, has mostly zero entries).

We describe each of these ideas in the following sections, using the BVP (5.28) as our model problem. We always assume that the coefficient $k(x)$ is positive, since it represents a positive physical parameter (stiffness or thermal conductivity, for example).

5.4.1 The principle of virtual work and the weak form of a BVP

When an elastic material is deformed, it stores potential energy due to internal elastic forces. It is not obvious from first principles how to measure (quantitatively) this *elastic potential energy*; however, we can deduce the correct definition from the equations of motion and the principle of conservation of energy.

5.4. Finite element methods for BVPs

Suppose an elastic bar, with its ends fixed, is in motion, and its displacement function $u(x,t)$ satisfies the homogeneous wave equation

$$A\rho(x)\frac{\partial^2 u}{\partial t^2} - A\frac{\partial}{\partial x}\left(k(x)\frac{\partial u}{\partial x}\right) = 0, \; 0 < x < \ell, \; t > 0,$$
$$u(0,t) = 0, \; t > 0,$$
$$u(\ell,t) = 0, \; t > 0.$$
(5.30)

We now perform the following calculation (here is a trick: multiply both sides of the wave equation by $\partial u/\partial t$ and integrate):

$$A\rho(x)\frac{\partial^2 u}{\partial t^2} - A\frac{\partial}{\partial x}\left(k(x)\frac{\partial u}{\partial x}\right) = 0, \; 0 < x < \ell$$

$$\Rightarrow A\rho(x)\frac{\partial^2 u}{\partial t^2}\frac{\partial u}{\partial t} - A\frac{\partial}{\partial x}\left(k(x)\frac{\partial u}{\partial x}\right)\frac{\partial u}{\partial t} = 0, \; 0 < x < \ell$$

$$\Rightarrow \int_0^\ell A\rho(x)\frac{\partial^2 u}{\partial t^2}\frac{\partial u}{\partial t}dx - \int_0^\ell A\frac{\partial}{\partial x}\left(k(x)\frac{\partial u}{\partial x}\right)\frac{\partial u}{\partial t}dx = 0.$$

Applying integration by parts, we obtain

$$-\int_0^\ell \frac{\partial}{\partial x}\left(k(x)\frac{\partial u}{\partial x}\right)\frac{\partial u}{\partial t}dx = -\left[k(x)\frac{\partial u}{\partial x}\frac{\partial u}{\partial t}\right]_0^\ell + \int_0^\ell k(x)\frac{\partial u}{\partial x}\frac{\partial^2 u}{\partial x \partial t}dx$$
$$= \int_0^\ell k(x)\frac{\partial u}{\partial x}\frac{\partial^2 u}{\partial t \partial x}dx \; \left(\frac{\partial u}{\partial t}(0,t) = \frac{\partial u}{\partial t}(\ell,t) = 0\right).$$

Also, we notice that

$$\frac{\partial^2 u}{\partial t^2}\frac{\partial u}{\partial t} = \frac{\partial}{\partial t}\left[\frac{1}{2}\left(\frac{\partial u}{\partial t}\right)^2\right], \; \frac{\partial^2 u}{\partial t \partial x}\frac{\partial u}{\partial x} = \frac{\partial}{\partial t}\left[\frac{1}{2}\left(\frac{\partial u}{\partial x}\right)^2\right].$$

Therefore, we obtain

$$\int_0^\ell A\rho(x)\frac{\partial}{\partial t}\left[\frac{1}{2}\left(\frac{\partial u}{\partial t}\right)^2\right]dx + \int_0^\ell Ak(x)\frac{\partial}{\partial t}\left[\frac{1}{2}\left(\frac{\partial u}{\partial x}\right)^2\right]dx = 0. \quad (5.31)$$

Applying Theorem 2.1 to (5.31) yields

$$\frac{d}{dt}\left[\frac{1}{2}\int_0^\ell A\rho(x)\left(\frac{\partial u}{\partial t}\right)^2 dx + \frac{1}{2}\int_0^\ell Ak(x)\left(\frac{\partial u}{\partial x}\right)^2 dx\right] = 0, \; t > 0. \quad (5.32)$$

This equation implies that the sum of the two integrals is constant with respect to time, that is,

$$\frac{1}{2}\int_0^\ell A\rho(x)\left(\frac{\partial u}{\partial t}\right)^2 dx + \frac{1}{2}\int_0^\ell Ak(x)\left(\frac{\partial u}{\partial x}\right)^2 dx$$

is conserved. Obviously, the first integral represents the kinetic energy (one-half mass times velocity squared), and we define the second integral to be the elastic potential energy. The elastic potential energy is the internal energy arising from the strain $\partial u/\partial x$.

We now return to the bar in equilibrium, that is, to the bar satisfying the BVP (5.28). We have the following expression for the elastic potential energy of the bar:

$$\frac{1}{2}\int_0^\ell Ak(x)\left(\frac{du}{dx}(x)\right)^2 dx.$$

The bar possesses another form of potential energy, due to the external force f acting upon it. The proper name of this energy depends on the nature of the force; if it were the force due to gravity, for example, it would be called the gravitational potential energy. We will call this energy the *external potential energy*. This energy is different in a fundamental way from the elastic potential energy: it depends on the absolute position of the bar, not on the relative position of parts of the bar to other parts. For this reason, we do not know the external potential energy of the bar in its "reference" (zero displacement) position (whereas the bar has zero elastic potential energy in the reference position). We define $\mathcal{E}_0$ to be the external potential energy of the bar in the reference position, and we now compute the change of energy between the bar in its reference position and the bar in its equilibrium position. The change in potential energy due to the external force f is equal to the work done by f when the bar is deformed from its reference position to its equilibrium position.

The work done by f acting on the part P of the bar originally between x and $x+\Delta x$ is approximately $Af(x)u(x)\Delta x$—$f(x)\cdot A\Delta x$ represents the force (the force per unit volume times the volume of P) and $u(x)$ represents the distance that P moves. To add up the work due to the external force acting on all little parts of the bar, we integrate

$$\int_0^\ell Af(x)u(x)dx. \tag{5.33}$$

We now have an expression for the total potential energy of the bar in its equilibrium position:

$$\mathcal{E}_{pot}(u) = \frac{1}{2}\int_0^\ell Ak(x)\left(\frac{du}{dx}(x)\right)^2 dx - \int_0^\ell Af(x)u(x)dx + \mathcal{E}_0.$$

To account for the sign on the second integral, consider the example of gravity. The corresponding force f is positive, since gravity acts in the positive (downward) direction, and the displacement u is also positive, because the bar is pulled down. This decreases the gravitational potential energy, which shows that the second integral should be subtracted from the first.

Next we wish to prove the following fact: *The equilibrium displacement u of the bar is the displacement with the least potential energy.* That is, if $w \neq u$ is any other displacement satisfying the Dirichlet conditions, then

$$\mathcal{E}_{pot}(w) > \mathcal{E}_{pot}(u).$$

We can write any displacement w as $w = u+v$ (just take $v = w-u$), and the conditions $u(0)=0$, $w(0)=0$ imply that $v(0)=0$, and similarly at $x=\ell$. Thus v must belong to $C_D^2[0,\ell]$. For brevity, we will write $V = C_D^2[0,\ell]$, the set of allowable displacements v.

5.4. Finite element methods for BVPs

Therefore, we wish to prove that

$$\mathcal{E}_{pot}(u+v) > \mathcal{E}_{pot}(u) \text{ for all } v \in V.$$

We have

$$\mathcal{E}_{pot}(u+v) = \frac{1}{2}\int_0^\ell k(x)\left(\frac{du}{dx}(x)+\frac{dv}{dx}(x)\right)^2 dx - \int_0^\ell f(x)(u(x)+v(x))\,dx + \mathcal{E}_0$$

$$= \frac{1}{2}\int_0^\ell k(x)\left(\left(\frac{du}{dx}(x)\right)^2 + 2\frac{du}{dx}(x)\frac{dv}{dx}(x) + \left(\frac{dv}{dx}(x)\right)^2\right)dx$$

$$- \int_0^\ell f(x)u(x)\,dx - \int_0^\ell f(x)v(x)\,dx + \mathcal{E}_0$$

$$= \frac{1}{2}\int_0^\ell k(x)\left(\frac{du}{dx}(x)\right)^2 dx + \int_0^\ell k(x)\frac{du}{dx}(x)\frac{dv}{dx}(x)\,dx$$

$$+ \frac{1}{2}\int_0^\ell k(x)\left(\frac{dv}{dx}(x)\right)^2 dx - \int_0^\ell f(x)u(x)\,dx - \int_0^\ell f(x)v(x)\,dx + \mathcal{E}_0;$$

that is,

$$\mathcal{E}_{pot}(u+v) = \mathcal{E}_{pot}(u) + \int_0^\ell k(x)\frac{du}{dx}(x)\frac{dv}{dx}(x)\,dx - \int_0^\ell f(x)v(x)\,dx$$
$$+ \frac{1}{2}\int_0^\ell k(x)\left(\frac{dv}{dx}(x)\right)^2 dx. \qquad (5.34)$$

Using integration by parts, we have

$$\int_0^\ell k(x)\frac{du}{dx}(x)\frac{dv}{dx}(x)\,dx - \int_0^\ell f(x)v(x)\,dx$$

$$= \left[k(x)\frac{du}{dx}(x)v(x)\right]_0^\ell - \int_0^\ell \left(\frac{d}{dx}\left(k(x)\frac{du}{dx}(x)\right)\right)v(x)\,dx - \int_0^\ell f(x)v(x)\,dx$$

$$= -\int_0^\ell \left(\frac{d}{dx}\left(k(x)\frac{du}{dx}(x)\right) + f(x)\right)v(x)\,dx.$$

(The boundary term disappears in the above calculation since $v(0) = v(\ell) = 0$.) The equilibrium displacement u satisfies the differential equation

$$\frac{d}{dx}\left(k(x)\frac{du}{dx}(x)\right) + f(x) = 0$$

(see (5.28)), as we derived in Section 2.2. Therefore, the integral above vanishes, and we have

$$\mathcal{E}_{pot}(u+v) = \mathcal{E}_{pot}(u) + \frac{1}{2}\int_0^\ell k(x)\left(\frac{dv}{dx}(x)\right)^2 dx.$$

If v is not identically zero, then neither is dv/dx (only a constant function has a zero derivative, and v cannot be constant unless it is identically zero, because $v(0) = v(\ell) = 0$). Since

$$\frac{1}{2}\int_0^\ell k(x)\left(\frac{dv}{dx}(x)\right)^2 dx > 0$$

whenever $v \neq 0$ (the integral of a nonzero, nonnegative function is positive), we have obtained the desired result:

$$v \neq 0 \Rightarrow \mathcal{E}_{pot}(u+v) > \mathcal{E}_{pot}(u).$$

This result gives us a different understanding of the equilibrium state of the bar—the bar assumes the state of minimal potential energy. This complements our earlier understanding that the bar assumes a state in which all forces balance.

The potential energy $\mathcal{E}_{pot}$ defines a function mapping the space of physically meaningful displacements into $\mathbf{R}$, and the equilibrium displacement u is its minimizer. A standard result from calculus is that the derivative of a real-valued function must be zero at a minimizer. Above we computed

$$\mathcal{E}_{pot}(u+v) = \mathcal{E}_{pot}(u) + \text{ a term linear in } v + \text{ a term quadratic in } v$$

(see (5.34)). The linear term in this expression must be $D\mathcal{E}_{pot}(u)v$, the directional derivative of $\mathcal{E}_{pot}$ at u in the direction of v. The minimality of the potential energy then implies that $D\mathcal{E}_{pot}(u) = 0$, or, equivalently,

$$D\mathcal{E}_{pot}(u)v = 0 \text{ for all } v \in V.$$

This yields

$$\int_0^\ell k(x)\frac{du}{dx}(x)\frac{dv}{dx}(x)\,dx - \int_0^\ell f(x)v(x)\,dx = 0 \text{ for all } v \in V$$

(as we saw above), which is called the *principle of virtual work*. (The term linear in u dominates the work $\mathcal{E}_{pot}(u+v) - \mathcal{E}_{pot}(u)$ when the displacement v is arbitrarily small ("virtual") and is called the *virtual work*. The principle says that the virtual work is zero when the bar is at equilibrium.)

The principle of virtual work is also called the *weak form* of the BVP; it is the form that the BVP would take if our basic principle were the minimality of potential energy rather than the balance of forces at equilibrium. By contrast, we will call the original BVP,

$$-\frac{d}{dx}\left(k(x)\frac{du}{dx}(x)\right) = f(x),\ 0 < x < \ell,$$
$$u(0) = 0,$$
$$u(\ell) = 0,$$
(5.35)

the *strong form*. We can show directly that the two forms of the BVP are equivalent, in the sense that, for $f \in C[0,\ell]$, u satisfies the strong form if and only if it satisfies the weak form.

5.4.2 The equivalence of the strong and weak forms of the BVP

The precise statement of the weak form of the BVP is

$$\text{find } u \in V \text{ such that } \int_0^\ell k(x)\frac{du}{dx}(x)\frac{dv}{dx}(x)\,dx = \int_0^\ell f(x)v(x)\,dx \text{ for all } v \in V. \quad (5.36)$$

5.4. Finite element methods for BVPs

We have already seen the proof of the following fact: If u satisfies (5.35), the strong form of the BVP, then u also satisfies (5.36), the weak form. This was proved just below (5.34), using integration by parts. Indeed, the simplest way of deriving the weak form is to start with the differential equation

$$-\frac{d}{dx}\left(k(x)\frac{du}{dx}(x)\right) = f(x),\ 0 < x < \ell,$$

multiply by an arbitrary "test function" $v \in V$ to get

$$-\frac{d}{dx}\left(k(x)\frac{du}{dx}(x)\right)v(x) = f(x)v(x),\ 0 < x < \ell,$$

and then integrate both sides from 0 to ℓ:

$$-\int_0^\ell \frac{d}{dx}\left(k(x)\frac{du}{dx}(x)\right)v(x)\,dx = \int_0^\ell f(x)v(x)\,dx.$$

Integrating by parts on the left and applying the boundary conditions yield the weak form.

Now suppose u satisfies (5.36), the weak form of the BVP. Then, by definition of V, $u(0) = u(\ell) = 0$. We now use integration by parts to show that u satisfies the strong form of the differential equation. We have

$$-\int_0^\ell k(x)\frac{du}{dx}(x)\frac{dv}{dx}(x)\,dx - \int_0^\ell f(x)v(x)\,dx = 0 \text{ for all } v \in V$$

$$\Rightarrow \left[k(x)\frac{du}{dx}(x)v(x)\right]_0^\ell - \int_0^\ell \frac{d}{dx}\left(k(x)\frac{du}{dx}(x)\right)v(x)\,dx$$

$$-\int_0^\ell f(x)v(x)\,dx = 0 \text{ for all } v \in V$$

$$\Rightarrow -\int_0^\ell \left\{\frac{d}{dx}\left(k(x)\frac{du}{dx}(x)\right) + f(x)\right\}v(x)\,dx = 0 \text{ for all } v \in V.$$

The boundary terms vanish because of the boundary conditions $v(0) = v(\ell) = 0$.

We have

$$\int_0^\ell \left\{\frac{d}{dx}\left(k(x)\frac{du}{dx}(x)\right) + f(x)\right\}v(x)\,dx = 0 \text{ for all } v \in V. \tag{5.37}$$

This condition can only be true if

$$\frac{d}{dx}\left(k(x)\frac{du}{dx}(x)\right) + f(x) = 0 \tag{5.38}$$

on the interval $[0,\ell]$. The proof is simple, given the following fact: If $0 < c < d < \ell$, then there exists a function $v_{[c,d]} \in V$ such that $v_{[c,d]}(x) > 0$ for all $x \in (c,d)$ and $v_{[c,d]}(x) = 0$ for all x in $[0,c]$ or $[d,\ell]$. (Exercise 3 asks the reader to construct such a function $v_{[c,d]}$.) If

$$\frac{d}{dx}\left(k(x)\frac{du}{dx}(x)\right) + f(x)$$

were positive on a small interval $[c,d] \subset [0,\ell]$, then

$$\left[\frac{d}{dx}\left(k(x)\frac{du}{dx}(x)\right) + f(x)\right]v_{[c,d]}(x)$$

would be positive on $[c,d]$ and zero on the rest of $[0,\ell]$, and hence

$$\int_0^\ell \left\{\frac{d}{dx}\left(k(x)\frac{du}{dx}(x)\right) + f(x)\right\} v_{[c,d]}(x)\,dx$$

would be positive. Since this integral is zero for all $v \in V$, this shows that

$$\frac{d}{dx}\left(k(x)\frac{du}{dx}(x)\right) + f(x)$$

cannot be positive on any small interval. By the same reasoning, it cannot be negative on any small interval. It follows that (5.38) must hold.

Thus, to solve (5.35), it suffices to solve the weak form (5.36). The weak form consists of infinitely many equations, since there are infinitely many "test functions" $v \in V$. The original differential equation also implies an infinite number of equations, since the equation must hold for each of the infinitely many values of $x \in [0,\ell]$. In either case, we cannot find the solution directly, since any practical computational procedure must reduce to a finite number of steps. The weak form admits the use of the Galerkin method, which reduces the infinitely many equations to a finite collection of equations whose solution provides an approximate solution to the BVP.

Exercises

1. Show that the differential operator

   $$K : C_D^2[0,\ell] \to C[0,\ell]$$

   defined by

   $$Ku = -\frac{d}{dx}\left(k(x)\frac{du}{dx}\right)$$

   is symmetric:

   $$(Ku,v) = (u,Kv) \text{ for all } u,v \in C_D^2[0,\ell].$$

2. Suppose that $k(x) > 0$ for all $x \in [0,\ell]$. Show that if the operator K, defined in the previous exercise, has an eigenvalue, then it must be positive. (Hint: Let λ be an eigenvalue and u be a corresponding eigenfunction. Compute (Ku,u) two ways, once using the fact that u is an eigenfunction and again using integration by parts once.)

3. Let $0 < c < d < \ell$ hold. Find a function $v_{[c,d]}$ that is twice-continuously differentiable on $[0,\ell]$ and satisfies the conditions that

   $$v_{[c,d]}(x) > 0 \text{ for all } x \in (c,d)$$

   and

   $$v_{[c,d]}(x) = 0 \text{ for all } x \in [0,c] \cup [d,\ell].$$

5.4. Finite element methods for BVPs

Hint: Let $p(x)$ be a polynomial such that

$$p(c) = \frac{dp}{dx}(c) = \frac{d^2p}{dx^2}(c) = 0, \; p(d) = \frac{dp}{dx}(d) = \frac{d^2p}{dx^2}(d) = 0.$$

Then define

$$v_{[c,d]}(x) = \begin{cases} p(x), & c < x < d, \\ 0, & 0 \leq x \leq c \text{ or } d \leq x \leq \ell. \end{cases}$$

4. Show that both of the integrals

$$\frac{1}{2}\int_0^\ell A\rho(x)\left(\frac{\partial u}{\partial t}\right)^2 dx, \; \frac{1}{2}\int_0^\ell Ak(x)\left(\frac{\partial u}{\partial x}\right)^2 dx$$

have units of energy (force times distance).

5. Consider the BVP

$$-\frac{d}{dx}\left(k(x)\frac{du}{dx}\right) + p(x)u = f(x), \; 0 < x < \ell,$$

$$u(0) = 0,$$

$$u(\ell) = 0.$$

Derive the weak form. Use the method described on page 171 (that is, multiply both sides of the differential equation by a test function and integrate by parts).

6. Find an energy function (similar to $\mathcal{E}_{pot}$) for the BVP in the previous exercise and minimize it to derive the weak form of the BVP.

7. Consider the BVP

$$-\frac{d}{dx}\left(k(x)\frac{du}{dx}\right) + c(x)\frac{du}{dx} + p(x)u = f(x), \; 0 < x < \ell,$$

$$u(0) = 0,$$

$$u(\ell) = 0.$$

Derive the weak form. (The weak form will not be symmetric in the solution u and the test function v.)

8. Explain why there cannot be an energy function for the BVP of the previous exercise whose minimization leads to the weak form.

9. Consider the damped wave equation

$$A\rho(x)\frac{\partial^2 u}{\partial t^2} + c\frac{\partial u}{\partial t} - A\frac{\partial}{\partial x}\left(k(x)\frac{\partial u}{\partial x}\right) = 0, \; 0 < x < \ell, \, t > 0,$$

$$u(0,t) = 0, \, t > 0,$$

$$u(\ell,t) = 0, \, t > 0,$$

where $c > 0$ is a constant. By modifying the calculation beginning on page 167, show that the total energy (kinetic energy plus elastic potential energy) is decreasing with time.

5.5 The Galerkin method

As we stated earlier, the Galerkin method defines an approximate solution to the weak form of the BVP by restricting the problem to a finite-dimensional subspace. This has the effect of reducing the infinitely many equations contained in (5.36) to a finite system of equations. To describe the Galerkin method, it is convenient to introduce the following notation. We define the symmetric *bilinear form* $a(\cdot,\cdot)$ by

$$a(u,v) = \int_0^\ell k(x)\frac{du}{dx}(x)\frac{dv}{dx}(x)\,dx.$$

The function is called bilinear because, holding one argument fixed, the function is linear in the other:

$$a(\alpha_1 u_1 + \alpha_2 u_2, v) = \alpha_1 a(u_1,v) + \alpha_2 a(u_2,v) \text{ for all } \alpha_1,\alpha_2 \in \mathbf{R},\ u_1,u_2,v,$$
$$a(u,\alpha_1 v_1 + \alpha_2 v_2) = \alpha_1 a(u,v_1) + \alpha_2 a(u,v_2) \text{ for all } \alpha_1,\alpha_2 \in \mathbf{R},\ u,v_1,v_2.$$

The bilinear form is called *symmetric* because

$$a(u,v) = a(v,u) \text{ for all } u,v.$$

Bilinearity and symmetry are two of the properties of an inner product (see Section 3.4). The third property also holds if we restrict $a(\cdot,\cdot)$ to vectors in $V = C_D^2[0,\ell]$:

$$a(u,u) \geq 0 \text{ for all } u \in V$$

and

$$a(u,u) = 0 \Leftrightarrow u = 0.$$

To see this, notice that

$$a(u,u) = \int_0^\ell k(x)\left(\frac{du}{dx}(x)\right)^2 dx.$$

Since the integrand is nonnegative, we clearly have $a(u,u) \geq 0$ for all u. Moreover, the integral of a nonnegative function can only be zero when the function (the integrand) is the zero function. Thus,

$$a(u,u) = 0 \Rightarrow k(x)\left(\frac{du}{dx}(x)\right)^2 = 0,\ 0 < x < \ell,$$
$$\Rightarrow \frac{du}{dx}(x) = 0,\ 0 < x < \ell \text{ (since } k(x) > 0),$$
$$\Rightarrow u(x) = C,\ 0 < x < \ell,$$

where C is a constant. But we already know that $u(0) = 0$ and u is continuous on $[0,\ell]$. Therefore, C must be zero, which is what we wanted to prove. Thus $a(\cdot,\cdot)$ defines an inner product on V.

We recall that

$$\int_0^\ell f(x)v(x)\,dx = (f,v),$$

5.5. The Galerkin method

where $(\cdot,\cdot)$ is the L^2 inner product. Thus, the weak form (5.36) of the model BVP (5.35) can be written as

$$\text{find } u \in V \text{ satisfying } a(u,v) = (f,v) \text{ for all } v \in V. \tag{5.39}$$

The Galerkin idea is simple. Choose a finite-dimensional subspace V_n of V, and reduce (5.39) to the subspace:

$$\text{find } v_n \in V_n \text{ satisfying } a(v_n,v) = (f,v) \text{ for all } v \in V_n. \tag{5.40}$$

Below we show that (5.40) has a unique solution that can be found by solving a matrix-vector equation (that is, a system of linear algebraic equations). First we show that solving (5.40) is useful: *the Galerkin approximation is the best approximation, from V_n, to the true solution.*

Suppose $u \in V$ is the solution to (5.39), V_n is a finite-dimensional subspace of V, and $v_n \in V_n$ is the solution to (5.40). Then we have

$$u \in V, \; a(u,v) = (f,v) \text{ for all } v \in V,$$
$$v_n \in V_n, \; a(v_n,v) = (f,v) \text{ for all } v \in V_n.$$

Subtracting the second equation from the first, we see that

$$a(u,v) - a(v_n,v) = 0 \text{ for all } v \in V_n$$

or, using the bilinearity of $a(\cdot,\cdot)$,

$$a(u - v_n, v) = 0 \text{ for all } v \in V_n.$$

Since $a(\cdot,\cdot)$ defines an inner product on V, the projection theorem (see Theorem 3.36 in Section 3.4) shows that v_n is the best approximation to u from V_n, when "best" is defined by the norm induced by $a(\cdot,\cdot)$:

$$\|u - v_n\|_E \leq \|u - w\|_E \text{ for all } w \in V_n,$$

where

$$\|v\|_E = \sqrt{a(v,v)} = \sqrt{\int_0^\ell k(x) \left(\frac{dv}{dx}(x)\right)^2 dx} \text{ for all } v \in V.$$

We usually refer to $\|\cdot\|_E$ as the *energy norm* and to $a(\cdot,\cdot)$ as the *energy inner product*.

To compute v_n, suppose $\{\phi_1, \phi_2, \ldots, \phi_n\}$ is a basis for V_n. Then (5.40) is equivalent to

$$\text{find } v_n \in V_n \text{ satisfying } a(v_n, \phi_i) = (f, \phi_i), \; i = 1, 2, \ldots, n. \tag{5.41}$$

Since v_n belongs to V_n, it can be written as a linear combination of the basis vectors:

$$v_n = \sum_{j=1}^n u_j \phi_j. \tag{5.42}$$

(We now have two meanings for the symbol "u." The function u is the exact solution to the BVP, while the components of the vector **u** are the weights in the representation of the approximation v_n in terms of the basis $\phi_1, \phi_2, \ldots, \phi_n$. We will live with this ambiguity in order to adhere to standard notation.)

Finding v_n is now equivalent to determining $u_1, u_2, \ldots, u_n$. Substituting (5.42) into (5.41) yields

$$a\left(\sum_{j=1}^{n} u_j \phi_j, \phi_i\right) = (f, \phi_i), \; i = 1, 2, \ldots, n,$$

or, using the bilinearity of $a(\cdot, \cdot)$,

$$\sum_{j=1}^{n} a(\phi_j, \phi_i) u_j = (f, \phi_i), \; i = 1, 2, \ldots, n.$$

This is a system of n linear equations for the unknowns $u_1, u_2, \ldots, u_n$. We define a matrix $\mathbf{K} \in \mathbf{R}^{n \times n}$ and a vector $\mathbf{f} \in \mathbf{R}^n$ by

$$K_{ij} = a(\phi_j, \phi_i), \; f_i = (f, \phi_i), \; i, j = 1, 2, \ldots, n.$$

(We now also have two meanings for the symbol "f": the function f is the right-hand side of the differential equation, while the vector $\mathbf{f}$ has components $f_i = (f, \phi_i)$.) Finding v_n is now equivalent to solving the linear system

$$\mathbf{Ku} = \mathbf{f}.$$

The matrix $\mathbf{K}$ is usually called the *stiffness matrix*, since, when applied to problems in mechanics, the entries in $\mathbf{K}$ depend on the stiffness $k(x)$ of the material. For analogous reasons, the vector $\mathbf{f}$ is usually called the *load vector*.[25]

The reader will recall that, in computing a best approximation, it is advantageous to have an orthogonal basis, since then the stiffness (or Gram) matrix is diagonal, and the projection is computed by n inner products. However, the basis for V_n must be orthogonal with respect to the energy inner product, and with a nonconstant coefficient $k(x)$, it is usually difficult[26] to find an orthogonal basis. Therefore, in the finite element method, we will use a nonorthogonal basis.

We now illustrate the Galerkin method by applying it to two BVPs, one with constant coefficients and the other with nonconstant coefficients.

Example 5.16. Let F_N be the subspace of V spanned by the basis

$$\{\sin(\pi x), \sin(2\pi x), \ldots, \sin(N \pi x)\}.$$

This basis is orthogonal with respect to the ordinary L^2 inner product on the interval $[0, 1]$.

---

[25] The stiffness matrix is precisely the Gram matrix appearing in the projection theorem; see page 63. If we apply the projection theorem directly to compute v_n, we need to compute the right-hand-side vector $\mathbf{f}$ whose components are $a(u, \phi_i)$; the weak form allows us to compute these components without knowing u, since $a(u, \phi_i) = (f, \phi_i)$.

[26] That is, in the sense that it is computationally expensive to compute an orthogonal basis. Given any basis, it is simple *in principle* to apply the Gram–Schmidt procedure to obtain an orthogonal basis. The reader can consult any introductory book on linear algebra for an explanation of the Gram–Schmidt procedure.

5.5. The Galerkin method

Consider the BVP

$$-\frac{d^2u}{dx^2} = x, \ 0 < x < 1,$$
$$u(0) = 0,$$
$$u(1) = 0$$

(5.43)

(this is (5.35) with $k(x) = 1$). The exact solution is

$$u(x) = \frac{x}{6} - \frac{x^3}{6}.$$

Since $k(x) = 1$, we see that

$$a(u,v) = \int_0^1 \frac{du}{dx}(x)\frac{dv}{dx}(x)dx$$

and

$$a(\phi_j, \phi_i) = ij\pi^2 \int_0^1 \cos(i\pi x)\cos(j\pi x)dx.$$

Therefore,

$$a(\phi_j, \phi_i) = \begin{cases} \frac{i^2\pi^2}{2}, & i = j, \\ 0, & i \neq j, \end{cases}$$

and the basis turns out to be orthogonal with respect to the energy inner product as well. We also have

$$(x, \phi_i) = \int_0^1 x\sin(i\pi x)dx = \frac{(-1)^{i+1}}{i\pi}.$$

These form the entries of the matrix $\mathbf{K}$ and the vector $\mathbf{f}$. Since $\mathbf{K}$ is diagonal, we can solve the system $\mathbf{Ku} = \mathbf{f}$ immediately to obtain

$$u_i = \frac{2(-1)^{i+1}}{i^3\pi^3}.$$

The resulting approximation, for $N = 20$, is displayed in Figure 5.9. The energy inner product, for a constant function $k(x)$, is just the L^2 inner product applied to the derivatives, and so the calculations are intimately related to those we perform in computing a Fourier series solution; indeed, the final result is the same! ∎

Example 5.17. Consider the BVP

$$-\frac{d}{dx}\left((1+x)\frac{du}{dx}\right) = x, \ 0 < x < 1,$$
$$u(0) = 0,$$
$$u(1) = 0.$$

(5.44)

The exact solution is

$$u(x) = \frac{x}{2} - \frac{x^2}{4} - \frac{\ln(1+x)}{4\ln 2}.$$

We will apply the Galerkin method with the same approximating subspace and basis as in the previous example. The calculations are similar, but the bilinear form changes to

$$a(u,v) = \int_0^1 (1+x)\frac{du}{dx}(x)\frac{dv}{dx}(x)\,dx,$$

and the basis functions are no longer orthogonal in the energy inner product.

We have

$$a(\phi_j,\phi_i) = ij\pi^2 \int_0^1 (1+x)\cos(i\pi x)\cos(j\pi x)\,dx$$

$$= \begin{cases} \frac{3i^2\pi^2}{4}, & i = j, \\ \frac{ij((-1)^{i+j}j^2+(-1)^{i+j}i^2-i^2-j^2)}{(i+j)^2(i-j)^2}, & i \neq j, \end{cases}$$

and

$$(x,\phi_i) = \int_0^1 x\sin(i\pi x)\,dx = \frac{(-1)^{i+1}}{i\pi}.$$

These form the entries of the matrix $\mathbf{K}$ and the vector $\mathbf{f}$. We cannot solve $\mathbf{Ku} = \mathbf{f}$ explicitly; that is, we cannot derive a useful formula for the coefficients $u_i, i = 1, 2, \ldots, N$, in the formula

$$v_N(x) = \sum_{i=1}^N u_i \sin(i\pi x).$$

However, we can solve for the unknowns numerically using Gaussian elimination, and thereby produce the approximation w from F_N. The result, for $N = 20$, is shown in Figure 5.9. ∎

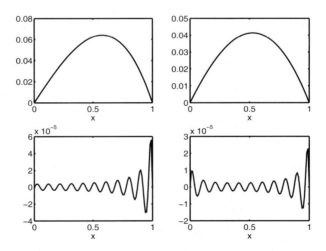

Figure 5.9. *Solving BVPs (5.43) (left) and (5.44) (right) using the Galerkin method. The solutions are graphed in the top figures and the errors in the solutions are shown in the bottom figures.*

5.5. The Galerkin method

The results in these two examples are of roughly the same quality, showing that the Galerkin method can be as effective as the Fourier method. The most significant difference in these two methods is the need to form the matrix **K** in the second example and solve the $N \times N$ system $\mathbf{Ku} = \mathbf{f}$. This is very time consuming compared to the computations required in the first example (where the system of linear equations is diagonal). This question of efficiency is particularly important when we have two or three spatial dimensions; in a problem of realistic size, it may take impossibly long to solve the resulting linear system if the coefficient matrix is *dense*. A dense matrix is a matrix in which most or all of the entries are nonzero. A *sparse matrix*, on the other hand, has mostly zero entries.

The finite element method is simply the Galerkin method with a special choice for the subspace and its basis; the basis leads to a sparse coefficient matrix. The ultimate sparse, nonsingular matrix is a diagonal matrix. Obtaining a diagonal matrix requires that the basis for the approximating subspace be chosen to be orthogonal with respect to the energy inner product. As mentioned earlier, it is too difficult, for a problem with variable coefficients, to find an orthogonal basis. The finite element method uses a basis in which *most* pairs of functions are orthogonal; the resulting matrix is not diagonal, but it is quite sparse.

Exercises

1. Determine whether the bilinear form

$$a(u,v) = \int_0^\ell \frac{du}{dx}(x) \frac{dv}{dx}(x) \, dx$$

   defines an inner product on each of the following subspaces of $C^2[0,\ell]$. If it does not, show why.

   (a) $\{v \in C^2[0,\ell] : v(\ell) = 0\}$;
   (b) $\{v \in C^2[0,\ell] : v(0) = v(\ell)\}$;
   (c) $C^2[0,\ell]$ (the entire space).

2. Repeat the previous exercise for the bilinear form

$$a(u,v) = \int_0^\ell \left\{ \frac{du}{dx}(x) \frac{dv}{dx}(x) + u(x)v(x) \right\} dx.$$

3. Show that if F_N is the subspace defined in Example 5.16, and the Galerkin method is applied to the weak form of

$$-k \frac{d^2 u}{dx^2} = f(x), \ 0 < x < \ell,$$
$$u(0) = 0,$$
$$u(\ell) = 0,$$

   with F_N as the approximating subspace, the result will always be the partial Fourier sine series (with N terms) of the exact solution u.

4. Does the result of the previous exercise still hold if the differential equation is changed to
$$-k\frac{d^2u}{dx^2} + pu = f(x),\ 0 < x < \ell?$$

5. Define S to be the set of all polynomials of the form $ax + bx^2$, considered as functions defined on the interval $[0,1]$.

    (a) Explain why S is a subspace of $C^2[0,1]$.

    (b) Explain why the bilinear form
    $$a(u,v) = \int_0^1 \frac{du}{dx}(x)\frac{dv}{dx}(x)\,dx$$
    defines an inner product on S.

    (c) Let $f(x) = e^x$. Compute the best approximation from S, in the energy norm, to f.

    (d) Explain why the best approximation from $\mathcal{P}_2$, in the energy norm, to $f(x) = e^x$ is not unique. (The subspace $\mathcal{P}_2$ is defined in Exercise 3.1.5(d).)

6. Repeat Exercise 5 with
$$S = \{v \in C^2[0,1]\ :\ v(x) = ax + bx^2 + cx^3 \text{ for some } a,b,c \in \mathbf{R}\}.$$

7. Define
$$V_2 = \text{span}\left\{x(1-x), x\left(\frac{1}{2}-x\right)(1-x)\right\}$$
and regard V_2 as a subspace of $C_D^2[0,1]$. Apply the Galerkin method, using V_2 as the approximating subspace, to estimate the solution of
$$-\frac{d}{dx}\left((1+x)\frac{du}{dx}\right) = x,\ 0 < x < 1,$$
$$u(0) = 0,$$
$$u(1) = 0.$$
Find the exact solution and graph the exact and approximate solutions together.

8. Repeat Exercise 7 using the subspace
$$V_3 = \text{span}\left\{x(1-x), x\left(\frac{1}{2}-x\right)(1-x), x\left(\frac{1}{3}-x\right)\left(\frac{2}{3}-x\right)(1-x)\right\}.$$

9. The standard Gaussian elimination algorithm for solving $\mathbf{Ax} = \mathbf{b}$ requires
$$O\left(\frac{2}{3}n^3\right)$$
arithmetic operations when $\mathbf{A} \in \mathbf{R}^{n \times n}$. (The exact number is a cubic polynomial in n, and $2n^3/3$ is the leading term. When n is large, the lower degree terms are negligible in comparison to the leading term.)

(a) Suppose that, on a certain computer, it takes 1 s (second) to solve a 100×100 dense linear system by Gaussian elimination. How long will it take to solve a 1000×1000 system? A 10000×10000 system?

(b) Now suppose $A \in \mathbf{R}^{n \times n}$ is *tridiagonal*; that is, suppose $A_{ij} = 0$ whenever $|i - j| > 1$. About how many arithmetic operations does it take to solve $Ax = b$ by Gaussian elimination, assuming that the algorithm takes advantage of the fact that most of the entries of A are already zero? (The reader should work out a small example by hand and count the number of operations taken at each step. A short calculation gives this operation count as a function of n.)

(c) Finally, suppose that, on a certain computer, it takes 0.01 s to solve a 100×100 tridiagonal linear system by Gaussian elimination. How long will it take to solve a 1000×1000 system? A 10000×10000 system?

5.6 Piecewise polynomials and the finite element method

To apply the Galerkin method to solve a BVP accurately and efficiently, we must choose a subspace V_n of $C^2[0, \ell]$ and a basis for V_n with the following properties:

1. The stiffness matrix **K** and the load vector **f** can be assembled efficiently. This means that the basis for V_n should consist of functions that are easy to manipulate, in particular, differentiate and integrate.

2. The basis for V_n should be as close to orthogonal as possible so that **K** will be sparse. Although we do not expect every pair of basis functions to be orthogonal (which would lead to a diagonal stiffness matrix), we want as many pairs as possible to be orthogonal so that **K** will be as close to diagonal as possible.

3. The true solution u of the BVP should be well approximated from subspace V_n, with the approximation becoming arbitrarily good as $n \to \infty$.

We will continue to use the BVP

$$-\frac{d}{dx}\left(k(x)\frac{du}{dx}\right) = f(x), \; 0 < x < \ell,$$
$$u(0) = 0, \quad (5.45)$$
$$u(\ell) = 0$$

as our model problem.

Finite element methods use subspaces of piecewise polynomials; for simplicity, we will concentrate on piecewise linear functions. To define a space of piecewise linear functions, we begin by creating a *mesh* (or *grid*) on the interval $[0, \ell]$:

$$0 = x_0 < x_1 < \cdots < x_n = \ell.$$

The points $x_0, x_1, \ldots, x_n$ are called the *nodes* of the mesh. Often we choose a regular mesh, with $x_i = ih$, $h = \ell/n$. A function $p : [0, \ell] \to \mathbf{R}$ is *piecewise linear* (relative to the given mesh) if, for each $i = 1, 2, \ldots, n$, there exist constants a_i, b_i, with

$$p(x) = a_i x + b_i \text{ for all } x \in (x_{i-1}, x_i).$$

See Figure 5.10 for an example of a continuous piecewise linear function.

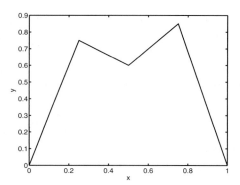

Figure 5.10. *A piecewise linear function relative to the mesh* $x_0 = 0.0$, $x_1 = 0.25$, $x_2 = 0.5$, $x_3 = 0.75$, $x_4 = 1.0$.

We define, for a fixed mesh on $[0, \ell]$,

$$S_n = \{p : [0, \ell] \to \mathbf{R} \ : \ p \text{ is continuous and piecewise linear}, p(0) = p(\ell) = 0\}. \quad (5.46)$$

We now argue that the subspace S_n satisfies the three requirements for an approximating subspace that are listed above. Two of the properties are almost obvious.

1. Since the functions belonging to S_n are piecewise polynomials, they are easy to manipulate—it is easy to differentiate or integrate a polynomial.

3. Smooth functions can be well approximated by piecewise linear functions. Indeed, in the days before handheld calculators, the elementary transcendental functions like $\sin(x)$ or e^x were given in tables. Only a finite number of function values could be listed in a table; the user was expected to use piecewise linear interpolation to estimate values not listed in the table.

The fact that smooth functions can be well approximated by piecewise linear functions can also be illustrated in a graph. Figure 5.11 shows a smooth function on $[0, 1]$ with two piecewise linear approximations, the first corresponding to $n = 10$ and the second to $n = 20$. Clearly the approximation can be made arbitrarily good by choosing n large enough.

In order to discuss property 2, that the stiffness matrix should be sparse, we must first define a basis for S_n. The idea of piecewise linear interpolation suggests the natural basis: a piecewise linear function is completely determined by its *nodal values* (that is, the values of the function at the nodes of the mesh). For $n = 1, 2, \ldots, n-1$, define $\phi_i \in S_n$ to be that piecewise linear function satisfying

$$\phi_i(x_j) = \begin{cases} 1, & j = i, \\ 0, & j \neq i. \end{cases}$$

Then each $p \in S_n$ satisfies

$$p(x) = \sum_{i=1}^{n-1} p(x_i) \phi_i(x). \quad (5.47)$$

5.6. Piecewise polynomials and the finite element method

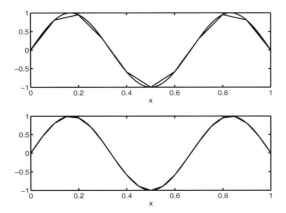

Figure 5.11. *A smooth function and two piecewise linear approximations. The top approximation is based on* 11 *nodes, while the bottom is based on* 21 *nodes.*

To prove this, we merely need to show that p and

$$\sum_{i=1}^{n-1} p(x_i)\phi_i \qquad (5.48)$$

have the same nodal values, since two continuous piecewise linear functions are equal if and only if they have the same nodal values. Therefore, we substitute $x = x_j$ into (5.48):

$$\sum_{i=1}^{n-1} p(x_i)\phi_i(x_j) = p(x_1)\cdot 0 + \cdots + p(x_j)\cdot 1 + \cdots + p(x_{n-1})\cdot 0 = p(x_j).$$

This shows that (5.47) holds.

Thus every $p \in S_n$ can be written as a linear combination of $\{\phi_1, \phi_2, \ldots, \phi_{n-1}\}$; that is, this set spans S_n. To show that $\{\phi_1, \phi_2, \ldots, \phi_{n-1}\}$ is linearly independent, suppose

$$\sum_{i=1}^{n-1} c_i \phi_i = 0.$$

Then, in particular,

$$\sum_{i=1}^{n-1} c_i \phi_i(x_j) = c_j = 0, \ j = 1, 2, \ldots, n-1.$$

This implies that all of the coefficients $c_1, c_2, \ldots, c_{n-1}$ are zero, so $\{\phi_1, \phi_2, \ldots, \phi_{n-1}\}$ is a linearly independent set. A typical basis function $\phi_i(x)$ is displayed in Figure 5.12.

The entries of the stiffness matrix are

$$K_{ij} = a(\phi_j, \phi_i), \ i, j = 1, 2, \ldots, n-1.$$

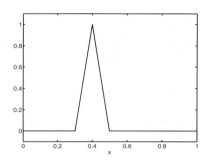

Figure 5.12. *A typical basis function for S_n.*

Here is the fundamental observation concerning the finite element method: *Since each ϕ_i is zero on most of the interval $[0,\ell]$, most of the inner products $a(\phi_i,\phi_j)$ are zero because the product*

$$k(x)\frac{d\phi_i}{dx}(x)\frac{d\phi_j}{dx}(x)$$

turns out to be zero for all $x \in [0,\ell]$.

To be specific, ϕ_i is zero except on the interval $[x_{i-1}, x_{i+1}]$. (We express this concisely by saying that the *support*[27] of ϕ_i is the interval $[x_{i-1}, x_{i+1}]$.) From this we see that

$$a(\phi_{i-1},\phi_i),\ a(\phi_i,\phi_i),\ a(\phi_{i+1},\phi_i)$$

can be nonzero, but

$$|j-i| > 1 \Rightarrow a(\phi_j,\phi_i) = 0,$$

since in that case, for any $x \in [0,\ell]$, either

$$\frac{d\phi_i}{dx}(x) \text{ or } \frac{d\phi_j}{dx}(x)$$

is zero. Therefore, the matrix **K** turns out to be *tridiagonal*, that is, all of its entries are zero except (possibly) those on three diagonals (see Figure 5.13).

We now see that the subspace S_n of continuous piecewise linear functions satisfies all three requirements for a good approximating subspace. However, the attentive reader may have been troubled by one apparent shortcoming of S_n: it is not a subset of $C_D^2[0,\ell]$, because most of the functions in S_n are not continuously differentiable, much less twice-continuously differentiable. This would seem to invalidate the entire context on which the Galerkin method depends. In fact, however, this presents no difficulty—the important fact is that equations defining the weak form are well defined, that is, that we can compute

$$\int_0^\ell k(x)\frac{d\phi_j}{dx}(x)\frac{d\phi_i}{dx}(x)dx \text{ and } \int_0^\ell f(x)\phi_i(x)dx.$$

---

[27] The support of a continuous function is the closure of the set on which the function is nonzero, that is, the set on which the function is nonzero, together with its boundary. In the case of ϕ_i, the function is nonzero on (x_{i-1},x_{i+1}). The closure of this open interval is the closed interval.

5.6. Piecewise polynomials and the finite element method

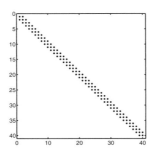

Figure 5.13. *A schematic view of a tridiagonal matrix; the only nonzeros are on the main diagonal and the first sub- and superdiagonal.*

Since ϕ_i is piecewise linear, its derivative $d\phi_i/dx$ is piecewise constant and hence integrable.

The key to a mathematically consistent treatment of finite element methods is to use a larger space of functions, one that includes both $C_D^2[0,\ell]$ and S_n as subspaces. This will show that the basic property of the Galerkin method still holds when we use S_n as the approximating subspace: the Galerkin method produces the best approximation from S_n, in the energy norm, to the true solution u. We sketch this theory in Section 13.3. For a more complete exposition of the theory, we refer the reader to more advanced texts, for example, Brenner and Scott [8].

5.6.1 Examples using piecewise linear finite elements

Let us apply the Galerkin method, with the subspace S_n defined in the previous section, to (5.45). As we showed in Section 5.5, the Galerkin approximation is

$$v_n(x) = \sum_{i=1}^{n-1} u_i \phi_i(x),$$

where the coefficients $u_1, u_2, \ldots, u_{n-1}$ satisfy

$$\mathbf{Ku} = \mathbf{f}, \ K_{ij} = a(\phi_j, \phi_i), \ f_i = (f, \phi_i).$$

Example 5.18. We apply the finite element method to the BVP

$$\begin{aligned} -\frac{d^2 u}{dx^2} &= e^x, \ 0 < x < 1, \\ u(0) &= 0, \\ u(1) &= 0. \end{aligned} \quad (5.49)$$

The exact solution is $u(x) = -e^x + (e-1)x + 1$. We use a regular mesh with n subintervals and $h = 1/n$. We then have

$$\phi_i(x) = \begin{cases} \frac{1}{h}(x - (i-1)h), & x_{i-1} < x < x_i, \\ -\frac{1}{h}(x - (i+1)h), & x_i < x < x_{i+1}, \\ 0 & \text{otherwise} \end{cases}$$

and

$$\frac{d\phi_i}{dx}(x) = \begin{cases} \frac{1}{h}, & x_{i-1} < x < x_i, \\ -\frac{1}{h}, & x_i < x < x_{i+1}, \\ 0 & \text{otherwise.} \end{cases}$$

Therefore,

$$K_{ii} = a(\phi_i, \phi_i) = \int_{(i-1)h}^{(i+1)h} \frac{dx}{h^2} = \frac{2}{h}$$

and

$$K_{i,i+1} = a(\phi_{i+1}, \phi_i) = \int_{ih}^{(i+1)h} \frac{1}{h}\left(-\frac{1}{h}\right) dx = -\frac{1}{h}.$$

(These calculations are critical, and the reader should make sure he or she thoroughly understands them. The basis function ϕ_i is nonzero only on the interval $[(i-1)h, (i+1)h]$, so the interval of integration reduces to this subinterval. The square of the derivative of ϕ_i is $1/h^2$ on this entire subinterval. This gives the result for K_{ii}. As for the computation of $K_{i,i+1}$, the only part of $[0,1]$ on which both ϕ_i and ϕ_{i+1} are nonzero is $[ih, (i+1)h]$. On this subinterval, the derivative of ϕ_i is $-1/h$ and the derivative of ϕ_{i+1} is $1/h$.)

This gives us the entries on the main diagonal and first superdiagonal of **K**; since **K** is symmetric and tridiagonal, we can deduce the rest of the entries. We also have

$$(f, \phi_i) = \int_{(i-1)h}^{ih} \left(\frac{1}{h}(x - (i-1)h)\right) e^x \, dx + \int_{ih}^{(i+1)h} \left(-\frac{1}{h}(x - (i+1)h)\right) e^x \, dx$$

$$= \frac{e^{ih}}{h}\left(e^h + e^{-h} - 2\right).$$

It remains only to assemble the tridiagonal matrix **K** and the right-hand-side vector **f**, and solve **Ku** = **f** using Gaussian elimination.

For example, for $n = 5$, we have $\mathbf{K} \in \mathbf{R}^{4 \times 4}$,

$$\mathbf{K} = \begin{bmatrix} 10 & -5 & 0 & 0 \\ -5 & 10 & -5 & 0 \\ 0 & -5 & 10 & -5 \\ 0 & 0 & -5 & 10 \end{bmatrix},$$

$\mathbf{f} \in \mathbf{R}^4$,

$$\mathbf{f} \doteq \begin{bmatrix} 0.2451 \\ 0.2994 \\ 0.3656 \\ 0.4466 \end{bmatrix},$$

and

$$\mathbf{u} \doteq \begin{bmatrix} 0.1223 \\ 0.1955 \\ 0.2089 \\ 0.1491 \end{bmatrix}.$$

5.6. Piecewise polynomials and the finite element method

The corresponding piecewise linear approximation, together with the exact solution, is displayed in Figure 5.14. ∎

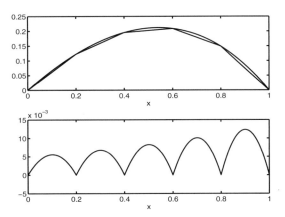

Figure 5.14. *Top: Exact solution and piecewise linear finite element approximation for $-\frac{d^2u}{dx^2} = e^x, u(0) = u(1) = 0$. Bottom: Error in piecewise linear approximation.*

Example 5.19. We now find an approximate solution to the BVP

$$-\frac{d}{dx}\left((1+x)\frac{du}{dx}(x)\right) = 1, \; 0 < x < 1,$$
$$u(0) = 0, \quad (5.50)$$
$$u(1) = 0.$$

The exact solution is $u(x) = \ln(1+x)/\ln 2 - x$. We again use a regular mesh with n subintervals and $h = 1/n$.

The bilinear form $a(\cdot, \cdot)$ now takes the form

$$a(u,v) = \int_0^1 (1+x)\frac{du}{dx}(x)\frac{dv}{dx}(x)\,dx.$$

We obtain

$$a(\phi_i, \phi_i) = \int_{(i-1)h}^{(i+1)h} \left((1+x)\frac{1}{h^2}\right) dx = \frac{2(ih+1)}{h}$$

and

$$a(\phi_{i+1}, \phi_i) = \int_{ih}^{(i+1)h} \left((1+x)\frac{1}{h}\left(-\frac{1}{h}\right)\right) dx = -\frac{h+2ih+2}{2h}.$$

As in the previous example, these calculations suffice to determine **K**. We also have

$$(f, \phi_i) = \int_{(i-1)h}^{ih} \left(\frac{1}{h}(x-(i-1)h)\right) dx + \int_{ih}^{(i+1)h} \left(-\frac{1}{h}(x-(i+1)h)\right) dx$$
$$= h.$$

It remains only to assemble the tridiagonal matrix **K** and the right-hand-side vector **f**, and solve **Ku** = **f** using Gaussian elimination.

For example, if $n = 5$, the matrix **K** is 4×4,

$$\mathbf{K} = \begin{bmatrix} 12 & -\frac{13}{2} & 0 & 0 \\ -\frac{13}{2} & 14 & -\frac{15}{2} & 0 \\ 0 & -\frac{15}{2} & 16 & -\frac{17}{2} \\ 0 & 0 & -\frac{17}{2} & 18 \end{bmatrix},$$

and **f** is a 4-vector,

$$\mathbf{f} = \begin{bmatrix} 1/5 \\ 1/5 \\ 1/5 \\ 1/5 \end{bmatrix}.$$

The resulting **u** is

$$\mathbf{u} \doteq \begin{bmatrix} 0.0628 \\ 0.0851 \\ 0.0778 \\ 0.0479 \end{bmatrix}.$$

The piecewise linear approximation, together with the exact solution, is shown in Figure 5.15. ∎

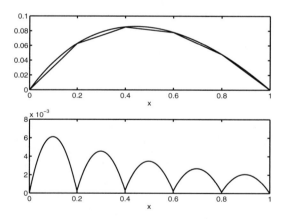

Figure 5.15. *Piecewise linear finite element approximation to the solution of* $-\frac{d}{dx}\left((x+1)\frac{du}{dx}\right) = 1, u(0) = u(1) = 0.$

The following are fundamental issues for the finite element method:

- How fast does the approximate solution converge to the true solution as $n \to \infty$?

- How can the stiffness matrix **K** and the load vector **f** be computed efficiently and automatically on a computer?

5.6. Piecewise polynomials and the finite element method

- How can the sparse system $\mathbf{Ku} = \mathbf{f}$ be solved efficiently? (This is primarily an issue when the problem involves two or three spatial dimensions, since then $\mathbf{K}$ can be very large.)

Another issue that will be important for higher-dimensional problems is the description of the mesh—special data structures are needed to allow efficient computation. A careful treatment of these questions is beyond the scope of this book, but we will provide some answers in Chapter 13.

5.6.2 Inhomogeneous Dirichlet conditions

The inhomogeneous Dirichlet problem,

$$-\frac{d}{dx}\left(k(x)\frac{du}{dx}\right) = f(x), \; 0 < x < \ell,$$
$$u(0) = \alpha,$$
$$u(\ell) = \beta$$
(5.51)

can be solved using the method of "shifting the data" introduced earlier. In this method, we find a function $g(x)$ satisfying the boundary conditions and define a new BVP for the unknown $w(x) = u(x) - g(x)$. We can apply this idea to the weak form of the BVP.

A calculation similar to that of Section 5.4.1 shows that the weak form of (5.51) is

$$a(u,v) = (f,v) \text{ for all } v \in V.$$

Here, however, we do not look for the solution u in V, but rather in $g + V = \{g + v : v \in V\}$. We now write $u = g + w$, and the problem becomes

$$\text{find } w \in V \text{ such that } a(w+g,v) = (f,v) \text{ for all } v \in V. \tag{5.52}$$

But $a(w+g,v) = a(w,v) + a(g,v)$, so we can write (5.52) as

$$\text{find } w \in V \text{ such that } a(w,v) = (f,v) - a(g,v) \text{ for all } v \in V. \tag{5.53}$$

It turns out that solving (5.53) is not much harder than solving the weak formulation of the homogeneous Dirichlet problem; the only difference is that we modify the right-hand-side vector $\mathbf{f}$, as we explain below.

Shifting the data is easy because we can delay the choice of the function g until we apply the Galerkin method; it is always simple to find a finite element function g_n which satisfies the boundary conditions on the boundary nodes (actually, for a one-dimensional problem like this, it is trivial to find such a function g in any case; however, this becomes difficult in higher dimensions, while finding a finite element function to satisfy the boundary conditions, at least approximately, is still easy in higher dimensions).

The Galerkin problem is

$$\text{find } w_n \in S_n \text{ such that } a(w_n,v) = (f,v) - a(g_n,v) \text{ for all } v \in S_n. \tag{5.54}$$

We choose g_n to be piecewise linear and satisfy $g_n(x_0) = \alpha$, $g_n(x_n) = \beta$; the simplest such function has value zero at the other nodes of the mesh, namely $g_n = \alpha\phi_0 + \beta\phi_n$. Substituting

$w_n = \sum_{i=1}^{n-1} \alpha_i \phi_i$ into (5.54), we obtain, as before, the linear system $\mathbf{Ku} = \mathbf{f}$. The matrix $\mathbf{K}$ does not change, but $\mathbf{f}$ becomes

$$f_i = (f, \phi_i) - a(g_n, \phi_i)$$
$$= \begin{cases} (f, \phi_i), & i = 2, 3, \ldots, n-2 \text{ (since } a(g_n, \phi_i) = 0), \\ (f, \phi_1) - \alpha a(\phi_0, \phi_1), & i = 1, \\ (f, \phi_{n-1}) - \beta a(\phi_n, \phi_{n-1}), & i = n-1. \end{cases} \quad (5.55)$$

Example 5.20. We now solve an inhomogeneous version of Example 5.19; specifically, we apply the finite element method to

$$-\frac{d}{dx}\left((1+x)\frac{du}{dx}(x)\right) = 1, \ 0 < x < 1,$$
$$u(0) = 2, \quad (5.56)$$
$$u(1) = 1.$$

As we have just explained, the approximate solution is given by

$$v_n(x) = \sum_{i=1}^{n-1} u_i \phi_i(x),$$

where the coefficients $u_1, u_2, \ldots, u_{n-1}$ satisfy $\mathbf{Ku} = \mathbf{f}$. The stiffness matrix $\mathbf{K}$ is identical to the one calculated in Example 5.19, while the load vector $\mathbf{f}$ is modified as indicated in (5.55). We obtain

$$f_i = (f, \phi_i) - a(g_n, \phi_i)$$
$$= \begin{cases} h, & i = 2, 3, \ldots, n-2 \text{ (since } a(g_n, \phi_i) = 0), \\ h - 2a(\phi_0, \phi_1), & i = 1, \\ h - a(\phi_n, \phi_{n-1}), & i = n-1. \end{cases}$$

Now,

$$a(\phi_0, \phi_1) = \int_0^h \left((1+x)\frac{1}{h}\left(-\frac{1}{h}\right)\right) dx = -\frac{h+2}{2h},$$
$$a(\phi_n, \phi_{n-1}) = \int_{1-h}^1 \left((1+x)\frac{1}{h}\left(-\frac{1}{h}\right)\right) dx = \frac{h-4}{2h},$$

so

$$f_1 = h + 2\frac{h+2}{2h}, \ f_{n-1} = h - \frac{h-4}{2h}.$$

With $n = 5$, we obtain

$$\mathbf{f} = \begin{bmatrix} 56/5 \\ 1/5 \\ 1/5 \\ 97/10 \end{bmatrix}.$$

5.6. Piecewise polynomials and the finite element method

The results are shown in Figure 5.16. In this case, the computed solution equals the exact solution (the error shown in Figure 5.16 is only due to roundoff error in the computer calculations); see Exercise 5.6.4. ∎

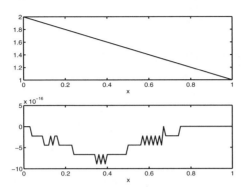

Figure 5.16. *Piecewise linear finite element approximation, with $n = 5$ subintervals in the mesh, to the solution of $-\frac{d}{dx}\left((x+1)\frac{du}{dx}\right) = 1, u(0) = 2, u(1) = 1$. The error shown in the bottom graph is due only to roundoff error—the "approximation" is in fact equal to the exact solution.*

Exercises

1. Using a uniform mesh with four elements, compute the piecewise linear finite element approximation to the solution of

$$-\frac{d^2u}{dx^2} = x, \; 0 < x < 1,$$
$$u(0) = 0,$$
$$u(\ell) = 0.$$

   Compute the exact solution by integration and compare with the approximate solution.

2. Using a uniform mesh with four elements, compute the piecewise linear finite element approximation to the solution of

$$-\frac{d}{dx}\left(e^x \frac{du}{dx}\right) = x, \; 0 < x < 1,$$
$$u(0) = 0,$$
$$u(\ell) = 0.$$

   Compute the exact solution by integration and compare with the approximate solution.

3. (a) Using direct integration, find the exact solution to the BVP from Example 5.19.
   (b) Repeat the calculations from Example 5.19, using an increasing sequence of values of n. Produce graphs similar to Figure 5.15, or otherwise measure the

errors in the approximation. Try to identify the size of the error as a function of h.[28]

4. (a) Compute the exact solution to the BVP from Example 5.20.

   (b) Explain why, for this particular BVP, the finite element method computes the exact, rather than an approximate, solution.

5. Use piecewise linear finite elements and a regular mesh to solve the following problem:
$$-\frac{d^2u}{dx^2} = x^2,$$
$$u(0) = 0,$$
$$u(1) = 0.$$

   Use $n = 10, 20, 40, \ldots$ elements, and determine (experimentally) an exponent p so that the error is
$$O\left(\frac{1}{n^p}\right).$$

   Here the error is defined to be the maximum absolute difference between the exact and approximate solutions. (Determine the exact solution by integration; you will need it to compute the error in the approximations.)

6. Repeat Exercise 5 for the BVP
$$-\frac{d}{dx}\left((1+x^2)\frac{du}{dx}\right) = 1, \ 0 < x < 1,$$
$$u(0) = 0,$$
$$u(1) = 0.$$

   The exact solution is
$$u(x) = \frac{2\ln 2}{\pi}\tan^{-1}(x) - \frac{1}{2}\ln(1+x^2).$$

7. Derive the weak form of the BVP
$$-\frac{d}{dx}\left(k(x)\frac{du}{dx}\right) + p(x)u = f(x), \ 0 < x < \ell,$$
$$u(0) = 0,$$
$$u(\ell) = 0,$$

   where k and p satisfy $k(x) > 0$, $p(x) > 0$ for $x \in [0, \ell]$. What is the bilinear form for this BVP?

---

[28] Performing these calculations for n much larger than 5 will require the use of a computer. One could, of course, write a program, in a language such as Fortran or C, to do the calculations. However, there exist powerful interactive software programs that integrate numerical calculations, programming, and graphics, and these are much more convenient to use. MATLAB is particularly suitable for the finite element calculations needed for this book, since it is designed to facilitate matrix computations. *Mathematica* and *Maple* are other possibilities.

5.6. Piecewise polynomials and the finite element method

8. Using a uniform mesh with four elements, compute the piecewise linear finite element approximation to the solution of

$$-\frac{d^2u}{dx^2} + u = x, \ 0 < x < 1,$$
$$u(0) = 0,$$
$$u(\ell) = 0.$$

The results of the previous exercise will be required. Compute the exact solution and compare with the approximate solution.

9. Repeat Exercise 5 for the BVP

$$-\frac{d^2u}{dx^2} + 2u = \frac{1}{2} - x,$$
$$u(0) = 0,$$
$$u(1) = 0.$$

The results of Exercise 7 will be required. The exact solution is

$$u(x) = c_1 e^{\sqrt{2}x} + c_2 e^{-\sqrt{2}x} + \frac{1}{4} - \frac{x}{2},$$

where

$$c_1 = -\frac{e^{-\sqrt{2}} + 1}{4\left(e^{-\sqrt{2}} - e^{\sqrt{2}}\right)},$$

$$c_2 = \frac{e^{\sqrt{2}} + 1}{4\left(e^{-\sqrt{2}} - e^{\sqrt{2}}\right)}.$$

10. Suppose we are given a finite element mesh on $[0, \ell]$ defined by the nodes $0 = x_0 < x_1 < x_2 < \cdots < x_n = \ell$. The continuous piecewise linear interpolant (relative to the given mesh) of a continuous function $u : [0, \ell] \to \mathbf{R}$ is the continuous piecewise linear function v that agrees with u at the nodes; that is, v is defined by $v(x_i) = u(x_i)$, $i = 0, 1, 2, \ldots, n$. Suppose u is twice-continuously differentiable and satisfies

$$-k\frac{d^2u}{dx^2} = f(x), \ 0 < x < \ell,$$
$$u(0) = 0,$$
$$u(\ell) = 0.$$

Then the Galerkin finite element method defines another continuous piecewise linear approximation of u, say w. Prove that $w = v$. (This result holds only for this simple BVP with constant coefficients; it is not a general result for finite element solutions.) (Hint: First derive the system of equations that the nodal values of w must satisfy. Then, using the fact that $f = -kd^2u/dx^2$, show that the nodal values of u satisfy this system of equations by manipulating $\int_0^\ell f(x)\phi_i(x)\,dx$.)

5.7 Suggestions for further reading

We have now introduced the two primary topics of this book, Fourier series and finite elements. Fourier analysis has been important in applied mathematics for 200 years, and there are many books on the subject. A classic reference to the theory of Fourier series is Zygmund [72]; more modern treatments include Folland [20] and Kammler [36]. A classic introductory textbook on Fourier series methods for BVPs is Churchill and Brown [12], which has been in print for over 60 years! Many, more modern, textbooks cover similar material; two are Bick [5] and Hanna and Rowland [32].

One can also look to introductory books on PDEs for details about Fourier series, in addition to other analytic and numeric techniques. Strauss [62] is a well-written text that concisely surveys analytic methods. It contains a range of topics, from elementary to relatively advanced. Another introduction to PDEs that covers Fourier series methods, as well as many other topics, is Haberman [30].

The finite element method is much more recent, having been popularized in the 1960s. The author has written an introductory text on Galerkin finite element methods for stationary problems [23]. Most textbooks on the subject are written at a more demanding level. Brenner and Scott [8] is a careful introduction to the theory of finite elements for steady-state problems, while Johnson [35] and Strang and Fix [61] treat both steady-state and time-dependent equations. Ciarlet [13] is a careful treatment, written at a demanding level, of the theory of finite elements for steady-state problems. There is also the six-volume *Texas Finite Element Series* by Becker, Carey, and Oden [4], which starts at an elementary level.

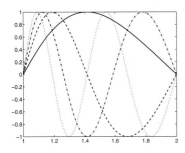

Chapter 6
Heat Flow and Diffusion

We now turn our attention to time-dependent problems, still restricting ourselves to problems in one spatial dimension. Since both time (t) and space (x) are independent variables, the differential equations will be PDEs, and we will need initial conditions as well as boundary conditions.

We begin with the heat equation,

$$\rho c \frac{\partial u}{\partial t} - \kappa \frac{\partial^2 u}{\partial x^2} = f(x,t),\ 0 < x < \ell,\ t > t_0.$$

This PDE models the temperature distribution $u(x,t)$ in a bar (see Section 2.1) and a similar equation models the concentration $u(x,t)$ of a chemical in solution (see Section 2.1.3). Since the first time derivative of the unknown appears in the equation, we need a single initial condition.

Our first model problem will be the following initial-boundary value problem (IBVP):

$$\begin{aligned}
\rho c \frac{\partial u}{\partial t} - \kappa \frac{\partial^2 u}{\partial x^2} &= f(x,t),\ 0 < x < \ell,\ t > t_0, \\
u(x,t_0) &= \psi(x),\ 0 < x < \ell, \\
u(0,t) &= 0,\ t > t_0, \\
u(\ell,t) &= 0,\ t > t_0.
\end{aligned} \quad (6.1)$$

We apply the Fourier series method first and then turn to the finite element method. Along the way we will introduce new combinations of boundary conditions.

6.1 Fourier series methods for the heat equation

We can solve (6.1) using Fourier series. Since the unknown $u(x,t)$ is a function of both time and space, it can be represented as a Fourier sine series in which the Fourier coefficients depend on t. In other words, for each t, we have a Fourier sine series representation of

$u(x,t)$ (regarded as a function of x). The series takes the form

$$u(x,t) = \sum_{n=1}^{\infty} a_n(t)\sin\left(\frac{n\pi x}{\ell}\right), \quad a_n(t) = \frac{2}{\ell}\int_0^{\ell} u(x,t)\sin\left(\frac{n\pi x}{\ell}\right)dx.$$

The function u will then automatically satisfy the boundary conditions in (6.1). We must choose the coefficients $a_n(t)$ so that the PDE and the initial condition are satisfied.

We represent the right-hand side $f(x,t)$ in the same way:

$$f(x,t) = \sum_{n=1}^{\infty} c_n(t)\sin\left(\frac{n\pi x}{\ell}\right), \quad c_n(t) = \frac{2}{\ell}\int_0^{\ell} f(x,t)\sin\left(\frac{n\pi x}{\ell}\right)dx.$$

We then express the left-hand side of the PDE as a Fourier series, in which the Fourier coefficients are all expressed in terms of the unknowns $a_1(t), a_2(t), \ldots$. To do this, we have to compute the Fourier coefficients of the functions

$$\rho c \frac{\partial u}{\partial t}(x,t)$$

and

$$-\kappa \frac{\partial^2 u}{\partial x^2}(x,t).$$

Just as with the function u itself, we are using the Fourier series to represent the spatial variation (i.e., the dependence on x) of these functions. Since they also depend on t, the resulting Fourier coefficients will be functions of t.

To compute the Fourier coefficients of

$$-\kappa \frac{\partial^2 u}{\partial x^2}(x,t),$$

we integrate by parts twice, just as we did in the previous chapter, and use the boundary conditions to eliminate the boundary terms:

$$-\frac{2\kappa}{\ell}\int_0^{\ell} \frac{\partial^2 u}{\partial x^2}(x,t)\sin\left(\frac{n\pi x}{\ell}\right)dx$$

$$= -\frac{2\kappa}{\ell}\left(\left[\frac{\partial u}{\partial x}(x,t)\sin\left(\frac{n\pi x}{\ell}\right)\right]_0^{\ell} - \frac{n\pi}{\ell}\int_0^{\ell}\frac{\partial u}{\partial x}(x,t)\cos\left(\frac{n\pi x}{\ell}\right)dx\right)$$

$$= \frac{2\kappa n\pi}{\ell^2}\int_0^{\ell}\frac{\partial u}{\partial x}(x,t)\cos\left(\frac{n\pi x}{\ell}\right)dx$$

$$= \frac{2\kappa n\pi}{\ell^2}\left(\left[u(x,t)\cos\left(\frac{n\pi x}{\ell}\right)\right]_0^{\ell} + \frac{n\pi}{\ell}\int_0^{\ell} u(x,t)\sin\left(\frac{n\pi x}{\ell}\right)dx\right)$$

$$= \frac{2\kappa n^2\pi^2}{\ell^3}\int_0^{\ell} u(x,t)\sin\left(\frac{n\pi x}{\ell}\right)dx$$

$$= \frac{\kappa n^2\pi^2}{\ell^2}a_n(t).$$

6.1. Fourier series methods for the heat equation

We used the conditions $\sin(0) = \sin(n\pi) = 0$ and $u(0,t) = u(\ell,t) = 0$ in canceling the boundary terms in the above calculation.

We now compute the Fourier sine coefficient of $\rho c \partial u/\partial t$. Using Theorem 2.1, we have

$$\frac{2}{\ell}\int_0^\ell \frac{\partial u}{\partial t}(x,t)\sin\left(\frac{n\pi x}{\ell}\right)dx = \frac{d}{dt}\left[\frac{2}{\ell}\int_0^\ell u(x,t)\sin\left(\frac{n\pi x}{\ell}\right)dx\right] = \frac{d}{dt}[a_n(t)]$$
$$= \frac{da_n}{dt}(t).$$

Thus the nth Fourier coefficient of $\rho c \partial u/\partial t$ is

$$\rho c \frac{da_n}{dt}(t).$$

We now see that

$$\rho c \frac{\partial u}{\partial t}(x,t) - \kappa \frac{\partial^2 u}{\partial x^2}(x,t) = \sum_{n=1}^\infty \left(\rho c \frac{da_n}{dt}(t) + \frac{\kappa n^2 \pi^2}{\ell^2} a_n(t)\right)\sin\left(\frac{n\pi x}{\ell}\right).$$

Since

$$f(x,t) = \sum_{n=1}^\infty c_n(t)\sin\left(\frac{n\pi x}{\ell}\right),$$

the PDE implies

$$\rho c \frac{da_n}{dt}(t) + \frac{\kappa n^2 \pi^2}{\ell^2} a_n(t) = c_n(t),\ n = 1,2,3,\ldots.$$

This is a sequence of ODEs for the coefficients $a_1(t), a_2(t), a_3(t), \ldots$. Moreover, we have

$$u(x,t_0) = \psi(x),\ 0 < x < \ell$$
$$\Rightarrow \sum_{n=1}^\infty a_n(t_0)\sin\left(\frac{n\pi x}{\ell}\right) = \sum_{n=1}^\infty b_n \sin\left(\frac{n\pi x}{\ell}\right),$$

where $b_1, b_2, \ldots$ are the Fourier sine coefficients of $\psi(x)$. Therefore, we obtain

$$a_n(t_0) = b_n,\ n = 1,2,3,\ldots.$$

These conditions provide initial conditions for the ODEs. To find $a_n(t)$, we solve the IVP

$$\rho c \frac{da_n}{dt} + \frac{\kappa n^2 \pi^2}{\ell^2} a_n = c_n(t), \tag{6.2}$$
$$a_n(t_0) = b_n.$$

The coefficients $c_n(t)$ and b_n are computable from the given functions $f(x,t)$ and $\psi(x)$.

The IVP (6.2) is of the type considered in Section 4.2.4, and we have a formula for the solution:

$$a_n(t) = b_n e^{-\frac{\kappa n^2 \pi^2}{\rho c \ell^2}(t-t_0)} + \frac{1}{\rho c}\int_{t_0}^t e^{-\frac{\kappa n^2 \pi^2}{\rho c \ell^2}(t-s)} c_n(s)\,ds. \tag{6.3}$$

6.1.1 The homogeneous heat equation

We will begin with examples of the homogeneous heat equation.

Example 6.1. We consider a 50 cm iron bar, with specific heat $c = 0.437 \, \text{J}/(\text{g K})$, density $\rho = 7.88 \, \text{g/cm}^3$, and thermal conductivity $\kappa = 0.836 \, \text{W}/(\text{cm K})$. We assume that the bar is insulated except at the ends and that it is (somehow) given the initial temperature

$$\psi(x) = 5 - \frac{1}{5}|x - 25|,$$

where $\psi(x)$ is given in degrees Celsius. Finally, we assume that, at time $t = 0$, the ends of the bar are placed in an ice bath (0 degrees Celsius). We will compute the temperature distribution after 20, 60, and 300 seconds.

We must solve the IBVP

$$\rho c \frac{\partial u}{\partial t} - \kappa \frac{\partial^2 u}{\partial x^2} = 0, \; 0 < x < 50, \; t > 0,$$
$$u(x,0) = \psi(x), \; 0 < x < 50,$$
$$u(0,t) = 0, \; t > 0,$$
$$u(50,t) = 0, \; t > 0.$$

The solution is

$$u(x,t) = \sum_{n=1}^{\infty} a_n(t) \sin\left(\frac{n\pi x}{50}\right),$$

where the coefficient $a_n(t)$ satisfies the IVP

$$\frac{da_n}{dt} + \frac{\kappa n^2 \pi^2}{50^2 \rho c} a_n = 0,$$
$$a_n(0) = b_n$$

and

$$b_n = \frac{2}{50} \int_0^{50} \psi(x) \sin\left(\frac{n\pi x}{50}\right) dx = \frac{40 \sin(n\pi/2)}{\pi^2 n^2}.$$

The coefficient a_n is given by

$$a_n(t) = b_n e^{-\frac{\kappa n^2 \pi^2}{50^2 \rho c} t},$$

where κ, ρ, and c have the values given above. This gives us an explicit formula for the solution u,

$$u(x,t) = \sum_{n=1}^{\infty} b_n e^{-\frac{\kappa n^2 \pi^2}{50^2 \rho c} t} \sin\left(\frac{n\pi x}{50}\right), \tag{6.4}$$

which can be approximated by a finite series of the form

$$u_N(x,t) = \sum_{n=1}^{N} b_n e^{-\frac{\kappa n^2 \pi^2}{50^2 \rho c} t} \sin\left(\frac{n\pi x}{50}\right).$$

In Figure 6.1, we display "snapshots" of the temperature distributions at the times $t = 20$, $t = 60$, and $t = 300$; that is, we show the graphs of the function $u(x,20)$, $u(x,60)$, and $u(x,300)$. This is often the preferred way to visualize a function of space and time. ∎

6.1. Fourier series methods for the heat equation

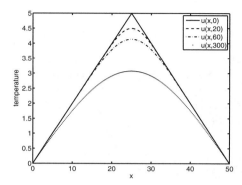

Figure 6.1. *The solution $u(x,t)$ from Example 6.1 at times 0, 20, 60, and 300 (seconds). Ten terms of the Fourier series were used to create these curves.*

We now make several observations concerning the homogeneous heat equation and its solution.

1. For a fixed $t > 0$, the Fourier coefficients $a_n(t)$ decay *exponentially* as $n \to \infty$ (cf. equation (6.3), in which the second term is zero for the homogeneous heat equation). Since larger values of n correspond to higher frequencies, this shows that the solution $u(x,t)$ is very smooth as a function of x for any $t > 0$. This can be seen in Figure 6.1; at $t = 0$, the temperature distribution has a singularity (a point of nondifferentiability) at $x = 25$. This singularity is not seen even after 10 seconds. In fact, it can be shown mathematically that the solution $u(x,t)$ is infinitely differentiable as a function of x for *every* $t > 0$, and this holds for every initial temperature distribution $\psi \in C[0,\ell]$.

   The relationship between the smoothness of a function and the rate of decay of its Fourier will be explained thoroughly in Section 12.5.1. For now, we will content ourselves with a qualitative description of this relationship. A "rough" function (one that varies rapidly with x) must have a lot of high-frequency content (because high-frequency modes—$\sin(n\pi x/\ell)$ with n large—change rapidly with x), and therefore the Fourier coefficients must decay to zero slowly as $n \to \infty$. On the other hand, a smooth function, one that changes slowly with x, must be made up mostly of low-frequency waves, and so the Fourier coefficients decay rapidly with n.

   For example, a function with a discontinuity is "rough" in the sense that we are discussing, and its Fourier coefficients go to zero only as fast as $1/n$. A function that is continuous but whose derivative is discontinuous has Fourier coefficients that decay to zero like $1/n^2$ (assuming the function satisfies the same boundary conditions as the eigenfunctions). This pattern continues: Each additional degree of smoothness in the function corresponds to an additional factor of n in the denominator of the Fourier coefficients. For examples, the reader is referred to Exercise 6.1.13 and, for a complete explanation, to Section 12.5.1.

   The fact that the solution $u(x,t)$ of the homogeneous heat equation is very smooth can be understood physically as follows: If somehow a temperature distribution that has a lot of high-frequency content (meaning that the temperature changes rapidly with x) is induced, and then the temperature distribution is allowed to evolve without external influence, the heat energy will quickly flow from high-temperature regions to low-temperature regions and smooth out the temperature distribution.

2. For each fixed n, the Fourier coefficient $a_n(t)$ decays exponentially as $t \to \infty$. Moreover, the rate of decay is faster for larger n. This is another implication of the fact that the temperature distribution u becomes smoother as t increases, and means that, as t grows, fewer and fewer terms in the Fourier series are required to produce an accurate approximation to the solution $u(x,t)$. *When t is large enough, the first term from the Fourier series provides an excellent approximation to the solution.*

As an illustration of this, we graph, in Figure 6.2, the difference between $u(x,t)$ and the first term of its Fourier series for $t = 10$, $t = 20$, and $t = 300$. (We used 10 terms of the Fourier series to approximate the exact solution. Figure 6.2 suggests that this is plenty.)

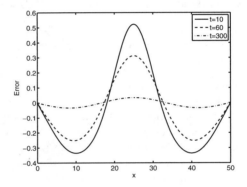

Figure 6.2. *The error in approximating, at times 0, 20, 60, and 300 (seconds), the solution $u(x,t)$ from Example 6.1 with only the first term in its Fourier series.*

3. The material characteristics of the bar appear in the formula for the solution only in the combination
$$\frac{\kappa}{\rho c \ell^2}.$$
For example, in the case of the iron bar of Example 6.1, we have
$$\frac{\kappa}{\rho c \ell^2} \doteq 9.711 \cdot 10^{-5}.$$
If the bar were made of aluminum instead, with density $\rho = 2.70\,\text{g/cm}^3$, specific heat $c = 0.875\,\text{J/(g K)}$, and thermal conductivity $\kappa = 2.36\,\text{W/(cm K)}$, and still had length 50 cm, we would have
$$\frac{\kappa}{\ell^2 \rho c} \doteq 3.996 \cdot 10^{-4}.$$
An examination of the solution (formula (6.4)) shows that the temperature $u(x,t)$ will decay to zero faster for a bar made of aluminum than for a bar made of iron. Figure 6.3 demonstrates this; it is exactly analogous to Figure 6.1, except the solution is for an aluminum bar instead of an iron bar.

6.1. Fourier series methods for the heat equation

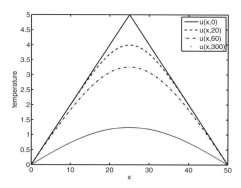

Figure 6.3. *The temperature distribution $u(x,t)$ for an aluminum bar at times 0, 20, 60, and 300 (seconds). Ten terms of the Fourier series were used to create these curves. The experiment is exactly as in Example 6.1, except the iron bar is replaced by an aluminum bar.*

6.1.2 Nondimensionalization

Point 3 above raises the following question. Suppose we cannot solve a PDE explicitly. Can we still identify the critical parameters (or combination of parameters) in the problem? In the case of the homogeneous heat equation, the parameters ρ, c, κ, and ℓ are not significant individually, but rather in the combination $\kappa/(\rho c \ell^2)$. Can we deduce this from the equation itself, rather than from its solution?

It turns out that this is often possible through *nondimensionalization*, which is the process of replacing the independent variables (and sometimes the dependent variable) with nondimensional variables. We continue to consider heat flow in a bar. The independent variables are x and t. The spatial variable has dimensions of centimeters, and it is rather obvious how to nondimensionalize x—we describe the spatial location in terms of the overall length ℓ of the bar. That is, we replace x with

$$y = \frac{x}{\ell}.$$

Since both x and ℓ have units of centimeters, y is dimensionless.

It is more difficult to nondimensionalize the time variable, but there is a general technique for creating nondimensional variables: List all parameters appearing in the problem, together with their units, and find a combination of the parameters which have the same units as the variable you wish to nondimensionalize. In this problem, the parameters and their units are as follows:

Parameter	Units
ρ	g/cm^3
c	J/(g K)
κ	J/(s cm K)
ℓ	cm

Then the ratio
$$\frac{\rho c \ell^2}{\kappa}$$
is seen to have units of seconds, so we define a dimensionless time variable s by
$$s = \frac{t}{\rho c \ell^2 / \kappa} = \frac{\kappa t}{\rho c \ell^2}.$$

We then define $v(y,s) = u(x,t)$, where (y,s) and (x,t) are related by
$$y = \frac{x}{\ell}, \quad s = \frac{\kappa t}{\rho c \ell^2}.$$

Then, by the chain rule,
$$\frac{\partial u}{\partial t} = \frac{\partial v}{\partial s}\frac{\partial s}{\partial t} = \frac{\kappa}{\rho c \ell^2}\frac{\partial v}{\partial s}$$

and
$$\frac{\partial^2 u}{\partial x^2} = \frac{\partial}{\partial x}\left(\frac{\partial u}{\partial x}\right) = \frac{\partial}{\partial x}\left(\frac{\partial v}{\partial y}\frac{\partial y}{\partial x}\right) = \frac{\partial}{\partial x}\left(\frac{1}{\ell}\frac{\partial v}{\partial y}\right) = \frac{1}{\ell}\frac{\partial^2 v}{\partial y^2}\frac{\partial y}{\partial x} = \frac{1}{\ell^2}\frac{\partial^2 v}{\partial y^2}.$$

Therefore, the PDE
$$\rho c \frac{\partial u}{\partial t} - \kappa \frac{\partial^2 u}{\partial x^2} = 0$$

is equivalent to
$$\rho c \frac{\kappa}{\rho c \ell^2}\frac{\partial v}{\partial s} - \frac{\kappa}{\ell^2}\frac{\partial^2 v}{\partial y^2} = 0,$$

which simplifies to
$$\frac{\partial v}{\partial s} - \frac{\partial^2 v}{\partial y^2} = 0.$$

Since $x = \ell y$, and $x = \ell$ corresponds to $y = 1$, the IBVP (6.1), with $f(x,t) = 0$, becomes
$$\begin{aligned}
\frac{\partial v}{\partial s} - \frac{\partial^2 v}{\partial y^2} &= 0, \; 0 < y < 1, \; s > s_0, \\
v(y,s_0) &= \psi(\ell y), \; 0 < y < 1, \\
v(0,s) &= 0, \; s > s_0, \\
v(1,s) &= 0, \; s > s_0,
\end{aligned} \qquad (6.5)$$

where $s_0 = \kappa t_0/(\rho c \ell^2)$. The solution of (6.5) is computed as above; the result is
$$v(s,y) = \sum_{n=1}^{\infty} b_n e^{-n^2 \pi^2 s} \sin(n\pi y).$$

The Fourier sine coefficients of $\psi(\ell y)$, considered as a function of y on the interval $[0,1]$, are the same as the Fourier sine coefficients of $\psi(x)$ considered as a function of x on the interval $[0,\ell]$ (see Exercise 6.1.6).

6.1. Fourier series methods for the heat equation

What did we gain from nondimensionalization? We identified the natural time scale, namely, $\rho c \ell^2 / \kappa$, for the experiment, and we did this without solving the equation. Whether we have an iron or aluminum bar, the problem can be written as (6.5). This shows that the evolution of the temperature will be exactly the same for the two bars, except that the temperature evolves on different time scales, depending on the material properties of the bars. Moreover, we see exactly how this time scale depends on the various parameters, and this can give some information that is not immediately obvious from the equation. For example, it is obvious from the equation that, holding ρ constant, a bar with specific heat c and thermal conductivity κ will behave the same as a bar with specific heat $2c$ and thermal conductivity 2κ. However, it is not at all obvious that, holding ρ and c constant, a bar of length ℓ with thermal conductivity κ will behave just as a bar of length 2ℓ with thermal conductivity 4κ.[29]

6.1.3 The inhomogeneous heat equation

We now consider a problem involving the inhomogeneous diffusion equation (which is mathematically equivalent to the heat equation).

Example 6.2. The diffusion coefficient for carbon monoxide (CO) diffusing in air is $D = 0.208 \, \text{cm}^2/\text{s}$. We consider a pipe of length 100 cm, filled with air, that initially contains no CO. We assume that the two ends of the pipe open onto large reservoirs of air, also containing no CO, and that CO is produced inside the pipe at a constant rate of $10^{-7} \, \text{g/cm}^3$ per second. We write $u(x,t)$, $0 < x < 100$, for the concentration of CO in air, in units of g/cm^3, and model u as the solution of the IBVP

$$\frac{\partial u}{\partial t} - D \frac{\partial^2 u}{\partial x^2} = 10^{-7}, \; 0 < x < 100, \; t > 0,$$
$$u(x,0) = 0, \; 0 < x < 100,$$
$$u(0,t) = 0, \; t > 0,$$
$$u(100,t) = 0, \; t > 0.$$

We wish to determine the evolution of the concentration of CO in the pipe.

We write the solution u in the form

$$u(x,t) = \sum_{n=1}^{\infty} a_n(t) \sin\left(\frac{n\pi x}{100}\right),$$

so that

$$\frac{\partial u}{\partial t}(x,t) - D \frac{\partial^2 u}{\partial x^2}(x,t) = \sum_{n=1}^{\infty} \left(\frac{da_n}{dt}(t) + \frac{Dn^2\pi^2}{100^2} a_n(t) \right) \sin\left(\frac{n\pi x}{100}\right).$$

We have

$$10^{-7} = \sum_{n=1}^{\infty} c_n \sin\left(\frac{n\pi x}{100}\right),$$

---

[29] For a thorough discussion of nondimensionalization and scale models, see [45, Chapter 6].

where

$$c_n = \frac{2}{100}\int_0^{100} 10^{-7}\sin\left(\frac{n\pi x}{100}\right)dx = \frac{2\cdot 10^{-7}(1-(-1)^n)}{n\pi}, \ n=1,2,3,\ldots.$$

To find the coefficients $a_n(t)$ of $u(x,t)$, we must solve the IVPs

$$\frac{da_n}{dt} + \frac{Dn^2\pi^2}{100^2}a_n = c_n,$$
$$a_n(0) = 0.$$

Using (6.3), we obtain

$$a_n(t) = c_n\int_0^t e^{-\frac{Dn^2\pi^2}{100^2}(t-s)}ds$$
$$= -\frac{100^2 c_n}{Dn^2\pi^2}\left(e^{-\frac{Dn^2\pi^2}{100^2}t} - 1\right)$$
$$= \frac{2\cdot 10^{-3}((-1)^n - 1)}{Dn^3\pi^3}\left(e^{-\frac{Dn^2\pi^2}{100^2}t} - 1\right).$$

In Figure 6.4, we show the concentration of CO in the pipe at times $t = 3600$ (1 hour), $t = 7200$ (2 hours), $t = 14400$ (4 hours), $t = 21600$ (6 hours), and $t = 360000$ (100 hours). The graph suggests that the concentration approaches steady state (the reader should notice how little the concentration changes between 4 hours and 100 hours of elapsed time). We discuss the approach to steady state in Section 6.1.5. ∎

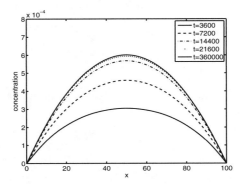

Figure 6.4. *The concentration of CO in the pipe of Example 6.2 after* 1, 2, 4, 6, *and* 100 *hours. Ten terms of the Fourier series were used to create these curves.*

6.1.4 Inhomogeneous boundary conditions

Inhomogeneous boundary conditions in a time-dependent problem can be handled by the method of shifting the data, much as in the steady-state problems discussed in the last chapter. We illustrate with an example.

6.1. Fourier series methods for the heat equation

Example 6.3. We suppose that the iron bar of Example 6.1 is heated to a constant temperature of 4 degrees Celsius, and that one end ($x = 0$) is placed in an ice bath (0 degree Celsius), while the other end is maintained at 4 degrees. What is the temperature distribution after 5 minutes?

The IBVP is

$$\rho c \frac{\partial u}{\partial t} - \kappa \frac{\partial^2 u}{\partial x^2} = 0, \ 0 < x < 50, \ t > 0,$$

$$u(x, 0) = 4, \ 0 < x < 50,$$

$$u(0, t) = 0, \ t > 0,$$

$$u(50, t) = 4, \ t > 0.$$

The function $p(x) = 4x/50$ satisfies the boundary conditions, and thus we define $v(x,t) = u(x,t) - p(x)$. Then, as is easily verified, v satisfies

$$\rho c \frac{\partial v}{\partial t} - \kappa \frac{\partial^2 v}{\partial x^2} = 0, \ 0 < x < 50, \ t > 0,$$

$$v(x, 0) = 4 - \frac{4}{50}x, \ 0 < x < 50,$$

$$v(0, t) = 0, \ t > 0,$$

$$v(50, t) = 0, \ t > 0.$$

The initial temperature satisfies

$$4 - \frac{4}{50}x = \sum_{n=1}^{\infty} b_n \sin\left(\frac{n\pi x}{50}\right),$$

where

$$b_n = \frac{8}{n\pi}, \ n = 1, 2, 3, \ldots.$$

The solution

$$v(x,t) = \sum_{n=1}^{\infty} a_n(t) \sin\left(\frac{n\pi x}{50}\right)$$

is determined by the IVPs

$$\frac{da_n}{dt} + \frac{\kappa n^2 \pi^2}{\rho c 50^2} a_n = 0,$$

$$a_n(0) = b_n,$$

so the solution is

$$v(x,t) = \sum_{n=1}^{\infty} \frac{8}{n\pi} e^{-\frac{\kappa n^2 \pi^2}{\rho c 50^2} t} \sin\left(\frac{n\pi x}{50}\right).$$

The solution u is then given by

$$u(x,t) = \frac{4}{50}x + \sum_{n=1}^{\infty} \frac{8}{n\pi} e^{-\frac{\kappa n^2 \pi^2}{\rho c 50^2}t} \sin\left(\frac{n\pi x}{50}\right).$$

The temperature distribution at $t = 300$ is shown in Figure 6.5. ∎

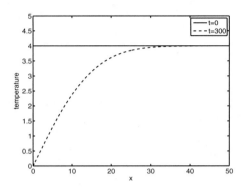

Figure 6.5. *The temperature distribution after* 300 *seconds (see Example* 6.3*).*

In a time-dependent problem, inhomogeneous boundary conditions can themselves be time dependent; for example,

$$u(0,t) = g(t), \ u(\ell,t) = h(t), \ t > 0,$$

are valid inhomogeneous Dirichlet conditions. The function

$$p(x,t) = g(t) + \frac{x}{\ell}(h(t) - g(t))$$

satisfies these conditions. The reader should notice that the function $v = u - p$ will *not* satisfy the same PDE as does u; the right-hand side will be different. See Exercise 6.1.5.

6.1.5 Steady-state heat flow and diffusion

We now discuss the special case of heat flow or diffusion in which the source function $f(x,t)$ appearing on the right-hand side of the differential equation is independent of t (this is the case in Example 6.2). In this case, the Fourier coefficients of f are constant with respect to t, and formula (6.3) can be simplified as follows:

$$\begin{aligned}
a_n(t) &= b_n e^{-\frac{\kappa n^2 \pi^2}{\rho c \ell^2}(t-t_0)} + \frac{1}{\rho c} \int_{t_0}^{t} e^{-\frac{\kappa n^2 \pi^2}{\rho c \ell^2}(t-s)} c_n \, ds \\
&= b_n e^{-\frac{\kappa n^2 \pi^2}{\rho c \ell^2}(t-t_0)} + \frac{c_n \ell^2}{\kappa n^2 \pi^2}\left(1 - e^{-\frac{\kappa n^2 \pi^2}{\rho c \ell^2}(t-t_0)}\right) \\
&= \left(b_n - \frac{c_n \ell^2}{\kappa n^2 \pi^2}\right) e^{-\frac{\kappa n^2 \pi^2}{\rho c \ell^2}(t-t_0)} + \frac{c_n \ell^2}{\kappa n^2 \pi^2}.
\end{aligned}$$

6.1. Fourier series methods for the heat equation

As $t \to \infty$, the first term tends to zero for each n, and so

$$a_n(t) \to \frac{c_n \ell^2}{\kappa n^2 \pi^2} \text{ as } t \to \infty.$$

This implies that

$$u(x,t) \to u_s(x) = \sum_{n=1}^{\infty} \frac{c_n \ell^2}{\kappa n^2 \pi^2} \sin\left(\frac{n\pi x}{\ell}\right).$$

The limit $u_s(x)$ is the *steady-state* temperature distribution. It is not difficult to show that u_s is the solution to the BVP

$$\begin{aligned} -\kappa \frac{d^2 u}{dx^2} &= f(x),\ 0 < x < \ell, \\ u(0) &= 0, \\ u(\ell) &= 0 \end{aligned} \tag{6.6}$$

(see Exercise 6.1.2). It should be noted that the initial temperature distribution is irrelevant in determining the steady-state temperature distribution.

Similar results hold for the diffusion equation, which is really no different from the heat equation except in the meaning of the parameters. Indeed, Example 6.2 is an illustration; the concentration u approaches a steady-state concentration, as suggested by Figure 6.4 (see Exercise 6.1.1).

In the next section, we will turn our attention to two new sets of boundary conditions for the heat equation and derive Fourier series methods that apply to the new boundary conditions. First, however, we briefly discuss another derivation of the Fourier series method.

6.1.6 Separation of variables

There is another way to derive Fourier series solutions of the heat equation, which is called the method of *separation of variables*. It is in some ways more elementary than the point of view we have presented, but it is less general.

To introduce the method of separation of variables, we will use the following IBVP:

$$\begin{aligned} \rho c \frac{\partial u}{\partial t} - \kappa \frac{\partial^2 u}{\partial x^2} &= 0,\ 0 < x < \ell,\ t > t_0, \\ u(x, t_0) &= \psi(x),\ 0 < x < \ell, \\ u(0, t) &= 0,\ t > t_0, \\ u(\ell, t) &= 0,\ t > t_0. \end{aligned} \tag{6.7}$$

The reader should notice that the PDE is homogeneous; this is necessary and is a significant limitation of this technique.

Here is the key idea of the method of separation of variables: we look for *separated* solutions: $u(x,t) = X(x)T(t)$ (X and T are functions of a single variable). Substituting u

into the PDE yields

$$u = XT, \quad \rho c \frac{\partial u}{\partial t} - \kappa \frac{\partial^2 u}{\partial x^2} = 0$$

$$\Rightarrow \rho c X \frac{dT}{dt} - \kappa T \frac{d^2 X}{dx^2} = 0$$

$$\Rightarrow -\kappa T \frac{d^2 X}{dx^2} = -\rho c X \frac{dT}{dt}$$

$$\Rightarrow -\kappa X^{-1} \frac{d^2 X}{dx^2} = -\rho c T^{-1} \frac{dT}{dt}.$$

The last step follows upon dividing both sides of the equation by XT. The crucial observation is the following:

$$-\kappa X^{-1} \frac{d^2 X}{dx^2}$$

is a function of x alone, while

$$-\rho c T^{-1} \frac{dT}{dt}$$

is a function of t alone. The only way these two functions can be equal is if they are both constants:

$$-\kappa X^{-1} \frac{d^2 X}{dx^2} = -\rho c T^{-1} \frac{dT}{dt} = \lambda.$$

We now have ODEs for $X(x)$ and $T(t)$:

$$-\kappa \frac{d^2 X}{dx^2} = \lambda X, \quad -\rho c \frac{dT}{dt} = \lambda T.$$

Moreover, the boundary conditions for u imply boundary conditions for X:

$$u(0,t) = 0 \Rightarrow X(0)T(t) = 0 \Rightarrow X(0) = 0,$$
$$u(\ell,t) = 0 \Rightarrow X(\ell)T(t) = 0 \Rightarrow X(\ell) = 0$$

(the function $T(t)$ cannot be zero, for otherwise u is the trivial solution).

We see that the function X must satisfy the eigenvalue problem

$$-\kappa \frac{d^2 X}{dx^2} = \lambda X, \quad 0 < x < \ell,$$
$$X(0) = 0,$$
$$X(\ell) = 0.$$

In Section 5.2.1, we showed that this problem has a sequence of solutions:

$$\lambda_n = \frac{\kappa n^2 \pi^2}{\ell^2}, \quad X_n(x) = \sin\left(\frac{n\pi x}{\ell}\right), \quad n = 1, 2, 3, \ldots.$$

For each value of λ_n, we can solve the ODE

$$-\rho c \frac{dT}{dt} = \lambda T$$

6.1. Fourier series methods for the heat equation

to find a corresponding function T_n:

$$T_n(t) = b_n e^{-\frac{\lambda_n(t-t_0)}{\rho c}}, \; n = 1, 2, 3, \ldots$$

(b_n is an arbitrary constant).

We have now found a sequence of separated solutions to the homogeneous heat equation, namely,

$$u_n(x,t) = b_n e^{-\frac{\lambda_n(t-t_0)}{\rho c}} \sin\left(\frac{n\pi x}{\ell}\right), \; n = 1, 2, 3, \ldots.$$

By construction, each of these functions satisfies the homogeneous Dirichlet conditions.

To satisfy the initial condition, we use the principle of superposition (which applies because the heat equation is linear) to form the solution

$$u(x,t) = \sum_{n=1}^{\infty} b_n e^{-\frac{\lambda_n(t-t_0)}{\rho c}} \sin\left(\frac{n\pi x}{\ell}\right).$$

We then have

$$u(x,t_0) = \sum_{n=1}^{\infty} b_n \sin\left(\frac{n\pi x}{\ell}\right).$$

If we choose $b_1, b_2, b_3, \ldots$ to be the Fourier sine coefficients of ψ,

$$b_n = \frac{2}{\ell} \int_0^\ell \psi(x) \sin\left(\frac{n\pi x}{\ell}\right) dx,$$

then $u(x,t_0) = \psi(x)$ will hold.

We thus see that the method of separation of variables yields the solution we obtained earlier (see Section 6.1.1). However, the method requires that the PDE and the boundary conditions be homogeneous, and this is a serious limitation. For example, one cannot apply separation of variables to a problem with inhomogeneous, time-dependent boundary conditions. The method does not apply directly, and one cannot use the trick of shifting the data, since it would result in an inhomogeneous PDE.

We will find separation of variables useful in Section 11.2, when we derive the eigenvalues and eigenfunctions for a differential operator defined on a rectangle.

Exercises

1. What is the steady-state solution of Example 6.2? What BVP does it satisfy?

2. Find the Fourier sine series of the solution of (6.6) and show that it equals $u_s(x)$.

3. Solve the following IBVP using the Fourier series method:

$$\frac{\partial u}{\partial t} - \frac{\partial^2 u}{\partial x^2} = 0, \; 0 < x < 1, \; t > 0,$$
$$u(x,0) = x, \; 0 < x < 1,$$
$$u(0,t) = 0, \; t > 0,$$
$$u(1,t) = 0, \; t > 0.$$

Produce an accurate graph of the "snapshot" $u(\cdot, 0.1)$.

4. Solve the following IBVP using the Fourier series method:

$$\frac{\partial u}{\partial t} - \frac{\partial^2 u}{\partial x^2} = \sin(t), \ 0 < x < 1, \ t > 0,$$
$$u(x,0) = 0, \ 0 < x < 1,$$
$$u(0,t) = 0, \ t > 0,$$
$$u(1,t) = 0, \ t > 0.$$

Produce an accurate graph of the "snapshot" $u(\cdot, 0.1)$.

5. Consider the heat equation with inhomogeneous boundary conditions:

$$\rho c \frac{\partial u}{\partial t} - \kappa \frac{\partial^2 u}{\partial x^2} = f(x,t), \ 0 < x < \ell, \ t > 0,$$
$$u(x,t_0) = \psi(x), \ 0 < x < \ell,$$
$$u(0,t) = g(t), \ t > t_0,$$
$$u(\ell,t) = h(t), \ t > t_0.$$

Define

$$p(x,t) = g(t) + \frac{x}{\ell}(h(t) - g(t))$$

and $v(x,t) = u(x,t) - p(x,t)$. What IBVP does v satisfy?

6. Suppose $\psi \in C[0,\ell]$, and $\phi \in C[0,1]$ is defined by $\phi(y) = \psi(\ell y)$. Show that the sine coefficients of ϕ on $[0,1]$ are the same as the sine coefficients of ψ on $[0,\ell]$.

7. Use the method of shifting the data to solve the following IBVP with inhomogeneous Dirichlet conditions:

$$\frac{\partial u}{\partial t} - \frac{\partial^2 u}{\partial x^2} = 0, \ 0 < x < 1, \ t > 0,$$
$$u(x,0) = x, \ 0 < x < 1,$$
$$u(0,t) = 0, \ t > 0,$$
$$u(1,t) = \cos(t), \ t > 0.$$

Produce an accurate graph of the "snapshot" $u(\cdot, 1.0)$.

8. Consider an iron bar, of diameter 4 cm and length 1 m, with specific heat $c = 0.437\,\text{J}/(\text{g K})$, density $\rho = 7.88\,\text{g/cm}^3$, and thermal conductivity $\kappa = 0.836\,\text{W}/(\text{cm K})$. Suppose that the bar is insulated except at the ends, it is heated to a constant temperature of 5 degrees Celsius, and the ends are placed in an ice bath (0 degrees Celsius). Compute the temperature (accurate to 3 digits) at the midpoint of the bar after 20 minutes. (Warning: Be sure to use consistent units.)

9. Repeat Exercise 8 with a copper bar: specific heat $c = 0.379\,\text{J}/(\text{g K})$, density $\rho = 8.97\,\text{g/cm}^3$, thermal conductivity $\kappa = 4.04\,\text{W}/(\text{cm K})$.

6.2. Pure Neumann conditions and the Fourier cosine series

10. Consider the bar of Exercise 8. Suppose the bar is completely insulated except at the ends. The bar is heated to a constant temperature of 5 degrees Celsius, and then the right end is placed in an ice bath (0 degrees Celsius), while the other end is maintained at 5 degrees Celsius.

    (a) What is the steady-state temperature distribution of the bar?

    (b) How long does it take the bar to reach steady-state (to within 1%)?

11. Repeat the preceding exercise with the bar of Exercise 9.

12. Consider a 100 cm circular bar, with radius 4 cm, designed so that the thermal conductivity is nonconstant and satisfies $\kappa(x) = 1 + \alpha x$ W/(cm K) for some $\alpha > 0$. Assume that the sides of the bar are completely insulated, one end ($x = 0$) is kept at 0 degrees Celsius, and heat is added to the other end at a rate of 4 W. The temperature of the bar reaches a steady state, $u = u(x)$, and the temperature at the end $x = 100$ is measured to be

    $$u(100) \doteq 7.6 \text{ degrees Celsius}.$$

    Estimate α.

13. Compute the Fourier sine coefficients of each of the following functions, and verify the statements on page 199 concerning the rate of decay of the Fourier coefficients. (Take the interval to be [0, 1].)

    (a)
    $$f(x) = \begin{cases} x, & 0 \leq x < 1/2, \\ 2 - 2x, & 1/2 \leq x \leq 1. \end{cases}$$

    (b)
    $$f(x) = \begin{cases} x, & 0 \leq x < 1/2, \\ 1 - x, & 1/2 \leq x \leq 1. \end{cases}$$

6.2 Pure Neumann conditions and the Fourier cosine series

In this section we will consider the effect of insulating one or both ends of a bar in which heat is flowing.[30] With the new boundary conditions that result, we must use different eigenfunctions in the Fourier series method. Of particular interest is the case in which both ends of the bar are insulated, as this raises some mathematical questions that we have not yet encountered.

6.2.1 One end insulated; mixed boundary conditions

We begin with the case in which one end of the bar is insulated, but the other is not. In Section 2.1, we saw that an insulated boundary corresponds to a homogeneous Neumann condition, which indicates that no heat energy is flowing through that end of the bar. We

---

[30]Or closing one or both ends of a pipe in which a chemical is diffusing.

consider a bar of length ℓ, perfectly insulated on the sides, and assume that one end ($x = \ell$) is perfectly insulated while the other end ($x = 0$) is uninsulated and placed in an ice bath. If the initial temperature distribution is $\psi(x)$, then the temperature distribution $u(x,t)$ satisfies the IBVP

$$\rho c \frac{\partial u}{\partial t} - \kappa \frac{\partial^2 u}{\partial x^2} = f(x,t), \ 0 < x < \ell, \ t > t_0,$$
$$u(x,t_0) = \psi(x), \ 0 < x < \ell,$$
$$u(0,t) = 0, \ t > t_0,$$
$$\frac{\partial u}{\partial x}(\ell,t) = 0, \ t > t_0.$$
(6.8)

The eigenvalue/eigenfunctions pairs for L_m, the negative second derivative operator under the mixed boundary conditions (Dirichlet at the left, Neumann at the right), are

$$\frac{(2n-1)^2 \pi^2}{4\ell^2}, \ \sin\left(\frac{(2n-1)\pi x}{2\ell}\right), \ n = 1, 2, 3, \ldots$$

(see Section 5.2.3). If we represent the solution as

$$u(x,t) = \sum_{n=1}^{\infty} a_n(t) \sin\left(\frac{(2n-1)\pi x}{2\ell}\right),$$

then u will satisfy the boundary conditions. Just as in the previous section, we can derive an ODE with an initial condition for each coefficient $a_n(t)$ and thereby determine u. We will illustrate with an example.

Example 6.4. We consider the temperature distribution in the iron bar of Example 6.1, with the experimental conditions unchanged except that the right end ($x = \ell$) of the bar is perfectly insulated. The temperature distribution $u(x,t)$ then satisfies the IBVP

$$\rho c \frac{\partial u}{\partial t} - \kappa \frac{\partial^2 u}{\partial x^2} = 0, \ 0 < x < \ell, \ t > 0,$$
$$u(x,0) = \psi(x), \ 0 < x < \ell,$$
$$u(0,t) = 0, \ t > 0,$$
$$\frac{du}{dx}(\ell,t) = 0, \ t > 0,$$

where $\rho = 7.88\,\text{g/cm}^3$, $c = 0.437\,\text{J/(g K)}$, $\kappa = 0.836\,\text{W/(cm K)}$, $\ell = 50\,\text{cm}$, and

$$\psi(x) = 5 - \frac{1}{5}|x - 25|.$$

We have

$$\psi(x) = \sum_{n=1}^{\infty} c_n \sin\left(\frac{(2n-1)\pi}{2\ell}\right),$$

6.2. Pure Neumann conditions and the Fourier cosine series

where

$$c_n = \frac{2}{\ell}\int_0^\ell \psi(x)\sin\left(\frac{(2n-1)\pi x}{2\ell}\right)dx$$
$$= -\frac{80\left(-\sqrt{2}\sin(n\pi/2) + \sqrt{2}\cos(n\pi/2) - (-1)^n\right)}{\pi^2(2n-1)^2}, \quad n = 1, 2, 3, \ldots.$$

We write the solution u in the form

$$u(x,t) = \sum_{n=1}^\infty a_n(t) \sin\left(\frac{(2n-1)\pi x}{2\ell}\right).$$

Then, by almost exactly the same computation as on pages 196–197, we have

$$\rho c \frac{\partial u}{\partial t}(x,t) - \kappa \frac{\partial^2 u}{\partial x^2}(x,t) = \sum_{n=1}^\infty \left(\rho c \frac{da_n}{dt}(t) + \frac{\kappa(2n-1)^2\pi^2}{4\ell^2} a_n(t)\right) \sin\left(\frac{(2n-1)\pi x}{2\ell}\right).$$

The PDE and the initial condition for u then imply that the coefficient $a_n(t)$ must satisfy the following IVP:

$$\rho c \frac{da_n}{dt} + \frac{\kappa(2n-1)^2\pi^2}{4\ell^2} a_n = 0,$$
$$a_n(0) = c_n.$$

The solution is

$$a_n(t) = c_n e^{-\frac{\kappa(2n-1)^2\pi^2}{4\ell^2 \rho c}t},$$

which determines $u(x,t)$.

In Figure 6.6, we show the temperature distributions at times $t = 20$, $t = 60$, and $t = 300$; this figure should be compared with Figure 6.1. The effect of insulating the right end is clearly seen. ∎

6.2.2 Both ends insulated; Neumann boundary conditions

In order to treat the case of heat flow in a bar with both ends insulated, we must first analyze the negative second derivative operator under Neumann conditions. We define

$$C_N^2[0,\ell] = \left\{u \in C^2[0,\ell] : \frac{du}{dx}(0) = \frac{du}{dx}(\ell) = 0\right\},$$

and define $L_N : C_N^2[0,\ell] \to C[0,\ell]$ by

$$L_N u = -\frac{d^2 u}{dx^2}.$$

Unlike the negative second derivative operator under the other sets of boundary conditions we have considered, L_N has a nontrivial null space; indeed, $L_N u = 0$ if and only if u is

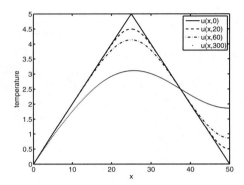

Figure 6.6. *The solution $u(x,t)$ from Example 6.4 at times 0, 20, 60, and 300 (seconds). Ten terms of the Fourier series were used to create these curves.*

a constant function (see Example 3.17). Therefore, $\lambda_0 = 0$ is an eigenvalue of L_N with eigenfunction $\gamma_0(x) = 1$. It is straightforward to show that L_N is symmetric, so that its eigenvalues are real and eigenfunctions corresponding to distinct eigenvalues are orthogonal. Moreover, apart from λ_0, the other eigenvalues of L_N are positive (see Exercise 6.2.3). We now determine these other eigenvalues and the corresponding eigenfunctions.

Since we are looking for positive eigenvalues, we write $\lambda = \theta^2$ and solve

$$-\frac{d^2 u}{dx^2} = \theta^2 u, \ 0 < x < \ell,$$
$$\frac{du}{dx}(0) = 0, \qquad (6.9)$$
$$\frac{du}{dx}(\ell) = 0.$$

The general solution of the ODE is

$$u(x) = c_1 \cos(\theta x) + c_2 \sin(\theta x).$$

Therefore,

$$\frac{du}{dx}(x) = -c_1 \theta \sin(\theta x) + c_2 \theta \cos(\theta x)$$

and

$$\frac{du}{dx}(0) = c_2 \theta,$$

so the first boundary condition implies $c_2 = 0$. We then have $u(x) = c_1 \cos(\theta x)$, and the second boundary condition becomes $c_1 \theta \sin(\theta \ell) = 0$. When $\lambda > 0$, this holds only if

$$\theta \ell = \pm n\pi, \ n = 1, 2, 3, \ldots,$$

that is, if

$$\lambda = \frac{n^2 \pi^2}{\ell^2}, \ n = 1, 2, 3, \ldots$$

6.2. Pure Neumann conditions and the Fourier cosine series

(we disregard the possibility that $c_1 = 0$, as this does not lead to an eigenfunction). The value of the constant c_1 is irrelevant.

We thus see that the operator $-d^2/dx^2$, under Neumann conditions, has the following eigenpairs:

$$\lambda_n = \frac{n^2\pi^2}{\ell^2}, \; \gamma_n(x) = \cos\left(\frac{n\pi x}{\ell}\right), \; n = 1, 2, 3, \ldots, \quad (6.10)$$

in addition to the eigenpair $\lambda_0 = 0$, $\gamma_0(x) = 1$.

A direct calculation shows that

$$(\gamma_m, \gamma_n) = \begin{cases} 0, & m \neq n, \\ \frac{\ell}{2}, & m = n > 0, \\ \ell, & m = n = 0. \end{cases}$$

The eigenfunctions are orthogonal, as we knew they would be due to the symmetry of L_N.

By the projection theorem, the best approximation to $f \in C[0, \ell]$ from the subspace spanned by

$$\left\{1, \cos\left(\frac{\pi x}{\ell}\right), \ldots, \cos\left(\frac{N\pi x}{\ell}\right)\right\}$$

is

$$f_N(x) = \sum_{n=0}^{N} \frac{(f, \gamma_n)}{(\gamma_n, \gamma_n)} \gamma_n(x) = b_0 + \sum_{n=1}^{N} b_n \cos\left(\frac{n\pi x}{\ell}\right),$$

where

$$b_0 = \frac{1}{\ell} \int_0^\ell f(x) dx, \; b_n = \frac{2}{\ell} \int_0^\ell f(x) \cos\left(\frac{n\pi x}{\ell}\right) dx, \; n = 1, 2, \ldots.$$

It can be shown that $\|f - f_N\| \to 0$ as $N \to \infty$; that is,

$$f(x) = b_0 + \sum_{n=1}^{\infty} b_n \cos\left(\frac{n\pi x}{\ell}\right),$$

where the convergence is in the mean-square sense (not necessarily in the pointwise sense—the distinction will be discussed in detail in Section 12.4.1). This is the *Fourier cosine series* of the function f, and the coefficients $b_0, b_1, b_2, \ldots$ are the *Fourier cosine coefficients* of f.

We can now solve the following IBVP for the heat equation:

$$\begin{aligned} \rho c \frac{\partial u}{\partial t} - \kappa \frac{\partial^2 u}{\partial x^2} &= f(x, t), \; 0 < x < \ell, \; t > t_0, \\ u(x, 0) &= \psi(x), \; 0 < x < \ell, \\ \frac{\partial u}{\partial x}(0, t) &= 0, \; t > t_0, \\ \frac{\partial u}{\partial x}(\ell, t) &= 0, \; t > t_0. \end{aligned} \quad (6.11)$$

We write the unknown solution $u(x,t)$ in a Fourier cosine series with time-dependent Fourier coefficients:

$$u(x,t) = b_0(t) + \sum_{n=1}^{\infty} b_n(t) \cos\left(\frac{n\pi x}{\ell}\right),$$

$$b_0(t) = \frac{1}{\ell} \int_0^{\ell} u(x,t)\,dx,$$

$$b_n(t) = \frac{2}{\ell} \int_0^{\ell} u(x,t) \cos\left(\frac{n\pi x}{\ell}\right) dx, \; n = 1, 2, \ldots.$$

The problem of determining $u(x,t)$ now reduces to the problem of solving for $b_0(t)$, $b_1(t), b_2(t), \ldots$. We write the PDE

$$\rho c \frac{\partial u}{\partial t} - \kappa \frac{\partial^2 u}{\partial x^2} = f(x,t)$$

in terms of Fourier cosine series; this requires determining formulas for the Fourier cosine coefficients of $\partial u/\partial t(x,t)$ and $-\partial^2 u/\partial x^2(x,t)$ in terms of the Fourier coefficients of $u(x,t)$. By Theorem 2.1, we have

$$\frac{2}{\ell} \int_0^{\ell} \frac{\partial u}{\partial t}(x,t) \cos\left(\frac{n\pi x}{\ell}\right) dx = \frac{d}{dt}\left[\frac{2}{\ell} \int_0^{\ell} u(x,t) \cos\left(\frac{n\pi x}{\ell}\right) dx\right] = \frac{db_n}{dt}(t)$$

(and similarly for $n = 0$). Therefore,

$$\frac{\partial u}{\partial t}(x,t) = \frac{db_0}{dt}(t) + \sum_{n=1}^{\infty} \frac{db_n}{dt}(t) \cos\left(\frac{n\pi x}{\ell}\right).$$

Since L_N is a symmetric operator, we can compute the Fourier coefficients of $-\partial^2 u/\partial x^2(x,t)$ just as in Section 5.3.5:

$$\frac{(L_N u, \gamma_n)}{(\gamma_n, \gamma_n)} = \frac{(u, L_N \gamma_n)}{(\gamma_n, \gamma_n)} = \frac{(u, \lambda_n \gamma_n)}{(\gamma_n, \gamma_n)} = \lambda_n \frac{(u, \gamma_n)}{(\gamma_n, \gamma_n)} = \lambda_n b_n(t).$$

Note that, in particular, the $n = 0$ Fourier coefficient is $\lambda_0 b_0(t) = 0 \cdot b_0(t) = 0$.
We obtain

$$-\frac{\partial^2 u}{\partial x^2}(x,t) = \sum_{n=1}^{\infty} \lambda_n b_n(t) \gamma_n(x) = \sum_{n=1}^{\infty} \frac{n^2 \pi^2}{\ell^2} b_n(t) \cos\left(\frac{n\pi x}{\ell}\right).$$

Therefore,

$$\rho c \frac{\partial u}{\partial t} - \kappa \frac{\partial^2 u}{\partial x^2} = \rho c \frac{db_0}{dt}(t) + \sum_{n=1}^{\infty} \left(\rho c \frac{db_n}{dt}(t) + \frac{\kappa \pi^2 n^2}{\ell^2} b_n(t)\right) \cos\left(\frac{n\pi x}{\ell}\right). \quad (6.12)$$

Now, since any function in $C[0,\ell]$ can be represented by a cosine series, we have

$$f(x,t) = c_0(t) + \sum_{n=1}^{\infty} c_n(t) \cos\left(\frac{n\pi x}{\ell}\right), \quad (6.13)$$

6.2. Pure Neumann conditions and the Fourier cosine series

where the coefficients $c_0(t), c_1(t), c_2(t), \ldots$ can be computed explicitly because f is given:

$$c_0(t) = \frac{1}{\ell} \int_0^\ell f(x,t)\,dx, \quad c_n(t) = \frac{2}{\ell} \int_0^\ell f(x,t) \cos\left(\frac{n\pi x}{\ell}\right) dx, \quad n = 1, 2, \ldots.$$

We equate the two series, given in (6.12) and (6.13), and obtain

$$\rho c \frac{db_0}{dt}(t) = c_0(t), \quad \rho c \frac{db_n}{dt}(t) + \frac{\kappa \pi^2 n^2}{\ell^2} b_n(t) = c_n(t), \quad n = 1, 2, \ldots.$$

We can then compute $b_0(t), b_1(t), b_2(t), \ldots$ by solving these ODEs, together with the initial values obtained from the initial condition $u(x, t_0) = \psi(x)$, $0 < x < \ell$.

Example 6.5. We continue to consider the iron bar of Example 6.1, now with both ends insulated. We must solve the IBVP

$$\begin{aligned}
\rho c \frac{\partial u}{\partial t} - \kappa \frac{\partial^2 u}{\partial x^2} &= 0, \ 0 < x < \ell, \ t > 0, \\
u(x, 0) &= \psi(x), \ 0 < x < \ell, \\
\frac{\partial u}{\partial x}(0, t) &= 0, \ t > 0, \\
\frac{\partial u}{\partial x}(\ell, t) &= 0, \ t > 0,
\end{aligned} \tag{6.14}$$

where ρ, c, κ, ℓ, and ψ are as given before.

Before solving for u, we note that it is easy to predict the long-time behavior of u, the temperature. Since the bar is completely insulated and there is no internal source of heat, the heat energy contained in the bar at time 0 (represented by the initial value of u) will flow from hot regions to colder regions and eventually reach equilibrium at a constant temperature. The purpose of modeling this phenomenon with the heat equation is to obtain *quantitative* information about this process—the process is already understood qualitatively.

We write

$$u(x,t) = b_0(t) + \sum_{n=1}^\infty b_n(t) \cos\left(\frac{n\pi x}{\ell}\right)$$

and then

$$\rho c \frac{\partial u}{\partial t}(x,t) - \kappa \frac{\partial^2 u}{\partial x^2}(x,t) = \rho c \frac{db_0}{dt}(t) + \sum_{n=1}^\infty \left(\rho c \frac{db_n}{dt}(t) + \frac{\kappa n^2 \pi^2}{\ell^2} b_n(t)\right) \cos\left(\frac{n\pi x}{\ell}\right).$$

Since the right-hand side of the PDE is zero, we obtain

$$\rho c \frac{db_0}{dt}(t) = 0, \quad \rho c \frac{db_n}{dt}(t) + \frac{\kappa n^2 \pi^2}{\ell^2} b_n(t) = 0.$$

Also,

$$\psi(x) = d_0 + \sum_{n=1}^\infty d_n \cos\left(\frac{n\pi x}{\ell}\right),$$

where
$$d_0 = \frac{1}{\ell}\int_0^\ell \psi(x)\,dx = \frac{5}{2},$$
$$d_n = \frac{2}{\ell}\int_0^\ell \psi(x)\cos\left(\frac{n\pi x}{\ell}\right)dx$$
$$= \frac{20(2\cos(n\pi/2) - 1 - (-1)^n)}{n^2\pi^2},\ n = 1,2,3,\ldots.$$

Since $u(x,0) = \psi(x)$, we have
$$b_0(0) + \sum_{n=1}^\infty b_n(0)\cos\left(\frac{n\pi x}{\ell}\right) = d_0 + \sum_{n=1}^\infty d_n \cos\left(\frac{n\pi x}{\ell}\right);$$
that is,
$$b_0(0) = d_0,\ b_n(0) = d_n,\ n = 1,2,3,\ldots.$$

Thus we solve
$$\rho c \frac{db_0}{dt}(t) = 0,\ b_0(0) = d_0$$
to get
$$b_0(t) = d_0,$$
and we solve
$$\rho c \frac{db_n}{dt} + \frac{\kappa n^2 \pi^2}{\ell^2} b_n = 0,\ b_n(0) = d_n$$
to get
$$b_n(t) = d_n e^{-\mu_n t},$$
where
$$\mu_n = \frac{\kappa n^2 \pi^2}{\ell^2 \rho c}.$$

Therefore,
$$u(x,t) = d_0 + \sum_{n=1}^\infty d_n e^{-\mu_n t} \cos\left(\frac{n\pi x}{\ell}\right)$$

with $d_0, d_1, d_2, \ldots$ given above. Figure 6.7 is analogous to Figures 6.1 and 6.6, showing snapshots of the temperature distribution at the same times as before. It also shows the equilibrium temperature; an inspection of the formula for $u(x,t)$ makes it clear that $u(x,t) \to d_0 = 5/2$ as $t \to \infty$. ∎

6.2.3 Pure Neumann conditions in a steady-state BVP

The BVP
$$-\kappa \frac{d^2 u}{dx^2} = f(x),\ 0 < x < \ell,$$
$$\frac{du}{dx}(0) = 0,$$
$$\frac{du}{dx}(\ell) = 0$$
(6.15)

6.2. Pure Neumann conditions and the Fourier cosine series

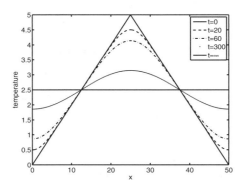

Figure 6.7. *The solution $u(x,t)$ from Example* 6.5 *at times* 0, 20, 60, *and* 300 *(seconds). Ten terms of the Fourier series were used to create these curves.*

describes the steady-state temperature distribution $u(x)$ in a completely insulated bar with a (time-independent) heat source. Up to this point, each BVP or IBVP in Chapters 5 and 6 has had a unique solution for any reasonable choice of the "ingredients" of the PDE: the coefficients (such as the stiffness or the heat conductivity), the initial and boundary conditions, and the right-hand-side function. We have not emphasized this point; rather, we have concentrated on developing solution techniques. However, we must now address the issues of *existence* (does the problem have a solution?) and *uniqueness* (is there only one solution?). The class of PDEs for which we can demonstrate existence and uniqueness by explicitly producing the solution is very small. That is, there are many PDEs for which we cannot derive a formula for the solution; we can only approximate the solution by numerical methods. For those problems, we need to know that a unique solution exists before we try to approximate it.

Indeed, this is an issue with the finite element method, although we did not acknowledge it when we first introduced the technique. We showed in Section 5.5 that the Galerkin method, on which the finite element method is based, produces the best approximation (in the energy norm, and from a certain subspace) to the true solution. However, we *assumed* in that argument that a true solution existed, and that there was only one!

The question of existence and uniqueness is not merely academic. The steady-state BVP (6.15) demonstrates this—it is a physically meaningful problem that a scientist may wish to solve, and yet, as we are about to show, either it does not have a solution or the solution is not unique! We already discussed this fact in Examples 3.14 and 3.17 in Section 3.2. For emphasis, we will repeat the justification from several points of view.

Physical reasoning

The source function $f(x)$ in (6.15) represents the rate at which (heat) energy is added to the bar (in units of energy per time per volume). It depends on x (the position in the bar) because we allow the possibility that different parts of the bar receive different amounts of heat. However, it is independent of time, indicating that the rate at which energy is added is constant. Clearly, for a steady-state solution to exist, the total rate at which energy is added must be zero; if energy is added to one part of the bar, it must be taken away from another part. Otherwise, the total heat energy in the bar would be constantly growing or shrinking, and the temperature could not be constant with respect to time. All of this implies that (6.15) may not have a solution; whether it does depends on the properties of $f(x)$.

On the other hand, if (6.15) does have a solution, it cannot be unique. Knowing only that no net energy is being added to or taken from the bar does not tell us how much energy is in the bar. That is, the energy, and hence the temperature, is not uniquely determined by the BVP.

Mathematical reasoning

Recall that $f(x)$ has units of energy per unit volume per unit time; the total rate at which energy is added to the bar is

$$\int_0^\ell f(x)A\,dx,$$

where A is the cross-sectional area of the bar. By physical reasoning, we concluded that

$$\int_0^\ell f(x)A\,dx = 0,$$

or simply

$$\int_0^\ell f(x)\,dx = 0, \tag{6.16}$$

must hold in order for a solution to exist. But we can see this directly from the BVP. Suppose u is a solution to (6.15). Then

$$\begin{aligned}
\int_0^\ell f(x)\,dx &= -\kappa \int_0^\ell \frac{d^2 u}{dx^2}(x)\,dx \\
&= -\kappa \frac{du}{dx}(x)\Big|_0^\ell \\
&= \kappa \frac{du}{dx}(0) - \kappa \frac{du}{dx}(\ell) \\
&= 0 - 0 = 0.
\end{aligned}$$

Thus, if a solution exists, then the *compatibility condition* (6.16) holds.

Now suppose that a solution u of (6.15) does exist. Let C be any constant, and define a function v by $v(x) = u(x) + C$. Then we have

$$-\kappa \frac{d^2 v}{dx^2}(x) = -\kappa \frac{d^2 u}{dx^2}(x) + 0 = -\kappa \frac{d^2 u}{dx^2}(x) = f(x)$$

and

$$\frac{dv}{dx}(0) = \frac{du}{dx}(0) + 0 = \frac{du}{dx}(0) = 0, \quad \frac{dv}{dx}(\ell) = \frac{du}{dx}(\ell) + 0 = \frac{du}{dx}(\ell) = 0.$$

Therefore, v is another solution of (6.15), and this is true for any real number C. This shows that if (6.15) has a solution, then it has infinitely many solutions.

6.2. Pure Neumann conditions and the Fourier cosine series

The Fourier series method

Suppose that, rather than engage in the above reasoning, we just set out to solve (6.15) by the Fourier series method. Because of the boundary conditions, we write the solution as a cosine series:

$$u(x) = b_0 + \sum_{n=1}^{\infty} b_n \cos\left(\frac{n\pi x}{\ell}\right).$$

Then, computing the cosine coefficients of $-d^2u/dx^2(x)$ in the usual way, we have

$$-\kappa \frac{d^2u}{dx^2}(x) = \sum_{n=1}^{\infty} \frac{\kappa n^2 \pi^2}{\ell^2} b_n \cos\left(\frac{n\pi x}{\ell}\right)$$

(note that the constant term b_0 vanishes when $u(x)$ is differentiated). Also,

$$f(x) = c_0 + \sum_{n=1}^{\infty} c_n \cos\left(\frac{n\pi x}{\ell}\right),$$

where

$$c_0 = \frac{1}{\ell}\int_0^\ell f(x)\,dx, \quad c_n = \frac{2}{\ell}\int_0^\ell f(x)\cos\left(\frac{n\pi x}{\ell}\right)dx, \; n=1,2,\ldots.$$

Equating the series for $-\kappa d^2u/dx^2(x)$ and $f(x)$, we find that the following equations must hold:

$$0 = c_0, \quad \frac{\kappa n^2 \pi^2}{\ell^2} b_n = c_n, \; n = 1, 2, \ldots.$$

Since the coefficients $c_0, c_1, c_2, \ldots$ are given and the coefficients $b_0, b_1, b_2, \ldots$ are to be determined, we can choose

$$b_n = \frac{\ell^2 c_n}{\kappa n^2 \pi^2}, \; n = 1, 2, \ldots,$$

and thereby satisfy all but the first equation. However, the equation

$$c_0 = 0$$

is not a condition on the solution u, but rather on the right-hand-side function f. Indeed, no choice of u can satisfy this equation unless

$$c_0 = \frac{1}{\ell}\int_0^\ell f(x)\,dx = 0.$$

Thus we see again that the compatibility condition (6.16) must hold in order for (6.15) to have a solution.

Moreover, assuming that $c_0 = 0$ does hold, we see that the value of b_0 is undetermined by the BVP. Indeed, with $c_0 = 0$, the general solution of the BVP is

$$u(x) = b_0 + \sum_{n=1}^{\infty} \frac{\ell^2 c_n}{\kappa n^2 \pi^2} \cos\left(\frac{n\pi x}{\ell}\right),$$

where b_0 can take on any value. Thus, when a solution exists, it is not unique.

Relationship to the Fredholm alternative for symmetric linear systems

If $\mathbf{A} \in \mathbf{R}^{n \times n}$ is symmetric and singular (so that $\mathcal{N}(\mathbf{A})$ contains nonzero vectors), then the Fredholm alternative (Theorem 3.19) implies that $\mathbf{Ax} = \mathbf{b}$ has a solution if and only if $\mathbf{b}$ is orthogonal to $\mathcal{N}(\mathbf{A})$. This is the compatibility condition for a symmetric, singular linear system. The compatibility condition (6.16) is precisely analogous, as we now show.

The operator L_N (the negative second derivative operator, restricted to functions satisfying homogeneous Neumann conditions) is symmetric. Moreover, the null space of the operator contains nonzero functions, namely, all nonzero solutions to

$$-\kappa \frac{d^2 u}{dx^2} = 0,\ 0 < x < \ell,$$
$$\frac{du}{dx}(0) = 0, \quad (6.17)$$
$$\frac{du}{dx}(\ell) = 0.$$

It is easy to show, by direct integration, that the solution set consists of all constant functions defined on $[0, \ell]$. Therefore, the compatibility condition suggested by the finite-dimensional situation is that (6.15) has a solution if and only if f is orthogonal to every constant function. But

$$(c, f) = 0 \text{ for all } c \in \mathbf{R}$$

is equivalent to

$$\int_0^\ell c f(x)\, dx = 0 \text{ for all } c \in \mathbf{R},$$

or simply to

$$\int_0^\ell f(x)\, dx = 0.$$

This is precisely (6.16).

Summary

As the above example shows, a steady-state (time-independent) BVP with pure Neumann conditions can have the property that either there is no solution or there is more than one solution. It is important to note, however, that this situation depends on the differential operator as well as on the boundary conditions. For example, the BVP

$$-\kappa \frac{d^2 u}{dx^2} + au = f(x),\ 0 < x < \ell,$$
$$\frac{du}{dx}(0) = 0, \quad (6.18)$$
$$\frac{du}{dx}(\ell) = 0$$

($\kappa, a > 0$) has a unique solution, and there is no compatibility condition imposed on f. The difference between (6.18) and (6.15) is that the differential operator in (6.18) involves the function u itself, not just the derivatives of u. Therefore, unlike in problem (6.15), the addition of a constant to u is not "invisible" to the PDE.

6.2. Pure Neumann conditions and the Fourier cosine series

Exercises

1. Solve the IBVP

$$\frac{\partial u}{\partial t} - \frac{\partial^2 u}{\partial x^2} = 0,\ 0 < x < 1,\ t > 0,$$
$$u(x,0) = x(1-x),\ 0 < x < 1,$$
$$\frac{\partial u}{\partial x}(0,t) = 0,\ t > 0,$$
$$\frac{\partial u}{\partial x}(1,t) = 0,\ t > 0,$$

   and graph the solution at times $0, 0.02, 0.04, 0.06$, along with the steady-state solution.

2. Solve the IBVP

$$\frac{\partial u}{\partial t} - \frac{\partial^2 u}{\partial x^2} = \frac{1}{2} - x,\ 0 < x < 1,\ t > 0,$$
$$u(x,0) = x(1-x),\ 0 < x < 1,$$
$$\frac{\partial u}{\partial x}(0,t) = 0,\ t > 0,$$
$$\frac{\partial u}{\partial x}(1,t) = 0,\ t > 0$$

   and graph the solution at times $0, 0.02, 0.04, 0.06$, along with the steady-state solution.

3. (a) Show that L_N is symmetric.
   (b) Show that the eigenvalues of L_N are nonnegative.

4. Following Section 6.1.4, show how to solve the following IBVP with inhomogeneous Neumann conditions:

$$\rho c \frac{\partial u}{\partial t} - \kappa \frac{\partial^2 u}{\partial x^2} = f(x,t),\ 0 < x < \ell,\ t > t_0,$$
$$u(x,t_0) = \psi(x),\ 0 < x < \ell,$$
$$\frac{\partial u}{\partial x}(0,t) = a(t),\ t > t_0,$$
$$\frac{\partial u}{\partial x}(\ell,t) = b(t),\ t > t_0.$$

5. (a) Consider the following BVP (with *inhomogeneous Neumann conditions*):

$$-\kappa \frac{d^2 u}{dx^2} = f(x),\ 0 < x < \ell,$$
$$\frac{du}{dx}(0) = a,$$
$$\frac{du}{dx}(\ell) = b.$$

Just as in the case of homogeneous Neumann conditions, if there is one solution $u(x)$, there are in fact infinitely many solutions $u(x)+C$, where C is a constant. Also as in the case of homogeneous Neumann conditions, there is a compatibility condition that the right-hand side $f(x)$ must satisfy in order for a solution to exist. What is it?

(b) Does the BVP

$$-\kappa \frac{d^2u}{dx^2} + u = f(x), \ 0 < x < \ell,$$

$$\frac{du}{dx}(0) = a,$$

$$\frac{du}{dx}(\ell) = b$$

(note the difference in the differential equation) require a compatibility condition on $f(x)$, or is there a solution for every $f \in C[0,\ell]$? Justify your answer.

6. Consider a copper bar ($\rho = 8.96\,\text{g/cm}^3$, $c = 0.385\,\text{J/(g K)}$, $\kappa = 4.01\,\text{W/(cm K)}$) of length 1 m and cross-sectional area $2\,\text{cm}^2$. Suppose the sides and the left end ($x = 0$) are perfectly insulated and that heat is added to (or taken from) the other end ($x = 100$, x in centimeters) at a rate of $f(t) = 0.1\sin(60\pi t)\,\text{W}$. If the initial ($t = 0$) temperature of the bar is a constant 25 degrees Celsius, find the temperature distribution $u(x,t)$ of the bar by formulating and solving an IBVP.

7. Consider the IBVP

$$\frac{\partial u}{\partial t} - \frac{\partial^2 u}{\partial x^2} = 0, \ 0 < x < \ell, \ t > 0,$$

$$u(x,0) = \psi(x), \ 0 < x < \ell,$$

$$\frac{\partial u}{\partial x}(0,t) = 0, \ t > 0,$$

$$\frac{\partial u}{\partial x}(\ell,t) = 0, \ t > 0.$$

Show that the solution $u(x,t)$ satisfies

$$\lim_{t \to \infty} u(x,t) = C \text{ for all } x \in (0,\ell)$$

for some constant C, and compute C.

8. Consider the experiment described in Exercise 6.1.8. Suppose that, after the 20 minutes are up, the ends of the bar are removed from the ice bath and are insulated. Compute the eventual steady-state temperature distribution in the bar.

9. Repeat Exercise 8 for the copper bar described in Exercise 6.1.9.

10. Consider an iron bar ($\rho = 7.87\,\text{g/cm}^3$, $c = 0.449\,\text{J/(g K)}$, $\kappa = 0.802\,\text{W/(cm K)}$) of length 1 m and cross-sectional area $2\,\text{cm}^2$. Suppose that the bar is heated to a constant temperature of 30 degrees Celsius, the left end ($x = 0$) is perfectly insulated, and

heat energy is added to (or taken from) the right end ($x = 100$) at the rate of 1 W (the variable x is given in centimeters). Suppose further that heat energy is added to the interior of the bar at a rate of $0.1\,\text{W}/\text{cm}^3$.

(a) Explain why the temperature will not reach steady state.

(b) Compute the temperature distribution $u(x,t)$ by formulating and solving the appropriate IBVP, and verify that $u(x,t)$ does not approach a steady state.

(c) Suppose that, instead of the left end being insulated, heat energy is removed from the left end of the bar. At what rate must energy be removed if the temperature distribution is to reach a steady state?

(d) Verify that your answer to part (c) is correct by formulating and solving the appropriate IBVP and showing that $u(x,t)$ approaches a limit.

(e) Suppose that a steady state is to be achieved by holding the temperature at the left end at a fixed value. What must the value be? Explain.

11. The purpose of this exercise is to show that if $u(x,t)$ satisfies homogeneous Neumann conditions, and $u(x,t)$ is represented by a Fourier sine series, then a formal calculation of the Fourier sine series of $-\partial^2 u/\partial x^2(x,t)$ is invalid. That is, suppose

$$u(x,t) = \sum_{n=1}^{\infty} a_n(t) \sin\left(\frac{n\pi x}{\ell}\right), \quad a_n(t) = \frac{2}{\ell} \int_0^{\ell} u(x,t) \sin\left(\frac{n\pi x}{\ell}\right) dx$$

and that u satisfies

$$\frac{\partial u}{\partial x}(0,t) = \frac{\partial u}{\partial x}(\ell,t) = 0 \text{ for all } t > 0.$$

We wish to show that

$$\sum_{n=1}^{\infty} \frac{n^2 \pi^2}{\ell^2} a_n(t) \sin\left(\frac{n\pi x}{\ell}\right) \tag{6.19}$$

does *not* represent $-\partial^2 u/\partial x^2(x,t)$.

(a) Show that the method of Sections 6.1 and 6.2.2 (two applications of integration by parts) cannot be used to compute the Fourier sine coefficients of $-\partial^2 u/\partial x^2(x,t)$ in terms of $a_1(t), a_2(t), \ldots$.

(b) Let $u(x,t) = t$ and $\ell = 1$. Compute the Fourier sine coefficients

$$a_1(t), a_2(t), a_3(t), \ldots$$

of u. Then compute the series (6.19), and explain why it does not equal $\frac{-\partial^2 u}{\partial x^2(x,t)}$.

6.3 Periodic boundary conditions and the full Fourier series

We now consider the case of a circular (instead of straight) bar, specifically, a set of the form

$$\left\{(x,y,z) \ : \ R - \delta \leq \sqrt{x^2 + y^2} \leq R + \delta, |z| \leq \sqrt{\delta^2 - (\sqrt{x^2+y^2} - R)^2}\right\}$$

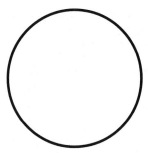

Figure 6.8. *A circular bar.*

(see Figure 6.8). We can model the flow of heat in the ring by the heat equation

$$\rho c \frac{\partial u}{\partial t} - \kappa \frac{\partial^2 u}{\partial x^2} = 0;$$

however, the variable x, representing the cross section of the bar, is now related to the polar angle θ. Purely for mathematical convenience, we will assume that the length of the bar is 2ℓ, and we will label the cross sections by $x \in (-\ell, \ell]$, where $x = \theta \ell/\pi$. In particular, $x = -\ell$ and $x = \ell$ now represent the same cross section of the bar.

There is an additional level of approximation involved in modeling heat flow in a circular bar (as compared to a straight bar) by the one-dimensional heat equation. We mentioned earlier that, in the case of the straight bar, if the initial temperature distribution and the heat source depend only on the longitudinal coordinate, then so does the temperature distribution for any subsequent time. This is not the case for a circular bar, as the geometry suggests (for example, the distance around the ring varies from $2\ell - 2\pi\delta$ to $2\ell + 2\pi\delta$, depending on the path). Nevertheless, the modeling error involved in using the one-dimensional heat equation is not large when the bar is thin (i.e., when δ is small compared to ℓ).

We will consider an initial condition as before:

$$u(x, t_0) = \psi(x), \ -\ell < x \leq \ell.$$

A ring does not have physical boundaries as a straight bar does (a ring has a lateral boundary, which we still assume to be insulated, but it does not have "ends"). However, since $x = -\ell$ represents the same cross section as does $x = \ell$, we still have boundary conditions:

$$u(-\ell, t) = u(\ell, t), \ \frac{\partial u}{\partial x}(-\ell, t) = \frac{\partial u}{\partial x}(\ell, t), \ t > t_0$$

(the temperature and temperature gradient must be the same whether we identify the cross section by $x = -\ell$ or by $x = \ell$). These equations are referred to as *periodic boundary conditions*.

6.3. Periodic boundary conditions and the full Fourier series

We therefore wish to solve the following IBVP:

$$\rho c \frac{\partial u}{\partial t} - \kappa \frac{\partial^2 u}{\partial x^2} = f(x,t), \ -\ell < x < \ell, \ t > t_0,$$
$$u(x,t_0) = \psi(x), \ -\ell < x < \ell,$$
$$u(-\ell,t) = u(\ell,t), \ t > t_0, \quad (6.20)$$
$$\frac{\partial u}{\partial x}(-\ell,t) = \frac{\partial u}{\partial x}(\ell,t), \ t > t_0.$$

As we have seen several times now, the first step is to develop a Fourier series method for the related BVP

$$-\kappa \frac{d^2 u}{dx^2} = f(x), \ -\ell \leq x < \ell,$$
$$u(-\ell) = u(\ell), \quad (6.21)$$
$$\frac{du}{dx}(-\ell) = \frac{du}{dx}(\ell).$$

6.3.1 Eigenpairs of $-\frac{d^2}{dx^2}$ under periodic boundary conditions

We define $L_p : C_p^2[-\ell,\ell] \to C[-\ell,\ell]$ by

$$L_p u = -\frac{d^2 u}{dx^2},$$

where

$$C_p^2[-\ell,\ell] = \left\{ u \in C^2[-\ell,\ell] \ : \ u(-\ell) = u(\ell), \ \frac{du}{dx}(-\ell) = \frac{du}{dx}(\ell) \right\}.$$

Using techniques that are by now familiar, the reader should be able to demonstrate that

- L_p is symmetric, and

- L_p has no negative eigenvalues

(see Exercise 6.3.5).

We first observe that $\lambda_0 = 0$ is an eigenvalue of L_p, and a corresponding eigenfunction is $g_0(x) = 1$. Indeed, it is straightforward to show that the only solutions to

$$-\frac{d^2 u}{dx^2} = 0, \ -\ell < x < \ell,$$
$$u(-\ell) = u(\ell),$$
$$\frac{du}{dx}(-\ell) = \frac{du}{dx}(\ell)$$

are the constant functions.

We next consider positive eigenvalues. Suppose $\lambda = \theta^2$, $\theta > 0$. We must solve

$$-\frac{d^2 u}{dx^2} = \theta^2 u, \quad -\ell \leq x < \ell,$$
$$u(-\ell) = u(\ell), \qquad (6.22)$$
$$\frac{du}{dx}(-\ell) = \frac{du}{dx}(\ell).$$

The general solution to the differential equation is

$$u(x) = c_1 \cos(\theta x) + c_2 \sin(\theta x).$$

The first boundary condition yields the equation

$$c_1 \cos(\theta \ell) + c_2 \sin(\theta \ell) = c_1 \cos(\theta \ell) - c_2 \sin(\theta \ell),$$

while the second yields

$$-c_1 \theta \sin(\theta \ell) + c_2 \theta \cos(\theta \ell) = c_1 \theta \sin(\theta \ell) + c_2 \theta \cos(\theta \ell).$$

Since $\theta > 0$, these equations reduce to

$$c_2 = 0 \text{ or } \sin(\theta \ell) = 0$$

and

$$c_1 = 0 \text{ or } \sin(\theta \ell) = 0,$$

respectively. It follows that the only nonzero solutions of (6.22) correspond to

$$\lambda = \frac{n^2 \pi^2}{\ell^2}, \; n = 1, 2, 3, \ldots,$$

and, moreover, that with one of these values for λ, every function of the form

$$\cos(\theta x) \text{ or } \sin(\theta x)$$

satisfies (6.22).

Now,
$$\cos\left(\frac{n\pi x}{\ell}\right) \text{ and } \sin\left(\frac{n\pi x}{\ell}\right)$$
are linearly independent, which means that there are *two* independent eigenfunctions for each positive eigenvalue, a situation we have not seen before. The complete list of eigenpairs for the negative second derivative operator, subject to periodic boundary conditions, is

$$\lambda_0 = 0, \; \gamma_0(x) = 1;$$
$$\lambda_n = \frac{n^2 \pi^2}{\ell^2}, \; \gamma_n(x) = \cos\left(\frac{n\pi x}{\ell}\right), \; n = 1, 2, 3, \ldots; \qquad (6.23)$$
$$\lambda_n = \frac{n^2 \pi^2}{\ell^2}, \; \psi_n(x) = \sin\left(\frac{n\pi x}{\ell}\right), \; n = 1, 2, 3, \ldots.$$

6.3. Periodic boundary conditions and the full Fourier series

We know that, since L_p is symmetric, eigenfunctions corresponding to *distinct* eigenvalues are orthogonal. That is, if $m \neq n$, then

$$\cos\left(\frac{m\pi x}{\ell}\right)$$

is orthogonal to both

$$\cos\left(\frac{n\pi x}{\ell}\right) \text{ and } \sin\left(\frac{n\pi x}{\ell}\right).$$

Similarly,

$$\sin\left(\frac{m\pi x}{\ell}\right)$$

is orthogonal to both

$$\cos\left(\frac{n\pi x}{\ell}\right) \text{ and } \sin\left(\frac{n\pi x}{\ell}\right).$$

There is no a priori guarantee that the two eigenfunctions corresponding to λ_m are orthogonal to each other. However, a direct calculation shows this to be true. We can also calculate

$$(1,1) = 2\ell, \ (\gamma_n, \gamma_n) = (\psi_n, \psi_n) = \ell, \ n = 1, 2, 3, \ldots. \tag{6.24}$$

The *full Fourier series* of a function $f \in C[-\ell, \ell]$ is given by

$$a_0 + \sum_{n=1}^{\infty} \left\{ a_n \cos\left(\frac{n\pi x}{\ell}\right) + b_n \sin\left(\frac{n\pi x}{\ell}\right) \right\},$$

where

$$\begin{aligned} a_0 &= \frac{1}{2\ell} \int_{-\ell}^{\ell} f(x)\, dx, \\ a_n &= \frac{1}{\ell} \int_{-\ell}^{\ell} f(x) \cos\left(\frac{n\pi x}{\ell}\right) dx, \ n = 1, 2, 3, \ldots, \\ b_n &= \frac{1}{\ell} \int_{-\ell}^{\ell} f(x) \sin\left(\frac{n\pi x}{\ell}\right) dx, \ n = 1, 2, 3, \ldots. \end{aligned} \tag{6.25}$$

We will show in Section 12.6 that the full Fourier series of a $C[-\ell, \ell]$ function converges to it in the mean-square sense (that is, the sequence of partial Fourier series converges to the function in the L^2 norm).

6.3.2 Solving the BVP using the full Fourier series

We now show how to solve the BVP (6.21) using the full Fourier series. Since the method should by now be familiar, we leave the intermediate steps to the exercises.

We write the (unknown) solution of (6.21) as

$$u(x) = a_0 + \sum_{n=1}^{\infty} \left\{ a_n \cos\left(\frac{n\pi x}{\ell}\right) + b_n \sin\left(\frac{n\pi x}{\ell}\right) \right\}, \tag{6.26}$$

where the coefficients $a_0, a_1, a_2, \ldots, b_1, b_2, \ldots$ are to be determined. Then u automatically satisfies the periodic boundary conditions. The full Fourier series for $-d^2u/dx^2$ is given by

$$-\kappa \frac{d^2u}{dx^2}(x) = \sum_{n=1}^{\infty} \left\{ a_n \frac{\kappa n^2 \pi^2}{\ell^2} \cos\left(\frac{n\pi x}{\ell}\right) + b_n \frac{\kappa n^2 \pi^2}{\ell^2} \sin\left(\frac{n\pi x}{\ell}\right) \right\} \quad (6.27)$$

(see Exercise 6.3.6).

We now assume that the source function f has full Fourier series

$$f(x) = c_0 + \sum_{n=1}^{\infty} \left\{ c_n \cos\left(\frac{n\pi x}{\ell}\right) + d_n \sin\left(\frac{n\pi x}{\ell}\right) \right\},$$

where the coefficients $c_0, c_1, c_2, \ldots, d_1, d_2, \ldots$ can be determined explicitly, since f is known. Then the differential equation

$$-\kappa \frac{d^2u}{dx^2} = f(x) \quad (6.28)$$

implies that the Fourier series (6.27) and (6.28) are equal and hence that

$$0 = c_0,$$
$$\frac{\kappa n^2 \pi^2}{\ell^2} a_n = c_n,$$
$$\frac{\kappa n^2 \pi^2}{\ell^2} b_n = d_n.$$

From these equations, we deduce first of all that the source function f must satisfy the compatibility condition

$$\int_{-\ell}^{\ell} f(x)\, dx = 0 \quad (6.29)$$

in order for a solution to exist. If (6.29) holds, then we have

$$a_n = \frac{\ell^2 c_n}{\kappa n^2 \pi^2}, \quad n = 1, 2, 3, \ldots,$$
$$b_n = \frac{\ell^2 d_n}{\kappa n^2 \pi^2}, \quad n = 1, 2, 3, \ldots,$$

but a_0 is not determined by the differential equation or by the boundary conditions. Therefore, for any a_0,

$$u(x) = a_0 + \sum_{n=1}^{\infty} \left\{ \frac{\ell^2 c_n}{\kappa n^2 \pi^2} \cos\left(\frac{n\pi x}{\ell}\right) + \frac{\ell^2 d_n}{\kappa n^2 \pi^2} \sin\left(\frac{n\pi x}{\ell}\right) \right\}$$

is a solution to (6.21) (again, assuming that (6.29) holds).

6.3. Periodic boundary conditions and the full Fourier series

Example 6.6. We consider a thin gold ring, of radius 1 cm. In the above notation, then, $\ell = \pi$. The material properties of gold are

$$\rho = 19.3\,\text{g/cm}^3,\ c = 0.129\,\text{J/(g K)},\ \kappa = 3.17\,\text{W/(cm K)}.$$

We assume that heat energy is being added to the ring at the rate of

$$f(x) = \frac{1}{10}x(x^2 - \pi^2)\,\text{W/cm}^3.$$

Since

$$\int_{-\pi}^{\pi} f(x)\,dx = 0,$$

the BVP

$$-\kappa \frac{d^2 u}{dx^2} = f(x),\ -\pi < x < \pi,$$
$$u(-\pi) = u(\pi),$$
$$\frac{du}{dx}(-\pi) = \frac{du}{dx}(\pi)$$

has a solution. We write

$$u(x) = \sum_{n=1}^{\infty}(a_n \cos(nx) + b_n \sin(nx))$$

(we take $a_0 = 0$, thus choosing one of the infinitely many solutions). We then have

$$-\kappa \frac{d^2 u}{dx^2}(x) = \sum_{n=1}^{\infty}\left(\kappa n^2 a_n \cos(nx) + \kappa n^2 b_n \sin(nx)\right).$$

Also,

$$f(x) = \sum_{n=1}^{\infty}(c_n \cos(nx) + d_n \sin(nx)),$$

where

$$c_n = \frac{1}{\pi}\int_{-\pi}^{\pi} f(x)\cos(nx)\,dx = 0,\ n = 1,2,3,\ldots,$$
$$d_n = \frac{1}{\pi}\int_{-\pi}^{\pi} f(x)\sin(nx)\,dx = \frac{6(-1)^n}{5n^3},\ n = 1,2,3,\ldots.$$

(The value of c_0 is already known to be zero since f satisfies the compatibility condition.) Equating the last two series yields

$$\kappa n^2 a_n = c_n \Rightarrow a_n = \frac{c_n}{\kappa n^2} = 0,\ n = 1,2,3,\ldots,$$
$$\kappa n^2 b_n = d_n \Rightarrow b_n = \frac{d_n}{\kappa n^2} = \frac{12(-1)^n}{10\kappa n^5},\ n = 1,2,3,\ldots.$$

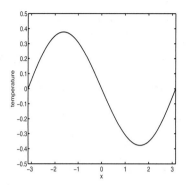

Figure 6.9. *The temperature distribution in Example* 6.6. *This graph was computed using* 10 *terms of the Fourier series.*

Thus

$$u(x) = \sum_{n=1}^{\infty} \frac{12(-1)^n}{10\kappa n^5} \sin(nx).$$

A graph of u is shown in Figure 6.9. We emphasize that this solution really only shows the variation in the temperature—the true temperature is $u(x) + C$ for some real number C. The assumptions above do not give enough information to determine C; we would have to know the amount of heat energy in the ring before the heat source f is applied. ∎

6.3.3 Solving the IBVP using the full Fourier series

Now we solve the IVP (6.20) for the heat equation, using a full Fourier series representation of the solution. As in the previous section, the technique should by now be familiar, and we will leave many of the supporting computations as exercises.

Let $u(x,t)$ be the solution to (6.20). We express u as a full Fourier series with time-varying Fourier coefficients:

$$u(x,t) = a_0(t) + \sum_{n=1}^{\infty} \left\{ a_n(t) \cos\left(\frac{n\pi x}{\ell}\right) + b_n(t) \sin\left(\frac{n\pi x}{\ell}\right) \right\}. \tag{6.30}$$

The series for $-\kappa d^2 u/dx^2$ is

$$-\kappa \frac{\partial^2 u}{\partial x^2}(x,t) = \sum_{n=1}^{\infty} \left\{ \frac{\kappa n^2 \pi^2 a_n(t)}{\ell^2} \cos\left(\frac{n\pi x}{\ell}\right) + \frac{\kappa n^2 \pi^2 b_n(t)}{\ell^2} \sin\left(\frac{n\pi x}{\ell}\right) \right\}. \tag{6.31}$$

As for the time derivative, we can apply Theorem 2.1 to obtain

$$\frac{\partial u}{\partial t}(x,t) = \frac{da_0}{dt}(t) + \sum_{n=1}^{\infty} \left\{ \frac{da_n}{dt}(t) \cos\left(\frac{n\pi x}{\ell}\right) + \frac{db_n}{dt}(t) \sin\left(\frac{n\pi x}{\ell}\right) \right\} \tag{6.32}$$

(see Exercise 6.3.8).

6.3. Periodic boundary conditions and the full Fourier series

We then write the source term $f(x,t)$ as the (time-varying) full Fourier series,

$$f(x,t) = c_0(t) + \sum_{n=1}^{\infty} \left\{ c_n(t) \cos\left(\frac{n\pi x}{\ell}\right) + d_n(t) \sin\left(\frac{n\pi x}{\ell}\right) \right\}, \quad (6.33)$$

where the Fourier coefficients can be computed explicitly from

$$c_0(t) = \frac{1}{2\ell} \int_{-\ell}^{\ell} f(x,t)\,dx,$$

$$c_n(t) = \frac{1}{\ell} \int_{-\ell}^{\ell} f(x,t) \cos\left(\frac{n\pi x}{\ell}\right) dx, \; n = 1,2,3,\ldots,$$

$$d_n(t) = \frac{1}{\ell} \int_{-\ell}^{\ell} f(x,t) \sin\left(\frac{n\pi x}{\ell}\right) dx, \; n = 1,2,3,\ldots.$$

Substituting (6.31), (6.32), and (6.33) into the heat equation and equating coefficients yield

$$\rho c \frac{da_0}{dt}(t) = c_0(t),$$

$$\rho c \frac{da_n}{dt}(t) + \frac{\kappa n^2 \pi^2}{\ell^2} a_n(t) = c_n(t), \; n = 1,2,3,\ldots, \quad (6.34)$$

$$\rho c \frac{db_n}{dt}(t) + \frac{\kappa n^2 \pi^2}{\ell^2} b_n(t) = d_n(t), \; n = 1,2,3,\ldots.$$

We can solve these ODEs for the unknown coefficients once we have derived initial conditions. These come from the initial condition for the PDE, just as in Section 6.1. We write the initial value $\psi(x)$ in a full Fourier series, say

$$\psi(x) = p_0 + \sum_{n=1}^{\infty} \left\{ p_n \cos\left(\frac{n\pi x}{\ell}\right) + q_n \sin\left(\frac{n\pi x}{\ell}\right) \right\}. \quad (6.35)$$

We also have

$$\psi(x) = u(x,0) = a_0(0) + \sum_{n=1}^{\infty} \left\{ a_n(0) \cos\left(\frac{n\pi x}{\ell}\right) + b_n(0) \sin\left(\frac{n\pi x}{\ell}\right) \right\}.$$

This yields the initial conditions

$$a_0(0) = p_0,$$
$$a_n(0) = p_n, \; n = 1,2,3,\ldots, \quad (6.36)$$
$$b_n(0) = q_n, \; n = 1,2,3,\ldots.$$

Solving the IVPs defined by (6.34) and (6.36) yields

$$a_0(t), a_1(t), a_2(t), \ldots, b_1(t), b_2(t), \ldots,$$

and then $u(x,t)$ is given by (6.30).

Example 6.7. We consider again the ring of Example 6.6. We will assume that the temperature in the ring is initially 25 degrees Celsius, and that at time $t = 0$ (t measured in seconds), heat energy is added according to the function f given in that example. We wish to find $u(x,t)$, which describes the evolution of temperature in the ring. To do this, we solve the IBVP

$$\begin{aligned} \rho c \frac{\partial u}{\partial t} - \kappa \frac{\partial^2 u}{\partial x^2} &= f(x), \ -\pi < x < \pi, \ t > 0, \\ u(x,0) &= 25, \ -\pi < x < \pi, \\ u(-\pi,t) &= u(\pi,t), \ t > 0, \\ \frac{\partial u}{\partial t}(-\pi,t) &= \frac{\partial u}{\partial t}(\pi,t), \ t > 0. \end{aligned} \quad (6.37)$$

We write the solution as

$$u(x,t) = a_0(t) + \sum_{n=1}^{\infty} \{a_n(t)\cos(nx) + b_n(t)\sin(nx)\}.$$

We already have the full Fourier series of f, and the initial temperature distribution is given by a full Fourier series with exactly one nonzero term, namely, the constant term 25. We must therefore solve the IVPs

$$\begin{aligned} \rho c \frac{da_0}{dt} &= 0, \ a_0(0) = 25, \\ \rho c \frac{da_n}{dt} + \kappa n^2 a_n &= 0, \ a_n(0) = 0, \ n = 1,2,3,\ldots, \\ \rho c \frac{db_n}{dt} + \kappa n^2 b_n &= d_n, \ b_n(0) = 0, \ n = 1,2,3,\ldots. \end{aligned}$$

The solutions are

$$\begin{aligned} a_0(t) &= 25, \\ a_n(t) &= 0, \ n = 1,2,3,\ldots, \\ b_n(t) &= \frac{12(-1)^n \rho c}{10 \kappa n^5}\left(1 - e^{-\frac{\kappa n^2}{\rho c}t}\right), \ n = 1,2,3,\ldots. \end{aligned}$$

Therefore, the solution is

$$u(x,t) = 25 + \sum_{n=1}^{\infty} \frac{12(-1)^n}{10 \kappa n^5}\left(1 - e^{-\frac{\kappa n^2}{\rho c}t}\right)\sin(nx).$$

Snapshots of this temperature distribution are shown in Figure 6.10. The reader should notice the relationship between the steady-state temperature of the ring and the solution to the previous example. ∎

6.3. Periodic boundary conditions and the full Fourier series

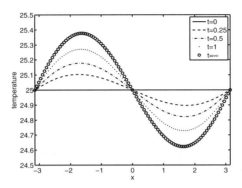

Figure 6.10. *The solution $u(x,t)$ to Example 6.7 at times 0, 0.25, 0.5, and 1 (seconds), along with the steady-state solution. These solutions were estimated using 10 terms in the Fourier series.*

Exercises

1. Repeat Example 6.6, assuming the ring is made of silver instead of gold: $\rho = 10.5\,\text{g/cm}^3$, $c = 0.235\,\text{J/(g K)}$, $\kappa = 4.29\,\text{W/(cm K)}$. How does the steady-state temperature compare to that of the gold ring?

2. Repeat Example 6.7, assuming the ring is made of silver instead of gold: $\rho = 10.5\,\text{g/cm}^3$, $c = 0.235\,\text{J/(g K)}$, $\kappa = 4.29\,\text{W/(cm K)}$.

3. Consider a ring, 5 cm in radius, made of lead (which has physical properties $\rho = 11.3\,\text{g/cm}^3$, $c = 0.129\,\text{J/(g K)}$, $\kappa = 0.353\,\text{W/(cm K)}$). Suppose, by heating one side of the ring, the temperature distribution

$$\psi(x) = 25 + \frac{(5\pi - x)^2(5\pi + x)^2}{3000}$$

is induced. Suppose the heat source is then removed and the ring is allowed to cool.

   (a) Write down an IBVP describing the cooling of the ring.
   (b) Solve the IBVP.
   (c) Find the steady-state temperature of the ring.
   (d) How long does it take the ring to reach steady state (within 1%)?

4. Consider the lead ring of the previous exercise. Suppose the temperature is a constant 25 degrees Celsius, and an (uneven) heat source is applied to the ring. If the heat source delivers heat energy to the ring at a rate of

$$f(x) = 1 - \frac{|x|}{5\pi}\,\text{W/cm}^3,$$

how long does it take for the temperature at the hottest part of the ring to reach 30 degrees Celsius?

5. (a) Show that L_p is symmetric.

   (b) Show that L_p does not have any negative eigenvalues.

6. Assuming that u is a smooth function defined on $[-\ell,\ell]$, the full Fourier series of u is given by (6.26), and u satisfies periodic boundary conditions, show that the full Fourier series of $-d^2u/dx^2$ is given by (6.27).

7. Justify the compatibility condition (6.29)

   (a) by physical reasoning (assume that (6.21) models a steady-state temperature distribution in a circular ring);

   (b) by using the differential equation
   $$-\frac{d^2u}{dx^2}(x) = f(x), \quad -\ell < x < \ell,$$
   and the periodic boundary conditions to compute
   $$\int_{-\ell}^{\ell} f(x)\,dx;$$

   (c) by analogy to the compatibility condition given in the Fredholm alternative.

8. Show that the Fourier series representation (6.32) follows from (6.30) and Theorem 2.1.

9. Consider the IBVP
   $$\rho c \frac{\partial u}{\partial t} - \kappa \frac{\partial^2 u}{\partial x^2} + pu = f(x,t), \quad -\ell < x < \ell, \ t > t_0,$$
   $$u(x,t_0) = \psi(x), \quad -\ell < x < \ell,$$
   $$u(-\ell,t) = u(\ell,t), \ t > t_0,$$
   $$\frac{\partial u}{\partial x}(-\ell,t) = \frac{\partial u}{\partial x}(\ell,t), \ t > t_0.$$

   Assume that p is a constant.

   (a) Derive the solution to the IBVP using the Fourier series method.

   (b) For what values of p does the associated spatial operator have only positive eigenvalues?

10. Let $f : (-\ell,\ell) \to \mathbf{R}$ be an odd function (that is, $f(-x) = -f(x)$ for all $x \in (0,\ell)$). Prove that the full Fourier series of f reduces to the sine series of f, regarded as a function on $(0,\ell)$.

11. Let $f : (-\ell,\ell) \to \mathbf{R}$ be an even function (that is, $f(-x) = f(x)$ for all $x \in (0,\ell)$). Prove that the full Fourier series of f reduces to the cosine series of f, regarded as a function on $(0,\ell)$.

6.4 Finite element methods for the heat equation

To apply the Fourier series method to the heat equation, we used the familiar eigenfunctions to represent the spatial variation of the solution, while allowing the Fourier coefficients to depend on time. We then found the values of these Fourier coefficients by solving ODEs. We can use the finite element method in an analogous fashion. We use finite element functions to approximate the spatial variation of the solution, while the coefficients in the representation depend on time. We end up with a system of ODEs whose solution yields the unknown coefficients.

We will consider again the following IBVP:

$$\rho c \frac{\partial u}{\partial t} - \kappa \frac{\partial^2 u}{\partial x^2} = f(x,t), \ 0 < x < \ell, \ t > t_0,$$
$$u(x,t_0) = \psi(x), \ 0 < x < \ell, \qquad (6.38)$$
$$u(0,t) = 0, \ t > t_0,$$
$$u(\ell,t) = 0, \ t > t_0.$$

We begin by deriving the weak form of the IBVP, following the pattern of Section 5.4.2 (multiply the differential equation by a test function and integrate by parts on the left). In the following calculation, V is the same space of test functions used for the problem in statics: $V = C_D^2[0,\ell]$. We have

$$\rho c \frac{\partial u}{\partial t} - \kappa \frac{\partial^2 u}{\partial x^2} = f(x,t), \ 0 < x < \ell, \ t > t_0$$
$$\Rightarrow \rho c \frac{\partial u}{\partial t}(x,t)v(x) - \kappa \frac{\partial^2 u}{\partial x^2}(x,t)v(x) = f(x,t)v(x), \ 0 < x < \ell, \ t > t_0, \ v \in V$$
$$\Rightarrow \int_0^\ell \left\{ \rho c \frac{\partial u}{\partial t}(x,t)v(x) - \kappa \frac{\partial^2 u}{\partial x^2}(x,t)v(x) \right\} dx = \int_0^\ell f(x,t)v(x)\,dx, \ t > t_0, \ v \in V.$$

We integrate by parts in the second term on the left; the boundary term from integration by parts vanishes because of the boundary conditions on the test function $v(x)$, and we obtain

$$\int_0^\ell \left\{ \rho c \frac{\partial u}{\partial t}(x,t)v(x) + \kappa \frac{\partial u}{\partial x}(x,t) \frac{dv}{dx}(x) \right\} dx$$
$$= \int_0^\ell f(x,t)v(x)\,dx, \ t > t_0 \text{ for all } v \in V.$$

This is the weak form of the IBVP. To get an approximate solution, we apply the Galerkin method with approximating subspace

$$S_n = \{p : [0,\ell] \to \mathbf{R} \ : \ p \text{ is continuous and piecewise linear, } p(0) = p(\ell) = 0\}$$
$$= \text{span}\{\phi_1, \phi_2, \ldots, \phi_{n-1}\}.$$

The piecewise linear functions are defined on the grid $0 = x_0 < x_1 < \cdots < x_n = \ell$, and $\{\phi_1, \phi_2, \ldots, \phi_{n-1}\}$ is the standard basis defined in the previous chapter. For each t, the function $u(\cdot,t)$ must lie in S_n, that is,

$$u_n(x,t) = \sum_{i=1}^{n-1} \alpha_i(t)\phi_i(x). \qquad (6.39)$$

Galerkin's method leads to

$$\int_0^\ell \left\{ \rho c \frac{\partial u_n}{\partial t}(x,t)v(x) + \kappa \frac{\partial u_n}{\partial x}(x,t)\frac{dv}{dx}(x) \right\} dx$$
$$= \int_0^\ell f(x,t)v(x)dx \text{ for all } v \in S_n, \ t > t_0,$$

or

$$\int_0^\ell \left\{ \rho c \frac{\partial u_n}{\partial t}(x,t)\phi_i(x) + \kappa \frac{\partial u_n}{\partial x}(x,t)\frac{d\phi_i}{dx}(x) \right\} dx$$
$$= \int_0^\ell f(x,t)\phi_i(x)dx, \ t > t_0, \ i = 1,2,\ldots,n-1. \quad (6.40)$$

Substituting (6.39) into (6.40) yields, for $t > t_0$ and $i = 1,2,\ldots,n-1$,

$$\sum_{j=1}^{n-1} \frac{d\alpha_j}{dt}(t) \int_0^\ell \rho c \phi_j(x)\phi_i(x)dx + \sum_{j=1}^{n-1} \alpha_j(t) \int_0^\ell \kappa \frac{d\phi_j}{dx}(x)\frac{d\phi_i}{dx}(x)dx$$
$$= \int_0^\ell f(x,t)\phi_i(x)dx. \quad (6.41)$$

If we now define the *mass matrix* **M** and the stiffness matrix **K** by

$$M_{ij} = \int_0^\ell \rho c \phi_j(x)\phi_i(x)dx, \ K_{ij} = \int_0^\ell \kappa \frac{d\phi_j}{dx}(x)\frac{d\phi_i}{dx}(x)dx,$$

and the vector-valued functions $\mathbf{f}(t)$ and $\alpha(t)$ by

$$\mathbf{f}(t) = \begin{bmatrix} \int_0^\ell f(x,t)\phi_1(x)dx \\ \int_0^\ell f(x,t)\phi_2(x)dx \\ \vdots \\ \int_0^\ell f(x,t)\phi_{n-1}(x)dx \end{bmatrix}, \ \alpha(t) = \begin{bmatrix} \alpha_1(t) \\ \alpha_2(t) \\ \vdots \\ \alpha_{n-1}(t) \end{bmatrix},$$

then we can write (6.41) as

$$\mathbf{M}\frac{d\alpha}{dt} + \mathbf{K}\alpha = \mathbf{f}(t).$$

(The names "mass matrix" and "stiffness matrix" for **M** and **K**, respectively, arise from the interpretation for similar matrices appearing in mechanical models. We have already seen the stiffness matrix in Chapter 5; we will see the mass matrix again in Chapter 7. In the context of the heat equation, these names are not particularly meaningful, but the usage is well established.)

The initial condition $u(x,t_0) = \psi(x), \ 0 < x < \ell$ can be approximately implemented as

$$u_n(x,t_0) = \sum_{i=1}^{n-1} \alpha_i(t_0)\phi_i(x) = \sum_{i=1}^{n-1} \psi(x_i)\phi_i(x),$$

6.4. Finite element methods for the heat equation

that is, as $\alpha_i(t_0) = \psi(x_i)$. This is reasonable, because the function

$$\tilde{\psi}(x) = \sum_{i=1}^{n-1} \psi(x_i)\phi_i(x)$$

satisfies

$$\tilde{\psi}(x_j) = \sum_{i=1}^{n-1} \psi(x_i)\phi_i(x_j) = \psi(x_j).$$

We call $\tilde{\psi}$ the *piecewise linear interpolant* of $\psi(x)$ (see Figure 6.11 for an example).

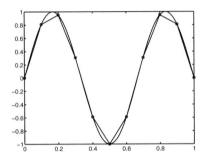

Figure 6.11. $\psi(x) = \sin(3\pi x)$ *and its piecewise linear interpolant.*

We therefore arrive at the following system of ODEs and initial conditions:

$$\mathbf{M}\frac{d\alpha}{dt} + \mathbf{K}\alpha = \mathbf{f}(t), \ t > t_0,$$

$$\alpha(t_0) = \alpha_0,$$

where

$$\alpha_0 = \begin{bmatrix} \psi(x_1) \\ \psi(x_2) \\ \vdots \\ \psi(x_{n-1}) \end{bmatrix}.$$

To solve this, we can multiply the ODE on both sides by $\mathbf{M}^{-1}$ to obtain

$$\frac{d\alpha}{dt} = -\mathbf{M}^{-1}\mathbf{K}\alpha + \mathbf{M}^{-1}\mathbf{f}(t);$$

that is,

$$\frac{d\alpha}{dt} = \mathbf{A}\alpha + \mathbf{g}(t), \ \mathbf{A} = -\mathbf{M}^{-1}\mathbf{K}, \ \mathbf{g}(t) = \mathbf{M}^{-1}\mathbf{f}(t).$$

We recognize this as an inhomogeneous system of linear, constant-coefficient first-order ODEs for the unknown $\alpha(t)$ (if the coefficients in the PDE are nonconstant, then so are the matrices **M** and **K** and hence **A**; in that case, the system of ODEs will not have constant coefficients).[31]

We have now discretized the spatial variation of the solution (a process referred to as *semidiscretization in space*) to obtain a system of ODEs. We can now apply a numerical method to integrate the ODEs. This general technique of solving a time-dependent PDE by integrating the system of (semidiscrete) ODEs is called the *method of lines*.

6.4.1 The method of lines for the heat equation

We now illustrate the method of lines via an example, which will show that the system of ODEs arising from the heat equation is stiff.

Example 6.8. Suppose an iron bar ($\rho = 7.88$, $c = 0.437$, $\kappa = 0.836$) is chilled to a constant temperature of 0 degrees Celsius and then heated internally with both ends maintained at 0 degrees Celsius. Suppose further that the bar is 100 cm in length and heat energy is added at the rate of

$$f(x,t) = 10^{-8} t x (100-x)^2 \text{ W/cm}^3.$$

We wish to find the temperature distribution in the bar after 3 minutes.

The temperature distribution $u(x,t)$ is the solution of the IBVP

$$\rho c \frac{\partial u}{\partial t} - \kappa \frac{\partial^2 u}{\partial x^2} = f(x,t),\ 0 < x < 100,\ t > 0,$$
$$u(x,0) = 0,\ 0 < x < 100,$$
$$u(0,t) = 0,\ t > 0,$$
$$u(100,t) = 0,\ t > 0.$$
(6.42)

We will approximate the solution using piecewise linear finite elements with a regular mesh of 100 subintervals. We write $n = 100$, $h = 100/n$, $x_i = ih$, $i = 0,1,2,\ldots,n$. As usual, $\{\phi_1,\phi_2,\ldots,\phi_{n-1}\}$ will be the standard basis for the subspace S_n of continuous piecewise linear finite elements. It is straightforward to compute the mass and stiffness matrices:

$$M_{ii} = \int_0^{100} \rho c (\phi_i(x))^2 \, dx = \frac{2h\rho c}{3},\ i = 1,2,\ldots,n-1,$$

$$M_{i,i+1} = \int_0^{100} \rho c \phi_i(x) \phi_{i+1}(x) \, dx = \frac{h\rho c}{6},\ i = 1,2,\ldots,n-2,$$

$$K_{ii} = \int_0^{100} \kappa \left(\frac{d\phi_i}{dx}(x)\right)^2 dx = \frac{2\kappa}{h},\ i = 1,2,\ldots,n-1,$$

$$K_{i,i+1} = \int_0^{100} \kappa \frac{d\phi_i}{dx}(x) \frac{d\phi_{i+1}}{dx}(x) \, dx = -\frac{\kappa}{h},\ i = 1,2,\ldots,n-2$$

---

[31] In practice, we do not actually compute either $\mathbf{M}^{-1}$ or **A**. When implementing numerical algorithms, it is rarely efficient to compute an inverse matrix, particularly when the matrix is sparse, as in this case. Instead, the presence of $\mathbf{M}^{-1}$ is a signal that a linear system with coefficient matrix **M** must be solved. This is explained below when we discuss the implementation of Euler's method and the backward Euler method.

6.4. Finite element methods for the heat equation

(both **M** and **K** are tridiagonal and symmetric). We also have

$$f_i(t) = \int_0^{100} f(x,t)\phi_i(x)\,dx$$
$$= t\frac{h^2(60000i - 200h - 1200i^2h + 6h^2i^3 + 3ih^2)}{6 \cdot 10^8}, \; i = 1,2,\ldots,n-1.$$

We first try taking $N = 180$ steps ($\Delta t = 1\,\text{s}$) of Euler's method; however, the result is meaningless, with temperatures on the order of 10^{40} degrees Celsius! Clearly Euler's method is unstable with this choice of Δt. A little experimentation shows that Δt cannot be much more than 0.7 seconds, or instability will result. The temperature distribution at $t = 180$ seconds, computed using 260 time steps ($\Delta t \doteq 0.69$), is shown in Figure 6.12.

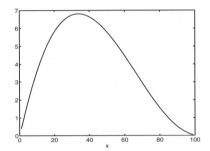

Figure 6.12. *The temperature distribution in Example* 6.8 *after* 180 *seconds (computed using the forward Euler method, with* $\Delta t \doteq 0.69$*).*

Before leaving this example, we should explain how Euler's method is implemented in practice. The system of ODEs is

$$\frac{d\alpha}{dt} = \mathbf{M}^{-1}(-\mathbf{K}\alpha + \mathbf{f}(t)),$$

and so Euler's method takes the form

$$\alpha^{(i+1)} = \alpha^{(i)} + \Delta t \mathbf{M}^{-1}\left(-\mathbf{K}\alpha^{(i)} + \mathbf{f}(t_i)\right).$$

As we mentioned before, it is not efficient to actually compute $\mathbf{M}^{-1}$, since $\mathbf{M}$ is tridiagonal and $\mathbf{M}^{-1}$ is completely dense. Instead, we implement the above iteration as

$$\alpha^{(i+1)} = \alpha^{(i)} + \Delta t \mathbf{s}^{(i)},$$

where $\mathbf{s}^{(i)}$ is found by solving

$$\mathbf{M}\mathbf{s}^{(i)} = -\mathbf{K}\alpha^{(i)} + \mathbf{f}(t_i).$$

Therefore, one tridiagonal solve is required during each iteration, at a cost of $O(8n)$ operations, instead of a multiplication by a dense matrix, which costs $O(2n^2)$ operations.[32] ∎

---

[32]We can even do a bit better, by *factoring* the matrix **M** once (outside the main loop) and then using the factors in the loop to solve for $\mathbf{s}^{(i)}$. This reduces the cost to $O(6n)$ per iteration. Factorization of matrices is discussed briefly in Section 13.2.

Although the time step of $\Delta t = 0.7$ seconds, required in Example 6.8 for stability, does not seem excessively small, it is easy to show by numerical experimentation that a time step Δt on the order of h^2 is required for stability. (Exercise 6.4.10 asks the reader to verify this for Example 6.8.) That is, if we wish to refine the spatial grid by a factor of 2, we must reduce the time step by a factor of 4. This is the typical situation when we apply the method of lines to the heat equation: The resulting system of ODEs is stiff, with the degree of stiffness increasing as the spatial grid is refined.

The reason for the stiffness in the heat equation is easy to understand on an intuitive level. The reader should recall from Section 4.5 that stiffness arises when a (physical) system has components that decay at very different rates. In heat flow, the components of interest are the spatial frequencies in the temperature distribution. High-frequency components are damped out very quickly, as we saw from the Fourier series solution (see Section 6.1.1). Low-frequency components, on the other hand, decay more slowly. Moreover, as we refine the spatial grid, we can represent higher frequencies, and so the degree of stiffness worsens.

Example 6.9. We now apply the backward Euler method to the previous example. Using $n = 100$ subintervals in space and $\Delta t = 2$ (seconds), we obtain the result shown in Figure 6.13. Using a time step for which Euler's method is unstable, we produced a result which is qualitatively correct.

The backward Euler method takes the form

$$\alpha^{(i+1)} = \alpha^{(i)} + \Delta t \mathbf{M}^{-1}\left(-\mathbf{K}\alpha^{(i+1)} + \mathbf{f}(t_{i+1})\right).$$

Multiplying through by $\mathbf{M}$ and manipulating, we obtain

$$(\mathbf{M} + \Delta t \mathbf{K})\alpha^{(i+1)} = \mathbf{M}\alpha^i + \Delta t \mathbf{f}(t_{i+1}).$$

We therefore compute $\alpha^{(i+1)}$ by solving a system with the (tridiagonal) matrix $\mathbf{M} + \Delta t \mathbf{K}$ as the coefficient matrix. ■

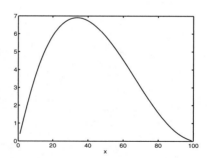

Figure 6.13. *The temperature distribution in Example 6.8 after 180 seconds (computed using the backward Euler method, with $\Delta t = 2$).*

Exercises

1. Show that the mass matrix $\mathbf{M}$ is nonsingular. (Hint: This follows from the general result that if $\{\mathbf{u}_1, \mathbf{u}_2, \ldots, \mathbf{u}_n\}$ is a linearly independent set in an inner product space,

6.4. Finite element methods for the heat equation

then the *Gram* matrix $\mathbf{G}$ defined by $G_{ij} = (\mathbf{u}_j, \mathbf{u}_i)$ is nonsingular. To prove this, suppose $\mathbf{Gx} = 0$. Then $(\mathbf{x}, \mathbf{Gx}) = 0$, and by expanding $(\mathbf{x}, \mathbf{Gx})$ in terms of the inner products $(\mathbf{u}_j, \mathbf{u}_i)$, it is possible to show that $(\mathbf{x}, \mathbf{Gx}) = 0$ is only possible if $\mathbf{x} = 0$. This suffices to show that $\mathbf{G}$ is nonsingular.)

2. Consider a function f in $C[0, \ell]$.

   (a) Show how to compute the projection of f onto the subspace S_n. Be sure to observe how the mass matrix $\mathbf{M}$ and load vector $\mathbf{f}$ arise naturally in this problem.

   (b) We can also find an approximation to f from S_n by computing the piecewise linear interpolant of f. Show by a specific example that the projection of f onto S_n and the piecewise linear interpolant of f from S_n are not the same. (The number n can be chosen small to make the computations simple.)

3. Apply the method of lines to the IBVP

$$\frac{\partial u}{\partial t} - \frac{\partial^2 u}{\partial x^2} = x(1-x)\cos(t),\ 0 < x < 1,\ t > 0,$$
$$u(x,0) = 1,\ 0 < x < 1,$$
$$u(0,t) = 0,\ t > t_0,$$
$$u(1,t) = 0,\ t > t_0.$$

Use the finite element method, with the approximating subspace S_3 (and a regular grid), to do the discretization in space.

   (a) Explicitly compute the mass matrix $\mathbf{M}$, the stiffness matrix $\mathbf{K}$, and the load vector $\mathbf{f}(t)$.

   (b) Explicitly set up the system of ODEs resulting from applying the method of lines.

4. Consider a heterogeneous bar, that is, a bar in which the material properties ρ, c, and κ are not constants but rather functions of space: $\rho = \rho(x)$, $c = c(x)$, $\kappa = \kappa(x)$.

   (a) What is the appropriate form of the heat equation for this bar?

   (b) What is the appropriate weak form of the resulting IBVP (assuming homogeneous Dirichlet conditions)?

   (c) How does the system of ODEs change?

5. An advantage of the weak formulation of a BVP (and therefore of the finite element method) is that the equation makes sense for discontinuous coefficients. Consider a metal bar of length 1 m and diameter 4 cm. Assume that half of the bar is made of copper ($c = 0.379\,\text{J}/(\text{g K})$, $\rho = 8.97\,\text{g/cm}^3$, $\kappa = 4.04\,\text{W}/(\text{cm K})$), and the other half is made of iron (specific heat $c = 0.437\,\text{J}/(\text{g K})$, density $\rho = 7.88\,\text{g/cm}^3$, thermal conductivity $\kappa = 0.836\,\text{W}/(\text{cm K})$). Suppose the bar is initially heated to 5 degrees Celsius and then (at time 0) its ends are placed in an ice bath (0 degrees Celsius).

   (a) Formulate the IBVP describing this experiment.

(b) Apply the finite element method and the method of lines to obtain the resulting system of ODEs. Use only a uniform grid with an even number of subintervals (so that the midpoint of the bar is always a grid point). Give the mass matrix **M** and the stiffness matrix **K** explicitly.

6. Consider the IBVP with inhomogeneous Dirichlet conditions:

$$\rho c \frac{\partial u}{\partial t} - \kappa \frac{\partial^2 u}{\partial x^2} = f(x,t),\ 0 < x < \ell,\ t > t_0,$$
$$u(x,t_0) = \psi(x),\ 0 < x < \ell, \qquad (6.43)$$
$$u(0,t) = a(t),\ t > t_0,$$
$$u(\ell,t) = b(t),\ t > t_0.$$

(a) Formulate the weak form of the IBVP.

(b) Show how to apply the finite element method by representing the approximate solution in the form $u_n(x,t) = v_n(x,t) + g_n(x,t)$, where

$$v_n(x,t) = \sum_{i=1}^{n-1} \alpha_i(t) \phi_i(x)$$

and

$$g_n(x,t) = a(t)\phi_0(x) + b(t)\phi_n(x).$$

(c) Illustrate by solving (6.43) with ρ, c, κ equal to the material constants for iron, $\ell = 100$ cm, $\psi(x) = 0$ degrees Celsius, $a(t) = 0$, $b(t) = \sin(60\pi t)$.

7. Consider the IBVP from Examples 6.8 and 6.9.

(a) Using the method of Fourier series, find the exact solution $u(x,t)$.

(b) Using enough terms in the Fourier series to obtain a highly accurate solution, evaluate $u(x, 180)$ on the regular grid with $n = 100$ used in Examples 6.8 and 6.9.

(c) Reproduce the numerical results in Examples 6.8 and 6.9, and, by comparing to the Fourier series result, determine the accuracy of each result.

8. Repeat Example 6.8, with the following changes. First, assume that the bar is made of copper ($\rho = 8.96$ g/cm^3, $c = 0.385$ J/(g K), $\kappa = 4.01$ W/(cm K)) instead of iron. Second, assume that the heat "source" is given by

$$f(x,t) = 10^{-7} t x (60 - x)(100 - x)$$

(note that f adds energy over part of the interval and takes it away over another part). Graph the temperature after 180 seconds. Does the temperature approach a steady state as $t \to \infty$?

9. Consider a heterogeneous bar of length 100 cm whose material properties are given by the following formulas:

$$\rho(x) = 7.5 + 0.01x \text{ g/cm}^3, \ 0 < x < 100,$$
$$c(x) = 0.45 + 0.0001x \text{ J/(g K)}, \ 0 < x < 100,$$
$$\kappa(x) = 2.5 + 0.05x \text{ g/(cm K)}, \ 0 < x < 100.$$

Suppose that the initial temperature in the bar is a uniform 5 degrees Celsius, and that at $t = 0$ both ends are placed in ice baths (while the lateral boundary of the bar is perfectly insulated).

(a) Formulate the IBVP describing this experiment.

(b) Formulate the weak form of the IBVP.

(c) Use the finite element method with backward Euler integration to estimate the temperature after 2 minutes.

10. Consider Example 6.8, and suppose that the number of elements in the mesh is increased from 100 to 200, so that the mesh size h is cut in half. Show, by numerical experimentation, that the time step Δt in the forward Euler method must be reduced by a factor of approximately four to preserve stability.

6.5 Finite elements and Neumann conditions

So far we have only used finite element methods for problems with Dirichlet boundary conditions. The weak form of the BVP or IBVP, on which the finite element method is based, incorporates the Dirichlet conditions in the definition of V, the space of test functions. When the weak form is discretized via the Galerkin method, the boundary conditions form part of the definition of S_n, the approximating subspace (see (5.46)).

It turns out that Neumann conditions are even easier to handle in the finite element method. As we show below, a Neumann condition does not appear explicitly in the weak form or in the definition of the approximating subspace (the analogue of S_n). For this reason, a Neumann condition is often called a *natural boundary condition* (since it is satisfied automatically by a solution of the weak form), while a Dirichlet condition is referred to as an *essential boundary condition* (since it is essential to include the condition explicitly in the weak form).

6.5.1 The weak form of a BVP with Neumann conditions

We will first consider the (time-independent) BVP

$$-\frac{d}{dx}\left(k(x)\frac{du}{dx}\right) = f(x), \ 0 < x < \ell,$$
$$\frac{du}{dx}(0) = 0, \qquad \qquad \qquad (6.44)$$
$$\frac{du}{dx}(\ell) = 0.$$

To derive the weak form, we multiply both sides of the differential equation by a test function and integrate by parts. We take for our space of test functions
$$\tilde{V} = C^2[0, \ell]$$
(note that the boundary conditions do not appear in this definition).

Assuming that u solves (6.44), we have, for each $v \in \tilde{V}$,

$$-\frac{d}{dx}\left(k(x)\frac{du}{dx}(x)\right)v(x) = f(x)v(x), \ 0 < x < \ell$$

$$\Rightarrow -\int_0^\ell \frac{d}{dx}\left(k(x)\frac{du}{dx}(x)\right)v(x)\,dx = \int_0^\ell f(x)v(x)\,dx$$

$$\Rightarrow -\left[k(x)\frac{du}{dx}(x)v(x)\right]_0^\ell + \int_0^\ell k(x)\frac{du}{dx}(x)\frac{dv}{dx}(x)\,dx = \int_0^\ell f(x)v(x)\,dx$$

$$\Rightarrow \int_0^\ell k(x)\frac{du}{dx}(x)\frac{dv}{dx}(x)\,dx = \int_0^\ell f(x)v(x)\,dx.$$

In the last step, we use the fact that
$$\frac{du}{dx}(0) = \frac{du}{dx}(\ell) = 0.$$

The weak form of (6.44) is

find $u \in \tilde{V}$ such that $\int_0^\ell k(x)\frac{du}{dx}(x)\frac{dv}{dx}(x)\,dx = \int_0^\ell f(x)v(x)\,dx$ for all $v \in \tilde{V}$. (6.45)

In the context of (6.45), we refer to (6.44) as the strong form of the BVP.

6.5.2 Equivalence of the strong and weak forms of a BVP with Neumann conditions

The derivation of the weak form (6.45) shows that if u satisfies the original BVP (6.44) (the strong form), then u also satisfies the weak form. We now demonstrate the converse: If u satisfies the weak form (6.45) of the BVP, then u also satisfies (6.44)—including the boundary conditions! This is quite surprising at first sight, since the boundary conditions mentioned in (6.44) are not mentioned in (6.45).

We assume that $u \in \tilde{V}$ and that
$$\int_0^\ell k(x)\frac{du}{dx}(x)\frac{dv}{dx}(x)\,dx = \int_0^\ell f(x)v(x)\,dx \text{ for all } v \in \tilde{V}.$$

Integrating by parts on the left side of this equation yields

$$\left[k(x)\frac{du}{dx}(x)v(x)\right]_0^\ell - \int_0^\ell \frac{d}{dx}\left[k(x)\frac{du}{dx}(x)\right]v(x)\,dx = \int_0^\ell f(x)v(x)\,dx \text{ for all } v \in \tilde{V}. \quad (6.46)$$

Now, $V \subset \tilde{V}$, where V is defined as the space of test functions used for a Dirichlet problem:
$$V = C_D^2[0, \ell] = \left\{v \in \tilde{V} \ : \ v(0) = v(\ell) = 0\right\}.$$

6.5. Finite elements and Neumann conditions

Therefore, in particular, (6.46) holds for all $v \in V$. Since the boundary term in (6.46) vanishes when $v(0) = v(\ell) = 0$, we obtain

$$-\int_0^\ell \frac{d}{dx}\left[k(x)\frac{du}{dx}(x)\right] v(x)\,dx = \int_0^\ell f(x)v(x)\,dx \text{ for all } v \in V,$$

or

$$\int_0^\ell \left\{\frac{d}{dx}\left[k(x)\frac{du}{dx}(x)\right] + f(x)\right\} v(x)\,dx = 0 \text{ for all } v \in V.$$

Using the same argument as in Section 5.4.2, we see that

$$\frac{d}{dx}\left[k(x)\frac{du}{dx}(x)\right] + f(x) = 0, \ 0 < x < \ell,$$

must hold, or equivalently, that

$$-\frac{d}{dx}\left[k(x)\frac{du}{dx}(x)\right] = f(x), \ 0 < x < \ell.$$

Thus, if u satisfies the weak form of the BVP, it must satisfy at least the differential equation appearing in the strong form.

It is now easy to show that the boundary conditions also hold. The condition (6.46) is equivalent to

$$-\left[k(x)\frac{du}{dx}(x)v(x)\right]_0^\ell + \int_0^\ell \left\{\frac{d}{dx}\left[k(x)\frac{du}{dx}(x)\right] + f(x)\right\} v(x)\,dx = 0 \text{ for all } v \in \tilde{V},$$

and, since we have already shown that

$$\frac{d}{dx}\left[k(x)\frac{du}{dx}(x)\right] + f(x) = 0, \ 0 < x < \ell,$$

we see that

$$\left[k(x)\frac{du}{dx}(x)v(x)\right]_0^\ell = 0 \text{ for all } v \in \tilde{V}. \tag{6.47}$$

We now take, for example, $v(x) = 1 - x/\ell$, which certainly belongs to $\tilde{V}$. With this choice of v, (6.47) becomes

$$-k(0)\frac{du}{dx}(0) = 0$$

(since $v(0) = 1$, $v(\ell) = 0$). Since $k(0) > 0$ by assumption, this can hold only if

$$\frac{du}{dx}(0) = 0.$$

Similarly, choosing $v(x) = x/\ell$ shows that

$$\frac{du}{dx}(\ell) = 0$$

must hold. Thus, when u satisfies the weak form (6.45), it must necessarily satisfy the Neumann boundary conditions. This completes the proof that the strong and weak forms of BVP (6.44) are equivalent.

6.5.3 Piecewise linear finite elements with Neumann conditions

Now that we have the weak form of the BVP, we choose the appropriate subspace of piecewise linear functions and apply the Galerkin technique to obtain a finite element method. Since no boundary conditions are imposed in the weak form, we augment the space S_n defined in Section 5.6 by adding the basis functions ϕ_0 and ϕ_n (see Figure 6.14). We denote the resulting subspace by $\tilde{S}_n$:

$$\tilde{S}_n = \{p : [0, \ell] \to \mathbf{R} \ : \ p \text{ is continuous and piecewise linear}\}$$
$$= \text{span}\{\phi_0, \phi_1, \ldots, \phi_n\}.$$

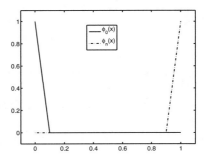

Figure 6.14. *The piecewise linear basis functions ϕ_0 (solid curve) and ϕ_n (dashed curve).*

Having defined the approximating subspace $\tilde{S}_n$, we apply the Galerkin method to the weak form (6.45), yielding the problem

$$\text{find } v_n \in \tilde{S}_n \text{ such that } \int_0^\ell k(x)\frac{dv_n}{dx}(x)\frac{dv}{dx}(x)\,dx = \int_0^\ell f(x)v(x)\,dx \text{ for all } v \in \tilde{S}_n,$$

or, equivalently,

$$\text{find } v_n \in \tilde{S}_n \text{ such that } \int_0^\ell k(x)\frac{dv_n}{dx}(x)\frac{d\phi_i}{dx}(x)\,dx = \int_0^\ell f(x)\phi_i(x)\,dx, \ i = 0, 1, \ldots, n. \tag{6.48}$$

For simplicity of notation, we will again write

$$a(v, w) = \int_0^\ell k(x)\frac{dv}{dx}(x)\frac{dw}{dx}(x)\,dx,$$

so that (6.48) can be written as

$$\text{find } v_n \in \tilde{S}_n \text{ such that } a(v_n, \phi_i) = (f, \phi_i), \ i = 0, 1, \ldots, n. \tag{6.49}$$

We then write the unknown v_n as

$$v_n = \sum_{j=0}^n \tilde{u}_j \phi_j.$$

6.5. Finite elements and Neumann conditions

and substitute into (6.49). The result is

$$\sum_{i=0}^{n} a(\phi_j, \phi_i)\tilde{u}_j = (f, \phi_i), \ i = 0, 1, \ldots, n.$$

This is a system of $n+1$ equations in the $n+1$ unknowns $\tilde{u}_0, \tilde{u}_1, \ldots, \tilde{u}_n$. We can write the system as the matrix-vector equation $\tilde{\mathbf{K}}\tilde{\mathbf{u}} = \tilde{\mathbf{f}}$ by defining

$$\tilde{K}_{ij} = a(\phi_j, \phi_i), \ i, j = 0, 1, \ldots, n,$$
$$\tilde{f}_i = (f, \phi_i), \ i = 0, 1, \ldots, n.$$

The calculations involved in forming the matrix $\tilde{\mathbf{K}}$ and the vector $\tilde{\mathbf{f}}$ are exactly the same as for a Dirichlet problem, except that ϕ_0 and ϕ_n are qualitatively different from $\phi_1, \phi_2, \ldots, \phi_{n-1}$ (ϕ_0 and ϕ_n are each nonzero on a single subinterval, namely $[x_0, x_1]$ and $[x_{n-1}, x_n]$, respectively, while each ϕ_i, $i = 1, 2, \ldots, n-1$, is nonzero on $[x_{i-1}, x_{i+1}]$).

Having now reduced the problem to the linear system $\tilde{\mathbf{K}}\tilde{\mathbf{u}} = \tilde{\mathbf{f}}$, where now $\tilde{\mathbf{K}} \in \mathbf{R}^{(n+1)\times(n+1)}$ and $\tilde{\mathbf{u}}, \tilde{\mathbf{f}} \in \mathbf{R}^{n+1}$, we face the difficulty that $\tilde{\mathbf{K}}$ is a singular matrix. This is not surprising, since we know that the original BVP (6.44) does not have a unique solution. We can show directly that $\tilde{\mathbf{K}}$ is singular, and understand the nature of the singularity, as follows. If

$$\mathbf{u}_c = \begin{bmatrix} 1 \\ 1 \\ \vdots \\ 1 \end{bmatrix},$$

then

$$(\tilde{\mathbf{K}}\mathbf{u}_c)_i = \sum_{j=0}^{n} \tilde{K}_{ij}(\mathbf{u}_c)_j = \sum_{j=0}^{n} a(\phi_j, \phi_i) = a\left(\sum_{j=0}^{n} \phi_j, \phi_i\right)$$
$$= \int_0^\ell k(x) \frac{d}{dx}\left[\sum_{j=0}^{n} \phi_j(x)\right] \frac{d\phi_i}{dx}(x)\, dx.$$

But $\sum_{j=0}^{n} \phi_j(x)$ is piecewise linear and has value one at each mesh node. It follows that $\sum_{j=0}^{n} \phi_j(x)$ is the constant function one, and hence has derivative zero. Therefore $(\tilde{\mathbf{K}}\mathbf{u}_c)_i = 0$ for each i, which shows that $\tilde{\mathbf{K}}\mathbf{u}_c = 0$ and hence that $\tilde{\mathbf{K}}$ is singular. Moreover, it can be shown that $\mathbf{u}_c$ spans the null space of $\tilde{\mathbf{K}}$; that is, if $\tilde{\mathbf{K}}\tilde{\mathbf{u}} = 0$, then $\tilde{\mathbf{u}}$ is a multiple of $\mathbf{u}_c$ (see Exercise 6.5.5).

The fact that $\tilde{\mathbf{K}}$ is singular means that we must give special attention to the process of solving $\tilde{\mathbf{K}}\tilde{\mathbf{u}} = \tilde{\mathbf{f}}$. In particular, if we ignore the singularity of $\tilde{\mathbf{K}}$ and solve $\tilde{\mathbf{K}}\tilde{\mathbf{u}} = \tilde{\mathbf{f}}$ using computer software, we will get either a meaningless solution or an error message. To solve this singular system correctly, we must add another equation (one additional equation, correctly chosen, will be sufficient, because the null space of $\tilde{\mathbf{K}}$ is one dimensional). A simple choice is the equation $\tilde{u}_n = 0$; this is equivalent to choosing, out of the infinitely many solutions to (6.44), the one with $u(\ell) = 0$. Moreover, we can impose this additional equation by simply removing the last row and column from $\tilde{\mathbf{K}}$, and the last entry from $\tilde{\mathbf{f}}$. The last column of $\tilde{\mathbf{K}}$ consists of the coefficients of the terms in the equations involving $\tilde{u}_n$; if we insist that $\tilde{u}_n = 0$, then we can remove these terms from all $n+1$ equations. We then have $n+1$ equations in the n unknowns $\tilde{u}_0, \tilde{u}_1, \ldots, \tilde{u}_{n-1}$. This is one more equation than we need, so we remove one equation, the last, to obtain a square system. It can be proved that the resulting $n \times n$ system is nonsingular (see Exercise 6.5.11).[33]

We could have removed a different row and column. If we remove the ith row and column, we are selecting the approximate solution v_n satisfying $v_n(x_i) = 0$.

Example 6.10. Consider the Neumann problem

$$-\frac{d^2 u}{dx^2} = x - \frac{1}{2}, \ 0 < x < 1,$$
$$\frac{du}{dx}(0) = 0, \qquad (6.50)$$
$$\frac{du}{dx}(1) = 0.$$

It is easy to show (by direct integration) that $u(x) = -x^3/6 + x^2/4 + C$ is a solution for any constant C, and that $u(x) = -x^3/6 + x^2/4 - 1/12$ is the solution satisfying $u(1) = 0$.

We will use a regular grid with the finite element method. We define $x_i = ih$, $i = 0, 1, \ldots, n$, $h = 1/n$. With piecewise linear finite elements, as we have seen before, the stiffness matrix $\tilde{\mathbf{K}}$ is tridiagonal since

$$|j - i| > 1 \Rightarrow \frac{d\phi_i}{dx}(x) \frac{d\phi_j}{dx}(x) = 0 \text{ on } (0, \ell).$$

We compute as follows (as suggested above, the computations involving ϕ_0 and ϕ_n must be handled separately from those involving $\phi_1, \phi_2, \ldots, \phi_{n-1}$):

$$\tilde{K}_{ii} = \int_{(i-1)h}^{(i+1)h} \left(\frac{d\phi_i}{dx}(x)\right)^2 dx = \int_{(i-1)h}^{(i+1)h} \frac{1}{h^2} dx = \frac{2}{h}, \ i = 1, 2, \ldots, n-1,$$

$$\tilde{K}_{00} = \int_0^h \left(\frac{d\phi_0}{dx}(x)\right)^2 dx = \int_0^h \frac{1}{h^2} dx = \frac{1}{h},$$

$$\tilde{K}_{nn} = \int_{1-h}^1 \left(\frac{d\phi_n}{dx}(x)\right)^2 dx = \int_{1-h}^1 \frac{1}{h^2} dx = \frac{1}{h},$$

$$\tilde{K}_{i,i+1} = \int_{ih}^{(i+1)h} \frac{d\phi_i}{dx}(x) \frac{d\phi_{i+1}}{dx}(x) dx = \int_{ih}^{(i+1)h} \left(-\frac{1}{h^2}\right) dx$$
$$= -\frac{1}{h}, \ i = 0, 1, \ldots, n-1.$$

---

[33] The method described here for solving the singular linear system has the advantage of simplicity, but it makes the resulting nonsingular matrix more *ill-conditioned* than necessary. This increases the error due to roundoff when the system is solved on a computer in finite-precision arithmetic. The conditioning of a matrix is a quantitative measure of how close the matrix is to being singular, and it is always desirable to have a well-conditioned matrix (that is, a matrix that is far from being singular). For methods that do not lead to any unnecessary ill-conditioning, the interested reader can consult [6] or Section 11.5 of [23].

6.5. Finite elements and Neumann conditions

Since $\tilde{\mathbf{K}}$ is symmetric and tridiagonal, this completes the computation of $\tilde{\mathbf{K}}$. Next, we have

$$\tilde{f}_i = \int_{(i-1)h}^{(i+1)h} \left(x - \frac{1}{2}\right) \phi_i(x)\,dx$$

$$= \int_{(i-1)h}^{ih} \left(x - \frac{1}{2}\right) \frac{(x - (i-1)h)}{h}\,dx + \int_{ih}^{(i+1)h} \left(x - \frac{1}{2}\right) \left(1 - \frac{x - ih}{h}\right)\,dx$$

$$= ih^2 - \frac{h}{2},$$

$$\tilde{f}_0 = \int_0^h \left(x - \frac{1}{2}\right) \left(1 - \frac{x}{h}\right)\,dx$$

$$= \frac{h^2}{6} - \frac{h}{4},$$

$$\tilde{f}_n = \int_{1-h}^1 \left(x - \frac{1}{2}\right) \frac{x - (1-h)}{h}\,dx$$

$$= \frac{h}{4} - \frac{h^2}{6}.$$

As suggested above, we compute a solution $\tilde{\mathbf{u}}$ to $\tilde{\mathbf{K}}\tilde{\mathbf{u}} = \tilde{\mathbf{f}}$ by removing the last row and column from $\tilde{\mathbf{K}}$ and the last component from $\tilde{\mathbf{f}}$, and solving the resulting square, nonsingular system. The result, for $n = 10$, is shown in Figure 6.15. ∎

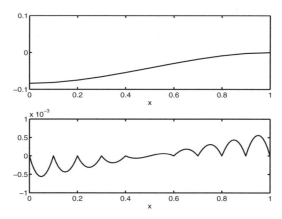

Figure 6.15. *Top: The exact solution (solid curve) and the computed solution (dashed curve). Bottom: The error in the computed solution for (6.50). The computed solution was found using piecewise linear finite elements with a regular grid of 10 subintervals.*

6.5.4 Inhomogeneous Neumann conditions

The finite element method requires only minor modifications to handle the inhomogeneous Neumann problem

$$-\frac{d}{dx}\left(k(x)\frac{du}{dx}\right) = f(x),\ 0 < x < \ell,$$
$$\frac{du}{dx}(0) = a, \tag{6.51}$$
$$\frac{du}{dx}(\ell) = b.$$

In deriving the weak form of the BVP, the boundary terms from the integration by parts do not vanish:

$$-\int_0^\ell \frac{d}{dx}\left[k(x)\frac{du}{dx}(x)\right] v(x)\,dx$$
$$= \left[-k(x)\frac{du}{dx}(x)v(x)\right]_0^\ell + \int_0^\ell k(x)\frac{du}{dx}(x)\frac{dv}{dx}(x)\,dx$$
$$= -k(\ell)\frac{du}{dx}(\ell)v(\ell) + k(0)\frac{du}{dx}(0)v(0) + \int_0^\ell k(x)\frac{du}{dx}(x)\frac{dv}{dx}(x)\,dx$$
$$= -k(\ell)v(\ell)b + k(0)v(0)a + \int_0^\ell k(x)\frac{du}{dx}(x)\frac{dv}{dx}(x)\,dx.$$

The weak form becomes

$$\text{find } u \in \tilde{V} \text{ such that } a(u,v) = (f,v) + k(\ell)v(\ell)b - k(0)v(0)a \text{ for all } v \in \tilde{V}. \tag{6.52}$$

When we apply the Galerkin method, we end up with a slightly different right-hand-side vector $\tilde{\mathbf{f}}$. We have

$$\tilde{f}_i = (f,\phi_i) + k(\ell)\phi_i(\ell)b - k(0)\phi_i(0)a,$$

or

$$\tilde{f}_i = \begin{cases} (f,\phi_i), & i = 1,2,\ldots,n-1 \text{ (since } \phi_i(\ell) = \phi_i(0) = 0), \\ (f,\phi_i) - k(0)a, & i = 0 \text{ (since } \phi_0(\ell) = 0, \phi_0(0) = 1), \\ (f,\phi_i) + k(\ell)b, & i = n \text{ (since } \phi_n(\ell) = 1, \phi_n(0) = 0). \end{cases} \tag{6.53}$$

Example 6.11. We now apply the finite element method to the inhomogeneous Neumann problem

$$-\frac{d^2u}{dx^2} = x - \frac{1}{2},\ 0 < x < 1,$$
$$\frac{du}{dx}(0) = 1, \tag{6.54}$$
$$\frac{du}{dx}(1) = 1.$$

6.5. Finite elements and Neumann conditions

The exact solution is

$$u(x) = -\frac{x^3}{6} + \frac{x^2}{4} + x - \frac{13}{12}$$

(again choosing the solution with $u(1) = 0$). The calculations are exactly the same as for Example 6.10, except that the first and last components of $\tilde{\mathbf{f}}$ are altered as given in (6.53). We therefore have

$$\tilde{f}_0 = \frac{h^2}{6} - \frac{h}{4} - 1, \ \tilde{f}_n = \frac{h}{4} - \frac{h^2}{6} + 1.$$

The results are shown, for $n = 10$, in Figure 6.16. ∎

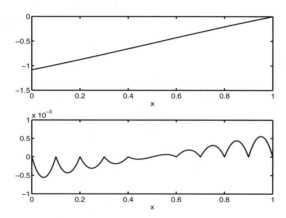

Figure 6.16. *The exact and computed solution (top) and the error in the computed solution (bottom) for (6.54). The computed solution was found using piecewise linear finite elements with a regular grid of 10 subintervals.*

6.5.5 The finite element method for an IBVP with Neumann conditions

We now briefly consider the following IBVP:

$$\begin{aligned}
\rho c \frac{\partial u}{\partial t} - \kappa \frac{\partial^2 u}{\partial x^2} &= f(x,t), \ 0 < x < \ell, \ t > t_0, \\
u(x,t_0) &= \psi(x), \ 0 < x < \ell, \\
\frac{du}{dx}(0,t) &= 0, \ t > t_0, \\
\frac{du}{dx}(\ell,t) &= 0, \ t > t_0.
\end{aligned} \quad (6.55)$$

Finite element methods for the heat equation with Neumann conditions are derived as in Section 6.4. The following differences arise.

- In the derivation of the weak form of the IBVP, the space V of test functions is replaced by $\tilde{V}$.

- When the Galerkin method is applied, the approximating subspace S_n is replaced by $\tilde{S}_n$.

- The mass matrix $\mathbf{M} \in \mathbf{R}^{(n-1)\times(n-1)}$, stiffness matrix $\mathbf{K} \in \mathbf{R}^{(n-1)\times(n-1)}$, and load vector $\mathbf{f}(t) \in \mathbf{R}^{n-1}$ are replaced by $\tilde{\mathbf{M}} \in \mathbf{R}^{(n+1)\times(n+1)}$, $\tilde{\mathbf{K}} \in \mathbf{R}^{(n+1)\times(n+1)}$, and $\tilde{\mathbf{f}}(t) \in \mathbf{R}^{n+1}$, respectively. The new stiffness matrix $\tilde{\mathbf{K}}$ was derived in Section 6.5.3, and the new form of the mass matrix, $\tilde{\mathbf{M}}$, is derived analogously.

In contrast to the case of a steady-state BVP, the singularity of the matrix $\tilde{\mathbf{K}}$ causes no particular problem in an IBVP, since we are not required to solve a linear system whose coefficient matrix is $\tilde{\mathbf{K}}$.

Moreover, the source function $f(x,t)$ is not required to satisfy a compatibility condition in order for (6.55) to have a solution. Physically, this makes sense. Thinking of heat flow, for example, the temperature distribution is, in general, changing, so there is no inconsistency in having a net gain or loss of total heat energy in the bar. On the other hand, for the steady-state case, the temperature could not be independent of time if the total amount of heat energy in the bar were changing with time.

This can also be seen directly. If u is the solution of (6.55), then

$$\int_0^\ell f(x,t)\,dx = \int_0^\ell \rho c \frac{\partial u}{\partial t}(x,t)\,dx - \int_0^\ell \kappa \frac{\partial^2 u}{\partial x^2}(x,t)\,dx$$

$$= \int_0^\ell \rho c \frac{\partial u}{\partial t}(x,t)\,dx - \left[\kappa \frac{\partial u}{\partial x}(x,t)\right]_{x=0}^\ell$$

$$= \int_0^\ell \rho c \frac{\partial u}{\partial t}(x,t)\,dx - \kappa\left(\frac{\partial u}{\partial x}(\ell,t) - \frac{\partial u}{\partial x}(0,t)\right)$$

$$= \int_0^\ell \rho c \frac{\partial u}{\partial t}(x,t)\,dx \text{ (by the Neumann conditions)}$$

$$= \frac{\partial}{\partial t}\left[\int_0^\ell \rho c u(x,t)\,dx\right].$$

This last expression is (proportional to) the time rate of change of the total heat energy contained in the bar (see (2.2) in Section 2.1).

We leave the detailed formulation of a piecewise linear finite element method for (6.55) to the exercises (see Exercise 6 below).

Exercises

1. Consider an aluminum bar of length 1 m and radius 1 cm. Suppose that the sides and ends of the bar are perfectly insulated, and heat energy is added to the interior of the bar at a rate of

$$f(x) = 10^{-7}x(25-x)(100-x) + \frac{1}{240} \text{ W/cm}^3$$

6.5. Finite elements and Neumann conditions

(x is given in centimeters, and the usual coordinate system is used). The thermal conductivity of the aluminum alloy is $1.5\,\text{W}/(\text{cm}\,\text{K})$. Find and graph the steady-state temperature of the bar. Use the finite element method.

2. Repeat the previous exercise, except now assume that the bar is heterogeneous, with thermal conductivity
$$\kappa(x) = 1.0 + \left(\frac{x}{100}\right)^2.$$

3. Consider a copper bar (thermal conductivity $4.04\,\text{W}/(\text{cm}\,\text{K})$) of length 100 cm. Suppose that heat energy is added to the interior of the bar at the constant rate of $f(x) = 0.005\,\text{W}/\text{cm}^3$, and that heat energy is added to the left end ($x = 0$) of the bar at the rate of $0.01\,\text{W}/\text{cm}^2$.

    (a) At what rate must heat energy be removed from the right end ($x = 100$) in order that a steady-state solution exist? (See Exercise 6.2.5(a), or just use common sense.)

    (b) Formulate the BVP leading to a steady-state solution, and approximate the solution using the finite element method.

4. Suppose that, in the Neumann problem (6.44), f does not satisfy the compatibility condition
$$\int_0^\ell f(x)\,dx = 0.$$

    Suppose we just ignore this fact and try to apply the finite element method anyway. Where does the method break down? Illustrate with the Neumann problem
$$-\frac{d^2u}{dx^2} = x,\ 0 < x < 1,$$
$$\frac{du}{dx}(0) = 0,$$
$$\frac{du}{dx}(1) = 0.$$

5. The purpose of this exercise is to prove that the null space of $\tilde{\mathbf{K}}$ is spanned by $\mathbf{u}_c$, where $\tilde{\mathbf{K}}$ is the stiffness matrix arising in a Neumann problem and $\mathbf{u}_c$ is the vector with each component equal to one.

    (a) Suppose $\tilde{\mathbf{K}}\mathbf{u} = 0$, where the components of $\mathbf{u}$ are $u_0, u_1, \ldots, u_n$. Show that this implies that $a(v,v) = 0$, where
$$v = \sum_{i=0}^n u_i \phi_i.$$

    (Hint: Compute $\mathbf{u} \cdot \tilde{\mathbf{K}}\mathbf{u}$.)

    (b) Explain why $a(v,v) = 0$ implies that v is a constant function.

    (c) Now explain why $\tilde{\mathbf{K}}\mathbf{u} = 0$ implies that $\mathbf{u}$ is a multiple of $\mathbf{u}_c$.

6. (a) Formulate the weak form of (6.55).

   (b) Show how to use piecewise linear finite elements and the method of lines to reduce the weak form to a system of ODEs.

   (c) Using the backward Euler method for the time integration, estimate the solution of the IBVP

   $$\frac{\partial u}{\partial t} - \frac{\partial^2 u}{\partial x^2} = 0, \ 0 < x < 1, \ t > 0,$$
   $$u(x,0) = x(1-x), \ 0 < x < 1,$$
   $$\frac{\partial u}{\partial x}(0,t) = 0, \ t > 0,$$
   $$\frac{\partial u}{\partial x}(1,t) = 0, \ t > 0$$

   and graph the solution at times $0, 0.02, 0.04, 0.06$, along with the steady-state solution. (See Exercise 6.2.1.)

7. (a) Formulate the weak form of the following BVP with mixed boundary conditions:

   $$-\frac{d}{dx}\left(k(x)\frac{du}{dx}\right) = f(x), \ 0 < x < \ell,$$
   $$u(0) = 0, \qquad (6.56)$$
   $$\frac{du}{dx}(\ell) = 0.$$

   (b) Show that the weak and strong forms of (6.56) are equivalent.

8. Repeat Exercise 7 for the BVP

   $$-\frac{d}{dx}\left(k(x)\frac{du}{dx}\right) = f(x), \ 0 < x < \ell,$$
   $$\frac{du}{dx}(0) = 0,$$
   $$u(\ell) = 0.$$

9. Consider a copper bar ($\rho = 8.97 \, \text{g/cm}^3$, $c = 0.379 \, \text{J/(g K)}$, $\kappa = 4.04 \, \text{W/(cm K)}$) of length 1 m and cross-sectional area $2 \, \text{cm}^2$. Suppose the bar is heated to a uniform temperature of 8 degrees Celsius, the sides and the left end ($x = 0$) are perfectly insulated, and the right end ($x = 100$) is placed in an ice bath. Find the temperature distribution $u(x,t)$ of the bar by formulating and solving an IBVP. Graph $u(x,300)$ (t in seconds). Use the finite element method with backward Euler time integration.

10. Consider a chromium bar of length 25 cm. The material constants for chromium are

    $$\rho = 7.15 \, \text{g/cm}^3, \ c = 0.439 \, \text{J/(g K)}, \ \kappa = 0.937 \, \text{W/(cm K)}.$$

Suppose that the left end of the bar is placed in an ice bath, the right end is held fixed at 8 degrees Celsius, and the bar is allowed to reach a steady-state temperature. Then (at $t = 0$) the right end is insulated. Find the temperature distribution using the finite element method and backward Euler integration. Graph $u(x, 100)$.

11. Let $\tilde{\mathbf{K}} \in \mathbf{R}^{(n+1) \times (n+1)}$ be the stiffness matrix for the Neumann problem, and let $\hat{\mathbf{K}} \in \mathbf{R}^{n \times n}$ be the matrix obtained by removing the last row and column of $\tilde{\mathbf{K}}$. Prove that $\hat{\mathbf{K}}$ is nonsingular. (Hint: $\hat{\mathbf{K}}$ is the stiffness matrix for the BVP with mixed boundary conditions, Neumann at the left and Dirichlet at the right.)

6.6 Suggestions for further reading

In this chapter, we have continued the development, begun in the previous chapter, of both Fourier series and finite element methods. The references noted in Section 5.7 are, for the most part, relevant for this chapter as well. Most books introduce Fourier series via the method of separation of variables briefly described in Section 6.1.6. Our approach is perhaps unique among introductory texts.

To gain a deep understanding of solution techniques, it is essential to know the qualitative theory of PDEs, which is developed in several texts. A basic understanding can be gained from introductory books such as Strauss [62] or Haberman [30]. An alternative is the older text by Weinberger [68]. More advanced references include McOwen [50], Folland [19], and Rennardy and Rogers [54].

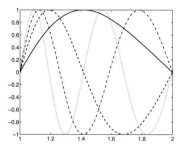

Chapter 7

Waves

We now treat the one-dimensional wave equation, which models the transverse vibrations of an elastic string or the longitudinal vibrations of a metal bar. We will concentrate on modeling a homogeneous medium, in which case the wave equation takes the form

$$\frac{\partial^2 u}{\partial t^2} - c^2 \frac{\partial^2 u}{\partial x^2} = f(x,t).$$

Our first order of business will be to understand the meaning of the parameter c. We then derive Fourier series and finite element methods for solving the wave equation.

7.1 The homogeneous wave equation without boundaries

We begin our study of the wave equation by supposing that there are no boundaries—that the wave equation holds for $-\infty < x < \infty$. Although this may not seem to be a very realistic problem, it will provide some useful information about wave motion.

We therefore consider the IVP

$$\frac{\partial^2 u}{\partial t^2} - c^2 \frac{\partial^2 u}{\partial x^2} = 0, \quad -\infty < x < \infty, \ t > 0,$$
$$u(x,0) = \psi(x), \quad -\infty < x < \infty, \tag{7.1}$$
$$\frac{\partial u}{\partial t}(x,0) = \gamma(x), \quad -\infty < x < \infty.$$

(To simplify the algebra that follows, we assume in this section that the initial time is $t_0 = 0$.) With a little cleverness, it is possible to derive an explicit formula for the solution in terms of the initial conditions $\psi(x)$ and $\gamma(x)$. The key point in deriving this formula is to notice that the wave operator

$$\frac{\partial^2}{\partial t^2} - c^2 \frac{\partial^2}{\partial x^2}$$

can be factored:

$$\frac{\partial^2}{\partial t^2} - c^2 \frac{\partial^2}{\partial x^2} = \left(\frac{\partial}{\partial t} - c\frac{\partial}{\partial x}\right)\left(\frac{\partial}{\partial t} + c\frac{\partial}{\partial x}\right) = \left(\frac{\partial}{\partial t} + c\frac{\partial}{\partial x}\right)\left(\frac{\partial}{\partial t} - c\frac{\partial}{\partial x}\right).$$

For example,
$$\left(\frac{\partial}{\partial t} - c\frac{\partial}{\partial x}\right)\left(\frac{\partial}{\partial t} + c\frac{\partial}{\partial x}\right)u(x,t)$$
$$= \left(\frac{\partial}{\partial t} - c\frac{\partial}{\partial x}\right)\left(\frac{\partial u}{\partial t}(x,t) + c\frac{\partial u}{\partial x}(x,t)\right)$$
$$= \frac{\partial^2 u}{\partial t^2}(x,t) + c\frac{\partial^2 u}{\partial t \partial x}(x,t) - c\frac{\partial^2 u}{\partial x \partial t}(x,t) - c^2\frac{\partial^2 u}{\partial x^2}(x,t)$$
$$= \frac{\partial^2 u}{\partial t^2}(x,t) - c^2\frac{\partial^2 u}{\partial x^2}(x,t).$$

The mixed partial derivatives cancel because, according to a theorem of calculus, if the second derivatives of a function $u(x,t)$ are continuous, then
$$\frac{\partial^2 u}{\partial t \partial x} = \frac{\partial^2 u}{\partial x \partial t}.$$

It follows that any solution $u(x,t)$ of either
$$\frac{\partial u}{\partial t} + c\frac{\partial u}{\partial x} = 0$$
or
$$\frac{\partial u}{\partial t} - c\frac{\partial u}{\partial x} = 0$$

will also solve the homogeneous wave equation. If $f : \mathbf{R} \to \mathbf{R}$ is any twice differentiable function and $u(x,t) = f(x-ct)$, then
$$\frac{\partial u}{\partial t}(x,t) + c\frac{\partial u}{\partial x}(x,t) = -cf'(x-ct) + cf'(x-ct) = 0.$$

Similarly, if $g : \mathbf{R} \to \mathbf{R}$ is twice differentiable, then $u(x,t) = g(x+ct)$ satisfies
$$\frac{\partial u}{\partial t}(x,t) - c\frac{\partial u}{\partial x}(x,t) = cg'(x+ct) - cg'(x+ct) = 0.$$

Thus, by linearity, every function of the form
$$u(x,t) = f(x-ct) + g(x+ct) \tag{7.2}$$

is a solution of the homogeneous wave equation. We now show that (7.2) is the general solution.

To prove that (7.2) really represents all possible solutions of the homogeneous wave equation, it suffices to show that we can always solve (7.1) with a function of the form (7.2) (since any solution of the wave equation satisfies (7.1) for some ψ and γ). Moreover, by linearity (the principle of superposition), it suffices to solve
$$\frac{\partial^2 u}{\partial t^2} - c^2\frac{\partial^2 u}{\partial x^2} = 0, \ -\infty < x < \infty, \ t > 0,$$
$$u(x,0) = \psi(x), \ -\infty < x < \infty, \tag{7.3}$$
$$\frac{\partial u}{\partial t}(x,0) = 0, \ -\infty < x < \infty,$$

7.1. The homogeneous wave equation without boundaries

and

$$\frac{\partial^2 u}{\partial t^2} - c^2 \frac{\partial^2 u}{\partial x^2} = 0, \quad -\infty < x < \infty, \ t > 0,$$
$$u(x,0) = 0, \quad -\infty < x < \infty, \tag{7.4}$$
$$\frac{\partial u}{\partial t}(x,0) = \gamma(x), \quad -\infty < x < \infty,$$

separately and add the solutions.

To find a solution to (7.3), we assume that $u(x,t) = f(x-ct) + g(x+ct)$. Then u solves the PDE. We want to choose f and g so that $u(x,0) = \psi(x)$ and $\partial u/\partial t(x,0) = 0$. But

$$u(x,0) = f(x) + g(x)$$

and

$$\frac{\partial u}{\partial t}(x,0) = c(g'(x) - f'(x)),$$

so if we take

$$f(x) = \frac{1}{2}\psi(x), \ g(x) = \frac{1}{2}\psi(x),$$

the required conditions are satisfied. That is,

$$u(x,t) = \frac{1}{2}(\psi(x-ct) + \psi(x+ct))$$

is the solution of (7.3).

Finding a solution to (7.4) is a bit harder. With $u(x,t) = f(x-ct) + g(x+ct)$, we want $u(x,0) = 0$ and $\partial u/\partial t(x,0) = \gamma(x)$. This yields two equations:

$$f(x) + g(x) = 0,$$
$$-cf'(x) + cg'(x) = \gamma(x).$$

The first equation implies that $f(x) = -g(x)$; substituting this into the second equation yields

$$2cg'(x) = \gamma(x).$$

A solution to this is

$$g(x) = \frac{1}{2c} \int_0^x \gamma(s) ds.$$

With $f(x) = -g(x)$, we obtain

$$u(x,t) = -\frac{1}{2c} \int_0^{x-ct} \gamma(s) ds + \frac{1}{2c} \int_0^{x+ct} \gamma(s) ds = \frac{1}{2c} \int_{x-ct}^{x+ct} \gamma(s) ds.$$

This is the solution to (7.4). Adding the two, we obtain

$$u(x,t) = \frac{1}{2}(\psi(x-ct) + \psi(x+ct)) + \frac{1}{2c} \int_{x-ct}^{x+ct} \gamma(s) ds. \tag{7.5}$$

This is *d'Alembert's* solution to (7.1).

We have only shown that (7.5) is *a* solution to (7.1). It can be shown (see Exercise 7.1.8) that this is the only solution to (7.1), so we are justified in calling it *the* solution.

We can now understand the significance of the constant c, and also why the PDE is called the *wave* equation. We first consider a function of the form $f(x-ct)$. Regarding $u(x,t) = f(x-ct)$ as a function of x for each fixed t, we see that each $u(x,t)$ is a translation of $u(x,0) = f(x)$. That is, the "time snapshots" of the function $u(x,t)$ all have the same shape; as time goes on, this shape moves to the right (since $c > 0$). We call $u(x,t) = f(x-ct)$ a *right-moving wave*; an example is shown in Figure 7.1. Similarly, $u(x,t) = g(x+ct)$ is a left-moving wave. Moreover, c is just the wave speed. It is therefore easy to understand d'Alembert's solution to the wave equation—it is the sum of a right-moving wave and a left-moving wave, both moving at speed c.

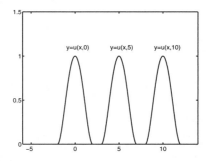

Figure 7.1. *A right-moving wave $u(x,t) = f(x-ct)$ (with $c = 1$).*

Example 7.1. The solution to

$$\frac{\partial^2 u}{\partial t^2} - c^2 \frac{\partial^2 u}{\partial x^2} = 0, \ -\infty < x < \infty, \ t > 0,$$
$$u(x,0) = e^{-x^2}, \ -\infty < x < \infty, \quad (7.6)$$
$$\frac{\partial u}{\partial t}(x,0) = 0, \ -\infty < x < \infty,$$

is

$$u(x,t) = \frac{1}{2}\left(e^{-(x-t)^2} + e^{-(x+t)^2}\right).$$

Several snapshots of this solution are graphed in Figure 7.2. Notice how the initial "blip" splits into two parts which move to the right and left. ∎

D'Alembert's solution to the wave equation shows that disturbances in a medium modeled by the wave equation travel at a finite speed. We can state this precisely as follows. Given any point $\bar{x}$ in space and any positive time $\bar{t}$, the interval $[\bar{x} - c\bar{t}, \bar{x} + c\bar{t}]$ consists of those points from which a signal, traveling at a speed of c, would reach $\bar{x}$ in time $\bar{t}$ or less.

Theorem 7.2. *Suppose $u(x,t)$ solves the IVP (7.1), $\bar{x}$ is any real number, and $\bar{t} > 0$. If $\psi(x)$ and $\gamma(x)$ are both zero for all x satisfying $\bar{x} - c\bar{t} \leq x \leq \bar{x} + c\bar{t}$, then $u(\bar{x}, \bar{t}) = 0$.*

7.1. The homogeneous wave equation without boundaries

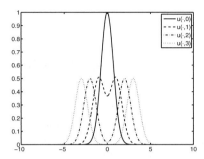

Figure 7.2. *Snapshots of the solution to (7.6) at times $t = 0, 1, 2, 3$.*

Proof. We have

$$u(\overline{x}, \overline{t}) = \frac{1}{2}[\psi(\overline{x} - c\overline{t}) + \psi(\overline{x} + c\overline{t})] + \frac{1}{2c}\int_{\overline{x} - c\overline{t}}^{\overline{x} + c\overline{t}} \gamma(s)\,ds.$$

If $\psi(x)$ and $\gamma(x)$ are both zero for all x satisfying $\overline{x} - c\overline{t} \le x \le \overline{x} + c\overline{t}$, then all three terms in this formula for $u(\overline{x}, \overline{t})$ are zero, and hence $u(\overline{x}, \overline{t}) = 0$. □

We call the interval $[\overline{x} - c\overline{t}, \overline{x} + c\overline{t}]$ the *domain of dependence* of the space-time point $(\overline{x}, \overline{t})$. We illustrate the domain of dependence in Figure 7.3.

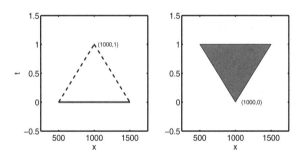

Figure 7.3. *Left: The domain of dependence of the point $(\overline{x}, \overline{t}) = (1000, 1)$ with $c = 500$. The domain of dependence is the interval shaded on the x-axis. Right: The domain of influence of the point $(x_0, 0) = (1000, 0)$.*

We can look at this result from another point of view. The initial data at a point x_0, $\psi(x_0)$ and $\gamma(x_0)$, can only affect the solution $u(x,t)$ if t is large enough, specifically, if

$$t \ge \left|\frac{x - x_0}{c}\right|.$$

The set

$$\{(x,t) \,:\, ct \ge |x - x_0|\}$$

is therefore called the *domain of influence*[34] of the point $(x_0, 0)$. An example of a domain of influence is given in Figure 7.3.

From the preceding discussion, we see that if a disturbance is initially confined to the interior of a bounded region, the boundary cannot affect the solution until sufficient time has passed. For this reason, the solution to the wave equation in an infinite medium provides a realistic idea of what happens (in certain experiments) in a bounded region, at least for a finite period of time (namely, until enough time has elapsed for the disturbance to reach the boundary). However, there is no simple formula for the solution of the wave equation subject to boundary conditions. To attack that problem, we must use Fourier series, finite elements, or other less elementary methods.

Exercises

1. Solve (7.1) with

$$\psi(x) = \begin{cases} 0, & x < -2 \text{ or } x > 2, \\ 5 \cdot 10^{-4} x + 10^{-3}, & -2 \leq x \leq 0, \\ -5 \cdot 10^{-4} x + 10^{-3}, & 0 < x \leq 2, \end{cases}$$

   $\gamma(x) = 0$, and $c = 500$. Graph several snapshots of the solutions on a common plot.

2. Solve (7.1) with $\psi(x) = 0$,

$$\gamma(x) = \begin{cases} 0, & x < -0.5 \text{ or } x > 0.5, \\ -1, & |x| \leq 0.5, \end{cases}$$

   and $c = 200$. Graph several snapshots of the solutions on a common plot.

3. Consider the IVP (7.1) with $\gamma(x) = 0$ and

$$\psi(x) = e^{-(x+5)^2} + e^{-(x-5)^2}.$$

   The initial displacement consists of two "blips," which, according to d'Alembert's formula, will each split into a right-moving and a left-moving wave. What happens when the right-moving wave from the first blip meets the left-moving wave from the second blip?

4. Repeat the previous exercise with

$$\psi(x) = e^{-(x+5)^2} - e^{-(x-5)^2}.$$

5. Consider the IVP

$$\frac{\partial^2 u}{\partial t^2} - c^2 \frac{\partial^2 u}{\partial x^2} = 0, \quad -\infty < x < \infty, \ t > 0,$$
$$u(x, 0) = \psi(x), \quad -\infty < x < \infty,$$
$$\frac{\partial u}{\partial t} = 0, \quad -\infty < x < \infty,$$

---

[34] Also called the *range of influence* by some authors.

7.2. Fourier series methods for the wave equation

where the support of ψ is $[a,b]$. (Recall that the support of a continuous function is the closure of the set on which the function is nonzero. Thus $\psi(x) = 0$ for all $x \notin [a,b]$.) Let $x_1 > b$ be given. Name all intervals of time during which $u(x_1, t) = 0$ holds.

6. Repeat the previous exercise, changing the initial conditions to

$$u(x,0) = 0, \ \frac{\partial u}{\partial t}(x,0) = \gamma(x), \ -\infty < x < \infty.$$

Assume that the support of γ is $[a,b]$.

7. Consider the IBVP

$$\frac{\partial^2 u}{\partial t^2} - c^2 \frac{\partial^2 u}{\partial x^2} = 0, \ 0 < x < \ell, \ t > 0,$$
$$u(x,0) = \psi(x), \ 0 < x < \ell,$$
$$\frac{\partial u}{\partial t} = 0, \ 0 < x < \ell,$$
$$u(0,t) = 0, \ t > 0,$$
$$u(\ell,t) = 0, \ t > 0,$$

where the support of ψ is $[a,b]$ and $0 < a < b < \ell$. Define

$$u(x,t) = \frac{1}{2}(\psi(x - ct) + \psi(x + ct)).$$

Prove that if t_f is sufficiently small, then u is a solution of the IBVP for $0 < t < t_f$. What is the largest such t_f?

8. Consider the IVP (7.1), where ψ and γ have bounded support. Show that the IVP has at most one solution as follows.

   (a) Assume that two solutions, say u and v, exist. Show that the difference $w = u - v$ satisfies the homogeneous wave equation with zero initial conditions.

   (b) Show that the total energy of any solution to the homogeneous wave equation (without boundaries) is conserved. (Hint: See Section 5.4.1.)

   (c) Use the conservation of energy to show that $w = 0$, that is, that $u = v$.

7.2 Fourier series methods for the wave equation

Fourier series methods for the wave equation are developed just as for the heat equation—the key is to represent the solution in a Fourier series in which the Fourier coefficients are functions of t. We begin with the IBVP

$$\frac{\partial^2 u}{\partial t^2} - c^2 \frac{\partial^2 u}{\partial x^2} = f(x,t),\ 0 < x < \ell,\ t > t_0,$$
$$u(x,t_0) = \psi(x),\ 0 < x < \ell,$$
$$\frac{\partial u}{\partial t}(x,t_0) = \gamma(x),\ 0 < x < \ell, \quad (7.7)$$
$$u(0,t) = 0,\ t > t_0,$$
$$u(\ell,t) = 0,\ t > t_0,$$

which models the small displacements of a vibrating string. We write the solution in the form

$$u(x,t) = \sum_{n=1}^{\infty} a_n(t) \sin\left(\frac{n\pi x}{\ell}\right),$$

where

$$a_n(t) = \frac{2}{\ell} \int_0^\ell u(x,t) \sin\left(\frac{n\pi x}{\ell}\right) dx,\ n = 1, 2, 3, \ldots.$$

We also expand the right-hand side $f(x,t)$ in a sine series:

$$f(x,t) = \sum_{n=1}^{\infty} c_n(t) \sin\left(\frac{n\pi x}{\ell}\right),$$

where

$$c_n(t) = \frac{2}{\ell} \int_0^\ell f(x,t) \sin\left(\frac{n\pi x}{\ell}\right) dx,\ n = 1, 2, 3, \ldots.$$

Using the same reasoning as in Section 6.1, we can express the left-hand side of the PDE in a sine series:

$$\frac{\partial^2 u}{\partial t^2}(x,t) - c^2 \frac{\partial^2 u}{\partial x^2}(x,t) = \sum_{n=1}^{\infty} \left(\frac{d^2 a_n}{dt^2}(t) + \frac{c^2 n^2 \pi^2}{\ell^2} a_n(t)\right) \sin\left(\frac{n\pi x}{\ell}\right).$$

Setting this equal to the series for $f(x,t)$, we obtain the ODEs

$$\frac{d^2 a_n}{dt^2} + \frac{c^2 n^2 \pi^2}{\ell^2} a_n = c_n(t),\ n = 1, 2, 3, \ldots.$$

Initial conditions for these ODEs are obtained from the initial conditions for the wave equation in (7.7). If

$$\psi(x) = \sum_{n=1}^{\infty} b_n \sin\left(\frac{n\pi x}{\ell}\right),\ \gamma(x) = \sum_{n=1}^{\infty} d_n \sin\left(\frac{n\pi x}{\ell}\right),$$

then we have

$$a_n(t_0) = b_n,\ \frac{da_n}{dt}(t_0) = d_n.$$

7.2. Fourier series methods for the wave equation

We thus find the Fourier sine coefficients of the solution $u(x,t)$ by solving the IVPs

$$\frac{d^2 a_n}{dt^2} + \frac{c^2 n^2 \pi^2}{\ell^2} a_n = c_n(t),$$
$$a_n(t_0) = b_n, \quad (7.8)$$
$$\frac{da_n}{dt}(t_0) = d_n$$

for $n = 1, 2, 3, \ldots$. In Section 4.2.3, we presented an explicit formula for the solution of (7.8). The use of this formula is illustrated in Example 7.4.

7.2.1 Fourier series solutions of the homogeneous wave equation

When the wave equation is homogeneous (that is, when $f \equiv 0$), the Fourier series solution is especially instructive. We then must solve

$$\frac{d^2 a_n}{dt^2} + \frac{c^2 n^2 \pi^2}{\ell^2} a_n = 0,$$
$$a_n(t_0) = b_n,$$
$$\frac{da_n}{dt}(t_0) = d_n,$$

which yields

$$a_n(t) = b_n \cos\left(\frac{cn\pi}{\ell}(t - t_0)\right) + \frac{d_n \ell}{cn\pi} \sin\left(\frac{cn\pi}{\ell}(t - t_0)\right).$$

Therefore, the solution to

$$\frac{\partial^2 u}{\partial t^2} - c^2 \frac{\partial^2 u}{\partial x^2} = 0, \ 0 < x < \ell, \ t > t_0,$$
$$u(x, t_0) = \psi(x), \ 0 < x < \ell,$$
$$\frac{\partial u}{\partial t}(x, t_0) = \gamma(x), \ 0 < x < \ell, \quad (7.9)$$
$$u(0, t) = 0, \ t > t_0,$$
$$u(\ell, t) = 0, \ t > t_0,$$

is

$$u(x,t) = \sum_{n=1}^{\infty} \left\{ b_n \cos\left(\frac{cn\pi}{\ell}(t - t_0)\right) + \frac{d_n \ell}{cn\pi} \sin\left(\frac{cn\pi}{\ell}(t - t_0)\right) \right\} \sin\left(\frac{n\pi x}{\ell}\right).$$

This formula shows the essentially oscillatory nature of solutions to the wave equation; indeed, the Fourier sine coefficients of the solution are periodic:

$$a_n\left(t + \frac{2\ell}{cn}\right) = a_n(t) \text{ for all } t.$$

The frequencies $cn/(2\ell)$ are called the *natural frequencies* of the string.[35] These frequencies are called natural because any solution of the homogeneous wave equation is an (infinite) combination of the *normal modes* of the string,

$$\left(A_n \cos\left(\frac{cn\pi}{\ell}(t-t_0)\right) + B_n \sin\left(\frac{cn\pi}{\ell}(t-t_0)\right)\right) \sin\left(\frac{n\pi x}{\ell}\right).$$

Each normal mode is a *standing wave* with temporal frequency $cn/(2\ell)$ cycles per second. Several examples are displayed in Figure 7.4.

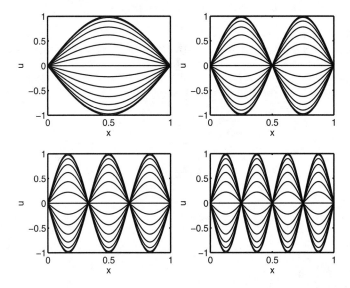

Figure 7.4. *The first four normal modes ($c = \ell = 1$, $t_0 = 0$). Displayed are multiple time snapshots of each standing wave.*

In particular, since $2\ell/c$ is an integer multiple of $2\ell/(cn)$, we have

$$a_n\left(t + \frac{2\ell}{c}\right) = a_n(t) \text{ for all } t, \text{ for all } n = 1,2,3,\ldots.$$

It follows that the solution $u(x,t)$ satisfies

$$u\left(x, t + \frac{2\ell}{c}\right) = u(x,t) \text{ for all } x \in (0, \ell), \text{ for all } t > 0;$$

that is, the solution is periodic with period $2\ell/c$. The frequency $c/(2\ell)$ is called the *fundamental frequency* of the string—it is the frequency at which the plucked string vibrates.

---

[35]The units of frequency are literally 1/second, which is interpreted as cycles per second, or Hertz. Sometimes the natural frequencies are given as $2\pi \frac{cn}{2\ell} = \frac{cn\pi}{\ell}$, in which case the units can be thought of as radians/seconds.

7.2. Fourier series methods for the wave equation

An interesting question is the following: What happens to a string that is subjected to an oscillatory transverse pressure with a frequency equal to one of the natural frequencies of the string? The answer is that *resonance* occurs. This is illustrated in Exercise 7.2.8 and also in Section 7.4.

We now give an example of using the Fourier series method to solve an IBVP for the wave equation.

Example 7.3. Consider (7.7) with $\ell = 1$ m, $c = 522$ m/s, $f(x,t) = 0$, $\gamma(x) = 0$, and

$$\psi(x) = \begin{cases} x - 0.4, & 0.4 < x < 0.5, \\ -(x - 0.6), & 0.5 < x < 0.6, \\ 0, & 0 < x < 0.4 \text{ or } 0.6 < x < 1. \end{cases} \quad (7.10)$$

The fundamental frequency of the string is then

$$\frac{c}{2\ell} = \frac{522}{2} = 261 \text{ Hz},$$

which is the frequency of the musical note called "middle C." The initial conditions indicate that the string is "plucked" at the center. Using the notation introduced above, we have $c_n(t) = 0$ and $d_n = 0$ for all n, and

$$b_n = 2\int_0^1 \psi(x)\sin(n\pi x)\,dx$$
$$= -2\frac{-2\sin(1/2 n\pi) + \sin(2/5 n\pi) + \sin(3/5 n\pi)}{n^2\pi^2}.$$

We must solve

$$\frac{d^2 a_n}{dt^2} + c^2 n^2 \pi^2 a_n = 0,$$
$$a_n(0) = b_n,$$
$$\frac{da_n}{dt}(0) = 0.$$

The solution is easily seen to be

$$a_n(t) = b_n \cos(cn\pi t), \quad n = 1, 2, 3, \ldots,$$

and therefore

$$u(x,t) = \sum_{n=1}^{\infty} b_n \cos(cn\pi t)\sin(n\pi x).$$

Several snapshots of the solution are graphed in Figure 7.5. The reader should observe that when a wave hits a boundary where there is a Dirichlet condition, the wave inverts and reflects. Therefore, for example, $u(x, 3t_f/8) = -u(x, t_f/8)$ and $u(x, t_f/2) = -u(x, 0)$,

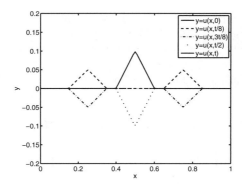

Figure 7.5. *The solution $u(x,t)$ to (7.7) at times 0, $t_f/8$, $3t_f/8$, $t_f/2$, and t_f. For this example, we took $\ell = 1$, $c = 522$, $f(x,t) = 0$, $\gamma(x) = 0$, and the initial condition ψ given in (7.10). One hundred terms of the Fourier series were used. The solutions at $t = 0$ and $t = t_f$ are identical.*

where t_f is the period corresponding to the fundamental frequency: $t_f = 2/522 = 1/261\,\text{s}$. Also, $u(x,t_f) = u(x,0)$, as expected. ∎

The reader should recall, from Section 2.3, that the wave speed c is determined by the tension T and the density ρ of the string: $c^2 = T/\rho$. The fundamental frequency of the string is

$$\frac{c}{2\ell},$$

which shows that if the tension and density of the string stay the same, but the string is shortened, then the fundamental frequency is increased. This explains how a guitar works—the strings are shortened by the pressing of the strings against the "frets," and thus different notes are sounded.

7.2.2 Fourier series solutions of the inhomogeneous wave equation

We now show two examples of the inhomogeneous wave equation.

Example 7.4. Consider a metal string, such as a guitar string, that can be attracted by a magnet. Suppose the string in question is 25 cm in length, with $c = 2500$ cm/s in the wave equation, its ends are fixed, and it is "plucked" so that its initial displacement is

$$\psi(x) = \frac{1}{10}\left(1 - \frac{2}{25}\left|x - \frac{25}{2}\right|\right)$$

and released (so that its initial velocity is zero). Suppose further that a magnet exerts a constant upward force of 1000 dynes. We wish to find the motion of the string.

7.2. Fourier series methods for the wave equation

We must solve the IBVP

$$\frac{\partial^2 u}{\partial t^2} - c^2 \frac{\partial^2 u}{\partial x^2} = 1000, \ 0 < x < 25, \ t > 0,$$

$$u(x,0) = \psi(x), \ 0 < x < 25,$$

$$\frac{\partial u}{\partial t}(x,0) = 0, \ 0 < x < 25,$$

$$u(0,t) = 0, \ t > 0,$$

$$u(25,t) = 0, \ t > 0,$$

where $c = 2500$ and ψ is as given above.

We write the solution as

$$u(x,t) = \sum_{n=1}^{\infty} a_n(t) \sin\left(\frac{n\pi x}{25}\right)$$

and determine the coefficients $a_1(t), a_2(t), a_3(t), \ldots$ by solving the IVPs (7.8). We have

$$1000 = \sum_{n=1}^{\infty} c_n \sin\left(\frac{n\pi x}{25}\right),$$

with

$$c_n = \frac{2}{25} \int_0^{25} 1000 \sin\left(\frac{n\pi x}{25}\right) dx = \frac{2000(1-(-1)^n)}{n\pi}, \ n = 1,2,3,\ldots,$$

and

$$\psi(x) = \sum_{n=1}^{\infty} b_n \sin\left(\frac{n\pi x}{25}\right),$$

with

$$b_n = \frac{2}{25} \int_0^{25} \psi(x) \sin\left(\frac{n\pi x}{25}\right) dx = \frac{4\sin(n\pi/2)}{5n^2\pi^2}, \ n = 1,2,3,\ldots.$$

To find $a_n(t)$, therefore, we must solve

$$\frac{d^2 a_n}{dt^2} + \frac{c^2 n^2 \pi^2}{25^2} a_n = c_n,$$

$$a_n(0) = b_n,$$

$$\frac{da_n}{dt}(0) = 0.$$

Using the results of Section 4.2.3, we see that the solution is

$$a_n(t) = b_n \cos\left(\frac{cn\pi t}{25}\right) + \frac{25}{cn\pi} \int_0^t \sin\left(\frac{cn\pi}{25}(t-s)\right) c_n \, ds$$

$$= \left(b_n - \frac{625 c_n}{c^2 n^2 \pi^2}\right) \cos\left(\frac{cn\pi t}{25}\right) + \frac{625 c_n}{c^2 n^2 \pi^2}.$$

The solution is periodic with period

$$\frac{2\pi}{c\pi/25} = \frac{50}{c}.$$

Twenty-five snapshots of the solution are shown in Figure 7.6. The effect of the external force on the motion of the string is clearly seen. (The reader should solve Exercise 7.2.1 if it is not clear how the string would move in the absence of the external force.) ∎

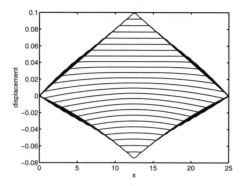

Figure 7.6. *Twenty-five snapshots of the vibrating string from Example 7.4.*

We now consider an example in which one of the boundary conditions is inhomogeneous. As usual, we use the method of shifting the data.

Example 7.5. Consider a string of length 100 cm that has one end fixed at $x = 0$, with the other end free to move in the vertical direction only (tied to a frictionless pole, for example). Let the wave speed in the string be $c = 2000$ cm/s.[36] Suppose the free end is "flicked," that is, moved up and down rapidly, according to the formula $u(100, t) = f(t)$, where

$$f(t) = \begin{cases} \epsilon^{-1}t - 3\epsilon^{-2}t^2 + 3\epsilon^{-3}t^3 - \epsilon^{-4}t^4, & 0 < t < \epsilon, \\ 0, & t > \epsilon, \end{cases}$$

where $\epsilon = 0.01$ (see Figure 7.7).

To determine the motion of the string, we must solve the IBVP

$$\frac{\partial^2 u}{\partial t^2} - c^2 \frac{\partial^2 u}{\partial x^2} = 0, \ 0 < x < 100, \ t > 0,$$
$$u(x, 0) = 0, \ 0 < x < 100,$$
$$\frac{\partial u}{\partial t}(x, 0) = 0, \ 0 < x < 100, \quad (7.11)$$
$$u(0, t) = 0, \ t > 0,$$
$$u(100, t) = f(t), \ t > 0.$$

---

[36]The fundamental frequency is then $\frac{c}{2 \cdot 100} = 10$ Hz. By comparison, the lowest tone of a piano has frequency 27.5 Hz.

7.2. Fourier series methods for the wave equation

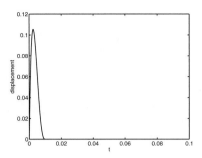

Figure 7.7. *The "flick" of the free end in Example 7.5.*

Since the boundary condition at the right end of the string is inhomogeneous, we will shift the data to obtain a problem with homogeneous boundary conditions. Define $v(x,t) = u(x,t) - p(x,t)$, where

$$p(x,t) = \frac{xf(t)}{100}.$$

Then v satisfies the IBVP

$$\frac{\partial^2 v}{\partial t^2} - c^2 \frac{\partial^2 v}{\partial x^2} = -\frac{x}{100} \frac{d^2 f}{dt^2}(t), \ 0 < x < 100, \ t > 0,$$

$$v(x,0) = 0, \ 0 < x < 100,$$

$$\frac{\partial v}{\partial t}(x,0) = -\frac{x}{100\epsilon}, \ 0 < x < 100,$$

$$v(0,t) = 0, \ t > 0,$$

$$v(100,t) = 0, \ t > 0.$$

In computing the initial conditions for v, we used

$$f(0) = 0, \ \frac{df}{dt}(0) = \frac{1}{\epsilon}.$$

The new right-hand side of the PDE is

$$g(x,t) = -\frac{x}{100} \frac{d^2 f}{dt^2}(t) = \begin{cases} -\frac{x}{100}\left(-6\epsilon^{-2} + 18\epsilon^{-3}t - 12\epsilon^{-4}t^2\right), & 0 \leq t \leq \epsilon, \\ 0, & t > \epsilon. \end{cases}$$

Since g is defined piecewise, so are its Fourier coefficients:

$$g(x,t) = \sum_{n=1}^{\infty} c_n(t) \sin\left(\frac{n\pi x}{100}\right),$$

$$c_n(t) = \frac{2}{100} \int_0^{100} g(x,t) \sin\left(\frac{n\pi x}{100}\right) dx$$

$$= \begin{cases} \frac{2(-1)^n}{n\pi} \frac{d^2 f}{dt^2}(t), & 0 \leq t \leq \epsilon, \\ 0, & t > \epsilon. \end{cases}$$

The initial velocity for v is given by

$$-\frac{x}{100\epsilon} = \sum_{n=1}^{\infty} d_n \sin\left(\frac{n\pi x}{100}\right),$$

$$d_n = \frac{2}{100} \int_0^{100} \left(-\frac{x}{100\epsilon}\right) \sin\left(\frac{n\pi x}{100}\right) dx$$

$$= \frac{2(-1)^n}{n\pi\epsilon}.$$

If we write

$$v(x,t) = \sum_{n=1}^{\infty} a_n(t) \sin\left(\frac{n\pi x}{100}\right),$$

then the coefficient $a_n(t)$ is determined by the IVP

$$\frac{d^2 a_n}{dt^2} + \frac{c^2 n^2 \pi^2}{100^2} a_n = c_n(t),$$

$$a_n(0) = 0,$$

$$\frac{da_n}{dt}(0) = d_n.$$

Again using the results of Section 4.2.3, we see that

$$a_n(t) = \frac{100}{cn\pi} \left(d_n \sin\left(\frac{cn\pi t}{100}\right) + \int_0^t \sin\left(\frac{cn\pi}{100}(t-s)\right) c_n(s) ds \right).$$

We note that

$$c_n(t) = \epsilon d_n \frac{d^2 f}{dt^2}(t),$$

which allows us to simplify the formulas slightly. We will write

$$\theta = \frac{cn\pi}{100}.$$

Then we have

$$a_n(t) = \frac{d_n}{\theta} \left(\sin(\theta t) + \epsilon \int_0^t \sin(\theta(t-s)) \left(-6\epsilon^{-2} + 18\epsilon^{-3} s - 12\epsilon^{-4} s^2 \right) ds \right).$$

For $t \in [0, \epsilon]$,

$$a_n(t) = \frac{d_n}{\theta} \left(\sin(\theta t) + \epsilon \int_0^t \sin(\theta(t-s)) \left(-6\epsilon^{-2} + 18\epsilon^{-3} s - 12\epsilon^{-4} s^2 \right) ds \right)$$

$$= \frac{d_n}{\theta} \left(\sin(\theta t) - 6\epsilon^{-1} \left(\frac{1 - \cos(\theta t)}{\theta} - 3\epsilon^{-1} \frac{\theta t - \sin(\theta t)}{\theta^2} \right.\right.$$

$$\left.\left. + 2\epsilon^{-2} \frac{\theta^2 t^2 - 2 + 2\cos(\theta t)}{\theta^3} \right) \right),$$

7.2. Fourier series methods for the wave equation

while for $t > \epsilon$,

$$
\begin{aligned}
a_n(t) &= \frac{d_n}{\theta} \left(\sin(\theta t) + \epsilon \int_0^\epsilon \sin(\theta(t-s)) \left(-6\epsilon^{-2} + 18\epsilon^{-3}s - 12\epsilon^{-4}s^2 \right) ds \right) \\
&= \frac{d_n}{\theta} \left(\sin(\theta t) - 6\epsilon^{-1} \left(\frac{\cos(\theta(t-\epsilon)) - \cos(\theta t)}{\theta} \right. \right. \\
&\quad - 3\epsilon^{-1} \frac{\sin(\theta(t-\epsilon)) + \epsilon\theta \cos(\theta(t-s)) - \sin(\theta t)}{\theta^2} \\
&\quad \left. \left. + 2\epsilon^{-2} \frac{\epsilon^2\theta^2 \cos(\theta(t-\epsilon)) - 2\cos(\theta(t-\epsilon)) + 2\epsilon\theta \sin(\theta(t-\epsilon)) + 2\cos(\theta t)}{\theta^3} \right) \right).
\end{aligned}
$$

Several snapshots of the solution are shown in Figure 7.8. For each t, the Fourier coefficients of $u(x,t)$ decay to zero relatively slowly in n, so a good approximation requires many terms in the Fourier series. To produce Figure 7.8, we used 200 terms. The reader should observe the behavior of the wave motion: the wave produced by the flick of the end travels from right to left, reflects, travels from left to right, reflects again, and so on. At each reflection, the wave inverts.[37] ∎

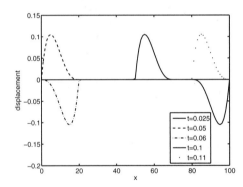

Figure 7.8. *Snapshots of the string from Example 7.5.*

7.2.3 Other boundary conditions

Using Fourier series to solve the wave equation with a different set of boundary conditions is not difficult, provided the eigenvalues and eigenfunctions are known. We now give an example with mixed boundary conditions.

Example 7.6. Consider a steel bar of length 1 m, fixed at the top ($x = 0$) with the bottom end ($x = 1$) free to move. Suppose the bar is stretched, by means of a downward pressure applied to the free end, to a length of 1.002 m, and then released. We wish to describe the vibrations of the bar, given that the type of steel has stiffness $k = 195\,\text{GPa}$ and density $\rho = 7.9\,\text{g/cm}^3$.

---

[37] The reader can approximate this experiment with a long rope.

The longitudinal displacement $u(x,t)$ of the bar satisfies the IBVP

$$\rho \frac{\partial^2 u}{\partial t^2} - k \frac{\partial^2 u}{\partial x^2} = 0, \ 0 < x < 1, \ t > 0,$$
$$u(x,0) = 0.002x, \ 0 < x < 1,$$
$$\frac{\partial u}{\partial t}(x,0) = 0, \ 0 < x < 1, \qquad (7.12)$$
$$u(0,t) = 0, \ t > 0,$$
$$\frac{\partial u}{\partial x}(1,t) = 0, \ t > 0$$

(see Section 2.2). The eigenpairs of the negative second derivative operator, under these mixed boundary conditions, are

$$\lambda_n = \frac{(2n-1)^2 \pi^2}{4\ell^2}, \ \phi_n(x) = \sin\left(\frac{(2n-1)\pi x}{2\ell}\right), \ n = 1, 2, 3, \ldots$$

(see Section 5.2.3), with $\ell = 1$. We therefore represent the solution of (7.12) as

$$u(x,t) = \sum_{n=1}^{\infty} a_n(t) \sin\left(\frac{(2n-1)\pi x}{2}\right).$$

The left side of the PDE is

$$\rho \frac{\partial^2 u}{\partial t^2}(x,t) - k \frac{\partial^2 u}{\partial x^2}(x,t) = \sum_{n=1}^{\infty} \left\{ \rho \frac{d^2 a_n}{dt^2}(t) + \frac{k(2n-1)^2 \pi^2}{4} a_n(t) \right\} \sin\left(\frac{(2n-1)\pi x}{2}\right);$$

since the right side is zero, we obtain the ODEs

$$\rho \frac{d^2 a_n}{dt^2} + \frac{k(2n-1)^2 \pi^2}{4} a_n = 0, \ n = 1, 2, 3, \ldots.$$

The initial displacement is given by

$$0.002x = \sum_{n=1}^{\infty} b_n \sin\left(\frac{(2n-1)\pi x}{2}\right)$$

with

$$b_n = 2 \int_0^1 0.002x \sin\left(\frac{(2n-1)\pi x}{2}\right) dx = \frac{2(-1)^{n+1}}{125(2n-1)^2 \pi^2}, \ n = 1, 2, 3, \ldots.$$

This, together with the fact that the initial velocity of the bar is zero, yields initial conditions for the ODEs:

$$\rho \frac{d^2 a_n}{dt^2} + \frac{k(2n-1)^2 \pi^2}{4} a_n = 0,$$
$$a_n(0) = b_n,$$
$$\frac{da_n}{dt}(0) = 0.$$

7.2. Fourier series methods for the wave equation

The solution is

$$a_n(t) = b_n \cos\left(\frac{c(2n-1)\pi t}{2}\right), \quad n = 1, 2, 3, \ldots$$

(with $c = \sqrt{k/\rho}$), so

$$u(x,t) = \sum_{n=1}^{\infty} b_n \cos\left(\frac{c(2n-1)\pi t}{2}\right) \sin\left(\frac{(2n-1)\pi x}{2}\right).$$

We see that the wave speed in the bar is

$$c = \sqrt{\frac{k}{\rho}} = \sqrt{\frac{1.95 \cdot 10^{11}}{7.9 \cdot 10^3}} \doteq 4970 \, \text{m/s}$$

(we convert k and ρ to consistent units before computing c), and the fundamental frequency is

$$\frac{c}{4} \doteq \frac{4970}{4} \doteq 1240 \, \text{Hz}.$$

Several snapshots of the displacement of the bar are shown in Figure 7.9. ∎

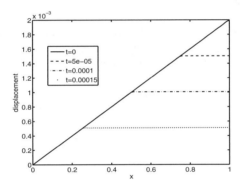

Figure 7.9. *Snapshots of the displacement of the bar in Example 7.6.*

Exercises

1. Find the motion of the string in Example 7.4 in the case that no external force is applied, and produce a graph analogous to Figure 7.6. Compare.

2. Consider a string of length 50 cm, with $c = 400$ cm/s. Suppose that the string is initially at rest, with both ends fixed, and that it is struck by a hammer at its center so as to induce an initial velocity of

$$\gamma(x) = \begin{cases} -20, & 24 < x < 26, \\ 0, & 0 \le x \le 24 \text{ or } 26 \le x \le 50. \end{cases}$$

(a) How long does it take for the resulting wave to reach the ends of the string?

(b) Formulate and solve the IBVP describing the motion of the string.

(c) Plot several snapshots of the string, showing the wave first reaching the ends of the string. Verify your answer to 2(a).

3. Find the motion of the string in Example 7.4 if the only change in the experimental conditions is that the right end ($x = 25$) is free to move vertically. (Imagine that the right end is held at $u = 0$ and then released, along with the rest of the string, at $t = 0$.)

4. Repeat Exercise 2, assuming that the right end of the string is free to move vertically.

5. Find the motion of the string in Example 7.4 in the case that both ends of the string are free to move vertically. (All other experimental conditions remain the same.)

6. Solve the IBVP from Example 7.4 with $c = 5000$ in place of $c = 2500$, and produce a graph analogous to Figure 7.6. How does the solution change? Explain this change in terms of the vibrating string.

7. How does the motion of the bar in Example 7.6 change if the force due to gravity is taken into account?

8. Consider a string that has one end fixed, with the other end free to move in the vertical direction. Suppose the string is put into motion by virtue of the free end's being manually moved up and down periodically. An IBVP describing this motion is

$$\frac{\partial^2 u}{\partial t^2} - c^2 \frac{\partial^2 u}{\partial x^2} = 0, \; 0 < x < \ell, \; t > t_0,$$
$$u(x, t_0) = 0, \; 0 < x < \ell,$$
$$\frac{\partial u}{\partial t}(x, t_0) = 0, \; 0 < x < \ell,$$
$$u(0, t) = 0, \; t > t_0,$$
$$u(\ell, t) = \epsilon \sin(2\pi \omega t), \; t > t_0.$$

Take $\ell = 1$, $c = 522$, $t_0 = 0$, and $\epsilon = 10^{-4}$.

(a) Solve the IBVP with ω not equal to a natural frequency of the string. Graph the motion of the string with ω equal to half the fundamental frequency.

(b) Solve the IBVP with ω equal to a natural frequency of the string. Show that resonance occurs. Graph the motion of the string with ω equal to the fundamental frequency.

9. Consider a string whose (linear) density (in the unstretched state) is 0.25 g/cm and whose longitudinal stiffness is 6500 N. If the unstretched (zero tension) length of the string is 40 cm, to what length must the string be stretched in order that its fundamental frequency be 261 Hz (middle C)?

10. Although a discontinuous initial displacement does not make sense for the vibrating string, we can explore such a situation purely from a mathematical point of view. Consider the IBVP

$$\frac{\partial^2 v}{\partial t^2} - c^2 \frac{\partial^2 v}{\partial x^2} = 0, \ 0 < x < 50, \ t > 0,$$
$$v(x,0) = \psi(x), \ 0 < x < 50,$$
$$\frac{\partial v}{\partial t}(x,0) = 0, \ 0 < x < 50,$$
$$v(0,t) = 0, \ t > 0,$$
$$v(50,t) = 0, \ t > 0,$$

where $c = 1\,\text{cm/s}$ and

$$\psi(x) = \begin{cases} -1, & 24 < x < 26, \\ 0, & 0 \leq x \leq 24 \text{ or } 26 \leq x \leq 50. \end{cases}$$

Solve the IBVP using the Fourier series method and graph several snapshots of the solutions. Focus on the time just before and after the disturbance reaches the boundary. Describe the effect of the boundary on the solution.

7.3 Finite element methods for the wave equation

The wave equation presents special difficulties for the design of numerical methods. These difficulties are largely caused by the fact that waves with abrupt changes (even singularities, such as jump discontinuities) will propagate in accordance with the wave equation, with no smoothing of the waves. (Of course, if $u(x,t)$ has a singularity, that is, it fails to be twice differentiable, then it cannot be a solution of the wave equation unless we expand our notion of what constitutes a solution. But there are mathematically consistent ways to do this, and they are important for modeling physical phenomena exhibiting discontinuous behavior.)

For example, the function $\psi : [0,1] \to \mathbf{R}$ defined by

$$\psi(x) = \begin{cases} 1, & |x - \tfrac{1}{2}| < 0.05, \\ 0, & |x - \tfrac{1}{2}| \geq 0.05 \end{cases}$$

is discontinuous. We solve both the heat equation,

$$\begin{aligned}
\frac{\partial u}{\partial t} - \frac{\partial^2 u}{\partial x^2} &= 0, \ 0 < x < 1, \ t > 0, \\
u(x,0) &= \psi(x), \ 0 < x < 1, \\
u(0,t) &= 0, \ t > 0, \\
u(1,t) &= 0, \ t > 0,
\end{aligned} \qquad (7.13)$$

and the wave equation,

$$\frac{\partial^2 u}{\partial t^2} - \frac{\partial^2 u}{\partial x^2} = 0,\ 0 < x < 1,\ t > 0,$$
$$u(x,0) = \psi(x),\ 0 < x < 1,$$
$$\frac{\partial u}{\partial t}(x,0) = 0,\ 0 < x < 1,$$
$$u(0,t) = 0,\ t > 0,$$
$$u(1,t) = 0,\ t > 0,$$

(7.14)

with $\psi(x)$ as the initial condition (we need a second initial condition in the wave equation; we take the initial velocity equal to zero). The results are shown in Figures 7.10 and 7.11. The graphs show that the discontinuity is immediately "smoothed away" by the heat equation, but preserved by the wave equation.

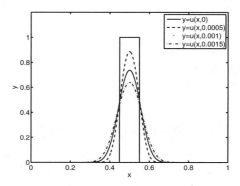

Figure 7.10. *The solution to the heat equation with a discontinuous initial condition (see (7.13)). Graphed are four time snapshots, including $t = 0$.*

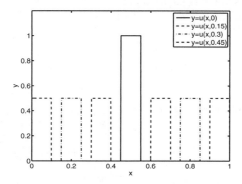

Figure 7.11. *The solution to the wave equation with a discontinuous initial condition (see (7.14)). Graphed are four time snapshots, including $t = 0$.*

7.3. Finite element methods for the wave equation

The standard method for applying finite elements to the wave equation, which we now present, works well if the true solution is smooth. Methods that are effective when the solution has singularities are beyond the scope of this book.

7.3.1 The wave equation with Dirichlet conditions

We proceed as we did for the heat equation. We suppose that $u(x,t)$ solves

$$\begin{aligned}
\frac{\partial^2 u}{\partial t^2} - c^2 \frac{\partial^2 u}{\partial x^2} &= f(x,t),\ 0 < x < \ell,\ t > t_0, \\
u(x,t_0) &= \psi(x),\ 0 < x < \ell, \\
\frac{\partial u}{\partial t}(x,t_0) &= \gamma(x),\ 0 < x < \ell, \\
u(0,t) &= 0,\ t > t_0, \\
u(\ell,t) &= 0,\ t > t_0.
\end{aligned} \qquad (7.15)$$

We then multiply the PDE on both sides by a test function $v \in V = C_D^2[0,\ell]$ and integrate by parts in the second term on the left to obtain

$$\int_0^\ell \left\{ \frac{\partial^2 u}{\partial t^2}(x,t)v(x) + c^2 \frac{\partial u}{\partial x}(x,t)\frac{dv}{dx}(x) \right\} dx = \int_0^\ell f(x,t)v(x)\,dx,\ t > t_0,\ v \in V. \qquad (7.16)$$

This is the weak form. We now apply the Galerkin technique, approximating $u(x,t)$ by

$$v_n(x,t) = \sum_{i=1}^{n-1} u_i(t)\phi_i(x), \qquad (7.17)$$

where $\phi_1, \phi_2, \ldots, \phi_{n-1}$ are the standard piecewise linear finite element basis functions, and requiring that the variational equation (7.16) holds for all continuous piecewise linear test functions:

$$\begin{aligned}
\int_0^\ell \left\{ \frac{\partial^2 v_n}{\partial t^2}(x,t)\phi_i(x) + c^2 \frac{\partial v_n}{\partial x}(x,t)\frac{d\phi_i}{dx}(x) \right\} dx \\
= \int_0^\ell f(x,t)\phi_i(x)\,dx,\ t > t_0\ \text{for all}\ i = 1,2,\ldots,n-1.
\end{aligned} \qquad (7.18)$$

Substituting (7.17) into (7.18) yields

$$\begin{aligned}
\sum_{j=1}^{n-1} \frac{d^2 u_j}{dt^2}(t) \int_0^\ell \phi_j(x)\phi_i(x)\,dx + \sum_{j=1}^{n-1} u_j(t) \int_0^\ell c^2 \frac{d\phi_j}{dx}(x)\frac{d\phi_i}{dx}(x)\,dx \\
= \int_0^\ell f(x,t)\phi_i(x)\,dx,\ t \geq t_0,\ i = 1,2,\ldots,n-1.
\end{aligned} \qquad (7.19)$$

If we now define matrices $\mathbf{M}$ and $\mathbf{K}$ (the mass and stiffness matrices, respectively) by

$$M_{ij} = \int_0^\ell \phi_j(x)\phi_i(x)\,dx,\quad K_{ij} = \int_0^\ell c^2 \frac{d\phi_j}{dx}(x)\frac{d\phi_i}{dx}(x)\,dx,$$

and the vector-valued functions $\mathbf{f}(t)$ and $\mathbf{u}(t)$ by

$$\mathbf{f}(t) = \begin{bmatrix} \int_0^\ell f(x,t)\phi_1(x)dx \\ \int_0^\ell f(x,t)\phi_2(x)dx \\ \vdots \\ \int_0^\ell f(x,t)\phi_{n-1}(x)dx \end{bmatrix}, \quad \mathbf{u}(t) = \begin{bmatrix} u_1(t) \\ u_2(t) \\ \vdots \\ u_{n-1}(t) \end{bmatrix},$$

we can write (7.19) as

$$\mathbf{M}\frac{d^2\mathbf{u}}{dt^2} + \mathbf{K}\mathbf{u} = \mathbf{f}(t).$$

We now have a system of second-order linear ODEs with constant coefficients. We need two initial conditions, and they are easily obtained from the initial conditions in (7.15). We have

$$u(x,t_0) = \psi(x), \; 0 < x < \ell,$$

and $\psi(x)$ can be approximated by its linear interpolant:

$$\psi(x) \doteq \sum_{i=1}^{n-1} \psi(x_i)\phi_i(x).$$

We therefore require that

$$u_i(t_0) = \psi(x_i).$$

Similarly, we require

$$\frac{du_i}{dt}(t_0) = \gamma(x_i).$$

These lead to the initial conditions

$$\mathbf{u}(t_0) = \mathbf{y}_0 = \begin{bmatrix} \psi(x_1) \\ \psi(x_2) \\ \vdots \\ \psi(x_{n-1}) \end{bmatrix}, \quad \frac{d\mathbf{u}}{dt}(t_0) = \mathbf{z}_0 = \begin{bmatrix} \gamma(x_1) \\ \gamma(x_2) \\ \vdots \\ \gamma(x_{n-1}) \end{bmatrix}.$$

We therefore obtain the IVP

$$\begin{aligned} \mathbf{M}\frac{d^2\mathbf{u}}{dt^2} + \mathbf{K}\mathbf{u} &= \mathbf{f}(t), \; t > t_0, \\ \mathbf{u}(t_0) &= \mathbf{y}_0, \\ \mathbf{u}'(t_0) &= \mathbf{z}_0. \end{aligned} \quad (7.20)$$

To apply one of the numerical methods for ODEs that we studied in Chapter 4, we must first convert (7.20) to a first-order system. We will define

$$\begin{aligned} \mathbf{y}(t) &= \mathbf{u}(t), \\ \mathbf{z}(t) &= \frac{d\mathbf{u}}{dt}(t) \end{aligned}$$

7.3. Finite element methods for the wave equation

$(\mathbf{y}, \mathbf{z} \in \mathbf{R}^{n-1})$. Then

$$\frac{d\mathbf{y}}{dt} = \mathbf{z},$$
$$\frac{d\mathbf{z}}{dt} = \frac{d^2\mathbf{u}}{dt^2}$$
$$= \mathbf{M}^{-1}(-\mathbf{K}\mathbf{u} + \mathbf{f}(t))$$
$$= -\mathbf{M}^{-1}\mathbf{K}\mathbf{y} + \mathbf{M}^{-1}\mathbf{f}(t)$$
$$= \mathbf{A}\mathbf{y} + \mathbf{g}(t),$$

where[38] $\mathbf{A} = -\mathbf{M}^{-1}\mathbf{K}$ and $\mathbf{g}(t) = \mathbf{M}^{-1}\mathbf{f}(t)$. We also have initial conditions:

$$\mathbf{y}(t_0) = \mathbf{u}(t_0) = \mathbf{y}_0,$$
$$\mathbf{z}(t_0) = \frac{d\mathbf{u}}{dt}(t_0) = \mathbf{z}_0.$$

We have thus obtained the $(2n - 2) \times (2n - 2)$ system of IVPs

$$\frac{d\mathbf{y}}{dt} = \mathbf{z}, \ \mathbf{y}(t_0) = \mathbf{y}_0,$$
$$\frac{d\mathbf{z}}{dt} = \mathbf{A}\mathbf{y} + \mathbf{g}(t), \ \mathbf{z}(t_0) = \mathbf{z}_0.$$

We can now apply any of the numerical methods that we have previously learned such as Euler's method, RK4, or an adaptive scheme.

Just as in the case of the heat equation, stability is an issue. If the time step Δt is chosen to be too large relative to the spatial mesh size h, then the approximate solution computed by the time-stepping method can "blow up" (i.e., increase without bound). For the heat equation, using Euler's method, we needed $\Delta t = O(h^2)$. This required such a large number of time steps that it was advantageous to use special methods (such as the backward Euler method) for the sake of efficiency. The requirement on Δt is not so stringent when solving the wave equation; we just need $\Delta t = O(h/c)$. For this reason, we usually use explicit methods to solve the system of ODEs arising from the wave equation; the improved stability properties of implicit methods are not really needed. We must keep the stability requirement in mind, though. If, in applying (7.20) to approximate a solution to the wave equation, we notice that the approximate solution is growing unreasonably in amplitude, then the time step must be decreased.

Example 7.7. We will solve the IBVP

$$\frac{\partial^2 u}{\partial t^2} - c^2 \frac{\partial^2 u}{\partial x^2} = 0, \ 0 < x < 1, \ t > 0,$$
$$u(x, 0) = \psi(x), \ 0 < x < 1,$$
$$\frac{\partial u}{\partial t}(x, 0) = 0, \ 0 < x < 1, \quad (7.21)$$
$$u(0, t) = 0, \ t > 0,$$
$$u(1, t) = 0, \ t > 0,$$

---

[38]As we discussed in Section 6.4, we do not actually form the inverse matrix $\mathbf{M}^{-1}$. Instead, when the action of $\mathbf{M}^{-1}$ is needed, we solve the appropriate linear system.

with $c = 522$ and
$$\psi(x) = 0.01x(1-x).$$

The resulting solution is quite smooth, so the finite element method should work well. We use a regular grid with $n = 20$ subintervals ($h = 1/n$) and apply the RK4 method to integrate the resulting ODEs. The fundamental frequency of the string is $c/(2\ell) = 261$, making the fundamental period
$$t_f = \frac{1}{261} \doteq 0.003814.$$
We compute the motion of the string over one period, using 50 steps ($\Delta t = t_f/50$). The results are shown in Figure 7.12. ∎

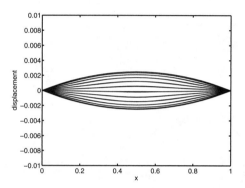

Figure 7.12. *The solution to the wave equation in Example 7.7. Shown are 25 snapshots taken over one period of the motion.*

The time step taken in the previous example,
$$\Delta t = \frac{t_f}{50} \doteq 7.663 \cdot 10^{-5},$$
seems rather small. Moreover, it can be shown that it cannot be much larger without instability appearing in the computation. On the other hand, as mentioned above, the time step Δt must be decreased only in proportion to h to preserve stability. This is in contrast to the situation with the heat equation, where $\Delta t = O(h^2)$ is required.

Example 7.8. We will now solve (7.21) with
$$\psi(x) = \begin{cases} 0.01(x-0.4), & 0.4 < x < 0.5, \\ -0.01(x-0.6), & 0.5 < x < 0.6, \\ 0, & 0 < x < 0.4 \text{ or } 0.6 < x < 1 \end{cases}$$

(compare Example 7.3). Since the initial displacement ψ is not very smooth (ψ is continuous but its derivative has singularities), our comments at the beginning of the section suggest that it may be difficult to accurately compute the wave motion using finite elements.

We use the same spatial mesh ($h = 1/20$), time-stepping algorithm (RK4), and time step ($\Delta t = t_f/50$) as in the previous example. In Figure 7.13, we show three snapshots of

7.3. Finite element methods for the wave equation

the solution. We already know how the solution should look: The initial "blip" splits into two pieces, one moving to the right and the other to the left, preserving the original shape (see Figure 7.5). The computed solution, while demonstrating the basic motion, is not very accurate—the shapes are not well represented.

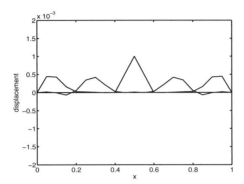

Figure 7.13. *The solution to the wave equation in Example 7.8, computed with a mesh size of $h = 1/20$ and a time step of $\Delta t = t_f/50$. Shown are 3 snapshots corresponding to $t = 0$, $t = t_f/50$, $t = 2t_f/50$.*

To get an accurate computed solution requires a fine grid. We repeat the calculations using $h = 1/100$ and $\Delta t = t_f/250$ and obtain the results shown in Figure 7.14. Even with this finer grid, there is a distortion in the waves. ∎

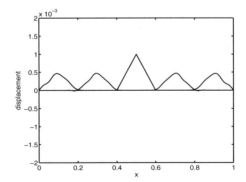

Figure 7.14. *The solution to the wave equation in Example 7.8, computed with a mesh size of $h = 1/100$ and a time step of $\Delta t = t_f/250$. Shown are 3 snapshots corresponding to $t = 0$, $t = t_f/50$, $t = 2t_f/50$.*

7.3.2 The wave equation under other boundary conditions

If an IBVP for the wave equation involves a Neumann condition, we can apply the finite element method by adjusting the weak form and the approximating subspace, as explained in

Section 6.5. By way of example, we will consider the following IBVP with mixed boundary conditions:

$$\frac{\partial^2 u}{\partial t^2} - c^2 \frac{\partial^2 u}{\partial x^2} = f(x,t),\ 0 < x < \ell,\ t > t_0,$$
$$u(x,t_0) = \psi(x),\ 0 < x < \ell,$$
$$\frac{\partial u}{\partial t}(x,t_0) = \gamma(x),\ 0 < x < \ell, \qquad (7.22)$$
$$u(0,t) = 0,\ t > t_0,$$
$$\frac{\partial u}{\partial x}(\ell,t) = 0,\ t > t_0.$$

Since the Dirichlet condition is essential, it appears in the definition of the space of test functions; however, the Neumann condition is natural and does not appear explicitly. Therefore, the weak form is

$$\int_0^\ell \left\{ \frac{\partial^2 u}{\partial t^2}(x,t)v(x) + c^2 \frac{\partial u}{\partial x}(x,t)\frac{dv}{dx}(x) \right\} dx = \int_0^\ell f(x,t)v(x)\,dx,\ t > t_0,\ v \in \tilde{V}, \quad (7.23)$$

where

$$\tilde{V} = \left\{ v \in C^2[0,\ell] \ :\ v(0) = 0 \right\}.$$

The reader should note that this is exactly the same as the weak form derived for the Dirichlet problem, except for the space of test functions used.

To apply the Galerkin method, we establish a grid

$$0 = x_0 < x_1 < x_2 < \cdots < x_n = \ell$$

and choose the finite element subspace consisting of all continuous piecewise linear functions (relative to the given mesh) satisfying the Dirichlet condition at the left endpoint. This subspace can be represented as

$$\tilde{S}_n = \text{span}\{\phi_1, \phi_2, \ldots, \phi_n\}.$$

The only difference between $\tilde{S}_n$ and S_n, the subspace used for the Dirichlet problem, is the inclusion of the basis function ϕ_n corresponding to the right endpoint. For simplicity, we will use a regular grid in the following calculations.

The calculation then proceeds just as for the Dirichlet problem, except that there is an extra basis function and hence an extra unknown and corresponding equation. We write

$$v_n(x,t) = \sum_{i=1}^n u_i(t)\phi_i(x)$$

for the approximation to the solution $u(x,t)$. Substituting into the weak form yields the equations

$$\tilde{\mathbf{M}} \frac{d^2 \mathbf{u}}{dt^2} + \tilde{\mathbf{K}} \mathbf{u} = \tilde{\mathbf{f}}(t),$$

7.3. Finite element methods for the wave equation

where $\tilde{\mathbf{M}} \in \mathbf{R}^{n\times n}$, $\tilde{\mathbf{K}} \in \mathbf{R}^{n\times n}$, and $\tilde{\mathbf{f}}(t) \in \mathbf{R}^n$. The mass matrix $\tilde{\mathbf{M}}$ is tridiagonal, with the following nonzero entries:

$$\tilde{M}_{ii} = \int_0^\ell (\phi_i(x))^2 \, dx = \frac{2h}{3}, \; i = 1,2,\ldots,n-1,$$

$$\tilde{M}_{nn} = \int_0^\ell (\phi_n(x))^2 \, dx = \frac{h}{3},$$

$$\tilde{M}_{i,i+1} = \int_0^\ell \phi_i(x)\phi_{i+1}(x) \, dx = \frac{h}{6}, \; i = 1,2,\ldots,n-1,$$

$$\tilde{M}_{i+1,i} = \tilde{M}_{i,i+1}, \; i = 1,2,\ldots,n-1.$$

The stiffness matrix $\tilde{\mathbf{K}}$ is also tridiagonal, with the following nonzero entries:

$$\tilde{K}_{ii} = \int_0^\ell c^2 \left(\frac{d\phi_i}{dx}(x)\right)^2 dx = \frac{2c^2}{h}, \; i = 1,2,\ldots,n-1,$$

$$\tilde{K}_{nn} = \int_0^\ell c^2 \left(\frac{d\phi_n}{dx}(x)\right)^2 dx = \frac{c^2}{h},$$

$$\tilde{K}_{i,i+1} = \int_0^\ell \frac{d\phi_i}{dx}(x) \frac{d\phi_{i+1}}{dx}(x) \, dx = -\frac{c^2}{h}, \; i = 1,2,\ldots,n-1,$$

$$\tilde{K}_{i+1,i} = \tilde{K}_{i,i+1}, \; i = 1,2,\ldots,n-1.$$

Finally, the load vector $\tilde{\mathbf{f}}(t)$ is given by

$$\tilde{f}_i(t) = \int_0^\ell f(x,t)\phi_i(x) \, dx, \; i = 1,2,\ldots,n.$$

Having derived the system of ODEs, the formulation of the initial conditions is just as before:

$$\mathbf{u}(t_0) = \mathbf{y}_0 = \begin{bmatrix} \psi(x_1) \\ \psi(x_2) \\ \vdots \\ \psi(x_n) \end{bmatrix},$$

$$\frac{d\mathbf{u}}{dt}(t_0) = \mathbf{z}_0 = \begin{bmatrix} \gamma(x_1) \\ \gamma(x_2) \\ \vdots \\ \gamma(x_n) \end{bmatrix}.$$

The IVP

$$\tilde{\mathbf{M}}\frac{d^2\mathbf{u}}{dt^2} + \tilde{\mathbf{K}}\mathbf{u} = \tilde{\mathbf{f}}(t), \; t > t_0,$$

$$\mathbf{u}(t_0) = \mathbf{y}_0,$$

$$\mathbf{u}'(t_0) = \mathbf{z}_0$$

is then converted to the first-order system

$$\frac{d\mathbf{y}}{dt} = \mathbf{z}, \ \mathbf{y}(t_0) = \mathbf{y}_0,$$
$$\frac{d\mathbf{z}}{dt} = \tilde{\mathbf{A}}\mathbf{y} + \tilde{\mathbf{g}}(t), \ \mathbf{z}(t_0) = \mathbf{z}_0,$$

where $\tilde{\mathbf{A}} = -\tilde{\mathbf{M}}^{-1}\tilde{\mathbf{K}}$ and $\tilde{\mathbf{g}}(t) = \tilde{\mathbf{M}}^{-1}\tilde{\mathbf{f}}(t)$.

Example 7.9. We will compute the motion of the string in Example 7.7, with the only change being that the right end of the string is now assumed to be free to move vertically. The IBVP is

$$\frac{\partial^2 u}{\partial t^2} - c^2 \frac{\partial^2 u}{\partial x^2} = 0, \ 0 < x < 1, \ t > 0,$$
$$u(x, 0) = x(1-x), \ 0 < x < 1,$$
$$\frac{\partial u}{\partial t}(x, 0) = 0, \ 0 < x < 1,$$
$$u(0, t) = 0, \ t > 0,$$
$$\frac{\partial u}{\partial x}(1, t) = 0, \ t > 0,$$

with $c = 522$. The only change from the earlier calculations is that there is an extra row and column in the mass and stiffness matrices. We will again use $n = 20$ subintervals in the spatial mesh ($h = 1/20$). The fundamental period is now

$$T = \frac{4}{c} \doteq 0.007663.$$

We use the RK4 algorithm and 100 time steps to compute the motion over one period ($\Delta t = T/100$). The results are shown in Figure 7.15. ∎

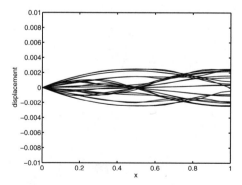

Figure 7.15. *The solution to the wave equation in Example 7.9. Shown are 25 snapshots taken over one period of the motion.*

7.3. Finite element methods for the wave equation

We pointed out above that, in solving the wave equation by the method of lines, implicit methods are not typically used, because the stability requirement ($\Delta t = O(h/c)$) for an explicit method is not too onerous. However, in using the finite element method, the mass matrix couples the time derivatives, and therefore, to advance a time-stepping method, it is necessary to solve a linear system. This means that an explicit method is no more efficient than an implicit method. The same was true in the case of the heat equation, with, however, a key difference. For the heat equation, the stability requirement is such that it is not advantageous to use explicit methods anyway, and so the coupling introduced by the finite element method is not important. For the wave equation, explicit *finite difference* methods do not couple the derivatives, and there is no need to solve a linear system to take a time step. For this reason, finite difference methods are often preferred for the wave equation. We present finite difference methods for the wave equation in Section 7.5.

Exercises

1. Use piecewise linear finite elements to solve the IBVP

$$\frac{\partial^2 u}{\partial t^2} - c^2 \frac{\partial^2 u}{\partial x^2} = 0, \ 0 < x < 100, \ t > 0,$$
$$u(x,0) = 10^{-6} x(100-x)^2, \ 0 < x < 100,$$
$$\frac{\partial u}{\partial t}(x,0) = 0, \ 0 < x < 100,$$
$$u(0,t) = 0, \ t > 0,$$
$$u(100,t) = 0, \ t > 0,$$

   on the interval $0 < t < T$, where T is the fundamental period. Take $c = 1000$. Graph several snapshots.

2. Repeat the previous exercise, assuming the PDE is inhomogeneous with right-hand side equal to the constant -100.

3. (Cf. Exercise 7.2.2.) Consider a string of length 50 cm, with $c = 400$ cm/s. Suppose that the string is initially at rest, with both ends fixed, and that it is struck by a hammer at its center so as to induce an initial velocity of

$$\gamma(x) = \begin{cases} -20, & 24 < x < 26, \\ 0, & 0 \le x \le 24 \text{ or } 26 \le x \le 50. \end{cases}$$

   (a) How long does it take for the resulting wave to reach the ends of the string?

   (b) Formulate and solve the IBVP describing the motion of the string. Use the finite element method with RK4 or some other numerical method for solving the system of ODEs.

   (c) Plot several snapshots of the string, showing the wave first reaching the ends of the string. Verify your answer to 3(a).

4. Repeat Exercise 3, assuming that the right end of the string is free to move vertically.

5. Solve (7.14) using the finite element method, and try to reproduce Figure 7.11. How small must h (the length of the subintervals in the mesh) be to obtain a good graph?

6. (Cf. Example 7.6.) Consider a steel bar of length 1 m, fixed at the top ($x = 0$) with the bottom end ($x = 1$) free to move. Suppose the bar is stretched, by means of a downward pressure applied to the free end, to a length of 1.002 m, and then released. Using the finite element method, compute the motion of the bar, given that the type of steel has stiffness $k = 195$ GPa and density $\rho = 7.9\,\text{g/cm}^3$. Produce a graph analogous to Figure 7.9.

7. Consider a heterogeneous bar with the top end fixed and the bottom end free. According to the derivation in Section 2.2, the displacement $u(x,t)$ of the bar satisfies the IBVP

$$\rho(x)\frac{\partial^2 u}{\partial t^2} - \frac{\partial}{\partial x}\left(k(x)\frac{\partial u}{\partial x}\right) = f(x,t),\ 0 < x < \ell,\ t > t_0,$$
$$u(x,t_0) = \psi(x),\ 0 < x < \ell,$$
$$\frac{\partial u}{\partial t}(x,t_0) = \gamma(x),\ 0 < x < \ell,\qquad (7.24)$$
$$u(0,t) = 0,\ t > t_0,$$
$$\frac{\partial u}{\partial x}(\ell,t) = 0,\ t > t_0.$$

(a) Formulate the weak form of (7.24).

(b) Apply the Galerkin method to obtain a system of ODEs with initial conditions. How do the mass and stiffness matrices change from the case of a homogeneous bar?

8. Repeat Exercise 6, assuming now that the bar is heterogeneous with density $\rho(x) = 7500 + 1000x\,\text{kg/m}^3$ and stiffness $k(x) = 1.9 \cdot 10^{11} + 10^{10}x\,\text{N/m}^2$. (The coordinate x is measured in meters: $0 < x < 1$.)

7.4 Resonance

We now consider two special experiments that can lead to resonance.

- A string has one end fixed, and the other is mechanically moved up and down: $u(\ell,t) = \sin(2\pi\omega t)$. When the frequency of the applied motion equals one of the natural frequencies of the string, *resonance* occurs. This means that the amplitudes of the vibration of the string increase without bound.

- A metal string (such as a guitar string), with both ends fixed, is subjected to an oscillatory force generated by a small electromagnet. Again, resonance occurs when the applied frequency equals one of the natural frequencies of the string.

7.4.1 The wave equation with a periodic boundary condition

The following IBVP describes an elastic string with the left endpoint held fixed and the right endpoint moved up and down in a periodic motion (cf. Exercise 7.2.8):

$$\frac{\partial^2 u}{\partial t^2} - c^2 \frac{\partial^2 u}{\partial x^2} = 0,\ 0 < x < \ell,\ t > t_0,$$
$$u(x, t_0) = 0,\ 0 < x < \ell,$$
$$\frac{\partial u}{\partial t}(x, t_0) = 0,\ 0 < x < \ell, \quad (7.25)$$
$$u(0, t) = 0,\ t > t_0,$$
$$u(\ell, t) = \epsilon \sin(2\pi \omega t),\ t > t_0.$$

For simplicity, we take $\ell = 1$ and $t_0 = 0$.

Since the Dirichlet condition at the right endpoint is inhomogeneous, we will shift the data. We define

$$p(x,t) = \epsilon x \sin(2\pi \omega t)$$

and $v(x,t) = u(x,t) - p(x,t)$. Then a straightforward calculation shows that v satisfies the IBVP

$$\frac{\partial^2 v}{\partial t^2} - c^2 \frac{\partial^2 v}{\partial x^2} = 4\pi^2 \omega^2 \epsilon x \sin(2\pi \omega t),\ 0 < x < 1,\ t > 0,$$
$$v(x, 0) = 0,\ 0 < x < 1,$$
$$\frac{\partial v}{\partial t}(x, 0) = -2\pi \omega \epsilon x,\ 0 < x < 1, \quad (7.26)$$
$$v(0, t) = 0,\ t > 0,$$
$$v(1, t) = 0,\ t > 0.$$

We write the solution as

$$v(x,t) = \sum_{n=1}^{\infty} a_n(t) \sin(n\pi x).$$

Then, by the usual calculation, the coefficient $a_n(t)$ satisfies the IVP

$$\frac{d^2 a_n}{dt^2} + c^2 n^2 \pi^2 a_n = c_n(t),$$
$$a_n(0) = 0,$$
$$\frac{da_n}{dt}(0) = b_n,$$

where

$$c_n(t) = 2 \int_0^1 4\pi^2 \omega^2 \epsilon x \sin(2\pi \omega t) \sin(n\pi x)\, dx$$
$$= \frac{8\pi \omega^2 \epsilon (-1)^{n+1}}{n} \sin(2\pi \omega t)$$
$$= e_n \sin(2\pi \omega t) \quad \left(e_n = \frac{8\pi \omega^2 \epsilon (-1)^{n+1}}{n} \right)$$

and
$$b_n = 2\int_0^1 (-2\pi\omega\epsilon x)\sin(n\pi x)\,dx = \frac{4\omega\epsilon(-1)^n}{n}.$$

By the results of Section 4.2.3, we have
$$a_n(t) = \frac{b_n}{cn\pi}\sin(cn\pi t) + \frac{1}{cn\pi}\int_0^t \sin(cn\pi(t-s))c_n(s)\,ds$$

or
$$a_n(t) = \frac{4\omega\epsilon(-1)^n}{cn^2\pi}\sin(cn\pi t) + \frac{8\omega^2\epsilon(-1)^{n+1}}{cn^2}\int_0^1 \sin(cn\pi(t-s))\sin(2\pi\omega s)\,ds. \quad (7.27)$$

If $\omega \neq cn/2$ for all n (that is, if ω is not one of the natural frequencies of the string), then we obtain
$$\int_0^t \sin(cn\pi(t-s))\sin(2\pi\omega s)\,ds = \frac{2}{\pi\left(c^2n^2 - 4\omega^2\right)}\left(\frac{cn}{2}\sin(2\pi\omega t) - \omega\sin(cn\pi t)\right). \quad (7.28)$$

On the other hand, if $\omega = cn/2$ for some positive integer n, then ω is a natural frequency of the string and we obtain
$$\int_0^t \sin(cn\pi(t-s))\sin(2\pi\omega s)\,ds = \frac{\sin(cn\pi t)}{2cn\pi} - \frac{t}{2}\cos(cn\pi t). \quad (7.29)$$

In the second case, there is a factor of t multiplying one of the sinusoids. This indicates resonance—as t increases, the magnitude of a_n increases without bound. This is true for the value of n for which $\omega = cn/2$. As a result, the vibration of the string will resemble the nth normal mode of the string, but with an ever-increasing amplitude.

If ω differs from all the natural frequencies, but $\omega \doteq cn/2$ for some positive integer n, then (7.28) shows that the corresponding a_n will be large (though bounded), due to the small denominator $c^2n^2 - 4\omega^2$. In this case, the vibration of the string will be dominated by the nth normal mode.

The following examples illustrate these conclusions.

Example 7.10. Suppose $c = 522$ and $\omega = c/2 = 261$, the fundamental frequency of the string. With $\epsilon = 0.001$, we obtain the motion shown in Figure 7.16. The reader should notice how motion of the string resembles the fundamental mode, and also how the amplitude increases as time progresses. ∎

Example 7.11. This example is the same as the previous one, except the frequency of the forcing motion is $\omega = 1040$, which is very close to the fourth natural frequency of 1044. Figure 7.17 shows several snapshots of the solution, which resembles the fourth normal mode of the string. Unlike the previous example, the amplitude of the vibrations remains bounded. ∎

7.4. Resonance

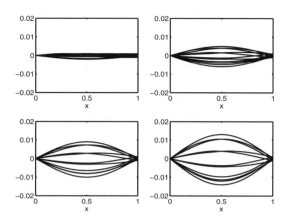

Figure 7.16. *Solution to (7.25) with an oscillatory Dirichlet condition (see Example 7.10). The frequency of the forcing term is 261, the fundamental frequency of the string. Shown are 10 snapshots from the first fundamental period (upper left), 10 from the third fundamental period (upper right), 10 from the fifth (lower left), and 10 from the seventh (lower right). The increase in the amplitude is due to resonance.*

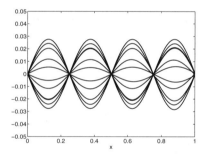

Figure 7.17. *Solution to (7.25) with an oscillatory Dirichlet condition (see Example 7.11). The frequency of the forcing term is 1040, close to but not equal to the fourth natural frequency of the spring. This example does not exhibit resonance.*

7.4.2 The wave equation with a localized source

We now present an example in which a string is driven by a force localized on a very small interval. We begin with the following IBVP for the wave equation:

$$\frac{\partial^2 u}{\partial t^2} - c^2 \frac{\partial^2 u}{\partial x^2} = f(x,t),\ 0 < x < \ell,\ t > t_0,$$
$$u(x,t_0) = 0,\ 0 < x < \ell,$$
$$\frac{\partial u}{\partial t}(x,t_0) = 0,\ 0 < x < \ell, \quad (7.30)$$
$$u(0,t) = 0,\ t > t_0,$$
$$u(\ell,t) = 0,\ t > t_0,$$

where $f(x,t)$ is defined by

$$f(x,t) = \begin{cases} \phi(t), & a < x < b, \\ 0 & \text{otherwise}, \end{cases}$$

and $0 < a < b < \ell$. This models the vibrating string, where the string is assumed to vibrate due to a time-varying force applied on the interval $[a,b]$.

We can solve (7.30) exactly as in Section 7.2. The Fourier sine coefficients of $f(x,t)$ are computed as follows:

$$\begin{aligned} c_n(t) &= \frac{2}{\ell} \int_0^\ell f(x,t) \sin\left(\frac{n\pi x}{\ell}\right) dx \\ &= \frac{2}{\ell} \int_a^b \phi(t) \sin\left(\frac{n\pi x}{\ell}\right) dx \\ &= \frac{2\phi(t)}{n\pi} \left(\cos\left(\frac{n\pi a}{\ell}\right) - \cos\left(\frac{n\pi b}{\ell}\right) \right) \\ &= d_n \phi(t), \; d_n = \frac{2}{n\pi} \left(\cos\left(\frac{n\pi a}{\ell}\right) - \cos\left(\frac{n\pi b}{\ell}\right) \right). \end{aligned}$$

We then must solve the following IVP to find the unknown coefficients $a_n(t)$:

$$\begin{aligned} \frac{d^2 a_n}{dt^2} + \frac{c^2 n^2 \pi^2}{\ell^2} a_n &= d_n \phi(t), \; t > t_0, \\ a_n(t_0) &= 0, \\ \frac{da_n}{dt}(t_0) &= 0. \end{aligned}$$

According to the results derived in Section 4.2.3, the solution to this IVP is

$$a_n(t) = \frac{d_n \ell}{c n \pi} \int_{t_0}^t \sin\left(\frac{c n \pi}{\ell}(t-s)\right) \phi(s) \, ds.$$

Oscillatory forcing and resonance

We are interested in the case in which $f(t)$ is oscillatory. For convenience, we take $\ell = 1$ and $t_0 = 0$, and let $f(t) = \sin(2\pi \omega t)$. Then the solution to the IBVP is

$$u(x,t) = \sum_{n=1}^\infty a_n(t) \sin(n\pi x),$$

where

$$a_n(t) = \frac{d_n}{c n \pi} \int_0^t \sin(c n \pi (t-s)) \sin(2\pi \omega s) \, ds.$$

7.4. Resonance

A lengthy but elementary calculation shows that, if $\omega \neq cn/2$, then

$$a_n(t) = \frac{d_n (cn \sin(2\pi \omega t) - 2\omega \sin(cn\pi t))}{cn\pi^2(c^2n^2 - 4\omega^2)}. \tag{7.31}$$

One the other hand, if $\omega = cm/2$, one of the natural frequencies of the string, then

$$a_m(t) = \frac{d_n (\sin(cm\pi t) - cm\pi t \cos(cm\pi t))}{2c^2m^2\pi^2} \tag{7.32}$$

(formula (7.31) still holds, with $\omega = cm/2$, for the coefficients $a_n(t)$ with $n \neq m$).

In the case that the external frequency ω is a natural frequency of the string, the result (7.32) shows that the string oscillates with an ever-increasing amplitude. (The reader should notice the factor of t, which shows that the amplitude increases without bound.) This phenomenon is called *resonance*. When the external frequency is close to but not equal to a natural frequency, formula (7.31) shows that the nearby normal mode is amplified in the solution (though it remains bounded)—this is because the denominator in (7.31) is very small.

Example 7.12. We now solve (7.30) with $c = 522$ and $\ell = 1$ (this corresponds to a fundamental frequency of middle C, as we saw in Example 7.3), $t_0 = 0$, and $\phi(t) = \sin(2\pi \omega t)$ for various values of ω. We localize the source on the interval $[0.35, 0.36]$ (that is, $f(x,t) = \sin(2\pi \omega t)$ for $0.35 \leq x \leq 0.36$, $f(x,t) = 0$ otherwise). Figure 7.18 shows the solutions corresponding to $\omega = 260$, $\omega = 520$, $\omega = 780$, and $\omega = 1040$, close to the natural frequencies of 261, 522, 783, and 1044, respectively. The reader will notice how the solutions resemble the first four normal modes of the string. In Figure 7.19, we display some snapshots of the solution for $\omega = 1044$, the fourth natural frequency. The effect of resonance is clearly seen in the increasing amplitude. ∎

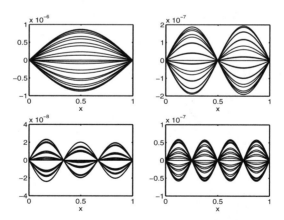

Figure 7.18. *Solutions to (7.30) with an oscillatory forcing term (see Example 7.12). The four solutions shown correspond to $\omega = 260$, $\omega = 520$, $\omega = 780$, and $\omega = 1040$; since these frequencies are close to the natural frequencies of 261, 522, 783, and 1044, the solutions are almost equal to the corresponding normal modes.*

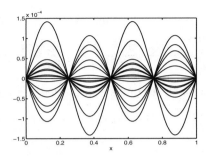

Figure 7.19. *Solution to (7.30) with an oscillatory forcing term (see Example 7.12). The frequency of the forcing term is 1044, the fourth natural frequency of the string, and the solution exhibits resonance.*

Exercises

1. Solve the IBVP

$$\frac{\partial^2 u}{\partial t^2} - c^2 \frac{\partial^2 u}{\partial x^2} = \sin(2\pi\omega t),\ 0 < x < \ell,\ t > 0,$$
$$u(x,0) = 0,\ 0 < x < \ell,$$
$$\frac{\partial u}{\partial t}(x,0) = 0,\ 0 < x < \ell,$$
$$u(0,t) = 0,\ t > 0,$$
$$u(\ell,t) = 0,\ t > 0$$

   in two different cases: ω is equal to a natural frequency, and ω does not equal any natural frequency. Describe the difference in the results. If ω is close to but not equal to a natural frequency, do the results indicate near-resonance (that is, large but bounded amplitudes)?

2. Suppose a string with total mass of 10 g is stretched to a length of 40 cm. By varying the frequency of an oscillatory forcing term until resonance occurs, it is determined that the fundamental frequency of the string is 500 Hz.

   (a) How fast do waves travel along the string?

   (b) What is the tension in the string? (Hint: See the derivation of the wave equation in Section 2.3.)

   (c) At what other frequencies will the string resonate?

3. Consider an iron bar of length 1 m, with a stiffness of $k = 90$ GPa and a density of $\rho = 7.2$ g/cm^3. Suppose that the bottom end ($x = 1$) of the bar is fixed, and the top end is subjected to an oscillatory pressure:

$$k \frac{\partial u}{\partial x}(0,t) = B \sin(2\pi\omega t),\ t > 0.$$

   (a) What is the smallest value of ω that causes resonance? Call this frequency ω_r.

(b) Find and graph the displacement of the bar for $\omega = \omega_r$.

(c) Find and graph the displacement of the bar for $\omega = \omega_r/4$.

4. If a string is driven by an external force whose frequency equals a natural frequency, does resonance always occur? Consider the string of Example 7.12, and suppose $\omega = 522$, $a = 1/2 - \delta$, $b = 1/2 + \delta$, where δ is a small positive number. Show analytically that there is no resonance. Can you explain this example?

5. Solve the IBVP

$$\frac{\partial^2 u}{\partial t^2} - c^2 \frac{\partial^2 u}{\partial x^2} = f(x,t),\ 0 < x < 1,\ t > 0,$$
$$u(x,0) = 0,\ 0 < x < 1,$$
$$\frac{\partial u}{\partial t}(x,0) = 0,\ 0 < x < 1,$$
$$u(0,t) = 0,\ t > 0,$$
$$u(1,t) = 0,\ t > 0,$$

where $c = 55$, $\omega = 120$, and

$$f(x,t) = \begin{cases} \sin(2\pi \omega t), & 0.6 < x < 0.65, \\ 0 & \text{otherwise.} \end{cases}$$

Graph several snapshots of the solution.

7.5 Finite difference methods for the wave equation

To this point, we have presented two methods for solving PDEs, the method of Fourier series and the finite element method. For stationary problems and for the heat equation, it can be argued that the finite element method is the most powerful and effective numerical method available. However, the wave equation presents special challenges, as we have discussed in this chapter. One issue is that singularities in the initial conditions are preserved as time proceeds, and it is difficult to design a numerical method that can approximate such a solution faithfully (see Example 7.8). Another issue is that the finite element method couples the time derivatives when the method of lines is applied, and therefore each iteration of an explicit integration scheme is as costly as an iteration of a comparable implicit scheme (see the discussion on page 289). For this reason, standard finite element methods are not very efficient for the wave equation.[39]

In this section, we will discuss *finite difference* methods for the wave equation, which address the second difficulty described above. The problem of accurately computing solutions with singularities is beyond the scope of this book.

---

[39]There are modifications of the finite element method, notably *mass lumping*, that address this issue. Mass lumping produces a diagonal mass matrix, which eliminates the coupling of the time derivatives.

7.5.1 Finite difference approximation of derivatives

The finite difference method works with the strong form of the PDE (rather than the weak form, as does the finite element method), and it is based on replacing the derivatives in the PDE with *finite difference* approximations. The simplest finite difference approximation to a derivative is the *forward difference* estimate of the first derivative:

$$\frac{d\phi}{dx}(x) \doteq \frac{\phi(x+\Delta x) - \phi(x)}{\Delta x}. \tag{7.33}$$

This estimate obviously arises from the definition of the derivative,

$$\frac{d\phi}{dx}(x) = \lim_{\Delta x \to 0} \frac{\phi(x+\Delta x) - \phi(x)}{\Delta x},$$

by taking a positive value for Δx. It can be shown that the error in the forward difference approximation of $d\phi/dx(x)$ is approximately proportional to Δx. The backward difference,

$$\frac{d\phi}{dx}(x) \doteq \frac{\phi(x) - \phi(x-\Delta x)}{\Delta x}, \tag{7.34}$$

where Δx is positive, also has an error that is approximately proportional to Δx. These error estimates can be derived from a Taylor expansion of ϕ (see Exercise 7.5.1).

Centered finite difference approximations are more accurate than forward or backward differences. The *central difference* estimate of $d\phi/dx(x)$ is

$$\frac{d\phi}{dx}(x) \doteq \frac{\phi(x+\Delta x) - \phi(x-\Delta x)}{2\Delta x}. \tag{7.35}$$

It can be shown that the error in (7.35) is approximately proportional to Δx^2 (see Exercise 7.5.2), which is much smaller than Δx when Δx itself is small.

To estimate the second derivative accurately, we can use central differences as follows:

$$\begin{aligned}\frac{d^2\phi}{dx^2}(x) &\doteq \frac{\frac{d\phi}{dx}(x+\Delta x/2) - \frac{d\phi}{dx}(x-\Delta x/2)}{\Delta x} \\ &\doteq \frac{\frac{\phi(x+\Delta x)-\phi(x)}{\Delta x} - \frac{\phi(x)-\phi(x-\Delta x)}{\Delta x}}{\Delta x} \\ &= \frac{\phi(x-\Delta x) - 2\phi(x) + \phi(x+\Delta x)}{\Delta x^2}.\end{aligned}$$

In the first step, we used a central difference approximation to $d^2\phi/dx^2(x)$, using values of $d\phi/dx$ at $x+\Delta x/2$ and $x-\Delta x/2$. We then approximated each of the first derivatives using central differences.

It can be shown that the error in

$$\frac{d^2\phi}{dx^2}(x) \doteq \frac{\phi(x-\Delta x) - 2\phi(x) + \phi(x+\Delta x)}{\Delta x^2} \tag{7.36}$$

is approximately proportional to Δx^2 (see Exercise 7.5.3).

7.5. Finite difference methods for the wave equation

Example 7.13. Let us consider approximating the derivative of $f(x) = e^x$ at $x = 0$. The exact value is $f'(0) = 1$. Using $\Delta x = 0.1$, we obtain

$$\frac{f(x+\Delta x) - f(x)}{\Delta x} \doteq 1.05170918,$$

$$\frac{f(x) - f(x-\Delta x)}{\Delta x} \doteq 0.95162582,$$

$$\frac{f(x+\Delta x) - f(x-\Delta x)}{2\Delta x} \doteq 1.00166750.$$

We see that the central difference formula is much more accurate than the forward or backward difference. If we reduce Δx to 0.01, we obtain

$$\frac{f(x+\Delta x) - f(x)}{\Delta x} \doteq 1.00501671,$$

$$\frac{f(x) - f(x-\Delta x)}{\Delta x} \doteq 0.99501663,$$

$$\frac{f(x+\Delta x) - f(x-\Delta x)}{2\Delta x} \doteq 1.00001667.$$

Reducing Δx by a factor of 10 reduces the error in the forward and backward differences by a factor of approximately 10, but the error in the central difference formula is reduced by a factor of approximately 100.

If we approximate $f''(0) = 1$ using the central difference formula (7.36) and $\Delta x = 0.1$, we obtain

$$\frac{f(x-\Delta x) - 2f(x) + f(x+\Delta x)}{\Delta x^2} \doteq 1.00083361,$$

while if $\Delta x = 0.01$, we have

$$\frac{f(x-\Delta x) - 2f(x) + f(x+\Delta x)}{\Delta x^2} \doteq 1.00000833.$$

Once again, if Δx is reduced by a factor of 10, the error is reduced by a factor of approximately 100. This is consistent with our earlier statement that the error is approximately proportional to Δx^2. ∎

7.5.2 The wave equation

The finite difference method for a PDE estimates the values of the solution at the nodes of a *grid*. As a concrete example, let us consider the following wave equation with Dirichlet boundary conditions:

$$\begin{aligned}
\frac{\partial^2 u}{\partial t^2} - c^2 \frac{\partial^2 u}{\partial x^2} &= f(x,t),\ 0 < x < \ell,\ t > 0, \\
u(x,0) &= \psi(x),\ 0 < x < \ell, \\
\frac{\partial u}{\partial t}(x,0) &= \gamma(x),\ 0 < x < \ell, \\
u(0,t) &= 0,\ t > 0, \\
u(\ell,t) &= 0,\ t > 0.
\end{aligned} \quad (7.37)$$

We choose a time interval of interest, say $[0, t_f]$, and define a uniform grid on $[0, \ell] \times [0, t_f]$ to be the collection of points (x_i, t_j), $i = 0, 1, \ldots, k$, $j = 0, 1, \ldots, n$, where

$$x_i = i \Delta x, \; i = 0, 1, \ldots, k, \; \Delta x = \frac{\ell}{k}, \; t_j = j \Delta t, \; j = 0, 1, \ldots, n, \; \Delta t = \frac{t_f}{n}.$$

A typical grid is shown in Figure 7.20.

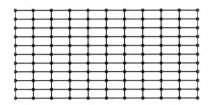

Figure 7.20. *A computational grid for the finite difference method.*

The goal is to compute estimates $u_i^{(j)}$ of $u(x_i, t_j)$. To do this, we replace each of the equations

$$\frac{\partial^2 u}{\partial t^2}(x_i, t_j) - c^2 \frac{\partial^2 u}{\partial x^2}(x_i, t_j) = f(x_i, t_j)$$

by an equation that arises from approximating each partial derivative using only values of u at the grid points. We will use the central difference scheme (7.36) presented above. We have

$$\frac{\partial^2 u}{\partial t^2}(x_i, t_j) \doteq \frac{u(x_i, t_{j-1}) - 2u(x_i, t_j) + u(x_i, t_{j+1})}{\Delta t^2}$$

and

$$\frac{\partial^2 u}{\partial x^2}(x_i, t_j) \doteq \frac{u(x_{i-1}, t_j) - 2u(x_i, t_j) + u(x_{i+1}, t_j)}{\Delta x^2}.$$

Therefore, the true solution of the IBVP (7.37) satisfies

$$\frac{u(x_i, t_{j-1}) - 2u(x_i, t_j) + u(x_i, t_{j+1})}{\Delta t^2}$$
$$- c^2 \frac{u(x_{i-1}, t_j) - 2u(x_i, t_j) + u(x_{i+1}, t_j)}{\Delta x^2} \doteq f(x_i, t_j)$$

for $i = 1, 2, \ldots, k-1$, $j = 1, 2, \ldots, n-1$. To obtain computable estimates, we replace $u(x_i, t_j)$ by $u_i^{(j)}$ and similarly for $u(x_{i-1}, t_j)$, $u(x_i, t_{j-1})$, and so forth:

$$\frac{u_i^{(j-1)} - 2u_i^{(j)} + u_i^{(j+1)}}{\Delta t^2} - c^2 \frac{u_{i-1}^{(j)} - 2u_i^{(j)} + u_{i+1}^{(j)}}{\Delta x^2} = f(x_i, t_j), \quad (7.38)$$

again for $i = 1, 2, \ldots, k-1$, $j = 1, 2, \ldots, n-1$. As we will now show, equations (7.38), together with the initial and boundary conditions, determine the estimates $u_i^{(j)}$ at all grid points.

7.5. Finite difference methods for the wave equation

We first note that the boundary conditions and the first initial condition yield the following values:

$$u_0^{(j)} = u_k^{(j)} = 0, \; j = 1, 2, \ldots, n, \tag{7.39}$$

$$u_i^{(0)} = \psi(x_i), \; i = 0, 1, \ldots, k. \tag{7.40}$$

Solving (7.38) for $u_i^{(j+1)}$ yields

$$u_i^{(j+1)} = 2\left(1 - \frac{c^2 \Delta t^2}{\Delta x^2}\right) u_i^{(j)} - u_i^{(j-1)} + \frac{c^2 \Delta t^2}{\Delta x^2} \left(u_{i-1}^{(j)} + u_{i+1}^{(j)}\right) + \Delta t^2 f(x_i, t_j), \tag{7.41}$$

$i = 1, 2, \ldots, k-1, j = 1, 2, \ldots, n-1$. Equation (7.41) gives an explicit formula for advancing the calculation one time step. When applying (7.41) with $i = 1$, we use the fact that $u_{i-1}^{(j)} = u_0^{(j)} = 0$, and similarly, for $i = k-1$, we use $u_{i+1}^{(j)} = u_k^{(j)} = 0$.

The only difficulty in using (7.41) is on the first step ($j = 1$), when we need the values of $u_i^{(j)}$ for $j = 0$ and $j = 1$. The values for $j = 0$ are provided by the initial condition $u(x, 0) = \psi(x)$, as noted above. To obtain the values for $j = 1$, we wish to use (7.41) with $j = 0$, which in turn requires that we know $u_i^{(-1)}$ (that is, estimates of $u(x_i, -\Delta t)$). Using a central difference approximation to $\partial u / \partial t$, we have

$$\frac{u(x_i, \Delta t) - u(x_i, -\Delta t)}{2\Delta t} \doteq \frac{\partial u}{\partial t}(x_i, 0) = \gamma(x_i).$$

We therefore impose the equation

$$\frac{u_i^{(1)} - u_i^{(-1)}}{2\Delta t} = \gamma(x_i),$$

which, upon solving for $u_i^{(-1)}$, yields

$$u_i^{(-1)} = u_i^{(1)} - 2\Delta t \gamma(x_i). \tag{7.42}$$

Equation (7.41), with $j = 0$, takes the form

$$u_i^{(1)} = 2\left(1 - \frac{c^2 \Delta t^2}{\Delta x^2}\right) u_i^{(0)} - u_i^{(-1)} + \frac{c^2 \Delta t^2}{\Delta x^2}\left(u_{i-1}^{(0)} + u_{i+1}^{(0)}\right) + \Delta t^2 f(x_i, 0). \tag{7.43}$$

Replacing $u_i^{(-1)}$ by the value determined by (7.42) and $u_i^{(0)}$ by $\psi(x_i)$, we obtain

$$u_i^{(1)} = \left(1 - \frac{c^2 \Delta t^2}{\Delta t^2}\right)\psi(x_i) + \Delta t \gamma(x_i) + \frac{c^2 \Delta t^2}{2\Delta t^2}(\psi(x_{i-1}) + \psi(x_i)) + \frac{\Delta t^2}{2} f(x_i, 0). \tag{7.44}$$

This allows us to determine $u_i^{(1)}$ for $i = 1, 2, \ldots, k-1$.

The explicit finite difference scheme defined by (7.41) and the side conditions (7.39), (7.40), and (7.44) is called the *2–2 scheme for the wave equation* because the error is second order in both space and time. In other words, the error is roughly proportional to $\Delta x^2 + \Delta t^2$.

However, the scheme is numerically unstable unless the time step is small enough compared to Δx; specifically, the condition

$$\frac{c\Delta t}{\Delta x} \leq 1 \qquad (7.45)$$

must hold. Inequality (7.45) is called the *CFL condition* (after mathematicians Courant, Friedrichs, and Lewy, who discovered it).

Example 7.14. We define f, ψ, and γ in (7.37) so that the exact solution is the function $u(x,t) = e^{\cos(10t)} \sin(2\pi x)$ for $0 < x < 1$ (that is, $\ell = 1$), $t > 0$. The necessary functions are

$$f(x,t) = 4e^{\cos(10t)} \left(c^2 \pi^2 - 25\cos(10t) + 25\sin^2(10t) \right) \sin(2\pi x),$$
$$\psi(x) = e\sin(2\pi x),$$
$$\gamma(x) = 0.$$

We take $c = 225$ and apply the 2–2 scheme to approximate u for $0 < t < 1$.

Figure 7.21 shows ten snapshots of the computed solution corresponding to $k = 20$ and $n = 5000$, on the time interval $[0, 1]$. The crucial ratio is

$$\frac{c\Delta t}{\Delta x} = 0.9,$$

and thus the CFL condition is satisfied. The maximum error at any grid point is approximately 0.04493. Doubling both k and n (that is, reducing both Δx and Δt by a factor of 2) should reduce the error by a factor of approximately 4. Indeed, if we double k and n to 40 and 10,000, respectively, we obtain a maximum error of approximately 0.01119, consistent with the predicted reduction in error. ∎

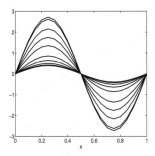

Figure 7.21. *The computed solution in Example 7.14.*

The reader should note that while the finite element method defines the approximate solution as a *function*, the finite difference method estimates the solution only at the grid points. (To be precise, for a time-dependent PDE like the wave equation, the finite element method produces a function of x approximating the solution $u(x,t_j)$ at each time step.)

7.5. Finite difference methods for the wave equation

7.5.3 Neumann boundary conditions

We can easily handle the case in which the Dirichlet conditions $u(0,t) = u(\ell,t) = 0$ are replaced by the Neumann conditions

$$\frac{\partial u}{\partial x}(0,t) = \frac{\partial u}{\partial t}(\ell,t) = 0.$$

With these boundary conditions, we must compute $u_0^{(j)}$ and $u_k^{(j)}$, since these values are no longer given. We impose the Neumann condition at the left endpoint using the central difference:

$$\frac{u_1^{(j)} - u_{-1}(j)}{2\Delta t} = 0 \tag{7.46}$$

(the value $u_{-1}^{(j)}$ is an estimate of $u(-\Delta t, t_j)$). Equation (7.46) implies that $u_{-1}^{(j)} = u_1^{(j)}$. We can now apply (7.41) with $i = 0$, replacing $u_{-1}^{(j)}$ by $u_1^{(j)}$, to obtain

$$u_0^{(j+1)} = 2\left(1 - \frac{c^2\Delta t^2}{\Delta x^2}\right)u_0^{(j)} - u_0^{(j-1)} + \frac{2c^2\Delta t^2}{\Delta x^2}u_1^{(j)} + \Delta t^2 f(x_0, t_j). \tag{7.47}$$

Treating the Neumann condition at the right-hand boundary in the same fashion, we obtain $u_{k+1}^{(j)} = u_{k-1}^{(j)}$, and (7.41) (with $i = k$) becomes

$$u_k^{(j+1)} = 2\left(1 - \frac{c^2\Delta t^2}{\Delta x^2}\right)u_k^{(j)} - u_k^{(j-1)} + \frac{2c^2\Delta t^2}{\Delta x^2}u_{k-1}^{(j)} + \Delta t^2 f(x_k, t_j). \tag{7.48}$$

The 2–2 scheme for the wave equation under Neumann condition is then defined by (7.41) together with the initial conditions (7.40), (7.44) and the boundary cases (7.47), (7.48). Equation (7.44), which defines $u_i^{(1)}$, must be modified for $i = 0$ and $i = k$ in the same way that (7.41) was modified to obtain (7.47) and (7.48). The details are left to the reader.

Example 7.15. We will apply the 2–2 scheme to the IBVP

$$\frac{\partial^2 u}{\partial t^2} - c^2 \frac{\partial^2 u}{\partial x^2} = f(x,t),\ 0 < x < \ell,\ t > 0,$$

$$u(x,0) = \psi(x),\ 0 < x < \ell,$$

$$\frac{\partial u}{\partial t}(x,0) = \gamma(x),\ 0 < x < \ell,$$

$$\frac{\partial u}{\partial x}(0,t) = 0,\ t > 0,$$

$$\frac{\partial u}{\partial x}(\ell,t) = 0,\ t > 0.$$

We choose $c = 90$, $\ell = 1$ and define ψ, γ, and f so that the exact solution is $u(x,t) = t(2x^3 - 2x^2)\cos(\pi x)$. We will approximate the solution on the time interval $[0, t_f]$, where $t_f = 1$.

The CFL condition requires

$$\frac{c\Delta t}{\Delta x} \leq 1 \Rightarrow \Delta t \leq \frac{\Delta x}{90},$$

and therefore we will take $n = 100k$. With $k = 100$ (that is, $\Delta x = 10^{-2}$ and $\Delta t = 10^{-4}$), we obtain a maximum error of approximately 0.2702. Increasing k and n by a factor of 2 each yields a maximum error of approximately 0.06754; the error is reduced by a factor of approximately four, as expected. ∎

Exercises

1. Recall that Taylor's theorem states that

$$\phi(x+\Delta x) = \phi(x) + \frac{d\phi}{dx}(x)\Delta x + \frac{1}{2!}\frac{d^2\phi}{dx^2}(x)\Delta x^2 + \cdots + \frac{1}{n!}\frac{d^n\phi}{dx^n}(x)\Delta x^n$$
$$+ \frac{1}{(n+1)!}\frac{d^{(n+1)}\phi}{dx^{(n+1)}}(c)\Delta x^{n+1},$$

   where c is some number between x and $x + \Delta x$ (c depends on x and Δx). This result holds provided ϕ is smooth enough, namely, that it has $n+1$ continuous derivatives. Use the above formula with $n = 1$ to prove that the error in the forward difference (7.33) is approximately proportional to Δx.

2. Use Taylor's theorem with $n = 2$ (see Exercise 1) to expand both $\phi(x+\Delta x)$ and $\phi(x-\Delta x)$, and use these expressions to prove that the error in the central difference (7.35) is approximately proportional to Δx^2.

3. Use Taylor's theorem with $n = 3$ (see Exercise 1) to expand both $\phi(x+\Delta x)$ and $\phi(x-\Delta x)$, and use these expressions to prove that the error in the central difference (7.36) is approximately proportional to Δx^2.

4. As discussed above in Section 7.5.3, equation (7.44) must be modified in the case of Neumann boundary conditions. Derive the correct equations for $u_0^{(1)}$ and $u_k^{(1)}$.

5. Modify the finite difference schemes presented in this section for the following IBVP with mixed boundary conditions:

$$\frac{\partial^2 u}{\partial t^2} - c^2 \frac{\partial^2 u}{\partial x^2} = f(x,t),\ 0 < x < \ell,\ t > 0,$$
$$u(x,0) = \psi(x),\ 0 < x < \ell,$$
$$\frac{\partial u}{\partial t}(x,0) = \gamma(x),\ 0 < x < \ell,$$
$$u(0,t) = 0,\ t > 0,$$
$$\frac{\partial u}{\partial x}(\ell,t) = 0,\ t > 0.$$

   Apply your scheme to the IBVP defined by $\ell = 1, c = 45, \psi(x) = x(1-x)^2, \gamma(x) = 0$, and

$$f(x,t) = -2(2\pi^2 x(1-x)^2 + c^2(3x-2))\cos(2\pi t).$$

   The exact solution is $u(x,t) = \cos(2\pi t)x(1-x)^2$. Verify that your scheme is second order in both Δt and Δx.

6. Derive a second-order finite-difference scheme for solving the BVP

$$-\kappa \frac{d^2 u}{dx^2} = f(x), \; 0 < x < \ell,$$
$$u(0) = a,$$
$$u(\ell) = b.$$

Compare the finite difference scheme to the standard piecewise linear finite element method for the same problem. How do the two methods differ?

7.6 Comparison of the heat and wave equations

We now wish to point out a fundamental difference between the heat equation

$$\rho c \frac{\partial u}{\partial t} - \kappa \frac{\partial^2 u}{\partial x^2} = f(x,t), \; 0 < x < \ell, t > 0 \qquad (7.49)$$

and the wave equation

$$\frac{\partial^2 u}{\partial t^2} - c^2 \frac{\partial^2 u}{\partial x^2} = f(x,t), \; 0 < x < \ell, t > 0. \qquad (7.50)$$

We have seen that, as time advances, solutions of the heat equation become smoother, while solutions of the wave equation maintain the same degree of smoothness (or lack thereof). This is related to the fact that it is relatively easy to solve the heat equation numerically, but difficult to accurately solve the wave equation. Any small errors that arise in the course of solving the heat equation are damped out as time advances; on the other hand, when solving the wave equation, such errors are retained and continue to corrupt the solution with increasing time.

Also related to the above fact is that it is possible to solve the wave equation backward in time, but not the heat equation. To explain this, we begin with the wave equation and assume that we wish to solve the "final value problem"

$$\begin{aligned}
\frac{\partial^2 u}{\partial t^2} - c^2 \frac{\partial^2 u}{\partial x^2} &= f(x,t), \; 0 < x < \ell, \; 0 < t < T, \\
u(x,T) &= \phi(x), \; 0 < x < \ell, \\
\frac{\partial u}{\partial t}(x,T) &= \psi(x), \; 0 < x < \ell, \\
u(0,t) &= 0, \; 0 < t < T, \\
u(\ell,t) &= 0, \; 0 < t < T.
\end{aligned} \qquad (7.51)$$

Let us suppose that the solution u exists, and define $v(x,t) = u(x, T - t)$. We have

$$\frac{\partial v}{\partial t}(x,t) = -\frac{\partial u}{\partial t}(x, T - t), \; \frac{\partial^2 v}{\partial t^2}(x,t) = \frac{\partial^2 u}{\partial t^2}(x, T - t)$$

and

$$\frac{\partial v}{\partial x}(x,t) = \frac{\partial u}{\partial x}(x, T - t), \; \frac{\partial^2 v}{\partial x^2}(x,t) = \frac{\partial^2 u}{\partial x^2}(x, T - t).$$

It follows that

$$\frac{\partial^2 v}{\partial t^2}(x,t) - c^2 \frac{\partial^2 v}{\partial x^2}(x,t) = \frac{\partial^2 u}{\partial t^2}(x,T-t) - c^2 \frac{\partial^2 u}{\partial x^2}(x,T-t) = f(x,T-t).$$

Also, we see that

$$v(x,0) = u(x,T) = \phi(x), \quad \frac{\partial v}{\partial t}(x,0) = -\frac{\partial u}{\partial t}(x,T) = -\psi(x), \; 0 < x < \ell,$$

and that

$$v(0,t) = u(0,T-t) = 0, \; v(\ell,t) = u(\ell,T-t) = 0, \; 0 < t < T.$$

Therefore, v is the solution of

$$\begin{aligned} \frac{\partial^2 v}{\partial t^2} - c^2 \frac{\partial^2 v}{\partial x^2} &= f(x,T-t), \; 0 < x < \ell, \; 0 < t < T, \\ v(x,0) &= \phi(x), \; 0 < x < \ell, \\ \frac{\partial v}{\partial t}(x,0) &= -\psi(x), \; 0 < x < \ell, \\ v(0,t) &= 0, \; 0 < t < T, \\ v(\ell,t) &= 0, \; 0 < t < T. \end{aligned} \quad (7.52)$$

Conversely, it is straightforward to show that if v solves (7.52), then the function $u(x,t) = v(x,T-t)$ solves (7.51). Since we know how to solve (7.52) (by the method of Fourier series, for example) and therefore the solution exists and is unique, this justifies our statement that it is possible to solve the wave equation backward in time.

On the other hand, it is not possible to solve the heat equation backward in time, at least in most cases. An argument similar to that given above shows that

$$\begin{aligned} \rho c \frac{\partial u}{\partial t} - \kappa \frac{\partial^2 u}{\partial x^2} &= f(x,t), \; 0 < x < \ell, \; 0 < t < T, \\ u(x,T) &= \phi(x), \; 0 < x < \ell, \\ u(0,t) &= 0, \; 0 < t < T, \\ u(\ell,t) &= 0, \; 0 < t < T \end{aligned} \quad (7.53)$$

if and only if $v(x,t) = u(x,T-t)$ solves

$$\begin{aligned} -\rho c \frac{\partial v}{\partial t} - \kappa \frac{\partial^2 v}{\partial x^2} &= f(x,T-t), \; 0 < x < \ell, \; 0 < t < T, \\ v(x,T) &= \phi(x), \; 0 < x < \ell, \\ v(0,t) &= 0, \; 0 < t < T, \\ v(\ell,t) &= 0, \; 0 < t < T. \end{aligned} \quad (7.54)$$

The reader should notice the change in sign on the $\partial v/\partial t$ terms, which means that (7.54) involves a different PDE than does (7.53); this new PDE is called the *backward heat equation*. We wish to determine whether (7.54) has a solution.

7.6. Comparison of the heat and wave equations

We proceed by the method of Fourier series. We write the proposed solution of (7.54) in the form

$$v(x,t) = \sum_{n=1}^{\infty} a_n(t) \sin\left(\frac{n\pi x}{\ell}\right),$$

where the coefficients $a_n(t)$, $n = 1, 2, 3, \ldots$, are to be determined. The calculation proceeds almost exactly as in Section 6.1, and we will not include all the details. We have

$$f(x, T-t) = \sum_{n=1}^{\infty} c_n(t) \sin\left(\frac{n\pi x}{\ell}\right),$$

$$\phi(x) = \sum_{n=1}^{\infty} d_n \sin\left(\frac{n\pi x}{\ell}\right),$$

where $c_n(t)$, d_n, $n = 1, 2, 3, \ldots$, are known. The backward heat equation is then equivalent to

$$\sum_{n=1}^{\infty} \left\{ -\rho c \frac{da_n}{dt}(t) + \frac{\kappa n^2 \pi^2}{\ell^2} a_n(t) \right\} \sin\left(\frac{n\pi x}{\ell}\right) = \sum_{n=1}^{\infty} c_n(t) \sin\left(\frac{n\pi x}{\ell}\right),$$

while the initial condition becomes

$$\sum_{n=1}^{\infty} a_n(0) \sin\left(\frac{n\pi x}{\ell}\right) = \sum_{n=1}^{\infty} d_n \sin\left(\frac{n\pi x}{\ell}\right).$$

Therefore, the unknown coefficients a_n are found by solving

$$-\rho c \frac{da_n}{dt}(t) + \frac{\kappa n^2 \pi^2}{\ell^2} a_n(t) = c_n(t), \quad t > 0,$$

$$a_n(0) = d_n.$$

The result is

$$a_n(t) = \left(d_n - \int_0^t e^{-\kappa n^2 \pi^2 s/(\rho c \ell^2)} c_n(s) \, ds \right) e^{\kappa n^2 \pi^2 t/(\rho c \ell^2)}, \quad n = 1, 2, 3, \ldots,$$

and, with this value for $a_n(t)$, the solution to (7.54) is, formally,

$$v(x,t) = \sum_{n=1}^{\infty} a_n(t) \sin\left(\frac{n\pi x}{\ell}\right). \tag{7.55}$$

However, in most cases, this series does not converge for any $t > 0$. For example, if $f = 0$ (and therefore $c_n(t) = 0$ for all n), we obtain the series

$$v(x,t) = \sum_{n=1}^{\infty} d_n e^{\kappa n^2 \pi^2 t/(\rho c \ell^2)} \sin\left(\frac{n\pi x}{\ell}\right).$$

For most choices of ϕ, this series does not converge because the coefficients grow exponentially with n.

A little thought shows that this is consistent with what we know about the (forward) heat equation. Even if the initial value has singularities, the solution is smooth for every $t > 0$ because the Fourier coefficients decay to zero exponentially with n (see the discussion on page 199). Solving the backward heat equation is equivalent to solving the final value problem (7.53), which is therefore possible only if the final value is very smooth (that is, only if the coefficients d_n decay to zero very rapidly).

7.7 Suggestions for further reading

The reader can consult Sections 5.7 and 6.6 for references that discuss PDEs, Fourier series, and finite element methods. Most of the books described there cover the wave equation.

Morton and Mayers [52] provides a good introduction to the finite difference methods, while Strikwerda [63] is a more advanced treatment. Also recommended is the book by LeVeque [43]. Celia and Gray [11] covers both finite differences and finite elements (including finite element methods that are not Galerkin methods).

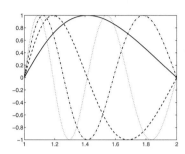

Chapter 8
First-Order PDEs and the Method of Characteristics

Previous chapters have focused on the most important second-order PDEs. We now turn our attention to first-order PDEs, such as the advection equation and other conservation laws introduced in Section 2.4.

8.1 The simplest PDE and the method of characteristics

We begin with the equation

$$\frac{\partial u}{\partial y} = 0, \qquad (8.1)$$

where the unknown is $u = u(x, y)$. Formally, (8.1) is a PDE; however, since only one partial derivative is involved, we can solve the equation by regarding the variable x as a parameter and the differential equation as an ODE. Integrating both sides of (8.1) with respect to y yields

$$u(x, y) = \phi(x),$$

where $\phi(x)$ is the "constant" of integration. Auxiliary conditions (that is, initial and/or boundary conditions) can be used to determine the function ϕ.

Many first-order PDEs that are not as simple as (8.1) can be solved by changing the variables to produce an equation like (8.1). Let us consider the PDE

$$a\frac{\partial u}{\partial x} + b\frac{\partial u}{\partial y} = 0, \qquad (8.2)$$

where a and b are constants (the advection equation discussed in Section 2.4 is a special case of (8.2)). The key to understanding (8.2) is to recognize the expression

$$a\frac{\partial u}{\partial x} + b\frac{\partial u}{\partial y}$$

309

as (proportional to) a directional derivative of u. In general, if $\mathbf{d} \in \mathbf{R}^2$ is a vector of length 1, then the directional derivative of u at (x,y) in the direction of $\mathbf{d}$ is[40]

$$\nabla u(x,y) \cdot \mathbf{d},$$

where $\nabla u(x,y)$ is the gradient vector:

$$\nabla u(x,y) = \begin{bmatrix} \frac{\partial u}{\partial x}(x,y) \\ \frac{\partial u}{\partial y}(x,y) \end{bmatrix}.$$

Thus we see that

$$a\frac{\partial u}{\partial x}(x,y) + b\frac{\partial u}{\partial y}(x,y) = \nabla u(x,y) \cdot (a,b),$$

which is a multiple of the directional derivative of u in the direction of (a,b) (it is not exactly equal to the directional derivative unless (a,b) happens to be a unit vector). Therefore, the PDE (8.2) simply states that this directional derivative is zero and hence that u *is constant on lines parallel to the vector* (a,b), that is, on the lines $bx - ay = c$, where c is a constant. These lines are called the *characteristics* of the PDE (8.2).

Any auxiliary condition that allows us to determine the value of u on each of the characteristics leads to a well-posed problem (that is, a problem that has a unique solution). For instance, suppose we are given

$$u(x,0) = u_0(x) \text{ for all } x \in \mathbf{R}.$$

We can write each characteristic $bx - ay = c$ in the form

$$y = \frac{b}{a}(x - x_0)$$

for some x_0. Given any $(x_1, y_1) \in \mathbf{R}^2$, there exists a unique x_0 such that $(x_0, 0)$ and (x_1, y_1) both lie on the line $y = (b/a)(x - x_0)$, namely, $x_0 = x_1 - (a/b)y_1$ (see Figure 8.1). Therefore, since u is constant on the line $y = (b/a)(x - x_0)$,

$$u(x_1, y_1) = u(x_0, 0) = u\left(x_1 - \frac{a}{b}y_1, 0\right) = u_0\left(x_1 - \frac{a}{b}y_1\right).$$

This holds for every (x_1, y_1), and therefore the solution of

$$a\frac{\partial u}{\partial x} + b\frac{\partial u}{\partial y} = 0, \ (x,y) \in \mathbf{R}^2,$$

$$u(x,0) = u_0(x), \ x \in \mathbf{R},$$

is

$$u(x,y) = u_0\left(x - \frac{a}{b}y\right).$$

It can be verified directly that this function satisfies both the PDE and the auxiliary condition. The reader should notice that if we regard y as time, then u is a right-moving wave when $a/b > 0$ and a left-moving wave when $a/b < 0$.

---

[40]Given u and $\mathbf{d} = (d_1, d_2)$, we can define $\phi(s) = u(x + sd_1, y + sd_2)$, which is simply the function u restricted to the line through (x,y) in the direction of $\mathbf{d}$. By the directional derivative of u at (x,y) in the direction of $\mathbf{d}$, we mean $d\phi/ds$, evaluated at $s = 0$. The formula for this derivative follows from the chain rule.

8.1. The simplest PDE and the method of characteristics

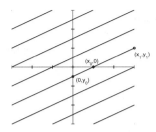

Figure 8.1. *The characteristics for the PDE (8.2) with $a = 2$, $b = 1$.*

The reader should appreciate that there is nothing special about the condition $u(x,0) = u_0(x)$; any condition that determines the value of u on every characteristic will do. For instance, let us suppose $u(0, y) = v_0(y)$ for all y. The characteristics can be expressed in the form $x = (a/b)(y - y_0)$. For all $(x_1, y_1) \in \mathbf{R}^2$, there exists a unique y_0 such that $(0, y_0)$ and (x_1, y_1) both lie on the line $x = (a/b)(y - y_0)$, namely, $y_0 = y_1 - (b/a)x_1$ (see Figure 8.1). Therefore,

$$u(x_1, y_1) = u(0, y_0) = u\left(0, y_1 - \frac{b}{a}x_1\right) = v_0\left(y_1 - \frac{b}{a}x_1\right),$$

and the solution u is given by

$$u(x, y) = v_0\left(y - \frac{b}{a}x\right) \text{ for all } (x, y) \in \mathbf{R}^2.$$

Still other auxiliary conditions are possible. The most general condition would specify the value of u on a curve specified in parametric form as $(f(s), g(s))$, $s \in \mathbf{R}$. We will refer to such a condition as an initial condition even though there may be no time variable in the problem (the reader should notice that $u(x, 0) = u_0(x)$ and $u(0, y) = v_0(y)$ are special cases). If, for each $(x_1, y_1) \in \mathbf{R}^2$, there exists a unique s_0 such that $(f(s_0), g(s_0))$ and (x_1, y_1) lie on the same characteristic, then a condition of the form

$$u(f(s), g(s)) = h(s), \ s \in \mathbf{R},$$

would determine a unique solution u. Certain curves would fail to satisfy the necessary condition; the most striking failure would be if the curve described by $(f(s), g(s))$ is itself a characteristic. For instance, suppose we try to solve, for a given function h,

$$a\frac{\partial u}{\partial x} + b\frac{\partial u}{\partial y} = 0, \ (x, y) \in \mathbf{R}^2, \tag{8.3}$$

$$u(as, bs) = h(s), \ s \in \mathbf{R}.$$

For a given $(x_1, y_1) \in \mathbf{R}^2$, we must find s_0 such that (as_0, bs_0) and (x_1, y_1) both lie on the characteristic $y = (b/a)(x - x_0)$. However, (as, bs), $s \in \mathbf{R}$, is a parametrization of the characteristic $y = (b/a)x$. Therefore, (x_1, y_1) lies on the same characteristic as some (as_0, bs_0) if and only if $y_1 = (b/a)x_1$.

There are two consequences of the fact that the initial data is specified on a characteristic. First of all, since the PDE shows that u is constant on the characteristic $y = (b/a)x$, the condition $u(as, bs) = h(s)$ is consistent with the PDE only if h is a constant function. If this is not the case, then there is no solution to (8.3). Second, if h is constant, then there exist infinitely many solutions to (8.3), namely, any function of the form $u(x,y) = u_0(x - (a/b)y)$ with u_0 satisfying $u_0(0) = h$. We already know that such a u satisfies the PDE; we also have

$$u(as, bs) = u_0\left(as - \frac{a}{b}bs\right) = u_0(as - as) = u_0(0),$$

which shows that u satisfies the initial condition provided $u_0(0) = h$.

8.1.1 Changing variables

We can show more directly the role of the characteristics by transforming the independent variables in the PDE. Still referring to the PDE (8.2) and imposing the condition $u(x,0) = u_0(x)$, we change variables from (x,y) to (s,t), where

$$\begin{aligned} \frac{\partial x}{\partial t} &= a, \; x(s,0) = s, \\ \frac{\partial y}{\partial t} &= b, \; y(s,0) = 0. \end{aligned} \tag{8.4}$$

These equations define x, y as functions of s, t ($x = x(s,t)$, $y = y(s,t)$) and impose the condition that t increases in the direction of the characteristics. To show this explicitly, we define $v(s,t) = u(x(s,t), y(s,t))$ and notice that

$$\frac{\partial v}{\partial t} = \frac{\partial u}{\partial x}\frac{\partial x}{\partial t} + \frac{\partial u}{\partial y}\frac{\partial y}{\partial t} = a\frac{\partial u}{\partial x} + b\frac{\partial u}{\partial y}.$$

Thus differentiating v with respect to t is equivalent to differentiating u in the direction of (a,b), and the PDE (8.2) reduces to

$$\frac{\partial v}{\partial t} = 0.$$

Moreover, the condition $u(s,0) = u_0(s)$ is equivalent to

$$u(x(s,0), y(s,0)) = u_0(s) \Rightarrow v(s,0) = u_0(s).$$

Therefore,

$$\begin{aligned} a\frac{\partial u}{\partial x} + b\frac{\partial u}{\partial y} &= 0, \; (x,y) \in \mathbf{R}^2, \\ u(x,0) &= u_0(x), \; x \in \mathbf{R}, \end{aligned} \tag{8.5}$$

is equivalent to

$$\begin{aligned} \frac{\partial v}{\partial t} &= 0, \; (s,t) \in \mathbf{R}^2, \\ v(s,0) &= u_0(s), \; s \in \mathbf{R}. \end{aligned} \tag{8.6}$$

8.1. The simplest PDE and the method of characteristics

The solution to (8.6) is $v(s,t) = u_0(s)$ (since the PDE implies that v is constant with respect to t). To find u, we must change variables from (s,t) back to (x,y). Equations (8.4) imply

$$x = s + at,$$
$$y = bt$$

and it is easy to solve for (s,t) to obtain

$$s = x - \frac{a}{b}y,$$
$$t = \frac{1}{b}y.$$

Then

$$u(x,y) = v(s,t) = v\left(x - \frac{a}{b}y, \frac{1}{b}y\right) = u_0\left(x - \frac{a}{b}y\right).$$

This is the same solution we found on page 310.

The technique of solving a PDE by reducing it to an ODE along the characteristics is referred to as the *method of characteristics*.

Example 8.1. We use the method of characteristics to solve the initial value problem

$$\frac{\partial u}{\partial x} - 2\frac{\partial u}{\partial y} = 0, \ (x,y) \in \mathbf{R}^2,$$
$$u(x,0) = \frac{1}{1+x^2}, \ x \in \mathbf{R}.$$

The changes of variables is $x = x(s,t)$, $y = y(s,t)$, defined by

$$\frac{\partial x}{\partial t} = 1, \ x(s,0) = s,$$
$$\frac{\partial y}{\partial t} = -2, \ y(s,0) = 0,$$

which yields

$$x(s,t) = s + t, \ y(s,t) = -2t.$$

We find $v(s,t) = u(x,y)$ by solving

$$\frac{\partial v}{\partial t} = 0, \ v(s,0) = \frac{1}{1+s^2}$$

to obtain

$$v(s,t) = \frac{1}{1+s^2}.$$

Solving $x = s+t$, $y = -2t$ for (s,t), we obtain

$$s = x + \frac{1}{2}y, \ t = -\frac{1}{2}y,$$

and then

$$u(x,y) = v(s(x,y), t(x,y)) = v\left(x + \frac{1}{2}y, -\frac{1}{2}y\right) = \frac{1}{1 + \left(x + \frac{1}{2}y\right)^2}.$$

The solution is shown in Figure 8.2, where it is seen that the solution is a left-moving wave (that is, a function of x, $u(x,y)$ has the shape of the initial function $1/(1+x^2)$, and this shape moves to the left as y increases). ■

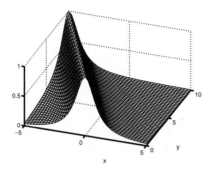

Figure 8.2. *The solution u for Example* 8.1.

In the case of the general initial condition $u(f(s), g(s)) = h(s)$, we replace (8.4) by

$$\begin{aligned}\frac{\partial x}{\partial t} &= a, \; x(s,0) = f(s), \\ \frac{\partial y}{\partial t} &= b, \; y(s,0) = g(s).\end{aligned} \tag{8.7}$$

The initial condition for v is $v(s,0) = h(s)$, and the solution is $v(s,t) = h(s)$. The only difficult part of finding u might be solving for (s,t) in terms of (x,y). Equations (8.7) yield

$$\begin{aligned}x &= f(s) + at, \\ y &= g(s) + bt,\end{aligned}$$

and, for given f and g, it may or may not be possible to solve algebraically for s and t.

For future reference, we point out that we can pose the equations for x, y, and v (as functions of s and t) as a system of three ODEs with initial conditions:

$$\begin{aligned}\frac{\partial x}{\partial t} &= a, \; x(s,0) = f(s), \\ \frac{\partial y}{\partial t} &= b, \; y(s,0) = g(s), \\ \frac{\partial v}{\partial t} &= 0, \; v(s,0) = h(s).\end{aligned} \tag{8.8}$$

It so happens that, in this simple case, we can solve the first two equations for x, y and solve the third equation (independently) for v. However, we will see that, for more complicated PDEs, the three equations are not decoupled, and they must be solved simultaneously.

8.1.2 An inhomogeneous PDE

The inhomogeneous version of (8.2) is

$$a\frac{\partial u}{\partial x} + b\frac{\partial u}{\partial y} = p(x,y), \tag{8.9}$$

where p is a given function, and a corresponding initial value problem is

$$\begin{aligned} a\frac{\partial u}{\partial x} + b\frac{\partial u}{\partial y} &= p(x,y),\ (x,y) \in \mathbf{R}^2, \\ u(f(s),g(s)) &= h(s),\ s \in \mathbf{R}. \end{aligned} \tag{8.10}$$

If we make the change of variables defined by (8.7), the PDE becomes the ODE

$$\frac{\partial v}{\partial t} = q(s,t),$$

where $q(s,t) = p(x(s,t),y(s,t))$. We still have the initial condition $v(s,0) = h(s)$, and thus we can solve the resulting initial value to find $v(s,t)$. As before, the challenge is to find (s,t) in terms of (x,y) so that we can express u explicitly as a function of x and y.

Example 8.2. This is an inhomogeneous version of the problem from Example 8.1:

$$\frac{\partial u}{\partial x} - 2\frac{\partial u}{\partial y} = e^{-y},\ (x,y) \in \mathbf{R}^2,$$

$$u(x,0) = \frac{1}{1+x^2},\ x \in \mathbf{R}.$$

The change of variables is the same as in the earlier example:

$$x = s+t,\ y = -2t \leftrightarrow s = x + \frac{1}{2}y,\ t = -\frac{1}{2}y.$$

With $p(x,y) = e^{-y}$, we have $q(s,t) = p(x,y) = e^{2t}$, and we must solve the IVP

$$\frac{\partial v}{\partial t} = e^{2t},\ v(s,0) = \frac{1}{1+s^2}.$$

The solution is easily seen to be

$$v(s,t) = \frac{1}{2}\left(e^{2t} - 1\right) + \frac{1}{1+s^2}.$$

Changing variables back to (x,y) yields

$$u(x,y) = \frac{1}{2}\left(e^{-y} - 1\right) + \frac{1}{1+\left(x+\frac{1}{2}y\right)^2}$$

(see Figure 8.3). ∎

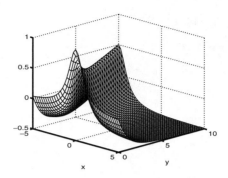

Figure 8.3. *The solution u for Example* 8.2.

Often we wish to solve the advection equation on a finite spatial interval, say $0 < x < \ell$. As discussed in Section 2.4, a boundary condition is needed at only one of the endpoints. We will illustrate with the following example.

Example 8.3. We will solve the IBVP

$$\frac{\partial u}{\partial t} + c\frac{\partial u}{\partial x} = 0,\ 0 < x < 1,\ t > 0,$$
$$u(x,0) = u_0(x),\ 0 < x < 1,$$
$$u(0,t) = \phi(t),\ t > 0,$$
(8.11)

where $u_0(x) = 1$, $\phi(t) = 1 + t^2$, and $c = 2$. The characteristics for this problem can be written as $x = x_0 + ct$ or $t = t_0 + c^{-1}x$, and each characteristic intersects either the positive x-axis (where the value $u(x_0,0) = u_0(x_0)$ is specified) or the positive t-axis (where the value $u(0,t_0) = \phi(t_0)$ is given). The characteristic $t = c^{-1}x$, which passes through the origin, divides the space-time region into two subregions (see Figure 8.4). If (x,t) satisfies

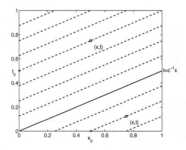

Figure 8.4. *The characteristics for the IBVP in Example* 8.3. *Since the solution is determined by an initial condition (on the x-axis) and a boundary condition (on the t-axis), we divide the characteristics into two families. Those below the line $t = c^{-1}x$ intersect the positive x-axis. If (x,t) lies on one of these characteristics, then $u(x,t) = u_0(x_0)$, $x_0 = x - ct$. The characteristics that lie above $t = c^{-1}x$ intersect the positive t-axis, and if (x,t) lies on one of these characteristics, then $u(x,t) = \phi(t_0)$, $t_0 = t - c^{-1}x$.*

8.1. The simplest PDE and the method of characteristics

$t < c^{-1}x$, then (x,t) lies on the characteristic $x = x_0 + ct$, $x_0 = x - ct$, and the value of the solution at (x,t) is $u(x,t) = u_0(x_0) = u_0(x - ct)$. On the other hand, if $t > c^{-1}x$, then (x,t) lies on the characteristic $t = t_0 + c^{-1}x$, $t_0 = t - c^{-1}x$. In this case, the value of the solution at (x,t) is $u(x,t) = \phi(t_0) = \phi(t - c^{-1}x)$.

We have thus found the solution of the IBVP:

$$u(x,t) = \begin{cases} u_0(x - ct), & t < c^{-1}x, \\ \phi(t - c^{-1}x), & t > c^{-1}x. \end{cases}$$

The solution for this problem is displayed in Figure 8.5. In this case ($u_0(x) = 1$, $\phi(t) = 1 + t^2$), the solution is continuous with continuous first derivatives. Exercise 8.1.5 asks the reader to determine conditions on u_0 and ϕ that guarantee that u, $\partial u/\partial x$, and $\partial u/\partial t$ are all continuous. ∎

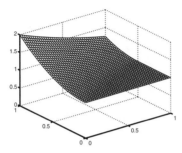

Figure 8.5. *The solution u for Example* 8.3.

Exercises

1. Consider the simple PDE

$$\frac{\partial u}{\partial y} = 0,$$

   where the unknown is $u = u(x,y)$.

   (a) What are the characteristics of this PDE?

   (b) Explain why the PDE, together with the condition $u(0,y) = \phi(y)$, is not a well-posed problem (that is, either it has no solution, or the solution is not unique).

2. Solve the PDE

$$2\frac{\partial u}{\partial x} - 3\frac{\partial u}{\partial y} = 0, \quad -\infty < x < \infty, \quad -\infty < y < \infty,$$

   under each of the following conditions:

   (a) $u(x,0) = \phi(x)$, $-\infty < x < \infty$;
   (b) $u(0,y) = \psi(y)$, $-\infty < y < \infty$;
   (c) $u(x,-x) = \theta(x)$, $-\infty < x < \infty$.

3. Solve the PDE
$$4\frac{\partial u}{\partial x} + 3\frac{\partial u}{\partial t} = 0, \quad 0 < x < 1, \ t > 0,$$
under each of the following sets of conditions:

   (a) $u(x,0) = u_0(x)$, $0 < x < 1$, $u(0,t) = \phi(t)$, $t > 0$;

   (b) $u(x,0) = u_0(x)$, $0 < x < 1$, $u(1,t) = \psi(t)$, $t > 0$.

4. Solve the PDE
$$\frac{\partial u}{\partial x} + \frac{\partial u}{\partial y} = 0, \quad -\infty < x < \infty, \ -\infty < y < \infty,$$
under each of the following conditions:

   (a) $u(x,x) = \phi(x)$, $-\infty < x < \infty$;

   (b) $u(x,-x) = \psi(x)$, $-\infty < x < \infty$.

5. Consider the IBVP (8.11), which has the solution
$$u(x,t) = \begin{cases} u_0(x - ct), & t < c^{-1}x, \\ \phi(t - c^{-1}x), & t > c^{-1}x \end{cases}$$
(see Example 8.3).

   (a) What conditions on u_0 and ϕ guarantee that u is continuous?

   (b) What conditions on u_0 and ϕ guarantee that $\partial u/\partial x$ and $\partial u/\partial t$ are continuous?

6. Let ϕ be defined by
$$\phi(t) = \begin{cases} 1 - 3t^2 + 2t^3, & 0 < t < 1, \\ 0, & t \geq 1, \end{cases}$$
and consider the following IBVP for the advection equation:
$$\frac{\partial u}{\partial t} + 2\frac{\partial u}{\partial x} = \frac{1}{2}, \quad 0 < x < 1, \ t > 0,$$
$$u(x,0) = 1, \quad 0 < x < 1,$$
$$u(0,t) = \phi(t), \quad t > 0.$$

This IBVP models the concentration of a chemical solution in a pipe. Initially the concentration is 1 g/cm throughout the pipe, the chemical is added at a rate of $1/2$ g/cm/s throughout the pipe, and after one second of transition, pure water enters the left end of the pipe.

   (a) Find the solution u.

   (b) Does the concentration eventually reach a steady state?

7. Solve the following IVP for $u = u(x,y)$:

$$\frac{\partial u}{\partial y} + \frac{\partial u}{\partial x} = x+y, \ (x,y) \in \mathbf{R}^2,$$

$$u(x,-x) = x^2, \ x \in \mathbf{R}.$$

8. Consider the IVP

$$\frac{\partial u}{\partial t} + c\frac{\partial u}{\partial x} = f(x,t), \ -\infty < x < \infty, \ t > 0,$$

$$u(x,0) = 0, \ -\infty < x < \infty.$$

Suppose there exists an interval $[a,b]$ such that, for all $t > 0$, $f(x,t) = 0$ for $x \notin [a,b]$. Find the set of points (x,t) such that $u(x,t) = 0$ is guaranteed.

8.2 First-order quasi-linear PDEs

In moving beyond the simple PDE discussed in the last section, we distinguish several classes of first-order equations.

1. A first-order *linear* PDE has the form

$$a(x,y)\frac{\partial u}{\partial x} + b(x,y)\frac{\partial u}{\partial y} + c(x,y)u = p(x,y). \tag{8.12}$$

The PDEs considered in Section 8.1 were first-order linear equations.

2. A first-order *semilinear* PDE has the form

$$a(x,y)\frac{\partial u}{\partial x} + b(x,y)\frac{\partial u}{\partial y} = c(x,y,u). \tag{8.13}$$

This equation is called semilinear because it is linear in the leading (highest-order) terms $\partial u/\partial x$ and $\partial u/\partial y$. However, it need not be linear in u.

3. A first-order *quasi-linear* PDE has the form

$$a(x,y,u)\frac{\partial u}{\partial x} + b(x,y,u)\frac{\partial u}{\partial y} = c(x,y,u). \tag{8.14}$$

This equation is linear in the leading terms $\partial u/\partial x$ and $\partial u/\partial y$, but the coefficients of these terms can depend on u.

4. A general (nonlinear) first-order PDE is written as

$$F\left(x,y,u,\frac{\partial u}{\partial x},\frac{\partial u}{\partial y}\right) = 0. \tag{8.15}$$

8.2.1 Linear equations

In this section, we will show how to solve linear equations. We can apply the method of characteristics, as in Section 8.1, with the sole difference that the characteristics are curves instead of lines. This is because

$$a(x,y)\frac{\partial u}{\partial x} + b(x,y)\frac{\partial u}{\partial y} = \nabla u \cdot (a(x,y), b(x,y)),$$

and therefore the PDE (8.12) still states that the directional derivative of u in the direction of $(a(x,y), b(x,y))$ is $f(x,y)$. However, the direction $(a(x,y), b(x,y))$ is not constant.

Example 8.4. We will solve the initial value problem

$$\frac{\partial u}{\partial x} + x\frac{\partial u}{\partial y} + yu = y, \ (x,y) \in \mathbf{R}^2,$$

$$u(0,y) = y, \ y \in \mathbf{R}.$$

The first step is to find the characteristics by solving the following system:

$$\frac{\partial x}{\partial t} = 1, \ x(s,0) = 0,$$

$$\frac{\partial y}{\partial t} = x, \ y(s,0) = s.$$

Since the first equation is independent of y, it can be solved immediately to yield $x(s,t) = t$. Substituting this into the second equation and solving, we obtain $y(s,t) = s + t^2/2$. We notice that, along the characteristics, we have $y = s + x^2/2$. Since s indicates a point on the y-axis (the original initial condition $u(0,y) = y$ leads to $x(s,0) = 0$, $y(s,0) = s$), we can write the characteristics as $y = y_0 + x^2/2$, which shows that the characteristics for this problem are parabolas.

We also note, for use below, that the equations

$$x = t, \ y = s + \frac{1}{2}t^2$$

are easily solved for (s,t) to yield

$$s = y - \frac{1}{2}x^2, \ t = x.$$

With the given $x(s,t)$, $y(s,t)$, and $v(s,t) = u(x(s,t), y(s,t))$, we have

$$\frac{\partial v}{\partial t} = \frac{\partial u}{\partial x}\frac{\partial x}{\partial t} + \frac{\partial u}{\partial y}\frac{\partial x}{\partial t} = \frac{\partial u}{\partial x} \cdot 1 + \frac{\partial u}{\partial y} \cdot x = \frac{\partial u}{\partial x} + x\frac{\partial u}{\partial y}.$$

Using this result and $y = s + t^2/2$, the original PDE, expressed in the new variables, is

$$\frac{\partial v}{\partial t} + \left(s + \frac{1}{2}t^2\right)v = s + \frac{1}{2}t^2.$$

8.2. First-order quasi-linear PDEs

The initial condition, in the new variables, is $v(s,0) = s$. Regarding s as a constant, we have a first-order linear ODE, with an initial condition, to solve for v. Using the method of integrating factors, as explained in Section 4.2.4, we obtain the solution

$$v(s,t) = 1 + (s-1)e^{-st-t^3/6}.$$

Substituting $s = y - x^2/2$, $t = x$, we obtain

$$u(x,y) = 1 + \left(y - \frac{1}{2}x^2 - 1\right)e^{-xy+x^3/3}. \quad \blacksquare$$

The preceding example illustrates the method of characteristics for a first-order linear IVP

$$a(x,y)\frac{\partial u}{\partial x} + b(x,y)\frac{\partial u}{\partial y} + c(x,y)u = p(x,y), \; u(f(s),g(s)) = h(s).$$

The technique can be summarized as follows.

1. Solve the system

$$\frac{\partial x}{\partial t} = a(x,y), \; x(s,0) = f(s),$$
$$\frac{\partial y}{\partial t} = b(x,y), \; x(s,0) = g(s),$$

to find $x = x(s,t)$, $y = y(s,t)$.

2. Define new functions $v(s,t) = u(x(s,t), y(s,t))$, $r(s,t) = c(x(s,t), y(s,t))$, and $q(s,t) = p(x(s,t), y(s,t))$. Then the original IVP becomes

$$\frac{\partial v}{\partial t} + r(s,t)v = q(s,t), \; v(s,0) = h(s),$$

which, at least in principle, can be solved for v. The key point is that the differential equation, while formally a PDE, is an ODE when s is regarded as a constant. Moreover, it is a first-order linear ODE, and the method of integrating factors is applicable.

3. Solve the equations $x = x(s,t)$, $y = y(s,t)$ for (s,t) to obtain $s = s(x,y)$, $t = t(x,y)$, and then obtain the final solution u by $u(x,y) = v(s(x,y), t(x,y))$.

The process is quite straightforward; however, it may not be possible to carry out all three steps explicitly, depending on how complicated a, b, c, and p are. For instance, it may not be possible to solve the ODEs in the first two steps explicitly, or it may not be possible to solve explicitly for $s = s(x,y)$, $t = t(x,y)$.

8.2.2 Noncharacteristic initial curves

In the previous section, we discussed the fact that the initial curve, parametrized as $(f(s), g(s))$, cannot be a characteristic curve (otherwise, the auxiliary condition $u(f(s), g(s)) = h(s)$ does not determine a unique solution, and there may be no solution satisfying this condition). In the case discussed in Section 8.1, the characteristics were straight lines, and so it was easy to determine if the initial curve was characteristic.

In the case of characteristic curves determined by

$$\frac{\partial x}{\partial t} = a(x,y), \, x(s,0) = f(s),$$

$$\frac{\partial y}{\partial t} = b(x,y), \, y(s,0) = g(s),$$

there is a simple test to check whether the initial curve is a characteristic: The tangents to the initial curve and to the characteristic must be independent (that is, point in different directions) at every point. These tangents are, respectively, the vectors

$$\left(\frac{\partial x}{\partial s}(s,0), \frac{\partial y}{\partial s}(s,0)\right) = \left(\frac{df}{ds}(s), \frac{dg}{ds}(s)\right),$$

$$\left(\frac{\partial x}{\partial t}(s,0), \frac{\partial y}{\partial t}(s,0)\right) = (a(f(s),g(s),),b(f(s),g(s)))).$$

These vectors are independent if and only if the *Jacobian determinant* is nonzero:

$$\begin{vmatrix} \frac{df}{ds}(s) & a(f(s),g(s)) \\ \frac{dg}{ds}(s) & b(f(s),g(s)) \end{vmatrix} = b(f(s),g(s))\frac{df}{ds}(s) - a(f(s),g(s))\frac{dg}{ds}(s) \neq 0. \qquad (8.16)$$

Example 8.5. Consider the IVP

$$\frac{\partial u}{\partial x} + x\frac{\partial u}{\partial y} = 0, \, (x,y) \in \mathbf{R}^2,$$

$$u(s,s^2/2) = 1. \qquad (8.17)$$

In terms of the earlier notation, we have

$$a(x,y) = 1, \, b(x,y) = x, \, f(s) = s, \, g(s) = \frac{1}{2}s^2, \, h(s) = 1,$$

and

$$\begin{vmatrix} \frac{df}{ds}(s) & a(f(s),g(s)) \\ \frac{dg}{ds}(s) & b(f(s),g(s)) \end{vmatrix} = \begin{vmatrix} 1 & 1 \\ s & s \end{vmatrix} = 1 \cdot s - s \cdot 1 = 0.$$

This result shows that the initial curve is a characteristic curve, and therefore the IVP is not well-posed. Indeed, the characteristics are parabolas (the same as in Example 8.4), as is the initial curve. ∎

By the inverse function theorem (see [37]), the condition

$$\begin{vmatrix} \frac{\partial x}{\partial s}(s,0) & \frac{\partial x}{\partial t}(s,0) \\ \frac{\partial y}{\partial s}(s,0) & \frac{\partial y}{\partial t}(s,0) \end{vmatrix} \neq 0$$

8.2. First-order quasi-linear PDEs

(which is equivalent to (8.16)) implies that the system

$$x = x(s,t),$$
$$y = y(s,t)$$

has a unique solution $s = s(x,y)$, $t = t(x,y)$ that exists on some neighborhood of the initial curve $(f(s), g(s))$. However, this neighborhood might be quite small.

At an arbitrary point $(\bar{s}, \bar{t})$, the condition

$$\begin{vmatrix} \frac{\partial x}{\partial s}(\bar{s},\bar{t}) & \frac{\partial x}{\partial t}(\bar{s},\bar{t}) \\ \frac{\partial y}{\partial s}(\bar{s},\bar{t}) & \frac{\partial y}{\partial t}(\bar{s},\bar{t}) \end{vmatrix} \neq 0$$

implies that the change of variables $(s,t) \mapsto (x,y)$ is well defined, at least locally: For each (x,y) sufficiently close to

$$(\bar{x}, \bar{y}) = (x(\bar{s},\bar{t}), y(\bar{s},\bar{t})),$$

there is one and only one point (s,t) such that

$$(x,y) = (x(s,t), y(s,t)).$$

The conclusion of these considerations is that if the initial curve is noncharacteristic, then the IVP has a unique solution that exists on some neighborhood of the initial curve. As explained above, though, it may not be possible to compute this solution explicitly. For instance, although the inverse function theorem guarantees that a unique solution exists, it does not guarantee that it can be computed explicitly.

8.2.3 Semilinear equations

Solving a semilinear equation is no different from solving a linear equation, except that the PDE is transformed into an ODE that is not linear. Specifically, the PDE

$$a(x,y)\frac{\partial u}{\partial x} + b(x,y)\frac{\partial u}{\partial y} = c(x,y,u)$$

becomes

$$\frac{\partial v}{\partial t} = d(s,t,v), \qquad (8.18)$$

where $d(s,t,v) = c(x(s,t), y(s,t), v)$. We illustrate with the following example.

Example 8.6. We will solve the IVP

$$2y\frac{\partial u}{\partial x} + \frac{\partial u}{\partial y} = u^2, \; u(x,0) = \frac{1}{1+x^2}.$$

The characteristics are found by solving

$$\frac{\partial x}{\partial t} = 2y, \; x(s,0) = s,$$
$$\frac{\partial y}{\partial t} = 1, \; y(s,0) = 0.$$

We can solve the second IVP to get $y(s,t) = t$; substituting this into the first and solving yields $x(s,t) = s + t^2$. We can solve $x = s + t^2$, $y = t$ to get $s = x - y^2$, $t = y$. The parameter s refers to the initial point on the x-axis, so $x = s + t^2$, $y = t$ also implies $x = s + y^2 = x_0 + y^2$, which shows that the characteristics are parabolas.

We now change variables by defining $v(s,t) = u(x(s,t), y(s,t))$, which yields

$$2y\frac{\partial u}{\partial x} + \frac{\partial u}{\partial y} = \frac{\partial v}{\partial t} \text{ and } u^2 = v^2.$$

Therefore, in the new variables, the IVP is

$$\frac{\partial v}{\partial t} = v^2, \ v(s,0) = \frac{1}{1+s^2}.$$

This ODE is nonlinear, but it is of the type called *separable*[41] and can be solved by direct integration. The solution is

$$v(s,t) = \frac{1}{1+s^2-t}.$$

Changing variables back to (x,y) yields

$$u(x,y) = \frac{1}{1+(x-y^2)^2 - y}.$$

The reader should note that the solution u is not defined everywhere, which is not surprising since the PDE is nonlinear. Nonlinear differential equations admit the possibility that solutions "blow up" on a finite domain. In this case, as t increases from 0 to $1+s^2$, the solution v goes from $1/(1+s^2)$ to ∞. Expressed in terms of the original variables, as (x,y) moves from $(x_0, 0)$ along the parabola $x = x_0 + y^2$ to the point of intersection with $1 + (x-y^2)^2 - y = 0$, u goes from $1/(1+x^2)$ to ∞. ∎

8.2.4 Quasi-linear equations

We now consider a quasi-linear equation of the form

$$a(x,y,u)\frac{\partial u}{\partial x} + b(x,y,u)\frac{\partial u}{\partial y} = c(x,y,u).$$

The left side of the PDE can still be regarded as (proportional to) a directional derivative of u, but now the direction $(a(x,y,u), b(x,y,u))$ depends not only on (x,y) but also on the value of u at that point. This means that we have to solve for u (or actually v) simultaneously with solving for the characteristics. Assuming an initial condition of the form $u(f(s), g(s)) = h(s)$, the characteristics $(x(s,t), y(s,t))$ and the solution $v(s,t)$ are defined by the system

$$\frac{\partial x}{\partial t} = a(x,y,v), \ x(s,0) = f(s),$$

$$\frac{\partial y}{\partial t} = b(x,y,v), \ y(s,0) = g(s),$$

$$\frac{\partial v}{\partial t} = c(x,y,v), \ v(s,0) = h(s).$$

---

[41] Every introductory text on differential equations explains how to solve separable ODEs. See, for example, Section 2.2 of [71].

8.2. First-order quasi-linear PDEs

Assuming x, y, v satisfy these equations and $v(s,t) = u(x(s,t), y(s,t))$, the fact that

$$\frac{\partial v}{\partial t} = \frac{\partial u}{\partial x}\frac{\partial x}{\partial t} + \frac{\partial u}{\partial y}\frac{\partial y}{\partial t},$$

together with the above equations for $\partial x/\partial t$, $\partial y/\partial t$, $\partial v/\partial t$, shows that u satisfies the original PDE.

We now present an example.

Example 8.7. We will solve the IVP

$$u\frac{\partial u}{\partial x} + y\frac{\partial u}{\partial y} = x, \ (x,y) \in \mathbf{R}^2, \tag{8.19}$$

$$u(x,1) = 2x, \ x \in \mathbf{R}.$$

This is an example of the general IVP described above, with

$$a(x,y,u) = u, \ b(x,y,u) = y, \ c(x,y,u) = x, \ f(s) = s, \ g(s) = 1, \ h(s) = 2s.$$

The solution (in the characteristic variables) is determined by the following system:

$$\frac{\partial x}{\partial t} = v, \ x(s,0) = s,$$

$$\frac{\partial y}{\partial t} = y, \ y(s,0) = 1,$$

$$\frac{\partial v}{\partial t} = x, \ v(s,0) = 2s.$$

This is a system of constant-coefficient, linear ODEs, and it can be solved by the methods of Section 4.3 to yield

$$x(s,t) = \frac{s}{2}\left(3e^t - e^{-t}\right),$$

$$y(s,t) = e^t,$$

$$v(s,t) = \frac{s}{2}\left(3e^t + e^{-t}\right).$$

Solving for (s,t) in terms of (x,y), we obtain

$$s(x,y) = \frac{2xy}{3y^2 - 1},$$

$$t(x,y) = \ln(y),$$

and then

$$u(x,y) = v(s(x,y), t(x,y)) = \frac{(3y^2 + 1)x}{3y^2 - 1}.$$

The reader can verify that u is a solution of the IVP.

It is interesting to note that the characteristic variables seem to imply that y must be positive ($y = e^t$). In fact, the computed u is a valid solution only for $3y^2 - 1 > 0$, that is, for

$$y > \frac{1}{\sqrt{3}} > 0.$$

This can be seen in the characteristic variables from the fact that the change of variables is valid only where

$$\begin{vmatrix} \frac{\partial x}{\partial s}(s,t) & \frac{\partial x}{\partial t}(s,t) \\ \frac{\partial y}{\partial s}(s,t) & \frac{\partial y}{\partial t}(s,t) \end{vmatrix} \neq 0.$$

The reader can verify that this condition corresponds to $t \neq \ln(1/\sqrt{3})$ or $y \neq 1/\sqrt{3}$. ∎

Exercises

1. Consider the following IVP:

$$x\frac{\partial u}{\partial x} + y\frac{\partial u}{\partial y} - (x+y)u = 0,$$

$$u(1,y) = \phi(y).$$

   (a) What are the characteristics of the PDE?
   (b) Find the solution.
   (c) On what set does the solution u exist? (Assume that $\phi(y)$ is defined and differentiable for all real numbers y.)

2. Consider the following IVP:

$$x\frac{\partial u}{\partial x} + y\frac{\partial u}{\partial y} - u^2 = 0,$$

$$u(x,1) = \phi(x).$$

   (a) What are the characteristics of the PDE?
   (b) Find the solution in terms of ϕ.
   (c) Let $\phi(x) = 1$ for all x. On what set does the solution u exist?

3. Solve the IVP

$$u\frac{\partial u}{\partial x} + u\frac{\partial u}{\partial y} = 0,$$

$$u(x,0) = \phi(x).$$

   What must be true for the initial curve to be noncharacteristic at every point?

4. Consider the IVP

$$u^2\frac{\partial u}{\partial x} + u\frac{\partial u}{\partial y} = 0,$$

$$u(x,0) = \phi(x).$$

   Show that the characteristics are straight lines.

8.3. Burgers's equation

5. Consider the IVP
$$\frac{\partial u}{\partial x} + u\frac{\partial u}{\partial y} = 0, \ x > 0,$$
$$u(x,0) = x, \ x > 0.$$

   (a) Solve the IVP using the method of characteristics and identify the largest subset of the right half-plane on which the solution is defined.
   (b) Find the characteristics of the PDE.
   (c) Each characteristic is indexed by a point on the initial curve, which in this case is the positive x-axis. In other words, each characteristic is indexed by a value of $x_0 > 0$. The solution u is constant along each characteristic. Show that, starting at $(x_0, 0)$ and proceeding along the characteristics in the direction of increasing y, the solution is well defined until the characteristic intersects with another characteristic. Find the point $(x(x_0), y(x_0))$ of intersection in terms of x_0, and identify the curve determined by these points of intersection.

6. Consider the IVP
$$\frac{\partial u}{\partial x} + u\frac{\partial u}{\partial y} = 0, \ x > 0,$$
$$u(x,0) = \frac{1}{x}, \ x > 0$$

(cf. Exercise 5).

   (a) Solve the IVP using the method of characteristics and identify the largest subset of the right half-plane on which the solution is defined.
   (b) Find the characteristics of the PDE.
   (c) Show that all of the characteristics intersect at a common point. What is it?

   Compare with the results of Exercise 5.

8.3 Burgers's equation

The inviscid Burgers's equation,

$$\frac{\partial u}{\partial t} + u\frac{\partial u}{\partial x} = 0, \qquad (8.20)$$

is an example of a nonlinear conservation law; it is of the form

$$\frac{\partial u}{\partial t} + \frac{\partial}{\partial x}(f(u)) = 0$$

with $f(u) = u^2/2$. The equation models certain phenomena in fluid flow, but these are complicated even to describe, and we will study the equation purely as an illustration of some mathematical aspects of nonlinear PDEs. We will focus on the initial value problem posed on the entire real line; the initial condition is $u(x,0) = u_0(x), \ x \in \mathbf{R}$.

In terms of the classification given in the last section, the inviscid Burgers's equation is a first-order quasi-linear PDE. The characteristics for (8.20) are determined by the system

$$\frac{\partial x}{\partial \tau} = v, \ x(s,0) = s,$$
$$\frac{\partial t}{\partial \tau} = 1, \ t(s,0) = 0,$$
$$\frac{\partial v}{\partial \tau} = 0, \ v(s,0) = u_0(s).$$

The second and third equations are easily solved to yield $t(s,\tau) = \tau$, $v(s,t) = u_0(s)$. We then have

$$\frac{\partial x}{\partial \tau} = u_0(s), \ x(s,0) = s \ \Rightarrow \ x(s,\tau) = s + u_0(s)\tau.$$

Therefore, the solution (in the characteristic variables) is defined by

$$\begin{aligned} x(s,\tau) &= s + u_0(s)\tau, \\ t(s,\tau) &= \tau, \\ v(s,\tau) &= u_0(s). \end{aligned} \quad (8.21)$$

Unless u_0 is expressed by an extremely simple formula, we cannot solve for (s,τ) explicitly in terms of (x,t). However, $t = \tau$ and therefore $x = s + u_0(s)t$. The variable s describes the position on the initial line $t = 0$ (that is, the x-axis) and plays the role of x_0. Therefore, the characteristics are defined by

$$x = x_0 + u_0(x_0)t. \quad (8.22)$$

We see that the characteristics for the inviscid Burgers's equation are straight lines. Also, the initial line $t = 0$ is not a characteristic, so the given IVP is well-posed.

Although we cannot solve explicitly for (s,τ) in terms of (x,t), we can use the Jacobian determinant

$$J(s,\tau) = \begin{vmatrix} \frac{\partial x}{\partial s}(s,\tau) & \frac{\partial x}{\partial t}(s,\tau) \\ \frac{\partial y}{\partial s}(s,\tau) & \frac{\partial y}{\partial t}(s,\tau) \end{vmatrix}$$

to determine values of (s,τ) for which the change of variables is valid. We have

$$J(s,\tau) = \begin{vmatrix} 1 + \frac{du_0}{ds}(s)\tau & u_0(s) \\ 0 & 1 \end{vmatrix} = 1 + \frac{du_0}{ds}(s)\tau.$$

First of all, we notice that $J(s,0) = 1$, which confirms that the initial line $t = 0$ is not characteristic. We will now consider $\tau \geq 0$, which corresponds to $t \geq 0$.

If $du_0/ds \geq 0$ everywhere (that is, if u_0 is an increasing function), then $J(s,\tau) \geq 1$ for all $t \geq 0$, and therefore the solution (8.21) is defined for all $t \geq 0$. Every point $(\bar{x},\bar{t})$, with $\bar{t} \geq 0$, lies on a unique characteristic line $x = x_0 + u_0(x_0)t$, and the value of the solution, $u(\bar{x},\bar{t})$, is $u_0(x_0)$.

8.3. Burgers's equation

If, on the other hand,
$$\frac{du_0}{ds}(\bar{s}) < 0,$$
then $J(\bar{s}, \bar{\tau}) = 0$ for
$$\tau = \bar{t} = -\frac{1}{\frac{du_0}{ds}(\bar{s})} > 0,$$
and the change of variables breaks down for $\tau = \bar{\tau}$. The smallest value of τ (or t) at which this occurs is
$$t^* = -\frac{1}{\min\left\{\frac{du_0}{ds}(s) : s \in \mathbf{R}\right\}}. \tag{8.23}$$

In this case, the IVP has a unique solution that exists on the interval $[0, t^*)$. The reader should notice that $du_0/ds < 0$ at some point implies that u_0 is decreasing on some interval.

We can study the above conclusion geometrically by considering the characteristics of the PDE. If u_0 is always increasing, then the slopes of the characteristics in the xt-plane are decreasing as the initial point x_0 moves to the right. Figure 8.6 illustrates this situation; it is easy to see that, in this case, no two characteristics can intersect for $t > 0$.

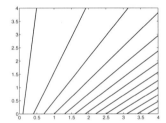

Figure 8.6. *The characteristics for the inviscid Burgers's equation when the initial function u_0 is always increasing.*

On the other hand, if u_0 is decreasing anywhere, then there exist initial points x_0, $\tilde{x}_0$ such that $\tilde{x}_0 > x_0$ and the characteristic through $\tilde{x}_0$ has a greater slope than the one through x_0. In this case, the second characteristic must intersect the first at some $t > 0$. This situation is illustrated in Figure 8.7.

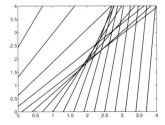

Figure 8.7. *The characteristics for the inviscid Burgers's equation when the initial function u_0 is decreasing.*

Example 8.8. We will solve the inviscid Burgers's equation with the initial condition

$$u(x,0) = u_0(x) = \frac{1}{1+x^2}, \ x \in \mathbf{R}.$$

We note that u_0 is increasing for $x < 0$ and decreasing for $x > 0$, so the solution will fail to be uniquely defined after some time for $x > 0$. By the above analysis, the solution is uniquely defined for all t between 0 and

$$t^* = -\frac{1}{\min\left\{\frac{du_0}{ds}(s) : s \in \mathbf{R}\right\}}.$$

For $u_0(x) = 1/(1+x^2)$, it is straightforward to determine that

$$t^* = 8/\left(3\sqrt{3}\right) \doteq 1.5396.$$

The solution to the IVP is uniquely defined on the interval $[0,t^*)$; Figure 8.8 shows three snapshots of this solution.

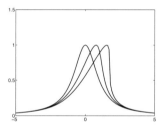

Figure 8.8. *Three snapshots of the solution of the IVP from Example 8.8, corresponding to $t = 0$ (the leftmost curve), $t = t^*/2$ (the middle curve), and $t = t^*$ (the rightmost curve).*

The reader should notice that

$$\frac{\partial u}{\partial x}(x,t^*)$$

appears to be infinite at a certain value of x ($x \doteq 1.75$). Moreover, it is easy to understand the reason for this. If we compare the inviscid Burgers's equation

$$\frac{\partial u}{\partial t} + u\frac{\partial u}{\partial x} = 0$$

with the simple advection equation

$$\frac{\partial u}{\partial t} + c\frac{\partial u}{\partial x} = 0,$$

we see that u itself plays the role of the velocity c. The velocity decreases to the right of the peak of the wave, so the left part of the wave actually travels faster than the right part, and eventually catches up to it. At this point, the wave "breaks," and the solution no longer exists, at least in the classical sense. However, it is possible to define meaningfully a discontinuous solution, in which case the discontinuity is called a *shock*. ∎

8.3. Burgers's equation

As we have seen above, if the initial curve is noncharacteristic, then the solution to the initial value problem for Burgers's equation exists for all times t less than a certain t^*. In other words, the solution $u = u(x,t)$ is uniquely defined on a "strip"

$$\{(x,t) : -\infty < x < \infty,\ 0 \leq t < t^*\}.$$

However, this conclusion does not mean that u fails to exist uniquely everywhere outside of this strip, only that there is some x^* such that (x^*, t^*) fails to be uniquely defined because two characteristics intersect at that point.

We can be more precise about the domain of existence of u by looking at each individual characteristic and determining the values of t for which u is uniquely defined along that characteristic. If $du_0/dx(x_0)$ is positive, then the characteristic through $(x_0, 0)$ never intersects another characteristic, and the solution is uniquely defined for all time along that characteristic. On the other hand, if $du_0/dx(x_0)$ is negative, then the solution is uniquely defined, on the characteristic through $(x_0, 0)$, only for

$$0 \leq t < -\frac{1}{\frac{du_0}{dx}(x_0)}.$$

The upper bound on t is easily derived as the time at which the change of variables from (s, τ) to (x,t) ceases to be well defined, and is found by solving

$$J(s,\tau) = 1 + \frac{du_0}{ds}(s)\tau = 0.$$

(The reader should recall that $t = \tau$ and s plays the role of x_0.) Moreover, from this value of t, the equations

$$x = s + u_0(s)\tau,\ t = \tau,$$

yield the following value of x:

$$x = x_0 - \frac{u_0(x_0)}{\frac{du_0}{dx}(x_0)}.$$

By these considerations, we have parametric equations for a certain curve:

$$x = x_0 - \frac{u_0(x_0)}{\frac{du_0}{dx}(x_0)},\ t = -\frac{1}{\frac{du_0}{dx}(x_0)}. \tag{8.24}$$

In these equations, x_0 ranges over all values for which $du_0/dx(x_0) < 0$. This curve is called a *caustic*, which refers to an "envelope" of curves.

The reader will recall that the change of variables from (s, τ) to (x,t) breaks down precisely when the direction of a characteristic, which is the vector

$$\left(\frac{\partial x}{\partial \tau}, \frac{\partial t}{\partial \tau}\right),$$

is parallel to the direction across the characteristic, which is defined by

$$\left(\frac{\partial x}{\partial s}, \frac{\partial t}{\partial s}\right);$$

the Jacobian determinant $J(s,\tau)$ is zero precisely when these two vectors are parallel. Solving $J(s,\tau) = 0$ yields a curve (parametrized by $x_0 = s$) which is tangent to each characteristic that it intersects. This curve is the caustic.

In Example 8.8, $u_0(x_0) = 1/(1+x_0^2)$, and equations (8.24) defining the caustic reduce to

$$x = \frac{1+3x_0^2}{2x_0},\ t = \frac{(1+x_0^2)^2}{2x_0},\ x_0 > 0.$$

This curve is displayed in Figure 8.9. Figure 8.10 shows the characteristics for the same problem, and the caustic can be clearly seen. From this graph, the reader should understand, at least intuitively, the meaning of the phrase "envelope of curves."

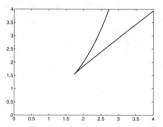

Figure 8.9. *The caustic from Example* 8.8.

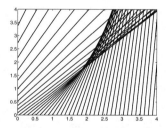

Figure 8.10. *The characteristics from Example 8.8. The caustic from Figure 8.9 can be clearly seen.*

Exercises

1. In this section we showed that the IVP for the inviscid Burgers's equation has a unique solution that exists on the interval $0 \le t < t^*$, where t^* is defined by (8.23). (This applies to the case in which the initial value u_0 is decreasing at least somewhere.) In other words, the solution u is guaranteed to exist and be unique on the set $\{(x,t) : x \in \mathbf{R},\ 0 \le t < t^*\}$. However, the solution will normally exist on a larger set; for most x, $u(x,t)$ is well defined for t larger than t^*. Consider the IVP in Example 8.8.

    (a) Show that the solution u exists and is unique for all $x < 0,\ t \ge 0$.

    (b) Describe the set of *all* (x,t) for which the solution is uniquely defined.

8.3. Burgers's equation

2. Consider the IVP

$$\frac{\partial u}{\partial t} + (1-u)\frac{\partial u}{\partial x} = 0, \ x > 1, \ t > 0,$$

$$u(x,0) = \frac{1}{x}, \ x > 1.$$

   (a) Find the equation of the characteristic passing through the point $(x_0, 0)$.

   (b) Prove that the solution u exists for all $t > 0$ by showing that the characteristics do not intersect for $t > 0$.

   (c) Find the solution u.

3. Consider the IVP

$$\frac{\partial u}{\partial t} + (1-u)\frac{\partial u}{\partial x} = 0, \ -\infty < x < 0, \ t > 0,$$

$$u(x,0) = \frac{1}{x}, \ -\infty < x < 0.$$

   (a) Find the equation of the characteristic passing through the point $(x_0, 0)$.

   (b) Show that there exists a point $(\bar{x}, \bar{t})$ such that any two characteristics intersect at that point.

   (c) Show that given any $x_0, x_1 < 0$, the characteristics through $(x_0, 0)$ and $(x_1, 0)$ intersect. Find the point of intersection.

4. The general conservation law (in one space dimension) takes the form

$$\frac{\partial u}{\partial t} + \frac{\partial}{\partial x}(f(u)) = 0,$$

   which is equivalent to

$$\frac{\partial u}{\partial t} + \frac{df}{du}(u)\frac{\partial u}{\partial x} = 0.$$

   We will write $a = df/du$ and assume that $a(u) > 0$ for all u.

   Consider the IVP

$$\frac{\partial u}{\partial t} + a(u)\frac{\partial u}{\partial x} = 0, \ -\infty < x < \infty, \ t > 0,$$

$$u(x,0) = \phi(x), \ x > 0.$$

   (a) Find the characteristic of the PDE passing through the point $(x_0, 0)$.

   (b) Show that, if the function $\psi(x) = a(\phi(x))$ is increasing as a function of x, then the characteristics do not intersect for $t > 0$, and therefore the IVP has a well-defined solution for all $t > 0$.

   (c) On the other hand, show that if ψ is decreasing somewhere, then characteristics will intersect for $t > 0$. Specifically, suppose $d\psi/dx(x_0) < 0$. Find the largest t^* such that the solution u is guaranteed to be uniquely defined, along the characteristic through $(x_0, 0)$, for $t < t^*$.

5. Consider the IVP from Exercise 4, and consider the case that the function $\psi(x) = a(\phi(x))$ is decreasing. We know that the solution will cease to be uniquely defined when characteristics intersect. On the other hand, we know that the method of characteristics breaks down when the change of variables $x = x(s,\tau)$, $t = t(s,\tau)$ is no longer valid (cf. the discussion on pages 331–332). Show that these two methods identify the same curve (the caustic).

8.4 Suggestions for further reading

Most introductory texts on PDEs present the method of characteristics; for example, the reader can consult the books by McOwen [50], Strauss [62], and Haberman [30]. Another introductory text that covers the method of characteristics is by Zachmanoglou and Thoe [69].

For more information about nonlinear PDEs and shocks, Haberman's text [29] on mathematical modeling covers nonlinear conservation laws in the context of modeling traffic flow. Numerical methods are covered by LeVeque in [42] and [43].

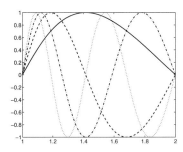

Chapter 9
Green's Functions

A square linear system, written in matrix-vector form as $\mathbf{A}\mathbf{x} = \mathbf{b}$, can be solved using the inverse of $\mathbf{A}$ (assuming it exists); the solution is $\mathbf{x} = \mathbf{A}^{-1}\mathbf{b}$. Although this is rarely the preferred numerical method for solving $\mathbf{A}\mathbf{x} = \mathbf{b}$, the formula is theoretically useful and often leads to insight about the nature of the solution, the sensitivity of the solution $\mathbf{x}$ to errors in $\mathbf{b}$, and other aspects of the problem.

There is a concept analogous to the inverse of a matrix that is useful for studying differential equations, namely, the concept of a Green's function. The analogy is strongest for boundary value problems (BVPs). For example, the solution to

$$-\frac{d}{dx}\left(k(x)\frac{du}{dx}\right) = f(x),\ a < x < b,$$
$$u(a) = 0,$$
$$u(b) = 0$$
(9.1)

can be written as

$$u(x) = \int_a^b G(x;y)f(y)\,dy,$$
(9.2)

where G is the *Green's function* for the BVP. We can make the analogy apparent by writing $\mathbf{x} = \mathbf{A}^{-1}\mathbf{x}$ in terms of components:

$$x_i = \sum_{j=1}^n \left(\mathbf{A}^{-1}\right)_{ij} b_j.$$
(9.3)

Bearing in mind that an integral is the limit of a sum, there is a direct correspondence between formulas (9.3) and (9.2):

$$i \leftrightarrow x,\ j \leftrightarrow y,\ x_i \leftrightarrow u(x),\ b_j \leftrightarrow f(y),\ \left(\mathbf{A}^{-1}\right)_{ij} \leftrightarrow G(x;y).$$

When we take into account the variety of problems in differential equations (ODEs, PDEs, IVPs, BVPs, IBVPs), the situation becomes more complicated, and there are a number of different kinds of Green's functions: *causal*, *free space*, and so forth. In this chapter, we will explain the different kinds of Green's functions and derive them explicitly in certain cases.

A word about notation: In previous chapters, we posed the differential equations on the interval $[0,\ell]$ because this convention led to simpler formulas for eigenvalues and eigenfunctions. For instance, under Dirichlet conditions, the eigenfunctions of $-d^2/dx^2$ on $[0,\ell]$ are

$$\sin\left(\frac{n\pi x}{\ell}\right), \ n = 1,2,3,\ldots.$$

If we worked on the general interval $[a,b]$, the formula would become

$$\sin\left(\frac{n\pi(x-a)}{b-a}\right), \ n = 1,2,3,\ldots.$$

Excluding Section 9.1, we will derive few explicit formulas in this chapter; therefore, beginning with Section 9.2, we will find it more convenient to work on a general interval $[a,b]$.

9.1 Green's functions for BVPs in ODEs: Special cases

As our first example, we consider a problem for which the Green's function can be identified by direct integration. The example is the BVP

$$-\frac{d}{dx}\left(k(x)\frac{du}{dx}\right) = f(x), \ 0 < x < \ell,$$
$$u(0) = 0, \qquad\qquad (9.4)$$
$$\frac{du}{dx}(\ell) = 0.$$

The "data" for (9.4) is the forcing function f, which is a function of the spatial variable x. By two integrations, we can determine a formula for u in terms of f:

$$-\frac{d}{dx}\left(k(x)\frac{du}{dx}(x)\right) = f(x)$$
$$\Rightarrow k(x)\frac{du}{dx}(x) = \int_x^\ell f(y)\,dy \ \left(\text{since } \frac{du}{dx}(\ell) = 0\right)$$
$$\Rightarrow \frac{du}{dx}(x) = \frac{1}{k(x)}\int_x^\ell f(y)\,dy$$
$$\Rightarrow u(x) = \int_0^x \frac{1}{k(z)}\int_z^\ell f(y)\,dy\,dz \ (\text{since } u(0) = 0)$$
$$\Rightarrow u(x) = \int_0^\ell \left\{\int_0^{\min\{x,y\}} \left(\frac{1}{k(z)}\right)dz\right\} f(y)\,dy.$$

(To verify the last step, the reader should sketch the domain of integration for the double integral and change the order of integration.) We can thus write the solution as

$$u(x) = \int_0^\ell G(x;y) f(y)\,dy, \qquad\qquad (9.5)$$

9.1. Green's functions for BVPs in ODEs: Special cases

where

$$G(x;y) = \int_0^{\min\{x,y\}} \left(\frac{1}{k(z)}\right) dz. \qquad (9.6)$$

By inspection of formula (9.5), we see that, for a particular x, the value $u(x)$ is a weighted sum of the data f over the interval $[0,\ell]$. Indeed, for a given x and y, $G(x;y)$ indicates the effect on the solution u at x of the data f at y. The function G is called the *Green's function* for BVP (9.4).

To expose the meaning of the Green's function more plainly, we consider a right-hand-side function f that is concentrated at a single point $x = \xi$. We define $d_{\Delta x}$ by

$$d_{\Delta x}(x) = \begin{cases} \frac{x + \Delta x}{\Delta x^2}, & -\Delta x < x < 0, \\ -\frac{x - \Delta x}{\Delta x^2}, & 0 < x < \Delta x, \\ 0 & \text{otherwise} \end{cases}$$

(see Figure 9.1) and let $f_{\Delta x}(x) = d_{\Delta x}(x - \xi)$. The solution of (9.4), with $f = f_{\Delta x}$, is

$$u(x) = \int_0^\ell G(x;y) d_{\Delta x}(y - \xi) dy$$
$$= \int_{\xi - \Delta x}^{\xi + \Delta x} G(x;y) d_{\Delta x}(y - \xi) dy.$$

Since

$$\int_{\xi - \Delta x}^{\xi + \Delta x} d_{\Delta x}(y - \xi) dy = 1$$

for all Δx, we see that $u(x)$ is a weighted average of $G(x;y)$ (considered as a function of y) over the interval $\xi - \Delta x \le y \le \xi + \Delta x$. As $\Delta x \to 0$, this weighted average converges to $G(x;\xi)$.

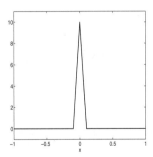

Figure 9.1. *The function $d_{\Delta x}$ ($\Delta x = 0.1$).*

We now interpret this result in terms of a specific application. We consider an elastic bar of length ℓ and stiffness $k(x)$ hanging with one end fixed at $x = 0$ and the other end (at $x = \ell$) free. The displacement u of the bar is modeled by the BVP (9.4), where f is the external force density (in units of force per volume). The quantity

$$\int_0^\ell A d_{\Delta x}(y) dy$$

is the total force exerted on the entire bar. Since

$$\int_0^\ell d_{\Delta x}(y)\,dy = 1 \text{ for all } \Delta x > 0$$

and $d_{\Delta x}$ is zero except on the interval $[\xi - \Delta x, \xi + \Delta x]$, we see that the total pressure exerted on the bar is A and it is concentrated more and more at $x = \xi$ as $\Delta x \to 0$. In the limit, the external force on the bar consists of a pressure of 1 unit acting on the cross section of the bar at $x = \xi$. The response of the bar, in the limit, is $u(x) = G(x;\xi)$.

Thus we see that, for a fixed ξ, $G(x;\xi)$ is a solution to (9.4) for a special right-hand side—a forcing function of magnitude 1 concentrated at $x = \xi$. However, this forcing function is not a true function in the usual sense. Indeed, the function $d_{\Delta x}$ has the following properties:

- $d_{\Delta x}(x) = 0$, $x \notin [-\Delta x, \Delta x]$.

- $d_{\Delta x}(0) \to \infty$ as $\Delta x \to 0$.

- $\int_a^b d_{\Delta x}(x)\,dx = \int_{-\Delta x}^{\Delta x} d_{\Delta x}(x)\,dx = 1$ for all $\Delta x > 0$, provided $[-\Delta x, \Delta x]$ is contained in $[a,b]$, and $\int_a^b d_{\Delta x}(x)\,dx = 0$ if $\Delta x \leq a$ or $-\Delta x \geq b$.

- If g is a continuous function and $[-\Delta x, \Delta x] \subset [a,b]$, then

$$\int_a^b d_{\Delta x}(x) g(x)\,dx$$

is a weighted average of g on the interval $[-\Delta x, \Delta x]$.

These properties suggest that the limit δ of $d_{\Delta x}$ as $\Delta x \to 0$ has the following properties.

- $\delta(x) = 0$ if $x \neq 0$.

- $\delta(0) = \infty$.

- $\int_a^b \delta(x)\,dx = 1$ if $0 \in [a,b]$, and $\int_a^b \delta(x)\,dx = 0$ if $0 \notin [a,b]$.

- If g is a continuous function and $0 \in [a,b]$, then $\int_a^b \delta(x)g(x)\,dx = g(0)$. If $0 \notin [a,b]$, then $\int_a^b \delta(x)g(x)\,dx = 0$.

The "function" δ is not a function at all—any ordinary function d has the property that if d is nonzero only at a single point on an interval, then the integral of d over that interval is zero. However, it is useful to regard δ as a *generalized function*. The particular generalized function δ is called the *Dirac delta function* (or simply the delta function) and is defined by the *sifting* property:

If g is a continuous function and $0 \in [a,b]$, then $\int_a^b \delta(x)g(x)\,dx = g(0)$.
If $0 \notin [a,b]$, then $\int_a^b \delta(x)g(x)\,dx = 0$.

Changing variables in the integral suggests that the following property also holds:

If g is a continuous function and $\xi \in [a,b]$, then $\int_a^b \delta(x)g(x-\xi)\,dx = g(\xi)$.

9.1. Green's functions for BVPs in ODEs: Special cases

Since δ is formally an even function ($\delta(-x) = \delta(x)$), we should also have the following property:

If g is a continuous function and $\xi \in [a,b]$, then $\int_a^b \delta(x) g(\xi - x) dx = g(\xi)$.

Since δ is not a function, but a generalized function, we regard the last two properties as assumptions.

From this discussion, we see that the Green's function G for (9.4) has the property that $u(x) = G(x;\xi)$ is the solution of the BVP

$$-\frac{d}{dx}\left(k(x)\frac{du}{dx}\right) = \delta(x-\xi), \ 0 < x < \ell,$$
$$u(0) = 0, \quad (9.7)$$
$$\frac{du}{dx}(\ell) = 0.$$

We can verify this result using physical reasoning; this is particularly simple in the case of a homogeneous bar. We consider the following thought experiment: We apply a unit pressure (in the positive direction) to the cross section at $x = \xi$, where $0 < \xi < \ell$. (It is not obvious how such a pressure could actually be applied in a physical experiment, which is why we term this a "thought experiment.") Since the bar is fixed at $x = 0$, and the end at $x = \ell$ is free, it is not hard to see what the resulting displacement will be. The part of the bar originally between $x = 0$ and $x = \xi$ will stretch from its original length of ξ to a length of $\xi + \Delta\xi$, where

$$k\frac{\Delta\xi}{\xi} = 1$$

(the pressure, 1, is proportional to the relative change in length ($\Delta\xi/\xi$)). Therefore, $\Delta\xi$, which is the displacement $u(\xi)$, is ξ/k. Since the bar is homogeneous, the displacement in the part of the bar above $x = \xi$ is

$$u(x) = \frac{x}{k}, \ 0 \le x \le \xi.$$

What about the part of the bar originally between $x = \xi$ and $x = \ell$? This part of the bar is not stretched at all; it moves merely because of the displacement in the part of the bar above it. We therefore obtain

$$u(x) = \frac{\xi}{k}, \ \xi < x < \ell.$$

Combining these two formulas gives

$$u(x) = \begin{cases} \frac{x}{k}, & 0 < x \le \xi, \\ \frac{\xi}{k}, & \xi \le x \le \ell, \end{cases}$$

or simply

$$u(x) = \frac{\min\{x,\xi\}}{k}.$$

Thus we see that $u(x) = G(x;\xi)$, where G is given by (9.6) (specialized to the case in which k is constant).

Example 9.1. Consider the BVP (9.4) with $\ell = 1$, $k(x) = 2 - x$, and $f(x) = 1$. We have

$$G(x;y) = \int_0^{\min\{x,y\}} \frac{dz}{2-z}$$
$$= -\log(2-z)\Big|_0^{\min\{x,y\}}$$
$$= \log 2 - \log(2 - \min\{x,y\})$$
$$= \begin{cases} \log 2 - \log(2-y), & 0 < y < x, \\ \log 2 - \log(2-x), & x < y < 1. \end{cases}$$

Therefore,

$$u(x) = \int_0^1 G(x;y)\,dy \quad (\text{since } f(y) = 1)$$
$$= \int_0^x \{\log 2 - \log(2-y)\}\,dy + \int_x^1 \{\log 2 - \log(2-x)\}\,dy$$
$$= x(1 + \log 2) + (2-x)\log(2-x) - \log 4 + (1-x)(\log 2 - \log(2-x))$$
$$= x + \log\left(1 - \frac{x}{2}\right). \quad \blacksquare$$

We can also verify mathematically that

$$u(x) = G(x;\xi) = \int_0^{\min\{x,\xi\}} \left(\frac{1}{k(z)}\right) dz$$

is the solution of (9.7). However, since δ is not a true function, but rather a generalized function, (9.7) must be interpreted as implying the weak form:

$$\int_0^\ell k(x)\frac{du}{dx}(x)\frac{dv}{dx}(x)\,dx = \int_0^\ell \delta(x - \xi)v(x)\,dx = v(\xi) \text{ for all } v \in V. \quad (9.8)$$

Here V is the space of test functions: $V = \{v \in C^2[0,\ell] : v(0) = 0\}$. Now,

$$G(x;\xi) = \int_0^{\min\{x,\xi\}} \frac{dz}{k(z)} = \begin{cases} \int_0^\xi \frac{dz}{k(z)}, & 0 \leq \xi \leq x, \\ \int_0^x \frac{dz}{k(z)}, & x \leq \xi \leq \ell, \end{cases}$$

and therefore

$$\frac{\partial G}{\partial x}(x;\xi) = \begin{cases} 0, & 0 \leq \xi < x, \\ \frac{1}{k(x)}, & x < \xi \leq \ell. \end{cases}$$

It follows that

$$\int_0^\ell k(x)\frac{\partial G}{\partial x}(x;\xi)\frac{dv}{dx}(x)\,dx = \int_0^\xi \frac{dv}{dx}(x)\,dx$$
$$= v(\xi) - v(0) = v(\xi)$$

as desired (since $v(0) = 0$ by assumption). It also follows immediately from the above formulas that

$$G(0;\xi) = 0, \quad \frac{\partial G}{\partial x}(\ell;\xi) = 0,$$

9.1. Green's functions for BVPs in ODEs: Special cases

and so the boundary conditions are satisfied as well. (Actually, we need only verify the essential boundary condition $G(0;\xi) = 0$; the Neumann condition at $x = \ell$ is a natural condition and must be true since $G(\cdot;\xi)$ satisfies the weak form of the BVP.)

9.1.1 The Green's function and the inverse of a differential operator

We write
$$C_m^2[0,\ell] = \left\{u \in C^2[0,\ell] : u(0) = \frac{du}{dx}(\ell) = 0\right\}$$

and define $K : C_m^2[0,\ell] \to C[0,\ell]$ by

$$Ku = -\frac{d}{dx}\left(k(x)\frac{du}{dx}\right).$$

Our results above show that, for each $f \in C[0,\ell]$, there is a unique solution to $Ku = f$. In other words, K is an invertible operator. Moreover, the solution to $Ku = f$ is $u = Mf$, where $M : C[0,\ell] \to C_m^2[0,\ell]$ is defined by

$$(Mf)(x) = \int_0^\ell G(x;y)f(y)\,dy.$$

Therefore, $KMf = f$, which shows that $M = K^{-1}$, the inverse operator of K. It follows that

$$MKu = u \text{ for all } u \in C_m^2[0,\ell] \tag{9.9}$$

must also hold. Thus the Green's function defines the inverse of the differential operator.

9.1.2 Symmetry of the Green's function; reciprocity

The Green's function (9.6) is symmetric in the sense that $G(x;y) = G(y;x)$ for all $x, y \in (0,\ell)$. We will see later that this property always holds for the Green's function of a symmetric differential operator. It is related to the fact that the operator M, defined above, is symmetric, which is analogous to the fact that the inverse of a symmetric matrix is also symmetric.

Physically, the fact that G is symmetric is the principle of *reciprocity*. Recalling the interpretation of G as the solution of (9.7), reciprocity implies that the solution at y induced by a point source at x is the same as the solution at x induced by a point source at y. This can be interpreted in terms of steady-state heat flow (or electrostatics, which we do not discuss in this text).

Exercises

1. Use the Green's function to solve (9.4) when $\ell = 1$, $f(x) = 1$, and $k(x) = e^x$.

2. Prove directly (by substituting u into the differential equation and boundary conditions) that the function u defined by (9.5) and (9.6) solves (9.4).

3. Consider the BVP

$$-\frac{d}{dx}\left(k(x)\frac{du}{dx}\right) = f(x), \ 0 < x < \ell,$$
$$\frac{du}{dx}(0) = 0, \quad (9.10)$$
$$u(\ell) = 0.$$

(a) Using physical reasoning (as on page 339), determine a formula for the Green's function in the case that k is constant.

(b) Derive the Green's function for the BVP (with a possibly nonconstant k) by integrating the differential equation twice and interchanging the order of integration in the resulting double integral.

(c) Simplify the Green's function found in (b) in the case that the stiffness k is constant. Compare to the result in part (a).

(d) Use the Green's function to solve (9.10) for $\ell = 1$, $f(x) = x$, $k(x) = 1 + x$.

4. Consider the BVP

$$-\frac{d}{dx}\left(k(x)\frac{du}{dx}\right) = f(x), \ 0 < x < \ell,$$
$$u(0) = 0,$$
$$u(\ell) = 0.$$

(a) Using physical reasoning (as in the text), determine a formula for the Green's function in the case that k is constant. (Hint: Since the bar is homogeneous, the displacement function $u(x) = G(x;y)$ (y fixed) will be piecewise linear, with $u(0) = u(\ell) = 0$. Determine the displacement of the cross section at $x = y$ by balancing the forces and using Hooke's law, and then the three values $u(0)$, $u(y)$, and $u(\ell)$ will determine the displacement function.)

(b) Derive the Green's function for the BVP, still assuming that k is constant, by integrating the differential equation twice and interchanging the order of integration in the resulting double integral. Verify that you obtain the same formula for the Green's function.

(c) **(Hard)** Derive the Green's function in the case of a nonconstant stiffness $k(x)$.

5. Since the BVP (9.4) is linear, the principle of superposition holds, and the solution to

$$-\frac{d}{dx}\left(k(x)\frac{du}{dx}\right) = f(x), \ 0 < x < \ell,$$
$$u(0) = 0,$$
$$k(\ell)\frac{du}{dx}(\ell) = p$$

9.1. Green's functions for BVPs in ODEs: Special cases

is $u(x) = w(x) + v(x)$, where w solves (9.4) (hence w is given by (9.5)) and v solves

$$-\frac{d}{dx}\left(k(x)\frac{du}{dx}\right) = 0,\ 0 < x < \ell,$$
$$u(0) = 0, \qquad (9.11)$$
$$k(\ell)\frac{du}{dx}(\ell) = p.$$

Solve (9.11) and show that the solution can be written in terms of the Green's function (9.6).

6. Solve the BVP

$$-\frac{d}{dx}\left(k(x)\frac{du}{dx}\right) = 0,\ 0 < x < \ell,$$
$$u(0) = a,$$
$$k(\ell)\frac{du}{dx}(\ell) = 0,$$

and show that the solution can be written in terms of the Green's function G defined by (9.6).

7. Prove directly that (9.9) holds. (Hint: Write

$$G(x;y) = \begin{cases} \int_0^y \frac{dz}{k(z)}, & y < x, \\ \int_0^x \frac{dz}{k(z)}, & x < y, \end{cases}$$

and rewrite

$$\int_0^\ell G(x;y)(Ku)(y)\,dy$$

as the sum of two integrals,

$$\int_0^x G(x;y)(Ku)(y)\,dy + \int_x^\ell G(x;y)(Ku)(y)\,dy.$$

Then apply integration by parts to the first integral, and simplify.)

8. Let $G(x;y)$ be the Green's function for the BVP

$$-k\frac{d^2u}{dx^2} = f(x),\ 0 < x < \ell,$$
$$u(0) = 0,$$
$$u(\ell) = 0.$$

The purpose of this problem is to derive the Fourier sine series of $G(x;y)$ in three different ways. Since G depends on the parameter y, its Fourier sine coefficients will be functions of y.

(a) First compute the Fourier sine series of $G(x; y)$ directly (that is, using the formula for the Fourier sine coefficients of a given function), using the formula for G as determined in Exercise 4.

(b) Next, derive the Fourier sine series of $G(x; y)$ as follows: Write down the Fourier series solution to the BVP. Pass the summation through the integral defining the Fourier coefficients of the right-hand side f. Identify the Fourier sine series of the Green's function $G(x; y)$.

(c) Find the Fourier sine series of $G(x; y)$ by solving the BVP

$$-k\frac{d^2u}{dx^2} = \delta(x-y),\ 0 < x < \ell,$$
$$u(0) = 0,$$
$$u(\ell) = 0$$

using the Fourier series method. Use the sifting property of the delta function to determine its Fourier coefficients.

Verify that all three methods give the same result.

9.2 Green's functions for BVPs in ODEs: The symmetric case

The examples in the previous section (including those in the exercises) could be handled directly by integration. In this section and the next, we discuss differential equations for which explicit integration is not possible. The general second-order linear ODE has the form

$$p(x)\frac{d^2u}{dx^2} + q(x)\frac{du}{dx} + r(x)u = f(x),\ a < x < b.$$

Such an ODE is called *regular* if $p(x) \neq 0$ for all $x \in [a,b]$; this implies in particular that p has a single sign on $[a,b]$. We will restrict ourselves to the case of regular ODEs, and since we will eventually be interested in eigenvalues and eigenvectors, we will introduce a negative sign and consider the equation in the form

$$-p(x)\frac{d^2u}{dx^2} + q(x)\frac{du}{dx} + r(x)u = f(x),\ a < x < b, \qquad (9.12)$$

where $p(x) > 0$ for all $x \in [a,b]$. Then the familiar equation

$$-k\frac{d^2u}{dx^2} = f(x),\ 0 < x < \ell,$$

$k > 0$, will fit naturally into the notation as a special case. We also assume that p, q, and r are all continuous on $[a,b]$.

In the special case that $q = -dp/dx$, we have

$$-p(x)\frac{d^2u}{dx^2} + q(x)\frac{du}{dx} + r(x)u = -\frac{d}{dx}\left(p(x)\frac{du}{dx}\right) + r(x)u.$$

9.2. Green's functions for BVPs in ODEs: The symmetric case

In such a case, the corresponding differential operator is symmetric. In the next section, we show how to transform the general equation (9.12) into

$$-\frac{d}{dx}\left(P(x)\frac{du}{dx}\right) + R(x)u = F(x), \; a < x < b, \quad (9.13)$$

where P is positive and belongs to $C^1[a,b]$. Therefore, in this section, we will find the Green's function for (9.13), subject to suitable boundary conditions.

The method presented here will not produce the Green's function explicitly in terms of P and R, as would be ideal, but rather in terms of solutions to the corresponding homogeneous ODE,

$$-\frac{d}{dx}\left(P(x)\frac{du}{dx}\right) + R(x)u = 0, \; a < x < b. \quad (9.14)$$

For this reason, we might as well deal with a general set of boundary conditions:

$$\alpha_1 u(a) + \alpha_2 \frac{du}{dx}(a) = 0,$$

$$\beta_1 u(b) + \beta_2 \frac{du}{dx}(b) = 0.$$

The only requirement is that α_1 and α_2 not be both zero, and similarly that β_1 and β_2 are not both zero. This set of boundary conditions then encompasses the three cases of most interest: If $\alpha_2 = 0$, the boundary condition at $x = a$ is a Dirichlet condition; if $\alpha_1 = 0$, it is a Neumann condition; and if α_1 and α_2 are both nonzero, then it is a Robin condition (see Exercise 5.1.11). Similar considerations hold at $x = b$. Moreover, if we define

$$V = \left\{ v \in C^2[a,b] \; : \; \alpha_1 v(a) + \alpha_2 \frac{dv}{dx}(a) = 0, \beta_1 v(b) + \beta_2 \frac{dv}{dx}(b) = 0 \right\} \quad (9.15)$$

and define $L : V \to C[a,b]$ by

$$Lu = -\frac{d}{dx}\left(P(x)\frac{du}{dx}\right) + R(x)u, \quad (9.16)$$

then L is a symmetric operator (see Exercise 9.2.4).

9.2.1 Derivation of the Green's function

For the reasons given above, we consider the BVP

$$-\frac{d}{dx}\left(P(x)\frac{du}{dx}\right) + R(x)u = F(x), \; a < x < b, \quad (9.17a)$$

$$\alpha_1 u(a) + \alpha_2 \frac{du}{dx}(a) = 0, \quad (9.17b)$$

$$\beta_1 u(b) + \beta_2 \frac{du}{dx}(b) = 0. \quad (9.17c)$$

Since P is assumed to be nonzero on $[a,b]$, we can apply standard theory of ODEs to conclude that any initial value problem for the ODE (9.17a) has a unique solution that exists on $[a,b]$; in this context, an initial value problem specifies the values of u and du/dx at the same point. We will use this fact below.

Let v_1, v_2 be two linearly independent solutions of the homogeneous equation (9.14). For the moment, we will not specify the initial conditions that v_1 and v_2 satisfy, but we will wait to see what conditions we need.

The reader may recall that $\{v_1, v_2\}$ is linearly independent if the Wronskian

$$W(x) = \begin{vmatrix} v_1(x) & v_2(x) \\ \frac{dv_1}{dx}(x) & \frac{dv_2}{dx}(x) \end{vmatrix} = v_1(x)\frac{dv_2}{dx}(x) - v_2(x)\frac{dv_1}{dx}(x)$$

is nonzero at any $x \in [a,b]$. Moreover, since v_1 and v_2 are solutions of the same linear homogeneous differential equation, $W(x) \neq 0$ at every $x \in [a,b]$ if and only if this is true for a single x (cf. Section 4.1.3). Thus we can say that $\{v_1, v_2\}$ is linearly independent if and only if $W(x) \neq 0$ for all $x \in [a,b]$.

We use the method of variation of parameters (cf. Section 4.2.2) to solve the inhomogeneous ODE (9.13) in terms of v_1 and v_2. We define u by

$$u(x) = c_1(x)v_1(x) + c_2(x)v_2(x)$$

and determine c_1, c_2 from the equations

$$v_1 \frac{dc_1}{dx} + v_2 \frac{dc_2}{dx} = 0,$$
$$\frac{dv_1}{dx}\frac{dc_1}{dx} + \frac{dv_2}{dx}\frac{dc_2}{dx} = -\frac{F}{P}.$$

By Cramer's rule, the solution is

$$\frac{dc_1}{dx} = \frac{\begin{vmatrix} 0 & v_2 \\ -\frac{F}{P} & \frac{dv_2}{dx} \end{vmatrix}}{\begin{vmatrix} v_1 & v_2 \\ \frac{dv_1}{dx} & \frac{dv_2}{dx} \end{vmatrix}} = \frac{v_2}{PW}F,$$

$$\frac{dc_2}{dx} = \frac{\begin{vmatrix} v_1 & 0 \\ \frac{dv_1}{dx} & -\frac{F}{P} \end{vmatrix}}{\begin{vmatrix} v_1 & v_2 \\ \frac{dv_1}{dx} & \frac{dv_2}{dx} \end{vmatrix}} = -\frac{v_1}{PW}F.$$

We now show that the function $w(x) = P(x)W(x)$ is constant by showing that dw/dx is identically zero:

$$\frac{dw}{dx}(x) = \frac{d}{dx}(P(x)W(x)) = \frac{d}{dx}\left(P(x)\frac{dv_2}{dx}(x)v_1(x) - P(x)\frac{dv_1}{dx}(x)v_2(x)\right)$$
$$= \frac{d}{dx}\left(P(x)\frac{dv_2}{dx}(x)\right)v_1(x) + P(x)\frac{dv_2}{dx}(x)\frac{dv_1}{dx}(x)$$
$$- \frac{d}{dx}\left(P(x)\frac{dv_1}{dx}(x)\right)v_2(x) - P(x)\frac{dv_1}{dx}(x)\frac{dv_2}{dx}(x)$$
$$= R(x)v_2(x)v_1(x) - R(x)v_1(x)v_2(x)$$
$$= 0.$$

9.2. Green's functions for BVPs in ODEs: The symmetric case

The reader should notice how the differential equation (9.17a) was used in the next-to-last step:

$$-\frac{d}{dx}\left(P(x)\frac{dv_1}{dx}\right) + R(x)v_1 = 0 \Rightarrow \frac{d}{dx}\left(P(x)\frac{dv_1}{dx}\right) = R(x)v_1,$$

and similarly for v_2. Since dw/dx is the zero function, w is a constant function, and since $P(x) > 0$ and $W(x) \neq 0$ for all $x \in [a,b]$, it follows that $k = 1/w(x)$ is a nonzero constant. We now obtain

$$\frac{dc_1}{dx} = kv_2 F, \quad \frac{dc_2}{dx} = -kv_1 F.$$

We must integrate dc_1/dx and dc_2/dx. It turns out that the boundary conditions are easier to impose if we integrate dc_1/dx from a to x and dc_2/dx from x to b:

$$c_1(x) = \int_a^x kv_2(y)F(y)\,dy + a_1,$$

$$c_2(x) = \int_x^b kv_1(y)F(y)\,dy + a_2.$$

Then

$$u(x) = v_1(x)\int_a^x kv_2(y)F(y)\,dy + v_2(x)\int_x^b kv_1(y)F(y)\,dy + a_1 v_1(x) + a_2 v_2(x).$$

Since u was constructed by the method of variation of parameters and hence satisfies

$$\frac{du}{dx}(x) = c_1(x)\frac{dv_1}{dx}(x) + c_2(x)\frac{dv_2}{dx}(x),$$

we also have

$$\frac{du}{dx}(x) = \frac{dv_1}{dx}(x)\left(a_1 + \int_a^x kv_2(y)F(y)\,dy\right) + \frac{dv_2}{dx}(x)\left(a_2 + \int_x^b kv_1(y)F(y)\,dy\right).$$

We now have a solution u to the inhomogeneous ODE (9.13), which is defined in terms of two solutions v_1 and v_2 of the corresponding homogeneous ODE (9.14). We still have freedom to impose initial conditions on v_1 and v_2 and to choose the constants a_1 and a_2 so that u will satisfy the desired boundary conditions. Let us consider the boundary condition at $x = a$. We have

$$u(a) = v_2(a)\int_a^b kv_1(y)F(y)\,dy + a_1 v_1(a) + a_2 v_2(a),$$

$$\frac{du}{dx}(a) = \frac{dv_2}{dx}(a)\int_a^b kv_1(y)F(y)\,dy + a_1 \frac{dv_1}{dx}(a) + a_2 \frac{dv_2}{dx}(a),$$

and thus

$$\alpha_1 u(a) + \alpha_2 \frac{du_2}{dx}(a) = \alpha_1 \left(v_2(a) \int_a^b k v_1(y) F(y) dy + a_1 v_1(a) + a_2 v_2(a) \right)$$
$$+ \alpha_2 \left(\frac{dv_2}{dx}(a) \int_a^b k v_1(y) F(y) dy + a_1 \frac{dv_1}{dx}(a) + a_2 \frac{dv_2}{dx}(a) \right)$$
$$= a_1 \left(\alpha_1 v_1(a) + \alpha_2 \frac{dv_1}{dx}(a) \right)$$
$$+ \left(a_2 + \int_a^b k v_1(y) F(y) dy \right) \left(\alpha_1 v_2(a) + \alpha_2 \frac{dv_2}{dx}(a) \right).$$

If we choose

$$a_2 = - \int_a^b k v_1(y) F(y) dy$$

and choose v_1 to be a solution of (9.14) that satisfies

$$\alpha_1 v_1(a) + \alpha_2 \frac{dv_1}{dx}(a) = 0,$$

then u satisfies the desired boundary condition at $x = a$. It is possible to impose this condition on v_1 by defining v_1 to be the solution of the IVP

$$-\frac{d}{dx}\left(P(x) \frac{dv}{dx} \right) + R(x)v = 0, \ v(a) = \alpha_2, \ \frac{dv}{dx}(a) = -\alpha_1.$$

Since α_1 and α_2 cannot both be zero, it follows that v_1 is a nonzero solution to the ODE, and

$$\alpha_1 v_1(a) + \alpha_2 \frac{dv_1}{dx}(a) = \alpha_1 \alpha_2 + \alpha_2 (-\alpha_1) = 0.$$

By similar reasoning, if we define

$$a_1 = - \int_a^b k v_2(y) F(y) dy,$$

and if v_2 satisfies

$$v_2(b) = \beta_2, \ \frac{dv_2}{dx}(b) = -\beta_1,$$

then u satisfies the desired boundary condition at $x = b$.

We can definitely choose v_1 and v_2 to satisfy the initial conditions given above, and therefore the boundary conditions

$$\alpha_1 v_1(a) + \alpha_2 \frac{dv_1}{dx}(a) = 0, \ \beta_1 v_2(b) + \beta_2 \frac{dv_2}{dx}(b) = 0.$$

However, we cannot ignore the crucial assumption that $\{v_1, v_2\}$ must be linearly independent. What happens if, in defining v_1, v_2 by the given initial conditions, $\{v_1, v_2\}$ turns out to be

9.2. Green's functions for BVPs in ODEs: The symmetric case

linearly dependent? This would imply that v_1 is just a multiple of v_2, from which it easily follows that v_1 satisfies both boundary conditions:

$$\alpha_1 v_1(a) + \alpha_2 \frac{dv_1}{dx}(a) = 0, \ \beta_1 v_1(b) + \beta_2 \frac{dv_1}{dx}(b) = 0.$$

Then v_1 is a nontrivial solution of the homogeneous equation $Lu = 0$ (that is, of the BVP (9.17) where $F = 0$). But this implies that the solution to $Lu = F$, if it exists, cannot be unique; if u is one solution, then $u + sv_1$ is another solution for every real number s (cf. the discussion in Section 3.2.2). In seeking the Green's function for (9.17), we are essentially looking for the inverse of L, which only exists if $Lu = F$ has a unique solution for every $F \in C[a,b]$.

We therefore assume that $Lu = 0$ has only the trivial solution, in which case the functions v_1, v_2 must be linearly independent. This is no restriction, since if it were not true, then the Green's function would not exist anyway.

Let us now manipulate the formula for the solution u to reveal the Green's function. Using the values of a_1, a_2 determined above, we have

$$u(x) = v_1(x)\left(a_1 + \int_a^x kv_2(y)F(y)dy\right) + v_2(x)\left(a_2 + \int_x^b kv_1(y)F(y)dy\right)$$

$$= v_1(x)\left(-\int_a^b kv_2(y)F(y)dy + \int_a^x kv_2(y)F(y)dy\right)$$

$$+ v_2(x)\left(-\int_a^b kv_1(y)F(y)dy + \int_x^b kv_1(y)F(y)dy\right)$$

$$= v_1(x)\left(-\int_x^b kv_2(y)F(y)dy\right) + v_2(x)\left(-\int_a^x kv_1(y)F(y)dy\right)$$

$$= \int_a^b G(x;y)F(y)dy,$$

where

$$G(x;y) = \begin{cases} -kv_1(y)v_2(x), & a \leq y \leq x, \\ -kv_1(x)v_2(y), & x \leq y \leq b. \end{cases} \quad (9.18)$$

Recalling that k is the constant $1/(P(x)W(x))$, we finally obtain

$$G(x;y) = \begin{cases} -\frac{v_1(y)v_2(x)}{P(x)W(x)}, & a \leq y \leq x, \\ -\frac{v_1(x)v_2(y)}{P(x)W(x)}, & x \leq y \leq b. \end{cases} \quad (9.19)$$

The reader should keep in mind that the quantity $P(x)W(x)$ is a constant. Also, v_1 is the solution of the homogeneous version of the ODE that satisfies the initial conditions $v_1(a) = \alpha_2$, $dv_1/dt(a) = -\alpha_1$, while v_2 is the solution satisfying $v_2(b) = \beta_2$, $dv_2/dt(b) = -\beta_1$.

The above derivation shows that G is the desired Green's function (since the resulting u is the unique solution of the BVP (9.17)). We now examine G from another point of view. The discussion in the last section suggests that, for a fixed y, $u(x) = G(x;y)$ should be the solution of

$$-\frac{d}{dx}\left(P(x)\frac{du}{dx}\right) + R(x)u = \delta(x-y), \quad a < x < b, \tag{9.20a}$$

$$\alpha_1 u(a) + \alpha_2 \frac{du}{dx}(a) = 0, \tag{9.20b}$$

$$\beta_1 u(b) + \beta_2 \frac{du}{dx}(b) = 0. \tag{9.20c}$$

By finding v_1 and v_2 that satisfy the homogeneous ODE (9.14) and such that v_1 and v_2 satisfy the boundary conditions on the left and the right, respectively, and "gluing" them together to obtain

$$g(x;y) = \begin{cases} v_1(y)v_2(x), & a \le y \le x, \\ v_1(x)v_2(y), & x \le y \le b, \end{cases}$$

we find that $u(x) = g(x;y)$ satisfies

$$-\frac{d}{dx}\left(P(x)\frac{du}{dx}\right) + R(x)u = 0, \quad a < x < b, \, x \ne y.$$

Also, $u(x) = g(x;y)$ satisfies both boundary conditions (9.20b) and (9.20c).

To see how the delta function arises, we note that while $g(x;y)$ is continuous at $x = y$, the first derivative $\partial g/\partial x(x;y)$ has a jump discontinuity at $x = y$. Exercise 9.2.7 asks the reader to show that the derivative of a function with a jump discontinuity is a delta function.

For the reasons given in the last two paragraphs, g would be a plausible guess for the Green's function. Substituting it into the ODE shows that it is off by the constant $-1/(P(x)W(x))$.

Example 9.2. Consider the BVP

$$-\frac{d}{dx}\left(x^2 \frac{du}{dx}\right) + 6u = F(x), \quad 1 < x < 3, \tag{9.21a}$$

$$u(1) = 0, \tag{9.21b}$$

$$u(3) = 0. \tag{9.21c}$$

In this example, $P(x) = x^2$, $R(x) = 6$, $a = 1$, and $b = 3$. The ODE is an Euler equation (see Section 4.2.5). By substituting $u(x) = x^m$ into the homogeneous ODE, we find two linearly independent solutions

$$u_1(x) = x^2, \quad u_2(x) = x^{-3}$$

of (9.21a). The general solution is therefore $v(x) = c_1 u_1(x) + c_2 u_2(x)$, from which we determine that

$$v_1(x) = \frac{1 - x^5}{5x^3}, \quad v_2(x) = \frac{243 - x^5}{15x^3}$$

solves the ODE subject to the initial conditions

$$v_1(1) = 0, \quad \frac{dv_1}{dx}(1) = -1$$

9.2. Green's functions for BVPs in ODEs: The symmetric case

and
$$v_2(3) = 0, \quad \frac{dv_2}{dx}(3) = -1,$$
respectively. A direct calculation shows that $P(x)W(x) = 242/15$, where W is the Wronskian of v_1 and v_2, and thus $k = 15/242$.

Therefore, the Green's function for the BVP (9.21) is
$$G(x;y) = \begin{cases} -kv_1(y)v_2(x), & 0 \leq y \leq x, \\ -kv_1(x)v_2(y), & x \leq y \leq 2 \end{cases}$$
$$= \begin{cases} -\frac{(1-y^5)(243-x^5)}{1210 y^3 x^3}, & 0 \leq y \leq x, \\ -\frac{(1-x^5)(243-y^5)}{1210 x^3 y^3}, & x \leq y \leq 2. \end{cases}$$

Figure 9.2 displays a graph of $z = G(x; 1.5)$. The reader should notice the corner at $x = 1.5$, which indicates the expected discontinuity in the first derivative.

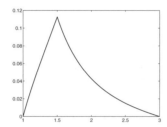

Figure 9.2. *The Green's function $z = G(x;y)$, for $y = 1.5$, from Example 9.2.*

The solution to (9.21) is
$$u(x) = \int_1^3 G(x;y) F(y) \, dy.$$

For example, if $F(x) = x$, then[42]
$$u(x) = \int_1^3 G(x;y) y \, dy = -\frac{81 - 121 x^4 + 40 x^5}{484 x^3}.$$

A graph of the solution is shown in Figure 9.3. ∎

One use of Green's functions is simply to solve inhomogeneous BVPs, as in the previous example. This assumes that we can find the Green's function explicitly, which is not always the case. Another, perhaps more important, use is to analyze the spectral properties (that is, facts about the eigenvalues and eigenfunctions) of the BVP. The operator M mapping F to u defined by
$$u(x) = \int_a^b G(x;y) F(y) \, dy$$

---
[42]The integrals were computed using *Mathematica*.

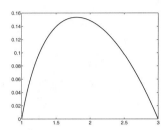

Figure 9.3. *The solution $u = u(x)$ to Example 9.2.*

is an example of an *integral operator*, and much is known about the eigenvalues and eigenfunctions of such operators. Moreover, this theory can be derived without knowing explicitly the Green's function for a given problem. For example, since $G(x;y)$ is symmetric (that is, $G(x;y) = G(y;x)$ for all x,y), it follows that M is a symmetric linear operator:

$$(Mf,g) = (f,Mg),$$

where $(\cdot,\cdot)$ represents the $L^2(a,b)$ inner product. It follows from the discussion in Section 5.1 that all eigenvalues of M are real, as are the corresponding eigenfunctions, and that eigenfunctions corresponding to distinct eigenvalues are orthogonal. Moreover, because M is an integral operator, it can be proved that M has an infinite sequence of eigenvalues converging to zero. Since M is the inverse of the differential operator L that defines the BVP, these and other results translate into information about the spectral properties of L.

The spectral theory of M and L will be presented in Chapter 10, which discusses the *Sturm–Liouville eigenvalue problem*.

9.2.2 Properties of the Green's function; inhomogeneous boundary conditions

We have an explicit formula (9.18) for the Green's function G of the BVP (9.17). Examination of this formula confirms that $G(y;x) = G(x;y)$ holds for all x, y in (a,b), which shows that the reciprocity property holds for G. The reader will recall from the previous section that this means that, for a system modeled by (9.17), the response at x to a concentrated source at y is the same as the response at y to a concentrated source at x.

We earlier remarked that the differential operator L (defined by (9.16)) is symmetric, meaning that

$$(u,Lv) = (Lu,v)$$

for all u, v satisfying the homogeneous boundary condition (as usual, $(\cdot,\cdot)$ represents the L^2 inner product). If we do not assume that u and v satisfy homogeneous boundary conditions,

9.2. Green's functions for BVPs in ODEs: The symmetric case

we have

$$(Lu,v) = P(a)\frac{du}{dx}(a)v(a) - P(b)\frac{du}{dx}(b)v(b)$$
$$+ \int_a^b \left\{ P(x)\frac{du}{dx}(x)\frac{dv}{dx}(x) + R(x)u(x)v(x) \right\} dx,$$

$$(u,Lv) = P(a)u(a)\frac{dv}{dx}(a) - P(b)u(b)\frac{dv}{dx}(b)$$
$$+ \int_a^b \left\{ P(x)\frac{du}{dx}(x)\frac{dv}{dx}(x) + R(x)u(x)v(x) \right\} dx,$$

and therefore

$$(u,Lv) - (Lu,v) = P(a)\left(u(a)\frac{dv}{dx}(a) - \frac{du}{dx}(a)v(a) \right) - P(b)\left(u(b)\frac{dv}{dx}(b) - \frac{du}{dx}(b)v(b) \right). \tag{9.22}$$

We now show that this formula allows us to solve a BVP with inhomogeneous boundary conditions.

For simplicity, we will examine the case of Dirichlet boundary conditions (which means that $\alpha_1 = 1$, $\alpha_2 = 0$, $\beta_1 = 1$, $\beta_2 = 0$). We let $v(x) = G(x;y)$ for a fixed $y \in (a,b)$, and we define u to be the solution of

$$-\frac{d}{dx}\left(P(x)\frac{du}{dx} \right) + R(x)u(x)v(x) = 0, \ a < x < b,$$
$$u(a) = u_a, \tag{9.23}$$
$$u(b) = u_b.$$

We now compute the various terms in (9.22). First, since $Lv = \delta(x-y)$, we see that

$$(u,Lv) = \int_a^b u(x)\delta(x-y)dx = u(y).$$

Since u solves the homogeneous ODE, $Lu = 0$, and therefore $(Lu,v) = 0$. We also have $u(a) = u_a$, $u(b) = u_b$, $v(a) = 0$, $v(b) = 0$, and therefore

$$P(a)\left(u(a)\frac{dv}{dx}(a) - \frac{du}{dx}(a)v(a) \right) - P(b)\left(u(b)\frac{dv}{dx}(b) - \frac{du}{dx}(b)v(b) \right)$$
$$= P(a)u_a\frac{dv}{dx}(a) - P(b)u_b\frac{dv}{dx}(b).$$

With $v(x) = G(x;y)$ and taking advantage of the symmetry of G, we see that

$$\frac{dv}{dx}(a) = \frac{\partial G}{\partial x}(a;y) = \frac{\partial G}{\partial y}(y;a), \ \frac{dv}{dx}(b) = \frac{\partial G}{\partial x}(b;y) = \frac{\partial G}{\partial y}(y;b).$$

Therefore, (9.22) simplifies to

$$u(y) = P(a)u_a\frac{\partial G}{\partial y}(y;a) - P(b)u_b\frac{\partial G}{\partial y}(y;b). \tag{9.24}$$

This is the solution to (9.23). With G defined by (9.18), we have

$$\frac{\partial G}{\partial y}(x;y) = \begin{cases} -k\frac{dv_1}{dx}(y)v_2(x), & a \le y < x, \\ -kv_1(x)\frac{dv_2}{dx}(y), & x < y \le b, \end{cases}$$

and therefore

$$\frac{\partial G}{\partial y}(x;a) = -k\frac{dv_1}{dx}(a)v_2(x), \quad \frac{\partial G}{\partial y}(x;b) = -kv_1(x)\frac{dv_2}{dx}(b).$$

Both of these functions satisfy the homogeneous ODE, and therefore so does the linear combination (9.24). Verification of the boundary conditions proceeds as follows:

$$u(a) = P(a)u_a \frac{\partial G}{\partial y}(a;a) - P(b)u_b \frac{\partial G}{\partial y}(a;b)$$

$$= P(a)u_a \left(-\frac{\frac{dv_1}{dx}(a)v_2(a)}{P(a)W(a)} \right) - P(b)u_b \left(-\frac{v_1(a)\frac{dv_2}{dx}(b)}{P(a)W(a)} \right).$$

Since $v_1(a) = 0$ and

$$W(a) = v_1(a)\frac{dv_2}{dx}(a) - \frac{dv_1}{dx}(a)v_2(a) = -\frac{dv_1}{dx}(a)v_2(a),$$

we see that $u(a)$ simplifies to

$$u(a) = P(a)u_a \left(-\frac{\frac{dv_1}{dx}(a)v_2(a)}{P(a)\left(-\frac{dv_1}{dx}(a)v_2(a)\right)} \right) - P(b)u_b \left(-\frac{0 \cdot \frac{dv_2}{dx}(b)}{P(a)W(a)} \right) = u_a.$$

The boundary condition $u(b) = u_b$ is verified in the same fashion.

In the exercises, the reader is asked to derive formulas for the solution of

$$-\frac{d}{dx}\left(P(x)\frac{du}{dx}\right) + R(x)u(x)v(x) = 0, \, a < x < b$$

subject to inhomogeneous Neumann conditions or Robin conditions. In these cases as well, the solution can be expressed in terms of the Green's function (9.18).

Incidentally, the reader may wonder about the validity of applying (9.22), which was derived for smooth u and v, with $v(x) = G(x;y)$ (since G is not a smooth function). However, the result is still valid since, in the case of $v(x) = G(x;y)$ (or any similar function v with a singularity in the first derivative), the expression (u, Lv) is *defined* by

$$(u, Lv) = P(a)u(a)\frac{dv}{dx}(a) - P(b)u(b)\frac{dv}{dx}(b)$$

$$+ \int_a^b \left\{ P(x)\frac{du}{dx}(x)\frac{dv}{dx}(x) + R(x)u(x)v(x) \right\} dx.$$

Verifying that the right-hand side of this expression equals $u(y)$ is essentially a part of Exercise 9.2.9.

Exercises

1. Find the Green's function for the BVP

$$-\frac{d^2u}{dx^2} - \theta^2 u = f(x), \ 0 < x < 1,$$
$$u(0) = 0,$$
$$u(1) = 0.$$

2. Find the Green's function for the BVP

$$-\frac{d^2u}{dx^2} - \pi^2 u = f(x), \ 0 < x < 1,$$
$$u(0) = 0,$$
$$\frac{du}{dx}(1) = 0.$$

3. Find the Green's function for the BVP

$$-\frac{d^2u}{dx^2} + \theta^2 u = f(x), \ 0 < x < 1,$$
$$u(0) = 0,$$
$$u(1) = 0.$$

4. Prove that the operator $L : V \to C[a,b]$ defined by (9.16) and (9.15) is symmetric, that is, prove that

$$(Lu, v) = (u, Lv) \text{ for all } u, v \in V,$$

where $(\cdot, \cdot)$ represents the L^2 inner product.

5. Let $G = G(x,y)$ be any continuous function of two variables, $a \leq x \leq b, a \leq y \leq b$, and define $M : C[a,b] \to C[a,b]$ by

$$(Mf)(x) = \int_a^b G(x,y) f(y) \, dy.$$

Prove that if G is symmetric in the sense that $G(y,x) = G(x,y)$ for all $x, y \in [a,b]$, then M is a symmetric operator with respect to the $L^2(a,b)$ inner product.

6. Using the methods of this section, derive the Green's function for the BVP (9.4) from Section 9.1. Show that the result is the same as was obtained by direct integration.

7. Suppose ψ, ϕ are smooth functions of a real variable, and $\phi(x) = 0$ for all x such that $|x|$ is sufficiently large (that is, there exists some $R > 0$ such that $\phi(x) = 0$ if $x > R$ or $x < -R$). We will denote the space of all such functions ϕ by $\mathcal{D}(\mathbf{R})$. Integration by parts shows that

$$\int_{-\infty}^{\infty} \frac{d\psi}{dx}(x) \phi(x) \, dx = -\int_{-\infty}^{\infty} \psi(x) \frac{d\phi}{dx}(x) \, dx$$

(notice how the boundary term vanishes because of the assumption on ϕ). Moreover, $d\psi/dx$ can be defined by this relationship, in the following sense: If ψ is differentiable and there exists a function q satisfying

$$\int_{-\infty}^{\infty} q(x)\phi(x)\,dx = -\int_{-\infty}^{\infty} \psi(x)\frac{d\phi}{dx}(x)\,dx \text{ for all } \phi \mathcal{D}(\mathbf{R}), \quad (9.25)$$

then $q = d\psi/dx$. We can extend this definition to functions for which the derivative does not exist in the classical sense (for example, to certain ψ not belonging to $C^1(\mathbf{R})$). If ψ and q are functions defined on $\mathbf{R}$ and satisfying (9.25), then we say that q is the *weak derivative* of ψ.

Define $H : \mathbf{R} \to \mathbf{R}$ by

$$H(x) = \begin{cases} 0, & x < 0, \\ 1, & 0 < x \end{cases}$$

(H is called the *Heaviside* function). Prove that the weak derivative of H is δ by proving that

$$\int_{-\infty}^{\infty} \delta(x)\phi(x)\,dx = -\int_{-1}^{1} H(x)\frac{d\phi}{dx}(x)\,dx$$

holds for all $\phi \in C^1[-1,1]$ satisfying $\phi(-1) = \phi(1) = 0$. Recall that the first integral in this equation is defined by the sifting property of δ.

8. Suppose that f is a smooth function and H is the Heaviside function (see the previous exercise). Prove that

$$\frac{d}{dx}(H(x)f(x)) = f(0)\delta(x) + H(x)\frac{df}{dx}(x),$$

where the derivative on the left is the weak derivative of Hf. (Since f is smooth by assumption, df/dx is unambiguous—the classical derivative and weak derivative of f are equal.)

9. Prove that $u(x) = G(x; y)$, where G is defined by (9.19), solves the BVP (9.20). Recall from the discussion on page 340 that the BVP must be interpreted in the weak sense.

10. By proceeding as in Section 9.2.2, find the solution of

$$-\frac{d}{dx}\left(P(x)\frac{du}{dx}\right) + R(x)u(x) = 0, \ a < x < b,$$

$$\frac{du}{dx}(a) = v_a,$$

$$\frac{du}{dx}(b) = v_b$$

in terms of the Green's function (9.18).

9.3. Green's functions for BVPs in ODEs: The general case

11. By proceeding as in Section 9.2.2, find the solution of

$$-\frac{d}{dx}\left(P(x)\frac{du}{dx}\right) + R(x)u(x) = 0, \ a < x < b,$$

$$\alpha_1 u(a) + \alpha_2 \frac{du}{dx}(a) = v_a,$$

$$\beta_1 u(b) + \beta_2 \frac{du}{dx}(b) = v_b,$$

in terms of the Green's function (9.18). Assume that $\alpha_1, \alpha_2, \beta_1, \beta_2$ are all nonzero.

9.3 Green's functions for BVPs in ODEs: The general case

We will now extend the results of the previous section to the general second-order BVP

$$-p(x)\frac{d^2 u}{dx^2} + q(x)\frac{du}{dx} + r(x)u = f(x), \ a < x < b, \tag{9.26a}$$

$$\alpha_1 u(a) + \alpha_2 \frac{du}{dx}(a) = 0, \tag{9.26b}$$

$$\beta_1 u(b) + \beta_2 \frac{du}{dx}(b) = 0. \tag{9.26c}$$

We assume that p, q, and r are continuous, and that $p(x) > 0$ for all $x \in [a,b]$.

Our approach will be to transform (9.26a) into the form (9.17a). We first divide through by the positive function p to obtain

$$-\frac{d^2 u}{dx^2} + \frac{q(x)}{p(x)}\frac{du}{dx} + \frac{r(x)}{p(x)}u = \frac{f(x)}{p(x)}, \ a < x < b.$$

We then multiply through by a positive function P, which remains to be determined, yielding

$$-P(x)\frac{d^2 u}{dx^2} + \frac{q(x)P(x)}{p(x)}\frac{du}{dx} + \frac{r(x)P(x)}{p(x)}u = \frac{f(x)P(x)}{p(x)}, \ a < x < b.$$

If this last equation is to have the desired form, then

$$\frac{dP}{dx}(x) = -\frac{q(x)}{p(x)}P(x)$$

must hold. This is a first-order linear ODE that can be solved for P; a solution is

$$P(x) = e^{g(x)},$$

where g is any antiderivative of $-q/p$; that is,

$$\frac{dg}{dx}(x) = -\frac{q(x)}{p(x)}.$$

With this choice of P, we see that (9.26a) is equivalent to

$$-\frac{d}{dx}\left(P(x)\frac{du}{dx}\right) + R(x) = F(x),\ a < x < b,$$

where

$$P(x) = e^{g(x)},\ R(x) = \frac{e^{g(x)}}{p(x)}r(x),\ F(x) = \frac{e^{g(x)}}{p(x)}f(x).$$

Moreover, the boundary conditions imposed on u are not affected by our manipulations, so (9.26) is equivalent to

$$-\frac{d}{dx}\left(P(x)\frac{du}{dx}\right) + R(x)u = F(x),\ a < x < b,$$
$$\alpha_1 u(a) + \alpha_2 \frac{du}{dx}(a) = 0, \qquad (9.27)$$
$$\beta_1 u(b) + \beta_2 \frac{du}{dx}(b) = 0.$$

For later use, we will write $w(x) = e^{g(x)}/p(x)$, and we note that $F(x) = w(x)f(x)$, and that $w(x) > 0$ for all $x \in [a,b]$.

In the last section, we saw how to determine the Green's function for (9.27), yielding a formula for the solution u in terms of F, and hence f:

$$u(x) = \int_a^b G(x;y)F(y)dy = \int_a^b G(x;y)f(y)w(y)dy.$$

The reader will recall that G is determined by two special solutions of the homogeneous differential equation. It should be noted that these solutions can be found from either of the equivalent equations,

$$-p(x)\frac{d^2u}{dx^2} + q(x)\frac{du}{dx} + r(x)u = 0$$

or

$$-\frac{d}{dx}\left(P(x)\frac{du}{dx}\right) + R(x)u = 0.$$

We will now present an example.

Example 9.3. Consider the BVP

$$-x^2\frac{d^2u}{dx^2} + x\frac{du}{dx} + u = f(x),\ 1 < x < 2, \qquad (9.28a)$$
$$u(1) = 0, \qquad (9.28b)$$
$$\frac{du}{dx}(2) = 0. \qquad (9.28c)$$

In terms of the notation adopted above, we have $p(x) = x^2$, $q(x) = x$, and $r(x) = 1$, with $a = 1$, $b = 2$. To transform (9.28) into an ODE defined by a symmetric differential operator, we must calculate an antiderivative g of

$$-\frac{q(x)}{p(x)} = -\frac{1}{x};$$

9.3. Green's functions for BVPs in ODEs: The general case

a suitable choice is $g(x) = -\ln x$. We then obtain

$$P(x) = e^{g(x)} = x^{-1}, \ R(x) = \frac{r(x)P(x)}{p(x)} = x^{-3}, \ w(x) = \frac{P(x)}{p(x)} = x^{-3}.$$

The ODE (9.28) is then equivalent to

$$-\frac{d}{dx}\left(P(x)\frac{du}{dx}\right) + R(x)u = w(x)f(x), \ 1 < x < 2. \tag{9.29}$$

The homogeneous version of (9.29) (or (9.28)) is an Euler equation (see Example 9.2 in Section 9.2). We solve by substituting $u(x) = x^m$ and find that two linearly independent solutions are

$$u_1(x) = x^{m_1}, \ u_2(x) = x^{m_2},$$

where $m_1 = -1 + \sqrt{2}, m_2 = 1 + \sqrt{2}$. Next, we solve the IVPs defining the special solutions v_1 and v_2, each of which can be written as a linear combination of u_1 and u_2:

$$v_1(1) = 0, \ \frac{dv_1}{dx}(1) = -1 \Rightarrow v_1(x) = \frac{1}{2\sqrt{2}}\left(x^{m_1} - x^{m_2}\right),$$

$$v_2(2) = 1, \ \frac{dv_2}{dx}(2) = 0 \Rightarrow v_2(x) = 2^{-5/2+\sqrt{2}}m_2 x^{m_1} - 2^{-5/2-\sqrt{2}}m_1 x^{m_2}.$$

Finally, we find that

$$P(x)W(x) = 2^{-5/2-\sqrt{2}}\left(\sqrt{2} + 2^{1/2+2\sqrt{2}} + 4^{\sqrt{2}} - 1\right) \doteq 1.1649.$$

We can then define

$$G(x;y) = \begin{cases} -\frac{v_1(y)v_2(x)}{P(x)W(x)}, & a \leq y \leq x, \\ -\frac{v_1(x)v_2(y)}{P(x)W(x)}, & x \leq y \leq b, \end{cases}$$

and we have an explicit, if somewhat messy, expression for the Green's function. The solution to (9.28) is then

$$u(x) = \int_1^2 G(x;y)f(y)w(y)dx. \quad \blacksquare$$

We can define operators $L : V \to C[a,b]$ and $M : C[a,b] \to V$ by

$$(Lu)(x) = -p(x)\frac{d^2u}{dx^2}(x) + q(x)\frac{du}{dx}(x) + r(x)u(x)$$

and

$$(Mf)(x) = \int_a^b G(x;y)f(y)w(y)dy,$$

respectively, where

$$V = \left\{v \in C^2[a,b] \ : \ \alpha_1 v(a) + \alpha_2 \frac{dv}{dx}(a) = 0, \beta_1 v(b) + \beta_2 \frac{dv}{dx}(b) = 0\right\},$$

$w(x) = e^{g(x)}/p(x)$, and g is an antiderivative of $-q/p$. The fact that, for each $f \in C[a,b]$, $u = Mf$ is the unique solution to $Lu = f$ means that M is the inverse of L. This is very similar to the situation in the last section. However, the operators L and M defined by the BVP (9.26) are not symmetric with respect to the $L^2(a,b)$ inner product. This would suggest that L might have complex eigenvalues, and that corresponding eigenfunctions could be nonorthogonal.

It turns out, though, that L and M *are* symmetric operators if we use an alternate inner product. Let us define

$$(f,g)_w = \int_a^b f(x)\overline{g(x)}w(x)\,dx.$$

In Exercises 9.3.6 and 9.3.7, the reader is asked to show that

$$(Lu,v)_w = (u,Lv)_2$$

for all u, v in complex $C^2[a,b]$ that satisfy the homogeneous boundary conditions, and that

$$(Mf,g)_w = (f,Mg)_w$$

for all f,g in complex $C[a,b]$. (The reader will recall that, in analyzing the eigenvalues of a symmetric operator, we initially assume that the eigenvalues and eigenfunctions might be complex, and therefore pose the problem in a complex inner product space. After proving that, in fact, the eigenvalues must be real numbers, we can restrict ourselves to the real number setting.) From this it follows that the eigenvalues of L are all real and eigenfunctions corresponding to distinct eigenvalues are orthogonal with respect to $(\cdot,\cdot)_w$.

Exercises

1. Find the Green's function for the BVP

$$-\frac{d^2u}{dx^2} + 3\frac{du}{dx} - 2u = f(x),\ 0 < x < 1,$$
$$u(0) = 0,$$
$$u(1) = 0.$$

2. Find the Green's function for the BVP

$$-\frac{d^2u}{dx^2} + 2\frac{du}{dx} - u = f(x),\ 0 < x < 1,$$
$$u(0) = 0,$$
$$u(1) = 0.$$

3. Find the Green's function for the BVP

$$-\frac{d^2u}{dx^2} + 2\frac{du}{dx} - 5u = f(x),\ 0 < x < 1,$$
$$u(0) = 0,$$
$$u(1) = 0.$$

9.3. Green's functions for BVPs in ODEs: The general case

4. Consider the BVP

$$-x^2 \frac{d^2u}{dx^2} - 3x\frac{du}{dx} - u = f(x),\ a < x < 1,$$
$$u(a) = 0,$$
$$u(1) = 0.$$

Find the Green's function for this BVP and solve the BVP for $f(x) = x^2$.

5. Consider the BVP

$$-x\frac{d^2u}{dx^2} + (x+3)\frac{du}{dx} - 2u = f(x),\ a < x < 1,$$
$$u(a) = 0,$$
$$\frac{du}{dx}(1) = 0.$$

Find the Green's function for this BVP and solve the BVP for $f(x) = x$. Two solutions to the homogeneous version of the ODE are

$$u_1(x) = e^{x+1},\ u_2(x) = (x^2 + 4x + 5)/2.$$

6. Define

$$S = \left\{ u \in C^2[a,b] : \alpha_1 u(a) + \alpha_2 \frac{du}{dx}(a) = 0,\ \beta_1 u(b) + \beta_2 \frac{du}{dx}(b) = 0 \right\},$$

and let $L : S \to C[a,b]$ be defined by

$$Lu = -p(x)\frac{d^2u}{dx^2} + q(x)\frac{du}{dx} + r(x)u$$

(here $C[a,b]$ and $C^2[a,b]$ are to be interpreted as spaces of complex-valued functions). Let $w(x) = e^{g(x)}/p(x)$, where g is an antiderivative of $-q/p$. Prove that

$$(Lu, v)_w = (u, Lv)_w \text{ for all } u, v \in S.$$

7. Let G by the Green's function derived in this section, and let M be defined by

$$(Mf)(x) = \int_a^b G(x;y) f(y) w(y)\, dy.$$

(Here M maps complex $C[a,b]$ into complex $C^2[a,b]$.) Prove that

$$(Mf, g)_w = (f, Mg)_w \text{ for all } f, g \in C[a,b].$$

9.4 Introduction to Green's functions for IVPs

The solution to an IVP for a linear ODE can be written in the form

$$u(t) = \int_0^\infty G(t;s)f(s)\,ds,$$

where f is the right-hand side of the ODE. The function G is called the *causal* Green's function for the problem; it satisfies $G(t;s) = 0$ for $s > t$, which explains the adjective *causal*—the value of u at t is caused by the values of $f(s)$ for $s \leq t$.[43]

In this section, we will introduce causal Green's functions for the simple ODEs that were considered in Section 4.2 and interpret it as a special solution of the differential equation.

9.4.1 The Green's function for first-order linear ODEs

The solution to

$$\frac{du}{dt} - au = f(t),\ t > 0,$$
$$u(0) = 0,$$
(9.30)

as derived in Section 4.2, is

$$u(t) = \int_0^t e^{a(t-s)} f(s)\,ds. \tag{9.31}$$

If we define

$$G(t;s) = \begin{cases} e^{a(t-s)}, & t > s, \\ 0, & t < s, \end{cases} \tag{9.32}$$

then we can write the formula for the solution u as

$$u(t) = \int_0^\infty G(t;s)f(s)\,ds. \tag{9.33}$$

The function G, as defined by (9.32), is the causal Green's function for (9.30). If we include a nonzero initial condition, say $u(0) = u_0$, then the solution becomes

$$u(t) = G(t;0)u_0 + \int_0^\infty G(t;s)f(s)\,ds. \tag{9.34}$$

Formula (9.34) indicates the significance of G. The effect of the initial datum u_0 on the solution u at time t is $G(t;0)u_0$, while the effect of the datum $f(s)$ on the solution u at

---

[43]Thus, we could define $G(t;s)$ only for $s \leq t$, and write $u(t) = \int_0^t G(t;s)f(s)\,ds$. However, as we will see, the Green's function, as it is usually presented (with a discontinuity at $s = t$), has an interpretation as a special solution of the given differential equation.

9.4. Introduction to Green's functions for IVPs

time t is $G(t;s)f(s)ds$.[44] We will expand on this interpretation below by showing that G itself satisfies the inhomogeneous ODE, where the right-hand side is a delta function.

The reader should recall that (9.31) was presented in Section 4.3.3 as an example of Duhamel's principle. As we will see, Duhamel's principle is invaluable for constructing Green's functions for IVPs and IBVPs.

Here is another example, this time involving a nonconstant-coefficient ODE. The solution to

$$\frac{du}{dt} - a(t)u = 0,$$
$$u(s) = u_0$$

is $u(t) = e^{A(t)-A(s)} u_0$, where A is an antiderivative of a (see Section 4.2.4). If we define

$$G(t;s) = \begin{cases} e^{A(t)-A(s)}, & t > s, \\ 0, & t < s, \end{cases}$$

then the solution of

$$\frac{du}{dt} - a(t)u = f(t),$$
$$u(0) = u_0$$

is

$$u(t) = G(t;0)u_0 + \int_0^\infty G(t;s)f(s)\,ds.$$

9.4.2 The Green's function for higher-order ODEs

As we discussed in Section 4.3.3, Duhamel's principle can be used to construct solutions for inhomogeneous ODEs of order greater than one, provided it is possible to solve the homogeneous version. If $u(t) = v(t;s)$ solves

$$\frac{d^{(k)}u}{dt} + a_{k-1}\frac{d^{(k-1)}u}{dt} + \cdots + a_1\frac{du}{dt} + a_0 u = 0,$$
$$u(0) = 0,$$
$$\frac{du}{dt}(0) = 0,$$
$$\vdots$$
$$\frac{d^{(k-2)}u}{dt}(0) = 0,$$
$$\frac{d^{(k-1)}u}{dt}(0) = f(s),$$

(9.35)

---

[44]The data represented by the function f have different units than that represented by u_0. Indeed, an examination of the differential equation shows that the units of f are the units of u_0 divided by time. Therefore, f is a rate, and it must be multiplied by the time "interval" ds prior to being multiplied by $G(t;s)$.

then the solution to

$$\frac{d^{(k)}u}{dt} + a_{k-1}\frac{d^{(k-1)}u}{dt} + \cdots + a_1 \frac{du}{dt} + a_0 u = f(t),$$
$$u(0) = 0,$$
$$\frac{du}{dt}(0) = 0,$$
$$\vdots$$
$$\frac{d^{(k-2)}u}{dt}(0) = 0,$$
$$\frac{d^{(k-1)}u}{dt}(0) = 0$$

(9.36)

is

$$u(t) = \int_0^t v(t-s; s) \, ds. \tag{9.37}$$

The reader should recall that this construction must be modified in the case of an ODE with nonconstant coefficients (specifically, the initial conditions in (9.35) must be specified at initial time $t = s$). From (9.37) (or the equivalent formula in the case of nonconstant coefficients), the Green's function can be identified. We will now present two examples.

A second-order, constant-coefficient example

In Section 4.2.3, we saw that the solution to

$$\frac{d^2 u}{dt^2} + \theta^2 u = f(t),$$
$$u(0) = 0, \tag{9.38}$$
$$\frac{du}{dt}(0) = 0$$

is

$$u(t) = \frac{1}{\theta} \int_0^t \sin(\theta(t-s)) f(s) \, ds.$$

Therefore, the (causal) Green's function for (9.38) is

$$G(t;s) = \begin{cases} \frac{\sin(\theta(t-s))}{\theta}, & t > s, \\ 0, & t < s. \end{cases} \tag{9.39}$$

The reader can verify that the solution to (9.38), with the initial conditions changed to

$$u(0) = u_0, \quad \frac{du}{dt}(0) = v_0,$$

is

$$u(t) = \frac{\partial G}{\partial t}(t;0) u_0 + G(t;0) v_0 + \int_0^\infty G(t;s) f(s) \, ds$$

(see Exercise 9.4.1).

9.4. Introduction to Green's functions for IVPs

A second-order, nonconstant-coefficient example

As an example of a problem with nonconstant coefficients, we consider the ODE

$$\frac{d^2u}{dt^2} + \frac{5}{t}\frac{du}{dt} + \frac{4}{t^2}u = f(t).$$

We must first solve the homogeneous IVP

$$\frac{d^2u}{dt^2} + \frac{5}{t}\frac{du}{dt} + \frac{4}{t^2}u = 0,$$
$$u(s) = 0, \quad (9.40)$$
$$\frac{du}{dt}(s) = c,$$

where we assume $s > 0$. The ODE is an Euler equation, which can be solved by standard methods (Section 4.2.5). The general solution is

$$u(t) = c_1 t^{-2} + c_2 t^{-2} \ln(t).$$

Applying the initial conditions yields the following solution to (9.40):

$$u(t) = \frac{s^3}{t^2} \ln\left(\frac{t}{s}\right) c.$$

We therefore define

$$G(t;s) = \begin{cases} \frac{s^3}{t^2} \ln\left(\frac{t}{s}\right), & t > s > 0, \\ 0, & 0 < t < s, \end{cases}$$

where now t and s are restricted to positive values. The solution of

$$\frac{d^2u}{dt^2} + \frac{5}{t}\frac{du}{dt} + \frac{4}{t^2}u = f(t),$$
$$u(t_0) = 0, \quad (9.41)$$
$$\frac{du}{dt}(t_0) = 0$$

is

$$u(t) = \int_{t_0}^{\infty} G(t;s) f(s) \, ds = \int_{t_0}^{\infty} \frac{s^3}{t^2} \ln\left(\frac{t}{s}\right) f(s) \, ds. \quad (9.42)$$

9.4.3 Interpretation of the causal Green's function

As in the case of a Green's function for a BVP, the causal Green's function is the solution of the inhomogeneous differential equation with a delta function as the right-hand side. As an example, we will show that the Green's function

$$G(t;s) = \begin{cases} e^{A(t)-A(s)}, & t > s, \\ 0, & t < s, \end{cases}$$

where $A'(t) = a(t)$, satisfies

$$\frac{\partial G}{\partial t}(t;s) - a(t)G(t;s) = \delta(t-s),$$

$$u(t_0) = 0,$$

provided $t_0 < s$. To prove this, it will be convenient to use the Heaviside function H defined by

$$H(t) = \begin{cases} 0, & t < 0, \\ 1, & t > 0. \end{cases}$$

We can then write $G(t;s) = H(t-s)S(t;s)$, where $S(t;s) = e^{A(t)-A(s)}$. We recall that S satisfies the homogeneous differential equation,

$$\frac{\partial S}{\partial t}(t;s) - a(t)S(t;s) = 0,$$

and $S(s;s) = 1$. It can be shown (see Exercise 9.2.8) that

$$\frac{\partial G}{\partial t}(t;s) = \frac{\partial}{\partial t}(H(t-s)S(t;s)) = S(s;s)\delta(t-s) + H(t-s)\frac{\partial S}{\partial t}(t;s)$$

$$= \delta(t-s) + H(t-s)\frac{\partial S}{\partial t}(t;s).$$

Therefore,

$$\frac{\partial G}{\partial t}(t;s) - a(t)G(t;s) = \delta(t-s) + H(t-s)\frac{\partial S}{\partial t}(t;s) - a(t)H(t-s)S(t;s)$$

$$= \delta(t-s) + H(t-s)\left(\frac{\partial S}{\partial t}(t;s) - a(t)S(t;s)\right)$$

$$= \delta(t-s),$$

as desired. The fact that $G(t_0;s) = 0$ for $t_0 < s$ follows immediately from the definition of G.

We can similarly show that the Green's function for a second-order (or higher-order) equation satisfies the ODE with a delta function for the right-hand side (see Exercises 9.4.9, 9.4.10).

Exercises

1. Let G be defined by (9.39). Prove that the solution of

$$\frac{d^2u}{dt^2} + \theta^2 u = f(t),$$

$$u(0) = u,$$

$$\frac{du}{dt}(0) = v_0$$

is

$$u(t) = \frac{\partial G}{\partial t}(t;0)u_0 + G(t;0)v_0 + \int_0^\infty G(t;s)f(s)\,ds.$$

9.4. Introduction to Green's functions for IVPs

2. Find the Green's function for the following IVP:
$$\frac{du}{dt} + 2u = f(t),$$
$$u(t_0) = u_0.$$

3. Find the Green's function for the IVP
$$\frac{d^2u}{dt^2} + 4u = f(t),$$
$$u(0) = 0,$$
$$\frac{du}{dt}(0) = 0.$$

4. Use the Green's function from the previous exercise to solve
$$\frac{d^2u}{dt^2} + 4u = \cos(t),$$
$$u(0) = 0,$$
$$\frac{du}{dt}(0) = 0.$$

5. Find the Green's function for the IVP
$$\frac{d^2u}{dt^2} + 3\frac{du}{dx} + 2u = f(t),$$
$$u(0) = 0,$$
$$\frac{du}{dt}(0) = 0.$$

6. Find the Green's function for the IVP
$$\frac{d^2u}{dt^2} + 2\frac{du}{dx} + 3u = f(t),$$
$$u(0) = 0,$$
$$\frac{du}{dt}(0) = 0.$$

7. Suppose a certain population grows according to the ODE
$$\frac{dP}{dt} = 0.02P + f(t),$$
where f is the immigration. If the population at the beginning of 1990 is 55.5 million and immigration during the next 10 years is given by the function $f(t) = 1 - 0.2t$ (in millions), where $t = 0$ represents the beginning of 1990, what will the population be at the beginning of 2000 ($t = 10$)? Use the Green's function to find $P(t)$ and then $P(10)$.

8. A certain radioactive isotope decays exponentially according to the law

$$\frac{dm}{dt} = -km,$$

where $k = \ln(2)/2$. Here $m(t)$ is the mass of the isotope at time t (seconds), and the differential equation indicates that a constant fraction of the atoms are disintegrating at each point in time. The above differential equation holds, of course, when none of the isotopes is being added or taken away by other means. Suppose that initially $10\,\mathrm{g}$ of the isotope are present, and mass is added from an external source at the constant rate of $0.1\,\mathrm{g/s}$. Solve the IVP modeling this situation, using the Green's function, and determine the long-time behavior of $m(t)$.

9. Let the solution of

$$\frac{d^2 u}{dt^2} + a_1(t)\frac{du}{dt} + a_0(t)u = 0,$$
$$u(s) = 0,$$
$$\frac{du}{dt}(s) = f(s)$$

be $u(t) = S(t;s)f(s)$, and define the Green's function $G(t;s)$ by

$$G(t;s) = \begin{cases} S(t;s), & t > s, \\ 0, & t < s. \end{cases}$$

Prove that G satisfies

$$\frac{\partial^2 G}{\partial t^2}(t;s) + a_1(t)\frac{\partial G}{\partial t}(t;s) + a_0(t)G(t;s) = \delta(t-s),$$
$$G(t_0;s) = 0,$$
$$\frac{\partial G}{\partial t}(t_0;s) = 0,$$

provided $t_0 < s$. (Hint: Since $S(s;s) = 0$, the weak derivative of $H(t-s)S(t;s)$ is simply $H(t-s)\frac{\partial S}{\partial t}(t;s)$. It is then possible to differentiate a second time.)

10. Generalize the previous exercise to the case of an nth-order linear differential equation with nonconstant coefficients by stating and proving the analogous result.

9.5 Green's functions for the heat equation

We will now discuss Green's functions for the heat equation, both on the entire real line and on a bounded interval. Before we present any specific Green's functions, we describe the form that the Green's function must take. We have so far seen two examples of Green's functions. In an IVP for an ODE, the solution $u = u(t)$ is of the form

$$u(t) = \int_{t_0}^{\infty} G(t;s)f(s)\,ds,$$

9.5. Green's functions for the heat equation

where f is the right-hand side of the ODE (see Section 9.4 for an example). This formula shows the contribution of the data at time s on the solution at time t—this contribution is $G(t;s)f(s)ds$. The reader should notice how the value $u(t)$ is found by "adding up" the contributions to $u(t)$ of the data from the time interval $[t_0, \infty)$. (The value of $G(t;s)$ is zero when $s > t$, so in fact only the data from the time interval $[t_0, t]$ can affect the value of $u(t)$.)

We have also seen examples of Green's function for a steady-state BVP (see Sections 9.1–9.3), where the data (the right-hand side of the differential equation) is given over a spatial interval such as $[a,b]$. In this case, the value $u(x)$ is determined by adding up the contributions from the data at all points in the interval. The solution formula takes the form

$$u(x) = \int_a^b G(x;y) f(y) \, dy.$$

In the case of a time-dependent PDE on the spatial interval $[a,b]$ (that is, in the case of an IBVP), the data $f(x,t)$ (the right-hand side of the PDE) is prescribed over both space and time. Therefore, we will have to add up the contributions over both space and time, and the formula for the solution will involve two integrals:

$$u(x,t) = \int_{t_0}^{\infty} \int_a^b G(x,t;y,s) f(y,s) \, dy \, ds.$$

If we consider the pure IVP on the spatial interval $(-\infty, \infty)$, then the spatial integral will extend over the entire real line:

$$u(x,t) = \int_{t_0}^{\infty} \int_{-\infty}^{\infty} G(x,t;y,s) f(y,s) \, dy \, ds.$$

9.5.1 The Gaussian kernel

The Green's function for the heat equation on the entire real line is best derived using the Fourier transform, which is beyond the scope of this book. Therefore, we will merely present the result.[45] We will write the heat equation in the form

$$\frac{\partial u}{\partial t} - k \frac{\partial^2 u}{\partial x^2} = f(x,t), \; -\infty < x < \infty, \; t > 0,$$

where $k = \kappa/(\rho c)$ and κ, ρ, and c are the thermal conductivity, the density, and the specific heat, respectively. (Thus the usual forcing function f, in units of J/(cm^3 s), say, must be replaced by $f/(\rho c)$, with units of K/s.) We wish to solve

$$\frac{\partial u}{\partial t} - k \frac{\partial^2 u}{\partial x^2} = f(x,t), \; -\infty < x < \infty, \; t > 0, \quad (9.43)$$
$$u(x,0) = 0, \; -\infty < x < \infty.$$

The discussion of Duhamel's principle in Section 4.3.3 suggests that it is sufficient to solve

$$\frac{\partial u}{\partial t} - k \frac{\partial^2 u}{\partial x^2} = 0, \; -\infty < x < \infty, \; t > 0, \quad (9.44)$$
$$u(x,0) = f(x,s), \; -\infty < x < \infty.$$

---
[45] For an elementary but lengthy derivation, the reader can consult [62, Section 2.4].

If the solution to (9.44) is $v(x,t;s)$, then the solution to (9.43) will be

$$u(x,t) = \int_0^t v(x,t-s;s)\,ds. \tag{9.45}$$

We can verify this directly. We have

$$\frac{\partial u}{\partial t}(x,t) = v(x,t-t;t) + \int_0^t \frac{\partial v}{\partial t}(x,t-s;s)\,ds.$$

Since $v(x,t-t;t) = v(x,0;t) = f(x,t)$, this reduces to

$$\frac{\partial u}{\partial t}(x,t) = f(x,t) + \int_0^t \frac{\partial v}{\partial t}(x,t-s;s)\,ds.$$

We also have

$$\frac{\partial^2 u}{\partial x^2}(x,t) = \int_0^t \frac{\partial^2 v}{\partial x^2}(x,t-s;s)\,ds.$$

Therefore,

$$\begin{aligned}\frac{\partial u}{\partial t}(x,t) - k\frac{\partial^2 u}{\partial x^2}(x,t) &= f(x,t) + \int_0^t \frac{\partial v}{\partial t}(x,t-s;s)\,ds \\ &\quad - k\int_0^t \frac{\partial^2 v}{\partial x^2}(x,t-s;s)\,ds \\ &= f(x,t) + \int_0^t \left\{\frac{\partial v}{\partial t}(x,t-s;s) - k\frac{\partial^2 v}{\partial x^2}(x,t-s;s)\right\}ds \\ &= f(x,t),\end{aligned}$$

since v satisfies the homogeneous heat equation by assumption. By the definition of u, it is obvious that $u(x,0) = 0$, and therefore we have verified that u is the desired solution.

As we stated above, we will not attempt to derive the solution of (9.44); it is

$$v(x,t;s) = \int_{-\infty}^{\infty} S(x-y,t)f(y,s)\,dy, \tag{9.46}$$

where

$$S(x,t) = \frac{1}{2\sqrt{\pi k t}} e^{-x^2/(4kt)}. \tag{9.47}$$

The function S is often called the *Gaussian kernel*. To verify that (9.46) is the solution to (9.44), we begin by showing that S itself is a solution to the heat equation. We have

$$\frac{\partial S}{\partial t}(x,t) = \frac{1}{8k^{3/2}\sqrt{\pi}}\left(\frac{x^2 - 2kt}{t^{5/2}}\right)e^{-4x^2/(4kt)}$$

9.5. Green's functions for the heat equation

and

$$\frac{\partial^2 S}{\partial x^2}(x,t) = \frac{1}{8k^{5/2}\sqrt{\pi}}\left(\frac{x^2 - 2kt}{t^{5/2}}\right)e^{-4x^2/(4kt)},$$

which together imply that

$$\frac{\partial S}{\partial t}(x,t) - k\frac{\partial^2 S}{\partial x^2}(x,t) = 0, \ -\infty < x < \infty, \ t > 0.$$

Since the PDE has constant coefficients, we also see that, for every $y \in \mathbf{R}$,

$$\frac{\partial S}{\partial t}(x-y,t) - k\frac{\partial^2 S}{\partial x^2}(x-y,t) = 0, \ -\infty < x < \infty, \ t > 0.$$

Therefore, with u defined by (9.46), we have

$$\frac{\partial u}{\partial t}(x,t) - k\frac{\partial^2 u}{\partial x^2}(x,t) = \int_{-\infty}^{\infty} \frac{\partial S}{\partial t}(x-y,t)f(y,s)\,dy - k\int_{-\infty}^{\infty} \frac{\partial^2 S}{\partial x^2}(x-y,t)f(y,s)\,dy$$
$$= \int_{-\infty}^{\infty} \left\{\frac{\partial S}{\partial t}(x-y,t) - k\frac{\partial^2 S}{\partial x^2}(x-y,t)\right\}f(y,s)\,dy = 0,$$

and thus u solves the homogeneous heat equation.

To verify the initial condition is more difficult, since, strictly speaking, u is undefined for $t = 0$. If the initial condition holds, it must hold in the sense that $u(x,t) \to f(x,s)$ as $t \to 0^+$. We begin by noting that

$$\int_{-\infty}^{\infty} S(x,t)\,dx = 1 \text{ for all } t > 0,$$

and hence also

$$\int_{-\infty}^{\infty} S(x-y,t)\,dy = 1 \text{ for all } x \in \mathbf{R}, \ t > 0$$

(see Exercise 9.5.1). This last result implies that, for all $t > 0$,

$$\int_{-\infty}^{\infty} S(x-y,t)f(y,s)\,dy$$

is a weighted average of the function $f(\cdot,s)$. Moreover, as Figure 9.4 shows, the weighting function $S(x-y,t)$ is concentrated more and more around $y = x$ as t decreases toward zero. It follows that $S(x-y,t)$ approximately equals $\delta(x-y)$ for small t, and it is reasonable to expect (although we do not prove it rigorously) that

$$u(x,t) = \int_{-\infty}^{\infty} S(x-y,t)f(y,s)\,dy \to f(x,s) \text{ as } t \to 0^+.$$

Having found the solution

$$v(x,t;s) = \int_{-\infty}^{\infty} S(x-y,t)f(y,s)\,dy$$

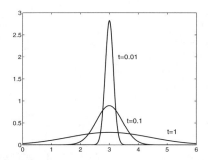

Figure 9.4. *The Gaussian kernel $S(x-3,t)$ for $t=1$, $t=0.1$, and $t=0.01$. The value $k=1$ was used to create these graphs.*

of (9.44), Duhamel's principle yields

$$u(x,t) = \int_0^t v(x,t-s;s)\,ds = \int_0^t \int_{-\infty}^\infty S(x-y,t-s)f(y,s)\,dy\,ds \qquad (9.48)$$

as the solution of (9.43). We can now define the causal Green's function for the heat equation:

$$G(x,t;y,s) = \begin{cases} S(x-y;t-s), & t>s, \\ 0, & t<s \end{cases}$$

$$= \begin{cases} \dfrac{1}{2\sqrt{\pi k(t-s)}} e^{-(x-y)^2/(4k(t-s))}, & t>s, \\ 0, & t<s. \end{cases} \qquad (9.49)$$

It can be shown that, for a fixed (y,s), G satisfies

$$\frac{\partial G}{\partial t} - k\frac{\partial^2 G}{\partial x^2} = \delta(x-y)\delta(t-s),\ -\infty<x<\infty,\ t>0, \qquad (9.50)$$

$$G(x,t;y,s) = 0,\ t<s$$

(see Exercise 9.5.2). The product $\delta(x-y)\delta(t-s)$ of delta functions is defined by

$$\int_0^\infty \int_{-\infty}^\infty \delta(x-y)\delta(t-s)f(x,t)\,dx\,dt = f(y,s)$$

for all smooth f.

9.5.2 The Green's function on a bounded interval

We now compute the Green's function for the IBVP

$$\begin{aligned}
\frac{\partial u}{\partial t} - k\frac{\partial^2 u}{\partial x^2} &= f(x,t),\ 0<x<\ell,\ t>0, \\
u(x,0) &= 0,\ 0<x<\ell, \\
u(0,t) &= 0,\ t>0, \\
u(\ell,t) &= 0,\ t>0
\end{aligned} \qquad (9.51)$$

9.5. Green's functions for the heat equation

(where $k = \kappa/(\rho c)$). The solution to (9.51) was derived in Section 6.1; it is

$$u(x,t) = \sum_{n=1}^{\infty} a_n(t)\sin\left(\frac{n\pi x}{\ell}\right),$$

where

$$a_n(t) = \int_0^t e^{-\frac{kn^2\pi^2}{\ell^2}(t-s)} c_n(s)\,ds$$

and

$$c_n(s) = \frac{2}{\ell}\int_0^\ell f(y,s)\sin\left(\frac{n\pi y}{\ell}\right)dy.$$

We can derive an expression for the Green's functions by manipulating the formula for $u(x,t)$:

$$u(x,t) = \sum_{n=1}^{\infty} a_n(t)\sin\left(\frac{n\pi x}{\ell}\right)$$

$$= \sum_{n=1}^{\infty} \frac{1}{\rho c} \int_0^t e^{-kn^2\pi^2(t-s)/\ell^2} c_n(s)\,ds \sin\left(\frac{n\pi x}{\ell}\right)$$

$$= \sum_{n=1}^{\infty} \int_0^t \int_0^\ell \frac{2}{\ell} e^{-kn^2\pi^2(t-s)/\ell^2} \sin\left(\frac{n\pi y}{\ell}\right)\sin\left(\frac{n\pi x}{\ell}\right) f(y,s)\,dy\,ds$$

$$= \int_0^t \int_0^\ell \left\{\frac{2}{\ell}\sum_{n=1}^{\infty} e^{-kn^2\pi^2(t-s)/\ell^2}\sin\left(\frac{n\pi y}{\ell}\right)\sin\left(\frac{n\pi x}{\ell}\right)\right\} f(y,s)\,dy\,ds.$$

If we define

$$G(x,t;y,s) = \begin{cases} \frac{2}{\ell}\sum_{n=1}^{\infty} e^{-kn^2\pi^2(t-s)/\ell^2}\sin\left(\frac{n\pi y}{\ell}\right)\sin\left(\frac{n\pi x}{\ell}\right), & 0 \leq s \leq t, \\ 0, & s > t, \end{cases} \quad (9.52)$$

then we have

$$u(x,t) = \int_0^\infty \int_0^\ell G(x,t;y,s)f(y,s)\,dy\,ds.$$

Therefore, $G(x,t;y,s)$ is the causal Green's function for (9.51).

The infinite series defining the Green's function converges rapidly if t is significantly larger than s; this is because the coefficients $e^{-kn^2\pi^2(t-s)/\ell^2}$ converge rapidly to zero as $n \to \infty$. Therefore, $G(x,t;y,s)$ can be accurately approximated by a few terms of the series when $t-s$ is large. However, if $t-s$ is small, then the series converges more slowly, and many terms of the series might be required for an accurate approximation. As in the case of the Gaussian kernel S, the infinite series above does not define $G(x,t;y,s)$ for $t = s$; when $t = s$, all of the coefficients equal 1, and the series does not converge. Nevertheless, again analogous to the case of the Gaussian kernel, it can be seen that $G(x,t;y,s)$ converges to $\delta(x-y)$ as $t \to s^+$. This is illustrated in Figure 9.5.

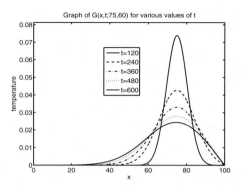

Figure 9.5. *Snapshots of the Green's function in Example 9.4 at times $t = 120, 240, 360, 480, 600$ seconds. Twenty terms of the Fourier series were used to create these graphs.*

As we should expect by now, the Green's function itself is the solution to the IBVP with a delta function as the right-hand side. If $s > 0$ and $0 < y < \ell$, then the solution to

$$\rho c \frac{\partial u}{\partial t} - \kappa \frac{\partial^2 u}{\partial x^2} = \delta(x-y)\delta(t-s), \ 0 < x < \ell, \ t > 0,$$
$$u(x,0) = 0, \ 0 < x < \ell,$$
$$u(0,t) = 0, \ t > 0,$$
$$u(\ell,t) = 0, \ t > 0,$$
(9.53)

is $u(x,t) = G(x,t;y,s)$. The right-hand side $\delta(x-y)\delta(t-s)$ represents a source concentrated entirely at the space-time point (y,s). The IBVP (9.53) describes the following experiment: A bar of length ℓ is initially at temperature 0, and at time $s > 0$, A Joules of energy, where A is the cross-sectional area of the bar,[46] are added to the cross section at $x = y$. The heat energy then flows in accordance with the heat equation. By examining snapshots of $G(x,t;y,s)$, we can get an idea of how the temperature changes with time.

Example 9.4. We consider an iron bar of length $\ell = 100$ cm ($\rho = 7.88$ g/cm^3, $c = 0.437$ J/(g K), $\kappa = 0.836$ W/(cm K)). We graph the solution to (9.53) with $s = 60$ and $y = 75$ (time measured in seconds) in Figure 9.5, which shows how the spike of heat energy diffuses over time. ∎

Formula (9.52) has its shortcomings; for example, if one wishes to compute a snapshot of $G(x,t;y,s)$ for t greater than but very close to s, then many terms of the Fourier series are required to obtain an accurate result (see Exercise 9.5.3). It is possible to obtain a more computationally efficient formula for use when t is very close to s. However, to do so would require a lengthy digression.[47]

---

[46] The reader should recall from Section 2.1 that each term in the heat equation, including the right-hand side, is multiplied by A in the original derivation.

[47] The interested reader can consult Haberman [30, Chapter 11].

9.5. Green's functions for the heat equation

9.5.3 Properties of the Green's function

An inspection of formulas (9.49) and (9.52) reveals that, when G denotes either the free-space or bounded-domain Green's function, the following properties are satisfied:

1. **Causality:** $G(x,t;y,s) = 0$ for $s > t$. The practical meaning of this is that the value of a heat source at time $s > t$ cannot affect the temperature at time t. In other words, the response at time t is caused only by what has happened before time t.

2. **Translation:** $G(x,t;y,s) = G(x,t-s;y,0)$. This implies that the response at time t to a source concentrated at time s depends only on the elapsed time $t-s$ and not on the times t and s themselves. The translation property holds only because the heat equation has constant coefficients.

3. **Reciprocity:** $G(x,t;y,s) = G(y,t;x,s)$. The temperature at the point x, given a heat source concentrated at the point y, is the same as the temperature at y, given a heat source at x.

9.5.4 Green's functions under other boundary conditions

The above method can be used to derive the Green's function for the one-dimensional heat equation under other boundary conditions, such as Neumann or mixed boundary conditions. The calculations are left to the exercises.

Exercises

1. Given the standard result
$$\int_{-\infty}^{\infty} e^{-x^2}\,dx = \sqrt{\pi},$$
compute
$$\int_{-\infty}^{\infty} S(x,t)\,dx$$
and show that the result is 1 for all $t > 0$. Use this result and a simple change of variables to prove that
$$\int_{-\infty}^{\infty} S(x-y,t)\,dy = 1 \text{ for all } y \in \mathbf{R},\ t > 0.$$

2. Using the fact that the causal Green's function for the heat equation on $(-\infty, \infty)$ satisfies $G(x,t;y,s) = H(t-s)S(x-y,t-s)$, and using reasoning similar to that in Section 9.4.3, prove that G satisfies (9.50).

3. Let $G(x,t;y,s)$ be the Green's function from Example 9.4 ($y = 75$, $s = 60$). Suppose one wishes to produce an accurate graph of $G(x,t;y,s)$ for various values of $t > s$. How many terms of the Fourier series are necessary to obtain an accurate graph for $t = 660$ (10 minutes after the heat energy is added)? For $t = 61$ (1 second afterward)? (Hint: Using trial and error, keep adding terms to the Fourier series until the graph no longer changes visibly.)

4. Compute the Green's function for the IBVP

$$\rho c \frac{\partial u}{\partial t} - \kappa \frac{\partial^2 u}{\partial x^2} = f(x,t),\ 0 < x < \ell,\ t > 0,$$
$$u(x,0) = 0,\ 0 < x < \ell,$$
$$\frac{\partial u}{\partial x}(0,t) = 0,\ t > 0,$$
$$\frac{\partial u}{\partial x}(\ell,t) = 0,\ t > 0.$$

Produce a graph similar to Figure 9.5 (use the values of ρ, c, κ, y, and s from Example 9.4).

5. Compute the Green's function for the IBVP

$$\rho c \frac{\partial u}{\partial t} - \kappa \frac{\partial^2 u}{\partial x^2} = f(x,t),\ 0 < x < \ell,\ t > 0,$$
$$u(x,0) = 0,\ 0 < x < \ell,$$
$$u(0,t) = 0,\ t > 0,$$
$$\frac{\partial u}{\partial x}(\ell,t) = 0,\ t > 0.$$

Produce a graph similar to Figure 9.5 (use the values of ρ, c, κ, y, and s from Example 9.4).

6. Compute the Green's function for the IBVP

$$\rho c \frac{\partial u}{\partial t} - \kappa \frac{\partial^2 u}{\partial x^2} = f(x,t),\ 0 < x < \ell,\ t > 0,$$
$$u(x,0) = 0,\ 0 < x < \ell,$$
$$\frac{\partial u}{\partial x}(0,t) = 0,\ t > 0,$$
$$u(\ell,t) = 0,\ t > 0.$$

Produce a graph similar to Figure 9.5 (use the values of ρ, c, κ, y, and s from Exercise 9.4).

7. Let G be the Green's function (9.52) for the IBVP (9.51), and define

$$u(x,t) = \int_0^\ell G(x,t;y,0)\phi(y)\,dy.$$

What IBVP does u satisfy? Use Fourier series to prove your answer.

9.6 Green's functions for the wave equation

In this section, we derive the Green's functions for the wave equation on the entire real line, and on a bounded interval.

9.6.1 The Green's function on the real line

We wish to derive the Green's function for the following IVP:

$$\frac{\partial^2 u}{\partial t^2} - c^2 \frac{\partial^2 u}{\partial x^2} = f(x,t), \quad -\infty < x < \infty, \ t > 0,$$
$$u(x,0) = 0, \quad -\infty < x < \infty, \tag{9.54}$$
$$\frac{\partial u}{\partial t}(x,0) = 0, \quad -\infty < x < \infty.$$

Applying Duhamel's principle, we start with the solution to

$$\frac{\partial^2 u}{\partial t^2} - c^2 \frac{\partial^2 u}{\partial x^2} = 0, \quad -\infty < x < \infty, \ t > 0,$$
$$u(x,0) = 0, \quad -\infty < x < \infty, \tag{9.55}$$
$$\frac{\partial u}{\partial t}(x,0) = f(x,s), \quad -\infty < x < \infty,$$

which we know from D'Alembert's formula (see Section 7.1):

$$v(x,t;s) = \frac{1}{2c} \int_{x-ct}^{x+ct} f(y,s)\,dy. \tag{9.56}$$

We wish to write (9.56) as an integral over $(-\infty, \infty)$, as the resulting form will make it easier to identify the Green's function. We notice that

$$x - ct < y < x + ct \iff -ct < y - x < ct \iff 0 < ct - |x - y|.$$

Using the Heaviside function H (defined by $H(x) = 1$ if $x > 0$ and $H(x) = 0$ otherwise), we see that $x - ct < y < x + ct$ if and only if $H(ct - |x - y|) = 1$, and therefore

$$v(x,t;s) = \frac{1}{2c} \int_{x-ct}^{x+ct} f(y,s)\,dy = \int_{-\infty}^{\infty} \frac{1}{2c} H(ct - |x - y|) f(y,s)\,dy.$$

Now, by Duhamel's principle, the solution of (9.54) is

$$u(x,t) = \int_0^t v(x, t-s; s)\,ds = \int_0^t \int_{-\infty}^{\infty} \frac{1}{2c} H(c(t-s) - |x-y|) f(y,s)\,dy\,ds.$$

Moreover, $c(t-s) - |x-y| < 0$ if $s > t$, and therefore we can also write

$$u(x,t) = \int_0^{\infty} \int_{-\infty}^{\infty} \frac{1}{2c} H(c(t-s) - |x-y|) f(y,s)\,dy\,ds. \tag{9.57}$$

From this, we see that the causal Green's function for (9.54) is

$$G(x,t;y,s) = \frac{1}{2c} H(c(t-s) - |x-y|). \tag{9.58}$$

If we wish to avoid the absolute value sign, we can also write

$$G(x,t;y,s) = \frac{1}{2c} H(c(t-s) - (x-y)) H(c(t-s) + (x-y)). \tag{9.59}$$

The function G is illustrated in Figure 9.6.

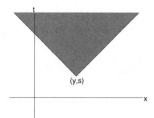

Figure 9.6. *The Green's function (9.58). The value of $G(x,t;y,s)$ is $1/(2c)$ in the shaded region, and 0 elsewhere.*

9.6.2 The Green's function on a bounded interval

Derivation using Fourier series

We will derive two representations of the Green's function for the problem

$$\frac{\partial^2 u}{\partial t^2} - c^2 \frac{\partial^2 u}{\partial x^2} = f(x,t),\ 0 < x < \ell,\ t > 0,$$
$$u(x,0) = 0,\ 0 < x < \ell,$$
$$\frac{\partial u}{\partial t}(x,0) = 0,\ 0 < x < \ell,\ \ \ \ \ \ (9.60)$$
$$u(0,t) = 0,\ t > 0,$$
$$u(\ell,t) = 0,\ t > 0.$$

First we derive a Fourier series representation, as we did in Section 9.5.2 for the heat equation. Using the methods of Section 7.2, the solution of (9.60) is

$$u(x,t) = \sum_{n=1}^{\infty} a_n(t) \sin\left(\frac{n\pi x}{\ell}\right),$$

where

$$a_n(t) = \frac{\ell}{cn\pi} \int_0^t \sin\left(\frac{cn\pi}{\ell}(t-s)\right) c_n(s)\,ds,\ c_n(s) = \frac{2}{\ell} \int_0^\ell f(y,s) \sin\left(\frac{n\pi y}{\ell}\right) dy,$$

or

$$a_n(t) = \frac{2}{cn\pi} \int_0^t \int_0^\ell \sin\left(\frac{n\pi y}{\ell}\right) \sin\left(\frac{cn\pi}{\ell}(t-s)\right) f(y,s)\,dy\,ds.$$

Therefore,

$$u(x,t) = \sum_{n=1}^{\infty} \frac{2}{cn\pi} \int_0^t \int_0^\ell \sin\left(\frac{n\pi y}{\ell}\right) \sin\left(\frac{cn\pi}{\ell}(t-s)\right) f(y,s)\,dy\,ds \sin\left(\frac{n\pi x}{\ell}\right)$$

$$= \int_0^t \int_0^\ell \left\{ \sum_{n=1}^{\infty} \frac{2}{cn\pi} \sin\left(\frac{cn\pi}{\ell}(t-s)\right) \sin\left(\frac{n\pi x}{\ell}\right) \sin\left(\frac{n\pi y}{\ell}\right) \right\} f(y,s)\,dy\,ds.$$

9.6. Green's functions for the wave equation

The causal Green's function for (9.60) is thus

$$G(x,t;y,s) = \begin{cases} \sum_{n=1}^{\infty} \frac{2}{cn\pi} \sin\left(\frac{cn\pi}{\ell}(t-s)\right) \sin\left(\frac{n\pi x}{\ell}\right) \sin\left(\frac{n\pi y}{\ell}\right), & t > s, \\ 0, & t < s. \end{cases} \quad (9.61)$$

The reader should notice that the series representing G does not converge rapidly; this is signaled by the fact that the coefficient is proportional to $1/n$ (as compared with the series representation of the Green's function for the heat equation, whose coefficients decay exponentially with n). For this reason, we expect to need many terms to get an accurate approximation of G (see Figure 9.7, where Gibb's phenomenon is evident).

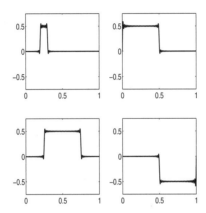

Figure 9.7. *Four snapshots of the causal Green's function $G(x,t;1/4,0)$ for the wave equation on $(0,1)$: $t = 0.05$ (top left), $t = 0.25$ (top right), $t = 0.5$ (bottom left), $t = 1.25$ (bottom right). The wave speed is $c = 1$. These graphs were computed using* 100 *terms of the Fourier series.*

Derivation using the method of images

To find a solution of the inhomogeneous wave equation on the bounded interval $[0, \ell]$ (that is, IBVP (9.60)), the method of images constructs an equivalent IVP that is defined on the entire real line—an instance of (9.54). This is done by extending the right-hand side f (with domain $0 < x < \ell$) to a function $\tilde{f}$ defined for $-\infty < x < \infty$ and satisfying $\tilde{f}(x) = f(x)$ for all $x \in (0, \ell)$.

To see how to define $\tilde{f}$, we consider what we know about the effect of boundaries on solutions to the wave equation. If we have a solution that is zero near the boundary of the interval $(0, \ell)$, then that solution solves the wave equation on the entire real line until it reaches the boundary, at which time it inverts and reflects back into $(0, \ell)$. This is illustrated in Figure 9.8 (see also Example 7.3 in Section 7.2).

We define $\tilde{f}$ to be the odd extension of f across the boundary $x = 0$, meaning that $\tilde{f}(-x, t) = -f(x, t)$ for all $x \in (-\ell, 0)$. This implies that if the source $f(x, t)$, $0 < x < \ell$, results in a left-moving wave reaching the boundary $x = 0$ from the right, then $\tilde{f}(x, t)$, $-\ell < x < 0$, will create a corresponding right-moving wave, of the opposite sign, approaching $x = 0$ from the left. We do the same thing across the boundary $x = \ell$, which amounts to defining $\tilde{f}(x,t) = -f(2\ell - x)$ for $\ell < x < 2\ell$. This implies that $\tilde{f}(x+2\ell,t) = \tilde{f}(x,t)$ for all $x \in (-\ell, 0)$ (see Figure 9.9).

Extending f from $(0,\ell)$ to $(-\ell, 2\ell)$ accounts for the primary reflections of waves starting in the interior of $(0,\ell)$. However, waves will reflect multiple times, as shown in Figure 9.8. We extend $\tilde{f}$ beyond $(-\ell, 2\ell)$ to account for these multiple reflections. A wave that reflects once at the left boundary and again at the right boundary will be inverted twice and hence have its original form (as in Figure 9.8); this can be produced by defining $\tilde{f}(x,t) = f(x - 2\ell, t)$ for $2\ell < x < 3\ell$. Similarly, a wave that reflects first at the right boundary and then at the left can be produced by defining the source $\tilde{f}(x,t) = f(x+2\ell, t)$. Continuing this reasoning to a higher number of reflections leads to a function $\tilde{f}$ like that shown in Figure 9.10. We can describe the desired $\tilde{f}$ as the *odd, periodic extension* of f, meaning that we first extend f from $0 < x < \ell$ to be an odd function defined on $-\ell < x < \ell$, and then extend to a periodic function with period 2ℓ.

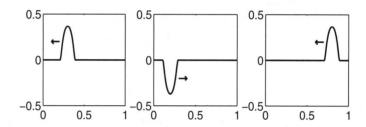

Figure 9.8. *A left-moving wave (left) is reflected at the boundary $x = 0$ (center) and then at the boundary $x = \ell$ (right). Each time the wave reflects, it inverts.*

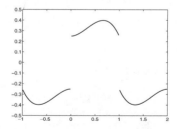

Figure 9.9. *A function f defined on $(0,1)$ and extended as an odd function across $x = 0$ and $x = 1$.*

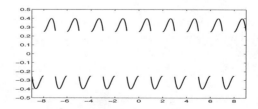

Figure 9.10. *The odd, periodic extension of a function f defined on $(0,1)$.*

9.6. Green's functions for the wave equation

The above discussion is intended to make the following result plausible: If $\tilde{f}$ is the odd, periodic extension of f from $0 < x < \ell$ to $-\infty < x < \infty$, then the solution u of

$$\frac{\partial^2 u}{\partial t^2} - c^2 \frac{\partial^2 u}{\partial x^2} = \tilde{f}(x,t), \quad -\infty < x < \infty, \; t > 0,$$
$$u(x,0) = 0, \quad -\infty < x < \infty, \tag{9.62}$$
$$\frac{\partial u}{\partial t}(x,0) = 0, \quad -\infty < x < \infty,$$

also solves

$$\frac{\partial^2 u}{\partial t^2} - c^2 \frac{\partial^2 u}{\partial x^2} = f(x,t), \; 0 < x < \ell, \; t > 0, \tag{9.63a}$$
$$u(x,0) = 0, \; 0 < x < \ell, \tag{9.63b}$$
$$\frac{\partial u}{\partial t}(x,0) = 0, \; 0 < x < \ell, \tag{9.63c}$$
$$u(0,t) = 0, \; t > 0, \tag{9.63d}$$
$$u(\ell,t) = 0, \; t > 0. \tag{9.63e}$$

We will verify this directly. Let u be the solution of (9.62). From our earlier results, we know that

$$u(x,t) = \int_0^t \int_{-\infty}^{\infty} \frac{1}{2c} H(c(t-s) - |x-y|) \tilde{f}(y,s) \, dy \, ds. \tag{9.64}$$

Since $\tilde{f}$ and f are identical for $0 < x < \ell$, u automatically satisfies the PDE (9.63a), and the initial conditions (9.63b) and (9.63c) are also satisfied by u. It remains only to verify that u satisfies the boundary conditions (9.63d) and (9.63e). We can do this directly from the formula for u. For example,

$$u(0,t) = \int_0^t \int_{-\infty}^{\infty} \frac{1}{2c} H(c(t-s) - |y|) \tilde{f}(y,s) \, dy \, ds$$
$$= \int_0^t \int_{-\infty}^{\infty} \frac{1}{2c} H(c(t-s) - |-y|) \tilde{f}(-y,s) \, dy \, ds$$
$$= -\int_0^t \int_{-\infty}^{\infty} \frac{1}{2c} H(c(t-s) - |y|) \tilde{f}(y,s) \, dy \, ds$$
$$= -u(0,t).$$

The key steps in the above calculation are the change of variables that replaces y by $-y$ and applying the fact that $\tilde{f}$ is an odd function: $\tilde{f}(-y,s) = -\tilde{f}(y,s)$. Since $u(0,t) = -u(0,t)$, it follows that $u(0,t) = 0$. A similar proof shows that $u(\ell,t) = 0$ (see Exercise 9.6.4).

We now manipulate (9.64) to determine the Green's function for (9.63). We begin by writing

$$\tilde{f}(x,t) = \sum_{n=-\infty}^{\infty} f_n(x,t),$$

where
$$f_n(x,t) = \begin{cases} \tilde{f}(x,t), & (2n-1)\ell < x < (2n+1)\ell, \\ 0 & \text{otherwise} \end{cases}$$
$$= \tilde{f}(x,t)H(\ell - |x - 2n\ell|).$$

Although $\tilde{f}$ is written in terms of an infinite series, we need not worry about convergence, since no two of the functions f_n are nonzero at a common point. The same is true for the following manipulations. We now have

$$u(x,t) = \int_0^t \int_{-\infty}^\infty \frac{1}{2c} H(c(t-s) - |x-y|)\tilde{f}(y,s)\,dy\,ds$$

$$= \int_0^t \int_{-\infty}^\infty \frac{1}{2c} H(c(t-s) - |x-y|) \sum_{n=-\infty}^\infty \tilde{f}(y,s)H(\ell - |y - 2n\ell|)\,dy\,ds$$

$$= \sum_{n=-\infty}^\infty \int_0^t \int_{-\infty}^\infty \frac{1}{2c} H(c(t-s) - |x-y|)\tilde{f}(y,s)H(\ell - |y - 2n\ell|)\,dy\,ds$$

$$= \sum_{n=-\infty}^\infty \int_0^t \int_{(2n-1)\ell}^{(2n+1)\ell} \frac{1}{2c} H(c(t-s) - |x-y|)\tilde{f}(y,s)\,dy\,ds.$$

Since $\tilde{f}$ is 2ℓ-periodic, we can rewrite

$$\int_{(2n-1)\ell}^{(2n+1)\ell} \frac{1}{2c} H(c(t-s) - |x-y|)\tilde{f}(y,s)\,dy$$

as an integral over $(-\ell, \ell)$; specifically, replacing y by $y + 2n\ell$ and using the fact that $\tilde{f}(y + 2n\ell, s) = \tilde{f}(y,s)$, we obtain

$$\int_{(2n-1)\ell}^{(2n+1)\ell} \frac{1}{2c} H(c(t-s) - |x-y|)\tilde{f}(y,s)\,dy = \int_{-\ell}^{\ell} \frac{1}{2c} H(c(t-s) - |x-y-2n\ell|)\tilde{f}(y,s)\,dy.$$

Also, since $\tilde{f}$ is odd, we have (replacing y by $-y$)

$$\int_{-\ell}^{0} \frac{1}{2c} H(c(t-s) - |x-y-2n\ell|)\tilde{f}(y,s)\,dy$$

$$= -\int_0^\ell \frac{1}{2c} H(c(t-s) - |x+y-2n\ell|)\tilde{f}(-y,s)\,dy$$

$$= -\int_0^\ell \frac{1}{2c} H(c(t-s) - |x+y-2n\ell|)f(y,s)\,dy$$

(since $\tilde{f}(-y,s) = f(y,s)$ for $-\ell < y < 0$). We can now write

$$u(x,t) = \sum_{n=-\infty}^\infty \Bigg\{ \int_0^t \int_0^\ell H(c(t-s) - |x-y-2nl|)f(y,s)\,dy\,ds$$

$$- \int_0^t \int_0^\ell H(c(t-s) - |x-(-y+2nl)|)f(y,s)\,dy\,ds \Bigg\}$$

$$= \int_0^t \int_0^\ell G(x,t;y,s)f(y,s)\,dy\,ds,$$

9.6. Green's functions for the wave equation

where

$$G(x,t;y,s) = \sum_{n=-\infty}^{\infty} \frac{1}{2c} \{H(c(t-s) - |x-y-2n\ell|) - H(c(t-s) - |x+y-2n\ell|)\}$$

(9.65)

is the causal Green's function for (9.63).

Figure 9.11 displays several snapshots of $G(x,t;y,s)$, as computed from (9.65). The reader should compare this to Figure 9.7.

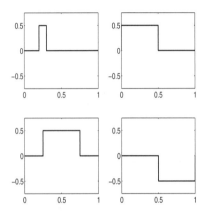

Figure 9.11. *Four snapshots of the causal Green's function $G(x,t;1/4,0)$ for the wave equation on $(0,1)$: $t = 0.05$ (top left), $t = 0.25$ (top right), $t = 0.5$ (bottom left), $t = 1.25$ (bottom right). The wave speed is $c = 1$. These graphs were computed using formula (9.65) derived by the method of images.*

The reader should notice from formulas (9.58) and (9.61) that the Green's functions for the wave equation satisfy the properties of causality, translation, and reciprocity, as do the Green's functions for the heat equation (see Section 9.5.3).

Exercises

1. Prove directly (that is, by computing $\partial^2 u/\partial t^2$ and $\partial^2 u/\partial x^2$ and substituting into the IVP) that

$$u(x,t) = \int_0^t \int_{x-c(t-s)}^{x+c(t-s)} f(y,s) \, dy \, ds$$

   solves (9.54).

2. Prove that (9.58) and (9.59) are equivalent.

3. Prove that the Green's function (9.58) can also be written as

$$G(x,t;y,s) = \frac{1}{2c} \{H((x-y) + c(t-s)) - H((x-y) - c(t-s))\} H(t-s).$$

4. Use formula (9.64) for the solution of (9.62) to prove that $u(\ell,t) = 0$.

5. Prove that the solution u of (9.62) is an odd function: $u(-x,t) = -u(x,t)$.

6. The causal Green's function (9.65) for (9.63) is defined by an infinite sum. Nevertheless, for fixed $y \in (0,\ell)$, s, and $t > s$, all but a finite number of the terms in the sum are identically zero. Given $y \in (0,\ell)$, s, and $t > s$, which terms in the sum (9.65) can be nonzero?

7. Use the method of images to find the causal Green's function for the IBVP

$$\frac{\partial^2 u}{\partial t^2} - c^2 \frac{\partial^2 u}{\partial x^2} = f(x,t),\ 0 < x < \ell,\ t > 0,$$
$$u(x,0) = 0,\ 0 < x < \ell,$$
$$\frac{\partial u}{\partial t}(x,0) = 0,\ 0 < x < \ell,$$
$$\frac{\partial u}{\partial x}(0,t) = 0,\ t > 0,$$
$$\frac{\partial u}{\partial x}(\ell,t) = 0,\ t > 0.$$

(Hint: Follow the procedure in Section 9.6.2, but define $\tilde{f}$ to be the even, periodic extension of f.)

8. Use the method of images to find the causal Green's function for the IBVP

$$\frac{\partial^2 u}{\partial t^2} - c^2 \frac{\partial^2 u}{\partial x^2} = f(x,t),\ 0 < x < \ell,\ t > 0,$$
$$u(x,0) = 0,\ 0 < x < \ell,$$
$$\frac{\partial u}{\partial t}(x,0) = 0,\ 0 < x < \ell,$$
$$\frac{\partial u}{\partial x}(0,t) = 0,\ t > 0,$$
$$u(\ell,t) = 0,\ t > 0.$$

9.7 Suggestions for further reading

Haberman [30] presents Green's functions for both stationary and time-dependent problems at an introductory level. More advanced references include the books by Stakgold [58, 59], Roach [55], and Duffy [18].

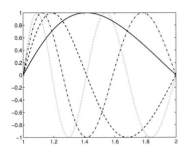

Chapter 10
Sturm–Liouville Eigenvalue Problems

10.1 Introduction

In this chapter, we will address problems of the form

$$-\frac{d}{dx}\left(P(x)\frac{du}{dx}\right) + R(x)u = \lambda w(x)u, \ a < x < b, \tag{10.1}$$

where P and w are positive functions on $[a,b]$, suitable boundary conditions are imposed, and both u and λ are unknown. Such a problem is called a *Sturm–Liouville* eigenvalue problem. It can be expressed in the form

$$Lu = \lambda u, \ Lu = \frac{1}{w(x)}\left(-\frac{d}{dx}\left(P(x)\frac{du}{dx}\right) + R(x)u\right),$$

and thus we are seeking the eigenvalues and eigenfunctions of the operator L. The simplest example of (10.1) is the equation

$$-k\frac{d^2u}{dx^2} = \lambda u, \ 0 < x < \ell, \tag{10.2}$$

which we have already studied extensively. Equation (10.2) is the special case of (10.1) in which $P(x) = k > 0$, $R(x) = 0$, and $w(x) = 1$. The operator appearing in (10.2) has a sequence of eigenvalues $\lambda_1 < \lambda_2 < \cdots$, with $\lambda_n \to \infty$ as $n \to \infty$, and corresponding eigenfunctions $\psi_1, \psi_2, \ldots$ (the values of λ_n and ψ_n depend on the particular boundary conditions imposed). A critical fact is that the eigenfunctions are mutually orthogonal, and any square integrable function f defined on $(0,\ell)$ can be expressed in terms of the eigenfunctions:

$$f(x) = \sum_{n=1}^{\infty} c_n \psi_n(x), \ c_n = \frac{(f, \psi_n)}{(\psi_n, \psi_n)}. \tag{10.3}$$

Such an expression is an *eigenfunction expansion* of the function f.

In this chapter, we will show that the operator L described above has similar properties to the much simpler operator defined by the negative second derivative operator.

In particular, there is an infinite sequence of eigenvalues, tending to infinity, and a corresponding orthogonal sequence of eigenfunctions. It is possible to represent an arbitrary square-integrable function by an eigenfunction expansion. As we will see, if w is not a constant, then orthogonality holds in a weighted inner product, but this detail does not cause any serious difficulties.

10.1.1 How Sturm–Liouville problems arise

Before we proceed to develop the properties of the operator L defined above, we will explain two sources of Sturm–Liouville eigenvalue problems. First, we consider the initial-boundary value problem (IBVP)

$$w(x)\frac{\partial u}{\partial t} - \frac{\partial}{\partial x}\left(P(x)\frac{\partial u}{\partial x}\right) + R(x)u = F(x,t),\ a < x < b,\ t > 0,$$

$$u(x,0) = u_0(x),\ a < x < b, \qquad (10.4)$$

$$u(a,t) = 0,\ t > 0,$$

$$u(b,t) = 0,\ t > 0.$$

If we take $w(x) = \rho(x)c(x)$, $P(x) = \kappa(x)$, and $R(x) = 0$, then the PDE is the heat equation

$$\rho(x)c(x)\frac{\partial u}{\partial t} - \frac{\partial}{\partial x}\left(\kappa(x)\frac{\partial u}{\partial x}\right) = F(x,t).$$

However, the reader should notice that this is the heat equation for a physically heterogeneous material, for example, a bar in which the density, heat capacity, and thermal conductivity vary along the length of the bar.

We wish to solve (10.4) using an eigenfunction expansion, analogous to the Fourier (sine) series expansion used in Section 2.1. We begin by dividing both sides of the PDE by $w(x)$, which is assumed to be a strictly positive function:

$$\frac{\partial u}{\partial t} - \frac{1}{w(x)}\frac{\partial}{\partial x}\left(P(x)\frac{\partial u}{\partial x}\right) + \frac{R(x)}{w(x)}u = \frac{F(x,t)}{w(x)},\ a < x < b,\ t > 0. \qquad (10.5)$$

In Section 6.1, we developed a Fourier series method for solving the simpler equation

$$\rho c\frac{\partial u}{\partial t} - \kappa\frac{d^2 u}{dx^2} = f(x,t),\ 0 \leq x \leq \ell,\ t > 0,$$

$$u(x,0) = u_0(x),\ 0 \leq x \leq \ell,$$

$$u(0,t) = 0,\ t > 0,$$

$$u(\ell,t) = 0,\ t > 0.$$

The solution was represented in the form

$$u(x,t) = \sum_{n=1}^{\infty} a_n(t)\sin\left(\frac{n\pi x}{\ell}\right).$$

The functions

$$\sin\left(\frac{n\pi x}{\ell}\right),\ n = 1,2,3,\ldots,$$

10.1. Introduction

are the eigenfunctions for the spatial part of the partial differential operator,

$$-\kappa \frac{d^2}{dx^2},$$

under the given boundary conditions (Dirichlet). To apply an analogous method to (10.5), we must find the eigenfunctions of the spatial operator

$$L(u) = \frac{1}{w(x)} \left(-\frac{d}{dx}\left(P(x)\frac{du}{dx} \right) + R(x)u \right), \tag{10.6}$$

under the given boundary conditions, $u(a) = u(b) = 0$. This is a Sturm–Liouville eigenvalue problem. The reader should notice that we are confronted with this problem when we try to solve certain PDEs with nonconstant coefficients by the method of eigenfunction expansion. The nonconstant coefficients reflect physical heterogeneity in the situation modeled by the PDE. This is a common source of Sturm–Liouville eigenvalue problems.

We also encounter Sturm–Liouville problems when solving a general BVP of the form

$$-p(x)\frac{d^2 u}{dx^2} + q(x)\frac{du}{dx} + r(x)u = f(x),\ a < x < b,$$
$$u(a) = 0, \tag{10.7}$$
$$u(b) = 0,$$

where p is assumed to be positive on the interval $[a,b]$. In Section 9.3, we showed how to transform this equation into one that leads to a symmetric operator. We now review this transformation. We find an antiderivative g of $-q/p$:

$$\frac{dg}{dx}(x) = -\frac{q(x)}{p(x)}.$$

We then define

$$w(x) = \frac{e^{g(x)}}{p(x)}$$

and multiply both sides of the ODE by $w(x)$ to get

$$-w(x)p(x)\frac{d^2 u}{dx^2} + q(x)w(x)\frac{du}{dx} + w(x)r(x)u = w(x)f(x).$$

The point of this transformation is that

$$\frac{d}{dx}(w(x)p(x)) = \frac{d}{dx}(e^{g(x)}) = g'(x)e^{g(x)} = -\frac{q(x)}{p(x)}e^{g(x)} = -w(x)q(x),$$

and therefore

$$-w(x)p(x)\frac{d^2 u}{dx^2} + q(x)w(x)\frac{du}{dx} = -\frac{d}{dx}\left(P(x)\frac{du}{dx} \right),$$

where $P(x) = w(x)p(x) = e^{g(x)}$. The ODE then can be written as

$$-\frac{d}{dx}\left(P(x)\frac{du}{dx} \right) + R(x)u = w(x)f(x),$$

where $R(x) = w(x)r(x)$, and the BVP becomes

$$-\frac{d}{dx}\left(P(x)\frac{du}{dx}\right) + R(x)u = w(x)f(x), \ a < x < b,$$
$$u(a) = 0,$$
$$u(b) = 0.$$

If we try to solve this BVP by the method of eigenfunction expansion, then we must solve a Sturm–Liouville eigenvalue problem to find the eigenvalues and eigenfunctions.

10.1.2 Boundary conditions for the Sturm–Liouville problem

For most nonconstant coefficients P, R, and w, it is not possible to compute the eigenvalues and eigenfunctions analytically. However, in this case, it is possible to approximate the smallest eigenvalues and the corresponding eigenfunctions numerically, and sometimes this is all that is needed. For instance, when solving the homogeneous heat equation by an eigenfunction expansion, the effect of the large eigenvalues decreases exponentially over time (see Section 6.1.1, page 200). Therefore an approximation using only the smallest eigenvalues and corresponding eigenfunctions can give an accurate approximation to the long-time behavior of the solution. Moreover, the knowledge that the eigenvalues and eigenfunctions exist, together with an understanding of their properties, is sometimes useful for theoretical purposes, even if explicit calculations cannot be carried out.

Since, as noted above, we will not be able to derive explicit formulas in most cases, we might as well develop the theory for a general set of boundary conditions,

$$\alpha_1 u(a) + \alpha_2 \frac{du}{dx}(a) = 0, \ \beta_1 u(b) + \beta_2 \frac{du}{dx}(b) = 0, \tag{10.8}$$

where α_1 and α_2 are not both zero, and similarly for β_1 and β_2. As we will see, these are "symmetric" boundary conditions, in the sense that they are consistent with the symmetry of the differential operator under consideration. Dirichlet conditions and Neumann conditions are included as special cases; for example, if $\alpha_1 = 1$ and $\alpha_2 = 0$, then the boundary condition at the left endpoint is a Dirichlet condition, while $\alpha_1 = 0$ and $\alpha_2 = 1$ corresponds to a Neumann condition. We also pose the problem on a general interval $[a,b]$ rather than the special interval $[0,\ell]$, which was used in earlier chapters because the Fourier sine and cosine series are slightly simpler if the left endpoint of the interval is 0.

When α_1 and α_2 are both nonzero, the boundary condition

$$\alpha_1 u(a) + \alpha_2 \frac{du}{dx}(a) = 0$$

is referred to as a *Robin* boundary condition (or sometimes as a boundary condition of the *third kind*). The reader should note that Robin is a French name (Victor Gustave Robin was a nineteenth century French mathematician) and is pronounced accordingly.

The main purpose of this chapter is to describe the spectral properties (that is, properties related to eigenvalues and eigenvectors) of the operator L defined by (10.6) under the boundary conditions (10.8). A complete development of these properties is beyond the scope of this book, so it will be necessary to state some results without proof. The key parts

10.2 Properties of the Sturm–Liouville operator

of the omitted proofs are based on the Green's functions derived in the previous chapter, and the theory will be outlined in Section 10.7. Apart from that section, this chapter is independent of Chapter 9.

10.2 Properties of the Sturm–Liouville operator

Let us consider the properties of the Sturm–Liouville operator

$$Lu = \frac{1}{w(x)}\left(-\frac{d}{dx}\left(P(x)\frac{du}{dx}\right) + R(x)u\right)$$

under the boundary conditions (10.8). To fully define L, we must specify its domain. We will write

$$S = \left\{v \in C^2[a,b] : \alpha_1 u(a) + \alpha_2 \frac{du}{dx}(a) = 0,\ \beta_1 u(b) + \beta_2 \frac{du}{dx}(b) = 0\right\}$$

and consider L as an operator mapping S into $C[a,b]$.

10.2.1 Symmetry

We recall from Section 5.1 that the essential property of an operator L that allows the development of a spectral method is that L be symmetric. The reader can verify that L is not symmetric with respect to the $L^2(a,b)$ inner product (unless $w(x) = 1$). However, we can define an alternate inner product with respect to which L is symmetric:

$$(f,g)_w = \int_a^b f(x)g(x)w(x)\,dx. \tag{10.9}$$

If $u,v \in S$, then

$$\begin{aligned}
(L(u),v)_w &= \int_a^b \frac{1}{w(x)}\left(-\frac{d}{dx}\left(P(x)\frac{du}{dx}(x)\right) + R(x)u(x)\right)v(x)w(x)\,dx \\
&= -\int_a^b \frac{d}{dx}\left(P(x)\frac{du}{dx}(x)\right)v(x)\,dx + \int_a^b R(x)u(x)v(x)\,dx \\
&= -P(x)\frac{du}{dx}(x)v(x)\Big|_a^b + \int_a^b P(x)\frac{du}{dx}(x)\frac{dv}{dx}(x)\,dx + \int_a^b R(x)u(x)v(x)\,dx.
\end{aligned}$$

The exact form of the boundary term depends on the boundary conditions. If the constants $\alpha_1,\alpha_2,\beta_1,\beta_2$ are all nonzero, then we have

$$\alpha_1 u(a) + \alpha_2 \frac{du}{dx}(a) = 0 \Rightarrow \frac{du}{dx}(a) = -\frac{\alpha_1}{\alpha_2}u(a),$$

$$\beta_1 u(b) + \beta_2 \frac{du}{dx}(b) = 0 \Rightarrow \frac{du}{dx}(b) = -\frac{\beta_1}{\beta_2}u(b),$$

and then

$$-P(x)\frac{du}{dx}(x)v(x)\Big|_a^b = -P(b)\frac{du}{dx}(b)v(b) + P(a)\frac{du}{dx}(a)v(a)$$
$$= \frac{\beta_1}{\beta_2}P(b)u(b)v(b) - \frac{\alpha_1}{\alpha_2}P(a)u(a)v(a).$$

In this case, we obtain

$$(L(u),v)_w = \frac{\beta_1}{\beta_2}P(b)u(b)v(b) - \frac{\alpha_1}{\alpha_2}P(a)u(a)v(a)$$
$$+ \int_a^b P(x)\frac{du}{dx}(x)\frac{dv}{dx}(x)\,dx + \int_a^b R(x)u(x)v(x)\,dx.$$

The expression on the right is symmetric in u and v, which shows that

$$(u,L(v))_w = (L(v),u)_w = (L(u),v)_w,$$

verifying that L is symmetric with respect to $(\cdot,\cdot)_w$ in this case. The reader can verify that, for every other valid choice of $\alpha_1, \alpha_2, \beta_1, \beta_2$, L remains symmetric. In particular, if the boundary condition imposed at each endpoint is a Dirichlet condition or a Neumann condition, then the boundary term in the expression for $(L(u),v)_w$ vanishes entirely.

The usual proof (see Sections 3.5.2 and 5.1) can be used to show that L has only real eigenvalues, and that the corresponding eigenfunctions can be chosen to be real. Eigenfunctions corresponding to distinct eigenvalues are orthogonal, again by the usual argument, but they are orthogonal with respect to $(\cdot,\cdot)_w$ rather than the usual L^2 inner product.

10.2.2 Existence of eigenvalues and eigenfunctions

Although we can prove that every eigenvalue of L is real, this does not show that L actually has any eigenvalues. When we studied the negative second derivative operator earlier in the text, this was not an issue because we could calculate all the eigenvalues and corresponding eigenfunctions directly. For most choices of the functions P, R, and w, this is not possible for L (in Example 10.1, we analyze a problem for which we can compute the eigenpairs explicitly). It is therefore desirable to prove by some theoretical means that L has eigenvalues and eigenfunctions. This is possible, but it requires quite a bit of mathematics that is beyond the scope of this book. Section 10.7 outlines this theory and gives references that the interested reader can consult to learn the details. Here we will only state the results.

Assuming that P is continuously differentiable, R, w are continuous on $[a,b]$, and P, w are positive, then

1. L has an infinite sequence of real eigenvalues $\lambda_1 < \lambda_2 < \lambda_3 < \cdots$, with $\lambda_n \to \infty$ as $n \to \infty$;

2. associated with each eigenvalue λ_n is a single (independent) eigenfunction ψ_n (in other words, if ψ_n and ϕ_n are both eigenfunctions corresponding to λ_n, then ϕ_n is just a multiple of ψ_n);

3. each eigenfunction ψ_n is smooth (at least twice continuously differentiable);

10.2. Properties of the Sturm–Liouville operator

4. eigenfunctions corresponding to distinct eigenvalues are orthogonal under the $(\cdot,\cdot)_w$ inner product;

5. every square-integrable function v on $[a,b]$ can be represented by an eigenfunction expansion:

$$v(x) = \sum_{n=1}^{\infty} a_n \psi_n(x), \; a_n = (\psi_n, v)_w = \int_a^b \psi_n(x) v(x) w(x) \, dx$$

(here we assume that the eigenfunctions have been normalized: $(\psi_n, \psi_n)_w = 1$ for all n);

6. the eigenfunction ψ_n has exactly $n-1$ zeros in the open interval (a,b), and each zero of ψ_n lies between two consecutive zeros of ψ_{n+1}.

We illustrate the above properties with the following example.

Example 10.1. We consider the Sturm–Liouville problem

$$-\frac{d}{dx}\left(x\frac{du}{dx}\right) = \frac{\lambda}{x} u, \; 1 < x < 2,$$
$$u(1) = 0,$$
$$u(2) = 0.$$

In this problem, we have $P(x) = x$, $R(x) = 0$, $w(x) = 1/x$, and $[a,b] = [1,2]$.

We begin by showing that the corresponding Sturm–Liouville operator

$$Lu = -x\frac{d}{dx}\left(x\frac{du}{dx}\right)$$

has only positive eigenvalues. If λ is an eigenvalue of L and u is a corresponding eigenfunction with $(u,u)_w = 1$, then

$$\lambda = \lambda(u,u)_w = (\lambda u, u)_w = (Lu, u)_w$$
$$= -\int_1^2 x\frac{d}{dx}\left(x\frac{du}{dx}(x)\right) u(x) \frac{1}{x} \, dx$$
$$= -\int_1^2 \frac{d}{dx}\left(x\frac{du}{dx}(x)\right) u(x) \, dx$$
$$= -x\frac{du}{dx}(x) u(x) \Big|_1^2 + \int_1^2 x\frac{du}{dx}(x) \frac{du}{dx}(x) \, dx$$
$$= \int_1^2 x\frac{du}{dx}(x) \frac{du}{dx}(x) \, dx.$$

This last integral is obviously nonnegative, and it is zero only if du/dx is the zero function. In that case, though, u is a constant function, in which case the boundary conditions on u imply that u is identically zero, which is impossible. Therefore, it follows that λ must be strictly positive.

The ODE we must solve can be written as

$$x^2 \frac{d^2 u}{dx^2} + x \frac{du}{dx} + \lambda u = 0;$$

equations of this form are called Euler equations (see Section 4.2.5). Solutions to Euler equations are of the form $u(x) = x^m$, where m must be determined by substitution into the equation. We have

$$u(x) = x^m, \quad x^2 \frac{d^2 u}{dx^2} + x \frac{du}{dx} + \lambda u = 0 \Rightarrow x^2 m(m-1) x^{m-2} + xm x^{m-1} + \lambda x^m = 0$$

$$\Rightarrow (m(m-1) + m + \lambda) x^m = 0$$

$$\Rightarrow m(m-1) + m + \lambda = 0$$

$$\Rightarrow m^2 + \lambda = 0.$$

Since we know that $\lambda > 0$, we see that m must be an imaginary number: $m = \pm i \sqrt{\lambda}$. For complex m and real x, x^m is interpreted as

$$x^m = e^{m \ln x}.$$

Therefore,

$$x^{-i\sqrt{\lambda}} = e^{-i\sqrt{\lambda} \ln x} = \cos\left(\sqrt{\lambda} \ln x\right) - i \sin\left(\sqrt{\lambda} \ln x\right),$$

$$x^{i\sqrt{\lambda}} = e^{i\sqrt{\lambda} \ln x} = \cos\left(\sqrt{\lambda} \ln x\right) + i \sin\left(\sqrt{\lambda} \ln x\right).$$

Since the ODE is linear, any linear combination of these solutions is again a solution, so we find real-valued solutions,

$$\frac{1}{2} x^{-i\sqrt{\lambda}} + \frac{1}{2} x^{i\sqrt{\lambda}} = \cos\left(\sqrt{\lambda} \ln x\right),$$

$$\frac{i}{2} x^{-i\sqrt{\lambda}} - \frac{i}{2} x^{i\sqrt{\lambda}} = \sin\left(\sqrt{\lambda} \ln x\right)$$

(see similar reasoning on page 86 of Section 4.2). The general solution of the ODE is therefore

$$u(x) = c_1 \cos\left(\sqrt{\lambda} \ln x\right) + c_2 \sin\left(\sqrt{\lambda} \ln x\right),$$

and λ is an eigenvalue if and only if there is a nonzero function of this form satisfying the Dirichlet boundary conditions. The first boundary condition yields

$$u(1) = 0 \Rightarrow c_1 \cdot 1 + c_2 \cdot 0 = 0 \Rightarrow c_1 = 0,$$

so u must be a multiple of $\sin(\sqrt{\lambda} \ln x)$. From the second boundary condition, we obtain

$$u(2) = 0 \Rightarrow u(2) = \sin(\sqrt{\lambda} \ln 2) = 0$$

$$\Rightarrow \sqrt{\lambda} \ln 2 = n\pi, \; n = 1, 2, 3, \ldots$$

$$\Rightarrow \lambda = \frac{n^2 \pi^2}{(\ln 2)^2}, \; n = 1, 2, 3, \ldots.$$

10.2. Properties of the Sturm–Liouville operator

Therefore, the eigenpairs are

$$\lambda_n = \frac{n^2\pi^2}{(\ln 2)^2}, \quad \psi_n(x) = \sin\left(\frac{n\pi \ln x}{\ln 2}\right), \quad n = 1,2,3,\ldots.$$

These eigenfunctions ψ_n have not been normalized; in fact,

$$(\psi_n, \psi_n)_w = \int_1^2 \sin^2\left(\frac{n\pi \ln x}{\ln 2}\right) \frac{1}{x} dx = \frac{\ln 2}{2}, \quad n = 1,2,3,\ldots.$$

Figure 10.1 shows the first four eigenfunctions.

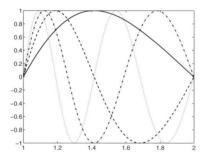

Figure 10.1. *The first four eigenfunctions in Example* 10.1.

It is now a straightforward matter to verify that these eigenvalues and eigenvectors satisfy the six properties described above. Obviously the eigenvalues λ_n are real and tend to infinity as $n \to \infty$, while the eigenfunctions are smooth, with a single eigenfunction for each eigenvalue. It can be verified directly that $(\psi_n, \psi_m)_w = 0$ if $m \neq n$. The interesting properties are the last two. To illustrate the representation of a given function by an eigenfunction expansion, we take the simple function $f(x) = \ln x (\ln(2/x))^2$. We have

$$f(x) = \sum_{n=1}^{\infty} c_n \psi_n(x), \quad c_n = \frac{(f, \psi_n)_w}{(\psi_n, \psi_n)_w}.$$

For each $n = 1, 2, 3, \ldots$,

$$\frac{(f, \psi_n)_w}{(\psi_n, \psi_n)_w} = \frac{2}{\ln 2} \int_1^2 \frac{f(x) \sin(n\pi \ln x / \ln 2)}{x} dx = \frac{4(2 + (-1)^n)(\ln 2)^3}{n^3 \pi^3}.$$

In Figure 10.2, we display the function f, together with approximations using two and four terms of the eigenfunction expansion (a small number of terms suffices for a reasonable approximation because f satisfies the same boundary conditions as the eigenfunctions).

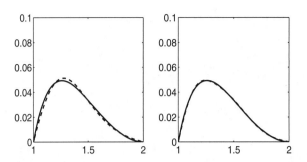

Figure 10.2. *The function $f(x) = \ln x (\ln(x/2))^2$ (solid curves) and its approximation by two (left) and four (right) terms in the eigenfunction expansion (cf. Example 10.1). The approximations are the dashed curves.*

Finally, we see that

$$\psi_n(x) = 0 \Rightarrow \sin\left(\frac{n\pi \ln x}{\ln 2}\right) = 0$$

$$\Rightarrow \frac{n\pi \ln x}{\ln 2} = k\pi, \ k = 1, 2, 3, \ldots$$

$$\Rightarrow \ln x = \frac{k}{n} \ln 2 = \ln\left(2^{k/n}\right), \ k = 1, 2, 3, \ldots$$

$$\Rightarrow x = 2^{k/n}, \ k = 1, 2, 3, \ldots.$$

The zeros of ψ_n that lie in the interval of interest, $(1, 2)$, satisfy

$$1 < 2^{k/n} < 2 \ \Rightarrow \ 0 < \frac{k}{n} < 1 \ \Rightarrow \ k = 1, 2, \ldots, n-1.$$

Thus we see that ψ_n has exactly $n - 1$ zeros in $(1, 2)$. ∎

We emphasize that, for most Sturm–Liouville problems, it is not possible to find formulas for the eigenvalues and eigenfunctions. Fortunately, it is possible to prove that the six properties listed above are always true.

Exercises

1. Let $L : S \to C[a, b]$ be the operator defined on page 389. Prove that L is symmetric in the case that $\alpha_2 = 0$ and $\alpha_1, \beta_1, \beta_2$ are all nonzero. (In this case, the boundary condition at the left endpoint is a Dirichlet condition.)

2. Repeat the previous exercise for the case that $\alpha_1 = 0$ and $\alpha_2, \beta_1, \beta_2$ are all nonzero. (In this case, the boundary condition at the left endpoint is a Neumann condition.)

3. Consider a heat conduction problem, in which $u(x,t)$ represents the temperature in an insulated bar. In this setting, a Robin boundary condition models Newton's law of heating/cooling at the ends of the bar, meaning that the rate of change of the temperature $u(a,t)$ is proportional to the difference between $u(a,t)$ and the temperature

10.2. Properties of the Sturm–Liouville operator

of the surroundings (and similarly for $u(b,t)$). Let $\alpha_2 = \beta_2 = \kappa$, where κ is the thermal conductivity in the bar. What sign should α_1 have to be physically meaningful? Similarly, what sign should β_1 have to be physically meaningful?

4. Consider the Sturm–Liouville operator L defined on page 389. Prove that, if the constants $\alpha_1, \alpha_2, \alpha_3, \alpha_4$ have the signs determined in the previous exercise, then all of the eigenvalues of L are positive.

5. Find the eigenvalues and eigenfunctions for the Sturm–Liouville problem

$$-\frac{d}{dx}\left(x\frac{du}{dx}\right) + \frac{3}{x}u = \frac{\lambda}{x}u, \ 1 < x < 2,$$
$$u(1) = 0,$$
$$u(2) = 0.$$

6. Find the eigenvalues and eigenfunctions for the Sturm–Liouville problem

$$-\frac{d}{dx}\left(x\frac{du}{dx}\right) = \frac{\lambda}{x}u, \ 1 < x < 2,$$
$$u(1) = 0,$$
$$\frac{du}{dx}(2) = 0.$$

7. Consider the Sturm–Liouville problem

$$-\frac{d}{dx}\left(\frac{1}{x^3}\frac{du}{dx}\right) + \frac{\beta}{x^5}u = \frac{\lambda}{x^5}u, \ 1 < x < 2,$$
$$u(1) = 0,$$
$$u(2) = 0.$$

   (a) Find the eigenvalues and eigenfunctions.
   (b) For what value(s) of β is 0 an eigenvalue?
   (c) For what value(s) of β does there exist exactly one negative eigenvalue? Exactly k negative eigenvalues?

8. Find the eigenvalues and eigenfunctions for the problem

$$-x^2\frac{d^2u}{dx^2} - 3x\frac{du}{dx} = \lambda u, \ 1 < x < 2,$$
$$u(1) = 0,$$
$$\frac{du}{dx}(2) = 0.$$

(Note: Although this is not the standard form for a Sturm–Liouville problem, the transformation described on page 387 can be used to write the problem in Sturm–Liouville form.)

9. Find the eigenvalues and eigenfunctions for the problem

$$-x^2 \frac{d^2u}{dx^2} - 3x \frac{du}{dx} + u = \lambda u, \ 1 < x < 2,$$
$$\frac{du}{dx}(1) = 0,$$
$$\frac{du}{dx}(2) = 0.$$

(Note: Although this is not the standard form for a Sturm–Liouville problem, the transformation described on page 387 can be used to write the problem in Sturm–Liouville form.)

10.3 Numerical methods for Sturm–Liouville problems

It is straightforward to apply the finite element method to address Sturm–Liouville problems. Just as the finite element method reduces a linear differential equation to a matrix-vector equation $\mathbf{Ku} = \mathbf{f}$, it reduces a Sturm–Liouville differential equation to a matrix eigenvalue problem $\mathbf{Au} = \lambda \mathbf{Wu}$, albeit a *generalized* eigenvalue problem (the standard eigenvalue problem has the form $\mathbf{Au} = \lambda \mathbf{u}$).

However, a (finite-dimensional) matrix has only finitely many eigenvalues, so clearly the finite element method cannot succeed in approximating all of the infinitely many eigenvalues of the Sturm–Liouville problem. Indeed, as we will see, this approach is effective for approximating some of the smallest eigenvalues and corresponding eigenfunctions, but it is quite time consuming if many eigenpairs are desired. Specialized methods have been developed that are more efficient for computing many eigenpairs.

10.3.1 The weak form

We begin by deriving the weak form of the Sturm–Liouville problem. We will focus on the Dirichlet problem, although these methods can be extended to the general boundary conditions considered in the previous sections (see Section 10.6). We assume that $u \neq 0$ and λ satisfy

$$-\frac{d}{dx}\left(P(x)\frac{du}{dx}\right) + R(x)u = \lambda w(x)u, \ a < x < b, \quad (10.10a)$$
$$u(a) = 0, \quad (10.10b)$$
$$u(b) = 0. \quad (10.10c)$$

Multiplying by the test function v, which is assumed to satisfy the boundary conditions $v(a) = v(b) = 0$, and integrating yield

$$-\int_a^b \left(P(x)\frac{du}{dx}(x)\right) v(x)\,dx + \int_a^b R(x)u(x)v(x)\,dx = \lambda \int_a^b w(x)u(x)v(x)\,dx.$$

Integrating the first term by parts and using the fact that $v(a) = v(b) = 0$, we obtain

$$\int_a^b P(x)\frac{du}{dx}(x)\frac{dv}{dx}(x)\,dx + \int_a^b R(x)u(x)v(x)\,dx = \lambda \int_a^b w(x)u(x)v(x)\,dx.$$

10.3. Numerical methods for Sturm–Liouville problems

This must hold for all test functions v. As in Section 5.6, we initially consider the space of test functions to be $C_D^2[a,b]$. However, the weak form depends only on the first derivatives of u and v, so we allow a larger space of test functions that includes all continuous piecewise linear functions (for more discussion, see pages 184–185 and Section 13.3). This allows the use of the finite element method.

We will use the notation established earlier (cf. Section 5.6):

$$a = x_0 < x_1 < \cdots < x_n = b$$

defines a uniform mesh on $[a,b]$, where $x_i = a+ih$, $i = 0, 1, \ldots, n$, $h = (b-a)/n$. The space of all continuous piecewise functions, relative to this mesh and satisfying the homogeneous Dirichlet boundary conditions, is denoted S_n. The standard basis for S_n is $\{\phi_1, \phi_2, \ldots, \phi_{n-1}\}$, where ϕ_i is defined by

$$\phi_i(x_j) = \begin{cases} 1, & j = i, \\ 0, & j \neq i \end{cases}$$

(see Figure 5.12).

The weak form of the Sturm–Liouville problem is to find u such that the following equation holds for all test functions v:

$$\int_a^b \left\{ P(x) \frac{du}{dx}(x) \frac{dv}{dx}(x) + R(x)u(x)v(x) \right\} dx = \lambda \int_a^b w(x)u(x)v(x)\,dx. \tag{10.11}$$

Applying the Galerkin method, we produce an approximation to u by finding v_n in S_n such that, for all $v \in S_n$,

$$\int_a^b \left\{ P(x) \frac{dv_n}{dx}(x) \frac{dv}{dx}(x) + R(x)v_n(x)v(x) \right\} dx = \lambda \int_a^b w(x)v_n(x)v(x)\,dx. \tag{10.12}$$

We represent v_n by the standard basis as

$$v_n(x) = \sum_{j=1}^{n-1} u_j \phi_j(x), \tag{10.13}$$

and the problem becomes to compute the vector $\mathbf{u} = (u_1, u_2, \ldots, u_{n-1})$. The reader should recall that the components of this vector are the nodal values of the approximate solution v_n.

To derive the system of equations defining v_n, we substitute (10.13),

$$\frac{dv_n}{dx}(x) = \sum_{j=1}^{n-1} u_j \frac{d\phi_j}{dx}(x),$$

and $v = \phi_i$ into (10.12) and manipulate as follows:

$$\int_a^b \left\{ P(x) \left(\sum_{j=1}^{n} u_j \frac{d\phi_j}{dx}(x) \right) \frac{d\phi_i}{dx}(x) + R(x) \left(\sum_{j=1}^{n-1} u_j \phi_j(x) \right) \phi_i(x) \right\} dx$$

$$= \lambda \int_a^b w(x) \left(\sum_{j=1}^{n-1} u_j \phi_j(x) \right) \phi_i(x) dx, \ i = 1, 2, \ldots, n-1$$

$$\Rightarrow \sum_{j=1}^{n-1} u_j \int_a^b \left\{ P(x) \frac{d\phi_j}{dx}(x) \frac{d\phi_i}{dx}(x) + R(x) \phi_j(x) \phi_i(x) \right\} dx$$

$$= \lambda \sum_{j=1}^{n-1} u_j \int_a^b w(x) \phi_j(x) \phi_i(x) dx, \ i = 1, 2, \ldots, n-1$$

$$\Rightarrow \sum_{j=1}^{n-1} A_{ij} u_j = \lambda \sum_{j=1}^{n-1} W_{ij} u_j, \ i = 1, 2, \ldots, n-1$$

$$\Rightarrow \mathbf{Au} = \lambda \mathbf{Wu}.$$

In this last equation, the $(n-1) \times (n-1)$ matrices $\mathbf{A}$ and $\mathbf{W}$ are defined by

$$A_{ij} = \int_a^b \left\{ P(x) \frac{d\phi_j}{dx}(x) \frac{d\phi_i}{dx}(x) + R(x) \phi_j(x) \phi_i(x) \right\} dx,$$

$$W_{ij} = \int_a^b w(x) \phi_j(x) \phi_i(x) dx.$$

These matrices should look familiar; $\mathbf{A}$ is the sum of the usual stiffness and mass matrices (with the nonconstant coefficients P and R, respectively), while $\mathbf{W}$ is another mass matrix (with nonconstant coefficient w). Thus the calculation of these matrices should be familiar from earlier chapters.

As mentioned above, the equation $\mathbf{Au} = \lambda \mathbf{Wu}$ is called a generalized eigenvalue problem. Algorithms for solving this problem are beyond the scope of this book, but software packages like MATLAB, *Mathematica*, and *Maple* have routines for finding the eigenvalues and eigenvectors numerically (that is, approximately). For our numerical examples, we will rely on such routines.

For each eigenvalue-eigenvector pair λ, $\mathbf{u}$ satisfying this equation, $\mathbf{u}$ defines an approximation to an eigenfunction of the Sturm–Liouville problem via (10.13), and λ approximates the corresponding Sturm–Liouville eigenvalue. Since $\mathbf{A}$ belongs to $\mathbf{R}^{(n-1) \times (n-1)}$, $\mathbf{Au} = \lambda \mathbf{Wu}$ yields $n-1$ eigenvalue-eigenvector pairs, approximating the $n-1$ smallest Sturm–Liouville eigenvalues and corresponding eigenfunctions. However, the largest of these $n-1$ approximate eigenvalues are typically quite inaccurate. Before we discuss the theoretical error estimates, we will present an example.

Example 10.2. We will apply the finite element method described above to the Sturm–Liouville problem from Example 10.1, for which we were able to derive exact formulas for the eigenvalues and eigenfunctions. In this way, we can judge the accuracy of the

10.3. Numerical methods for Sturm–Liouville problems

approximations. The eigenvalue problem is

$$-\frac{d}{dx}\left(x\frac{du}{dx}\right) = \frac{\lambda}{x}u, \quad 1 < x < 2,$$
$$u(1) = 0,$$
$$u(2) = 0.$$

We have

$$A_{ij} = \int_1^2 x \frac{d\phi_j}{dx}(x) \frac{d\phi_i}{dx}(x)\, dx = \begin{cases} -\frac{2+2ih-h}{2h}, & j = i-1, \\ \frac{2(1+ih)}{h}, & j = i, \\ -\frac{2+2ih+h}{2h}, & j = i+1, \\ 0 & \text{otherwise}, \end{cases}$$

and

$$W_{ij} = \int_1^2 \frac{d\phi_i}{dx}(x) \frac{d\phi_j}{dx}(x) \frac{1}{x}\, dx$$
$$= \begin{cases} \frac{1}{h} + i - \frac{1}{2} + \frac{(1+ih-h)^2 + h(1+ih-h)}{h^2} \ln\left(1 - \frac{h}{1+ih}\right), & j = i-1, \\ -\frac{2}{h} - 2i + \frac{(1+ih+h)^2}{h^2}\ln\left(1 - \frac{h}{1+ih}\right) - \frac{(1+ih-h)^2}{h^2}\ln\left(1 + \frac{h}{1+ih}\right), & j = i, \\ \frac{1}{h} + i + \frac{1}{2} - \frac{(1+ih)^2 + h(1+ih)}{h^2} \ln\left(1 + \frac{h}{1+ih}\right), & j = i+1. \end{cases}$$

Table 10.1 shows the nine eigenvalues satisfying $\mathbf{Au} = \lambda \mathbf{Wu}$ for $n = 10$, together with the exact eigenvalues (cf. Example 10.1) and the relative errors. We see that the smallest eigenvalues are fairly accurate (for example, the error in the three smallest eigenvalues is less than 10%), but the error in the computed λ_k grows with k.

Table 10.1. *Exact and computed eigenvalues for Example* 10.2 *($n = 10$), and the associated relative errors.*

k	Exact λ_k	Approx. λ_k	Relative error
1	$2.0542 \cdot 10^1$	$2.0752 \cdot 10^1$	$1.0205 \cdot 10^{-2}$
2	$8.2169 \cdot 10^1$	$8.5317 \cdot 10^1$	$3.8315 \cdot 10^{-2}$
3	$1.8488 \cdot 10^2$	$2.0072 \cdot 10^2$	$8.5654 \cdot 10^{-2}$
4	$3.2868 \cdot 10^2$	$3.7871 \cdot 10^2$	$1.5222 \cdot 10^{-1}$
5	$5.1356 \cdot 10^2$	$6.3475 \cdot 10^2$	$2.3599 \cdot 10^{-1}$
6	$7.3952 \cdot 10^2$	$9.8383 \cdot 10^2$	$3.3035 \cdot 10^{-1}$
7	$1.0066 \cdot 10^3$	$1.4368 \cdot 10^3$	$4.2743 \cdot 10^{-1}$
8	$1.3147 \cdot 10^3$	$2.0373 \cdot 10^3$	$5.4961 \cdot 10^{-1}$
9	$1.6639 \cdot 10^3$	$2.9833 \cdot 10^3$	$7.9291 \cdot 10^{-1}$

To demonstrate the effect of refining the mesh, Table 10.2 shows the relative errors in the nine smallest eigenvalues computed on meshes with $n = 10, 20, 40, 80$. We see that, for each eigenvalue, the error goes down by approximately a factor of 4 each time h is divided in half. In other words, the error is approximately proportional to h^2, which we express briefly by stating that the error is $O(h^2)$. In particular, focusing on λ_9, we see that the relative error is reduced from 79% on the coarsest mesh to just over 1% on the finest of the four meshes. On the other hand, the calculation on the finest mesh also yields an estimate of λ_{79}, which turns out to be in error by more than 100%! ∎

Table 10.2. *Relative errors in the approximate eigenvalues from Example* 10.2 *on successively finer meshes. Only the nine smallest eigenvalues are included.*

k	$n = 10$	$n = 20$	$n = 40$	$n = 80$
1	$1.0205 \cdot 10^{-2}$	$2.5433 \cdot 10^{-3}$	$6.3535 \cdot 10^{-4}$	$1.5881 \cdot 10^{-4}$
2	$3.8315 \cdot 10^{-2}$	$9.5094 \cdot 10^{-3}$	$2.3731 \cdot 10^{-3}$	$5.9301 \cdot 10^{-4}$
3	$8.5654 \cdot 10^{-2}$	$2.1158 \cdot 10^{-2}$	$5.2719 \cdot 10^{-3}$	$1.3168 \cdot 10^{-3}$
4	$1.5222 \cdot 10^{-1}$	$3.7538 \cdot 10^{-2}$	$9.3353 \cdot 10^{-3}$	$2.3305 \cdot 10^{-3}$
5	$2.3599 \cdot 10^{-1}$	$5.8696 \cdot 10^{-2}$	$1.4568 \cdot 10^{-2}$	$3.6344 \cdot 10^{-3}$
6	$3.3035 \cdot 10^{-1}$	$8.4653 \cdot 10^{-2}$	$2.0976 \cdot 10^{-2}$	$5.2289 \cdot 10^{-3}$
7	$4.2743 \cdot 10^{-1}$	$1.1536 \cdot 10^{-1}$	$2.8566 \cdot 10^{-2}$	$7.1145 \cdot 10^{-3}$
8	$5.4961 \cdot 10^{-1}$	$1.5061 \cdot 10^{-1}$	$3.7343 \cdot 10^{-2}$	$9.2918 \cdot 10^{-3}$
9	$7.9291 \cdot 10^{-1}$	$1.8995 \cdot 10^{-1}$	$4.7313 \cdot 10^{-2}$	$1.1761 \cdot 10^{-2}$

The pattern exhibited in Example 10.2 is typical. It can be proved that the error in each λ_k estimated by the finite element method, as described above, is $O(h^2)$, where h is the mesh size. As stated above, this means that the error is approximately proportional to h^2. However, it is important to note that the constant of proportionality depends on k, and, in fact, grows with k. Thus, given mesh size h and the corresponding $n-1$ approximate eigenvalues, some of the eigenvalues (the smaller ones) will be accurate and others (the larger ones) will be inaccurate.

The error in the computed eigenfunctions follows the same pattern if the error is measured in the L^2 norm: If $\psi_k^{(h)}$ is the finite element approximation to the eigenfunction ψ_k, then[48]

$$\left\| \psi_k - \psi_k^{(h)} \right\| = \left[\int_a^b \left(\psi_k(x) - \psi_k^{(h)}(x) \right)^2 dx \right]^{1/2} = O(h^2).$$

Once again, on any given mesh, the approximate eigenfunctions corresponding to the smallest k will generally be the most accurate. There is, however, another issue that affects the accuracy of the eigenfunction estimates: An eigenfunction is easier to estimate if the corresponding eigenvalue is well separated from other eigenvalues, and harder to estimate if the eigenvalue is almost equal to an adjacent eigenvalue. In the latter case, the error in the eigenfunction is still $O(h^2)$, but the constant of proportionality is larger.

---

[48]The same result holds if the error is measured in the weighted L^2 norm:

$$\left\| \psi_k - \psi_k^{(h)} \right\|_w = \left[\int_a^b \left(\psi_k(x) - \psi_k^{(h)}(x) \right)^2 w(x) dx \right]^{1/2} = O(h^2).$$

Exercises

1. Consider the Sturm–Liouville problem (10.10), but with (10.10c) changed to the Neumann condition $du/dx(b) = 0$. Derive the finite element method for estimating the eigenvalues and eigenfunctions.

2. Apply the method derived in the previous exercise to the Sturm–Liouville problem from Exercise 10.2.6. Compare the numerical estimates for the eigenvalues with the exact values derived in that exercise on several meshes. Does the error appear to be decreasing like $O(h^2)$?

3. Consider the Sturm–Liouville problem (10.10), but with both boundary conditions changed to homogeneous Neumann conditions. Derive the finite element method for estimating the eigenvalues and eigenvectors.

4. Apply the method derived in the previous exercise to the Sturm–Liouville problem from Exercise 10.2.9. Compare the numerical estimates for the eigenvalues with the exact values derived in that exercise on several meshes. Does the error appear to be decreasing like $O(h^2)$?

5. Suppose $\mathbf{A}, \mathbf{W} \in \mathbf{R}^{n \times n}$ are symmetric and that $\lambda \in \mathbf{C}$, $\mathbf{x} \in \mathbf{C}^n$ form a generalized eigenpair: $\mathbf{A}\mathbf{x} = \lambda \mathbf{W}\mathbf{x}$. Prove that λ is real. (Hint: $\mathbf{W}$ defines a weighted inner product on $\mathbf{C}^n$:

$$(\mathbf{x}, \mathbf{y})_\mathbf{W} = (\mathbf{W}\mathbf{x}) \cdot \overline{\mathbf{y}} = \sum_{i=1}^{n} (\mathbf{W}\mathbf{x})_i \overline{y}_i.$$

Assume that $\mathbf{x}$ is normalized with respect to the corresponding norm, and mimic the proof of Theorem 3.44 in Section 3.5.)

6. Suppose $\mathbf{A}, \mathbf{W} \in \mathbf{R}^{n \times n}$ are symmetric, and that $\lambda_1 \in \mathbf{R}$, $\mathbf{x}_1 \in \mathbf{R}^n$ and $\lambda_2 \in \mathbf{R}$, $\mathbf{x}_2 \in \mathbf{R}^n$ are generalized eigenpairs of $\mathbf{A}$ with respect to $\mathbf{W}$. Prove that $\mathbf{x}_1, \mathbf{x}_2$ are orthogonal with respect to the weighted inner product defined by $\mathbf{W}$ (see the previous exercise). (Hint: Mimic the proof of Theorem 3.45 of Section 3.5, but using the weighted inner product defined by $\mathbf{W}$ in place of the ordinary dot product.)

10.4 Examples of Sturm–Liouville problems

10.4.1 A guitar string with variable density

The reader will recall from Section 7.2.1 that the vibrations of a plucked string (such as a guitar string) can be expressed as the sum of *normal modes*, that is, standing waves vibrating at the *natural frequencies* $cn/(2\ell)$ of the string. The natural frequencies are simply the integer multiples of the *fundamental frequency* $c/(2\ell)$ of the string; these integer multiples of the fundamental frequency are often called *harmonics*. The fact that the fundamental frequencies of a guitar string are harmonics gives the instrument its pleasing sound; other instruments are often designed so that the first few fundamental frequencies are at least approximately equal to harmonics.

We will consider a string with variable density and see how the fundamental frequencies are affected. In particular, we wish to know if the fundamental frequencies are still harmonics.

Example 10.3. We consider an elastic string of length $\ell = 0.6$ m, (linear) density $\rho(x) = 6 \cdot 10^{-4} + b(x)$ kg/m, where

$$b(x) = \begin{cases} 6(x-0.2)^2(x-0.4)^2, & 0.2 < x < 0.4, \\ 0 & \text{otherwise,} \end{cases}$$

and tension $T = 58$ N. The wave equation is

$$\rho(x)\frac{\partial^2 u}{\partial t^2} - T\frac{\partial^2 u}{\partial x^2} = 0, \ 0 < x < \ell, \ t > 0,$$

or

$$\frac{\partial^2 u}{\partial t^2} - \frac{T}{\rho(x)}\frac{\partial^2 u}{\partial x^2} = 0, \ 0 < x < \ell, \ t > 0. \tag{10.14}$$

We can compare this string with one that has a constant density of $6 \cdot 10^{-4}$ kg/m. Such a string would have a wave speed of $c \doteq 310.9$ m/s ($c^2 = T/\rho$) and a fundamental frequency of

$$f_1 = \frac{c}{2\ell} \doteq 259.1 \text{ Hz}.$$

The other natural frequencies of the string would be $2f_1, 3f_1, \ldots$.

To find the natural frequencies of the string with variable density, we follow the same procedure as in Section 7.2. We consider the wave equation in the form (10.14), where the initial conditions are

$$u(x,0) = \psi(x), \ \frac{\partial u}{\partial t}(x,0) = \gamma(x), \ 0 < x < \ell,$$

and the boundary conditions are

$$u(0,t) = u(\ell,t) = 0, \ t > 0.$$

The spatial part of the differential operator is

$$Lu = -\frac{T}{\rho(x)}\frac{d^2 u}{dx^2}.$$

Finding the eigenvalues and eigenfunctions of L is equivalent to solving the Sturm–Liouville eigenvalue problem

$$\begin{aligned} -\frac{d^2 u}{dx^2} &= \lambda \frac{\rho(x)}{T} u, \ 0 < x < \ell, \\ u(0) &= 0, \\ u(\ell) &= 0. \end{aligned} \tag{10.15}$$

This problem has the form discussed in the preceding sections, with $P(x) = 1$, $R(x) = 0$, $w(x) = \rho(x)/T$. Assuming for the moment that we have found the eigenpairs

$$\lambda_k, \ \psi_k, \ k = 1, 2, 3, \ldots,$$

10.4. Examples of Sturm–Liouville problems

we solve the IBVP as follows. We write the solution as

$$u(x,t) = \sum_{k=1}^{\infty} a_k(t)\psi_k(x)$$

and substitute into the PDE to obtain

$$\sum_{k=1}^{\infty} \left\{ \frac{d^2 a_k}{dt^2}(t) + \lambda_k a_k(t) \right\} = 0.$$

This implies that a_k must satisfy the ODE

$$\frac{d^2 a_k}{dt^2} + \lambda_k a_k = 0, \ k = 1, 2, 3, \ldots. \tag{10.16}$$

If we write the initial values as

$$\psi(x) = \sum_{k=1}^{\infty} c_k \psi_k(x),$$

$$\gamma(x) = \sum_{k=1}^{\infty} d_k \psi_k(x),$$

then we obtain the following initial conditions for a_k:

$$a_k(0) = c_k, \ \frac{da_k}{dt}(0) = d_k, \ k = 1, 2, 3, \ldots. \tag{10.17}$$

The solution to (10.16) that satisfies (10.17) has the form

$$a_k(t) = A_k \cos\left(\sqrt{\lambda_k} t\right) + B_k \sin\left(\sqrt{\lambda_k} t\right).$$

The exact values of A_k and B_k are not important for our present purposes.[49] The key point is that $u(x,t)$ is the superposition of standing waves of the form

$$\left\{ A_k \cos\left(\sqrt{\lambda_k} t\right) + B_k \sin\left(\sqrt{\lambda_k} t\right) \right\} \psi_k(x).$$

The natural frequencies are therefore $\sqrt{\lambda_k}/(2\pi)$, $k = 1, 2, 3, \ldots$.

The answer to our question thus boils down to finding the eigenvalues by solving (10.15). We will estimate the smallest eigenvalues by applying the finite element method developed in the preceding section. The eigenvalues (at least the first $n-1$ of them) are

---

[49] To compute A_k and B_k, one must first find c_k and d_k, which are determined by the usual formula for generalized Fourier series coefficients, albeit in the weighted inner product:

$$c_k = \frac{(\psi, \psi_k)_w}{(\psi_k, \psi_k)_w}, \ d_k = \frac{(\gamma, \psi_k)_w}{(\psi_k, \psi_k)_w}, \ k = 1, 2, 3, \ldots.$$

Table 10.3. *Estimates of the five smallest eigenvalues in Example* 10.3.

	$n = 30$	$n = 60$	$n = 120$
λ_1	$1.9627 \cdot 10^6$	$1.9612 \cdot 10^6$	$1.9609 \cdot 10^6$
λ_2	$1.0104 \cdot 10^7$	$1.0076 \cdot 10^7$	$1.0069 \cdot 10^7$
λ_3	$1.9998 \cdot 10^7$	$1.9871 \cdot 10^7$	$1.9839 \cdot 10^7$
λ_4	$3.7353 \cdot 10^7$	$3.6922 \cdot 10^7$	$3.6815 \cdot 10^7$
λ_5	$5.9392 \cdot 10^7$	$5.8348 \cdot 10^7$	$5.8088 \cdot 10^7$

Table 10.4. *The five smallest natural frequencies in Example* 10.3.

k	1	2	3	4	5
f_k	$2.23 \cdot 10^2$	$5.05 \cdot 10^2$	$7.09 \cdot 10^2$	$9.66 \cdot 10^2$	$1.21 \cdot 10^3$

Table 10.5. *The ratios of the natural frequencies to the fundamental frequency in Example* 10.3.

k	2	3	4	5
f_k/f_1	2.27	3.18	4.33	5.44

estimated to be the values of λ that satisfy the generalized eigenvalue equation $\mathbf{Au} = \lambda \mathbf{Wu}$, where $\mathbf{A}$, $\mathbf{W}$ are the $(n-1) \times (n-1)$ matrices defined by

$$A_{ij} = \int_a^b \frac{d\phi_j}{dx}(x) \frac{d\phi_i}{dx}(x) \, dx, \quad W_{ij} = \int_a^b \frac{\rho(x)}{T} \phi_j(x) \phi_i(x) \, dx.$$

Here $\phi_1, \phi_2, \ldots, \phi_{n-1}$ are the standard nodal basis functions for a uniform mesh with n elements. The entries in $\mathbf{A}$ are the usual values ($2/h$ on the main diagonal, $-1/h$ on the first subdiagonal and first superdiagonal, 0 otherwise), but the entries in $\mathbf{W}$ must be computed from the above formula since ρ is not constant. We omit the details of the calculations (but, as usual, we recommend the use of a computer algebra system for calculations of this sort).

We compute the five smallest eigenvalues for $n = 30, 60,$ and 120, with the results shown in Table 10.3. The reader should recall that the error in each should reduce by a factor of approximately 4 each time n is doubled. Given these results, it is reasonable to assume that all five eigenvalues computed with $n = 120$ are accurate to three significant figures. The corresponding frequencies are given in Table 10.4.

Finally, we can answer our original question, Are the natural frequencies in the given string harmonics, that is, does $f_k = kf_1$ hold? The answer is no, as shown in Table 10.5. ∎

10.4.2 Heat flow with a variable thermal conductivity

As another example of Sturm–Liouville analysis, we consider the temperature in a one-dimensional bar with a variable thermal conductivity. Specifically, we will study the

10.4. Examples of Sturm–Liouville problems

following IBVP:

$$\rho(x)c(x)\frac{\partial u}{\partial t} - \frac{\partial}{\partial x}\left(\kappa(x)\frac{\partial u}{\partial x}\right) = 0, \ 0 < x < \ell, \ t > 0,$$
$$u(x,0) = u_0(x), \ 0 < x < \ell,$$
$$u(0,t) = 0, \ t > 0,$$
$$\frac{\partial u}{\partial x}(0,t) = 0, \ t > 0. \tag{10.18}$$

We recall from Section 6.2 that if ρ, c, and κ are all constant, then the solution of (10.18) is

$$u(x,t) = \sum_{k=1}^{\infty} c_k e^{-(2k-1)^2 \pi^2 \kappa t/(4\ell^2 \rho c)} \sin\left(\frac{(2k-1)\pi x}{2\ell}\right),$$

where c_k is the kth coefficient in the Fourier quarter-wave sine series of the initial temperature u_0. We can also express the solution u as

$$u(x,t) = \sum_{k=1}^{\infty} c_k e^{-\lambda_k t} \psi_k(x), \tag{10.19}$$

where

$$\lambda_k = \frac{(2k-1)^2 \pi^2 \kappa}{4\ell^2 \rho c}, \ \psi_k(x) = \sin\left(\frac{(2k-1)\pi x}{2\ell}\right), \ k = 1,2,3,\ldots,$$

are the eigenvalues and eigenfunctions of the spatial operator

$$Lu = -\frac{\kappa}{\rho c}\frac{d^2 u}{dx^2},$$

where the domain of L consists of u satisfying $u(0) = du/dx(\ell) = 0$. In solving (10.18), we will obtain a solution of the same form (10.19), in which the eigenvalues and eigenfunctions correspond to the spatial operator

$$Lu = -\frac{1}{\rho(x)c(x)}\frac{d}{dx}\left(\kappa(x)\frac{du}{dx}\right).$$

We now consider an example.

Example 10.4. We now consider a 1 m bar with constant density and heat capacity, $\rho = 7.88 \cdot 10^3 \text{ kg/m}^3$ and $c = 437 \text{ J/(kg K)}$, but variable thermal conductivity

$$\kappa(x) = \kappa_0 + 400\left(x - \frac{1}{2}\right)^2 - \frac{1}{3}, \ \kappa_0 = 83.6 \text{ W/(m K)}.$$

We notice that the mean value of $\kappa(x)$ on the interval $[0,1]$ is κ_0, and later we will compare heat flow in this bar with the results for the corresponding homogeneous bar.

As indicated above, the first step is to find the eigenvalues and eigenfunctions of the spatial operator

$$Lu = -\frac{1}{\rho c}\frac{d}{dx}\left(\kappa(x)\frac{du}{dx}\right) = -\frac{d}{dx}\left(\frac{\kappa(x)}{\rho c}\frac{du}{dx}\right)$$

under the given mixed boundary conditions. The reader should notice that, for convenience, we have moved the constant $1/(\rho c)$ inside the derivative. The problem is then to solve

$$-\frac{d}{dx}\left(\frac{\kappa(x)}{\rho c}\frac{du}{dx}\right) = \lambda u, \ 0 < x < \ell,$$

$$u(0) = 0,$$

$$\frac{du}{dx}(0) = 0,$$

which is a Sturm–Liouville eigenvalue problem with

$$P(x) = \frac{\kappa(x)}{\rho c}, \ R(x) = 0, \ w(x) = 1.$$

We will find the smallest eigenvalues and the corresponding eigenvalues using the finite element method. We must solve the generalized eigenvalue problem

$$\mathbf{Au} = \lambda \mathbf{Wu},$$

where

$$A_{ij} = \int_0^1 P(x)\frac{d\phi_j}{dx}(x)\frac{d\phi_i}{dx}(x)\,dx, \ W_{ij} = \int_0^1 \phi_j(x)\phi_i(x)\,dx, \ i,j = 1,2,\ldots,n-1,$$

and $\phi_1, \phi_2, \ldots, \phi_{n-1}$ are the standard nodal basis functions.

A little experimentation shows that $n = 80$ (that is, a mesh with 80 subintervals) yields three digits of accuracy for each of the four smallest eigenvalues. Their values are shown in Table 10.6, together with the four smallest eigenvalues of the constant-coefficient problem (in which the function $\kappa(x)$ is replaced by the constant κ_0). The eigenfunctions for both problems are shown in Figure 10.3.

Table 10.6. *The four smallest eigenvalues λ_k from Example 10.4, together with the eigenvalues $\lambda_k^{(0)}$ of the corresponding constant-coefficient problem.*

k	λ_k	$\lambda_k^{(0)}$
1	$5.2797 \cdot 10^{-5}$	$5.9904 \cdot 10^{-5}$
2	$5.0165 \cdot 10^{-4}$	$5.3927 \cdot 10^{-4}$
3	$1.3783 \cdot 10^{-3}$	$1.4987 \cdot 10^{-3}$
4	$2.6956 \cdot 10^{-3}$	$2.9398 \cdot 10^{-3}$

What conclusion can we draw from this Sturm–Liouville analysis? We see that the variable conductivity leads to smaller eigenvalues (for example, λ_1 is about 12% smaller than $\lambda_1^{(0)}$). Comparing formula (10.19) for the solution u, we expect that the temperature changes more slowly in the heterogeneous bar than in the homogeneous bar. In both cases, the temperature tends to a constant temperature of zero as $t \to \infty$. However, the terms in the eigenfunction expansion of the solution for the heterogeneous bar are damped more slowly than are the corresponding terms for the homogeneous bar.

10.4. Examples of Sturm–Liouville problems

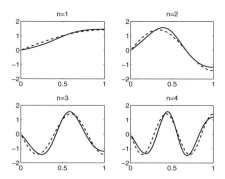

Figure 10.3. *The first four eigenfunctions in Example* 10.4 *(solid curves) and the corresponding eigenfunctions from the constant-coefficient problem (dashed curves).*

To verify this conclusion, we suppose the initial temperature in both bars is $10°$ Celsius, so that $u_0(x) = 10$ in (10.18). We will solve (10.18) for $u(x,t)$, and also the corresponding constant-coefficient problem to obtain $u_c(x,t)$. Both solutions have the form (10.19) in terms of the eigenvalues and eigenfunctions for the given problem. As we noted in Section 10.3, it is difficult to compute many eigenpairs accurately. However, we also know from Section 6.1.1 (see page 200) that, for large t, a few terms of the eigenfunction expansion provides an excellent approximation to the exact solution. We compare the temperatures in the two bars after five hours ($t = 18000$ seconds), by which time the first four terms of the eigenfunction expansion will provide an excellent approximation to each solution.

We have already computed the first four eigenvalues and eigenfunctions for each problem; it remains only to compute the coefficients c_1, c_2, c_3, c_4 in the eigenfunction expansion for u_0. Let α be the vector of nodal values of u_0 ($\alpha_i = u_0(x_i)$), where x_i is the ith node in the finite element mesh: $x_i = ih$). Then we can approximate u_0 accurately by the piecewise linear interpolant of u_0:

$$u_0 \doteq \sum_{i=1}^{n} \alpha_i \phi_i.$$

The coefficients c_k are defined by

$$c_k = \frac{(u_0, \psi_k)_w}{(\psi_k, \psi_k)_w}, \ k = 1, 2, 3, \ldots,$$

where $(\cdot, \cdot)$ represents the weighted $L^2(0, \ell)$ inner product determined by the function $w(x)$. Therefore,

$$\begin{aligned}
(u_0, \psi_k)_w &= \int_0^\ell u_0(x) \psi_k(x) \, dx \\
&\doteq \int_0^\ell \left(\sum_{i=1}^n \alpha_i \phi_i(x) \right) \psi_k(x) w(x) \, dx \\
&= \sum_{i=1}^n \alpha_i \int_0^\ell \phi_i(x) \psi_k(x) w(x) \, dx.
\end{aligned}$$

Now, assuming the eigenfunctions have been computed by the finite element method, we actually have, instead of ψ_k, a vector of (approximate) nodal values of ψ_k, say $\mathbf{v_k}$. Therefore,

$$(u_0, \psi_k)_w \doteq \sum_{i=1}^{n} \alpha_i \int_0^\ell \phi_i(x) \left(\sum_{j=1}^{n} (\mathbf{v_k})_j \phi_j(x) \right) w(x) dx$$

$$= \sum_{i=1}^{n} \sum_{j=1}^{n} \alpha_i (\mathbf{v_k})_j \int_0^\ell \phi_i(x) \phi_j(x) w(x) dx$$

$$= \sum_{i=1}^{n} \sum_{j=1}^{n} W_{ij} \alpha_i (\mathbf{v_k})_j$$

$$= \alpha \cdot \mathbf{W v_k},$$

where $\mathbf{W}$ is the mass matrix determined by the weight function w. Similarly,

$$(\psi_k, \psi_k)_w \doteq \mathbf{v_k} \cdot \mathbf{W v_k}.$$

Therefore, we can easily estimate the needed coefficients, using only information that we have already computed:

$$c_k \doteq \frac{\alpha \cdot \mathbf{W v_k}}{\mathbf{v_k} \cdot \mathbf{W v_k}}, \ k = 1, 2, 3, \ldots.$$

Using the above results, we obtain accurate estimates of $u(x, 18000)$ and $u_c(x, 18000)$, which are displayed in Figure 10.4. We see that the average temperature in the heterogeneous bar is greater than (that is, has decreased less than) the temperature in the homogeneous bar, consistent with the eigenvalues for the two problems. The average temperature in the heterogeneous bar at this time is approximately $2.906°$ Celsius, while in the homogeneous bar it is $2.784°$. (Exercise 1 asks the reader to determine a formula for these averages.) ∎

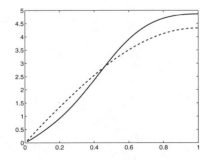

Figure 10.4. *The temperature at $t = 18000$ in the two bars of Example 10.4; the solid curve is the temperature in the heterogeneous bar, while the dashed curve is the temperature in the homogeneous bar.*

10.5. Robin boundary conditions

Exercises

1. Suppose u is a piecewise linear function defined on a regular mesh on $[0,\ell]$ (n subintervals, each of length $h = \ell/n$), and let **u** be the vector of nodal values of u. Explain how to compute the average value of u on ℓ.

2. Consider the temperature distribution in a 50 cm bar with the following physical parameters, where we assume that the bar occupies the interval $[0,50]$:

$$\rho(x) = 7 + 0.04x \text{ g/cm}^3,$$
$$c(x) = 0.4 - 0.005x \text{ J/g K},$$
$$\kappa(x) = 1 + 0.2x \text{ W/cm K}.$$

Let the left endpoint of the bar be completely insulated (along with the lateral boundary of the bar), and let the right endpoint be immersed in an ice bath.

   (a) Formulate the IBVP that describes $u(x,t)$, the temperature at point $x \in (0,50)$ in the bar and time $t > 0$.

   (b) Formulate the related Sturm–Liouville problem.

   (c) Estimate the five smallest eigenvalues and corresponding eigenfunctions for the Sturm–Liouville problem.

3. Consider a one-dimensional bar of length 25 cm with a variable stiffness. Assume that the left end of the bar ($x = 0$) is fixed, while the right end of the bar ($x = 25$) is free to move, and let the density and stiffness in the bar be

$$\rho(x) = 8 + 0.04x \text{ g/cm}^3,$$
$$k(x) = 10^{12} - 10^{10}x \text{ g/cm s}^2.$$

Estimate the fundamental frequency (that is, the smallest natural frequency) of the bar.

10.5 Robin boundary conditions

In presenting the basic theory of Sturm–Liouville problems, we introduced the boundary conditions

$$\alpha_1 u(a) + \alpha_2 \frac{du}{dx}(a) = 0, \; \beta_1 u(b) + \beta_2 \frac{du}{dx}(b) = 0,$$

which are called Robin conditions if the constants are all nonzero. We now investigate these boundary conditions further. Since it is not possible to obtain explicit formulas for the solution of an arbitrary Sturm–Liouville problem with nonconstant coefficients and Robin boundary conditions, we begin with a relatively simple problem that admits an analytic solution. Even in this case, as we will see, the eigenvalues must be estimated numerically. In the next section, we will discuss finite element methods for BVPs under Robin boundary conditions; these methods, of course, allow nonconstant coefficients in the differential equation.

10.5.1 Eigenvalues under Robin conditions

We will study the following BVP:

$$-\frac{\kappa}{\rho c}\frac{d^2u}{dx^2} = \lambda u, \ 0 < x < \ell,$$
$$\frac{du}{dx}(0) = 0, \quad (10.20)$$
$$\kappa\frac{du}{dx}(\ell) + \alpha u(\ell) = 0.$$

We have written the second boundary condition as it might appear in a steady-state heat flow problem. The expression

$$-\kappa\frac{du}{dx}(\ell)$$

is the heat flux (in the positive direction) at $x = \ell$, and the boundary condition

$$-\kappa\frac{du}{dx}(\ell) = \alpha u(\ell)$$

indicates that the heat flux is proportional to the temperature at the endpoint. This boundary condition more commonly appears in the inhomogeneous form

$$\kappa\frac{du}{dx}(\ell) + \alpha u(\ell) = \alpha T \ \Leftrightarrow \ -\kappa\frac{du}{dx}(\ell) = \alpha(u(\ell) - T).$$

This states that the heat flux is proportional to the difference between the temperature at the end of the bar and the surrounding temperature T. It is important to notice that $\alpha > 0$ in the physically meaningful case (if $u(\ell) > T$, then heat flows out of the bar at $x = \ell$).

The BVP (10.20) arises when the method of eigenfunction expansion is applied to the IBVP

$$\rho c\frac{\partial u}{\partial t} - \kappa\frac{\partial^2 u}{\partial x^2} = f(x,t), \ 0 < x < \ell, \ t > 0,$$
$$u(x,0) = u_0(x), \ 0 < x < \ell,$$
$$\frac{\partial u}{\partial x}(0,t) = 0, \ t > 0, \quad (10.21)$$
$$\kappa\frac{\partial u}{\partial x}(\ell,t) + \alpha u(\ell,t) = \alpha T, \ t > 0.$$

The boundary conditions indicate that the left end of the bar is perfectly insulated, and heat flows into or out of the bar through the right end according to the Robin boundary condition. As usual, to apply the method of eigenfunction expansions, we must find the eigenpairs of the related spatial operator under homogeneous boundary conditions.

For simplicity in what follows, we replace (10.20) with

$$-\frac{d^2u}{dx^2} = \lambda u, \ 0 < x < \ell,$$
$$\frac{du}{dx}(0) = 0, \quad (10.22)$$
$$\frac{du}{dx}(\ell) + \overline{\alpha} u(\ell) = 0,$$

10.5. Robin boundary conditions

where $\overline{\alpha} = \alpha/\kappa$. Having found the eigenvalues for this problem, one can merely multiply them by the constant $\kappa/(\rho c)$ to find the eigenvalues for (10.20). The eigenfunctions for (10.20) and (10.22) are the same. We will focus on the case in which $\overline{\alpha} > 0$, since this is the physically meaningful case; for the sake of completeness, we will briefly consider $\overline{\alpha} < 0$ at the end of the discussion.

The operator associated with the eigenvalue problem (10.22) is $L : S \to C[0, \ell]$,

$$Lu = -\frac{d^2 u}{dx^2},$$

where

$$S = \left\{ v \in C^2[0, \ell] : \frac{du}{dx}(0) = 0, \frac{du}{dx}(\ell) + \overline{\alpha} u(\ell) = 0 \right\}.$$

The reader should recall that we have already verified (in Section 10.2.1) that L is symmetric, and hence that L has only real eigenvalues. In fact, all the eigenvalues of L must be positive, as we now show. We suppose that λ is an eigenvalue of L with associated eigenfunction u, where $(u, u) = 1$ (as usual, $(\cdot, \cdot)$ represents the L^2 inner product). Then

$$\begin{aligned}
\lambda = \lambda(u, u) = (\lambda u, u) &= -\int_0^\ell \frac{d^2 u}{dx^2}(x) u(x) \, dx \\
&= -\frac{du}{dx}(x) u(x) \Big|_0^\ell + \int_0^\ell \left(\frac{du}{dx}(x) \right)^2 dx \\
&= -\frac{du}{dx}(\ell) u(\ell) + \int_0^\ell \left(\frac{du}{dx}(x) \right)^2 dx.
\end{aligned}$$

In the last step, we used the fact that u must satisfy the boundary condition

$$\frac{du}{dx}(0) = 0.$$

We also have

$$\frac{du}{dx}(\ell) + \overline{\alpha} u(\ell) = 0 \implies \frac{du}{dx}(\ell) = -\overline{\alpha} u(\ell),$$

and thus we obtain

$$\lambda = \overline{\alpha} u(\ell)^2 + \int_0^\ell \left(\frac{du}{dx}(x) \right)^2 dx.$$

Since $\overline{\alpha} > 0$ and $u \neq 0$ by assumption, we see that λ is positive (certainly the expression for λ shows that $\lambda \geq 0$, and if λ were zero, then $du/dx(x) = 0$ for all $x \in [0, \ell]$, $u(\ell) = 0$ would imply that u is the zero function).

With the knowledge that every eigenvalue of L is positive, we now derive the eigenfunctions. The differential equation is

$$\frac{d^2 u}{dx^2} + \lambda u = 0, \quad 0 < x < \ell,$$

and the general solution is

$$u(x) = a\cos\left(\sqrt{\lambda}x\right) + b\sin\left(\sqrt{\lambda}x\right).$$

Then

$$\frac{du}{dx}(x) = -a\sqrt{\lambda}\sin\left(\sqrt{\lambda}x\right) + b\sqrt{\lambda}\cos\left(\sqrt{\lambda}x\right),$$

and the Neumann boundary condition yields

$$-a\sqrt{\lambda}\cdot 0 + b\sqrt{\lambda}\cdot 1 = 0 \;\Rightarrow\; b = 0.$$

Therefore, $u(x)$ must be a multiple of $\cos(\sqrt{\lambda}x)$. We must now find the values of λ that cause the Robin boundary condition to be satisfied:

$$\frac{du}{dx}(\ell) + \overline{\alpha}u(\ell) = 0 \;\Rightarrow\; -\sqrt{\lambda}\sin\left(\sqrt{\lambda}\ell\right) + \overline{\alpha}\cos\left(\sqrt{\lambda}\ell\right) = 0$$

$$\Rightarrow\; \sqrt{\lambda}\sin\left(\sqrt{\lambda}\ell\right) = \overline{\alpha}\cos\left(\sqrt{\lambda}\ell\right)$$

$$\Rightarrow\; \tan\left(\sqrt{\lambda}\ell\right) = \frac{\overline{\alpha}}{\sqrt{\lambda}}.$$

This last equation must be solved for λ. To make the following analysis simpler, we define $s = \sqrt{\lambda}\ell$. Then we must solve

$$\tan(s) = \frac{\overline{\alpha}\ell}{s} \tag{10.23}$$

for $s > 0$. This equation cannot be solved in closed form, but a graph reveals the nature of the solutions. We graph $y = \tan(s)$ and $y = \overline{\alpha}\ell/s$ and look for the points of intersection. From Figure 10.5, we can draw the following conclusions:

- There are infinitely many solutions $0 < s_0 < s_1 < s_2 < \cdots$.

- The smallest solution s_0 satisfies $0 < s_0 < \pi/2$.

- The subsequent solutions $s_1, s_2, s_3, \ldots$ satisfy

$$k\pi < s_k < \frac{(2k+1)\pi}{2}, \; k = 1, 2, 3, \ldots,$$

with $s_k \doteq k\pi$ for k large.

Since $\lambda = s^2/\ell^2$, we see that there is an infinite sequence of positive eigenvalues $\lambda_0, \lambda_1, \lambda_2, \ldots$ with

$$0 < \lambda_0 < \frac{\pi^2}{4\ell^2}$$

and

$$\frac{k^2\pi^2}{\ell^2} < \lambda_k < \frac{(2k+1)^2\pi^2}{4\ell^2}, \; k = 1, 2, 3, \ldots.$$

10.5. Robin boundary conditions

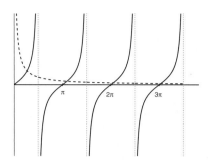

Figure 10.5. *The functions $y = \tan(s)$ (solid curves) and $y = \overline{\alpha}\ell/s$ (dashed curve). The intersection points are related to the eigenvalues of $-d^2/dx^2$ under a Robin boundary condition.*

Also,

$$\lambda_k \doteq \frac{k^2 \pi^2}{\ell^2} \text{ for } k \text{ large.}$$

With these values for the eigenvalues, the corresponding eigenfunctions are

$$\psi_k(x) = \cos\left(\sqrt{\lambda_k}\, x\right), \ k = 0, 1, 2, \ldots.$$

We have not given any precise values for the solutions s_k of (10.23) or the resulting eigenvalues, because they depend on the particular values of $\overline{\alpha}$ and ℓ. However, given $\overline{\alpha}$, ℓ, it is straightforward to solve the equation for as many s_k as desired using Newton's method or a similar technique. Knowledge of a good estimate of the solution makes it easy to solve an equation such as (10.23) accurately.

Example 10.5. We study a 100 cm iron bar that is perfectly insulated except at the right end; there we assume that heat is lost in accordance with the Robin boundary condition

$$\kappa \frac{\partial u}{\partial x}(100) + \alpha u(100) = 0.$$

We assume that the initial temperature in the bar is 10° Celsius, and we take $\alpha = 0.005$. The other physical parameters are $\rho = 7.88 \text{ g/cm}$, $c = 0.437 \text{ J/(g K)}$, and $\kappa = 0.836 \text{ W/(cm K)}$. We wish to solve the heat equation

$$\begin{aligned}
\rho c \frac{\partial u}{\partial t} - \kappa \frac{\partial^2 u}{\partial x^2} &= 0, \ 0 < x < \ell, \ t > 0, \\
u(x, 0) &= 10, \ 0 < x < \ell, \\
\frac{\partial u}{\partial x}(0) &= 0, \ t > 0, \\
\kappa \frac{\partial u}{\partial x}(\ell) + \alpha u(\ell) &= 0, \ t > 0.
\end{aligned} \quad (10.24)$$

The reader should notice that the Robin boundary condition models the situation in which the temperature of the surroundings is 0° Celsius, and the heat flux from the (warmer) bar to the surroundings is proportional to the temperature difference at the end. This modeling assumption should be compared to the assumption that the temperature at the end actually equals the surrounding temperature, which is modeled by the Dirichlet boundary condition $u(\ell) = 0$.

The first step in solving the above IBVP is to find the eigenvalues and eigenfunctions of the operator L considered above (that is, the negative second derivative operator). We have already done all of the necessary analysis, except that we must find accurate estimates of the solutions s_k of (10.23). We did this for $k = 0, 1, \ldots, 9$ using the MATLAB library routine **fsolve**; we then computed the corresponding eigenvalues

$$\lambda_k = \frac{s_k^2}{\ell^2}$$

of $-d^2/dx^2$. The results are shown in Table 10.7, along with the estimates $k\pi$ for s_k (which we expect to be accurate except for the smallest values of k), and the values of $\kappa/(\rho c)\lambda_k$, which are the eigenvalues of

$$-\frac{\kappa}{\rho c}\frac{d^2}{dx^2}.$$

Table 10.7. *The first* 10 *solutions s_k to* (10.23) *for Example* 10.5, *along with the estimates* $(k-1)\pi$ *of s_k and the resulting eigenvalues λ_k. Also shown are the values $\kappa/(\rho c)\lambda_k$, the scaled eigenvalues.*

k	s_k	$k\pi$	λ_k	$\frac{\kappa}{\rho c}\lambda_k$
0	0.70414	0	$4.9581 \cdot 10^{-5}$	$1.2037 \cdot 10^{-5}$
1	3.3198	3.1416	$1.1021 \cdot 10^{-3}$	$2.6757 \cdot 10^{-4}$
2	6.3767	6.2832	$4.0662 \cdot 10^{-3}$	$9.8717 \cdot 10^{-4}$
3	9.4877	9.4248	$9.0017 \cdot 10^{-3}$	$2.1854 \cdot 10^{-3}$
4	12.614	12.566	$1.5911 \cdot 10^{-2}$	$3.8627 \cdot 10^{-3}$
5	15.746	15.708	$2.4793 \cdot 10^{-2}$	$6.0192 \cdot 10^{-3}$
6	18.881	18.850	$3.5650 \cdot 10^{-2}$	$8.6548 \cdot 10^{-3}$
7	22.018	21.991	$4.8481 \cdot 10^{-2}$	$1.1770 \cdot 10^{-2}$
8	25.157	25.133	$6.3285 \cdot 10^{-2}$	$1.5364 \cdot 10^{-2}$
9	28.295	28.274	$8.0063 \cdot 10^{-2}$	$1.9437 \cdot 10^{-2}$

The eigenfunctions corresponding to the eigenvalues found above are given by $\psi_k(x) = \cos(\sqrt{\lambda_k}x)$. We can now solve the IBVP (10.24) using the method of eigenfunction expansion. The solution is

$$u(x,t) = \sum_{k=0}^{\infty} a_k(t)\psi_k(x),$$

where a_k satisfies the initial value problem (IVP)

$$\frac{da_k}{dt} + \frac{\kappa\lambda_k}{\rho c}a_k = 0, \; a_k(0) = c_k,$$

10.5. Robin boundary conditions

and $c_1, c_2, c_3, \ldots$ are the generalized Fourier coefficients for the initial temperature $u_0(x) = 10$:

$$c_k = \frac{(u_0, \psi_k)}{(\psi_k, \psi_k)} = \frac{\int_0^\ell 10 \psi_k(x) \, dx}{\int_0^\ell \psi_k(x)^2 \, dx} = \frac{40 \sin\left(\sqrt{\lambda_k} \ell\right)}{2\sqrt{\lambda_k} \ell + \sin\left(2\sqrt{\lambda_k} \ell\right)}, \; k = 0, 1, 2, \ldots.$$

The coefficients a_k are given by

$$a_k(t) = c_k e^{-\kappa \lambda_k t / (\rho c)}, \; k = 0, 1, 2, \ldots.$$

Figure 10.6 shows the temperature distribution in the bar after 30, 60, 90, 120, 150, and 180 minutes, computed using 10 terms in the eigenfunction expansion. ■

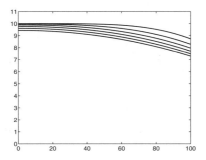

Figure 10.6. *The temperature in the bar of Example* 10.5 *after* 30, 60, 90, 12, 150, *and* 180 *minutes.*

10.5.2 The nonphysical case

Out of mathematical curiosity, let us briefly examine the case of a Robin condition

$$\frac{du}{dx}(\ell) + \overline{\alpha} u(\ell) = 0 \tag{10.25}$$

with $\overline{\alpha} < 0$. Much of the analysis we did above will carry over to this case. We used the assumption $\overline{\alpha} > 0$ in two ways: to conclude that all the eigenvalues must be positive, and to draw Figure 10.5. The counterpart of Figure 10.5 is Figure 10.7, in which $\overline{\alpha} < 0$ is assumed. This graph still determines the positive eigenvalues, with the difference that we now cannot assume that all the eigenvalues are positive.

Reasoning as before, we see that, with $\overline{\alpha} < 0$, the operator L defined on page 411 has positive eigenvalues $\lambda_1, \lambda_2, \lambda_3, \ldots$, with

$$\frac{(2k-1)^2 \pi^2}{4\ell^2} < \lambda_k < \frac{k^2 \pi^2}{\ell^2}, \; k = 1, 2, 3, \ldots.$$

We still have

$$\lambda_k \doteq \frac{k^2 \pi^2}{\ell^2} \text{ for } k \text{ large,}$$

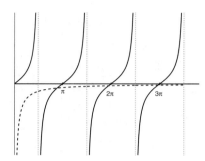

Figure 10.7. *The functions $y = \tan(s)$ (solid curves) and $y = \overline{\alpha}\ell/s$ (dashed curve) for $\overline{\alpha} < 0$. The intersection points are related to the positive eigenvalues of $-d^2/dx^2$ under a nonphysical Robin boundary condition.*

although now this estimate of λ_k is slightly too large rather than slightly too small. With these values for the eigenvalues, the corresponding eigenfunctions are

$$\psi_k(x) = \cos\left(\sqrt{\lambda_k}x\right), \ k = 0, 1, 2, \ldots.$$

We now look for nonpositive eigenvalues. If $\lambda = 0$ were an eigenvalue, we would have a nonzero function u satisfying

$$-\frac{d^2u}{dx^2} = 0, \ 0 < x < \ell,$$

and

$$\frac{du}{dx}(0) = 0, \ \frac{du}{dx}(\ell) + \overline{\alpha}u(\ell) = 0.$$

The differential equation implies that $u(x) = mx + b$ for some constants m and b, and the Neumann condition at the left implies that the slope m is zero. Then $u(x) = b$, and the Robin condition at the right endpoint forces b to be zero. Thus, in fact, there is no nonzero function satisfying the given conditions, and therefore 0 is not an eigenvalue.

If $\lambda < 0$, then we will write $\mu = -\lambda$, and the BVP becomes

$$\frac{d^2u}{dx^2} - \mu u = 0, \ 0 < x < \ell,$$

$$\frac{du}{dx}(0) = 0,$$

$$\frac{du}{dx}(\ell) + \overline{\alpha}u(\ell) = 0.$$

The general solution of the ODE is

$$u(x) = ae^{\sqrt{\mu}x} + be^{-\sqrt{\mu}x}$$

10.5. Robin boundary conditions

and

$$\frac{du}{dx}(x) = a\sqrt{\mu}e^{\sqrt{\mu}x} - b\sqrt{\mu}e^{-\sqrt{\mu}x}.$$

The Neumann condition at $x = 0$ implies $a = b$, and the Robin condition at $x = \ell$ can be manipulated to yield

$$\frac{e^{\sqrt{\mu}\ell} - e^{-\sqrt{\mu}\ell}}{e^{\sqrt{\mu}\ell} + e^{-\sqrt{\mu}\ell}} = -\frac{\overline{\alpha}}{\sqrt{\mu}} \Leftrightarrow \tanh\left(\sqrt{\mu}\ell\right) = -\frac{\overline{\alpha}}{\sqrt{\mu}}$$

(we leave the intermediate steps to the reader). For later reference, we point out that $a = b$ means that u can be represented conveniently as $u(x) = \cosh(\sqrt{\mu}x)$. With $s = \sqrt{\mu}\ell$, we must solve

$$\tanh(s) = -\frac{\overline{\alpha}\ell}{s}$$

for $s > 0$. Figure 10.8 shows that there is a single solution s_0. We therefore see that, if $\overline{\alpha} < 0$, there is a single negative eigenvalue $\lambda_0 = -s_0^2/\ell^2$ with corresponding eigenfunction $\psi_0(x) = \cosh(\sqrt{-\lambda_0}x)$.

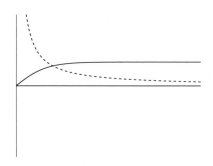

Figure 10.8. *The functions $y = \tanh(s)$ (solid curve) and $y = -\overline{\alpha}\ell/s$ (dashed curve) for $\overline{\alpha} < 0$. The intersection point determines the only negative eigenvalue of $-d^2/dx^2$ under a nonphysical Robin boundary condition.*

We can now summarize what we have learned about the eigenvalues of the operator L (the negative second derivative operator subject to a Neumann condition on the left and the Robin condition (10.25) on the right).

- If $\overline{\alpha} > 0$, then L has positive eigenvalues $0 < \lambda_0 < \lambda_1 < \lambda_2 < \cdots$ with $\lambda_k \doteq k^2\pi^2/\ell^2$ for large k.

- If $\overline{\alpha} < 0$, then L has a single negative eigenvalue, and the remaining eigenvalues are positive: $\lambda_0 < 0 < \lambda_1 < \lambda_2 < \cdots$. The estimate $\lambda_k \doteq k^2\pi^2/\ell^2$ for large k still holds.

Thus the smallest eigenvalue changes from positive to negative with $\overline{\alpha}$. One would naturally expect that λ_0 would be zero when $\overline{\alpha} = 0$, and this is true: $\overline{\alpha} = 0$ corresponds to a Neumann condition at $x = \ell$, and we already know that $-d^2/dx^2$ has a zero eigenvalue under Neumann conditions.

Exercises

1. Repeat Example 10.5, but with a homogeneous Dirichlet condition at the right endpoint instead of the Robin condition. Produce a plot like Figure 10.6.

2. Consider the following Sturm–Liouville problem:
$$-\frac{d^2u}{dx^2} = \lambda u, \ 0 < x < \ell,$$
$$u(0) = 0,$$
$$\kappa \frac{du}{dx}(\ell) + \alpha u(\ell) = 0.$$

   Analyze the eigenvalues and eigenfunctions in the case that $\alpha > 0$.

3. Repeat the previous problem, but for $\alpha < 0$.

4. Consider the following Sturm–Liouville problem with Robin conditions at both ends:
$$-\frac{d^2u}{dx^2} = \lambda u, \ 0 < x < \ell,$$
$$\frac{du}{dx}(0) - \alpha u(0) = 0,$$
$$\frac{du}{dx}(\ell) + \beta u(\ell) = 0$$

   (where $\alpha, \beta > 0$).

   (a) Prove that all the eigenvalues are positive.

   (b) Derive formulas for the eigenvalues and eigenfunctions. As above, an equation must be solved (numerically) to find the values of the eigenvalues. What is this equation?

   (c) Show graphically that, if the eigenvalues are labeled $\lambda_0, \lambda_1, \lambda_2, \ldots$, then $\lambda_k \doteq k^2\pi^2/\ell^2$ for large k.

10.6 Finite element methods for Robin boundary conditions

We can derive finite element methods for problems with Robin boundary conditions by the usual method of multiplying the differential equation by a test function and integrating by parts. In the process, we will see that a Robin condition is a natural boundary condition—it is not imposed explicitly on the solution or the test functions.

In this section, we will address two model problems, a boundary value problem (BVP) and a Sturm–Liouville eigenvalue problem, both with mixed boundary conditions.

10.6.1 A BVP with a Robin condition

For our first model problem, we take

$$-\frac{d}{dx}\left(P(x)\frac{du}{dx}\right) + R(x)u = F(x), \ a < x < b,$$
$$\frac{du}{dx}(a) = 0, \qquad (10.26)$$
$$\beta\frac{du}{dx}(b) + \alpha u(b) = 0.$$

The function P is assumed to be positive on the interval $[a,b]$.

If we multiply both sides of the differential equation by a test function v and integrate from a to b, we obtain

$$-\int_a^b \frac{d}{dx}\left(P(x)\frac{du}{dx}(x)\right)v(x)\,dx + \int_a^b R(x)u(x)v(x)\,dx = \int_a^b F(x)v(x)\,dx.$$

We integrate by parts in the first integral to obtain

$$-\int_a^b \frac{d}{dx}\left(P(x)\frac{du}{dx}(x)\right)v(x)\,dx = -\left.P(x)\frac{du}{dx}(x)v(x)\right|_a^b + \int_a^b P(x)\frac{du}{dx}(x)\frac{dv}{dx}(x)\,dx$$
$$= -P(b)\frac{du}{dx}(b)v(b) + \int_a^b P(x)\frac{du}{dx}(x)\frac{dv}{dx}(x)\,dx$$

(since $du/dx(0) = 0$). The Robin boundary condition yields

$$\frac{du}{dx}(b) = -\frac{\alpha}{\beta}u(b),$$

and therefore

$$-\int_a^b \frac{d}{dx}\left(P(x)\frac{du}{dx}(x)\right)v(x)\,dx = \frac{\alpha P(b)}{\beta}u(b)v(b) + \int_a^b P(x)\frac{du}{dx}(x)\frac{dv}{dx}(x)\,dx.$$

We thus obtain the weak form of the BVP (10.26): Find u such that

$$\frac{\alpha P(b)}{\beta}u(b)v(b) + \int_a^b P(x)\frac{du}{dx}(x)\frac{dv}{dx}(x)\,dx + \int_a^b R(x)u(x)v(x)\,dx = \int_a^b F(x)v(x)\,dx$$
(10.27)

for all test functions v. We derived this equation without any assumptions about the boundary values of v, and hence we impose no such conditions on v; both the Neumann condition at $x = a$ and the Robin condition at $x = b$ are natural boundary conditions.

Having derived the weak form of the BVP, we apply the Galerkin method with a space of finite element functions. We choose a positive integer n, define $h = (b-a)/n$, and establish a mesh with nodes $x_i = a + ih$, $i = 0, 1, \ldots, n$. We will use piecewise linear functions, for which the standard nodal basis functions are $\phi_0, \phi_1, \ldots, \phi_n$ satisfying

$$\phi_i(x_j) = \begin{cases} 1, & i = j, \\ 0, & i \neq j. \end{cases}$$

The Galerkin method, with the given approximating subspace, requires finding an approximation

$$v_n = \sum_{j=0}^{n} u_j \phi_j$$

to u that satisfies

$$\frac{\alpha P(b)}{\beta} v_n(b)\phi_i(b) + \int_a^b P(x)\frac{dv_n}{dx}(x)\frac{d\phi_i}{dx}(x)\,dx \\ + \int_a^b R(x)v_n(x)v(x)\,dx = \int_a^b F(x)\phi_i(x)\,dx \qquad (10.28)$$

for $i = 0, 1, \ldots, n$. Three of the expressions appearing in this equation are familiar:

$$\int_a^b P(x)\frac{dv_n}{dx}(x)\frac{d\phi_i}{dx}(x)\,dx = \sum_{j=0}^{n} K_{ij}u_j, \; K_{ij} = \int_a^b P(x)\frac{d\phi_j}{dx}(x)\frac{d\phi_i}{dx}(x)\,dx,$$

$$\int_a^b R(x)v_n(x)v(x)\,dx = \sum_{j=0}^{n} M_{ij}u_j, \; M_{ij} = \int_a^b R(x)\phi_j(x)\phi_i(x)\,dx,$$

$$\int_a^b F(x)\phi_i(x)\,dx = f_i,$$

where $\mathbf{K}$, $\mathbf{M}$, and $\mathbf{f}$ are the usual stiffness matrix, mass matrix, and load vector, respectively. To be specific, $\mathbf{K}$ and $\mathbf{M}$ are the usual stiffness and mass matrices corresponding to Neumann conditions at both endpoints.

The boundary term can be written as $\mathbf{Gu}$, but the matrix $\mathbf{G}$ is extremely simple. We have

$$\frac{\alpha P(b)}{\beta} v_n(b)\phi_i(b) = \frac{\alpha P(b)}{\beta} \sum_{j=0}^{n} u_j \phi_j(b)\phi_i(b) = \begin{cases} \frac{\alpha P(b)}{\beta} u_n, & i = n, \\ 0 & \text{otherwise} \end{cases}$$

since $\phi_i(b) = 0$ unless $i = n$, and similarly for $\phi_j(b)$. This shows that

$$G_{ij} = \begin{cases} \frac{\alpha P(b)}{\beta}, & i = j = n, \\ 0 & \text{otherwise.} \end{cases}$$

We then see that (10.28) is equivalent to

$$\mathbf{Au} = \mathbf{f}, \qquad (10.29)$$

where $\mathbf{A} = \mathbf{G} + \mathbf{K} + \mathbf{M}$ and $\mathbf{G}$, $\mathbf{K}$, $\mathbf{M}$, and $\mathbf{f}$ are as defined above.

10.6. Finite element methods for Robin boundary conditions

Example 10.6. We will solve the BVP

$$-\kappa \frac{du}{dx} = F(x),\ 0 < x < \ell,$$
$$\frac{du}{dx}(0) = 0, \quad (10.30)$$
$$\kappa \frac{du}{dx}(\ell) + \alpha u(\ell) = 0,$$

where $\ell = 100$, $\kappa = 0.836$, $\alpha = 1$,

$$F(x) = \begin{cases} cx^2(25-x)^2, & 0 < x < 25, \\ 0 & \text{otherwise,} \end{cases}$$

and $c = 3.26 \cdot 10^6$. This BVP models steady-state heat flow in a 100 cm bar, insulated on the sides and at the left end. Heat flows out of the right end in accordance with the given Robin condition, while heat is added to part of the bar, as described by the source F. The value of c was chosen so that the heat energy added to the bar is $0.1\,\text{J}/(\text{cm}^3\text{s})$.

In terms of the notation established above, we have $P(x) = \kappa$ and $R(x) = 0$. Since R is zero, so is the matrix $\mathbf{M}$. Since P is constant, $\mathbf{K}$ is the usual stiffness matrix, under Neumann conditions, multiplied by the constant κ:

$$\mathbf{K} = \kappa \begin{bmatrix} \frac{1}{h} & -\frac{1}{h} & & & & \\ -\frac{1}{h} & \frac{2}{h} & -\frac{1}{h} & & & \\ & -\frac{1}{h} & \frac{2}{h} & -\frac{1}{h} & & \\ & & \ddots & \ddots & \ddots & \\ & & & -\frac{1}{h} & \frac{2}{h} & -\frac{1}{h} \\ & & & & -\frac{1}{h} & \frac{1}{h} \end{bmatrix}$$

(cf. Example 6.10). The load vector $\mathbf{f}$ is given by

$$f_i = \frac{ch^3}{30}\left(30h^2i^4 + 30h^2i^2 + 2h^2 - 1500hi^3 - 750hi + 18750i^2 + 3125\right)$$

for $i = 1, 2, \ldots, n/4 - 1$, with

$$f_i = \begin{cases} \frac{ch^3}{60}\left(2h^2 - 150h + 3125\right), & i = 0, \\ \frac{ch^3}{60}\left(2h^2 - 150h + 3125\right), & i = \frac{n}{4}, \\ 0, & i = \frac{n}{4}+1, \ldots, n. \end{cases}$$

Because F is nonzero only on $(0, 25) = (0, \ell/4)$, the load vector $\mathbf{f}$ has nonzero components only in components $0, 1, \ldots, n/4$. Finally, the matrix $\mathbf{A} = \mathbf{K} + \mathbf{G}$ is simply $\mathbf{K}$ with $\alpha p(\ell)/\kappa = \alpha$ added to the n, n entry.

Solving the system $\mathbf{A}\mathbf{u} = \mathbf{f}$, we obtain the approximate solution shown in Figure 10.9. As a check on the solution, Exercise 10.6.5 shows that

$$\alpha u(\ell) = \int_0^\ell f(x)\,dx$$

must hold for the exact solution. We constructed f so that

$$\int_0^\ell f(x)\,dx = 0.1,$$

and $\alpha = 1$ in this problem. We should therefore have $u(100) = 0.1$; the computed solution has value 0.099853, which seems quite satisfactory. ∎

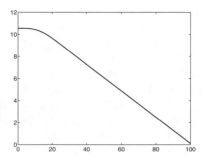

Figure 10.9. *The piecewise linear solution to Example* 10.6 *computed by the finite element method with* $n = 100$.

10.6.2 A Sturm–Liouville problem with a Robin condition

To handle a Sturm–Liouville eigenvalue problem with a Robin boundary condition, we combine our results from Section 10.3 and Section 10.6.1. We take as our model problem

$$-\frac{d}{dx}\left(P(x)\frac{du}{dx}\right) + R(x)u = \lambda w(x)u, \ a < x < b,$$

$$\frac{du}{dx}(a) = 0, \quad (10.31)$$

$$\beta\frac{du}{dx}(b) + \alpha u(b) = 0.$$

The weak form is derived just as above, and only the right-hand side is different:

$$\frac{\alpha P(b)}{\beta}u(b)v(b) + \int_a^b P(x)\frac{du}{dx}(x)\frac{dv}{dx}(x)\,dx$$
$$+ \int_a^b R(x)u(x)v(x)\,dx = \lambda \int_a^b w(x)u(x)v(x)\,dx.$$

When we apply the Galerkin method with the usual approximating subspace of piecewise linear functions, we obtain the generalized eigenvalue problem

$$\mathbf{Au} = \lambda \mathbf{Wu}, \quad (10.32)$$

10.6. Finite element methods for Robin boundary conditions

where

$$W_{ij} = \int_a^b w(x)\phi_j(x)\phi_i(x)\,dx$$

(cf. page 398) and $\mathbf{A} = \mathbf{G} + \mathbf{K} + \mathbf{M}$, as above.

Example 10.7. We will consider again Example 10.5, this time approximating the eigenvalues and eigenfunctions by the finite element method. The Sturm–Liouville problem is

$$-\frac{\kappa}{\rho c}\frac{d^2 u}{dx^2} = \lambda u, \ 0 < x < \ell,$$
$$\frac{du}{dx}(0) = 0, \quad (10.33)$$
$$\kappa \frac{du}{dx}(\ell) + \alpha u(\ell) = 0.$$

The parameters are $\ell = 100$, $\alpha = 0.005$, $\rho = 7.88$, $c = 0.437$, and $\kappa = 0.836$. In terms of the general formulas developed above, we have $P(x) = \kappa/(\rho c)$, $R(x) = 0$, and $w(x) = 1$. The matrix $\mathbf{M}$ (which corresponds to the $R(x)u$ term in the differential equation) vanishes here because $R(x) = 0$. The matrix $\mathbf{K}$ is the same as in Example 10.6, except multiplied by $\kappa/(\rho c)$ instead of κ. The matrix $\mathbf{W}$ is the usual mass matrix (under Neumann conditions):

$$\mathbf{W} = \begin{bmatrix} \frac{h}{3} & \frac{h}{6} & & & & & \\ \frac{h}{6} & \frac{2h}{3} & \frac{h}{6} & & & & \\ & \frac{h}{6} & \frac{2h}{3} & \frac{h}{6} & & & \\ & & \ddots & \ddots & \ddots & & \\ & & & \frac{h}{6} & \frac{2h}{3} & \frac{h}{6} \\ & & & & \frac{h}{6} & \frac{h}{3} \end{bmatrix}.$$

Finally, $\mathbf{A} = \mathbf{G} + \mathbf{K}$ (since $\mathbf{M} = 0$ for this problem). Since all the entries in $\mathbf{G}$ are zero except the one in the lower right-hand corner of the matrix, $\mathbf{A}$ is equal to $\mathbf{K}$ with $\alpha P(b)/\kappa = \alpha/(\rho c)$ added to the n, n entry.

We computed the matrices $\mathbf{A}$, $\mathbf{W}$ for $n = 10$ and $n = 20$ and solved the generalized eigenvalue problem $\mathbf{A}\mathbf{u} = \lambda \mathbf{W}\mathbf{u}$ using a library routine in MATLAB. The results are shown in Table 10.8, which compares the estimated eigenvalues with those computed using the methods of the previous section. The table shows that the finite element method succeeds in computing the smallest eigenvalues accurately, although a finer mesh would be required to compute all of these 10 eigenvalues accurately (on a mesh with 20 elements, for instance, the error in the tenth eigenvalue is about 17%). ∎

It should be straightforward to adapt the derivations of this section to BVPs or eigenvalue problems with the Robin condition at the left endpoint, or a Robin condition combined with a Dirichlet condition, or Robin conditions at both ends.

Table 10.8. *The smallest 10 eigenvalues for the Sturm–Liouville problem in Example 10.7, computed with meshes of $n = 10$ and $n = 20$ elements. Also shown are the relative differences between these estimates and those computed in Example 10.5 using the methods of Section 10.5.*

	$n = 10$			$n = 20$	
k	λ_k	Rel. diff.	k	λ_k	Rel. diff.
1	$1.2042 \cdot 10^{-5}$	$4.1336 \cdot 10^{-4}$	1	$1.2038 \cdot 10^{-5}$	$1.0331 \cdot 10^{-4}$
2	$2.7003 \cdot 10^{-4}$	$9.2247 \cdot 10^{-3}$	2	$2.6818 \cdot 10^{-4}$	$2.2986 \cdot 10^{-3}$
3	$1.0211 \cdot 10^{-3}$	$3.4354 \cdot 10^{-2}$	3	$9.9556 \cdot 10^{-4}$	$8.5014 \cdot 10^{-3}$
4	$2.3538 \cdot 10^{-3}$	$7.7098 \cdot 10^{-2}$	4	$2.2267 \cdot 10^{-3}$	$1.8895 \cdot 10^{-2}$
5	$4.3967 \cdot 10^{-3}$	$1.3826 \cdot 10^{-1}$	5	$3.9924 \cdot 10^{-3}$	$3.3575 \cdot 10^{-2}$
6	$7.3267 \cdot 10^{-3}$	$2.1723 \cdot 10^{-1}$	6	$6.3361 \cdot 10^{-3}$	$5.2657 \cdot 10^{-2}$
7	$1.1328 \cdot 10^{-2}$	$3.0881 \cdot 10^{-1}$	7	$9.3149 \cdot 10^{-3}$	$7.6260 \cdot 10^{-2}$
8	$1.6439 \cdot 10^{-2}$	$3.9672 \cdot 10^{-1}$	8	$1.2999 \cdot 10^{-2}$	$1.0448 \cdot 10^{-1}$
9	$2.2199 \cdot 10^{-2}$	$4.4485 \cdot 10^{-1}$	9	$1.7474 \cdot 10^{-2}$	$1.3734 \cdot 10^{-1}$
10	$2.7177 \cdot 10^{-2}$	$3.9822 \cdot 10^{-1}$	10	$2.2833 \cdot 10^{-2}$	$1.7471 \cdot 10^{-1}$

Exercises

1. Modify the finite element method presented above to apply to the BVP

$$-\frac{d}{dx}\left(P(x)\frac{du}{dx}\right) + R(x)u = F(x), \ a < x < b,$$

$$u(a) = 0,$$

$$\beta \frac{du}{dx}(b) + \alpha u(b) = 0,$$

where $P(x) > 0$ for all $x \in [a,b]$, $\alpha, \beta > 0$. The finite element solution is defined by the solution to $\mathbf{Au} = \mathbf{f}$. Specify carefully $\mathbf{A}$ and $\mathbf{f}$, and explain how they differ from the case presented in Section 10.6.1.

2. Using the result of the previous exercise, solve the BVP of Example 10.6 with the Neumann condition $du/dx(0) = 0$ changed to the Dirichlet condition $u(0) = 0$.

3. Modify the finite element method presented above to apply to the BVP

$$-\frac{d}{dx}\left(P(x)\frac{du}{dx}\right) + R(x)u = F(x), \ a < x < b,$$

$$-\alpha_1 u(a) + \alpha_2 \frac{du}{dx}(a) = 0,$$

$$\beta_1 u(b) + \beta_2 \frac{du}{dx}(b) = 0,$$

where $P(x) > 0$ for all $x \in [a,b]$, $\alpha_1, \alpha_2, \beta_1, \beta_2 > 0$. The finite element solution is defined by the solution to $\mathbf{Au} = \mathbf{f}$. Specify carefully $\mathbf{A}$ and $\mathbf{f}$, and explain how they differ from the case presented in Section 10.6.1.

4. Using the result of the previous exercise, solve the BVP of Example 10.6 with the Neumann condition $du/dx(0) = 0$ changed to the Robin condition

$$\kappa \frac{du}{dx}(0) - \alpha u(0) = 0.$$

5. Suppose u solves (10.30). Prove that

$$\alpha u(\ell) = \int_0^\ell f(x)\, dx.$$

10.7 The theory of Sturm–Liouville problems: An outline

The properties of Sturm–Liouville operators were described in Section 10.2. The purpose of this section is to show the reader, in outline form, how these are derived. Our discussion depends on the material in Chapter 9.

For convenience, we will take as our model problem

$$-\frac{d}{dx}\left(P(x)\frac{du}{dx}\right) + R(x)u = \lambda u, \; a < x < b,$$

$$\alpha_1 u(a) + \alpha_2 \frac{du}{dx}(a) = 0, \quad (10.34)$$

$$\beta_1 u(b) + \beta_2 \frac{du}{dx}(b) = 0,$$

where P is assumed to be positive on the interval $[a,b]$. The problem (10.34) differs from the general Sturm–Liouville problem discussed in Section 10.2 in that the weight function w is taken to be the constant function 1. If w is nonconstant, then a similar analysis can be performed using the weighted L^2 inner product

$$(f,g)_w = \int_a^b f(x)g(x)w(x)\, dx$$

in place of the ordinary L^2 inner product. We also assume that $\lambda = 0$ is not an eigenvalue of (10.34), that is, if $\lambda = 0$, then the only solution to (10.34) is $u = 0$.

We define the Sturm–Liouville operator $L : S \to C[a,b]$ by

$$Lu = -\frac{d}{dx}\left(P(x)\frac{du}{dx}\right) + R(x)u,$$

where

$$S = \left\{ v \in C^2[a,b] : \alpha_1 u(a) + \alpha_2 \frac{du}{dx}(a) = 0, \; \beta_1 u(b) + \beta_2 \frac{du}{dx}(b) = 0 \right\}.$$

We saw in Section 9.2 that there exists a function $G(x;y)$ such that the unique solution u to

$$-\frac{d}{dx}\left(P(x)\frac{du}{dx}\right) + R(x)u = F(x), \ a < x < b,$$
$$\alpha_1 u(a) + \alpha_2 \frac{du}{dx}(a) = 0, \qquad (10.35)$$
$$\beta_1 u(b) + \beta_2 \frac{du}{dx}(b) = 0$$

is

$$u(x) = \int_a^b G(x;y)F(x)\,dy. \qquad (10.36)$$

The function G is the Green's function for (10.35). We define a linear operator $M : C[a,b] \to C[a,b]$ by $MF = u$, where u is given by (10.36). For later reference, we note that M is an example of an *integral operator*.[50]

To say that MF is the solution to (10.35) is to say that $LMF = F$ for all $F \in C[a,b]$. On the other hand, given any $u \in S$, obviously u is the solution to (10.35) with $F = Lu$. In other words, $MLu = u$. This shows that L and M are inverse operators: $L = M^{-1}$.

A general fact about inverse operators is the following: If μ, ψ is an eigenvalue/eigenfunction pair for M, then

$$M\psi = \mu\psi \ \Rightarrow \ \psi = M^{-1}(\mu\psi) = \mu M\psi \ \Rightarrow \ M^{-1}\psi = \mu^{-1}\psi.$$

This shows that M and M^{-1} have the same eigenfunctions, and that the eigenvalues of one are the reciprocals of the other. The reader should notice that $\mu = 0$ cannot be an eigenvalue of M if M is invertible. This follows from the discussion of uniqueness in Section 3.2.2 (if M is invertible, then $MF = u$ has a unique solution for each u, which implies that $MF = 0$ has only the trivial solution $F = 0$).

Therefore, we can find the eigenvalues and eigenfunctions of L (our real interest) by studying the spectral properties of its inverse operator M. This is advantageous, since M has certain mathematical properties, to be described below, that allow far-reaching conclusions about its eigenvalues and eigenfunctions.

10.7.1 Facts about the eigenvalues

We begin by showing that M is symmetric, which follows from the symmetry of the Green's function ($G(x;y) = G(y;x)$). As usual, we use the complex inner product until we know

---

[50] An integral operator $T : C[a,b] \to C[a,b]$ has the form $(Tf)(x) = \int_a^b k(x;y)f(y)\,dy$, where the function k is called the *kernel* of the operator.

10.7. The theory of Sturm–Liouville problems: An outline

that the eigenvalues must be real:

$$
\begin{aligned}
(MF,H) &= \int_a^b (MF)(x)\overline{H(x)}\,dx = \int_a^b \left(\int_a^b G(x;y)F(y)\,dy \right) \overline{H(x)}\,dx \\
&= \int_a^b \int_a^b G(x;y)F(y)\overline{H(x)}\,dy\,dx \\
&= \int_a^b F(y)\left(\int_a^b G(x;y)\overline{H(x)}\,dx \right) dy \\
&= \int_a^b F(y)\left(\int_a^b \overline{G(x;y)H(x)}\,dx \right) dy \\
&= \int_a^b F(y)\left(\int_a^b \overline{G(y;x)H(x)}\,dx \right) dy \\
&= \int_a^b F(y)\overline{\left(\int_a^b G(y;x)H(x)\,dx \right)} dy \\
&= \int_a^b F(y)\overline{(MH)(y)}\,dy \\
&= (F, MH).
\end{aligned}
$$

The reader should notice how the above calculation was based on changing the order of integration in the double integral, on the symmetry of G, and on the fact that G is real (so that $\overline{G(x;y)} = G(x;y)$).

Since M is symmetric, we know that all the eigenvalues of M are real, and that eigenfunctions corresponding to distinct eigenvalues are orthogonal. The range of M is precisely the domain S of L; in other words, for each $F \in C[a,b]$, MF belongs to $S \subset C^2[a,b]$. If μ, ψ is an eigenpair of M, then

$$M\psi = \mu\psi \;\Rightarrow\; \psi = \mu^{-1}M\psi,$$

which shows that ψ lies in $C^2[a,b]$. Thus all eigenfunctions of M are smooth.

Next, we will show that the eigenspace of each eigenvalue of M is one dimensional; in other words, for a given eigenpair μ, ψ of M, every eigenfunction of M corresponding to μ is a multiple of ψ. We suppose that μ is an eigenvalue of M and that ϕ, ψ are two eigenfunctions corresponding to μ. Then ϕ, ψ are eigenfunctions of L ($L\phi = \mu^{-1}\phi$, $L\psi = \mu^{-1}\psi$), and hence ϕ, ψ are solutions of the BVP

$$
\begin{aligned}
-\frac{d}{dx}\left(P(x)\frac{du}{dx} \right) + (R(x) - \mu^{-1})u &= 0, \; a < x < b, \\
\alpha_1 u(a) + \alpha_2 \frac{du}{dx}(a) &= 0, \\
\beta_1 u(b) + \beta_2 \frac{du}{dx}(b) &= 0.
\end{aligned}
\quad (10.37)
$$

To show that ϕ and ψ are linearly dependent, it suffices to show that the Wronskian

$$W(x) = \begin{vmatrix} \phi(x) & \psi(x) \\ \frac{d\phi}{dx}(x) & \frac{d\psi}{dx}(x) \end{vmatrix} = \phi(x)\frac{d\psi}{dx}(x) - \frac{d\phi}{dx}(x)\psi(x)$$

is zero. We have already shown (see page 346) that $w(x) = P(x)W(x)$ is a constant function. Moreover, $P(x) > 0$ by assumption, so $W(x)$ is zero if and only if the constant $w(x)$ is zero. But it is straightforward to prove, from the fact that ϕ and ψ satisfy the same boundary condition at $x = a$, that $w(a) = 0$ (the details are left to the reader). Therefore $w(x) = 0$, and hence $W(x) = 0$, which shows that ϕ is just a multiple of ψ.

Compact sets and compact operators

The following ideas about *compactness* are the most complicated in this section, and they are the key to deriving the desired facts about the eigenvalues of M.

Definition 10.8. *Let V be any vector space with a norm, and let S be a subset of V. We say that S is (sequentially)* compact *if each sequence $\{v_n\}$ in S has a subsequence $\{v_{n_k}\}$ that converges to an element $v \in S$ ($\{v_n\}$ is concise notation for the sequence $v_1, v_2, v_3, \ldots$). We say that S is* precompact *if each sequence $\{v_n\}$ in S has a subsequence $\{v_{n_k}\}$ that converges to an element $v \in V$. (The distinction is that v may not belong to S.)*

There is a more general concept of compactness, which is equivalent to sequential compactness in a normed vector space. We will have no need for the more general concept, and we will use the word "compact" below, although "sequentially compact" would be more precise.

One of the basic facts about the set of real numbers, or more generally about any Euclidean space $\mathbf{R}^n$, is that every closed and bounded set is compact. For instance, any closed and bounded interval of real numbers, such as $[0,1]$, is compact. Therefore, no matter how one defines a sequence $\{x_n\}$ of numbers in $[0,1]$, there is guaranteed to be a subsequence $\{x_{n_k}\}$ and some $x \in [0,1]$ such that $x_{n_k} \to x$. Similarly, in any Euclidean space $\mathbf{R}^n$, the closed *unit ball* $\overline{B} = \{\mathbf{x} \in \mathbf{R}^n : \|\mathbf{x}\| \leq 1\}$ is compact.

On the other hand, in an infinite-dimensional space such as $C[a,b]$, the closed unit ball cannot be compact. This can be demonstrated as follows: In an infinite-dimensional inner product space, we can find an (infinite) orthonormal sequence $\{\psi_n\}$. Any orthonormal sequence lies in the closed unit ball. But, for $m \neq n$,

$$\|\psi_n - \psi_m\|^2 = (\psi_n - \psi_m, \psi_n - \psi_m) = (\psi_n, \psi_n) - 2(\psi_n, \psi_m) + (\psi_m, \psi_m)$$
$$= \|\psi_n\|^2 + \|\psi_m\|^2 \text{ (since } (\psi_n, \psi_m) = 0\text{)}$$
$$= 1 + 1 = 2,$$

which shows that $\|\psi_n - \psi_m\| = \sqrt{2}$ for all $m \neq n$. A little thought now shows that no subsequence of $\{\psi_n\}$ can converge; vectors getting closer and closer to a limit must get closer and closer to one another. This cannot occur with an infinite, orthonormal sequence. This argument holds only for spaces with an inner product, but a more complicated proof can be given in the case of a space whose norm is not derived from an inner product.

We also need the concept of a compact operator.

Definition 10.9. *Let U, V be normed vector spaces, and let $T : U \to V$ be a linear operator. We say that T is* compact *if the image under T of every bounded subset of U is a precompact subset of V (in other words, if $S \subset U$ is bounded, then $\{Tu \in V : v \in S\}$ is precompact).*

10.7. The theory of Sturm–Liouville problems: An outline

In particular, if $T : U \to V$ is a compact operator and $\{u_n\}$ is a bounded sequence in U, then $\{Tu_n\}$ has a subsequence that converges to some $v \in V$. This means that there is a subsequence $\{u_{n_k}\}$ of $\{u_n\}$ such that $Tu_{n_k} \to v$ as $k \to \infty$.

The above concepts are relevant to our discussion because any integral operator (see footnote 50 on page 426), such as M, is a compact operator. Before we can use this fact, we need one more concept, which is that of the norm of a linear operator.

Definition 10.10. *Let U and V be normed linear spaces with norms $\|\cdot\|_U$ and $\|\cdot\|_V$, respectively, and let $T : U \to V$ be linear. We define the* norm *of T by*

$$\|T\| = \sup\left\{ \frac{\|Tu\|_V}{\|u\|_U} : u \in U, u \neq 0 \right\}.$$

We say that T is bounded *if $\|T\| < \infty$ and* unbounded *otherwise.*

In this definition, "sup" denotes the *supremum* of the set, which is the same as the maximum except that the supremum need not be a member of the set. The following examples illustrate the distinction:

$$S = [0,1] \Rightarrow \max(S) = 1 \text{ (1 is the largest element of } S\text{)},$$
$$S = [0,1) \Rightarrow \max(S) \text{ does not exist (there is no largest element of } S\text{)},$$
$$S = [0,1) \Rightarrow \sup(S) = 1.$$

In the last example, 1 is the supremum of $S = [0,1)$ because every element of S is bounded by 1, and there are elements of S arbitrarily close to 1.

It can be shown that every compact operator is also bounded (see Exercise 10.7.1). Therefore, M, being an integral operator and thus compact, is also bounded.

We can now use the fact that M is a compact operator to derive an important fact about its eigenvalues. Let us suppose that M has an infinite sequence $\{\mu_n\}$ of eigenvalues, with corresponding eigenfunctions $\{\psi_n\}$. We can always assume that the eigenfunctions have been normalized ($\|\psi_n\| = 1$ for all n), and since M is symmetric, the eigenfunctions are mutually orthogonal. Thus, $\{\psi_n\}$ is an orthonormal sequence. Since M is compact, there must exist a subsequence $\{\psi_{n_k}\}$ such that $\{T\psi_{n_k}\}$ converges. However, $T\psi_{n_k} = \mu_{n_k}\psi_{n_k}$, and therefore $\{T\psi_{n_k}\} = \{\mu_{n_k}\psi_{n_k}\}$ is an orthogonal sequence. A little reasoning, analogous to the argument given above that an infinite orthonormal sequence cannot converge, shows that $\{\mu_{n_k}\psi_{n_k}\}$ converges only if $\mu_{n_k} \to 0$. In fact, we can conclude that the sequence of all eigenvalues must converge to zero: $\mu_n \to 0$ as $n \to \infty$ (the details of the argument are omitted). This is a fact about every compact operator: If it has an infinite sequence of eigenvalues, then that sequence must converge to zero.

The significance of this fact for us is the following: Since the eigenvalues of L are the reciprocals of the eigenvalues of M, if L has an infinite sequence $\{\lambda_n\}$ of eigenvalues, then $|\lambda_n| \to \infty$ as $n \to \infty$. Moreover, as we will discuss below, $\{\lambda_n\}$ must be bounded below, and therefore $\lambda_n \to \infty$ as $n \to \infty$ must hold.

More properties of the eigenvalues of L and M

Let us suppose that λ, ψ is an eigenpair of L with $(\psi, \psi) = 1$. We then have

$$\lambda = \lambda(\psi, \psi) = (\lambda \psi, \psi)$$
$$= \int_a^b \left(-\frac{d}{dx}\left(P(x)\frac{d\psi}{dx}(x)\right) + R(x)\psi(x) \right) \psi(x) dx$$
$$= -\left. P(x)\frac{d\psi}{dx}(x)\psi(x) \right|_a^b + \int_a^b \left\{ P(x)\left(\frac{d\psi}{dx}(x)\right)^2 + R(x)(\psi(x))^2 \right\} dx.$$

We will now assume that the boundary conditions at each end are either Dirichlet or Neumann conditions. In this case, the boundary term in the above expression vanishes. Moreover,

$$\int_a^b P(x)\left(\frac{d\psi}{dx}(x)\right)^2 dx \geq 0,$$
$$\int_a^b R(x)(\psi(x))^2 dx \geq R_{min} \int_a^b (\psi(x))^2 dx \geq R_{min}.$$

In these calculations, we have used the fact that $P(x) > 0$ on $[a,b]$, and $(\psi, \psi) = 1$. The value R_{min} is the minimum value of R on $[a,b]$, a value that is known to exist because R is assumed continuous.

Therefore, in the case of Dirichlet or Neumann conditions, we have shown that $\lambda \geq R_{min}$. In the case of Robin conditions, it is possible that $\lambda < R_{min}$, but a somewhat complicated argument (which we omit) shows that at most two eigenvalues of L can be strictly less than R_{min}. Thus, in any case, we know that the eigenvalues of L are bounded below. As explained above, a consequence of this is that if L has an infinite sequence $\{\lambda_n\}$ of eigenvalues, then $\lambda_n \to \infty$ as $n \to \infty$.

Finally, we must address the following issue: While a symmetric matrix is guaranteed to have eigenvalues (they are roots of the characteristic polynomial and therefore have to exist), a linear operator defined on an infinite-dimensional space need not have any eigenvalues. We have described above properties that any eigenvalues of M must have; however, we have not shown that M *has* any eigenvalues.

In fact, for any symmetric, compact operator $T : U \to U$, either $\|T\|$ or $-\|T\|$ must be an eigenvalue of T. Here is a sketch of the proof. First of all, it can be shown that, for a symmetric linear operator T, the following is an alternate formula for $\|T\|$:

$$\|T\| = \sup\{|(Tu,u)| : u \in U, \|u\| = 1\}. \tag{10.38}$$

It follows that

$$\|T\| = \sup\{(Tu,u) : u \in U, \|u\| = 1\} \text{ or } \|T\| = \sup\{-(Tu,u) : u \in U, \|u\| = 1\}.$$

We will assume that the first case holds; the proof in the second case is similar. There must be a sequence $\{\psi_n\}$ in U such that $\|\psi_n\| = 1$ for all n and

$$(T\psi_n, \psi_n) \to \|T\| \text{ as } n \to \infty.$$

10.7. The theory of Sturm–Liouville problems: An outline

Since T is compact, there must be a subsequence $\{\psi_{n_k}\}$ such that $\{T\psi_{n_k}\}$ converges to, say, $\phi \in U$. Now, we can show that $\|T\psi_{n_k} - \|T\|\psi_{n_k}\| \to 0$ as $k \to \infty$ as follows:

$$\begin{aligned}
0 \leq \|T\psi_{n_k} - \|T\|\psi_{n_k}\|^2 &= (T\psi_{n_k} - \|T\|\psi_{n_k}, T\psi_{n_k} - \|T\|\psi_{n_k}) \\
&= (T\psi_{n_k}, T\psi_{n_k}) - 2(T\psi_{n_k}, \|T\|\psi_{n_k}) + (\|T\|\psi_{n_k}, \|T\|\psi_{n_k}) \\
&= \|T\psi_{n_k}\|^2 - 2\|T\|(T\psi_{n_k}, \psi_{n_k}) + \|T\|^2(\psi_{n_k}, \psi_{n_k}) \\
&= \|T\psi_{n_k}\|^2 - 2\|T\|(T\psi_{n_k}, \psi_{n_k}) + \|T\|^2 \\
&\to \|T\|^2 - 2\|T\|^2 + \|T\|^2 = 0.
\end{aligned}$$

We then have
$$\|\phi - \|T\|\psi_{n_k}\| \leq \|\phi - T\psi_{n_k}\| + \|T\psi_{n_k} - \|T\|\psi_{n_k}\|,$$
and therefore $\|\phi - \|T\|\psi_{n_k}\| \to 0$ since both terms on the right go to zero. But then
$$\|T\phi - T(\|T\|\psi_{n_k})\| \to 0,$$
which implies, since $T(\psi_{n_k}) \to \phi$, that
$$T\phi = \|T\|\phi,$$
as desired. Thus we have shown that any symmetric, compact linear operator must have at least one eigenvalue. As we will see in the next section, when 0 is not an eigenvalue, such an operator (if defined on an infinite-dimensional space) actually has an infinite sequence of eigenvalues.

10.7.2 Facts about the eigenfunctions

We will now describe two facts about the eigenfunctions of M (and hence of L, since the inverse operators share the same eigenfunctions). One of these facts holds for the eigenvectors of any compact and symmetric linear operator, while the other is based on properties of the particular operator M that we are studying.

The sequence of eigenvectors of a compact, symmetric operator

We assume that $T : U \to U$ is compact and symmetric, where U is an infinite-dimensional inner product space, and that 0 is not an eigenvalue of T. As we argue below, T must have an infinite sequence of nonzero eigenvalues $\{\mu_n\}$ and a sequence of corresponding eigenvectors, $\{\psi_n\}$. The eigenvectors can be chosen to be an orthonormal set, and every element of U can be represented by an eigenvector expansion:[51]

$$u = \sum_{n=1}^{\infty} (\psi_n, u)\psi_n. \tag{10.39}$$

---

[51] For this conclusion to be true, we need a technical assumption about the inner product space: U must be *complete*. The concept of completeness is discussed in detail in Section 12.6.3, and we refer the interested reader to that section.

A full proof of these assertions is beyond the scope of this text, but we can concisely sketch the necessary reasoning.

First, we have already explained that T must have an eigenvalue μ_1 and a corresponding eigenvector ψ_1. We define the subspace U_2 to consist of all vectors in U orthogonal to ψ_1:
$$U_2 = \{u \in U : (\psi_1, u) = 0\}.$$
We then note that if $u \in U_2$, then
$$(\psi_1, Tu) = (T\psi_1, u) = (\mu_1 \psi_1, u) = \mu_1(\psi_1, u) = 0$$
(the reader should notice how the symmetry of T was used in an essential way). This shows that if $u \in U_2$, then Tu is also in U_2. We can then define T_2 to be the *restriction* of T to U_2:
$$T_2 : U_2 \to U_2, \; T_2 u = Tu.$$

The reader should find it easy to believe that T_2 is compact and symmetric, just like T, and therefore T_2 has an eigenpair μ_2, ψ_2: $T_2 \psi_2 = \mu_2 \psi_2$. But then, since T_2 is the restriction of T to U_2, this shows that $T\psi_2 = \mu_2 \psi_2$, and therefore μ_2, ψ_2 is an eigenpair of T.

We continue the above argument by defining U_3 to be the space of all vectors in U orthogonal to both ψ_1 and ψ_2:
$$U_3 = \{u \in U : (\psi_1, u) = (\psi_2, u) = 0\}.$$
By an argument similar to that given above, $Tu \in U_3$ for all $u \in U_3$, and therefore we can define T_3 to be the restriction of T to U_3. As in the case of T_2, T_3 inherits the properties of compactness and symmetry from T, and therefore we know that T_3 has an eigenpair μ_3, ψ_3, and by definition of T_3, this yields $T\psi_3 = \mu_3 \psi_3$.

We can continue this process to produce a sequence of eigenvalues $\{\mu_n\}$ and a sequence of corresponding eigenvectors $\{\psi_n\}$. All the eigenvalues are nonzero by assumption, and for any given eigenvalue μ_k, there can be only finitely many independent eigenvectors. This follows from the compactness of T, by an argument similar to that given on page 429 (where we argued that if a compact operator has a sequence of eigenvalues, then it must converge to zero). This means that in the sequence $\mu_1, \mu_2, \mu_3, \ldots$, any number is repeated at most a finite number of times. (The assumption that 0 is not an eigenvalue of T is essential, since otherwise T could have a finite number of distinct eigenvalues; the above construction would still be valid, but each μ_n could be zero beyond a certain value of n.)

Now we assume that $\{\mu_n\}$ is the sequence of all eigenvalues of T and show that every element of U can be expressed in terms of the associated eigenvectors. Given any $u \in U$, it can be shown that the series
$$\sum_{n=1}^{\infty} (\psi_n, u) \psi_n$$
converges to some vector $v \in U$. (The details of this argument are omitted, but here we must use the property of completeness of U mentioned in footnote 51 on page 431. We also need Bessel's inequality, which is derived in Section 12.4.2.) If $v \neq u$, then $u - v$ can be shown to be orthogonal to every eigenvector ψ_n. But then the subspace
$$U_\infty = \{u \in U : (\psi_n, u) = 0 \text{ for all } n = 1, 2, 3, \ldots\}$$

10.7. The theory of Sturm–Liouville problems: An outline

is nontrivial. We can then define T_∞ to be T restricted to U_∞, and just as above, T_∞ would be a compact, symmetric linear operator which must have an eigenvalue and a corresponding eigenvector. But this eigenpair could not be μ_n, ψ_n for any n, since the eigenvector lies in U_∞ and hence is orthogonal to every ψ_n. This contradicts the assumption that $\{\mu_n\}, \{\psi_n\}$ include every eigenvalue and eigenvector of T. The only unjustified assumption made in deriving this contradiction is that $v \neq u$, which must be false. Therefore,

$$u = \sum_{n=1}^{\infty} (\psi_n, u)\psi_n$$

must hold, which is what we wanted to show.

Properties of the eigenfunctions of L

We now know that M, and hence L, has an infinite sequence of eigenfunctions, which we will denote $\{\psi_n\}$. We have already shown above that each of these eigenfunctions is smooth (at least twice continuously differentiable), each corresponds to a different eigenvalue, and, since M is symmetric, they are mutually orthogonal. We assume that the eigenvalues $\{\mu_n\}$ and eigenfunctions $\{\psi_n\}$ of M have been ordered so that $\lambda_1 < \lambda_2 < \lambda_3 < \cdots$, where $\lambda_n = \mu_n^{-1}$. As explained in Section 10.2.2, we know something further:

- each ψ_n has exactly $n-1$ zeros in the interval (a,b);

- between any two consecutive zeros of ψ_n lies exactly one zero of ψ_{n-1} ($n \geq 3$).

To prove these properties requires rather involved reasoning that we do not give here. However, part of the second property can be proven using elementary reasoning about the BVP represented by the operator equation $Lu = \lambda u$. We will assume that λ is an eigenvalue of L (the reader should recall that $\lambda = \mu_n^{-1}$ for some n) with eigenfunction ψ, and that $\tilde{\lambda}, \tilde{\psi}$ is another eigenpair with $\tilde{\lambda} > \lambda$. Then

$$-\frac{d}{dx}\left(P(x)\frac{d\psi}{dx}\right) + R(x)\psi = \lambda \psi, \ a < x < b,$$

implies

$$-\frac{d}{dx}\left(P(x)\frac{d\psi}{dx}\right) + (R(x) - \lambda)\psi = 0, \ a < x < b,$$

and similarly

$$-\frac{d}{dx}\left(P(x)\frac{d\tilde{\psi}}{dx}\right) + (R(x) - \tilde{\lambda})\tilde{\psi} = 0, \ a < x < b.$$

Since $\tilde{\lambda} > \lambda$, $R(x) - \tilde{\lambda} < R(x) - \lambda$ for all $x \in [a,b]$. We will now prove the following theorem.

Theorem 10.11. *Suppose $P : [a,b] \to \mathbf{R}$ is continuously differentiable and positive, and let $R : [a,b] \to \mathbf{R}$, $\tilde{R} : [a,b] \to \mathbf{R}$ be two continuous functions satisfying $\tilde{R}(x) < R(x)$ for* all $x \in [a,b]$. Suppose ψ is a solution of

$$-\frac{d}{dx}\left(P(x)\frac{du}{dx}\right) + R(x)u = 0, \ a < x < b,$$

$$\alpha_1 u(a) + \alpha_2 \frac{du}{dx}(a) = 0, \quad (10.40)$$

$$\beta_1 u(b) + \beta_2 \frac{du}{dx}(b) = 0$$

and $\tilde{\psi}$ is a solution of the same BVP, with R replaced by $\tilde{R}$.

1. If $x_1, x_2 \in (a,b)$, $x_1 < x_2$, are two consecutive zeros of ψ ($\psi(x_1) = \psi(x_2) = 0$ and $\psi(x) \neq 0$ for $x \in (x_1, x_2)$), then $\tilde{\psi}$ has a zero in (x_1, x_2).

2. If $\tilde{x}$ is the largest zero of ψ in (a,b), then $\tilde{\psi}$ has at least one zero in $(\tilde{x}, b)$.

3. If $\tilde{x}$ is the smallest zero of ψ in (a,b), then $\tilde{\psi}$ has at least one zero in $(a, \tilde{x})$.

4. If ψ is nonzero on (a,b), then $\tilde{\psi}$ has at least one zero in (a,b).

This theorem is an example of a *Sturm comparison theorem*; it provides a partial proof of the second property described above if we think of $R(x) - \lambda$, $\mathbf{R}(x) - \tilde{\lambda}$ as the functions R, $\tilde{R}$ of the theorem.

Proof. We will prove the first conclusion; the proofs of the other conclusions use similar reasoning. We argue by contradiction and suppose $\tilde{\psi}(x) \neq 0$ for all $x \in (x_1, x_2)$. We are already assuming that $\psi(x) \neq 0$ for all $x \in (x_1, x_2)$, and we can assume that both ψ, $\tilde{\psi}$ are positive on (x_1, x_2). (If, for instance, ψ is negative on this interval, then $-\psi$ is positive on (x_1, x_2), and $-\psi$ is also a solution of the BVP. We could then apply the following argument to $-\psi$.) Since $\psi(x_1) = 0$ and $\psi(x) > 0$ for $x \in (x_1, x_2)$, it follows that $d\psi/dx(x_1) \geq 0$, and similar reasoning shows that $d\psi/dx(x_2) \leq 0$.

We now define

$$W(x) = \begin{vmatrix} \psi(x) & \tilde{\psi}(x) \\ \frac{d\psi}{dx}(x) & \frac{d\tilde{\psi}}{dx}(x) \end{vmatrix} = \psi(x)\frac{d\tilde{\psi}}{dx}(x) - \frac{d\psi}{dx}(x)\tilde{\psi}(x).$$

We have

$$W(x_1) = -\frac{d\psi}{dx}(x_1)\tilde{\psi}(x_1) \leq 0, \ W(x_2) = -\frac{d\psi}{dx}(x_2)\tilde{\psi}(x_2) \geq 0,$$

and hence

$$P(x_1)W(x_1) \leq 0, \ P(x_2)W(x_2) \geq 0.$$

10.7. The theory of Sturm–Liouville problems: An outline

But we also have

$$\begin{aligned}\frac{d}{dx}(P(x)W(x)) &= \frac{d}{dx}\left(P(x)\psi(x)\frac{d\tilde{\psi}}{dx}(x) - P(x)\frac{d\psi}{dx}(x)\tilde{\psi}(x)\right) \\
&= \frac{d}{dx}\left(P(x)\frac{d\tilde{\psi}}{dx}(x)\right)\psi(x) + P(x)\frac{d\tilde{\psi}}{dx}(x)\frac{d\psi}{dx}(x) \\
&\quad - \frac{d}{dx}\left(P(x)\frac{d\psi}{dx}(x)\right)\tilde{\psi}(x) - P(x)\frac{d\psi}{dx}(x)\frac{d\tilde{\psi}}{dx}(x) \\
&= \tilde{R}(x)\tilde{\psi}(x)\psi(x) - R(x)\psi(x)\tilde{\psi}(x) \\
&= \left(\tilde{R}(x) - R(x)\right)\tilde{\psi}(x)\psi(x) \\
&< 0 \text{ for all } x \in (x_1, x_2).\end{aligned}$$

The inequality in the last step follows from the fact that

$$\tilde{R}(x) < R(x) \text{ and } \psi(x), \tilde{\psi}(x) > 0 \text{ for all } x \in (x_1, x_2).$$

This calculation shows that $P(x)W(x)$ is strictly decreasing on the interval (x_1, x_2), which contradicts $P(x_1)W(x_1) \le 0$, $P(x_2)W(x_2) \ge 0$. This contradiction shows that $\tilde{\psi}(x) \ne 0$ for all $x \in (x_1, x_2)$ is impossible, that is, $\tilde{\psi}$ must have a zero in (x_1, x_2). □

Returning now to the operator L and its eigenfunctions, we have ordered the eigenvalues $\{\lambda_n\}$ so that $\lambda_1 < \lambda_2 < \lambda_3 < \cdots$. Theorem 10.11 shows that each eigenfunction ψ_{n+1} has at least one more zero than does ψ_n. Thus ψ_2 has at least one zero, ψ_3 has at least two, and so forth. As mentioned above, it can be shown that ψ_n has exactly $n-1$ zeros, but this proof is much more difficult.

Exercises

1. Let U and V be normed linear spaces and let $L : U \to V$ be a compact linear operator. Prove that L is bounded. (Hint: Use the alternate definition (10.38) of the norm. If L is unbounded, there must exist a sequence of vectors $\{u_n\}$ in U such that $\|u_n\|_U = 1$ for all n and $\|Lu_n\|_V \to \infty$. Prove that this is impossible if L is compact.)

2. Modify the proof of Theorem 10.11, part 1, to prove the other three parts of the theorem.

3. Consider the Sturm–Liouville problem

$$\begin{aligned}-\frac{d}{dx}\left(P(x)\frac{du}{dx}\right) + R(x)u &= \lambda w(x)u, \ a < x < b, \\
\alpha_1 u(a) + \alpha_2 \frac{du}{dx}(a) &= 0, \\
\beta_1 u(b) + \beta_2 \frac{du}{dx}(b) &= 0,\end{aligned} \quad (10.41)$$

where P and w are assumed to be positive on the interval $[a,b]$. Prove that, if λ is any eigenvalue for this problem, and ϕ, ψ are two eigenfunctions, both corresponding to the eigenvalue λ, then $\{\phi, \psi\}$ is linearly dependent.

4. Consider the Sturm–Liouville problem (10.41). Let the eigenvalues be

$$\lambda_1 < \lambda_2 < \lambda_3 < \cdots,$$

with corresponding eigenfunctions $\psi_1, \psi_2, \ldots$. Using Theorem 10.11, prove that ψ_{n+1} has at least one more zero in (a,b) than ψ_n has.

10.8 Suggestions for further reading

An excellent introduction to Sturm–Liouville problems, written at an introductory level, is Al-Gwaiz [1]. More advanced references include Zettl [70] and Reid [53]. In Section 10.2.2, we described the properties of the zeros of the eigenfunctions of a Sturm–Liouville operator (see page 391); a proof of these properties can be found in Chapter 8 of Coddington and Levinson [14].

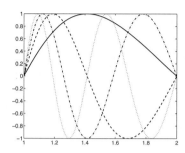

Chapter 11
Problems in Multiple Spatial Dimensions

Most practical applications involve multiple spatial dimensions, leading to PDEs involving two to four independent variables: x_1, x_2 or x_1, x_2, t or x_1, x_2, x_3, t. The purpose of this chapter is to extend the methods we have studied—Fourier series, finite elements, and Green's functions—to PDEs involving two or more spatial dimensions.

We begin by developing the fundamental physical models in two and three dimensions. We then present Fourier series methods; as we saw earlier, these techniques are applicable only in the case of constant-coefficient differential equations. Moreover, in higher dimensions, we can find the eigenfunctions explicitly only when the computational domain is simple; we treat the case of a rectangle and a circular disk in two dimensions.

We then turn to finite element methods, focusing on two-dimensional problems and piecewise linear finite elements defined on a *triangulation* of the computational domain. More details about the finite element method for problems in two spatial dimensions will be presented in Chapter 13.

We end the chapter by presenting Green's functions for the fundamental PDEs in two and three spatial dimensions.

11.1 Physical models in two or three spatial dimensions

The derivation of the physical models in Chapter 2 depended on the fundamental theorem of calculus. Using this result, we were able to relate a quantity defined on the boundary of an interval (force acting on a cross section or heat energy flowing across a cross section, for example) to another quantity in the interior of the interval.

In higher dimensions, the analogue of the fundamental theorem of calculus is the divergence theorem, which relates a vector field acting on the boundary of a domain to the *divergence* of the vector field in the interior. We begin by explaining the divergence theorem and proceed to apply it to a derivation of the heat equation. We must first establish some notation.

11.1.1 The divergence theorem

An advantage of using vector notation is that many results of calculus take the same form for two and three dimensions when expressed in vector form. Therefore, we will treat both two-dimensional and three-dimensional problems in this section, and much of the background calculus requires no distinction between $\mathbf{R}^2$ and $\mathbf{R}^3$.

Let Ω be a domain (a connected open set) in $\mathbf{R}^2$ or $\mathbf{R}^3$ with a piecewise smooth boundary $\partial \Omega$.[52] We will denote points in $\mathbf{R}^3$ using either vector notation or coordinate notation as is convenient, and similarly for points in $\mathbf{R}^2$. Thus a point in $\mathbf{R}^3$ can be denoted as $\mathbf{x}$ or as (x_1, x_2, x_3).

An important concept related to a domain Ω is that of the (outward) *unit normal*. At each point $\mathbf{x}$ of $\partial \Omega$ at which the boundary is smooth, there is a unique unit vector $\mathbf{n}(\mathbf{x})$ that is orthogonal to $\partial \Omega$ at $\mathbf{x}$ (to be precise, $\mathbf{n}(\mathbf{x})$ is orthogonal to the line or plane tangent to $\partial \Omega$ at $\mathbf{x}$) and points toward the exterior of Ω. It is customary to suppress the dependence of the normal vector $\mathbf{n}$ on $\mathbf{x}$, writing $\mathbf{n}$ instead of $\mathbf{n}(\mathbf{x})$, but it is important to keep this dependence in mind. Figure 11.1 shows an example in two dimensions.

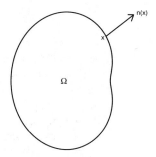

Figure 11.1. *A two-dimensional domain Ω with a sample unit normal vector.*

A (volume) integral over Ω in $\mathbf{R}^3$ will be denoted as

$$\int_\Omega f(\mathbf{x}) d\mathbf{x}$$

or simply as

$$\int_\Omega f,$$

where $f : \Omega \to \mathbf{R}$. If a domain Ω in $\mathbf{R}^3$ can be described as

$$\{(x_1, x_2, x_3) : a < x_1 < b, g_1(x_1) < x_2 < g_2(x_1), h_1(x_1, x_2) < x_3 < h_2(x_1, x_2)\},$$

for example, then the volume integral can be rewritten as an iterated integral:

$$\int_\Omega f(\mathbf{x}) d\mathbf{x} = \int_a^b \int_{g_1(x_1)}^{g_2(x_1)} \int_{h_1(x_1,x_2)}^{h_2(x_1,x_2)} f(x_1, x_2, x_3) dx_3 dx_2 dx_1.$$

---

[52]For our purposes, it does not seem worthwhile to define precise conditions on Ω and its boundary that allow the application of the divergence theorem. We will assume that the reader has an intuitive idea of the meanings of "smooth boundary" and "piecewise smooth boundary." For example, the unit ball, $S = \{(x_1, x_2, x_3) \in \mathbf{R}^3 : x^2 + y^2 + z^2 < 1\}$, has a smooth boundary, while the unit cube, $R = \{(x_1, x_2, x_3) \in \mathbf{R}^3 : 0 \leq x_1 \leq 1, 0 \leq x_2 \leq 1, 0 \leq x_3 \leq 1\}$, has a boundary that is piecewise smooth.

11.1. Physical models in two or three spatial dimensions

A (surface) integral over $\partial\Omega$ will be denoted by

$$\int_{\partial\Omega} g(\mathbf{x})\,d\sigma,$$

where $g : \partial\Omega \to \mathbf{R}$, and $d\sigma$ represents an infinitesimal surface area element, or simply

$$\int_{\partial\Omega} g.$$

Both volume and surface integrals can have vector-valued integrands, in which case each component is integrated.

If Ω is a domain in $\mathbf{R}^2$, then

$$\int_{\Omega} f$$

is an area integral that can, for many regions Ω, be rewritten as a doubly iterated integral. Since $\partial\Omega$ is a curve in this case,

$$\int_{\partial\Omega} g$$

is a line integral.[53]

A *vector field* defined on Ω is just a vector-valued function—a mapping of the form

$$\mathbf{f}(\mathbf{x}) = \begin{bmatrix} f_1(\mathbf{x}) \\ f_2(\mathbf{x}) \end{bmatrix}$$

if Ω is in $\mathbf{R}^2$ or

$$\mathbf{f}(\mathbf{x}) = \begin{bmatrix} f_1(\mathbf{x}) \\ f_2(\mathbf{x}) \\ f_3(\mathbf{x}) \end{bmatrix}$$

if Ω is in $\mathbf{R}^3$. The *divergence* of a vector field $\mathbf{f}$ is denoted by $\nabla \cdot \mathbf{f}$ and is defined by

$$\nabla \cdot \mathbf{f}(\mathbf{x}) = \frac{\partial f_1}{\partial x_1}(\mathbf{x}) + \frac{\partial f_2}{\partial x_2}(\mathbf{x})$$

in two dimensions and

$$\nabla \cdot \mathbf{f}(\mathbf{x}) = \frac{\partial f_1}{\partial x_1}(\mathbf{x}) + \frac{\partial f_2}{\partial x_2}(\mathbf{x}) + \frac{\partial f_3}{\partial x_3}(\mathbf{x})$$

in three dimensions. The following theorem, which holds in both two and three dimensions, explains why this combination of derivatives is so significant.

Theorem 11.1 (the divergence theorem). *Let Ω be a bounded open set with a piecewise smooth boundary $\partial\Omega$. Assume that $\mathbf{f}$ is a smooth vector field defined on $\Omega \cup \partial\Omega$, and that $\mathbf{n}$ is the (outward pointing) unit normal to $\partial\Omega$. Then*

$$\int_{\Omega} \nabla \cdot \mathbf{f} = \int_{\partial\Omega} \mathbf{f} \cdot \mathbf{n}.$$

---

[53]Line integral is a misnomer, since the domain of integration is generally a curve, not a (straight) line.

In the case of a rectangular domain in $\mathbf{R}^2$, the divergence theorem follows immediately from the fundamental theorem of calculus (see Exercise 11.1.1).

11.1.2 The heat equation for a three-dimensional domain

We now consider a three-dimensional solid occupying a domain Ω in $\mathbf{R}^3$. (We will often call the solid Ω, although, in fact, Ω is a mathematical description of the location of the solid.) For simplicity, assume that the solid is homogeneous, with density ρ, specific heat c, and thermal conductivity κ. Analogous to the one-dimensional case, if the domain ω is a subset of Ω, then the total heat energy contained in ω is

$$E_0 + \int_\omega \rho c(u(\mathbf{x},t) - T_0)\,d\mathbf{x},$$

where $u(\mathbf{x},t)$ is the temperature at $\mathbf{x} \in \Omega$ at time t, T_0 is the reference temperature, and E_0 is the thermal energy contained in the solid at temperature T_0.

The rate of change of the energy contained in ω is given by

$$\frac{d}{dt}\left[E_0 + \int_\omega \rho c(u(\mathbf{x},t) - T_0)\,d\mathbf{x}\right] = \frac{d}{dt}\int_\omega \rho c u(\mathbf{x},t)\,d\mathbf{x}.$$

Moving the derivative through the integral sign, we obtain the following expression for the rate of change of total heat energy contained in ω:

$$\int_\omega \rho c \frac{\partial u}{\partial t}(\mathbf{x},t)\,d\mathbf{x}. \tag{11.1}$$

We can also describe this rate of change by computing the rate at which heat energy flows across $\partial\omega$. Let $\mathbf{q}(\mathbf{x},t) \in \mathbf{R}^3$ be the heat flux at $\mathbf{x} \in \Omega$ at time t. The heat flux is a vector—its direction is the direction of the flow of energy, and its magnitude gives the rate at which energy is flowing across the plane at $\mathbf{x}$ to which $\mathbf{q}(\mathbf{x},t)$ is orthogonal, in units of energy/(time $\times$ area). At each point of $\partial\omega$, the rate at which energy is flowing into ω is $-\mathbf{q}\cdot\mathbf{n}$. (This follows from the fact that $\mathbf{q}$ is a vector quantity and hence can be decomposed into the component in the direction of $-\mathbf{n}$, which is $-\mathbf{q}\cdot\mathbf{n}$, and the component orthogonal to $-\mathbf{n}$.) The total amount of energy flowing into ω is then

$$-\int_{\partial\omega} \mathbf{q}\cdot\mathbf{n}.$$

We now make the assumption, called Fourier's law, that the heat flux is proportional to the temperature gradient:[54]

$$\mathbf{q}(\mathbf{x},t) = -\kappa \nabla u(\mathbf{x},t),$$

where

$$\nabla u = \begin{bmatrix} \frac{\partial u}{\partial x_1} \\ \frac{\partial u}{\partial x_2} \\ \frac{\partial u}{\partial x_3} \end{bmatrix}.$$

---

[54]The gradient of a scalar-valued function of several variables points in the direction in which the function increases most rapidly.

11.1. Physical models in two or three spatial dimensions

The constant of proportionality, $\kappa > 0$, is the thermal conductivity, just as in one spatial dimension. We therefore obtain the following expression for the rate of change of heat energy contained in ω:

$$\int_{\partial \omega} \kappa \nabla u \cdot \mathbf{n}.$$

We wish to equate this expression with (11.1) and derive a PDE. To do so, we must apply the divergence theorem to convert the boundary integral to a volume integral:

$$\int_{\partial \omega} \kappa \nabla u \cdot \mathbf{n} = \int_{\omega} \nabla \cdot (\kappa \nabla u) = \int_{\omega} \kappa \nabla \cdot \nabla u.$$

We thus have

$$\int_{\omega} \rho c \frac{\partial u}{\partial t}(\mathbf{x}, t) \, d\mathbf{x} = \int_{\omega} \kappa \nabla \cdot \nabla u(\mathbf{x}, t) \, d\mathbf{x},$$

or

$$\int_{\omega} \left(\rho c \frac{\partial u}{\partial t}(\mathbf{x}, t) - \kappa \nabla \cdot \nabla u(\mathbf{x}, t) \right) d\mathbf{x} = 0. \tag{11.2}$$

Since (11.2) holds for every subdomain ω of Ω, by exactly the same reasoning as in one dimension, the integrand must be identically zero. We therefore obtain the PDE

$$\rho c \frac{\partial u}{\partial t} - \kappa \nabla \cdot \nabla u = 0, \ \mathbf{x} \in \Omega, \ t > t_0.$$

The combination $\nabla \cdot \nabla$ of differential operators arises frequently due to just such an application of the divergence theorem, and it can be simplified as follows:

$$\nabla \cdot \nabla u = \frac{\partial}{\partial x_1} \frac{\partial u}{\partial x_1} + \frac{\partial}{\partial x_2} \frac{\partial u}{\partial x_2} + \frac{\partial}{\partial x_3} \frac{\partial u}{\partial x_3} = \frac{\partial^2 u}{\partial x_1^2} + \frac{\partial^2 u}{\partial x_2^2} + \frac{\partial^2 u}{\partial x_3^2} = \Delta u.$$

The differential operator

$$\Delta = \frac{\partial^2}{\partial x_1^2} + \frac{\partial^2}{\partial x_2^2} + \frac{\partial^2}{\partial x_3^2}$$

is called the *Laplacian*.[55] Using this notation, we obtain the three-dimensional heat equation

$$\rho c \frac{\partial u}{\partial t} - \kappa \Delta u = 0, \ \mathbf{x} \in \Omega, \ t > t_0. \tag{11.3}$$

If $f : \Omega \times (t_0, \infty) \to \mathbf{R}$ is a heat source (or sink), with units of energy per volume per time, then we obtain the inhomogeneous three-dimensional heat equation

$$\rho c \frac{\partial u}{\partial t} - \kappa \Delta u = f(\mathbf{x}, t), \ \mathbf{x} \in \Omega, \ t > t_0. \tag{11.4}$$

It should be obvious to the reader how to modify this equation if the material properties (ρ, c, κ) are not constants but rather functions of $\mathbf{x} \in \Omega$ (see Exercise 11.1.4).

---
[55] Another common symbol for the Laplacian is ∇^2.

11.1.3 Boundary conditions for the three-dimensional heat equation

Two common types of boundary conditions for the heat equation are Dirichlet and Neumann conditions. An equation of the form

$$u = 0 \text{ on } \partial\Omega$$

is called a homogeneous Dirichlet condition; it corresponds to an experiment in which the temperature of the boundary is held fixed at 0. Inhomogeneous Dirichlet conditions are also possible, and boundary data can be time dependent:

$$u(\mathbf{x},t) = g(\mathbf{x}),\ \mathbf{x} \in \partial\Omega,\ t > t_0,\ \text{or}\ u(\mathbf{x},t) = g(\mathbf{x},t),\ \mathbf{x} \in \partial\Omega,\ t > t_0.$$

It is also possible to insulate the boundary, which leads to the condition that the heat flux across the boundary is zero:

$$\nabla u(\mathbf{x},t) \cdot \mathbf{n} = 0,\ \mathbf{x} \in \partial\Omega,\ t > t_0.$$

The directional derivative $\nabla u \cdot \mathbf{n}$ is often referred to as the *normal derivative* and denoted

$$\frac{\partial u}{\partial \mathbf{n}}.$$

Just as in the one-dimensional case, we can have mixed boundary conditions, since it is possible to treat different parts of the boundary differently. Suppose $\partial\Omega = \Gamma_1 \cup \Gamma_2$ is a partition of the boundary into two disjoint sets. Then it is possible to pose a problem with the following mixed boundary conditions:

$$u(\mathbf{x},t) = 0,\ \mathbf{x} \in \Gamma_1,\ t > t_0,$$
$$\frac{\partial u}{\partial \mathbf{n}}(\mathbf{x},t) = 0,\ \mathbf{x} \in \Gamma_2,\ t > t_0.$$

11.1.4 The heat equation in a bar

In Section 2.1 and in subsequent sections, we discussed heat flow in a bar with insulated sides. We mentioned the following fact: If the initial temperature distribution in a bar depends only on the longitudinal coordinate (that is, if it is constant in each cross section), and if any heat source also depends only on the longitudinal coordinate, then the temperature at all subsequent times also depends only on a single spatial coordinate. We can now justify this statement.

Let the domain Ω in $\mathbf{R}^3$ be defined by

$$\Omega = \left\{ (x_1, x_2, x_3)\ :\ 0 < x_1 < \ell,\ x_2^2 + x_3^2 < r^2 \right\},$$

where ℓ and r are given positive constants. The set Ω is a circular cylinder centered on the interval $[0, \ell]$ on the x_1-axis. Consider the following initial-boundary value problem (IBVP):

$$\rho c \frac{\partial u}{\partial t} - \kappa \Delta u = f(x_1, t),\ \mathbf{x} \in \Omega,\ t > t_0,$$
$$u(\mathbf{x}, t_0) = \psi(x_1),\ \mathbf{x} \in \Omega,$$
$$\frac{\partial u}{\partial \mathbf{n}}(\mathbf{x}, t) = 0,\ \mathbf{x} \in \Gamma_t,\ t > t_0,$$
$$u(\mathbf{x}, t) = 0,\ \mathbf{x} \in \Gamma_e,\ t > t_0,$$

(11.5)

11.1. Physical models in two or three spatial dimensions

where Γ_e represents the ends of the bar and Γ_t the transverse part of the boundary (so that $\partial \Omega = \Gamma_e \cup \Gamma_t$):

$$\Gamma_e = \{(0, x_2, x_3) \,:\, x_2^2 + x_3^2 < r^2\} \cup \{(\ell, x_2, x_3) \,:\, x_2^2 + x_3^2 < r^2\},$$
$$\Gamma_t = \{(x_1, x_2, x_3) \,:\, 0 < x_1 < \ell,\ x_2^2 + x_3^2 = r^2\}.$$

The reader should notice that both the heat source f and the initial temperature distribution ψ are independent of x_2 and x_3.

The solution to (11.5) is $u(x_1, x_2, x_3, t) = v(x_1, t)$, where v is the solution of the IBVP

$$\begin{aligned}
\rho c \frac{\partial v}{\partial t} - \kappa \frac{\partial^2 v}{\partial x_1^2} &= f(x_1, t),\ 0 < x_1 < \ell,\ t > t_0, \\
v(x_1, t_0) &= \psi(x_1),\ 0 < x_1 < \ell, \\
v(0, t) &= 0,\ t > t_0, \\
v(\ell, t) &= 0,\ t > t_0.
\end{aligned} \tag{11.6}$$

The proof of this is a direct verification of the equations in (11.5) and is left as an exercise.

11.1.5 The heat equation in two dimensions

It is straightforward to restrict the heat equation to two dimensions. If heat flows in a solid, such as a thin plate, in such a way that the temperature is constant in one dimension, the derivative with respect to the corresponding spatial variable, say x_3, vanishes. The result is the two-dimensional heat equation, which is usually written exactly as in (11.3), since we will also use Δ to denote the Laplacian in two independent variables:

$$\Delta = \frac{\partial^2}{\partial x_1^2} + \frac{\partial^2}{\partial x_2^2}.$$

The distinction between the two- and three-dimensional Laplacian will be understood from the context.

11.1.6 The wave equation for a three-dimensional domain

The derivation of the wave equation in three dimensions is considerably more complicated than that of the heat equation. A complete treatment begins with an elastic solid, and applies Newton's law to the forces, both internal and external, acting on the solid. The result is a system of three (coupled) PDEs for the three components of displacement that describes the vibration of the solid. (Each point in the solid can move in three dimensions, so there are three dependent variables.) This system is one form of the wave equation.

Much useful physical modeling can be performed under apparently severe simplifying assumptions. Specifically, assuming that the solid is a fluid (meaning that the only stress supported is a pressure) and that the motion under consideration is a small perturbation of an equilibrium state, with the motion induced by a force density $\mathbf{F}(\mathbf{x}, t)$, the result is the *acoustic wave equation* for the pressure perturbation $u = u(\mathbf{x}, t)$:

$$\frac{\partial^2 u}{\partial t^2} - c^2 \Delta u = f(\mathbf{x}, t). \tag{11.7}$$

In this equation, the forcing function f is the negative divergence of the body force:

$$f(\mathbf{x},t) = -\nabla \cdot \mathbf{F}(\mathbf{x},t).$$

An example of a physical phenomenon modeled by the three-dimensional acoustic wave equation is the propagation of sound waves in air.

11.1.7 The wave equation in two dimensions

The two-dimensional (acoustic) wave equation models the small transverse vibrations of an elastic membrane. An elastic membrane is analogous to an elastic string, in that it does not resist bending. The form of the wave equation is exactly as in (11.7); however, the meaning is quite different. The dependent variable u is the vertical component of displacement, while the right-hand side f is the transverse pressure. Figure 11.2 illustrates the small deflection of a square membrane.

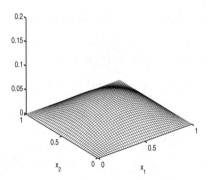

Figure 11.2. *The small vertical deflection of a square membrane under Dirichlet conditions.*

Dirichlet boundary conditions for the vibrating membrane indicate that the boundary is fixed (one can picture, for example, a circular drumhead with its boundary attached to the drum). Neumann conditions indicate that the boundary is free to move in a vertical direction.

11.1.8 Equilibrium problems and Laplace's equation

The differential operator Δ appears in both the heat equation and the wave equation, and therefore in the equilibrium versions of these equations. For example, steady-state heat flow is modeled by *Poisson's equation*,

$$-\kappa \Delta u = f(\mathbf{x}), \ x \in \Omega,$$

or by its counterpart for a heterogeneous material,

$$-\nabla \cdot (\kappa(\mathbf{x})\nabla u) = f(\mathbf{x}), \ \mathbf{x} \in \Omega.$$

This equation can be applied in either two or three dimensions.

11.1. Physical models in two or three spatial dimensions

An elastic membrane subject to a steady transverse pressure f has a vertical deflection u approximately satisfying

$$-k\Delta u = f(\mathbf{x}), \ \mathbf{x} \in \Omega.$$

The heterogeneous version of this equation is

$$-\nabla \cdot (k(\mathbf{x})\nabla u) = f(\mathbf{x}), \ \mathbf{x} \in \Omega.$$

The homogeneous version of Poisson's equation is referred to as *Laplace's equation*:

$$-\Delta u = 0, \ \mathbf{x} \in \Omega.$$

Of course, this equation is interesting only if the boundary conditions are inhomogeneous.

11.1.9 Advection and other first-order PDEs

If $u = u(\mathbf{x},t)$ represents the concentration of a chemical in solution, then we can model the advection of the chemical (that is, the change in $u(\mathbf{x},t)$ due to the flow of the fluid itself) with a first-order PDE. We will begin with a more general situation than simple advection, by modeling the flux of the chemical. We suppose that $\mathbf{F}(u)$ represents the flux of the chemical, in the sense that if P is a small planar surface passing through a point $\mathbf{x}$ and having normal vector $\mathbf{n}$, then the flux of the chemical across P is $\mathbf{F}(u(\mathbf{x},t)) \cdot \mathbf{n}$, in units of mass per area per time (for example, g/(cm^{2}s)). The reader should notice that we are assuming that the flux depends only on the concentration of the chemical at $\mathbf{x}$, and not directly on the location $\mathbf{x}$ or the time t. Simple advection would correspond to $\mathbf{F}(u) = u\mathbf{v}$, where $\mathbf{v}$ is a constant vector representing the velocity of the fluid.

Given the flux $\mathbf{F}$, we can derive the PDE modeling u as follows: Let ω be a small domain with the larger domain Ω of interest. The total mass of the chemical within ω at time t is

$$\int_\omega u(\mathbf{x},t)\,d\mathbf{x},$$

and therefore the rate of change of this mass is

$$\frac{d}{dt}\int_\omega u(\mathbf{x},t)\,d\mathbf{x} = \int_\omega \frac{\partial u}{\partial t}(\mathbf{x},t)\,d\mathbf{x}. \tag{11.8}$$

We can compute the same rate of change by measuring the amount of mass flowing across the boundary of ω. Using the flux $\mathbf{F}$ and applying the divergence theorem, this is

$$-\int_{\partial\omega} \mathbf{F}(u(\mathbf{x},t)) \cdot \mathbf{n} = -\int_\omega \nabla \cdot (\mathbf{F}(u(\mathbf{x},t)))\,d\mathbf{x}. \tag{11.9}$$

Equating (11.8) and (11.9) yields

$$\int_\omega \frac{\partial u}{\partial t}(\mathbf{x},t)\,d\mathbf{x} = -\int_\omega \nabla \cdot (\mathbf{F}(u(\mathbf{x},t)))\,d\mathbf{x}$$

or

$$\int_\omega \left\{\frac{\partial u}{\partial t}(\mathbf{x},t)\,d\mathbf{x} + \nabla \cdot (\mathbf{F}(u(\mathbf{x},t)))\right\}d\mathbf{x} = 0.$$

Since this holds for all domains ω within Ω (no matter how small), it follows that the integrand must be identically zero:

$$\frac{\partial u}{\partial t} + \nabla \cdot (\mathbf{F}(u)) = 0. \tag{11.10}$$

This is the general conservation law for a scalar quantity $u = u(\mathbf{x},t)$. Since $\mathbf{F}$ is a vector-valued function of a scalar variable, we have

$$\nabla \cdot (\mathbf{F}(u)) = \frac{d\mathbf{F}}{du}(u) \cdot \nabla u$$

(see Exercise 11.1.7), and (11.10) can be written as

$$\frac{\partial u}{\partial t} + \frac{d\mathbf{F}}{du}(u) \cdot \nabla u = 0. \tag{11.11}$$

(Here ∇u denotes the gradient of the function $u = u(\mathbf{x},t)$ with respect to the spatial variable $\mathbf{x}$; t is held constant.) In the case of simple advection, where the velocity of the fluid is given by the constant vector $\mathbf{v}$, we have $\mathbf{F}(u) = u\mathbf{v}$ and $\mathbf{F}'(u) = \mathbf{v}$. Therefore, the advection equation is

$$\frac{\partial u}{\partial t} + \mathbf{v} \cdot \nabla u = 0. \tag{11.12}$$

Equation (11.12) implies that any solution u is constant along the characteristics, which are lines in the direction of the vector $\mathbf{v}$. This reasoning results in solutions of the form $u(\mathbf{x},t) = \phi(\mathbf{x} - t\mathbf{v})$, where ϕ is a scalar-valued function of n variables. We can derive this family of solutions by applying the method of characteristics to the IVP

$$\begin{aligned}\frac{\partial u}{\partial t} + \mathbf{v} \cdot \nabla u &= 0, \ \mathbf{x} \in \mathbf{R}^n, \ t > 0, \\ u(\mathbf{x},0) &= \phi(\mathbf{x}), \ \mathbf{x} \in \mathbf{R}^n.\end{aligned} \tag{11.13}$$

The following system of ODEs defines the characteristic curves:

$$\begin{aligned}\frac{\partial x_1}{\partial \tau} &= v_1, \ x_1(\mathbf{s},0) = s_1, \\ \frac{\partial x_2}{\partial \tau} &= v_2, \ x_2(\mathbf{s},0) = s_2, \\ \frac{\partial x_3}{\partial \tau} &= v_3, \ x_3(\mathbf{s},0) = s_3, \\ \frac{\partial t}{\partial \tau} &= 1, \ t(\mathbf{s},0) = 0.\end{aligned}$$

We obtain $t(\mathbf{s},\tau) = \tau$ and $x_i(\mathbf{s},\tau) = s_i + \tau v_i$, $i = 1,2,3$; that is, $\mathbf{x}(\mathbf{s},\tau) = \mathbf{s} + \tau\mathbf{v}$. These equations can be inverted to obtain $\mathbf{s} = \mathbf{x} - t\mathbf{v}$, $\tau = t$. The solution, in the characteristic variables, is $w = w(\mathbf{s},\tau)$ as determined by the IVP

$$\frac{\partial w}{\partial \tau} = 0, \ w(\mathbf{s},0) = \phi(\mathbf{s}).$$

We obtain $w(\mathbf{s},\tau) = \phi(\mathbf{s})$. We then have $u(\mathbf{x},t) = w(\mathbf{s},\tau) = \phi(\mathbf{x} - t\mathbf{v})$, as expected.

11.1. Physical models in two or three spatial dimensions

The method of characteristics, as presented in Section 8.2.4, can be applied to quasilinear PDEs in the unknown $u = u(\mathbf{x},t)$; such PDEs have the form

$$a(\mathbf{x},t,u)\frac{\partial u}{\partial t} + \mathbf{b}(\mathbf{x},t,u) \cdot \nabla u = c(\mathbf{x},t,u).$$

However, we will not develop this method here. The interested reader is referred to [50] and other advanced texts on PDEs.

11.1.10 Green's identities and the symmetry of the Laplacian

We should expect to be able to extend the methods from the preceding chapters only if $-\Delta$ is a symmetric operator. The inner product is

$$(f,g) = \int_\Omega fg,$$

and we must show that $(-\Delta u, v) = (u, -\Delta v)$ for all u, v satisfying the desired boundary conditions. We define $C^2(\overline{\Omega})$ to be the set of all $u : \overline{\Omega} \to \mathbf{R}$ such that u and its partial derivatives up to order 2 are continuous ($\overline{\Omega}$ is the *closure* of Ω, that is, Ω together with its boundary: $\overline{\Omega} = \Omega \cup \partial\Omega$). We then define

$$C_D^2(\overline{\Omega}) = \left\{ u \in C^2(\overline{\Omega}) : \mathbf{x} \in \partial\Omega \Rightarrow u(\mathbf{x}) = 0 \right\}$$

and

$$L_D : C_D^2(\overline{\Omega}) \to C(\overline{\Omega}),$$
$$L_D u = -\Delta u.$$

In one dimension, the fundamental manipulation underlying the development of both Fourier series and finite element methods is integration by parts. The analogue in higher dimensions is Green's (first) identity, which we now derive. It is helpful to recall that integration by parts is based on the product rule for differentiation and the fundamental theorem of calculus, as follows:

$$\frac{d}{dx}[u(x)v(x)] = u(x)\frac{dv}{dx}(x) + \frac{du}{dx}(x)v(x)$$
$$\Rightarrow \int_a^b \frac{d}{dx}[u(x)v(x)]\,dx = \int_a^b u(x)\frac{dv}{dx}(x)\,dx + \int_a^b \frac{du}{dx}(x)v(x)\,dx$$
$$\Rightarrow u(x)v(x)\big|_a^b = \int_a^b u(x)\frac{dv}{dx}(x)\,dx + \int_a^b \frac{du}{dx}(x)v(x)\,dx$$
$$\Rightarrow \int_a^b u(x)\frac{dv}{dx}(x)\,dx = u(x)v(x)\big|_a^b - \int_a^b \frac{du}{dx}(x)v(x)\,dx.$$

To derive Green's identity, we need an analogous product rule for higher dimensions.

It is not completely obvious how to obtain the needed product rule, but we can be guided by the fact that we want to obtain a formula involving $v\Delta u = v\nabla \cdot \nabla u$. The desired product rule is

$$\nabla \cdot (v\nabla u) = \nabla u \cdot \nabla v + v\Delta u. \tag{11.14}$$

This is proved directly; see Exercise 11.1.9. Combining (11.14) with the divergence theorem gives the desired result:

$$\nabla \cdot (v \nabla u) = \nabla u \cdot \nabla v + v \Delta u$$

$$\Rightarrow \int_\Omega \nabla \cdot (v \nabla u) = \int_\Omega \nabla u \cdot \nabla v + \int_\Omega v \Delta u$$

$$\Rightarrow \int_{\partial \Omega} v \nabla u \cdot \mathbf{n} = \int_\Omega \nabla u \cdot \nabla v + \int_\Omega v \Delta u$$

$$\Rightarrow \int_\Omega v \Delta u = \int_{\partial \Omega} v \frac{\partial u}{\partial \mathbf{n}} - \int_\Omega \nabla u \cdot \nabla v.$$

This is *Green's first identity*:

$$\int_\Omega v \Delta u = \int_{\partial \Omega} v \frac{\partial u}{\partial \mathbf{n}} - \int_\Omega \nabla u \cdot \nabla v. \tag{11.15}$$

We can now prove the symmetry of the Laplacian under Dirichlet conditions. If $u, v \in C^2_D(\overline{\Omega})$, then

$$\begin{aligned}(L_D u, v) &= -\int_\Omega v \Delta u \\ &= \int_\Omega \nabla u \cdot \nabla v - \int_{\partial \Omega} v \frac{\partial u}{\partial \mathbf{n}} \text{ (Green's first identity)} \\ &= \int_\Omega \nabla u \cdot \nabla v \text{ (since } v = 0 \text{ on } \partial \Omega\text{)} \\ &= \int_{\partial \Omega} u \frac{\partial v}{\partial \mathbf{n}} - \int_\Omega u \Delta v \text{ (Green's first identity)} \\ &= -\int_\Omega u \Delta v \text{ (since } u = 0 \text{ on } \partial \Omega\text{)} \\ &= (u, L_D v).\end{aligned}$$

It is no more difficult to prove the symmetry of $-\Delta$ under Neumann conditions (see Exercise 11.1.10).

Exercises

1. Let $\Omega \subset \mathbf{R}^2$ be the rectangular domain

$$\Omega = \{\mathbf{x} \in \mathbf{R}^2 : a < x_1 < b, \ c < x_2 < d\},$$

and let $\mathbf{F} : \mathbf{R}^2 \to \mathbf{R}^2$ be a smooth vector field. Prove that

$$\int_\Omega \nabla \cdot \mathbf{F} = \int_{\partial \Omega} \mathbf{F} \cdot \mathbf{n}.$$

(Hint: Rewrite the area integral on the left as an iterated integral and apply the fundamental theorem of calculus. Compare the result to the boundary integral on the right.)

11.1. Physical models in two or three spatial dimensions

2. Let $\Omega \subset \mathbf{R}^2$ be the unit square: $\Omega = \{\mathbf{x} \in \mathbf{R}^2 : 0 < x_1 < 1, 0 < x_2 < 1\}$. Define $\mathbf{F} : \overline{\Omega} \to \mathbf{R}^2$ by
$$\mathbf{F}(\mathbf{x}) = \begin{bmatrix} x_2 \\ x_1 + x_2 \end{bmatrix}.$$
Verify that the divergence theorem holds for this domain Ω and this vector field $\mathbf{F}$.

3. Let $\Omega \subset \mathbf{R}^2$ be the unit disk: $\Omega = \{\mathbf{x} \in \mathbf{R}^2 : x_1^2 + x_2^2 < 1\}$. Let $\mathbf{F} : \overline{\Omega} \to \mathbf{R}^2$ be defined by
$$\mathbf{F}(\mathbf{x}) = \begin{bmatrix} x_2 \\ x_1 + x_2 \end{bmatrix}.$$
Verify that the divergence theorem holds for this domain Ω and this vector field $\mathbf{F}$.

4. How does the heat equation (11.3) change if ρ, c, and κ are functions of $\mathbf{x}$?

5. The nonconstant-coefficient version of the (negative) Laplace operator is
$$Lu = -\nabla \cdot (k(\mathbf{x})\nabla u),$$
where k is a real-valued function defined on the given domain Ω. Prove that L is symmetric under either Dirichlet or Neumann conditions.

6. Formulate Robin boundary conditions for a domain Ω in $\mathbf{R}^2$ or $\mathbf{R}^3$, and prove that $-\Delta$ is symmetric under homogeneous Robin conditions.

7. Let $\mathbf{F} : \mathbf{R} \to \mathbf{R}^3$ be a vector-valued function of a scalar variable,
$$\mathbf{F}(u) = (F_1(u), F_2(u), F_3(u)),$$
and let $\mathbf{F}'(u) = (F_1'(u), F_2'(u), F_3'(u))$. Prove that, if $u : \mathbf{R}^3 \to \mathbf{R}$, then
$$\nabla \cdot (\mathbf{F}(u)) = \mathbf{F}'(u) \cdot \nabla u.$$

8. Verify that $u(\mathbf{x}, t) = \phi(\mathbf{x} - t\mathbf{v})$ satisfies (11.13).

9. Prove (11.14) as follows: Write
$$\nabla \cdot (v \nabla u) = \sum_{i=1}^{3} \frac{\partial}{\partial x_i} \left[v \frac{\partial u}{\partial x_i} \right]$$
and apply the ordinary (scalar) product rule to each term on the right.

10. Define $C_N^2(\overline{\Omega}) = \left\{ u \in C^2(\overline{\Omega}) : \mathbf{x} \in \partial \Omega \Rightarrow \frac{\partial u}{\partial \mathbf{n}}(\mathbf{x}) = 0 \right\}$ and
$$L_N : C_N^2(\overline{\Omega}) \to C(\overline{\Omega}),$$
$$L_N u = -\Delta u.$$
Show that L_N is symmetric:
$$(L_N u, v) = (u, L_N v) \text{ for all } u, v \in C_N^2(\overline{\Omega}).$$

11. Suppose that the boundary of Ω is partitioned into two disjoint sets: $\partial\Omega = \Gamma_1 \cup \Gamma_2$. Define

$$C_m^2(\overline{\Omega}) = \left\{ u \in C^2(\overline{\Omega}) : \mathbf{x} \in \Gamma_1 \Rightarrow u(\mathbf{x}) = 0, \ \mathbf{x} \in \Gamma_2 \Rightarrow \frac{\partial u}{\partial \mathbf{n}}(\mathbf{x}) = 0 \right\}$$

and

$$L_m : C_m^2(\overline{\Omega}) \to C(\overline{\Omega}),$$
$$L_m u = -\Delta u.$$

Show that L_m is symmetric:

$$(L_m u, v) = (u, L_m v) \text{ for all } u, v \in C_m^2(\overline{\Omega}).$$

12. Verify that the solution to (11.6) is also the solution to (11.5).

11.2 Fourier series on a rectangular domain

We now develop Fourier series methods for the fundamental equations (Poisson's equation, the heat equation, and the wave equation) on the two-dimensional rectangular domain

$$\Omega = \left\{ \mathbf{x} \in \mathbf{R}^2 : 0 < x_1 < \ell_1, \ 0 < x_2 < \ell_2 \right\}. \tag{11.16}$$

We will begin by discussing Dirichlet conditions, so the operator is L_D, as defined at the end of the last section.

11.2.1 Dirichlet boundary conditions

As we should expect from the development in Chapters 5, 6, and 7, the crux of the matter is to determine the eigenvalues and eigenfunctions of L_D. We have already seen that L_D is symmetric, so we know that eigenfunctions corresponding to distinct eigenvalues must be orthogonal. Moreover, it is easy to show directly that L_D has only positive eigenvalues. For suppose λ is an eigenvalue of L_D and u is a corresponding eigenfunction, normalized to have norm one (the norm is derived from the inner product: $\|u\| = \sqrt{(u,u)}$). Then

$$\lambda = \lambda(u,u) = (\lambda u, u) = (L_D u, u) = -\int_\Omega \Delta u \, u = \int_\Omega \nabla u \cdot \nabla u,$$

with the last step following from Green's first identity and the fact that u vanishes on the boundary of Ω. Since $\nabla u \cdot \nabla u = \|\nabla u\|^2 \geq 0$, this certainly shows that $\lambda \geq 0$. Moreover,

$$\int_\Omega \|\nabla u\|^2 = 0$$

only if ∇u is identically equal to the zero vector. But this holds only if u is a constant function, and, due to the boundary conditions, the only constant function in $C_D^2(\overline{\Omega})$ is the zero function. By assumption, $\|u\| = 1$, so u is not the zero function. Thus we see that,

11.2. Fourier series on a rectangular domain

in fact, $\lambda > 0$ must hold. This proof did not use the particular form of Ω, and the result is therefore true for nonrectangular domains.

Thus we wish to find all positive values of λ such that the BVP

$$-\Delta u = \lambda u, \ \mathbf{x} \in \Omega,$$
$$u = 0, \ \mathbf{x} \in \partial\Omega, \tag{11.17}$$

has a nonzero solution. We are now faced with a PDE for which we have no general solution techniques. We fall back on a time-honored approach: make an inspired guess as to the general form of the solution, substitute into the equation, and try to determine specific solutions. We will look for solutions of the form

$$u(\mathbf{x}) = u_1(x_1)u_2(x_2),$$

that is, functions of two variables that can be written as the product of two functions, each with only one independent variable. Such functions are called *separated*, and this technique is called the method of *separation of variables*.

We therefore suppose that $u(\mathbf{x}) = u_1(x_1)u_2(x_2)$ satisfies

$$-\Delta u = \lambda u, \ \mathbf{x} \in \Omega.$$

We have

$$-\Delta u = -\frac{\partial^2 u}{\partial x_1^2} - \frac{\partial^2 u}{\partial x_2^2}$$
$$= -\frac{d^2 u_1}{dx_1^2} u_2 - u_1 \frac{d^2 u_2}{dx_2^2},$$

so the PDE becomes

$$-\frac{d^2 u_1}{dx_1^2} u_2 - u_1 \frac{d^2 u_2}{dx_2^2} = \lambda u_1 u_2, \ \mathbf{x} \in \Omega.$$

Dividing through by $u_1 u_2$ yields

$$-u_1^{-1}\frac{d^2 u_1}{dx_1^2} - u_2^{-1}\frac{d^2 u_2}{dx_2^2} = \lambda, \ (x_1, x_2) \in \Omega.$$

If we rewrite this as

$$-u_1^{-1}\frac{d^2 u_1}{dx_1^2} = \lambda + u_2^{-1}\frac{d^2 u_2}{dx_2^2}, \tag{11.18}$$

we obtain the conclusion that both

$$u_1^{-1}\frac{d^2 u_1}{dx_1^2} \text{ and } u_2^{-1}\frac{d^2 u_2}{dx_2^2}$$

must be constant functions. For on the left side of (11.18) is a function depending only on x_1, and on the right is a function depending only on x_2. If we differentiate with respect

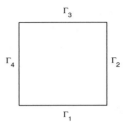

Figure 11.3. *The domain Ω and its boundary.*

to x_1, we see that the derivative of the first function must be zero, and hence that the function itself must be constant. Similarly, the second function must be constant.

We have therefore shown the following: If λ is positive and $-\Delta u = \lambda u$ has a solution of the form $u(x_1, x_2) = u_1(x_1)u_2(x_2)$, then u_1 and u_2 satisfy

$$-u_1^{-1}\frac{d^2 u_1}{dx_1^2} = \theta_1, \quad -u_2^{-1}\frac{d^2 u_2}{dx_2^2} = \theta_2,$$

where $\theta_1 + \theta_2 = \lambda$. These we can rewrite as

$$-\frac{d^2 u_1}{dx_1^2} = \theta_1 u_1, \quad -\frac{d^2 u_2}{dx_2^2} = \theta_2 u_2,$$

and so we obtain ODEs for u_1 and u_2.

Moreover, we can easily find boundary conditions for the ODEs, since the boundary conditions on the PDE also separate. The boundary of Ω consists of four line segments (see Figure 11.3):

$$\begin{aligned}
\Gamma_1 &= \{(x_1, x_2) : x_2 = 0, \ 0 \leq x_1 \leq \ell_1\} \text{ (bottom)}, \\
\Gamma_2 &= \{(x_1, x_2) : x_1 = \ell_1, \ 0 \leq x_2 \leq \ell_2\} \text{ (right)}, \\
\Gamma_3 &= \{(x_1, x_2) : x_2 = \ell_2, \ 0 \leq x_1 \leq \ell_1\} \text{ (top)}, \\
\Gamma_4 &= \{(x_1, x_2) : x_1 = 0, \ 0 \leq x_2 \leq \ell_2\} \text{ (left)}.
\end{aligned} \quad (11.19)$$

On Γ_1, for example, we have

$$u = 0 \Rightarrow u_1(x_1)u_2(0) = 0 \Rightarrow u_2(0) = 0.$$

(There is also the possibility that $u_1(x_1) \equiv 0$, but then u is the zero function, and we are only interested in nontrivial solutions.) By similar reasoning, we obtain

$$u_1(0) = u_1(\ell_1) = 0, \ u_2(0) = u_2(\ell_2) = 0.$$

Our problem now reduces to finding nonzero solutions to the BVPs

$$\begin{aligned}
-\frac{d^2 u_1}{dx_1^2} &= \theta_1 u_1, \ 0 < x_1 < \ell_1, \\
u_1(0) &= 0, \\
u_1(\ell_1) &= 0
\end{aligned} \quad (11.20)$$

11.2. Fourier series on a rectangular domain

and

$$-\frac{d^2 u_2}{dx_2^2} = \theta_2 u_2, \; 0 < x_2 < \ell_2,$$
$$u_2(0) = 0,$$
$$u_2(\ell_2) = 0,$$
(11.21)

where θ_1 and θ_2 can be any real numbers adding to λ. But we have already solved these problems (in Section 5.2). For (11.20), the permissible values of θ_1 are

$$\theta_1^{(m)} = \frac{m^2 \pi^2}{\ell_1^2}, \; m = 1, 2, 3, \ldots,$$

and the corresponding eigenfunctions are

$$\psi_1^{(m)}(x_1) = \sin\left(\frac{m \pi x_1}{\ell_1}\right), \; m = 1, 2, 3, \ldots.$$

For (11.21), we have

$$\theta_2^{(n)} = \frac{n^2 \pi^2}{\ell_2^2}, \; n = 1, 2, 3, \ldots,$$

and the corresponding eigenfunctions are

$$\psi_2^{(n)}(x_2) = \sin\left(\frac{n \pi x_2}{\ell_2}\right), \; n = 1, 2, 3, \ldots.$$

Any solution to (11.20) times any solution to (11.21) forms a solution to (11.17), so we obtained a doubly indexed sequence of solutions to (11.17):

$$\lambda_{mn} = \frac{m^2 \pi^2}{\ell_1^2} + \frac{n^2 \pi^2}{\ell_2^2},$$
$$\psi_{mn}(x_1, x_2) = \sin\left(\frac{m \pi x_1}{\ell_1}\right) \sin\left(\frac{n \pi x_2}{\ell_2}\right), \; m, n = 1, 2, 3, \ldots.$$
(11.22)

We have succeeded in computing all the eigenvalue-eigenfunction pairs of L_D in which the eigenfunctions are separated. We have no guarantee (at least, not from our analysis so far) that there are not other eigenpairs with nonseparated eigenfunctions. However, it turns out that this question is not particularly important, because we can show that the eigenpairs (11.22) are sufficient for our purposes. Indeed, it is not difficult to argue that every function $u \in C(\overline{\Omega})$ should be representable as

$$u(x_1, x_2) = \sum_{m=1}^{\infty} \sum_{n=1}^{\infty} a_{mn} \sin\left(\frac{m \pi x_1}{\ell_1}\right) \sin\left(\frac{n \pi x_2}{\ell_2}\right),$$

where the convergence is in the mean-square sense. We will call such a series a Fourier (double) sine series. We consider any $u = u(x_1, x_2)$ (not necessarily satisfying the Dirichlet boundary conditions). Regarding $x_2 \in (0, \ell_2)$ as a parameter, we can write

$$u(x_1, x_2) = \sum_{m=1}^{\infty} b_m(x_2) \sin\left(\frac{m \pi x_1}{\ell_1}\right),$$

where
$$b_m(x_2) = \frac{2}{\ell_1} \int_0^{\ell_1} u(x_1, x_2) \sin\left(\frac{m\pi x_1}{\ell_1}\right) dx_1.$$

But now b_m is a function of $x_2 \in (0, \ell_2)$ that can be expanded in a Fourier sine series as well:
$$b_m(x_2) = \sum_{n=1}^{\infty} a_{mn} \sin\left(\frac{n\pi x_2}{\ell_2}\right),$$
where
$$a_{mn} = \frac{2}{\ell_2} \int_0^{\ell_2} b_m(x_2) \sin\left(\frac{n\pi x_2}{\ell_2}\right) dx_2.$$

Putting these results together, we obtain
$$u(x_1, x_2) = \sum_{m=1}^{\infty} \sum_{n=1}^{\infty} a_{mn} \sin\left(\frac{m\pi x_1}{\ell_1}\right) \sin\left(\frac{n\pi x_2}{\ell_2}\right), \tag{11.23}$$

with
$$a_{mn} = \frac{4}{\ell_1 \ell_2} \int_0^{\ell_1} \int_0^{\ell_2} u(x_1, x_2) \sin\left(\frac{m\pi x_1}{\ell_1}\right) \sin\left(\frac{n\pi x_2}{\ell_2}\right) dx_2 dx_1. \tag{11.24}$$

Equation (11.23) is valid in the mean-square sense.

It is straightforward to show that formulas (11.23), (11.24) are exactly what the projection theorem produces; that is,
$$\sum_{m=1}^{M} \sum_{n=1}^{N} a_{mn} \sin\left(\frac{m\pi x_1}{\ell_1}\right) \sin\left(\frac{n\pi x_2}{\ell_2}\right)$$

is the best approximation to u, in the L^2 norm, from the subspace spanned by
$$\{\psi_{mn} : m = 1, 2, \ldots, M, n = 1, 2, \ldots, N\}$$

(see Exercise 11.2.6).

Example 11.2. Let $\Omega = \{(x_1, x_2) \in \mathbf{R}^2 : 0 < x_1 < 1, 0 < x_2 < 1\}$, and let
$$u(x_1, x_2) = x_1 x_2.$$

Then
$$a_{mn} = 4 \int_0^1 \int_0^1 x_1 x_2 \sin(m\pi x_1) \sin(n\pi x_2) = \frac{4(-1)^m (-1)^n}{mn\pi^2}.$$

In Figure 11.4, we graph the resulting partial series approximation to u, using $M = 20$ and $N = 20$ (for a total of 400 terms). Gibbs's phenomenon is clearly visible, as would be expected since u only satisfies the Dirichlet boundary conditions on part of the boundary. ∎

11.2. Fourier series on a rectangular domain

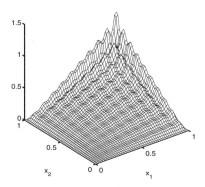

Figure 11.4. *A partial Fourier (double) sine series (400 terms) approximating* $u(x_1, x_2) = x_1 x_2$.

11.2.2 Solving a boundary value problem

It is no more difficult to apply the Fourier series method to a BVP in two or three dimensions than it was in one dimension, provided, of course, that we know the eigenvalues and eigenfunctions.

Example 11.3. We will solve the following BVP:

$$-\Delta u = 1 \text{ in } \Omega,$$
$$u = 0 \text{ on } \partial\Omega, \quad (11.25)$$

where Ω is the unit square:

$$\Omega = \{(x_1, x_2) \in \mathbf{R}^2 \,:\, 0 < x_1 < 1,\, 0 < x_2 < 1\}.$$

We first write the constant function $f(\mathbf{x}) = 1$ as a Fourier sine series. We have

$$1 = \sum_{m=1}^{\infty} \sum_{n=1}^{\infty} c_{mn} \sin(m\pi x_1) \sin(n\pi x_2),$$

where

$$c_{mn} = 4 \int_0^1 \int_0^1 \sin(m\pi x_1) \sin(n\pi x_2)\, dx_2\, dx_1$$
$$= \frac{4((-1)^n(-1)^m - (-1)^n - (-1)^m + 1)}{mn\pi^2}.$$

We next write the solution u as

$$u(x_1, x_2) = \sum_{m=1}^{\infty} \sum_{n=1}^{\infty} a_{mn} \sin(m\pi x_1) \sin(n\pi x_2).$$

It is straightforward to show that, since u satisfies homogeneous Dirichlet conditions,

$$-\Delta u(x_1, x_2) = \sum_{m=1}^{\infty} \sum_{n=1}^{\infty} \lambda_{mn} a_{mn} \sin(m\pi x_1) \sin(n\pi x_2),$$

where

$$\lambda_{mn} = (m^2 + n^2)\pi^2, \ m,n = 1,2,3,\ldots$$

(see Exercise 11.2.8). The PDE therefore implies that

$$\sum_{m=1}^{\infty} \sum_{n=1}^{\infty} \lambda_{mn} a_{mn} \sin(m\pi x_1) \sin(n\pi x_2) = \sum_{m=1}^{\infty} \sum_{n=1}^{\infty} c_{mn} \sin(m\pi x_1) \sin(n\pi x_2),$$

and therefore

$$\lambda_{mn} a_{mn} = c_{mn}, \ m,n = 1,2,3,\ldots.$$

This in turn implies that

$$a_{mn} = \frac{c_{mn}}{\lambda_{mn}} = \frac{4((-1)^n(-1)^m - (-1)^n - (-1)^m + 1)}{(m^2+n^2)mn\pi^4}, \ m,n = 1,2,3,\ldots.$$

We can approximate the solution by a partial Fourier series of the form

$$\sum_{m=1}^{M} \sum_{n=1}^{N} a_{mn} \sin(m\pi x_1) \sin(n\pi x_2).$$

In Figure 11.5, we graph the partial Fourier series with $M = 10$, $N = 10$. ∎

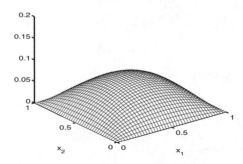

Figure 11.5. *The solution to the BVP* (11.25), *approximated by a Fourier series with* 100 *terms.*

11.2.3 Time-dependent problems

It is also straightforward to extend the one-dimensional Fourier series techniques for time-dependent problems to two- or three-dimensional problems, again assuming that eigenvalues and eigenfunctions are known. We illustrate with an example.

Example 11.4. We will solve the following IBVP for the wave equation on the unit square:

$$\begin{aligned}
\frac{\partial^2 u}{\partial t^2} - c^2 \Delta u &= 0, \ \mathbf{x} \in \Omega, \ t > 0, \\
u(\mathbf{x}, 0) &= 0, \ \mathbf{x} \in \Omega, \\
\frac{\partial u}{\partial t}(\mathbf{x}, 0) &= \gamma(\mathbf{x}), \ \mathbf{x} \in \Omega, \\
u(\mathbf{x}, t) &= 0, \ \mathbf{x} \in \partial\Omega, \ t > 0.
\end{aligned} \quad (11.26)$$

The initial velocity function will be taken to be the function

$$\gamma(\mathbf{x}) = \begin{cases} -1, & \frac{2}{5} < x_1 < \frac{3}{5}, \ \frac{2}{5} < x_2 < \frac{3}{5}, \\ 0 & \text{otherwise.} \end{cases}$$

We take $c = 261\sqrt{2}$. This IBVP models a (square) drum struck by a square hammer.

We write the solution u as

$$u(x_1, x_2, t) = \sum_{m=1}^{\infty} \sum_{n=1}^{\infty} a_{mn}(t) \sin(m\pi x_1) \sin(n\pi x_2),$$

where

$$a_{mn}(t) = 4 \int_0^1 \int_0^1 u(x_1, x_2, t) \sin(m\pi x_1) \sin(n\pi x_2) \, dx_2 \, dx_1.$$

The PDE then takes the form

$$\sum_{m=1}^{\infty} \sum_{n=1}^{\infty} \left\{ \frac{d^2 a_{mn}}{dt^2}(t) + c^2 \lambda_{mn} a_{mn}(t) \right\} \sin(m\pi x_1) \sin(n\pi x_2) = 0,$$

where $\lambda_{mn} = (m^2 + n^2)\pi^2$. The initial conditions become

$$u(x_1, x_2, 0) = \sum_{m=1}^{\infty} \sum_{n=1}^{\infty} a_{mn}(0) \sin(m\pi x_1) \sin(n\pi x_2) = 0$$

and

$$\frac{\partial u}{\partial t}(x_1, x_2, 0) = \sum_{m=1}^{\infty} \sum_{n=1}^{\infty} \frac{da_{mn}}{dt}(0) \sin(m\pi x_1) \sin(n\pi x_2) = \gamma(x_1, x_2).$$

We have

$$\gamma(x_1, x_2) = \sum_{m=1}^{\infty} \sum_{n=1}^{\infty} b_{mn} \sin(m\pi x_1) \sin(n\pi x_2),$$

where

$$b_{mn} = 4\int_0^1\int_0^1 \gamma(x_1,x_2)\sin(m\pi x_1)\sin(n\pi x_2)\,dx_2\,dx_1$$
$$= -4\int_{2/5}^{3/5}\int_{2/5}^{3/5} \sin(m\pi x_1)\sin(n\pi x_2)\,dx_2\,dx_1$$
$$= -\frac{4(\cos(3n\pi/5)-\cos(2n\pi/5))(\cos(3m\pi/5)-\cos(2m\pi/5))}{mn\pi^2}.$$

Putting together the PDE and the initial conditions, we obtain the following sequence of IVPs:

$$\frac{d^2 a_{mn}}{dt^2} + c^2\lambda_{mn}a_{mn} = 0,$$
$$a_{mn}(0) = 0,$$
$$\frac{da_{mn}}{dt}(0) = b_{mn},$$

$m,n = 1,2,3,\ldots$. The solutions are

$$a_{mn}(t) = d_{mn}\sin\left(c\sqrt{m^2+n^2}\pi t\right),\ m,n=1,2,3,\ldots,$$

where

$$d_{mn} = \frac{b_{mn}}{c\sqrt{m^2+n^2}\pi}.$$

We graph selected snapshots of u in Figure 11.6. ∎

11.2.4 Other boundary conditions for the rectangle

Using separation of variables, it is straightforward to find the eigenvalues and eigenfunctions for the negative Laplacian, on a rectangle, under Neumann conditions or some combination of Dirichlet and Neumann conditions. For example, if Ω is the rectangle (11.16) and $\partial\Omega = \Gamma_1\cup\Gamma_2\cup\Gamma_3\cup\Gamma_4$, with the Γ_i defined as in (11.19), we can consider the following IBVP:

$$\begin{aligned}\rho c\frac{\partial u}{\partial t} - \kappa\Delta u &= f(\mathbf{x},t),\ \mathbf{x}\in\Omega,\ t>0,\\ u(\mathbf{x},0) &= \psi(\mathbf{x}),\ \mathbf{x}\in\Omega,\\ u(\mathbf{x},t) &= 0,\ \mathbf{x}\in\Gamma_1,\ t>0,\\ \frac{\partial u}{\partial \mathbf{n}}(\mathbf{x},t) &= 0,\ \mathbf{x}\in\Gamma_2,\ t>0,\\ \frac{\partial u}{\partial \mathbf{n}}(\mathbf{x},t) &= 0,\ \mathbf{x}\in\Gamma_3,\ t>0,\\ u(\mathbf{x},t) &= 0,\ \mathbf{x}\in\Gamma_4,\ t>0.\end{aligned} \qquad(11.27)$$

11.2. Fourier series on a rectangular domain

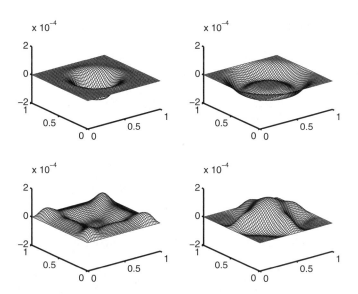

Figure 11.6. *Snapshots of the vibrating membrane of Example* 11.4: $t = 5 \cdot 10^{-4}$ *(upper left),* $t = 10^{-3}$ *(upper right),* $t = 2 \cdot 10^{-3}$ *(lower left),* $t = 3 \cdot 10^{-3}$ *(lower right).*

By applying the method of separation of variables, we can determine the appropriate Fourier series to represent the solution:

$$u(x_1, x_2, t) = \sum_{m=1}^{\infty} \sum_{n=1}^{\infty} a_{mn}(t) \sin\left(\frac{(2m-1)\pi x_1}{2\ell_1}\right) \sin\left(\frac{(2n-1)\pi x_2}{2\ell_2}\right)$$

(see Exercise 11.2.14). The determination of the coefficients $a_{mn}(t)$ follows the now-familiar pattern.

11.2.5 Neumann boundary conditions

Neumann conditions are slightly more difficult to handle than Dirichlet or mixed boundary conditions. This is simply because Neumann conditions lead to two different kinds of eigenfunctions: the constant function 1, and the cosines of increasing frequencies. To be specific, the reader will recall from Section 6.2 that the eigenpairs of the negative second derivative operator, under Neumann conditions, are

$$\lambda_0 = 0, \ \gamma_0(x) = 1,$$
$$\lambda_n = \frac{n^2 \pi^2}{\ell^2}, \ \gamma_n(x) = \cos\left(\frac{n\pi x}{\ell}\right), \ n = 1, 2, 3, \ldots.$$

Just as in the case of Dirichlet or mixed boundary conditions, the eigenfunctions of the negative Laplacian, under Neumann conditions and on the rectangle, are products of the one-dimensional eigenfunctions. Since there are two formulas for the one-dimensional eigenfunctions, this leads to four different formulas for the two-dimensional eigenfunctions (see Exercise 11.2.12):

$$\lambda_{00} = 0, \; \gamma_{00}(\mathbf{x}) = 1,$$
$$\lambda_{m0} = \frac{m^2\pi^2}{\ell_1^2}, \; \gamma_{m0}(\mathbf{x}) = \cos\left(\frac{m\pi x_1}{\ell_1}\right), \; m = 1,2,3,\ldots,$$
$$\lambda_{0n} = \frac{n^2\pi^2}{\ell_1^2}, \; \gamma_{0n}(\mathbf{x}) = \cos\left(\frac{n\pi x_2}{\ell_2}\right), \; n = 1,2,3,\ldots, \tag{11.28}$$
$$\lambda_{mn} = \frac{m^2\pi^2}{\ell_1^2} + \frac{n^2\pi^2}{\ell_2^2}, \; \gamma_{mn}(\mathbf{x}) = \cos\left(\frac{m\pi x_1}{\ell_1}\right)\cos\left(\frac{n\pi x_2}{\ell_1}\right), \; m,n = 1,2,3,\ldots.$$

By the same reasoning given in Section 11.2.1, this collection of eigenfunctions should be sufficient to represent any function in $C(\overline{\Omega})$; the series representation is

$$f(x_1,x_2) = a_{00} + \sum_{m=1}^{\infty} a_{m0}\cos\left(\frac{m\pi x_1}{\ell_1}\right) + \sum_{n=1}^{\infty} a_{0n}\cos\left(\frac{n\pi x_2}{\ell_2}\right)$$
$$+ \sum_{m=1}^{\infty}\sum_{n=1}^{\infty} a_{mn}\cos\left(\frac{m\pi x_1}{\ell_1}\right)\cos\left(\frac{n\pi x_2}{\ell_2}\right),$$

where

$$a_{00} = \frac{1}{\ell_1\ell_2}\int_0^{\ell_1}\int_0^{\ell_2} f(x_1,x_2)\,dx_2\,dx_1,$$
$$a_{m0} = \frac{2}{\ell_1\ell_2}\int_0^{\ell_1}\int_0^{\ell_2} f(x_1,x_2)\cos\left(\frac{m\pi x_1}{\ell_1}\right)dx_2\,dx_1,$$
$$a_{0n} = \frac{2}{\ell_1\ell_2}\int_0^{\ell_1}\int_0^{\ell_2} f(x_1,x_2)\cos\left(\frac{n\pi x_2}{\ell_2}\right)dx_2\,dx_1,$$
$$a_{mn} = \frac{4}{\ell_1\ell_2}\int_0^{\ell_1}\int_0^{\ell_2} f(x_1,x_2)\cos\left(\frac{m\pi x_1}{\ell_1}\right)\cos\left(\frac{n\pi x_2}{\ell_2}\right)dx_2\,dx_1.$$

A steady-state Neumann problem imposes a compatibility condition on the right-hand side of the PDE, as might be expected from the existence of a zero eigenvalue. Suppose u is a solution of the BVP

$$-\Delta u = f(\mathbf{x}) \text{ in } \Omega,$$
$$\frac{\partial u}{\partial \mathbf{n}} = 0 \text{ on } \partial\Omega. \tag{11.29}$$

The compatibility condition follows from the divergence theorem:

$$\int_\Omega f = -\int_\Omega \Delta u = -\int_\Omega \nabla\cdot\nabla u = -\int_{\partial\Omega}\nabla u\cdot\mathbf{n} = -\int_{\partial\Omega}\frac{\partial u}{\partial\mathbf{n}} = 0.$$

11.2. Fourier series on a rectangular domain

If f satisfies the compatibility condition

$$\int_\Omega f = 0,$$

then the Neumann problem (11.29) has infinitely many solutions, while if f does not satisfy the compatibility condition, then there is no solution.

Example 11.5. We will solve the Neumann problem

$$-\Delta u = x_1^2 x_2 - \frac{1}{6} \text{ in } \Omega,$$
$$\frac{\partial u}{\partial \mathbf{n}} = 0 \text{ on } \partial \Omega, \tag{11.30}$$

where Ω is the unit square. We write the solution as

$$u(x_1, x_2) = a_{00} + \sum_{m=1}^{\infty} a_{m0} \cos(m\pi x_1) + \sum_{n=1}^{\infty} a_{0n} \cos(n\pi x_2)$$
$$+ \sum_{m=1}^{\infty} \sum_{n=1}^{\infty} a_{mn} \cos(m\pi x_1) \cos(n\pi x_2).$$

Then

$$-\Delta u(x_1, x_2) = \sum_{m=1}^{\infty} m^2 \pi^2 a_{m0} \cos(m\pi x_1) + \sum_{n=1}^{\infty} n^2 \pi^2 a_{0n} \cos(n\pi x_2)$$
$$+ \sum_{m=1}^{\infty} \sum_{n=1}^{\infty} \left(m^2 \pi^2 + n^2 \pi^2 \right) a_{mn} \cos(m\pi x_1) \cos(n\pi x_2).$$

We also have

$$f(x_1, x_2) = c_{00} + \sum_{m=1}^{\infty} c_{m0} \cos(m\pi x_1) + \sum_{n=1}^{\infty} c_{0n} \cos(n\pi x_2)$$
$$+ \sum_{m=1}^{\infty} \sum_{n=1}^{\infty} c_{mn} \cos(m\pi x_1) \cos(n\pi x_2),$$

where

$$c_{00} = 0,$$
$$c_{m0} = \frac{2(-1)^m}{m^2 \pi^2},$$
$$c_{0n} = \frac{2((-1)^n - 1)}{3n^2 \pi^2},$$
$$c_{mn} = \frac{8\left((-1)^{m+n} + (-1)^{m+1}\right)}{m^2 n^2 \pi^4}.$$

The reader should notice that $c_{00} = 0$ is the compatibility condition, and so there are infinitely many solutions. Equating the series for $-\Delta u$ and f yields

$$a_{m0} = \frac{c_{m0}}{m^2\pi^2},$$
$$a_{0n} = \frac{c_{0n}}{n^2\pi^2},$$
$$a_{mn} = \frac{c_{mn}}{(m^2+n^2)\pi^2},$$

with a_{00} undetermined. The graph of u, with $a_{00} = 0$, is shown in Figure 11.7. ∎

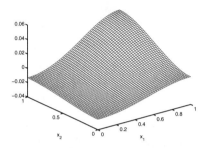

Figure 11.7. *The solution to the BVP in Example* 11.5. *This graph was produced using a total of* 120 *terms of the (double) Fourier cosine series.*

11.2.6 Dirichlet and Neumann problems for Laplace's equation

The PDE
$$-\Delta u = 0 \text{ in } \Omega \tag{11.31}$$
is called Laplace's equation. This equation, with inhomogeneous boundary conditions, is commonly encountered in applications. For example, a steady-state heat flow problem with no heat source, in a homogeneous domain, leads to Laplace's equation, which can be paired with inhomogeneous Dirichlet or Neumann conditions. The Dirichlet conditions indicate that the temperature is fixed on the boundary, while the Neumann conditions indicate that the heat flux is prescribed.

As another example, if a membrane is stretched on a frame described by a curve $(x, y, g(x_1, x_2))$, $(x_1, x_2) \in \partial\Omega$, and if no transverse force is applied, then the shape of the membrane is described by the solution of Laplace's equation with inhomogeneous Dirichlet conditions $u(x_1, x_2) = g(x_1, x_2)$, $(x_1, x_2) \in \partial\Omega$.

We will consider the following Dirichlet problem for Laplace's equation:
$$\begin{aligned} -\Delta u &= 0 \text{ in } \Omega, \\ u &= g \text{ on } \partial\Omega. \end{aligned} \tag{11.32}$$

If we wish to solve (11.32) using the method of Fourier series, then it is desirable to shift the data to obtain homogeneous boundary conditions. The reader should recall the method of shifting the data from Section 5.3: Given a smooth function p defined on $\overline{\Omega}$ satisfying $p(x_1,x_2) = g(x_1,x_2)$ on $\partial\Omega$, we define $v = u - p$, where u satisfies (11.32). Then, in Ω,

$$-\Delta v = -\Delta u + \Delta p = 0 + \Delta p = \Delta p,$$

and, on $\partial\Omega$,

$$v = u - p = g - g = 0.$$

Therefore, v satisfies the BVP

$$\begin{aligned} -\Delta v &= \Delta p \text{ in } \Omega, \\ v &= 0 \text{ on } \partial\Omega. \end{aligned} \quad (11.33)$$

For a general domain Ω, it may be difficult or impossible to find (explicitly) a smooth function p, defined on all of $\overline{\Omega}$ and satisfying $p(x_1,x_2) = g(x_1,x_2)$ for all $(x_1,x_2) \in \partial\Omega$. However, for such a general domain, it is probably impossible to find the eigenvalues and eigenfunction of the negative Laplacian, also. For the simple domains (rectangles and disks) for which we can explicitly find the eigenpairs of the Laplacian, we can also find the function p explicitly.

For the rectangle (11.16), the boundary data g will typically be described as follows:

$$g(x_1,x_2) = \begin{cases} g_1(x_1), & (x_1,x_2) \in \Gamma_1, \\ g_2(x_2), & (x_1,x_2) \in \Gamma_2, \\ g_3(x_1), & (x_1,x_2) \in \Gamma_3, \\ g_4(x_2), & (x_1,x_2) \in \Gamma_4, \end{cases}$$

where $\partial\Omega$ is partitioned as in (11.19). There is a completely mechanical technique for finding $p : \overline{\Omega} \to \mathbf{R}$ satisfying $p(x_1,x_2) = g(x_1,x_2)$ on $\partial\Omega$; however, this technique requires a lengthy explanation, which we relegate to Appendix B. In the following example, we use the results from Appendix B, which the interested reader can consult for the details.

Example 11.6. We assume that an elastic membrane of dimensions 10 cm by 15 cm occupies the set

$$\Omega = \left\{ \mathbf{x} \in \mathbf{R}^2 : 0 < x_1 < 10,\ 0 < x_2 < 15 \right\},$$

and the edges of the membrane are fixed to a frame, so that the vertical deflection of the membrane satisfies the following boundary conditions:

$$u(x_1,x_2) = g(x_1,x_2) = \begin{cases} \frac{x_1^2}{200}, & (x_1,x_2) \in \Gamma_1, \\ \frac{1}{2} - \frac{x_2}{30}, & (x_1,x_2) \in \Gamma_2, \\ \frac{1}{10} - \frac{x_1}{100}, & (x_1,x_2) \in \Gamma_3, \\ \frac{x_2}{15}, & (x_1,x_2) \in \Gamma_4. \end{cases}$$

Then u satisfies

$$\begin{aligned} -\Delta u &= 0 \text{ in } \Omega, \\ u &= g \text{ on } \partial\Omega. \end{aligned} \quad (11.34)$$

A smooth function p satisfying $p = g$ on $\partial \Omega$ is

$$p(x_1, x_2) = \frac{20x_2 - 2x_1 x_2 - x_1^2(x_2 - 15)}{3000},$$

and we have

$$\Delta p(x_1, x_2) = \frac{x_2 - 15}{1500}.$$

We therefore solve

$$-\Delta v = f(x_1, x_2) \text{ in } \Omega,$$
$$v = 0 \text{ on } \partial \Omega,$$

where $f(x_1, x_2) = (x_2 - 15)/1500$, by the Fourier series method. The eigenpairs of $-\Delta$ on Ω are

$$\lambda_{mn} = \frac{m^2 \pi^2}{100} + \frac{n^2 \pi^2}{225}, \quad \psi_{mn}(x_1, x_2) = \sin\left(\frac{m\pi x_1}{10}\right) \sin\left(\frac{n\pi x_2}{15}\right), \quad m, n = 1, 2, 3, \ldots.$$

The Fourier coefficients of f are

$$c_{mn} = \frac{4}{150} \int_0^{10} \int_0^{15} f(x_1, x_2) \sin\left(\frac{m\pi x_1}{10}\right) \sin\left(\frac{n\pi x_2}{15}\right) dx_2 \, dx_1$$

$$= \frac{(-1)^m - 1}{25 mn\pi^2}, \quad m, n = 1, 2, 3, \ldots,$$

and so the solution v is given by

$$v(x_1, x_2) = \sum_{m=1}^{\infty} \sum_{n=1}^{\infty} a_{mn} \sin\left(\frac{m\pi x_1}{10}\right) \sin\left(\frac{n\pi x_2}{15}\right),$$

where

$$a_{mn} = \frac{c_{mn}}{\lambda_{mn}}, \quad m, n = 1, 2, 3, \ldots.$$

The desired solution u is then given by

$$u(x_1, x_2) = p(x_1, x_2) + \sum_{m=1}^{\infty} \sum_{n=1}^{\infty} a_{mn} \sin\left(\frac{m\pi x_1}{10}\right) \sin\left(\frac{n\pi x_2}{15}\right).$$

The graph of u is shown in Figure 11.8. ∎

The Neumann problem for Laplace's equation can be handled in much the same way. An example is given in Appendix B (Example B.2).

11.2. Fourier series on a rectangular domain

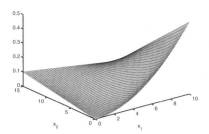

Figure 11.8. *The solution to the BVP in Example* 11.6. *This graph was produced using* 100 *terms of the (double) Fourier sine series.*

11.2.7 Fourier series methods for a rectangular box in three dimensions

It is straightforward, in principle, to extend the Fourier series method to problems posed on a rectangular box in three dimensions. For example, if

$$\Omega = \{(x_1, x_2, x_3) \in \mathbf{R}^3 : 0 < x_1 < \ell_1, \, 0 < x_2 < \ell_2, \, 0 < x_3 < \ell_3\},$$

then the eigenpairs of $-\Delta$ on Ω and under Dirichlet conditions are

$$\lambda_{kmn} = \frac{k^2 \pi^2}{\ell_1^2} + \frac{m^2 \pi^2}{\ell_2^2} + \frac{n^2 \pi^2}{\ell_3^2},$$

$$\psi_{kmn}(x_1, x_2, x_3) = \sin\left(\frac{k\pi x_1}{\ell_1}\right) \sin\left(\frac{m\pi x_2}{\ell_2}\right) \sin\left(\frac{n\pi x_3}{\ell_3}\right), \quad k, m, n = 1, 2, 3, \ldots.$$

Again, having determined these eigenpairs, it is straightforward to solve Poisson's equation, the heat equation, or the wave equation under Dirichlet conditions. The reader should note, however, that the resulting solutions are rather expensive to evaluate, because they are given as triply-indexed series. For example, to include the lowest 10 frequencies in each dimension means working with $10^3 = 1000$ terms! Fast transforms, such as the fast Fourier transform, can be used to reduce the cost (see Section 12.2).

Exercises

1. Let Ω be the unit square in $\mathbf{R}^2$,

$$\Omega = \{\mathbf{x} \in \mathbf{R}^2 : 0 < x_1 < 1, \, 0 < x_2 < 1\},$$

and define $f : \Omega \to \mathbf{R}$ by

$$f(\mathbf{x}) = x_1 x_2 \left(\frac{1}{4} - x_1\right)(1 - x_1)(1 - x_2).$$

(a) Compute the (double) Fourier sine series

$$\sum_{m=1}^{\infty}\sum_{n=1}^{\infty} c_{mn} \sin(m\pi x_1)\sin(n\pi x_2)$$

of f (that is, compute the coefficients c_{mn}).

(b) Graph the error in approximating f by

$$\sum_{m=1}^{M}\sum_{n=1}^{N} c_{mn} \sin(m\pi x_1)\sin(n\pi x_2)$$

for $M = N = 2$ and again for $M = N = 5$.

2. Repeat the previous exercise with

$$f(\mathbf{x}) = x_1(1-x_1)^2.$$

3. Solve the BVP

$$-\Delta u = f(\mathbf{x}) \text{ in } \Omega,$$
$$u = 0 \text{ on } \partial\Omega,$$

where Ω is the unit square in $\mathbf{R}^2$ and f is the function in Exercise 1.

4. Solve the BVP

$$-\Delta u = f(\mathbf{x}) \text{ in } \Omega,$$
$$u = 0 \text{ on } \partial\Omega,$$

where Ω is the unit square in $\mathbf{R}^2$ and f is the function in Exercise 2.

5. Suppose that a square iron plate of side length 50 cm initially has a constant temperature of 5 degrees Celsius and that its edges are placed in an ice bath (0 degrees Celsius) (the plate is perfectly insulated on the top and bottom). At the same time that the edges are placed in the ice bath, we begin adding heat energy everywhere in the interior of the plate at the (constant) rate of $0.02\,\text{W}/\text{cm}^3$. The material constants are $\rho = 7.88\,\text{g}/\text{cm}^3$, $c = 0.437\,\text{J}/(\text{g\,K})$, and $\kappa = 0.836\,\text{W}/(\text{cm\,K})$.

(a) Formulate an IBVP that describes the temperature $u(\mathbf{x},t)$.

(b) Find the (double) Fourier sine series of u.

(c) Find the Fourier series of the steady-state temperature u_s. What BVP does u_s satisfy?

(d) How close is u to u_s after 10 minutes? Graph the difference in the two temperature distributions.

6. Let $u \in C(\overline{\Omega})$ be given, and define F_{MN} to be the subspace of $C(\overline{\Omega})$ spanned by

$$\{\psi_{mn} : m = 1, 2, \ldots, M, \, n = 1, 2, \ldots, N\},$$

where

$$\psi_{mn}(x_1, x_2) = \sin\left(\frac{m\pi x_1}{\ell_1}\right)\sin\left(\frac{n\pi x_2}{\ell_2}\right).$$

Show that the best approximation to u (in the L^2 norm) from F_{MN} is given by (11.23), (11.24).

7. Consider the iron plate of Exercise 5. Suppose that the plate initially has a constant temperature of 2 degrees Celsius and that at time $t = 0$ the edges are instantly brought to a temperature of 5 degrees and held there. How long does it take until the entire plate has a temperature of at least 4 degrees? (Hint: To deal with the inhomogeneous boundary condition, just shift the data. It is trivial in this case, because the boundary data is constant.)

8. Suppose Ω is the rectangle $\{\mathbf{x} \in \mathbf{R}^2 : 0 < x_1 < \ell_1, \, 0 < x_2 < \ell_2\}$, and that u is a twice-continuously differentiable function defined on $\overline{\Omega}$. Let the Fourier sine series of u be

$$u(x_1, x_2) = \sum_{m=1}^{\infty}\sum_{n=1}^{\infty} a_{mn} \sin\left(\frac{m\pi x_1}{\ell_1}\right)\sin\left(\frac{n\pi x_2}{\ell_2}\right).$$

Suppose u satisfies $u(\mathbf{x}) = 0$ for $\mathbf{x} \in \partial\Omega$. Prove that the Fourier series of $-\Delta u$ is

$$-\Delta u(x_1, x_2) = \sum_{m=1}^{\infty}\sum_{n=1}^{\infty} \lambda_{mn} a_{mn} \sin\left(\frac{m\pi x_1}{\ell_1}\right)\sin\left(\frac{n\pi x_2}{\ell_2}\right),$$

where

$$\lambda_{mn} = \frac{m^2\pi^2}{\ell_1^2} + \frac{n^2\pi^2}{\ell_2^2}, \, m, n = 1, 2, 3, \ldots.$$

9. In Example 11.4, how long does it take for the leading edge of the wave to reach the boundary of Ω? Is the computed time consistent with Figure 11.6?

10. Suppose Ω is a domain in $\mathbf{R}^2$ or $\mathbf{R}^3$ with a piecewise smooth boundary $\partial\Omega$, and suppose $\partial\Omega$ is partitioned into two disjoint subsets: $\partial\Omega = \Gamma_1 \cup \Gamma_2$. Define

$$C_m^2(\overline{\Omega}) = \left\{u \in C^2(\overline{\Omega}) : u(\mathbf{x}) = 0, \, \mathbf{x} \in \Gamma_1, \, \frac{\partial u}{\partial \mathbf{n}}(\mathbf{x}) = 0, \, \mathbf{x} \in \Gamma_2\right\}.$$

Let $L_m : C_m^2(\overline{\Omega}) \to C(\overline{\Omega})$ be defined by $L_m u = -\Delta u$. Show that L_m is symmetric and has only positive eigenvalues.

11. Consider a square drumhead occupying the unit square,

$$\Omega = \{(x_1, x_2) \in \mathbf{R}^2 : 0 < x_1 < 1, \, 0 < x_2 < 1\},$$

when at rest, and suppose the drum is "plucked" so that its initial (vertical) displacement is the (piecewise linear) function

$$\psi(x_1,x_2) = \begin{cases} 2\beta x_2, & (x_1,x_2) \in \Omega_1, \\ 2\beta(1-x_1), & (x_1,x_2) \in \Omega_2, \\ 2\beta(1-x_2), & (x_1,x_2) \in \Omega_3, \\ 2\beta x_1, & (x_1,x_2) \in \Omega_4. \end{cases}$$

Here $\beta = 10^{-4}$ and

$$\Omega_1 = \left\{ (x_1,x_2) \in \mathbf{R}^2 \ : \ 0 < x_2 < \frac{1}{2}, \ x_2 < x_1 < 1 - x_2 \right\},$$

$$\Omega_2 = \left\{ (x_1,x_2) \in \mathbf{R}^2 \ : \ \frac{1}{2} < x_1 < 1, \ 1 - x_1 < x_2 < x_1 \right\},$$

$$\Omega_3 = \left\{ (x_1,x_2) \in \mathbf{R}^2 \ : \ \frac{1}{2} < x_2 < 1, \ 1 - x_2 < x_1 < x_2 \right\},$$

$$\Omega_4 = \left\{ (x_1,x_2) \in \mathbf{R}^2 \ : \ 0 < x_1 < \frac{1}{2}, \ x_1 < x_2 < 1 - x_1 \right\}$$

(see Figure 11.9). Find the resulting vibrations of the drumhead by solving

$$\frac{\partial^2 u}{\partial t^2} - c^2 \Delta u = 0, \ \mathbf{x} \in \Omega, \ t > 0,$$

$$u(\mathbf{x},0) = \psi(\mathbf{x}), \ \mathbf{x} \in \Omega,$$

$$\frac{\partial u}{\partial t}(\mathbf{x},0) = 0, \ \mathbf{x} \in \Omega,$$

$$u(\mathbf{x},t) = 0, \ \mathbf{x} \in \partial\Omega, \ t > 0.$$

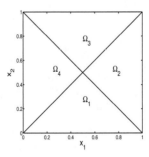

Figure 11.9. *The partition of the unit square used in Exercise* 11.

12. Let Ω be the rectangle (11.16). Define

$$C_N^2(\overline{\Omega}) = \left\{ u \in C^2(\overline{\Omega}) \ : \ \frac{\partial u}{\partial \mathbf{n}}(\mathbf{x}) = 0, \ \mathbf{x} \in \partial\Omega \right\}.$$

Let $L_N : C_N^2(\overline{\Omega}) \to C(\overline{\Omega})$ be defined by $L_N u = -\Delta u$. Use separation of variables to find the eigenpairs of L_N with separated eigenfunctions.

11.3. Fourier series on a disk

13. Suppose an elastic membrane occupies the unit square,

$$\Omega = \{(x_1, x_2) \in \mathbf{R}^2 \,:\, 0 < x_1 < 1,\, 0 < x_2 < 1\},$$

and something pushes up at the center of the membrane (a pin or the tip of a pencil, for example). The purpose of this exercise is to determine the resulting shape of the membrane. This can be done by solving the BVP

$$-\Delta u = \delta(x_1 - 1/2)\delta(x_2 - 1/2) \text{ in } \Omega,$$
$$u = 0 \text{ on } \partial\Omega.$$

Graph the solution. Note: The Fourier coefficients of the point source can be found using the sifting property of the Dirac delta function.[56]

14. Let Ω be the rectangle (11.16), and suppose $\partial\Omega$ is partitioned as in (11.19). Define

$$C_m^2(\overline{\Omega}) = \left\{ u \in C^2(\overline{\Omega}) \,:\, u(\mathbf{x}) = 0,\, \mathbf{x} \in \Gamma_1 \cup \Gamma_4,\, \frac{\partial u}{\partial \mathbf{n}}(\mathbf{x}) = 0,\, \mathbf{x} \in \Gamma_2 \cup \Gamma_3 \right\}.$$

Let $L_m : C_m^2(\overline{\Omega}) \to C(\overline{\Omega})$ be defined by $L_m u = -\Delta u$. Use separation of variables to find the eigenpairs of L_m with separated eigenfunctions.

15. Repeat Exercise 5, but assuming now that the edges Γ_2 and Γ_3 are insulated, while edges Γ_1 and Γ_4 are placed in an ice bath ($\partial\Omega$ is partitioned as in (11.19)). You will need the eigenpairs determined in Exercise 14.

11.3 Fourier series on a disk

We now discuss another simple two-dimensional domain for which we can derive Fourier series methods, namely, the case of a circular disk. We define

$$\Omega = \{(x_1, x_2) \in \mathbf{R}^2 \,:\, x_1^2 + x_2^2 < A\},$$

where $A > 0$ is a constant. We wish to develop Fourier series methods for the following Dirichlet problem:

$$-\Delta u = f(\mathbf{x}),\, \mathbf{x} \in \Omega,$$
$$u(\mathbf{x}) = 0,\, \mathbf{x} \in \partial\Omega.$$
(11.35)

The first step is to find the eigenvalues and eigenfunctions of the Laplacian, subject to the Dirichlet conditions on the circle.

We have only one technique for finding eigenfunctions of a differential operator in two variables, namely separation of variables. However, this technique is not very promising in rectangular coordinates, since the boundary condition does not separate. For this reason, we first change to polar coordinates. If we define

$$v(r, \theta) = u(x_1, x_2),$$

---
[56] The material on the delta function found in Section 9.1 is needed for this exercise.

where (r,θ) are the polar coordinates corresponding to the rectangular coordinates (x_1,x_2), then the Dirichlet condition becomes simply

$$v(A,\theta) = 0, \ -\pi \leq \theta < \pi.$$

If we apply separation of variables and write $v(r,\theta) = R(r)T(\theta)$, then the Dirichlet condition is $R(A)T(\theta) = 0$ for all θ, which implies (if v is nontrivial) that $R(A) = 0$. The use of polar coordinates introduces periodic boundary conditions for T:

$$T(-\pi) = T(\pi),$$
$$\frac{dT}{d\theta}(-\pi) = \frac{dT}{d\theta}(\pi).$$

There is also a boundary condition for R at $r = 0$, although it is not one we have seen before. We simply need $R(0)$ to be a finite number.

11.3.1 The Laplacian in polar coordinates

Before we can apply separation of variables to the PDE, we must change variables to determine the form of the (negative) Laplacian in polar coordinates. This is an exercise in the chain rule, as we now explain. The chain rule implies, for example, that

$$\frac{\partial u}{\partial x_1} = \frac{\partial v}{\partial r}\frac{\partial r}{\partial x_1} + \frac{\partial v}{\partial \theta}\frac{\partial \theta}{\partial x_1}.$$

This allows us to replace the partial derivative with respect to x_1 by partial derivatives with respect to the new variables r and θ, provided we can compute

$$\frac{\partial r}{\partial x_1}, \ \frac{\partial \theta}{\partial x_1}.$$

To deal with derivatives with respect to x_2, we will also need

$$\frac{\partial r}{\partial x_2}, \ \frac{\partial \theta}{\partial x_2}.$$

It is straightforward to compute the needed derivatives from the relationship between rectangular and polar coordinates:

$$x_1 = r\cos(\theta), \ x_2 = r\sin(\theta),$$
$$r = \sqrt{x_1^2 + x_2^2}, \ \tan(\theta) = \frac{x_2}{x_1}.$$

We have

$$\frac{\partial r}{\partial x_1} = \frac{x_1}{\sqrt{x_1^2 + x_2^2}} = \frac{r\cos(\theta)}{r} = \cos(\theta),$$

and, similarly,

$$\frac{\partial r}{\partial x_2} = \sin(\theta).$$

11.3. Fourier series on a disk

Also,
$$\tan(\theta) = \frac{x_2}{x_1}$$
$$\Rightarrow \sec^2(\theta)\frac{\partial \theta}{\partial x_1} = -\frac{x_2}{x_1^2} = -\frac{r\sin(\theta)}{r^2\cos^2(\theta)}$$
$$\Rightarrow \frac{\partial \theta}{\partial x_1} = -\frac{\sin(\theta)}{r},$$

and, similarly,
$$\frac{\partial \theta}{\partial x_2} = \frac{\cos(\theta)}{r}.$$

We collect these formulas for future reference:
$$\frac{\partial r}{\partial x_1} = \cos(\theta), \quad \frac{\partial r}{\partial x_2} = \sin(\theta),$$
$$\frac{\partial \theta}{\partial x_1} = -\frac{\sin(\theta)}{r}, \quad \frac{\partial \theta}{\partial x_2} = \frac{\cos(\theta)}{r}.$$

We can now compute the Laplacian in polar coordinates. We have
$$\frac{\partial u}{\partial x_1} = \frac{\partial v}{\partial r}\frac{\partial r}{\partial x_1} + \frac{\partial v}{\partial \theta}\frac{\partial \theta}{\partial x_1}$$
$$= \cos(\theta)\frac{\partial v}{\partial r} - \frac{\sin(\theta)}{r}\frac{\partial v}{\partial \theta},$$

and so
$$\frac{\partial^2 u}{\partial x_1^2} = \cos(\theta)\left(\frac{\partial^2 v}{\partial r^2}\frac{\partial r}{\partial x_1} + \frac{\partial^2 v}{\partial \theta \partial r}\frac{\partial \theta}{\partial x_1}\right) - \sin(\theta)\frac{\partial \theta}{\partial x_1}\frac{\partial v}{\partial r}$$
$$- \frac{\sin(\theta)}{r}\left(\frac{\partial^2 v}{\partial r \partial \theta}\frac{\partial r}{\partial x_1} + \frac{\partial^2 v}{\partial \theta^2}\frac{\partial \theta}{\partial x_1}\right) - \frac{r\cos(\theta)\frac{\partial \theta}{\partial x_1} - \sin(\theta)\frac{\partial r}{\partial x_1}}{r^2}\frac{\partial v}{\partial \theta}$$
$$= \cos(\theta)\left(\cos(\theta)\frac{\partial^2 v}{\partial r^2} - \frac{\sin(\theta)}{r}\frac{\partial^2 v}{\partial \theta \partial r}\right) + \frac{\sin^2(\theta)}{r}\frac{\partial v}{\partial r}$$
$$- \frac{\sin(\theta)}{r}\left(\cos(\theta)\frac{\partial^2 v}{\partial r \partial \theta} - \frac{\sin(\theta)}{r}\frac{\partial^2 v}{\partial \theta^2}\right) + \frac{2\sin(\theta)\cos(\theta)}{r^2}\frac{\partial v}{\partial \theta}$$
$$= \cos^2(\theta)\frac{\partial^2 v}{\partial r^2} - \frac{2\sin(\theta)\cos(\theta)}{r}\frac{\partial^2 v}{\partial \theta \partial r} + \frac{\sin^2(\theta)}{r^2}\frac{\partial^2 v}{\partial \theta^2} + \frac{\sin^2(\theta)}{r}\frac{\partial v}{\partial r}$$
$$+ \frac{2\sin(\theta)\cos(\theta)}{r^2}\frac{\partial v}{\partial \theta}.$$

A similar calculation shows that
$$\frac{\partial^2 u}{\partial x_2^2} = \sin^2(\theta)\frac{\partial^2 v}{\partial r^2} + \frac{2\sin(\theta)\cos(\theta)}{r}\frac{\partial^2 v}{\partial \theta \partial r} + \frac{\cos^2(\theta)}{r^2}\frac{\partial^2 v}{\partial \theta^2} + \frac{\cos^2(\theta)}{r}\frac{\partial v}{\partial r}$$
$$- \frac{2\sin(\theta)\cos(\theta)}{r^2}\frac{\partial v}{\partial \theta}.$$

Adding these two results and using the identity $\cos^2(\theta) + \sin^2(\theta) = 1$, we obtain the Laplacian in polar coordinates:

$$-\Delta v = -\frac{\partial^2 v}{\partial r^2} - \frac{1}{r^2}\frac{\partial^2 v}{\partial \theta^2} - \frac{1}{r}\frac{\partial v}{\partial r}. \tag{11.36}$$

11.3.2 Separation of variables in polar coordinates

We now apply separation of variables and look for eigenfunctions of $-\Delta$ (under Dirichlet conditions) of the form $v(r,\theta) = R(r)T(\theta)$. That is, we wish to find all solutions of

$$\begin{aligned}-\Delta v &= \lambda v \text{ in } \Omega, \\ v &= 0 \text{ on } \partial\Omega\end{aligned} \tag{11.37}$$

that have the form $v(r,\theta) = R(r)T(\theta)$. Substituting the separated function $v = RT$ into the PDE yields the following:

$$\begin{aligned}-\Delta v &= \lambda v \\ \Rightarrow -\frac{\partial^2 v}{\partial r^2} - \frac{1}{r^2}\frac{\partial^2 v}{\partial \theta^2} - \frac{1}{r}\frac{\partial v}{\partial r} &= \lambda v \\ \Rightarrow -\frac{d^2 R}{dr^2}T - \frac{1}{r^2}R\frac{d^2 T}{d\theta^2} - \frac{1}{r}\frac{dR}{dr}T &= \lambda RT \\ \Rightarrow -R^{-1}\frac{d^2 R}{dr^2} - \frac{1}{r^2}T^{-1}\frac{d^2 T}{d\theta^2} - \frac{1}{r}R^{-1}\frac{dR}{dr} &= \lambda \\ \Rightarrow -T^{-1}\frac{d^2 T}{d\theta^2} &= \lambda r^2 + R^{-1}r^2\frac{d^2 R}{dr^2} + R^{-1}r\frac{dR}{dr}.\end{aligned}$$

Since the left side of this equation is a function of θ alone, while the right side is a function of r alone, both sides must be constant. We will write

$$-T^{-1}\frac{d^2 T}{d\theta^2} = \gamma$$

or

$$-\frac{d^2 T}{d\theta^2} = \gamma T.$$

We then have

$$\lambda r^2 + R^{-1}r^2\frac{d^2 R}{dr^2} + R^{-1}r\frac{dR}{dr} = \gamma$$

or

$$\frac{d^2 R}{dr^2} + \frac{1}{r}\frac{dR}{dr} + \left(\lambda - \frac{\gamma}{r^2}\right)R = 0.$$

This last equation can be written as an eigenvalue problem for R, but it is not one we have studied before. We will study it in detail below, but first we deal with the easier eigenvalue problem for T.

11.3. Fourier series on a disk

As we mentioned above, the angular variable θ introduces periodic boundary conditions for T, so we must solve the eigenvalue problem

$$-\frac{d^2 T}{d\theta^2} = \gamma T, \quad -\pi < \theta < \pi,$$
$$T(-\pi) = T(\pi), \qquad (11.38)$$
$$\frac{dT}{d\theta}(-\pi) = \frac{dT}{d\theta}(\pi).$$

As we saw in Section 6.3, the eigenvalues are 0 and n^2, $n = 1, 2, 3, \ldots$. The constant function 1 is an eigenfunction corresponding to $\gamma = 0$, and each positive eigenvalue $\gamma = n^2$ has two independent eigenfunctions, $\cos(n\theta)$ and $\sin(n\theta)$.

11.3.3 Bessel's equation

We now consider the problem

$$\frac{d^2 R}{dr^2} + \frac{1}{r}\frac{dR}{dr} + \left(\lambda - \frac{\gamma}{r^2}\right) R = 0, \qquad (11.39a)$$
$$R(0) \in \mathbf{R}, \qquad (11.39b)$$
$$R(A) = 0. \qquad (11.39c)$$

We know that λ must be positive, since the Laplacian has only positive eigenvalues under Dirichlet conditions (see Section 11.2.1). We can considerably simplify the analysis by the change of variables

$$s = \sqrt{\lambda} r.$$

We define

$$S(s) = R\left(\frac{s}{\sqrt{\lambda}}\right) \Leftrightarrow R(r) = S\left(\sqrt{\lambda} r\right),$$

so that

$$\frac{dR}{dr}(r) = \sqrt{\lambda}\frac{dS}{ds}\left(\sqrt{\lambda} r\right) = \sqrt{\lambda}\frac{dS}{ds}(s)$$

and

$$\frac{d^2 R}{dr^2}(r) = \lambda \frac{d^2 S}{ds^2}\left(\sqrt{\lambda} r\right) = \lambda \frac{d^2 S}{ds^2}(s).$$

We then have

$$\frac{d^2 R}{dr^2} + \frac{1}{r}\frac{dR}{dr} + \left(\lambda - \frac{\gamma}{r^2}\right) R$$
$$= \lambda \frac{d^2 S}{ds^2} + \frac{\sqrt{\lambda}}{s/\sqrt{\lambda}}\frac{dS}{ds} + \left(\lambda - \frac{n^2}{s^2/\lambda}\right) S$$
$$= \lambda \left(\frac{d^2 S}{ds^2} + \frac{1}{s}\frac{dS}{ds} + \left(1 - \frac{n^2}{s^2}\right) S\right).$$

The ODE

$$\frac{d^2S}{ds^2} + \frac{1}{s}\frac{dS}{ds} + \left(1 - \frac{n^2}{s^2}\right)S = 0$$

is called *Bessel's equation of order n*. In the new variables, the BVP (11.39) becomes

$$\frac{d^2S}{ds^2} + \frac{1}{s}\frac{dS}{ds} + \left(1 - \frac{n^2}{s^2}\right)S = 0,$$
$$S(0) \in \mathbf{R}, \qquad (11.40)$$
$$S\left(\sqrt{\lambda}A\right) = 0.$$

Bessel's equation does not have solutions that can be expressed in terms of elementary functions. To find a solution of Bessel's equation, we can use a power series expansion of the solution. We suppose[57] that (11.40) has a solution of the form

$$S(s) = r^\alpha \sum_{k=0}^{\infty} a_k s^k = \sum_{k=0}^{\infty} a_k s^{k+\alpha}, \; a_0 \neq 0.$$

The value of α, as well as the coefficients $a_0, a_1, a_2, \ldots$, must be determined from the differential equation and boundary conditions. We have

$$\frac{dS}{ds}(s) = \sum_{k=0}^{\infty}(k+\alpha)a_k s^{k+\alpha-1},$$

and so

$$\frac{1}{s}\frac{dS}{ds}(s) = \sum_{k=0}^{\infty}(k+\alpha)a_k s^{k+\alpha-2}. \qquad (11.41)$$

Also,

$$\frac{d^2S}{ds^2}(s) = \sum_{k=0}^{\infty}(k+\alpha)(k+\alpha-1)a_k s^{k+\alpha-2}. \qquad (11.42)$$

It is helpful to write

$$S(s) = \sum_{k=0}^{\infty} a_k s^{k+\alpha} = \sum_{k=2}^{\infty} a_{k-2} s^{k+\alpha-2}. \qquad (11.43)$$

Substituting (11.43), (11.41), and (11.42) into Bessel's equation yields

$$\sum_{k=0}^{\infty}(k+\alpha)(k+\alpha-1)a_k s^{k+\alpha-2} + \sum_{k=0}^{\infty}(k+\alpha)a_k s^{k+\alpha-2}$$
$$+ \sum_{k=2}^{\infty} a_{k-2} s^{k+\alpha-2} - n^2 \sum_{k=0}^{\infty} a_k s^{k+\alpha-2} = 0$$

---

[57] This is another example of an inspired guess at the form of a solution. It is based on the fact that Euler's equation, $r^2 \frac{d^2R}{dr^2} + r\frac{dR}{dr} - \lambda R = 0$, has solutions of the form $R(r) = r^\alpha$.

11.3. Fourier series on a disk

or

$$(\alpha(\alpha-1)+\alpha-n^2)a_0 s^{\alpha-2} + ((\alpha+1)\alpha+(\alpha+1)-n^2)a_1 s^{\alpha-1}$$
$$+ \sum_{k=2}^{\infty} \left\{ ((k+\alpha)(k+\alpha-1)-n^2)a_k + a_{k-1} \right\} s^{k+\alpha-2} = 0.$$

This equation implies that the coefficient of each power of s must vanish, and so we obtain the equations

$$(\alpha(\alpha-1)+\alpha-n^2)a_0 = 0,$$
$$((\alpha+1)\alpha+(\alpha+1)-n^2)a_1 = 0,$$
$$((k+\alpha)(k+\alpha-1)+(k+\alpha)-n^2)a_k + a_{k-2} = 0, \ k=2,3,4,\ldots,$$

which simplify to

$$\left(\alpha^2 - n^2\right)a_0 = 0,$$
$$\left((\alpha+1)^2 - n^2\right)a_1 = 0, \tag{11.44}$$
$$\left((k+\alpha)^2 - n^2\right)a_k + a_{k-2} = 0, \ k=2,3,4,\ldots.$$

We have assumed that $a_0 \neq 0$, so the first equation in (11.44) implies that $\alpha = \pm n$. Moreover, the condition that $S(0)$ be finite rules out the possibility that $\alpha = -n$, so we conclude that $\alpha = n$ must hold. This, together with the second equation in (11.44), immediately implies that $a_1 = 0$. The third equation in (11.44) can be written as the recursion relation

$$a_k = -\frac{a_{k-2}}{(k+n)^2 - n^2} = -\frac{a_{k-2}}{k(k+2n)}, \ k=2,3,4,\ldots. \tag{11.45}$$

Since $a_1 = 0$, (11.45) implies that all of the odd coefficients are zero:

$$a_{2j+1} = 0, \ j=0,1,2,\ldots.$$

On the other hand, we have

$$a_2 = -\frac{a_0}{2(2+2n)} = -\frac{a_0}{2^2(n+1)},$$
$$a_4 = -\frac{a_2}{4(4+2n)} = \frac{a_0}{2!2^4(n+1)(n+2)},$$
$$a_6 = -\frac{a_4}{6(6+2n)} = -\frac{a_0}{3!2^6(n+1)(n+2)(n+3)},$$

and, in general,

$$a_{2j} = \frac{(-1)^j a_0}{j!2^{2j}(n+1)(n+2)\cdots(n+j)}, \ j=0,1,2,\ldots. \tag{11.46}$$

Since any multiple of a solution of (11.40) is again a solution, we may as well choose the value of a_0 to give as simple a formula as possible for a_{2j}. For this reason, we choose

$$a_0 = \frac{1}{2^n n!}$$

and obtain

$$a_{2j} = \frac{(-1)^j}{2^{2j+n} j!(n+j)!}, \ j = 0, 1, 2, \ldots.$$

We then obtain the solution

$$S(s) = \sum_{j=0}^{\infty} \frac{(-1)^j}{j!(n+j)!} \left(\frac{s}{2}\right)^{2j+n}.$$

This function is a solution of Bessel's equation of order n. It is usually written

$$J_n(s) = \sum_{j=0}^{\infty} \frac{(-1)^j}{j!(n+j)!} \left(\frac{s}{2}\right)^{2j+n} \tag{11.47}$$

and called the *Bessel function of order n*. The reader should note that the above calculations are valid for $n = 0$ as well as $n = 1, 2, 3, \ldots$.

11.3.4 Properties of the Bessel functions

The Bessel functions have been extensively studied because of their utility in applied mathematics, and their properties are well known.[58] We will need the following properties of the Bessel functions.

1. If $n > 0$, then $J_n(0) = 0$. This follows directly from (11.47).

2. The Bessel functions satisfy

$$J_{n+1}(s) = \frac{n}{s} J_n(s) - \frac{dJ_n}{ds}(s). \tag{11.48}$$

   This recursion relation can be verified directly by computing the power series of the right-hand side and simplifying it to show that it equals the left-hand side (see Exercise 11.2.10).

3. If, for a given value of α, $J_n(\alpha A) = 0$, then

$$\int_0^A (J_n(\alpha r))^2 r \, dr = \frac{A^2}{2} (J_{n+1}(\alpha A))^2. \tag{11.49}$$

   We will sketch the proof of this result. Since J_n satisfies Bessel's equation, we have

$$\frac{d^2 J_n}{ds^2} + \frac{1}{s} \frac{dJ_n}{ds} + \left(1 - \frac{n^2}{s^2}\right) J_n = 0,$$

---

[58] Standard software packages, including MATLAB, *Mathematica*, and *Maple*, implement Bessel functions just as they do elementary functions such as the natural logarithm or sine.

11.3. Fourier series on a disk

and multiplying through by s yields

$$s\frac{d^2 J_n}{ds^2} + \frac{dJ_n}{ds} + s^{-1}\left(s^2 - n^2\right)J_n = 0$$

or

$$\frac{d}{ds}\left[s\frac{dJ_n}{ds}(s)\right] + s^{-1}\left(s^2 - n^2\right)J_n(s) = 0.$$

Next, we multiply through by $2s\, dJ_n/ds$ and manipulate the result:

$$2s\frac{dJ_n}{ds}(s)\frac{d}{ds}\left[s\frac{dJ_n}{ds}(s)\right] + 2\left(s^2 - n^2\right)J_n(s)\frac{dJ_n}{ds}(s) = 0$$

$$\Rightarrow \frac{d}{ds}\left[\left(s\frac{dJ_n}{ds}(s)\right)^2\right] + \frac{d}{ds}\left[\left(s^2 - n^2\right)(J_n(s))^2\right] = 2s\,(J_n(s))^2.$$

Integrating both sides yields

$$2\int_0^B (J_n(s))^2 s\, ds = s^2\left(\frac{dJ_n}{ds}(s)\right)^2\bigg|_0^B + \left(s^2 - n^2\right)(J_n(s))^2\bigg|_0^B$$

$$= B^2\left(\frac{dJ_n}{ds}(B)\right)^2 + \left(B^2 - n^2\right)(J_n(B))^2 + n^2(J_n(0))^2$$

$$= B^2\left(\frac{dJ_n}{ds}(B)\right)^2 + \left(B^2 - n^2\right)(J_n(B))^2.$$

In the last step, we used the fact that $n^2 J_n(0) = 0$ since either $n = 0$ or $J_n(0) = 0$. We can now evaluate the integral

$$\int_0^A (J_n(\alpha r))^2 r\, dr$$

by making the change of variables $s = \alpha r$. The result is

$$\int_0^A (J_n(\alpha r))^2 r\, dr = \alpha^{-2}\int_0^{\alpha A} (J_n(s))^2 s\, ds$$

$$= \frac{A^2}{2}\left(\frac{dJ_n}{ds}(\alpha A)\right)^2 + \frac{1}{2\alpha^2}\left(\alpha^2 A^2 - n^2\right)(J_n(\alpha A))^2$$

$$= \frac{A^2}{2}\left(\frac{dJ_n}{ds}(\alpha A)\right)^2,$$

with the last step following from the assumption that $J_n(\alpha A) = 0$. Finally, applying (11.48), we obtain (11.49).

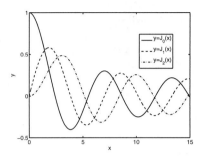

Figure 11.10. *The Bessel function J_0, J_1, and J_2.*

4. Each J_n, $n = 0, 1, 2, \ldots$, has an infinite number of positive roots, which we label $s_{n1} < s_{n2} < s_{n3} < \cdots$. (There is no simple formula for the roots s_{nm}.) We will not prove this fact. Figure 11.10 shows the graphs of J_0, J_1, and J_2. For future reference, we give here some of the roots of the Bessel functions:[59]

$$\begin{aligned}
s_{0m} &= 2.404825557690968, 5.520078109856846, \\
&\quad 8.65372791291017, 11.79153381314112, \ldots, \\
s_{1m} &= 3.831704472655219, 7.015586669057602, \\
&\quad 10.17346813505608, 13.32368935305259, \ldots, \\
s_{2m} &= 5.135622247449045, 8.41724413443633, \\
&\quad 11.61984117182147, 14.79595178232688, \ldots, \\
s_{3m} &= 6.380161894536216, 9.76102306245301, \\
&\quad 13.01520072163691, 16.22346616026436, \ldots.
\end{aligned} \qquad (11.50)$$

11.3.5 The eigenfunctions of the negative Laplacian on the disk

We originally set out to solve (11.39), and we have seen that $R(r) = J_n(\sqrt{\lambda}r)$ solves the ODE (11.39). Since J_n has an infinite sequence of positive roots, there are infinitely many positive solutions to $R(A) = J_n(\sqrt{\lambda}A) = 0$, namely,

$$\lambda_{mn} = \frac{s_{nm}^2}{A^2}, \ m = 1, 2, 3, \ldots.$$

Therefore, for each value of n, $n = 0, 1, 2, \ldots$, there are infinitely many solutions to (11.39). For each value of n, there are two independent solutions to (11.38):

$$\cos(n\theta), \ \sin(n\theta).$$

---

[59] These estimates were computed in *Mathematica* using the **BesselJ** function and the **FindRoot** command.

11.3. Fourier series on a disk

We therefore obtain the following doubly indexed sequence of eigenvalues for $-\Delta$ (on the disk and under Dirichlet conditions), with one or two independent eigenfunctions for each eigenvalue:

$$\begin{aligned}\lambda_{m0} &= \frac{s_{0m}^2}{A^2}, \; \varphi_{m0}^{(1)}(r,\theta) = J_0\left(\frac{s_{0m}r}{A}\right), \; m=1,2,3,\ldots, \\ \lambda_{mn} &= \frac{s_{nm}^2}{A^2}, \; \varphi_{mn}^{(1)}(r,\theta) = J_n\left(\frac{s_{nm}r}{A}\right)\cos(n\theta), \; m,n=1,2,3,\ldots, \\ & \qquad \varphi_{mn}^{(2)}(r,\theta) = J_n\left(\frac{s_{nm}r}{A}\right)\sin(n\theta), \; m,n=1,2,3,\ldots.\end{aligned} \quad (11.51)$$

Since $-\Delta$ is symmetric under Dirichlet conditions, we know that eigenfunctions corresponding to distinct eigenvalues are orthogonal. It turns out that $\varphi_{mn}^{(1)}$ and $\varphi_{mn}^{(2)}$ are orthogonal to each other as well, as a direct calculation shows:

$$\begin{aligned}\left(\varphi_{mn}^{(1)}, \varphi_{mn}^{(2)}\right) &= \int_{-\pi}^{\pi}\int_0^A J_n\left(\frac{s_{nm}r}{A}\right)\cos(n\theta) J_n\left(\frac{s_{nm}r}{A}\right)\sin(n\theta) r\, dr\, d\theta \\ &= \left(\int_{-\pi}^{\pi}\cos(n\theta)\sin(n\theta)\, d\theta\right)\left(\int_0^A J_n\left(\frac{s_{nm}r}{A}\right)^2 r\, dr\right) \\ &= 0 \cdot \left(\int_0^A J_n\left(\frac{s_{nm}r}{A}\right)^2 r\, dr\right) \\ &= 0.\end{aligned}$$

For convenience, we will write

$$\alpha_{mn} = \frac{s_{nm}}{A}, \; n=0,1,2,\ldots, \; m=1,2,3,\ldots. \quad (11.52)$$

We now have a sequence of eigenfunctions of $-\Delta$ on Ω (the disk of radius A centered at the origin), and these eigenfunctions are orthogonal on Ω. Just as in the case in which Ω was a rectangle, it is possible to represent any function in $C(\overline{\Omega})$ in terms of these eigenfunctions:

$$f(r,\theta) = \sum_{m=1}^{\infty} c_{m0} J_0(\alpha_{m0} r) + \sum_{m=1}^{\infty}\sum_{n=1}^{\infty}(c_{mn}\cos(n\theta) + d_{mn}\sin(n\theta)) J_n(\alpha_{mn} r). \quad (11.53)$$

The coefficients are determined by the usual formulas:

$$\begin{aligned}c_{mn} &= \frac{\left(f, \varphi_{mn}^{(1)}\right)}{\left(\varphi_{mn}^{(1)}, \varphi_{mn}^{(1)}\right)}, \; n=0,1,2,\ldots, \; m=1,2,3,\ldots, \\ d_{mn} &= \frac{\left(f, \varphi_{mn}^{(2)}\right)}{\left(\varphi_{mn}^{(2)}, \varphi_{mn}^{(2)}\right)}, \; m,n=1,2,3,\ldots.\end{aligned}$$

Using the properties of the Bessel functions developed above, we can simplify $\left(\varphi_{mn}^{(1)}, \varphi_{mn}^{(1)}\right)$ and $\left(\varphi_{mn}^{(2)}, \varphi_{mn}^{(2)}\right)$:

$$\left(\varphi_{m0}^{(1)}, \varphi_{m0}^{(1)}\right) = \int_{-\pi}^{\pi} \int_0^A (J_0(\alpha_{m0} r))^2 \, r \, dr \, d\theta$$

$$= 2\pi \left(\frac{A^2}{2} (J_1(\alpha_{m0} A))^2\right)$$

$$= \pi A^2 (J_1(\alpha_{m0} A))^2,$$

$$\left(\varphi_{mn}^{(1)}, \varphi_{mn}^{(1)}\right) = \int_{-\pi}^{\pi} \int_0^A \cos^2(n\theta)(J_n(\alpha_{mn} r))^2 \, r \, dr \, d\theta$$

$$= \left(\int_{-\pi}^{\pi} \cos^2(n\theta) \, d\theta\right) \left(\int_0^A (J_n(\alpha_{mn} r))^2 \, r \, dr\right)$$

$$= \frac{\pi A^2}{2} (J_{n+1}(\alpha_{mn} A))^2,$$

and, similarly,

$$\left(\varphi_{mn}^{(2)}, \varphi_{mn}^{(2)}\right) = \frac{\pi A^2}{2} (J_{n+1}(\alpha_{mn} A))^2.$$

We can now represent a function $f(r, \theta)$ as in (11.53), where

$$c_{m0} = \frac{\left(f, \varphi_{mn}^{(1)}\right)}{\pi A^2 (J_1(\alpha_{m0} A))^2}, \quad m = 1, 2, 3, \ldots,$$

$$c_{mn} = \frac{2\left(f, \varphi_{mn}^{(1)}\right)}{\pi A^2 (J_{n+1}(\alpha_{mn} A))^2}, \quad n = 1, 2, 3, \ldots, \, m = 1, 2, 3, \ldots, \quad (11.54)$$

$$d_{mn} = \frac{2\left(f, \varphi_{mn}^{(2)}\right)}{\pi A^2 (J_{n+1}(\alpha_{mn} A))^2}, \quad n = 1, 2, 3, \ldots, \, m = 1, 2, 3, \ldots.$$

Given that software routines implementing the Bessel functions are almost as readily available as routines for the trigonometric functions, it would appear that these formulas are as usable as the analogous formulas on the rectangle. However, we must keep in mind that we do not have formulas for the values of α_{mn}. Therefore, to actually use Bessel functions, we must do a certain amount of numerical work to obtain the needed values of α_{mn} and the eigenvalues $\lambda_{mn} = a_{mn}^2$.

Example 11.7. Consider the function $f \in C(\overline{\Omega})$ defined by

$$f(\mathbf{x}) = 1 - x_1^2 - x_2^2,$$

where Ω is the unit disk ($A = 1$). The corresponding function expressed in polar coordinates is

$$g(r, \theta) = 1 - r^2.$$

11.3. Fourier series on a disk

We wish to compute an approximation to g using the eigenfunctions of $-\Delta$. This is a particularly simple example, since, for any $n > 0$,

$$
\begin{aligned}
(g, \varphi_{mn}^{(1)}) &= \int_{-\pi}^{\pi} \int_0^1 g(r,\theta) \varphi_{mn}^{(1)}(r,\theta) r \, dr \, d\theta \\
&= \int_{-\pi}^{\pi} \int_0^1 \left(1 - r^2\right) J_n(\alpha_{mn} r) \cos(n\theta) r \, dr \, d\theta \\
&= \left(\int_{-\pi}^{\pi} \cos(n\theta) \, d\theta \right) \left(\int_0^1 \left(1 - r^2\right) J_n(\alpha_{mn} r) r \, dr \right) \\
&= 0 \cdot \left(\int_0^1 \left(1 - r^2\right) J_n(\alpha_{mn} r) r \, dr \right) \\
&= 0.
\end{aligned}
$$

Similarly,
$$(g, \varphi_{mn}^{(2)}) = 0$$

for every $n > 0$. It follows that g can be represented as follows:

$$g(r,\theta) = \sum_{m=1}^{\infty} c_{m0} J_0(\alpha_{m0} r).$$

This is not surprising; since g is independent of θ, it is reasonable that only the eigenfunctions independent of θ are needed to represent g.

Since we must compute numerical estimates of the eigenvalues λ_{mn}, we will content ourselves with using three eigenfunctions, those corresponding to the eigenvalues

$$\lambda_{10}, \lambda_{20}, \lambda_{30}.$$

To six digits, we have

$$s_{10} \doteq 2.40483, \; s_{20} \doteq 5.52008, \; s_{30} \doteq 8.65373$$

(see (11.50)). Since $A = 1$ in this example, we have $\alpha_{mn} = s_{mn}$ and $\lambda_{mn} = s_{mn}^2$. The needed coefficients are [60]

$$
\begin{aligned}
c_{10} &= \frac{1}{\pi (J_1(\alpha_{10}))^2} \int_{-\pi}^{\pi} \int_0^1 g(r,\theta) J_0(\alpha_{10} r) r \, dr \, d\theta \doteq 1.10802, \\
c_{20} &= \frac{1}{\pi (J_1(\alpha_{20}))^2} \int_{-\pi}^{\pi} \int_0^1 g(r,\theta) J_0(\alpha_{20} r) r \, dr \, d\theta \doteq -0.139778, \\
c_{30} &= \frac{1}{\pi (J_1(\alpha_{30}))^2} \int_{-\pi}^{\pi} \int_0^1 g(r,\theta) J_0(\alpha_{30} r) r \, dr \, d\theta \doteq 0.0454765.
\end{aligned}
$$

We now have the first three terms of the Fourier series for g. The graph of g is shown in Figure 11.11, while the approximation is shown in Figure 11.12. Exercise 11.2.1 asks the reader to improve the approximation by using more terms. ∎

---
[60] The integrals were computed in *Mathematica* using the **Integrate** command.

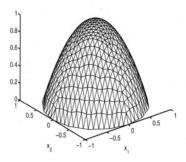

Figure 11.11. *The function $g(r,\theta) = 1 - r^2$.*

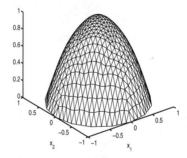

Figure 11.12. *The approximation to function $g(r,\theta) = 1 - r^2$, using three terms of the Fourier series (see Example 11.7).*

11.3.6 Solving PDEs on a disk

Having found the eigenvalues and eigenfunctions of $-\Delta$ on the disk and under Dirichlet conditions, we can now apply Fourier series methods to solve any of the familiar equations: the Poisson equation, the heat equation, and the wave equation.

Example 11.8. We will solve (approximately) the BVP

$$\begin{aligned} -\Delta u &= 1 - r^2 \text{ in } \Omega, \\ u &= 0 \text{ on } \partial\Omega, \end{aligned} \tag{11.55}$$

where Ω is the unit disk. We have already seen that

$$1 - r^2 = \sum_{m=1}^{\infty} c_{m0} J_0(\alpha_{m0} r),$$

11.3. Fourier series on a disk

where the coefficients c_{m0} can be computed as in the previous example. We now write

$$u(r,\theta) = \sum_{m=1}^{\infty} b_{m0} J_0(\alpha_{m0} r),$$

where the coefficients b_{m0} are to be determined. (Since the forcing function is radial (i.e., independent of θ), the solution will also be radial. Therefore, we do not need the eigenfunctions that depend on θ.) Since each $J_0(\alpha_{m0} r)$ is an eigenfunction of $-\Delta$ under the given boundary conditions, we have

$$-\Delta u(r,\theta) = \sum_{m=1}^{\infty} \lambda_{m0} b_{m0} J_0(\alpha_{m0} r),$$

where the eigenvalue λ_{m0} is given by $\lambda_{m0} = \alpha_{m0}^2$. Therefore, the PDE implies that

$$\lambda_{m0} b_{m0} = c_{m0}, \ m = 1, 2, 3, \ldots,$$

and so the solution is

$$u(r,\theta) = \sum_{m=1}^{\infty} \frac{c_{m0}}{\alpha_{m0}^2} J_0(\alpha_{m0} r).$$

We computed $c_{10}, c_{20}, c_{30}, \alpha_{10}, \alpha_{20}$, and α_{30} in the previous example, so we see that

$$u(r,\theta) \doteq 0.191594 J_0(\alpha_{10} r) - 0.00458719 J_0(\alpha_{20} r) + 0.000607268 J_0(\alpha_{30} r) + \cdots. \quad \blacksquare$$

Example 11.9. We will now solve the wave equation on a disk and compare the fundamental frequency of a circular drum to that of a square drum. We consider the IBVP

$$\begin{aligned} \frac{\partial^2 u}{\partial t^2} - c^2 \Delta u &= 0, \ (r,\theta) \in \Omega, \ t > 0, \\ u(r,\theta,0) &= \phi(r,\theta), \ (r,\theta) \in \Omega, \\ \frac{\partial u}{\partial t}(r,\theta,0) &= \gamma(r,\theta), \ (r,\theta) \in \Omega, \\ u(A,\theta,t) &= 0, \ -\pi \leq \theta < \pi, \ t > 0, \end{aligned} \quad (11.56)$$

where Ω is the disk of radius A, centered at the origin. We write the solution as

$$u(r,\theta,t) = \sum_{m=1}^{\infty} a_{m0}(t) J_0(\alpha_{m0} r) + \sum_{m=1}^{\infty} \sum_{n=1}^{\infty} (a_{mn}(t) \cos(n\theta) + b_{mn}(t) \sin(n\theta)) J_n(\alpha_{mn} r).$$

Then, by the usual argument,

$$\frac{\partial^2 u}{\partial t^2}(r,\theta,t) - c^2 \Delta u(r,\theta,t) = \sum_{m=1}^{\infty} \left(\frac{d^2 a_{m0}}{dt^2}(t) + c^2 \lambda_{m0} a_{m0}(t) \right) J_0(\alpha_{m0} r)$$

$$+ \sum_{m=1}^{\infty} \sum_{n=1}^{\infty} \left(\left(\frac{d^2 a_{mn}}{dt^2}(t) + c^2 \lambda_{mn} a_{mn}(t) \right) \cos(n\theta) \right.$$

$$+ \left. \left(\frac{d^2 b_{mn}}{dt^2}(t) + c^2 \lambda_{mn} b_{mn}(t) \right) \sin(n\theta) \right) J_n(\alpha_{mn} r).$$

The wave equation then implies the following ODEs:

$$\frac{d^2 a_{mn}}{dt^2} + c^2 \lambda_{mn} a_{mn} = 0, \; m = 1, 2, 3, \ldots, \; n = 0, 1, 2, \ldots,$$

$$\frac{d^2 b_{mn}}{dt^2} + c^2 \lambda_{mn} b_{mn} = 0, \; m, n = 1, 2, 3, \ldots.$$

From the initial conditions for the PDE, we obtain

$$a_{mn}(0) = c_{mn},$$

$$\frac{d a_{mn}}{dt}(0) = d_{mn}, \; m = 1, 2, 3, \ldots, \; n = 0, 1, 2, \ldots,$$

$$b_{mn}(0) = e_{mn},$$

$$\frac{d b_{mn}}{dt}(0) = f_{mn}, \; m, n = 1, 2, 3, \ldots,$$

where c_{mn}, d_{mn} are the (generalized) Fourier coefficients for ϕ, and e_{mn}, f_{mn} are the (generalized) Fourier coefficients for γ. By applying the results of Section 4.15 and using the fact that $\lambda_{mn} = \alpha_{mn}^2$, we obtain the following formulas for the coefficients of the solution:

$$a_{mn}(t) = c_{mn} \cos(c \alpha_{mn} t) + \frac{d_{mn}}{c \alpha_{mn}} \sin(c \alpha_{mn} t), \; m = 1, 2, 3, \ldots, \; n = 0, 1, 2, \ldots,$$

$$b_{mn}(t) = e_{mn} \cos(c \alpha_{mn} t) + \frac{f_{mn}}{c \alpha_{mn}} \sin(c \alpha_{mn} t), \; m, n = 1, 2, 3, \ldots.$$

The longest period of any of these coefficients is that of a_{10}:

$$T_{10} = \frac{2\pi}{c \alpha_{10}} = \frac{2\pi}{c s_{01}/A} = \frac{2 A \pi}{c s_{01}}.$$

The corresponding frequency, which is the fundamental frequency of this circular drum, is

$$F_{10} = \frac{c s_{01}}{2 A \pi} = \frac{c}{A} \frac{s_{01}}{2\pi} \doteq 0.382740 \frac{c}{A}.$$

It would be reasonable to compare a circular drum of radius A (diameter $2A$) with a square drum of side length $2A$. (Another possibility is to compare the circular drum with a square

11.3. Fourier series on a disk

drum of the same area. This is Exercise 11.3.9.) The fundamental frequency of such a square drum is

$$\frac{c\sqrt{\frac{\pi^2}{(2A)^2}+\frac{\pi^2}{(2A)^2}}}{2\pi} = \frac{c}{A}\frac{1}{2\sqrt{2}} \doteq 0.353553\frac{c}{A}.$$

The square drum sounds a lower frequency than the circular drum. ∎

Exercises

1. Let $g(r,\theta) = 1 - r^2$, as in Example 11.7. Compute the next three terms in the series for g.

2. Let $f(r,\theta) = 1 - r$. Compute the first nine terms in the generalized Fourier series for f (that is, those corresponding to the eigenvalues

$$\lambda_{10},\lambda_{20},\lambda_{30},\lambda_{11},\,\lambda_{21},\lambda_{31},\lambda_{12},\lambda_{22},\lambda_{32}).$$

   Take Ω to be the unit disk.

3. Solve the BVP
$$-\Delta u = 1 - \sqrt{x^2 + y^2} \text{ in } \Omega,$$
$$u = 0 \text{ on } \partial\Omega,$$

   where Ω is the unit disk. Graph the solution. (Hint: Change to polar coordinates and use the results of the previous exercise.)

4. Let $f(r,\theta) = r$. Compute the first nine terms in the generalized Fourier series for f; that is, those corresponding to the eigenvalues

$$\lambda_{10},\lambda_{20},\lambda_{30},\lambda_{11},\,\lambda_{21},\lambda_{31},\lambda_{12},\lambda_{22},\lambda_{32}.$$

   Take Ω to be the unit disk.

5. Solve the BVP
$$-\Delta u = \sqrt{x^2 + y^2} \text{ in } \Omega,$$
$$u = 0 \text{ on } \partial\Omega,$$

   where Ω is the unit disk. Graph the solution. (Hint: Change to polar coordinates and use the results of the previous exercise.)

6. Let $f(\mathbf{x}) = (1 - x_1^2 - x_2^2)(x_1 + x_2)$. Convert to polar coordinates and compute the first 16 terms in the (generalized) Fourier series (that is, those corresponding to λ_{mn}, $m = 1,2,3,4$, $n = 0,1,2,3$). Graph the approximation and its error.

7. Consider a disk, 10 cm in radius and made of copper (the material constants of copper are $\rho = 8.97\,\text{g/cm}^3$, $c = 0.379\,\text{J/(g K)}$, $\kappa = 4.04\,\text{W/(cm K)}$). Suppose that the temperature in the disk is initially $\phi(r,\theta) = r\cos(\theta)(10 - r)/5$. What is the temperature distribution after 30 seconds (assuming that the top and bottom of the disk are perfectly insulated, and the edge of the disk is held at 0 degrees Celsius)?

8. Consider a disk of radius 15 cm, made of iron (the material constants of iron are $\rho = 7.88\,\text{g/cm}^3$, $c = 0.437\,\text{J/(g K)}$, $\kappa = 0.836\,\text{W/(cm K)}$). Suppose that heat energy is added to the disk at the constant rate of

$$f(r,\theta) = \frac{1}{150} r(\sin(\theta) + \cos(\theta))\,\text{W/cm}^3,$$

where the disk occupies the set

$$\Omega = \{(r,\theta) : r < 15\}.$$

Assume that the top and bottom of the disk are perfectly insulated and that the edge of the disk is held at 0 degrees Celsius.

   (a) What is the steady-state temperature of the disk?

   (b) Assuming the disk is initially 0 degrees Celsius throughout, how long does it take for the disk to reach steady state (to within 1%)?

9. Which has a lower fundamental frequency, a square drum or a circular drum of equal area?

10. Show that (11.48) holds by deriving the power series for the right-hand side and showing that it simplifies to the power series for the left-hand side.

11. Compute s_{41}, s_{42}, s_{43}, s_{44} (cf. (11.51)).

11.4 Finite elements in two dimensions

We now turn our attention to finite element methods for multidimensional problems. For the sake of simplicity, we will restrict our attention to problems in two spatial dimensions and, as in one dimension, to piecewise linear finite elements.

The general outline of the finite element method does not change when we move to two-dimensional space. The finite element method is obtained via the following steps.

1. Derive the weak form of the given BVP.

2. Apply the Galerkin method to the weak form.

3. Choose a space of piecewise polynomials for the approximating subspace in the Galerkin method.

The first step is very similar to the derivation in one dimension; as one might expect, the main difference is that integration by parts is replaced by Green's first identity. The Galerkin method is unchanged when applied to a two-dimensional problem. The most significant difference in moving to two-dimensional problems is in defining the finite element spaces. We must create a mesh on the computational domain and define piecewise polynomials on the mesh. We will restrict ourselves to *triangulations* and piecewise linear functions.

11.4.1 The weak form of a BVP in multiple dimensions

We begin with the following Dirichlet problem:

$$-\nabla \cdot (k(\mathbf{x})\nabla u) = f(\mathbf{x}), \; \mathbf{x} \in \Omega, \\ u(\mathbf{x}) = 0, \; \mathbf{x} \in \partial\Omega. \tag{11.57}$$

Here Ω is a domain in $\mathbf{R}^2$ or $\mathbf{R}^3$; there is no difference in the following derivation.

We define the space V of test functions by

$$V = C_D^2(\overline{\Omega}) = \left\{ v \in C^2(\overline{\Omega}) : v(\mathbf{x}) = 0 \text{ for all } \mathbf{x} \in \partial\Omega \right\}.$$

We then multiply the original PDE by an arbitrary $v \in V$ and apply Green's identity:

$$-\nabla \cdot (k\nabla u) = f \text{ in } \Omega$$
$$\Rightarrow -\nabla \cdot (k\nabla u)v = fv \text{ in } \Omega, \; v \in V$$
$$\Rightarrow -\int_\Omega \nabla \cdot (k\nabla u)v = \int_\Omega fv \text{ for all } v \in V$$
$$\Rightarrow -\int_{\partial\Omega} kv \frac{\partial u}{\partial \mathbf{n}} + \int_\Omega k\nabla u \cdot \nabla v = \int_\Omega fv \text{ for all } v \in V$$
$$\Rightarrow \int_\Omega k\nabla u \cdot \nabla v = \int_\Omega fv \text{ for all } v \in V.$$

The last step follows from the fact that v vanishes on $\partial\Omega$.

We define the bilinear form $a(\cdot,\cdot)$ by

$$a(u,v) = \int_\Omega k\nabla u \cdot \nabla v.$$

We then obtain the weak form of the BVP (11.57):

$$\text{Find } u \in V \text{ such that } a(u,v) = (f,v) \text{ for all } v \in V. \tag{11.58}$$

The proof that a solution of the weak form also satisfies the original BVP follows the same pattern as in Section 5.4.2 (see Exercise 11.4.8).

11.4.2 Galerkin's method

To apply Galerkin's method, we choose a finite-dimensional subspace V_n of V and a basis $\{\phi_1, \phi_2, \ldots, \phi_n\}$ of V_n. We then pose the Galerkin problem:

$$\text{Find } u \in V_n \text{ such that } a(u,v) = (f,v) \text{ for all } v \in V_n. \tag{11.59}$$

We write the solution as

$$v_n = \sum_{i=1}^n u_i \phi_i \tag{11.60}$$

and note that (11.59) is equivalent to

$$\text{Find } u \in V_n \text{ such that } a(u,\phi_i) = (f,\phi_i), \; i = 1,2,\ldots,n. \tag{11.61}$$

Substituting (11.60) into (11.61), we obtain the following equations:

$$a\left(\sum_{j=1}^{n} u_j \phi_j, \phi_i\right) = (f, \phi_i), \ i = 1, 2, \ldots, n$$

$$\Rightarrow \sum_{j=1}^{n} a(\phi_j, \phi_i) u_j = (f, \phi_i), \ i = 1, 2, \ldots, n$$

$$\Rightarrow \mathbf{Ku} = \mathbf{f}.$$

The stiffness matrix $\mathbf{K}$ and the load vector $\mathbf{f}$ are defined by

$$K_{ij} = a(\phi_j, \phi_i), \ i, j = 1, 2, \ldots, n,$$
$$f_i = (f, \phi_i), \ i = 1, 2, \ldots, n.$$

The reader will notice that the derivation of the equation $\mathbf{Ku} = \mathbf{f}$ is exactly as in Section 5.5, since Galerkin's method is described in a completely abstract fashion.[61] The specific details, and the differences from the one-dimensional case, arise only when the approximating subspace V_n is chosen.

Moreover, just as in the one-dimensional case, the bilinear form $a(\cdot, \cdot)$ defines an inner product, called the energy inner product, and the Galerkin method produces the best approximation, in the energy norm, to the true solution u from the approximating subspace V_n.

11.4.3 Piecewise linear finite elements in two dimensions

The graph of a first-degree polynomial in two variables is a plane, and three (noncollinear) points determine a plane. Or, looking at it algebraically, the equation of a first-degree polynomial is determined by three parameters:

$$z = a + bx + cy.$$

The parameters a, b, and c can be determined by three points satisfying the equation. We can discretize a two-dimensional domain Ω by defining a *triangulation* on Ω—that is, by dividing Ω into triangular subdomains. (If Ω is not a polygonal domain, then Ω itself must be approximated by a polygonal domain at the cost of some approximation error.[62]) The three vertices of a triangle, together with function values specified there, yield three points that define a linear function on the triangle. See Figure 11.13 for examples of triangular meshes defined on various regions and Figure 11.14 for the graph of a piecewise linear function. The reader should notice how the graph of a piecewise linear function is made up of triangular "patches."

For a given triangulation $\mathcal{T}$ of Ω, we write n for the number of "free" nodes, that is, nodes that do not lie on the boundary and hence do not correspond to a Dirichlet boundary

---

[61] Just as in Section 5.5, each of the symbols "u" and "f" has two meanings: u is the true solution of the BVP, while $\mathbf{u} \in \mathbf{R}^n$ is the vector whose components are the unknown weights in the expression (11.60) for the approximate solution v_n. Similarly, f is the forcing function in the BVP, while $\mathbf{f} \in \mathbf{R}^n$ is the load vector in the matrix-vector equation $\mathbf{Ku} = \mathbf{f}$.

[62] Another way to handle this is to allow "triangles" with a curved edge.

11.4. Finite elements in two dimensions

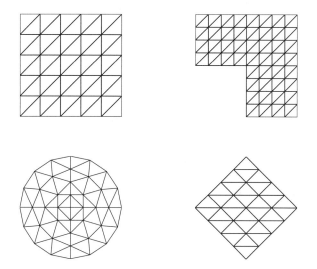

Figure 11.13. *Triangular meshes defined on (1) a square (upper left), (2) a union of three squares (upper right), (3) a disk (lower left), and (4) a rhombus (lower right). Since the disk is not polygonal, the triangulated domain is only an approximation to the original domain.*

Figure 11.14. *A piecewise linear function defined on the triangulation of the disk from Figure 11.13.*

condition. We will denote a typical triangular element of $\mathcal{T}$ by T, and a typical node by $\mathbf{z}$. Let V_n be the following approximating subspace of V:

$$V_n = \left\{ v \in C(\overline{\Omega}) \,:\, v \text{ is piecewise linear on } \mathcal{T},\ v(\mathbf{z}) = 0 \text{ for all nodes } \mathbf{z} \in \partial\Omega \right\}.$$

To apply the Galerkin method, we must choose a basis for V_n. This is done exactly as in one dimension. We number the free nodes of $\mathcal{T}$ as $\mathbf{z}_1, \mathbf{z}_2, \ldots, \mathbf{z}_n$. Then, since a piecewise linear function is determined by its values at the nodes of the mesh, we define $\phi_i \in V_n$ by the condition

$$\phi_i(\mathbf{z}_j) = \begin{cases} 1, & i = j, \\ 0, & i \neq j. \end{cases} \tag{11.62}$$

A typical ϕ_i is shown in Figure 11.15. Just as in the one-dimensional case, it is straightforward to use property (11.62) to show that $\{\phi_1, \phi_2, \ldots, \phi_n\}$ is a basis for V_n and that, for

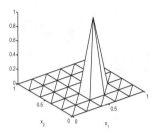

Figure 11.15. *One of the standard piecewise linear basis functions.*

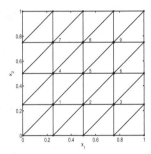

Figure 11.16. *The mesh for Example* 11.10.

any $v \in V_n$,

$$v(\mathbf{x}) = \sum_{i=1}^{M} v(\mathbf{z}_i)\phi_i(\mathbf{x}).$$

It is now straightforward (albeit quite tedious) to assemble the stiffness matrix **K** and the load vector **f**—it is just a matter of computing the quantities $a(\phi_j, \phi_i)$ and (f, ϕ_i). We will illustrate this calculation on an example.

Example 11.10. For simplicity, we take as our example the constant-coefficient problem

$$\begin{aligned}-\Delta u &= x_1, \ \mathbf{x} \in \Omega, \\ u(\mathbf{x}) &= 0, \ \mathbf{x} \in \partial\Omega,\end{aligned} \quad (11.63)$$

where Ω is the unit square:

$$\Omega = \left\{ (x_1, x_2) \in \mathbf{R}^2 \ : \ 0 < x_1 < 1, \ 0 < x_2 < 1 \right\}.$$

We choose a regular triangulation with 32 triangles, 25 nodes, and 9 free nodes. The mesh is shown in Figure 11.16, with the free nodes labeled from 1 to 9.

11.4. Finite elements in two dimensions

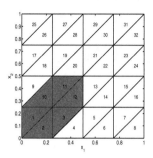

Figure 11.17. *The support of ϕ_1 (Example 11.10). The triangles of the mesh, $T_1, T_2, \ldots, T_{32}$ are also labeled.*

We begin with the computation of

$$K_{11} = a(\phi_1, \phi_1) = \int_\Omega \nabla \phi_1 \cdot \nabla \phi_1.$$

The support[63] of ϕ_1 is shown in Figure 11.17 (which also labels the triangles in the mesh, $T_1, T_2, \ldots, T_{32}$). This support is made up of six triangles; on each of these triangles, ϕ_1 has a different formula. We therefore compute K_{11} by adding up the contributions from each of the six triangles:

$$\int_\Omega \nabla \phi_1 \cdot \nabla \phi_1 = \int_{T_1} \nabla \phi_1 \cdot \nabla \phi_1 + \int_{T_2} \nabla \phi_1 \cdot \nabla \phi_1 + \int_{T_3} \nabla \phi_1 \cdot \nabla \phi_1$$
$$+ \int_{T_{10}} \nabla \phi_1 \cdot \nabla \phi_1 + \int_{T_{11}} \nabla \phi_1 \cdot \nabla \phi_1 + \int_{T_{12}} \nabla \phi_1 \cdot \nabla \phi_1.$$

On T_1, $\nabla \phi_1(\mathbf{x}) = (1/h, 0)$, where $h = 1/4$ (observing how ϕ_1 changes along the horizontal and vertical edges of T_1 leads to this conclusion). The area of T_1 (and of all the other triangles in the mesh) is $h^2/2$. Thus,

$$\int_{T_1} \nabla \phi_1 \cdot \nabla \phi_1 = \int_{T_1} \frac{1}{h^2} = \left(\frac{1}{h^2}\right)\left(\frac{h^2}{2}\right) = \frac{1}{2}.$$

On T_2, $\nabla \phi_1(\mathbf{x}) = (0, 1/h)$, and we obtain

$$\int_{T_2} \nabla \phi_1 \cdot \nabla \phi_1 = \int_{T_2} \frac{1}{h^2} = \left(\frac{1}{h^2}\right)\left(\frac{h^2}{2}\right) = \frac{1}{2}.$$

On T_3, $\nabla \phi_1(\mathbf{x}) = (-1/h, 1/h)$, and

$$\int_{T_3} \nabla \phi_1 \cdot \nabla \phi_1 = \int_{T_2} \frac{2}{h^2} = \left(\frac{2}{h^2}\right)\left(\frac{h^2}{2}\right) = 1.$$

---

[63]As we explained in Section 5.6, the support of a function is the set on which the function is nonzero, together with the boundary of that set.

It should be easy to believe, from symmetry, that

$$\int_{T_{12}} \nabla \phi_1 \cdot \nabla \phi_1 = \int_{T_1} \nabla \phi_1 \cdot \nabla \phi_1,$$

$$\int_{T_{11}} \nabla \phi_1 \cdot \nabla \phi_1 = \int_{T_2} \nabla \phi_1 \cdot \nabla \phi_1,$$

$$\int_{T_{10}} \nabla \phi_1 \cdot \nabla \phi_1 = \int_{T_3} \nabla \phi_1 \cdot \nabla \phi_1.$$

Adding up the contributions from the six triangles, we obtain

$$K_{11} = \int_{\Omega} \nabla \phi_1 \cdot \nabla \phi_1 = 4.$$

Since the PDE has constant coefficients (and the basis functions are all the same, up to translation), it is easy to see that

$$K_{ii} = 4, \; i = 1, 2, \ldots, 9.$$

We now turn our attention to the off-diagonal entries. For what values of $j \neq i$ will K_{ij} be nonzero? By examining the support of ϕ_1, we see that only ϕ_2, ϕ_4, and ϕ_5 have a support that overlaps that of ϕ_1 (in a set of positive area, that is, not just along the edge of a triangle). The reader should study Figure 11.16 until he or she is convinced of the following fundamental conclusion: *The support of ϕ_j and the support of ϕ_i have a nontrivial intersection if and only if nodes i and j belong to a common triangle.* By examining Figure 11.16, we see that the following entries of **K** might be nonzero:

$$K_{11}, K_{12}, K_{14}, K_{15},$$
$$K_{21}, K_{22}, K_{23}, K_{25}, K_{26},$$
$$K_{32}, K_{33}, K_{36},$$
$$K_{41}, K_{44}, K_{45}, K_{47}, K_{48},$$
$$K_{51}, K_{52}, K_{54}, K_{55}, K_{56}, K_{58}, K_{59},$$
$$K_{62}, K_{63}, K_{65}, K_{66}, K_{69},$$
$$K_{74}, K_{77}, K_{78},$$
$$K_{84}, K_{85}, K_{87}, K_{88}, K_{89},$$
$$K_{95}, K_{96}, K_{98}, K_{99}.$$

We now compute the first row of **K**. We already know that $K_{11} = 4$. Consulting Figure 11.17, we see that

$$K_{12} = \int_{\Omega} \nabla \phi_2 \cdot \nabla \phi_1 = \int_{T_3} \nabla \phi_2 \cdot \nabla \phi_1 + \int_{T_{12}} \nabla \phi_2 \cdot \nabla \phi_1.$$

On T_3,

$$\nabla \phi_1(\mathbf{x}) = (-1/h, 1/h), \; \nabla \phi_2(\mathbf{x}) = (1/h, 0),$$

so

$$\int_{T_3} \nabla \phi_2 \cdot \nabla \phi_1 = \left(-\frac{1}{h^2}\right)\left(\frac{h^2}{2}\right) = -\frac{1}{2}.$$

11.4. Finite elements in two dimensions

On T_{12},
$$\nabla\phi_1(\mathbf{x}) = (-1/h, 0), \ \nabla\phi_2(\mathbf{x}) = (1/h, -1/h),$$
so
$$\int_{T_{12}} \nabla\phi_2 \cdot \nabla\phi_1 = \left(-\frac{1}{h^2}\right)\left(\frac{h^2}{2}\right) = -\frac{1}{2}.$$

Thus $K_{12} = -1$.

The reader can verify the following calculations:
$$K_{14} = \int_{T_{10}} \nabla\phi_4 \cdot \nabla\phi_1 + \int_{T_{11}} \nabla\phi_4 \cdot \nabla\phi_1$$
$$= \left(-\frac{1}{h^2}\right)\left(\frac{h^2}{2}\right) + \left(-\frac{1}{h^2}\right)\left(\frac{h^2}{2}\right)$$
$$= -1,$$
$$K_{15} = \int_{T_{11}} \nabla\phi_5 \cdot \nabla\phi_1 + \int_{T_{12}} \nabla\phi_5 \cdot \nabla\phi_1$$
$$= (0)\left(\frac{h^2}{2}\right) + (0)\left(\frac{h^2}{2}\right)$$
$$= 0.$$

The rest of the calculations are similar, and the result is the following stiffness matrix:

$$\mathbf{K} = \begin{bmatrix} 4 & -1 & 0 & -1 & 0 & 0 & 0 & 0 & 0 \\ -1 & 4 & -1 & 0 & -1 & 0 & 0 & 0 & 0 \\ 0 & -1 & 4 & 0 & 0 & -1 & 0 & 0 & 0 \\ -1 & 0 & 0 & 4 & -1 & 0 & -1 & 0 & 0 \\ 0 & -1 & 0 & -1 & 4 & -1 & 0 & -1 & 0 \\ 0 & 0 & -1 & 0 & -1 & 4 & 0 & 0 & -1 \\ 0 & 0 & 0 & -1 & 0 & 0 & 4 & -1 & 0 \\ 0 & 0 & 0 & 0 & -1 & 0 & -1 & 4 & -1 \\ 0 & 0 & 0 & 0 & 0 & -1 & 0 & -1 & 4 \end{bmatrix}.$$

To compute the load vector $\mathbf{f}$, we must evaluate the integrals
$$\int_\Omega f(\mathbf{x})\phi_i(\mathbf{x})\,d\mathbf{x},$$
which is quite tedious when done by hand (unlike $\nabla\phi_i$, ϕ_i itself is not piecewise constant). We just show one representative calculation. We have

$$\phi_1(x_1, x_2) = \begin{cases} \frac{x_1}{h}, & \mathbf{x} \in T_1, \\ \frac{x_2}{h}, & \mathbf{x} \in T_2, \\ \frac{x_2 - x_1 + h}{h}, & \mathbf{x} \in T_3, \\ \frac{x_1 - x_2 + h}{h}, & \mathbf{x} \in T_{10}, \\ \frac{2h - x_2}{h}, & \mathbf{x} \in T_{11}, \\ \frac{2h - x_1}{h}, & \mathbf{x} \in T_{12}, \\ 0 & \text{otherwise.} \end{cases}$$

Therefore,

$$f_1 = \int_\Omega f(\mathbf{x})\phi_i(\mathbf{x})\,d\mathbf{x}$$

$$= \int_{T_1} \frac{x_1^2}{h}\,d\mathbf{x} + \int_{T_2} \frac{x_1 x_2}{h}\,d\mathbf{x} + \int_{T_3} \frac{x_1(x_2 - x_1 + h)}{h}\,d\mathbf{x}$$

$$+ \int_{T_{10}} \frac{x_1(x_1 - x_2 + h)}{h}\,d\mathbf{x} + \int_{T_{11}} \frac{x_1(2h - x_2)}{h}\,d\mathbf{x} + \int_{T_{12}} \frac{x_1(2h - x_1)}{h}\,d\mathbf{x}$$

$$= \frac{h^3}{12} + \frac{h^3}{8} + \frac{5h^3}{24} + \frac{h^3}{8} + \frac{5h^3}{24} + \frac{h^3}{4}$$

$$= h^3.$$

Continuing in this manner, we find

$$\mathbf{f} = \begin{bmatrix} h^3 \\ 2h^3 \\ 3h^3 \\ h^3 \\ 2h^3 \\ 3h^3 \\ h^3 \\ 2h^3 \\ 3h^3 \end{bmatrix}.$$

We can now solve[64] $\mathbf{Ku} = \mathbf{f}$ to obtain

$$\mathbf{u} \doteq \begin{bmatrix} 0.015904 \\ 0.027344 \\ 0.027065 \\ 0.020647 \\ 0.035156 \\ 0.034040 \\ 0.015904 \\ 0.027344 \\ 0.027065 \end{bmatrix}.$$

The resulting piecewise linear approximation is displayed in Figure 11.18. ∎

The calculations in the previous example are elementary but time consuming, and are ideal for implementation in a computer program. When these calculations are programmed, the operations are not organized in the same way as in the above hand calculation. For example, instead of computing one entry in $\mathbf{K}$ at a time, it is common to loop over the

---

[64] It is expected that a computer program will be used to solve any linear system larger than 2×2 or 3×3. We used MATLAB to solve $\mathbf{Ku} = \mathbf{f}$ for this example.

11.4. Finite elements in two dimensions

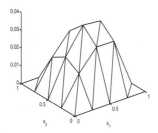

Figure 11.18. *The piecewise linear approximation to the solution of* (11.63).

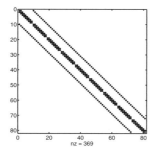

Figure 11.19. *The sparsity pattern of the discrete Laplacian (200 triangular elements).*

elements of the mesh and to compute all the contributions to the various entries in **K** and **f** from each element in turn. Among other things, this simplifies the data structure for describing the mesh. (To automate the calculations presented above, it would be necessary to know, given a certain node, which nodes are adjacent to it. This is avoided when one loops over the elements instead of over the nodes.) Also, the various integrations are rarely carried out exactly, but rather by using quadrature rules. These implementation details are discussed in Section 13.1.

As we have discussed before, an important aspect of the finite element method is the fact that the linear systems that must be solved are sparse. In Figure 11.19, we display the sparsity pattern of the stiffness matrix for Poisson's equation, as in Example 11.10, but with a finer grid. The matrix **K** is *banded*, that is, all of its entries are zero except those found in a band around the main diagonal. The efficient solution of banded and other sparse systems is discussed in Section 13.2.

11.4.4 Finite elements and Neumann conditions

We will close by describing briefly how Neumann conditions are handled in two-dimensional problems. For the sake of generality, we will consider a problem with possibly mixed boundary conditions. So suppose Ω is a domain in either $\mathbf{R}^2$ or $\mathbf{R}^3$, and assume that $\partial \Omega$

has been partitioned into two disjoint sets: $\partial\Omega = \Gamma_1 \cup \Gamma_2$. We consider the following BVP:

$$\begin{aligned}
-\nabla \cdot (k(\mathbf{x})\nabla u) &= f(\mathbf{x}), \ \mathbf{x} \in \Omega, \\
u(\mathbf{x}) &= 0, \ \mathbf{x} \in \Gamma_1, \\
\frac{\partial u}{\partial \mathbf{n}}(\mathbf{x}) &= 0, \ \mathbf{x} \in \Gamma_2.
\end{aligned} \qquad (11.64)$$

As we discussed in Section 6.5, Dirichlet conditions are termed *essential boundary conditions* because they must be explicitly imposed in the finite element method, while Neumann conditions are called *natural* and need not be mentioned. We therefore define the space of test functions by

$$\tilde{V} = \left\{ u \in C^2(\overline{\Omega}) : u(\mathbf{x}) = 0 \text{ for all } \mathbf{x} \in \Gamma_1 \right\}.$$

The weak form of (11.64) is now derived exactly as on page 487; the boundary integral

$$\int_{\partial\Omega} kv \frac{\partial u}{\partial \mathbf{n}} = \int_{\Gamma_1} kv \frac{\partial u}{\partial \mathbf{n}} + \int_{\Gamma_2} kv \frac{\partial u}{\partial \mathbf{n}}$$

now vanishes because $v = 0$ on Γ_1 and $\partial u/\partial \mathbf{n} = 0$ on Γ_2. Thus the weak form of (11.64) is

$$\text{Find } u \in \tilde{V} \text{ such that } a(u,v) = (f,v) \text{ for all } v \in \tilde{V}. \qquad (11.65)$$

The bilinear form $a(\cdot,\cdot)$ is defined exactly as before (the only difference from (11.58) is in the space of test functions).

We now restrict our discussion once more to two-dimensional polygonal domains. To apply the finite element method, we must choose an approximating subspace of $\tilde{V}$. Since the boundary conditions are mixed, there are at least two points where the boundary conditions change from Dirichlet to Neumann. We will make the assumption that the mesh is chosen so that all such points are nodes (and that all such nodes belong to Γ_1, that is, that Γ_1 includes its "endpoints"). We can then choose the approximating subspace of $\tilde{V}$ as follows:

$$\tilde{V}_n = \left\{ v \in C(\overline{\Omega}) : v \text{ is piecewise linear on } \mathcal{T}, v(\mathbf{z}) = 0 \text{ for all nodes } \mathbf{z} \in \Gamma_1 \right\}.$$

A basis for $\tilde{V}_n$ is formed by including all basis functions corresponding to interior nodes (as in the Dirichlet case), as well as the basis functions corresponding to boundary nodes that do not belong to Γ_1. For an example of a basis function corresponding to a boundary node, see Figure 11.20.

Example 11.11. We consider the BVP

$$\begin{aligned}
-\Delta u &= x_1, \ \mathbf{x} \in \Omega, \\
u(\mathbf{x}) &= 0, \ \mathbf{x} \in \Gamma_3, \\
\frac{\partial u}{\partial \mathbf{n}}(\mathbf{x}) &= 0, \ \mathbf{x} \in \Gamma_1 \cup \Gamma_2 \cup \Gamma_4,
\end{aligned} \qquad (11.66)$$

where Ω is the unit square, as in the previous example, and Γ_1, Γ_2, Γ_3, and Γ_4 are defined as in Figure 11.3. We use the same regular triangulation of Ω as in Example 11.10. Now,

11.4. Finite elements in two dimensions

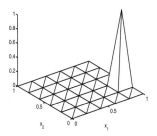

Figure 11.20. *A standard piecewise linear basis function corresponding to a boundary node.*

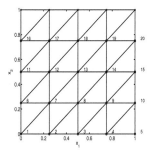

Figure 11.21. *The mesh for Example* 11.11.

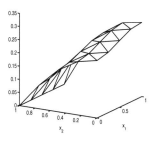

Figure 11.22. *The piecewise linear approximation to the solution of* (11.66).

however, the boundary nodes belonging to Γ_1, Γ_2, and Γ_4 are free nodes (but not the corners $(1, 1)$ and $(0, 1)$ of Ω). It follows that $\tilde{V}_n$ has dimension 20. The free nodes are labeled in Figure 11.21.

The calculation of the new stiffness matrix **K** and load vector **f** proceeds as before, and the resulting piecewise linear approximation to the true solution is displayed in Figure 11.22. ∎

If the BVP includes only Neumann conditions, then the stiffness matrix will be singular, reflecting the fact that BVP either does not have a solution or has infinitely many solutions. Special care must be taken to compute a meaningful solution to $\mathbf{Ku} = \mathbf{f}$.

11.4.5 Inhomogeneous boundary conditions

In a two-dimensional problem, inhomogeneous boundary conditions are handled just as in one dimension. Inhomogeneous Dirichlet conditions are addressed via the method of shifting the data (with a specially chosen piecewise linear function), while inhomogeneous Neumann conditions are taken into account directly when deriving the weak form. Both types of boundary conditions lead to a change in the load vector. Exercises 11.4.7 and 11.4.9 ask the reader to derive the load vectors in these cases.

Exercises

1. Consider the BVP

$$-\nabla \cdot (k(\mathbf{x})\nabla u) = f(\mathbf{x}), \ \mathbf{x} \in \Omega,$$
$$u(\mathbf{x}) = 0, \ \mathbf{x} \in \partial\Omega,$$

   where Ω is the unit square, $k(\mathbf{x}) = 1 + x_1 x_2$, and $f(\mathbf{x}) = 1$. Produce the piecewise linear finite element (approximate) solution using a regular grid with 18 triangles.

2. Repeat Exercise 1, but suppose that the boundary conditions are homogeneous Neumann conditions on the right side Γ_2 of the square and homogeneous Dirichlet conditions on the other three sides.

3. Repeat Exercise 1, but suppose that the boundary conditions are homogeneous Neumann conditions on every part of the boundary of Ω. Let the right-hand side be $f(\mathbf{x}) = x_1 - 1/2$ (so that the compatibility condition is satisfied and a solution exists).

4. If u does not belong to V_n, then

$$w = \sum_{i=1}^{n} u(\mathbf{z}_i)\phi_i$$

   is not equal to u. Instead, w is the piecewise linear interpolant of u (relative to the given mesh $\mathcal{T}$). The interpolant of u is one piecewise linear approximation to u; if u is the solution to a BVP, then the finite element method computes another. The purpose of this exercise is to compare the two.

   (a) Verify that $u(\mathbf{x}) = x_1 x_2 (1 - x_1)(1 - x_2)$ is the solution to the Dirichlet problem

$$-\Delta u = 2x_1(1-x_1) + 2x_2(1-x_2), \ \mathbf{x} \in \Omega,$$
$$u(\mathbf{x}) = 0, \ \mathbf{x} \in \partial\Omega,$$

   where Ω is the unit square.

11.4. Finite elements in two dimensions

(b) Let $\mathcal{T}$ be the regular triangulation of Ω with 18 triangles. Compute the finite element approximation to u using piecewise linear functions on $\mathcal{T}$.

(c) Compute the piecewise linear interpolant of u relative to $\mathcal{T}$.

(d) Which is a better approximation of u, the finite element approximation or the interpolant? Justify your answer both theoretically and numerically.

5. Given a triangulation of Ω, there is another way to construct a continuous piecewise linear approximation of a given function u, namely, by computing the best approximation to u from V_n in the $L^2(\Omega)$ norm (here V_n is the space of continuous piecewise linear functions relative to the given mesh). Explain how to calculate this approximation.

6. Let Ω be a (polygonal) domain in $\mathbf{R}^2$, and let $\mathcal{T}$ be a triangulation of Ω. Explain how to use continuous piecewise linear finite elements to approximate the solution of

$$\rho c \frac{\partial u}{\partial t} - \kappa \Delta u = f(\mathbf{x},t), \ \mathbf{x} \in \Omega, \ t > 0,$$
$$u(\mathbf{x},0) = \phi(\mathbf{x}), \ \mathbf{x} \in \Omega,$$
$$u(\mathbf{x},t) = 0, \ \mathbf{x} \in \partial\Omega, \ t > 0.$$

7. (a) Explain how to solve an inhomogeneous Dirichlet problem using piecewise linear finite elements. (Hint: Shift the data using a piecewise linear interpolant of the Dirichlet boundary data.)

(b) Illustrate the procedure on the following BVP:

$$-\Delta u = 0 \text{ in } \Omega,$$
$$u(\mathbf{x}) = x_1^2, \ \mathbf{x} \in \partial\Omega.$$

Let Ω be the unit square and use a regular grid with 18 triangles.

8. Suppose u is a solution to (11.58). Prove that u also solves (11.57).

9. (a) Explain how to solve an inhomogeneous Neumann problem using piecewise linear finite elements. (Hint: Do not shift the data; rather, incorporate the boundary data into the weak form.)

(b) What is the compatibility condition for an inhomogeneous Neumann problem?

(c) Illustrate the procedure on the following BVP:

$$-\Delta u = x_1 x_2 + \frac{3}{4} \text{ in } \Omega,$$
$$\frac{\partial u}{\partial \mathbf{n}}(\mathbf{x}) = -\frac{3x_1^2}{5}, \ \mathbf{x} \in \partial\Omega.$$

Let Ω be the unit square and use a regular grid with 18 triangles.

10. Let Ω be a (polygonal) domain in $\mathbf{R}^2$ and consider the following BVP:

$$-\kappa \Delta u = f(\mathbf{x}), \ \mathbf{x} \in \Omega,$$
$$\alpha u(\mathbf{x}) + \kappa \frac{\partial u}{\partial \mathbf{n}} = 0, \ \mathbf{x} \in \partial\Omega.$$

(a) Derive the weak form of this BVP.

(b) Let $\mathcal{T}$ be a triangulation of Ω. Derive the Galerkin finite element method for this problem. What changes from the case of Neumann conditions?

11.5 The free-space Green's function for the Laplacian

We will now consider Green's functions in multiple spatial dimensions, beginning with Green's functions for the Laplacian, both in space and on a bounded domain.

We will first find a function $G = G(\mathbf{x}, \mathbf{y})$ with the property that

$$u(\mathbf{x}) = \int_{\mathbf{R}^2} G(\mathbf{x}, \mathbf{y}) f(\mathbf{y}) d\mathbf{y}$$

satisfies $-\Delta u = f(\mathbf{x})$ in all of $\mathbf{R}^2$. Such a function G is called a *free-space Green's function*[65] (or a *fundamental solution* of Poisson's equation). There is not a unique free-space Green's function, since the PDE $-\Delta u = f(\mathbf{x})$ (unconstrained by any boundary conditions) does not have a unique solution. We need just one such G, and will directly construct one that has a simple formula.

We will then use the free-space Green's function to construct a Green's function for $-\Delta$, subject to boundary conditions, on a bounded domain Ω. We will call this bounded-domain Green's function G_Ω, and it will allow us to solve the BVP

$$\begin{aligned}
-\Delta u &= f(\mathbf{x}) \text{ in } \Omega, \\
\alpha_1 u + \alpha_2 \frac{\partial u}{\partial \mathbf{n}} &= 0 \text{ on } \partial\Omega.
\end{aligned} \quad (11.67)$$

We assume that $\alpha_1 \neq 0$, so the cases of Dirichlet and Robin conditions are included, but not the case of Neumann conditions. Since the BVP with $f = 0$ and Neumann boundary conditions has a nontrivial solution (namely, a nonzero constant function), we know that uniqueness fails and therefore no Green's function exists in that case.

We find the Green's function for (11.67) by solving the BVP

$$\begin{aligned}
-\Delta u &= 0 \text{ in } \Omega, \\
\alpha_1 u + \alpha_2 \frac{\partial u}{\partial \mathbf{n}} &= w \text{ on } \partial\Omega
\end{aligned}$$

for a certain choice of boundary function w and using the solution to modify the free-space Green's function. We explain how this works in the Section 11.6, and carry out the computations for a specific domain Ω.

11.5.1 The free-space Green's function in two dimensions

We begin by considering the problem in $\mathbf{R}^2$. To find a free-space Green's function, we recall that it should satisfy

$$-\Delta_\mathbf{x} G(\mathbf{x}; \mathbf{y}) = \delta(\mathbf{x} - \mathbf{y}),$$

---

[65] Another term is *infinite-space Green's function*.

11.5. The free-space Green's function for the Laplacian

where δ is the two-dimensional delta function, defined by the sifting property:

$$\int_{\mathbf{R}^2} \phi(\mathbf{x})\delta(\mathbf{x}-\mathbf{y})d\mathbf{x} = \phi(\mathbf{y}) \text{ for all smooth } \phi.$$

We will find a formula for G through some formal reasoning and then prove that the formula is correct.

The Laplace operator is invariant under a rotation of the coordinate system. To be precise, suppose we define coordinates (y_1, y_2) by rotating the $x_1 x_2$-axes by α radians. Some analytic geometry (see Exercise 11.5.1) shows that

$$\begin{aligned} y_1 &= \cos(\alpha)x_1 + \sin(\alpha)x_2, \\ y_2 &= -\sin(\alpha)x_1 + \cos(\alpha)x_2. \end{aligned} \quad (11.68)$$

We can then express

$$\Delta u = \frac{\partial^2 u}{\partial x_1^2} + \frac{\partial^2 u}{\partial x_2^2}$$

in terms of partial derivatives with respect to y_1, y_2, and it turns out that

$$\frac{\partial^2 u}{\partial x_1^2} + \frac{\partial^2 u}{\partial x_2^2} = \frac{\partial^2 u}{\partial y_1^2} + \frac{\partial^2 u}{\partial y_2^2} \quad (11.69)$$

(see Exercise 11.5.2). Thus the partial differential operator Δ is the same whether expressed in terms of x_1, x_2 or y_1, y_2. The Laplacian is also invariant under a translation of the coordinate system (as is any linear, constant-coefficient partial differential operator).

To see the practical implication of these two facts, consider the equation $-\Delta u = f(\mathbf{x})$, where f has the form $f(\mathbf{x}) = \phi(\|\mathbf{x}-\mathbf{y}\|)$ for some fixed point $\mathbf{y}$ in the plane. Here ϕ is a real-valued function of a single real variable, and f has the property that its value at $\mathbf{x}$ depends only on the distance of $\mathbf{x}$ from $\mathbf{y}$. Such an f is called a *radial function* (with respect to $\mathbf{y}$). Since the Laplacian is invariant under rotations and translations, it is natural to expect that $-\Delta u = f(\mathbf{x})$ has solutions that are radial with respect to $\mathbf{y}$. Indeed, if we define (r, θ) to be the polar coordinates of the point $(x_1 - y_1, x_2 - y_2)$, then the Laplacian in (r, θ) is

$$\Delta u = \frac{\partial^2 u}{\partial r^2} + \frac{1}{r}\frac{\partial u}{\partial r} + \frac{1}{r^2}\frac{\partial^2 u}{\partial \theta^2} \quad (11.70)$$

(see Section 11.3.1). If we assume u is radial with respect to $\mathbf{y}$, then u has the form $u = \psi(r)$, and

$$\Delta u = \frac{d^2 \psi}{dr^2} + \frac{1}{r}\frac{d\psi}{dr}.$$

For such a u, the equation $-\Delta u = f(\mathbf{x})$ reduces to

$$-\frac{d^2 \psi}{dr^2} - \frac{1}{r}\frac{d\psi}{dr} = \phi(r). \quad (11.71)$$

It can be shown directly that (11.71) has solutions ψ (see Exercise 11.5.3), and it follows that, for any such ψ, $u(\mathbf{x}) = \psi(\|\mathbf{x}-\mathbf{y}\|)$ is a solution of $-\Delta u = f(\mathbf{x})$ that is radial with respect to $\mathbf{y}$.

Since the delta function $\delta(\mathbf{x}-\mathbf{y})$ is (formally) radial with respect to $\mathbf{y}$, we can therefore expect that there is a free-space Green's function of the form

$$G(\mathbf{x};\mathbf{y}) = g(\|\mathbf{x}-\mathbf{y}\|) = g(r).$$

Recalling that $\delta(\mathbf{x}-\mathbf{y})$ is formally zero for $\mathbf{x} \neq \mathbf{y}$, it follows that g should satisfy

$$-\frac{d^2 g}{dr^2} - \frac{1}{r}\frac{dg}{dr} = 0,\ r > 0. \tag{11.72}$$

The general solution of (11.72) is

$$g(r) = c_1 \ln(r) + c_2,\ r > 0, \tag{11.73}$$

where c_1 and c_2 are constants (see Exercise 11.5.4). We will show that for $c_2 = 0$ and an appropriate choice of c_1, $g(r) = c_1 \ln(r)$ defines the desired free-space Green's function:

$$G(\mathbf{x};\mathbf{y}) = c_1 \ln(\|\mathbf{x}-\mathbf{y}\|).$$

To demonstrate this, we show that $u(\mathbf{x}) = G(\mathbf{x};\mathbf{y})$ satisfies the weak form of the PDE

$$-\Delta u = \delta(\mathbf{x}-\mathbf{y}).$$

We have not considered the weak form of a PDE without boundary conditions before. The idea is to ensure that the PDE holds on every region in $\mathbf{R}^2$ by using as test functions all smooth functions having *compact support*. The reader will recall that the support of a function is the closure of the set on which the function is nonzero; the support is compact if it is a closed and bounded set. For instance, by choosing as test functions all functions whose support is contained in a given set Ω, the weak form will guarantee that the PDE is satisfied on Ω. By doing this for every possible domain Ω, we ensure that the PDE is satisfied everywhere. We will write V for the space of all test functions (that is, all smooth functions with compact support).

The weak form of $-\Delta u = \delta(\mathbf{x}-\mathbf{y})$ is

$$\int_{\mathbf{R}^2} \nabla u(\mathbf{x}) \cdot \nabla v(\mathbf{x})\, d\mathbf{x} = \int_{\mathbf{R}^2} \delta(\mathbf{x}-\mathbf{y}) v(\mathbf{x})\, d\mathbf{x} = v(\mathbf{y}) \text{ for all } v \in V \tag{11.74}$$

in Cartesian coordinates, or

$$\int_0^{2\pi}\int_0^\infty \left(\frac{\partial u}{\partial r}\frac{\partial v}{\partial r} + \frac{1}{r^2}\frac{\partial u}{\partial \theta}\frac{\partial v}{\partial \theta}\right) v\, r\, dr\, d\theta = v(\mathbf{y}) \text{ for all } v \in V \tag{11.75}$$

in polar coordinates (see Exercise 11.5.5). Here polar coordinates correspond to the Cartesian coordinates $(x_1 - y_1, x_2 - y_2)$, so $v(\mathbf{y})$ is the value of v for $r = 0$ (and any value of θ). In either Cartesian or polar form, there are no boundary terms because every test function v has compact support, and is therefore zero outside a bounded set.

With $u = G(\mathbf{x};\mathbf{y}) = c_1 \ln(r)$, we have

$$\frac{\partial u}{\partial r} = \frac{c_1}{r},\ \frac{\partial u}{\partial \theta} = 0.$$

11.5. The free-space Green's function for the Laplacian

Therefore,

$$\int_0^{2\pi}\int_0^\infty \left(\frac{\partial u}{\partial r}\frac{\partial v}{\partial r} + \frac{1}{r^2}\frac{\partial u}{\partial \theta}\frac{\partial v}{\partial \theta}\right) vr\,dr\,d\theta = \int_0^{2\pi}\int_0^\infty \frac{c_1}{r}\frac{\partial v}{\partial r}r\,dr\,d\theta$$
$$= c_1 \int_0^{2\pi}\int_0^\infty \frac{\partial v}{\partial r}\,dr\,d\theta$$
$$= 2\pi c_1 \int_0^\infty \frac{\partial v}{\partial r}\,dr$$
$$= 2\pi c_1 \left.(-v)\right|_{r=0}$$
$$= -2\pi c_1 v(\mathbf{y}).$$

The last integration follows from the fact that v is zero for r sufficiently large. We now see that if $c_1 = -1/2\pi$, then the weak form of the PDE is satisfied. Thus the desired free-space Green's function is

$$G(\mathbf{x};\mathbf{y}) = -\frac{1}{2\pi}\ln(\|\mathbf{x}-\mathbf{y}\|). \tag{11.76}$$

We can now verify directly that

$$u(\mathbf{x}) = \int_\Omega G(\mathbf{x};\mathbf{y})f(\mathbf{y})\,d\mathbf{y} \tag{11.77}$$

satisfies $-\Delta u = f(\mathbf{x})$ in the weak sense:

$$\int_{\mathbf{R}^2} \nabla u(\mathbf{x}) \cdot \nabla v(\mathbf{x})\,d\mathbf{x} = \int_{\mathbf{R}^2} f(\mathbf{x})v(\mathbf{x})\,d\mathbf{x} \text{ for all } v \in V. \tag{11.78}$$

We assume that f is integrable and has compact support. (It is actually not necessary that the support of f be compact, provided $|f(\mathbf{x})|$ decays to zero sufficiently quickly as $\|\mathbf{x}\| \to \infty$. We will assume the simpler condition on f, however.) Before we can verify (11.78), we make some preliminary observations and definitions.

We will use the fact that, for each fixed $\mathbf{y}$, $G(\cdot;\mathbf{y})$ satisfies $\Delta G(\mathbf{x};\mathbf{y}) = 0$ for all $\mathbf{x} \neq \mathbf{y}$ (see Exercise 11.5.6). Specifically,

$$\nabla G(\mathbf{x};\mathbf{y}) = -\frac{\mathbf{x}-\mathbf{y}}{2\pi\|\mathbf{x}-\mathbf{y}\|^2} \tag{11.79}$$

and

$$\Delta G(\mathbf{x};\mathbf{y}) = \nabla \cdot \nabla G(\mathbf{x};\mathbf{y}) = \nabla \cdot \left(-\frac{\mathbf{x}-\mathbf{y}}{2\pi\|\mathbf{x}-\mathbf{y}\|^2}\right) = 0. \tag{11.80}$$

Given any v with compact support, we choose $R > 0$ sufficiently large such that $v(\mathbf{x}) = 0$ for all $\mathbf{x}$ satisfying $\|\mathbf{x}-\mathbf{y}\| \geq R$. Writing $\Omega = \{\mathbf{x} \in \mathbf{R}^2 : \|\mathbf{x}-\mathbf{y}\| < R\}$, we see that

$$\int_{\mathbf{R}^2} \nabla u(\mathbf{x}) \cdot \nabla v(\mathbf{x})\,d\mathbf{x} = \int_{\mathbf{R}^2} f(\mathbf{x})v(\mathbf{x})\,d\mathbf{x}$$

is equivalent to

$$\int_\Omega \nabla u(\mathbf{x}) \cdot \nabla v(\mathbf{x})\,d\mathbf{x} = \int_\Omega f(\mathbf{x})v(\mathbf{x})\,d\mathbf{x}.$$

We can now proceed with the proof. Since

$$u(\mathbf{x}) = \int_{\mathbf{R}^2} G(\mathbf{x};\mathbf{y}) f(\mathbf{y}) d\mathbf{y},$$

we see that

$$\nabla u(\mathbf{x}) = \int_{\mathbf{R}^2} \nabla G(\mathbf{x};\mathbf{y}) f(\mathbf{y}) d\mathbf{y} = -\frac{1}{2\pi} \int_{\mathbf{R}^2} f(\mathbf{y}) \frac{\mathbf{x}-\mathbf{y}}{\|\mathbf{x}-\mathbf{y}\|^2} d\mathbf{y}.$$

Thus,

$$\int_{\mathbf{R}^2} \nabla u(\mathbf{x}) \cdot \nabla v(\mathbf{y}) d\mathbf{y} = \int_{\Omega} \nabla u(\mathbf{x}) \cdot \nabla v(\mathbf{y}) d\mathbf{y}$$

$$= -\frac{1}{2\pi} \int_{\Omega} \int_{\mathbf{R}^2} f(\mathbf{y}) \frac{\mathbf{x}-\mathbf{y}}{\|\mathbf{x}-\mathbf{y}\|^2} \cdot \nabla v(\mathbf{x}) d\mathbf{y} d\mathbf{x}$$

$$= -\frac{1}{2\pi} \int_{\mathbf{R}^2} \left\{ \int_{\Omega} \frac{\mathbf{x}-\mathbf{y}}{\|\mathbf{x}-\mathbf{y}\|^2} \cdot \nabla v(\mathbf{x}) d\mathbf{x} \right\} f(\mathbf{y}) d\mathbf{y}.$$

It now suffices to prove that

$$-\frac{1}{2\pi} \int_{\Omega} \frac{\mathbf{x}-\mathbf{y}}{\|\mathbf{x}-\mathbf{y}\|^2} \cdot \nabla v(\mathbf{x}) d\mathbf{x} = v(\mathbf{y}).$$

We would like to apply Green's identity (integration by parts); however, we cannot do this directly since the integrand is singular at $\mathbf{x} = \mathbf{y}$. We therefore define

$$\Omega_\epsilon = \{\mathbf{x} \in \mathbf{R}^2 : \epsilon < \|\mathbf{x}-\mathbf{y}\| < R\}$$

and notice that

$$\int_{\Omega} \frac{\mathbf{x}-\mathbf{y}}{\|\mathbf{x}-\mathbf{y}\|^2} \cdot \nabla v(\mathbf{x}) d\mathbf{x} = \lim_{\epsilon \to 0^+} \int_{\Omega_\epsilon} \frac{\mathbf{x}-\mathbf{y}}{\|\mathbf{x}-\mathbf{y}\|^2} \cdot \nabla v(\mathbf{x}) d\mathbf{x}.$$

The boundary of Ω_ϵ consists of two circles centered at $\mathbf{y}$:

$$\partial \Omega_\epsilon = S_R \cup S_\epsilon, \ S_R = \{\mathbf{x} \in \mathbf{R}^2 : \|\mathbf{x}-\mathbf{y}\| = R\}, \ S_\epsilon = \{\mathbf{x} \in \mathbf{R}^2 : \|\mathbf{x}-\mathbf{y}\| = \epsilon\}.$$

We have

$$\int_{\Omega_\epsilon} \frac{\mathbf{x}-\mathbf{y}}{\|\mathbf{x}-\mathbf{y}\|^2} \cdot \nabla v(\mathbf{x}) d\mathbf{x} = \int_{\partial \Omega_\epsilon} v(\mathbf{x}) \frac{\mathbf{x}-\mathbf{y}}{\|\mathbf{x}-\mathbf{y}\|^2} \cdot \mathbf{n} d\sigma_\mathbf{x}$$

$$- \int_{\Omega_\epsilon} \nabla \cdot \left(\frac{\mathbf{x}-\mathbf{y}}{\|\mathbf{x}-\mathbf{y}\|^2} \right) v(\mathbf{x}) d\mathbf{x}$$

$$= \int_{S_R} v(\mathbf{x}) \frac{\mathbf{x}-\mathbf{y}}{\|\mathbf{x}-\mathbf{y}\|^2} \cdot \mathbf{n} d\sigma_\mathbf{x}$$

$$+ \int_{S_\epsilon} v(\mathbf{x}) \frac{\mathbf{x}-\mathbf{y}}{\|\mathbf{x}-\mathbf{y}\|^2} \cdot \mathbf{n} d\sigma_\mathbf{x}$$

$$- \int_{\Omega_\epsilon} \nabla \cdot \left(\frac{\mathbf{x}-\mathbf{y}}{\|\mathbf{x}-\mathbf{y}\|^2} \right) v(\mathbf{x}) d\mathbf{x}.$$

11.5. The free-space Green's function for the Laplacian

As pointed out above,
$$\nabla \cdot \left(\frac{\mathbf{x} - \mathbf{y}}{\|\mathbf{x} - \mathbf{y}\|^2} \right) = 0,$$
and Ω was chosen so that $v(\mathbf{x}) = 0$ for all $\mathbf{x} \in S_R$. Therefore,
$$\int_{\Omega_\epsilon} \frac{\mathbf{x} - \mathbf{y}}{\|\mathbf{x} - \mathbf{y}\|^2} \cdot \nabla v(\mathbf{x}) \, d\mathbf{x} = \int_{S_\epsilon} v(\mathbf{x}) \frac{\mathbf{x} - \mathbf{y}}{\|\mathbf{x} - \mathbf{y}\|^2} \cdot \mathbf{n} \, d\sigma_\mathbf{x}.$$

Moreover, on S_ϵ, the normal vector is
$$\mathbf{n} = -\frac{\mathbf{x} - \mathbf{y}}{\|\mathbf{x} - \mathbf{y}\|},$$
and therefore
$$\frac{\mathbf{x} - \mathbf{y}}{\|\mathbf{x} - \mathbf{y}\|^2} \cdot \mathbf{n} = -\frac{(\mathbf{x} - \mathbf{y}) \cdot (\mathbf{x} - \mathbf{y})}{\|\mathbf{x} - \mathbf{y}\|^3} = -\frac{1}{\|\mathbf{x} - \mathbf{y}\|} = -\frac{1}{\epsilon}.$$

Therefore,
$$-\frac{1}{2\pi} \int_{\Omega_\epsilon} \frac{\mathbf{x} - \mathbf{y}}{\|\mathbf{x} - \mathbf{y}\|^2} \cdot \nabla v(\mathbf{x}) \, d\mathbf{x} = -\frac{1}{2\pi} \int_{S_\epsilon} v(\mathbf{x}) \frac{\mathbf{x} - \mathbf{y}}{\|\mathbf{x} - \mathbf{y}\|^2} \cdot \mathbf{n} \, d\sigma_\mathbf{x} = \frac{1}{2\pi\epsilon} \int_{S_\epsilon} v(\mathbf{x}) \, d\mathbf{x}.$$

This last integral is the average value of v on S_ϵ, and the continuity of v therefore yields
$$-\frac{1}{2\pi} \int_{\Omega_\epsilon} \frac{\mathbf{x} - \mathbf{y}}{\|\mathbf{x} - \mathbf{y}\|^2} \cdot \nabla v(\mathbf{x}) \, d\mathbf{x} = \frac{1}{2\pi\epsilon} \int_{S_\epsilon} v(\mathbf{x}) \, d\mathbf{x} \to v(\mathbf{y}) \text{ as } \epsilon \to 0^+.$$

It follows that
$$-\frac{1}{2\pi} \int_{\Omega} \frac{\mathbf{x} - \mathbf{y}}{\|\mathbf{x} - \mathbf{y}\|^2} \cdot \nabla v(\mathbf{x}) \, d\mathbf{x} = v(\mathbf{y}),$$
as desired. This completes the proof that u satisfies (11.78), and therefore that G is indeed a free-space Green's function for the Laplacian.

11.5.2 The free-space Green's function in three dimensions

It is now straightforward to find a free-space Green's function in $\mathbf{R}^3$. We follow the same reasoning as before, taking advantage of the symmetry of the Laplace operator and now using spherical coordinates instead of polar coordinates. The spherical coordinates of a point in $\mathbf{R}^3$ are (ρ, ϕ, θ), where $\rho \geq 0$ is the distance of the point from the origin, $\phi \in [0, \pi]$ is the angle from the positive x_3-axis, and $\theta \in [0, 2\pi]$ is the angle, in the $x_1 x_2$-plane, from the positive x_1-axis. The relationship between Cartesian and spherical coordinates is defined by
$$x_1 = \rho \sin(\phi) \cos(\theta),$$
$$x_2 = \rho \sin(\phi) \sin(\theta),$$
$$x_3 = \rho \cos(\phi).$$

Exercise 11.5.7 asks the reader to verify that the Laplacian in spherical coordinates is given by
$$\Delta u = \frac{1}{\rho^2} \frac{\partial}{\partial \rho} \left(\rho^2 \frac{\partial u}{\partial \rho} \right) + \frac{1}{\rho^2 \sin(\phi)} \frac{\partial}{\partial \phi} \left(\sin(\phi) \frac{\partial u}{\partial \phi} \right) + \frac{1}{\rho^2 \sin^2(\phi)} \frac{\partial^2 u}{\partial \theta^2}. \tag{11.81}$$

If $u = u(\rho, \phi, \theta)$ is a radial function (that is, u is constant with respect to ϕ and θ), then Δu simplifies to

$$\Delta u = \frac{1}{\rho^2} \frac{\partial}{\partial \rho}\left(\rho^2 \frac{\partial u}{\partial \rho}\right).$$

As in the two-dimensional case, we assume that the desired free-space Green's function has the form $G(\mathbf{x};\mathbf{y}) = g(\|\mathbf{x}-\mathbf{y}\|) = g(\rho)$. We want $\Delta G(\mathbf{x};\mathbf{y}) = \delta(\mathbf{x}-\mathbf{y})$, and therefore, formally, $-\Delta g(\rho) = 0$ for $\rho > 0$ should hold. We therefore solve

$$\frac{1}{\rho^2}\frac{\partial}{\partial \rho}\left(\rho^2 \frac{\partial g}{\partial \rho}\right) = 0, \; \rho > 0,$$

to obtain

$$g(\rho) = \frac{c_1}{\rho} + c_2.$$

We next show that, for $c_2 = 0$ and a suitable value of c_1, $g(\rho) = c_1/g$ is a free-space Green's function. We determine c_1 by substituting g into the weak form of the equation $-\Delta g = \delta$. Exercise 11.5.8 asks the reader to show that the weak form of $-\Delta u = \delta$, expressed in spherical coordinates, is

$$\int_0^{2\pi}\int_0^{\pi}\int_0^{\infty}\left\{\frac{\partial u}{\partial \rho}\frac{\partial v}{\partial \rho} + \frac{1}{\rho^2}\frac{\partial u}{\partial \phi}\frac{\partial v}{\partial \phi} + \frac{1}{\rho^2 \sin^2(\phi)}\frac{\partial u}{\partial \theta}\frac{\partial v}{\partial \theta}\right\}\rho^2 \sin(\phi)\,d\rho\,d\phi\,d\theta \qquad (11.82)$$
$$= v(\mathbf{y}) \text{ for all } v \in V.$$

Here V represents the set of all smooth functions defined on $\mathbf{R}^3$ and having compact support, and $\mathbf{y}$ is the origin for the spherical coordinate system (so that $\rho = 0$ represents $\mathbf{x} = \mathbf{y}$).

Given that g is independent of ϕ and θ, we wish to show that $g(\rho) = c_1/\rho$ satisfies

$$\int_0^{2\pi}\int_0^{\pi}\int_0^{\infty}\frac{\partial g}{\partial \rho}\frac{\partial v}{\partial \rho}\rho^2 \sin(\phi)\,d\rho\,d\phi\,d\theta = v(\mathbf{y}) \text{ for all } v \in V. \qquad (11.83)$$

Exercise 11.5.9 asks the reader to show that g satisfies this variational equation provided $c_1 = 1/(4\pi)$. It follows that the free-space Green's function in three dimensions is

$$G(\mathbf{x};\mathbf{y}) = \frac{1}{4\pi \|\mathbf{x}-\mathbf{y}\|}. \qquad (11.84)$$

As in the case of two dimensions, we can verify that (11.84) defines a free-space Green's function for the Laplacian by showing that

$$u(\mathbf{x}) = \int_{\mathbf{R}^3} G(\mathbf{x};\mathbf{y}) f(\mathbf{y})\,d\mathbf{y} \qquad (11.85)$$

satisfies

$$\int_{\mathbf{R}^3} \nabla u(\mathbf{x}) \cdot \nabla v(\mathbf{x})\,d\mathbf{x} = \int_{\mathbf{R}^3} v(\mathbf{x}) f(\mathbf{x})\,d\mathbf{x} \text{ for all } v \in V, \qquad (11.86)$$

where V is the space of all smooth function defined on $\mathbf{R}^3$ and having compact support. The proof is similar to the two-dimensional case and is left to Exercise 11.5.10.

11.5. The free-space Green's function for the Laplacian

Exercises

1. Suppose Cartesian coordinates (x_1, x_2) are given, and the $x_1 x_2$-axes are rotated by α radians to give new coordinates (y_1, y_2). Prove that (11.68) holds. (Hint: A point in the plane with coordinates (x_1, x_2) in the original coordinate system has polar coordinates (r, θ). If polar coordinates are then defined with respect to the second coordinate systems—that is, if the polar angle is measured from the y_1-axis—then the polar coordinates of the same point are $(r, \theta - \alpha)$ (draw a picture if this is not clear). We then have $y_1 = r\cos(\theta - \alpha)$ and $y_2 = r\sin(\theta - \alpha)$, and applying trigonometric identities yields the desired result.)

2. Given Cartesian coordinates (x_1, x_2), define (y_1, y_2) by (11.68). Prove that (11.69) holds.

3. Show that (11.71) is equivalent to
$$-\frac{d}{dr}\left(r\frac{d\psi}{dr}\right) = r\phi(r),$$
and solve to find the general solution of (11.71).

4. Show that (11.72) is equivalent to
$$-\frac{d}{dr}\left(r\frac{dg}{dr}\right) = 0,\ r > 0,$$
and solve to find (11.73).

5. Show that the weak form (11.75) is equivalent to (11.74) by two different methods:
   (a) Start with (11.74) and change variables in the double integral to obtain (11.75).
   (b) Start with the Laplacian in polar coordinates (equation (11.70)) and derive the weak form in the usual way (multiply by a test function and then integrate by parts in the appropriate term).

6. Verify formulas (11.79) and (11.80) and thus verify that
$$G(\mathbf{x}; \mathbf{y}) = -\frac{1}{2\pi} \ln(\|\mathbf{x} - \mathbf{y}\|)$$
satisfies $\Delta G(\mathbf{x}; \mathbf{y}) = 0$ for all $\mathbf{x} \neq \mathbf{y}$.

7. Derive (11.81). (Hint: Mimic the derivation of the Laplacian in polar coordinates given in Section 11.3.1.)

8. Prove that (11.82) is the weak form of $-\Delta u = \delta$ in spherical coordinates.

9. Show that $g(\rho) = c_1/\rho$ satisfies (11.83) if and only if $c_1 = 1/(4\pi)$.

10. Prove that (11.84) is a free-space Green's function for the Laplacian by showing that (11.85) satisfies the variational equation (11.86).

11.6 The Green's function for the Laplacian on a bounded domain

We now wish to find the Green's function for the BVP

$$-\Delta u = f(\mathbf{x}) \text{ in } \Omega,$$
$$\alpha_1 u + \alpha_2 \frac{\partial u}{\partial \mathbf{n}} = 0 \text{ on } \partial\Omega. \qquad (11.87)$$

As we stated in the previous section, it is possible to use the free-space Green's function to find the Green's function for (11.87). The function f may be defined only on Ω, but we can extend it to all of $\mathbf{R}^2$ by defining $f(\mathbf{x}) = 0$ for $\mathbf{x} \notin \Omega$. Of course, f (thus extended) may not be continuous across $\partial\Omega$, but this does not matter; as pointed out in the previous section, f need only be integrable in order that the Green's function lead to a solution of the PDE (interpreted in the weak sense):

$$u_1(\mathbf{x}) = \int_\Omega G(\mathbf{x};\mathbf{y}) f(\mathbf{y}) d\mathbf{y} = \int_{\mathbf{R}^2} G(\mathbf{x};\mathbf{y}) f(\mathbf{y}) d\mathbf{y}.$$

Since u_1 is constructed by ignoring the boundary condition

$$\alpha_1 u + \alpha_2 \frac{\partial u}{\partial \mathbf{n}} = 0 \text{ on } \partial\Omega,$$

it is almost certain that u_1 does not satisfy this condition. We have

$$\alpha_1 u_1(\mathbf{x}) + \alpha_2 \frac{\partial u_1}{\partial \mathbf{n}}(\mathbf{x}) = \int_\Omega \left(\alpha_1 G(\mathbf{x};\mathbf{y}) + \alpha_2 \frac{\partial G}{\partial \mathbf{n}}(\mathbf{x};\mathbf{y}) \right) f(\mathbf{y}) d\mathbf{y}.$$

Here $\partial G/\partial \mathbf{n}$ represents the normal derivative of G with respect to the variable x:

$$\frac{\partial G}{\partial \mathbf{n}}(\mathbf{x};\mathbf{y}) = \nabla_\mathbf{x} G(\mathbf{x};\mathbf{y}) \cdot \mathbf{n}(\mathbf{x}).$$

Let us define

$$w(\mathbf{x};\mathbf{y}) = \alpha_1 G(\mathbf{x};\mathbf{y}) + \alpha_2 \frac{\partial G}{\partial \mathbf{n}}(\mathbf{x};\mathbf{y}), \ \mathbf{y} \in \Omega, \ \mathbf{x} \in \partial\Omega.$$

For each $\mathbf{y} \in \Omega$, the BVP

$$-\Delta u = 0 \text{ in } \Omega,$$
$$\alpha_1 u + \alpha_2 \frac{\partial u}{\partial \mathbf{n}} = w(\mathbf{x};\mathbf{y}) \text{ on } \partial\Omega \qquad (11.88)$$

has a solution; since it depends on both $\mathbf{x}$ and $\mathbf{y}$ (and on the domain Ω), we denote it by $v_\Omega(\mathbf{x};\mathbf{y})$. We then define

$$G_\Omega(\mathbf{x};\mathbf{y}) = G(\mathbf{x};\mathbf{y}) - v_\Omega(\mathbf{x};\mathbf{y}).$$

11.6. The Green's function for the Laplacian on a bounded domain

We will now sketch the verification that G_Ω is the desired Green's function, with the intermediate steps left to the exercises. We define

$$u(\mathbf{x}) = \int_\Omega G_\Omega(\mathbf{x};\mathbf{y}) f(\mathbf{y}) d\mathbf{y} = \int_\Omega G(\mathbf{x};\mathbf{y}) f(\mathbf{y}) d\mathbf{y} - \int_\Omega v_\Omega(\mathbf{x};\mathbf{y}) f(\mathbf{y}) d\mathbf{y}. \quad (11.89)$$

Exercise 11.6.1 asks the reader to show that the weak form of the BVP (11.87) is

$$\frac{\alpha_1}{\alpha_2} \int_{\partial\Omega} uv + \int_\Omega \nabla u \cdot \nabla v = \int_\Omega fv \text{ for all } v \in C^2(\overline{\Omega}), \quad (11.90)$$

where $C^2(\overline{\Omega})$ is the space of all functions that are twice continuously differentiable on $\overline{\Omega}$ (the open set Ω together with its boundary). On the other hand, the weak form of (11.88) is

$$\frac{\alpha_1}{\alpha_2} \int_{\partial\Omega} uv + \int_\Omega \nabla u \cdot \nabla v = \frac{1}{\alpha_2} \int_{\partial\Omega} wv \text{ for all } v \in C^2(\overline{\Omega}) \quad (11.91)$$

(see Exercise 11.6.2). If we write

$$u_1(\mathbf{x}) = \int_\Omega G(\mathbf{x};\mathbf{y}) f(\mathbf{y}) d\mathbf{y}, \ u_2(\mathbf{x}) = \int_\Omega v_\Omega(\mathbf{x};\mathbf{y}) f(\mathbf{y}) d\mathbf{y}, \quad (11.92)$$

then the proposed solution is $u = u_1 - u_2$ (cf. (11.89)). It can be shown (see Exercise 11.6.3) that u_2 satisfies the following variational equation:

$$\frac{\alpha_1}{\alpha_2} \int_{\partial\Omega} u_2 v + \int_\Omega \nabla u_2 \cdot \nabla v = \frac{1}{\alpha_2} \int_\Omega \left\{ \int_{\partial\Omega} w(\cdot;\mathbf{y}) v \right\} f(\mathbf{y}) d\mathbf{y}. \quad (11.93)$$

The reader should note that $u_2(\mathbf{x})$ is defined by an integral that is parametrized by $\mathbf{x}$, and derivatives of u_2 can be computed by differentiating under the integral sign. In particular,

$$\nabla u_2(\mathbf{x}) = \int_\Omega f(\mathbf{y}) \nabla v_\Omega(\mathbf{x};\mathbf{y}) d\mathbf{y}.$$

Similar remarks apply to u_1. (Even though G is singular at $\mathbf{x} = \mathbf{y}$, $\nabla G(\mathbf{x};\mathbf{y})$ is still integrable, and therefore it is still valid to differentiate under the integral sign.)

The function u_1 satisfies the variational equation

$$\frac{\alpha_1}{\alpha_2} \int_{\partial\Omega} u_1 v + \int_\Omega \nabla u_1 \cdot \nabla v = \frac{1}{\alpha_2} \int_\Omega \left\{ \int_{\partial\Omega} w(\cdot;\mathbf{y}) v \right\} f(\mathbf{y}) d\mathbf{y} + \int_\Omega f(\mathbf{x}) v(\mathbf{x}) d\mathbf{x}. \quad (11.94)$$

To derive (11.94), we use the fact that u_1 is a weak solution to $-\Delta u = f(\mathbf{x})$ over all of $\mathbf{R}^2$. This means that

$$\int_{\mathbf{R}^2} \nabla u_1(\mathbf{x}) \cdot \nabla v(\mathbf{x}) d\mathbf{x} = \int_{\mathbf{R}^2} f(\mathbf{x}) v(\mathbf{x}) d\mathbf{x} \text{ for all } v \in V.$$

Since f is zero outside of Ω, we have

$$\int_{\mathbf{R}^2} f(\mathbf{x}) v(\mathbf{x}) d\mathbf{x} = \int_\Omega f(\mathbf{x}) v(\mathbf{x}) d\mathbf{x} \text{ for all } v \in V.$$

On the other hand,

$$\int_{\mathbf{R}^2} \nabla u_1(\mathbf{x}) \cdot \nabla v(\mathbf{x}) d\mathbf{x} = \int_\Omega \nabla u_1(\mathbf{x}) \cdot \nabla v(\mathbf{x}) d\mathbf{x} + \int_{\mathbf{R}^2 \setminus \Omega} \nabla u_1(\mathbf{x}) \cdot \nabla v(\mathbf{x}) d\mathbf{x},$$

where $\mathbf{R}^2 \setminus \Omega$ denotes the set of all points in the plane $\mathbf{R}^2$ not lying in Ω. If we apply Green's first identity to

$$\int_{\mathbf{R}^2 \setminus \Omega} \nabla u_1(\mathbf{x}) \cdot \nabla v(\mathbf{x}) d\mathbf{x}$$

and use the fact that $G(\mathbf{x}; \mathbf{y})$ satisfies $\Delta_\mathbf{x} G(\mathbf{x}; \mathbf{y}) = 0$ for $\mathbf{x} \notin \Omega$, $\mathbf{y} \in \Omega$, we can show that

$$\int_{\mathbf{R}^2 \setminus \Omega} \nabla u_1(\mathbf{x}) \cdot \nabla v(\mathbf{x}) d\mathbf{x} = -\int_\Omega \left\{ \int_{\partial\Omega} \frac{\partial G}{\partial \mathbf{n}}(\cdot; \mathbf{y}) v \right\} f(\mathbf{y}) d\mathbf{y}.$$

Exercise 11.6.4 asks the reader to put these facts together to derive (11.94).

Finally, since u_1 and u_2 satisfy (11.94) and (11.93), it follows immediately that $u = u_1 - u_2$ satisfies (11.90), the weak form of the original BVP (11.87). Throughout this analysis, we have assumed that $\alpha_2 \neq 0$. If this is not the case, then the boundary condition is a Dirichlet condition, and the derivations simplify somewhat. The Dirichlet problem is addressed in Exercise 11.6.5.

11.6.1 Reciprocity

In Chapter 9, we derived explicit formulas for Green's functions for the standard differential equations in one spatial dimension. From these formulas, we could see directly that the Green's functions satisfied the appropriate reciprocity property. We can also see from the formulas for the free-space Green's function G for the Laplacian in two and three dimensions that $G(\mathbf{x}; \mathbf{y}) = G(\mathbf{y}; \mathbf{x})$ for all $\mathbf{x}$ and $\mathbf{y}$.

Without deriving an explicit formula for the Green's function G_Ω of the Laplacian on a bounded domain Ω, we can prove that G_Ω is symmetric. In fact, this follows from the fact that the Laplacian defines a symmetric linear operator (under any of the usual boundary conditions). Let us define $L : S \to C(\overline{\Omega})$, where

$$S = \left\{ u \in C^2(\overline{\Omega}) : \alpha_1 u + \alpha_2 \frac{\partial u}{\partial \mathbf{n}} = 0 \text{ on } \partial\Omega \right\}.$$

We already know that L is a symmetric operator under the L^2 inner product: $(Lu, v) = (u, Lv)$ for all $u, v \in S$. Also, the inverse $M : C(\overline{\Omega}) \to S$ of L exists and is defined by G_Ω:

$$(Mf)(\mathbf{x}) = \int_\Omega G_\Omega(\mathbf{x}; \mathbf{y}) f(\mathbf{y}) d\mathbf{y}.$$

We can show immediately that M is also symmetric. Let f, g be arbitrary elements of $C(\overline{\Omega})$, and define $u = Mf$, $v = Mg$. Then $Lu = f$, $Lv = g$, and

$$(Mf, g) = (u, Lv) = (Lu, v) = (f, Mg).$$

11.6. The Green's function for the Laplacian on a bounded domain

From this it follows easily that $G_\Omega(\mathbf{x};\mathbf{y})$ is symmetric in $\mathbf{x}$ and $\mathbf{y}$. We have

$$(Mf,g) = (f,Mg)$$
$$\Leftrightarrow \int_\Omega \left\{ \int_\Omega G(\mathbf{x};\mathbf{y}) f(\mathbf{y}) d\mathbf{y} \right\} g(\mathbf{x}) d\mathbf{x} = \int_\Omega f(\mathbf{x}) \left\{ \int_\Omega G(\mathbf{x};\mathbf{y}) g(\mathbf{y}) d\mathbf{y} \right\} d\mathbf{x}$$
$$\Leftrightarrow \int_\Omega \int_\Omega G(\mathbf{x};\mathbf{y}) f(\mathbf{y}) g(\mathbf{x}) d\mathbf{y} d\mathbf{x} = \int_\Omega \int_\Omega G(\mathbf{x};\mathbf{y}) f(\mathbf{x}) g(\mathbf{y}) d\mathbf{y} d\mathbf{x}$$
$$\Leftrightarrow \int_\Omega \int_\Omega G(\mathbf{x};\mathbf{y}) f(\mathbf{y}) g(\mathbf{x}) d\mathbf{y} d\mathbf{x} = \int_\Omega \int_\Omega G(\mathbf{y};\mathbf{x}) f(\mathbf{y}) g(\mathbf{x}) d\mathbf{x} d\mathbf{y}$$
$$\Leftrightarrow \int_\Omega \int_\Omega G(\mathbf{x};\mathbf{y}) f(\mathbf{y}) g(\mathbf{x}) d\mathbf{y} d\mathbf{x} = \int_\Omega \int_\Omega G(\mathbf{y};\mathbf{x}) f(\mathbf{y}) g(\mathbf{x}) d\mathbf{y} d\mathbf{x}.$$

(On the fourth line, we just interchanged the names of the variables in the integral on the right, and on the fifth line, we changed the order of integration.) This last equation is equivalent to

$$\int_\Omega \left\{ \int_\Omega (G(\mathbf{x};\mathbf{y}) - G(\mathbf{y};\mathbf{x})) f(\mathbf{y}) d\mathbf{y} \right\} g(\mathbf{x}) d\mathbf{x} = 0.$$

Since this holds for all continuous g, it follows that

$$\int_\Omega (G(\mathbf{x};\mathbf{y}) - G(\mathbf{y};\mathbf{x})) f(\mathbf{y}) d\mathbf{y} = 0 \text{ for all } \mathbf{x} \in \Omega,$$

and since this holds for all continuous f, it follows that

$$G(\mathbf{x};\mathbf{y}) - G(\mathbf{y};\mathbf{x}) = 0 \text{ for all } \mathbf{x}, \mathbf{y} \in \Omega,$$

as desired.

11.6.2 The Green's function for a disk

The results above show how to find the Green's function for a specific domain, provided one can find the solution v_Ω to BVP (11.88). For most domains Ω, it is not possible to find v_Ω explicitly. We will now show how to proceed in a special case in which this is possible: Ω is a disk and Dirichlet conditions are imposed. We assume that Ω is the disk of radius R centered at the origin. We wish to find the Green's function for the BVP

$$-\Delta u = f(\mathbf{x}) \text{ in } \Omega,$$
$$u = 0 \text{ on } \partial\Omega.$$

The free-space Green's function is

$$G(\mathbf{x};\mathbf{y}) = -\frac{1}{2\pi} \ln(\|\mathbf{x} - \mathbf{y}\|).$$

Since the boundary conditions are Dirichlet, the function v_Ω must satisfy

$$v_\Omega(\mathbf{x};\mathbf{y}) = G(\mathbf{x};\mathbf{y}) \text{ for all } y \in \Omega, \, x \in \partial\Omega.$$

We will show that $v_\Omega(\mathbf{x};\mathbf{y})$ can be constructed from $G(\mathbf{x};\mathbf{y}^*)$ for some $\mathbf{y}^*$, depending on $\mathbf{y}$ and lying outside of Ω. To be precise, we will define

$$v_\Omega(\mathbf{x};\mathbf{y}) = G(\mathbf{x};\mathbf{y}^*) + c(\mathbf{y}),$$

where c is a scalar function of $\mathbf{y}$ and $\mathbf{y}^*$ is a function of $\mathbf{y}$ (with the dependence on $\mathbf{y}$ suppressed). To see how $\mathbf{y}^*$ is determined, we consider the resulting function G_Ω:

$$\begin{aligned} G_\Omega(\mathbf{x};\mathbf{y}) &= G(\mathbf{x};\mathbf{y}) - G(\mathbf{x};\mathbf{y}^*) - c(\mathbf{y}) \\ &= -\frac{1}{2\pi} \ln(\|\mathbf{x}-\mathbf{y}\|) + \frac{1}{2\pi} \ln(\|\mathbf{x}-\mathbf{y}^*\|) - c(\mathbf{y}) \\ &= \frac{1}{2\pi} \ln\left(\frac{\|\mathbf{x}-\mathbf{y}^*\|}{\|\mathbf{x}-\mathbf{y}\|}\right) - c(\mathbf{y}). \end{aligned}$$

If $\mathbf{y}^*$ lies outside of Ω, then the proposed v_Ω satisfies $-\Delta v_\Omega = 0$ in Ω, so we need only choose $\mathbf{y}^* \notin \Omega$ so that $G_\Omega(\mathbf{x};\mathbf{y}) = 0$ for all $x \in \partial \Omega$. According to the above formula for $G_\Omega(\mathbf{x};\mathbf{y})$, if we choose $\mathbf{y}^*$ so that

$$\frac{\|\mathbf{x}-\mathbf{y}^*\|}{\|\mathbf{x}-\mathbf{y}\|} = m(\mathbf{y}) \text{ for all } \mathbf{x} \in \partial \Omega, \tag{11.95}$$

where m is a scalar-valued function of $\mathbf{y}$, then we obtain

$$G_\Omega(\mathbf{x};\mathbf{y}) = \frac{1}{2\pi} \ln(m(\mathbf{y})) - c(\mathbf{y}) \text{ for all } \mathbf{x} \in \partial \Omega.$$

Then, choosing $c(\mathbf{y}) = \ln(m(\mathbf{y}))/(2\pi)$, we obtain $G_\Omega(\mathbf{x};\mathbf{y}) = 0$ for all $\mathbf{x} \in \partial \Omega$, $\mathbf{y} \in \Omega$.

Equation (11.95) states that, for each fixed $\mathbf{y}$, $\mathbf{y}^*$ should be chosen so that $\|\mathbf{x}-\mathbf{y}^*\|$ is a fixed multiple of $\|\mathbf{x}-\mathbf{y}\|$ for all $\mathbf{x} \in \partial \Omega$:

$$\|\mathbf{x}-\mathbf{y}^*\| = m\|\mathbf{x}-\mathbf{y}\| \text{ for all } \mathbf{x} \in \partial \Omega. \tag{11.96}$$

Although it is not obvious, it is possible to find such a $\mathbf{y}^*$. We emphasize that both $\mathbf{y}^*$ and m depend on $\mathbf{y}$, although we suppress this dependence in the following calculations.

A consideration of symmetry makes it clear (or at least plausible) that $\mathbf{y}^*$ lies on the same radial line from the origin as does $\mathbf{y}$ (see Figure 11.23). The point y^* is called the *image* of y and this method of determining v_Ω is called the *method of images*.

Finding $\mathbf{y}^*$ now reduces to an exercise in geometry. We wish to find $m > 0$ and $\mathbf{y}^* \notin \Omega$ such that (11.96) holds. We begin by assuming that such m and $\mathbf{y}^*$ exist. We will use polar coordinates extensively and write (r_x, θ_x) for the polar coordinates of $\mathbf{x}$, and similarly for $\mathbf{y}$ and $\mathbf{y}^*$. We write $\phi = \theta_x - \theta_y$. Referring to Figure 11.23 and applying the law of cosines, we have

$$\begin{aligned} \|\mathbf{x}-\mathbf{y}\|^2 &= R^2 + r_y^2 - 2R r_y \cos(\phi), \\ \|\mathbf{x}-\mathbf{y}^*\|^2 &= R^2 + r_{y^*}^2 - 2R r_{y^*} \cos(\phi). \end{aligned}$$

Equation (11.96) must hold for all angles ϕ. Applying it for $\phi = 0$ and $\phi = \pi$ yields

$$\begin{aligned} R^2 + r_{y^*}^2 - 2R r_{y^*} &= m^2 \left(R^2 + r_y^2 - 2R r_y\right), \\ R^2 + r_{y^*}^2 + 2R r_{y^*} &= m^2 \left(R^2 + r_y^2 + 2R r_y\right). \end{aligned}$$

11.6. The Green's function for the Laplacian on a bounded domain

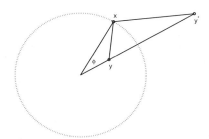

Figure 11.23. *The method of images for the disk.*

By subtracting the first equation from the second, we obtain

$$4Rr_{y^*} = 4m^2 Rr_y \implies r_{y^*} = m^2 r_y.$$

On the other hand, with $\phi = \pi/2$, (11.96) yields

$$R^2 + r_{y^*}^2 = m^2 \left(R^2 + r_y^2 \right).$$

We can solve these last two equations to find

$$m = \frac{R}{r_y}, \quad r_{y^*} = \frac{R^2}{r_y}.$$

Since $r_y = \|\mathbf{y}\|$, this last condition means that

$$\mathbf{y}^* = \frac{R^2}{\|\mathbf{y}\|} \frac{\mathbf{y}}{\|\mathbf{y}\|},$$

and the first implies that

$$\ln \left(\frac{\|\mathbf{x} - \mathbf{y}^*\|}{\|\mathbf{x} - \mathbf{y}\|} \right) = \ln \left(\frac{R}{\|\mathbf{y}\|} \right).$$

We therefore define

$$c(\mathbf{y}) = \frac{1}{2\pi} \ln \left(\frac{R}{\|\mathbf{y}\|} \right),$$

and obtain the Green's function

$$G_\Omega(\mathbf{x};\mathbf{y}) = \frac{1}{2\pi} \ln \left(\frac{\|\mathbf{x} - \mathbf{y}^*\|}{\|\mathbf{x} - \mathbf{y}\|} \right) - \frac{1}{2\pi} \ln \left(\frac{R}{\|\mathbf{y}\|} \right) = \frac{1}{2\pi} \ln \left(\frac{\|\mathbf{y}\| \|\mathbf{x} - \mathbf{y}^*\|}{R \|\mathbf{x} - \mathbf{y}\|} \right). \quad (11.97)$$

If we substitute the formula for $\mathbf{y}^*$, we can write G_Ω explicitly as

$$G_\Omega(\mathbf{x};\mathbf{y}) = \frac{1}{2\pi} \ln \left(\frac{\|\|\mathbf{y}\|^2 \mathbf{x} - R^2 \mathbf{y}\|}{R \|\mathbf{y}\| \|\mathbf{x} - \mathbf{y}\|} \right). \quad (11.98)$$

The reader will recall that we derived this formula by assuming that m and $\mathbf{y}^*$ exist. Exercise 11.6.6 asks the reader to verify that (11.98) has the desired properties.

11.6.3 Inhomogeneous boundary conditions

The Green's function G_Ω, as developed above, allows us to construct the solution of an inhomogeneous PDE subject to homogeneous boundary conditions. We now show that the same Green's function can be used to produce the solution to a homogeneous PDE with inhomogeneous boundary conditions. To demonstrate this, we first must derive Green's second identity. The first Green's identity states that

$$\int_\Omega v\Delta u = \int_{\partial\Omega} v\frac{\partial u}{\partial \mathbf{n}} - \int_\Omega \nabla u \cdot \nabla v. \tag{11.99}$$

If we interchange the roles of u and v in (11.99), we obtain

$$\int_\Omega u\Delta v = \int_{\partial\Omega} u\frac{\partial v}{\partial \mathbf{n}} - \int_\Omega \nabla v \cdot \nabla u.$$

Subtracting the second equation from the first yields

$$\int_\Omega (v\Delta u - u\Delta v) = \int_{\partial\Omega} \left(v\frac{\partial u}{\partial \mathbf{n}} - u\frac{\partial v}{\partial \mathbf{n}} \right). \tag{11.100}$$

This is *Green's second identity*. It applies to any $u,v \in C^2(\overline{\Omega})$ (provided that the boundary of Ω is sufficiently regular).

We will now construct the solution of

$$\begin{aligned} -\Delta u &= 0 \text{ in } \Omega, \\ u &= \phi(\mathbf{x}) \text{ on } \partial\Omega, \end{aligned} \tag{11.101}$$

where ϕ is a given function defined on $\partial\Omega$. To find a formula for u, we apply (11.100) with u equal to the solution of (11.101) and $v(\mathbf{x}) = G(\mathbf{x};\mathbf{y})$ for a fixed $\mathbf{y} \in \Omega$. Since v has a singularity at $\mathbf{x} = \mathbf{y}$, it is not obvious that Green's second identity is valid. We will proceed formally at first, and then outline a proof that our results are correct.

By construction, G_Ω satisfies

$$\begin{aligned} -\Delta G_\Omega(\mathbf{x};\mathbf{y}) &= \delta(\mathbf{x}-\mathbf{y}) \text{ for all } \mathbf{x} \in \Omega, \\ G_\Omega(\mathbf{x};\mathbf{y}) &= 0 \text{ for all } \mathbf{x} \in \partial\Omega, \end{aligned} \tag{11.102}$$

where $\mathbf{y}$ is any point in Ω. Using (11.101) and (11.102), Green's second identity,

$$\int_\Omega (G(\mathbf{x};\mathbf{y})\Delta u(\mathbf{x}) - u(\mathbf{x})\Delta G_\Omega)\,d\mathbf{x} = \int_{\partial\Omega} \left(G_\Omega(\mathbf{x};\mathbf{y})\frac{\partial u}{\partial n}(\mathbf{x}) - u(\mathbf{x})\frac{\partial G_\Omega}{\partial \mathbf{n_x}}(\mathbf{x};\mathbf{y}) \right) d\sigma_\mathbf{x},$$

becomes

$$\int_\Omega u(\mathbf{x})\delta(\mathbf{x}-\mathbf{y})\,d\mathbf{x} = -\int_{\partial\Omega} u(\mathbf{x})\frac{\partial G_\Omega}{\partial \mathbf{n_x}}(\mathbf{x};\mathbf{y})\,d\sigma_\mathbf{x},$$

which further reduces to

$$u(\mathbf{y}) = -\int_{\partial\Omega} \frac{\partial G_\Omega}{\partial \mathbf{n_x}}(\mathbf{x};\mathbf{y})\phi(\mathbf{x})\,d\sigma_\mathbf{x}.$$

11.6. The Green's function for the Laplacian on a bounded domain

Finally, since G_Ω is symmetric (that is, $G_\Omega(\mathbf{x};\mathbf{y}) = G_\Omega(\mathbf{y};\mathbf{x})$ for all $\mathbf{x},\mathbf{y} \in \Omega$—see Exercise 11.6.7), we can write the formula for the solution to (11.101) as

$$u(\mathbf{x}) = -\int_{\partial\Omega} \frac{\partial G_\Omega}{\partial \mathbf{n_y}}(\mathbf{x};\mathbf{y})\phi(\mathbf{y})\,d\sigma_\mathbf{y}. \tag{11.103}$$

This formula is valid for $\Omega \subset \mathbf{R}^2$ or $\Omega \subset \mathbf{R}^3$.

The preceding derivation relied on applying Green's second identity to G_Ω, even though this function has a singularity at $\mathbf{x} = \mathbf{y} \in \Omega$. We can prove that our derivation is correct by using a standard method for handling the singularity. We choose $\epsilon > 0$ sufficiently small that $B_\epsilon(\mathbf{y})$ is entirely contained within Ω, where $\mathbf{y} \in \Omega$ is fixed throughout this derivation, and we define

$$\Omega_\epsilon = \{\mathbf{x} \in \Omega : \|\mathbf{x} - \mathbf{y}\| > \epsilon\}. \tag{11.104}$$

Then $v(\mathbf{x}) = G_\Omega(\mathbf{x};\mathbf{y})$ and u (the solution of (11.101)) are both smooth on Ω_ϵ, and Green's second identity holds over Ω_ϵ. The functions u and v then satisfy the properties

$$\Delta u = 0 \text{ in } \Omega, \ \Delta_\mathbf{x} G_\Omega(\mathbf{x};\mathbf{y}) = 0 \text{ in } \Omega_\epsilon, \ G_\Omega(\mathbf{x};\mathbf{y}) = 0 \text{ for all } \mathbf{x} \in \partial\Omega,$$

and Green's second identity reduces to

$$0 = -\int_{\partial\Omega} u(\mathbf{x})\frac{\partial G_\Omega}{\partial \mathbf{n_x}}(\mathbf{x};\mathbf{y})\,d\sigma_\mathbf{x} + \int_{S_\epsilon(\mathbf{y})} \left\{ G_\Omega(\mathbf{x};\mathbf{y})\frac{\partial u}{\partial \mathbf{n}}(\mathbf{x})\,d\sigma_\mathbf{x} - \frac{\partial G_\Omega}{\partial \mathbf{n_x}}(\mathbf{x}) \right\} d\sigma_\mathbf{x}, \tag{11.105}$$

where $S_\epsilon(\mathbf{y}) = \partial B_\epsilon(\mathbf{y})$. By taking the limit as $\epsilon \to 0^+$, (11.103) then follows (see Exercise 11.6.9).

11.6.4 The Poisson integral formula

Since we know an explicit formula for the Green's function G_Ω for $\Omega = B_R(0)$, we can also express (11.103) explicitly in this case. The Green's function is

$$G_\Omega(\mathbf{x};\mathbf{y}) = \frac{1}{2\pi}\ln\left(\frac{\|\|\mathbf{y}\|^2\mathbf{x} - R^2\mathbf{y}\|}{R\|y\|\|\mathbf{x}-\mathbf{y}\|}\right).$$

A lengthy calculation shows that

$$\nabla_\mathbf{y} G_\Omega(\mathbf{x};\mathbf{y})\big|_{\|\mathbf{y}\|=R} = \frac{\|\mathbf{x}\|^2 - R^2}{2\pi R^2 \|\mathbf{x}-\mathbf{y}\|^2}\mathbf{y}. \tag{11.106}$$

On $\partial\Omega = S_R(0)$,

$$\mathbf{n} = \mathbf{n}(\mathbf{y}) = \frac{\mathbf{y}}{\|\mathbf{y}\|},$$

and hence

$$\frac{\partial G_\Omega}{\partial \mathbf{n_y}}(\mathbf{x};\mathbf{y}) = \nabla_\mathbf{y} G_\Omega(\mathbf{x};\mathbf{y})\big|_{\|\mathbf{y}\|=R} \cdot \mathbf{n} = \frac{\|\mathbf{x}\|^2 - R^2}{2\pi R\|\mathbf{x}-\mathbf{y}\|^2}.$$

Therefore, with $\Omega = B_R(0)$, the solution u of (11.101) is

$$u(\mathbf{x}) = \int_{S_R(0)} \frac{R^2 - \|\mathbf{x}\|^2}{2\pi R \|\mathbf{x} - \mathbf{y}\|^2} \phi(\mathbf{y}) \, d\sigma_{\mathbf{y}}. \tag{11.107}$$

With $\Omega = B_R(0)$, $\partial\Omega$ is the circle of radius R, which is naturally parametrized by the polar angle θ. It is therefore natural to specify the boundary data by a function $\phi(\theta)$. We can then express the solution u of (11.101) in terms of polar coordinates; that is, we can write $u = u(r, \theta)$ instead of $u = u(\mathbf{x})$. Exercise 11.6.11 asks the reader to show that (11.107) is equivalent to

$$u(r, \theta) = \frac{1}{2\pi R} \int_0^{2\pi} \frac{R^2 - r^2}{R^2 + r^2 - 2Rr\cos(\psi - \theta)} \phi(\psi) \, d\psi. \tag{11.108}$$

This is called the *Poisson integral formula*.

Exercises

1. Show that the weak form of (11.87) is (11.90).

2. Show that the weak form of (11.88) is (11.91).

3. Prove that u_2, as defined in (11.92), satisfies the variational equation (11.93). (Hint: Use the remark given in the text immediately following (11.93).)

4. Prove that u_1, as defined in (11.92), satisfies the variational equation (11.94). (Hint: Use the remarks given in the text immediately following (11.94).)

5. Use the free-space Green's function G for $-\Delta$ to derive the Green's function for the BVP

$$-\Delta u = f(\mathbf{x}) \text{ in } \Omega,$$
$$u = 0 \text{ on } \partial\Omega.$$

   (That is, repeat the derivation of $G_\Omega(\mathbf{x}; \mathbf{y})$ but with the Dirichlet condition in place of $\alpha_1 u + \alpha_2 \partial u/\partial \mathbf{n} = 0$.)

6. Consider the Green's function G_Ω for the Dirichlet problem on the disk, as given by (11.97) or (11.98). By construction, G_Ω satisfies $-\Delta G_\Omega(\mathbf{x}, \mathbf{y}) = \delta(\mathbf{x} - \mathbf{y})$ for each $\mathbf{y} \in \Omega$. Prove directly that G_Ω also satisfies $G_\Omega(\mathbf{x}, \mathbf{y}) = 0$ for all $\mathbf{x} \in \partial\Omega, \mathbf{y} \in \Omega$. (Hint: Use the law of cosines to express $\|\mathbf{x} - \mathbf{y}^*\|$ in terms of $\|\mathbf{x} - \mathbf{y}\|$.)

7. Let G_Ω be the Green's function for a given domain Ω. Prove that G_Ω is symmetric by applying Green's second identity to $u(\mathbf{x}) = G_\Omega(\mathbf{x}; \mathbf{y})$, $v(\mathbf{x}) = G_\Omega(\mathbf{x}; \mathbf{z})$.

8. Let $\Omega \subset \mathbf{R}^2$ be the disk of radius $R > 0$ centered at the origin. According to the previous exercise, the Green's function

$$G_\Omega(\mathbf{x}; \mathbf{y}) = \frac{1}{2\pi} \ln\left(\frac{\|\|\mathbf{y}\|^2 \mathbf{x} - R^2 \mathbf{y}\|}{R \|\mathbf{y}\| \|\mathbf{x} - \mathbf{y}\|}\right)$$

   should satisfy $G_\Omega(\mathbf{y}; \mathbf{x}) = G_\Omega(\mathbf{x}; \mathbf{y})$ for all $\mathbf{x}, \mathbf{y} \in \Omega$. Prove this directly.

11.6. The Green's function for the Laplacian on a bounded domain

9. Let Ω_ϵ be defined by (11.104), where $\epsilon > 0$ is small enough that $B_\epsilon(\mathbf{y}) \subset \Omega$. Let G_Ω be the Green's function for the Laplacian on Ω under homogeneous Dirichlet boundary conditions, and let u be the solution of (11.101).

   (a) Prove that Green's second identity, applied to u and $v(\mathbf{x}) = G_\Omega(\mathbf{x};\mathbf{y})$, simplifies to (11.105).

   (b) Prove (11.103) by taking the limit of (11.105) as $\epsilon \to 0^+$.

10. Derive (11.106) from (11.98).

11. Prove that (11.107) is equivalent to (11.108).

12. Let Ω be a domain in $\mathbf{R}^2$ or $\mathbf{R}^3$, and let G_Ω be the Green's function for the Laplacian under Robin boundary conditions. That is, assume that the solution to

    $$-\Delta u = f(\mathbf{x}) \text{ in } \Omega,$$
    $$\alpha_1 u + \alpha_2 \frac{\partial u}{\partial \mathbf{n}} = 0 \text{ on } \partial\Omega$$

    is

    $$u(\mathbf{x}) = \int_\Omega G_\Omega(\mathbf{x};\mathbf{y}) f(\mathbf{y}) d\mathbf{y}.$$

    For the purposes of this exercise, assume that α_1 and α_2 are both nonzero. Consider the BVP

    $$-\Delta u = 0 \text{ in } \Omega,$$
    $$\alpha_1 u + \alpha_2 \frac{\partial u}{\partial \mathbf{n}} = \phi(\mathbf{x}) \text{ on } \partial\Omega.$$

    Derive two equivalent formulas for the solution u, one involving G_Ω and the other involving $\partial G_\Omega/\partial \mathbf{n}$ (both involving an integral over $\partial\Omega$).

13. Show that the solution (11.103) to the Dirichlet problem (11.101) is analogous to the solution (9.24) of the BVP (9.23) in Section 9.2.2. More specifically, show that (9.24) is equivalent to (11.103) if we identify Ω as (a,b).

14. We have derived G_Ω for $\Omega = B_R(0)$, the disk of radius R centered at the origin in $\mathbf{R}^2$. Find the corresponding Green's function for $\Omega = B_R(\mathbf{z})$, where $\mathbf{z}$ is a fixed point in $\mathbf{R}^2$. (Hint: Perform a simple change of variables to transform the BVP

    $$-\Delta u = f(\mathbf{x}), \ \mathbf{x} \in B_R(\mathbf{z}),$$
    $$u(\mathbf{x}) = 0, \ \mathbf{x} \in \partial B_R(\mathbf{z}),$$

    into a BVP on $B_R(0)$.)

15. A function satisfying $\Delta u = 0$ is often called a *harmonic* function. Use the previous exercise and the Poisson integral formula to prove the mean value property of harmonic functions:

    $$\Delta u = 0 \text{ in } \Omega \implies u(\mathbf{y}) = \frac{1}{2\pi R} \int_{B_R(\mathbf{y})} u(\mathbf{x}) d\sigma_\mathbf{x} \text{ provided } B_R(\mathbf{y}) \subset \Omega.$$

    This equation states that, if u is harmonic, then $u(\mathbf{y})$ equals the average value of u on the circle of radius R centered at $\mathbf{y}$.

16. Use the mean value property of harmonic functions (see the previous exercise) to prove the *maximum principle*: If u is harmonic on Ω and continuous on $\overline{\Omega}$, then the maximum and minimum values of u on $\overline{\Omega}$ occur on $\partial \Omega$. Moreover, u cannot have a local maximum or local minimum in Ω unless u is constant on Ω.

11.7 Green's function for the wave equation

11.7.1 The free-space Green's function

We now consider the following IVP for the wave equation in three-dimensional space:

$$\frac{\partial^2 u}{\partial t^2} - c^2 \Delta u = f(\mathbf{x},t), \ \mathbf{x} \in \mathbf{R}^3, \ t > 0,$$
$$u(\mathbf{x},0) = 0, \ \mathbf{x} \in \mathbf{R}^3, \tag{11.109}$$
$$\frac{\partial u}{\partial t}(\mathbf{x},0) = 0, \ \mathbf{x} \in \mathbf{R}^3.$$

We wish to find the Green's function. We will apply Duhamel's principle (see Sections 4.3.3, 9.4, and 9.6) and start with the IVP

$$\frac{\partial^2 u}{\partial t^2} - c^2 \Delta u = 0, \ \mathbf{x} \in \mathbf{R}^3, \ t > 0,$$
$$u(\mathbf{x},0) = 0, \ \mathbf{x} \in \mathbf{R}^3, \tag{11.110}$$
$$\frac{\partial u}{\partial t}(\mathbf{x},0) = \psi(\mathbf{x}), \ \mathbf{x} \in \mathbf{R}^3.$$

The solution to (11.110) is

$$u(\mathbf{x},t) = \frac{t}{4\pi (ct)^2} \int_{\partial B_{ct}(\mathbf{x})} \psi(\mathbf{z}) d\sigma_{\mathbf{z}}. \tag{11.111}$$

The reader will recall that a signal governed by the wave equation travels with a velocity of c. The expression

$$\frac{1}{4\pi (ct)^2} \int_{\partial B_{ct}(\mathbf{x})} \psi(\mathbf{z}) d\sigma_{\mathbf{z}}$$

is the average value of the function ψ over the set of all points at distance ct from $\mathbf{x}$.

We will verify directly that u satisfies (11.110). Since it is difficult to differentiate (11.111) directly, we first perform a change of variables in the integral: $\mathbf{z} = \mathbf{x} + ct\mathbf{y}$ transforms the sphere $\partial B_{ct}(\mathbf{x})$ into the sphere of radius 1 centered at the origin, $\partial B_1(0)$. We have $d\sigma_{\mathbf{z}} = (ct)^2 d\sigma_{\mathbf{y}}$, and therefore u can be written as

$$u(\mathbf{x},t) = \frac{t}{4\pi} \int_{\partial B_1(0)} \psi(\mathbf{x}+ct\mathbf{y}) d\sigma_{\mathbf{y}}. \tag{11.112}$$

By the chain rule,

$$\frac{\partial}{\partial t}(\psi(\mathbf{x}+ct\mathbf{y})) = c\nabla \psi(\mathbf{x}+ct\mathbf{y}) \cdot \mathbf{y},$$
$$\frac{\partial^2}{\partial t^2}(\psi(\mathbf{x}+ct\mathbf{y})) = c^2 \mathbf{y} \cdot \nabla^2 \psi(\mathbf{x}+ct\mathbf{y})\mathbf{y}. \tag{11.113}$$

11.7. Green's function for the wave equation

Here $\nabla^2 \psi$ is the matrix of second partial derivatives of ψ:

$$\nabla^2 \psi(\mathbf{x}) = \begin{bmatrix} \frac{\partial^2 \psi}{\partial x_1^2}(\mathbf{x}) & \frac{\partial^2 \psi}{\partial x_2 \partial x_1}(\mathbf{x}) & \frac{\partial^2 \psi}{\partial x_3 \partial x_1}(\mathbf{x}) \\ \frac{\partial^2 \psi}{\partial x_1 \partial x_2}(\mathbf{x}) & \frac{\partial^2 \psi}{\partial x_2^2}(\mathbf{x}) & \frac{\partial^2 \psi}{\partial x_3 \partial x_2}(\mathbf{x}) \\ \frac{\partial^2 \psi}{\partial x_1 \partial x_3}(\mathbf{x}) & \frac{\partial^2 \psi}{\partial x_2 \partial x_3}(\mathbf{x}) & \frac{\partial^2 \psi}{\partial x_3^2}(\mathbf{x}) \end{bmatrix}.$$

Using (11.113), and recalling that we can differentiate under an integral, we obtain

$$\frac{\partial}{\partial t}\left(\int_{\partial B_1(0)} \psi(\mathbf{x}+ct\mathbf{y})\,d\sigma_{\mathbf{y}}\right) = c\int_{\partial B_1(0)} \nabla \psi(\mathbf{x}+ct\mathbf{y}) \cdot \mathbf{y}\,d\sigma_{\mathbf{y}},$$

$$\frac{\partial^2}{\partial t^2}\left(\int_{\partial B_1(0)} \psi(\mathbf{x}+ct\mathbf{y})\,d\sigma_{\mathbf{y}}\right) = c^2 \int_{\partial B_1(0)} \mathbf{y} \cdot \nabla^2 \psi(\mathbf{x}+ct\mathbf{y})\mathbf{y}\,d\sigma_{\mathbf{y}}, \quad (11.114)$$

$$\Delta\left(\int_{\partial B_1(0)} \psi(\mathbf{x}+ct\mathbf{y})\,d\sigma_{\mathbf{y}}\right) = \int_{\partial B_1(0)} \Delta \psi(\mathbf{x}+ct\mathbf{y})\,d\sigma_{\mathbf{y}}.$$

From these calculations, we obtain

$$\frac{\partial^2 u}{\partial t^2}(\mathbf{x},t) = \frac{c}{2\pi}\int_{\partial B_1(0)} \nabla \psi(\mathbf{x}+ct\mathbf{y}) \cdot \mathbf{y}\,d\sigma_{\mathbf{y}}$$
$$+ \frac{c^2 t}{4\pi}\int_{\partial B_1(0)} \mathbf{y} \cdot \nabla^2 \psi(\mathbf{x}+ct\mathbf{y})\mathbf{y}\,d\sigma_{\mathbf{y}}, \quad (11.115)$$

$$c^2 \Delta u(\mathbf{x},t) = \frac{c^2 t}{4\pi}\int_{\partial B_1(0)} \Delta \psi(\mathbf{x}+ct\mathbf{y})\,d\sigma_{\mathbf{y}}.$$

It is not obvious from these formulas that

$$\frac{\partial^2 u}{\partial t^2}(\mathbf{x},t) = c^2 \Delta u(\mathbf{x},t).$$

However, we can verify that this holds by using the divergence theorem to convert each of the surface integrals over $\partial B_1(0)$ to a volume integral over $B_1(0)$. This is possible because the unit normal vector to $\partial B_1(0)$ at $\mathbf{y} \in \partial B_1(0)$ is $\mathbf{y}$ itself. We obtain

$$\int_{\partial B_1(0)} \nabla \psi(\mathbf{x}+ct\mathbf{y}) \cdot \mathbf{y}\,d\sigma_{\mathbf{y}} = \int_{B_1(0)} \Delta \psi(\mathbf{x}+ct\mathbf{y})\,d\mathbf{y},$$

$$\int_{\partial B_1(0)} \mathbf{y} \cdot \nabla^2 \psi(\mathbf{x}+ct\mathbf{y})\mathbf{y}\,d\sigma_{\mathbf{y}} = \int_{B_1(0)} \nabla_{\mathbf{y}} \cdot \left(\nabla^2 \psi(\mathbf{x}+ct\mathbf{y})\mathbf{y}\right)\,d\mathbf{y},$$

$$\int_{\partial B_1(0)} \Delta \psi(\mathbf{x}+ct\mathbf{y})\,d\sigma_{\mathbf{y}} = \int_{\partial B_1(0)} \Delta \psi(\mathbf{x}+ct\mathbf{y})\mathbf{y} \cdot \mathbf{y}\,d\sigma_{\mathbf{y}} \quad (11.116)$$

$$= \int_{B_1(0)} \nabla_{\mathbf{y}} \cdot (\Delta \psi(\mathbf{x}+ct\mathbf{y})\mathbf{y})\,d\mathbf{y}.$$

Exercise 11.7.2 asks the reader to use (11.116) to prove that u satisfies the wave equation. A key part of the calculation is the following:

$$\nabla_{\mathbf{y}} \cdot \left(\nabla^2 \psi(\mathbf{x}+ct\mathbf{y})\mathbf{y} \right) = \sum_{i=1}^{3} \frac{\partial}{\partial y_i} \left(\sum_{j=1}^{3} \frac{\partial^2 \psi}{\partial x_j \partial x_i}(\mathbf{x}+ct\mathbf{y})y_j \right)$$

$$= \sum_{i=1}^{3} \left(\frac{\partial^2 \psi}{\partial x_i^2}(\mathbf{x}+ct\mathbf{y}) + \sum_{j=1}^{3} ct \frac{\partial^3 \psi}{\partial x_i \partial x_j \partial x_i}(\mathbf{x}+ct\mathbf{y}) \right)$$

$$= \Delta \psi(\mathbf{x}+ct\mathbf{y}) + ct \sum_{j=1}^{3} y_j \sum_{i=1}^{3} \frac{\partial^3 \psi}{\partial x_i^2 \partial x_j}(\mathbf{x}+ct\mathbf{y})$$

$$= \Delta \psi(\mathbf{x}+ct\mathbf{y}) + ct \sum_{j=1}^{3} y_j \Delta \left(\frac{\partial \psi}{\partial x_j} \right)(\mathbf{x}+ct\mathbf{y}).$$

Similar expressions arise when $\nabla_{\mathbf{y}} \cdot (\Delta \psi(\mathbf{x}+ct\mathbf{y})\mathbf{y})$ is simplified.

Since we have a formula for the solution of (11.110), we can now apply Duhamel's principle. The solution to (11.110), with $\psi(\mathbf{x})$ replaced by $f(\mathbf{x},s)$, is

$$v(\mathbf{x},t;s) = \frac{1}{4\pi c^2 t} \int_{\partial B_{ct}(\mathbf{x})} f(\mathbf{z},s) d\sigma_{\mathbf{z}}.$$

By Duhamel's principle, the solution of (11.109) is

$$u(\mathbf{x},t) = \int_0^t \int_{\partial B_{c(t-s)}(\mathbf{x})} \frac{1}{4\pi c^2(t-s)} f(\mathbf{z},s) d\sigma_{\mathbf{z}} ds. \tag{11.117}$$

We can write this in the following equivalent form (by applying Duhamel's principle to the alternate form (11.112) of the solution of (11.110)):

$$u(\mathbf{x},t) = \int_0^t \int_{\partial B_1(0)} \frac{t-s}{4\pi} f(\mathbf{x}+c(t-s)\mathbf{y},s) d\sigma_{\mathbf{y}} ds. \tag{11.118}$$

From formula (11.118) for the solution of the wave equation, we can identify the Green's function for the wave equation in three dimensions; however, we need some new notation to do so. The reader is familiar with the use of the Dirac delta function to represent the evaluation of a function at a point:

$$\int_{\mathbf{R}^3} F(\mathbf{x})\delta(\mathbf{x}-\mathbf{y}) d\mathbf{x} = F(\mathbf{y}) \text{ for all smooth } F.$$

The delta function δ is concentrated at the origin, and therefore $\delta(\mathbf{x}-\mathbf{y})$ is concentrated at $\mathbf{x}=\mathbf{y}$. If the equation $\phi(\mathbf{x}) = 0$ defines a surface S in $\mathbf{R}^3$, then we define, for any smooth function F,

$$\int_{\mathbf{R}^3} F(\mathbf{x})\delta(\phi(\mathbf{x})) d\mathbf{x} = \int_S F(\mathbf{x}) d\sigma.$$

11.7. Green's function for the wave equation

Thus integration against $\delta(\phi(\mathbf{x}))$ has the effect of picking out the (total) value of F on the surface S defined by $\phi(\mathbf{x}) = 0$.[66]

Using this notation, we can write (11.118) as

$$u(\mathbf{x},t) = \int_0^\infty \int_{\mathbf{R}^3} \frac{1}{4\pi c^2(t-s)} \delta(c(t-s) - \|\mathbf{x}-\mathbf{y}\|) f(\mathbf{y},s) \, d\mathbf{y} \, ds.$$

It follows that the Green's function for the wave equation in three dimensions is

$$G(\mathbf{x},t;\mathbf{y},s) = \frac{1}{4\pi c^2(t-s)} \delta(c(t-s) - \|\mathbf{x}-\mathbf{y}\|). \tag{11.119}$$

Formally, $G(\mathbf{x},t;\mathbf{y},s) = 0$ for $s > t$ (since then the argument to the delta function is strictly negative), and therefore G satisfies the causality property. By inspection, G also satisfies the translation and reciprocity properties.

11.7.2 The wave equation in two-dimensional space

We can derive the solution of the wave equation in $\mathbf{R}^2$ from the solution just derived for $\mathbf{R}^3$. The key observation is that if $f : \mathbf{R}^3 \times [0,\infty) \to \mathbf{R}$ is independent of x_3, then the solution u of (11.109) is also independent of x_3. This follows immediately from (11.118), which shows that if $\partial f/\partial x_3 = 0$, then also $\partial u/\partial x_3 = 0$. It then follows immediately that the solution of (11.109), if f is independent of x_3, solves

$$\frac{\partial^2 u}{\partial t^2} - c^2 \Delta u = f(\mathbf{x},t), \ \mathbf{x} \in \mathbf{R}^2, \ t > 0,$$
$$u(\mathbf{x},0) = 0, \ \mathbf{x} \in \mathbf{R}^2, \tag{11.120}$$
$$\frac{\partial u}{\partial t}(\mathbf{x},0) = 0, \ \mathbf{x} \in \mathbf{R}^2.$$

(Here we are abusing notation slightly, by using the same symbol f both for the function $f : \mathbf{R}^3 \times [0,\infty) \to \mathbf{R}$ that is assumed independent of x_3, and for the equivalent function $f : \mathbf{R}^2 \times [0,\infty) \to \mathbf{R}$. Similarly, we are writing u for the solution of (11.109) and the equivalent function that solves (11.120).)

It now remains to reduce formula (11.118) to an expression involving only x_1 and x_2 when f is independent of x_3; that is, we wish to write (11.118) in terms of an equivalent integral over a set in $\mathbf{R}^2$. This can be done by the standard method for writing a surface integral in $\mathbf{R}^3$ as an area integral in $\mathbf{R}^2$. In general, if a surface S in $\mathbf{R}^3$ is defined by the equation $x_3 = z(x_1,z_2)$ for $(x_1,x_2) \in D$, where D is a domain in $\mathbf{R}^2$, then

$$\int_S \psi(\mathbf{x}) \, d\sigma = \iint_D \psi(x_1,x_2,z(x_3)) \sqrt{1 + \left(\frac{\partial z}{\partial x_1}\right)^2 + \left(\frac{\partial z}{\partial x_2}\right)^2} \, dx_1 \, dx_2.$$

In order to apply this to (11.118), we write S for the upper hemisphere of $\partial B_1(0)$, and notice that

$$\int_{\partial B_1(0)} f(\mathbf{x} + c(t-s)\mathbf{y},s) \, d\sigma_\mathbf{y} = 2 \int_S f(\mathbf{x} + c(t-s)\mathbf{y},s) \, d\sigma_\mathbf{y}.$$

---

[66]Similarly, if $\phi(\mathbf{x}) = 0$ defines a curve in $\mathbf{R}^2$, then we define $\int_{\mathbf{R}^2} F(\mathbf{x}) \delta(\phi(\mathbf{x})) \, d\mathbf{x} = \int_S F(\mathbf{x}) \, d\sigma$ for all smooth F defined on $\mathbf{R}^2$.

The hemisphere S is defined by the equation

$$z = \sqrt{1 - x_1^2 - x_2^2}, \quad -\sqrt{1-x_1^2} \leq x_2 \leq \sqrt{1-x_1^2}, \quad -1 \leq x_1 \leq 1,$$

and we have

$$\sqrt{1 + \left(\frac{\partial z}{\partial x_1}\right)^2 + \left(\frac{\partial z}{\partial x_2}\right)^2} = \frac{1}{\sqrt{1-x_1^2-x_2^2}}.$$

Therefore,

$$\int_{\partial B_1(0)} f(\mathbf{x}+c(t-s)\mathbf{y},s)\,d\sigma_\mathbf{y}$$

$$= 2\int_{-1}^{1}\int_{-\sqrt{1-x^2}}^{\sqrt{1-x^2}} \frac{f(x_1+c(t-s)y_1, x_2+c(t-s)y_2, s)}{\sqrt{1-y_1^2-y_2^2}}\,dy_2\,dy_1.$$

We can write this result more simply as

$$2\int_{B_1(0)} \frac{f(\mathbf{x}+c(t-s)\mathbf{y},s)}{\sqrt{1-\|\mathbf{y}\|^2}}\,d\mathbf{y},$$

with the proviso that in this integral, $\mathbf{x}$ and $\mathbf{y}$ are vectors in $\mathbf{R}^2$, and $B_1(0)$ is the unit disk in $\mathbf{R}^2$. We therefore obtain the solution of (11.120):

$$u(\mathbf{x},t) = \int_0^t \int_{B_1(0)} \frac{t-s}{2\pi}\, \frac{f(\mathbf{x}+c(t-s)\mathbf{y},s)}{\sqrt{1-\|\mathbf{y}\|^2}}\,d\mathbf{y}\,ds. \tag{11.121}$$

If we perform the change of variables $\mathbf{z} = \mathbf{x}+c(t-s)\mathbf{y}$, we obtain the equivalent formula

$$u(\mathbf{x},t) = \int_0^t \int_{B_{c(t-s)}(\mathbf{x})} \frac{f(\mathbf{z},s)}{2\pi c\sqrt{c^2(t-s)^2 - \|\mathbf{z}-\mathbf{x}\|^2}}\,d\mathbf{z}\,ds. \tag{11.122}$$

From formula (11.122), it is easy to identify the Green's function for the wave equation in two dimensions. It is

$$G(\mathbf{x},t;\mathbf{y},s) = \begin{cases} \frac{1}{2\pi c\sqrt{c^2(t-s)^2 - \|\mathbf{y}-\mathbf{x}\|^2}}, & \|\mathbf{x}-\mathbf{y}\| < c(t-s), \\ 0 & \text{otherwise.} \end{cases} \tag{11.123}$$

As in the case of three dimensions, the Green's function satisfies the properties of causality, translation, and reciprocity.

Exercise 11.7.3 asks the reader to derive two equivalent formulas for the solution of the IVP

$$\frac{\partial^2 u}{\partial t^2} - c^2 \Delta u = 0, \; \mathbf{x} \in \mathbf{R}^2, \; t > 0,$$

$$u(\mathbf{x},0) = 0, \; \mathbf{x} \in \mathbf{R}^2, \tag{11.124}$$

$$\frac{\partial u}{\partial t}(\mathbf{x},0) = \psi(\mathbf{x}), \; \mathbf{x} \in \mathbf{R}^2.$$

11.7.3 Huygen's principle

Formulas (11.117) and (11.122) show that there is an essential difference in how waves propagate in three-dimensional space as opposed to two-dimensional space. We already know that (in any dimension) no signal can propagate with a velocity greater than c. In $\mathbf{R}^3$, the value of the forcing function at the point $\mathbf{y}$ and time $t = s$ affects the signal $u(\mathbf{x},t)$ if and only if the distance from $\mathbf{y}$ to $\mathbf{x}$ is exactly $c(t-s)$. This means that an impulsive source (such as a handclap) propagates as a sharp signal with velocity c; there is no trailing wave. On the other hand, in $\mathbf{R}^2$, the values of $f(\mathbf{y},s)$ on the entire disk $B_{ct}(\mathbf{x})$ influence the signal $u(\mathbf{x},t)$. Therefore, an impulsive source, originating at point $\mathbf{y}$ and time s, affects the value of $u(\mathbf{x},t)$ not only on the wavefront $B_{c(t-s)}(\mathbf{y})$, but also behind this wavefront. Thus signals that propagate in accordance with the two-dimensional wave equation are not sharp. (This can be seen by dropping a pebble into a still pond and watching how the waves propagate.)

The fact that signals propagate sharply in three dimensions is called *Huygens's principle*. Mathematically, Huygens's principle is valid in all odd dimensions greater than one, but not in even dimensions. For explicit solutions of the wave equation in dimensions greater than three, the reader can consult Chapter 5 of Folland [19].

11.7.4 The Green's function for the wave equation on a bounded domain

In principle, we can find the Green's function for the wave equation on a bounded domain Ω via an eigenfunction expansion, just as in Section 9.6.2. This, of course, assumes that we can find the eigenvalues and eigenfunctions of the Laplacian on the domain Ω of interest. As in the one-dimensional case, we would not expect the series to converge rapidly.

Without finding an explicit formula for the Green's function on Ω, we wish to show that the important reciprocity property is necessarily satisfied. For a time-dependent PDE, this takes the form $G(\mathbf{x},t;\mathbf{y},s) = G(\mathbf{y},t;\mathbf{x},s)$ for all $\mathbf{x},\mathbf{y} \in \Omega$, and it means that the response at $\mathbf{x}$ to a source at $\mathbf{y}$ is the same as the response at $\mathbf{y}$ to a source at $\mathbf{x}$. The following argument is valid for $\Omega \subset \mathbf{R}^2$ and $\Omega \subset \mathbf{R}^3$.

We would like to use symmetry; unfortunately, the wave operator is not quite symmetric. To be specific, if u and v satisfy homogeneous Dirichlet conditions on Ω, then

$$\int_0^T \int_\Omega u(\mathbf{x},t) \left(\frac{\partial^2 v}{\partial t^2}(\mathbf{x},t) - c^2 \Delta v(\mathbf{x},t) \right) d\mathbf{x}\, dt$$

$$= \int_\Omega \left(u(\mathbf{x},t) \frac{\partial v}{\partial t}(\mathbf{x},t) \right) \bigg|_0^T d\mathbf{x} - \int_0^T \int_\Omega \frac{\partial u}{\partial t}(\mathbf{x},t) \frac{\partial v}{\partial t}(\mathbf{x},t) d\mathbf{x}\, dt$$

$$- c^2 \int_0^T \int_{\partial\Omega} u(\mathbf{x},t) \frac{\partial v}{\partial \mathbf{n}}(\mathbf{x},t) d\sigma\, dt + c^2 \int_0^T \int_\Omega \nabla u(\mathbf{x},t) \cdot \nabla v(\mathbf{x},t) d\mathbf{x}\, dt$$

$$= \int_\Omega \left(u(\mathbf{x},t) \frac{\partial v}{\partial t}(\mathbf{x},t) \right) \bigg|_0^T d\mathbf{x} - \int_0^T \int_\Omega \frac{\partial u}{\partial t}(\mathbf{x},t) \frac{\partial v}{\partial t}(\mathbf{x},t) d\mathbf{x}\, dt$$

$$+ c^2 \int_0^T \int_\Omega \nabla u(\mathbf{x},t) \cdot \nabla v(\mathbf{x},t) d\mathbf{x}\, dt$$

The reader should notice that the spatial boundary integral vanishes because u is assumed to satisfy homogeneous Dirichlet conditions, but the temporal boundary integral does not vanish. We similarly have

$$\int_0^T \int_\Omega \left(\frac{\partial^2 u}{\partial t^2}(\mathbf{x},t) - c^2 \Delta u(\mathbf{x},t)\right) v(\mathbf{x},t) d\mathbf{x} dt$$
$$= \int_\Omega \left(\frac{\partial u}{\partial t}(\mathbf{x},t) v(\mathbf{x},t)\right)\bigg|_0^T d\mathbf{x} - \int_0^T \int_\Omega \frac{\partial u}{\partial t}(\mathbf{x},t) \frac{\partial v}{\partial t}(\mathbf{x},t) d\mathbf{x} dt$$
$$+ c^2 \int_0^T \int_\Omega \nabla u(\mathbf{x},t) \cdot \nabla v(\mathbf{x},t) d\mathbf{x} dt.$$

We notice that if

$$u(\mathbf{x},0) = \frac{\partial u}{\partial t}(\mathbf{x},0) = v(\mathbf{x},T) = \frac{\partial v}{\partial t}(\mathbf{x},T) = 0,$$

then

$$\int_\Omega \left(u(\mathbf{x},t) \frac{\partial v}{\partial t}(\mathbf{x},t)\right)\bigg|_0^T d\mathbf{x} = \int_\Omega \left(\frac{\partial u}{\partial t}(\mathbf{x},t) u(\mathbf{x},t)\right)\bigg|_0^T d\mathbf{x} = 0,$$

and hence

$$\int_0^T \int_\Omega u(\mathbf{x},t) \left(\frac{\partial^2 v}{\partial t^2}(\mathbf{x},t) - c^2 \Delta v(\mathbf{x},t)\right) d\mathbf{x} dt$$
$$= \int_0^T \int_\Omega \left(\frac{\partial^2 u}{\partial t^2}(\mathbf{x},t) - c^2 \Delta u(\mathbf{x},t)\right) v(\mathbf{x},t) d\mathbf{x} dt \qquad (11.125)$$

holds. We will now show how this can be arranged.

We first remind the reader that it is possible to solve the wave equation backwards in time, that is, to solve a final value problem such as

$$\begin{aligned}
\frac{\partial^2 v}{\partial t^2} - c^2 \Delta v &= f(\mathbf{x},t), \ \mathbf{x} \in \Omega, \ 0 < t < T, \\
v(\mathbf{x},T) &= 0, \ \mathbf{x} \in \Omega, \\
\frac{\partial v}{\partial t}(\mathbf{x},T) &= 0, \ \mathbf{x} \in \Omega, \\
v(\mathbf{x},t) &= 0, \ \mathbf{x} \in \partial\Omega, \ 0 < t < T.
\end{aligned} \qquad (11.126)$$

The solution is $v(\mathbf{x},t) = u(\mathbf{x}, T-t)$, where v solves

$$\begin{aligned}
\frac{\partial^2 u}{\partial t^2} - c^2 \Delta u &= f(\mathbf{x}, T-t), \ \mathbf{x} \in \Omega, \ 0 < t < T, \\
u(\mathbf{x},0) &= 0, \ \mathbf{x} \in \Omega, \\
\frac{\partial u}{\partial t}(\mathbf{x},0) &= 0, \ \mathbf{x} \in \Omega, \\
u(\mathbf{x},t) &= 0, \ \mathbf{x} \in \partial\Omega, \ 0 < t < T.
\end{aligned}$$

11.7. Green's function for the wave equation

Assuming G is the Green's function for the wave equation on Ω (subject to Dirichlet conditions), we have

$$u(\mathbf{x},t) = \int_0^T \int_\Omega G(\mathbf{x},t;\mathbf{y},s)f(\mathbf{y},T-s)\,d\mathbf{y}\,ds.$$

(For convenience, we have written the upper limit of the time integration as T rather than ∞; this is equivalent since $G(\mathbf{x},t;\mathbf{y},s) = 0$ for $s > t$.) It follows that the solution of (11.126) is

$$v(\mathbf{x},t) = \int_0^T \int_\Omega G(\mathbf{x},T-t;\mathbf{y},s)f(\mathbf{y},T-s)\,d\mathbf{y}\,ds.$$
$$= \int_0^T \int_\Omega G(\mathbf{x},T-t;\mathbf{y},T-s)f(\mathbf{y},s)\,d\mathbf{y}\,ds$$

(to get the second expression, we changed variables, replacing s by $T-s$ in the time integration).

Now, by the translation property, which holds because the wave equation has constant coefficients (see Exercise 11.7.4), we have

$$G(\mathbf{x},T-t;\mathbf{y},T-s) = G(\mathbf{x},T-t-(T-s);\mathbf{y},0) = G(\mathbf{x},s-t;\mathbf{y},0) = G(\mathbf{x},s;\mathbf{y},t).$$

Therefore, we can write the solution of (11.126) as

$$v(\mathbf{x},t) = \int_0^T \int_\Omega G(\mathbf{x},s;\mathbf{y},t)f(\mathbf{y},s)\,d\mathbf{y}\,ds. \tag{11.127}$$

This suggests that $G(\mathbf{x},s;\mathbf{y},t)$, for fixed $\mathbf{y}, s$, solves

$$\begin{aligned}\frac{\partial^2 u}{\partial t^2} - c^2 \Delta u &= \delta(\mathbf{x}-\mathbf{y})\delta(t-s),\ \mathbf{x} \in \Omega,\ 0 < t < T, \\ u(\mathbf{x},T) &= 0,\ \mathbf{x} \in \Omega, \\ \frac{\partial u}{\partial t}(\mathbf{x},T) &= 0,\ \mathbf{x} \in \Omega, \\ u(\mathbf{x},t) &= 0,\ \mathbf{x} \in \partial\Omega,\ 0 < t < T.\end{aligned} \tag{11.128}$$

We now consider

$$u(\mathbf{x},t) = \int_0^T \int_\Omega G(\mathbf{x},s;\mathbf{y},t)f(\mathbf{y},s)\,d\mathbf{y}\,ds,$$
$$v(\mathbf{x},t) = \int_0^T \int_\Omega G(\mathbf{x},t;\mathbf{y},s)g(\mathbf{y},s)\,d\mathbf{y}\,ds,$$

where f and g are any continuous functions. The function u solves an initial value problem with homogeneous initial conditions, while v solves a final value problem with homogeneous final conditions. It follows that u and v satisfy (11.125). Now suppose we take $\mathbf{y}_1, \mathbf{y}_s \in \Omega$ and $s_1, s_2 \in (0,T)$, and define

$$f(\mathbf{x},t) = \delta(\mathbf{x}-\mathbf{y}_1)\delta(t-s_1),\ g(\mathbf{x},t) = \delta(\mathbf{x}-\mathbf{y}_2)\delta(t-s_2).$$

Then $u(\mathbf{x},t) = G(\mathbf{x},t;\mathbf{y}_1,s_1)$ and $v(\mathbf{x},t) = G(\mathbf{x},s_2;\mathbf{y}_2,t)$. Moreover,

$$\frac{\partial^2 u}{\partial t^2}(\mathbf{x},t) - c^2 \Delta u(\mathbf{x},t) = \delta(\mathbf{x}-\mathbf{y}_1)\delta(t-s_1),$$

$$\frac{\partial^2 v}{\partial t^2}(\mathbf{x},t) - c^2 \Delta v(\mathbf{x},t) = \delta(\mathbf{x}-\mathbf{y}_2)\delta(t-s_2),$$

and hence

$$\int_0^T \int_\Omega u(\mathbf{x},t)\left(\frac{\partial^2 v}{\partial t^2}(\mathbf{x},t) - c^2 \Delta v(\mathbf{x},t)\right) d\mathbf{x}\, dt = u(\mathbf{y}_2,s_2) = G(\mathbf{y}_2,s_2;\mathbf{y}_1,s_1),$$

$$\int_0^T \int_\Omega \left(\frac{\partial^2 u}{\partial t^2}(\mathbf{x},t) - c^2 \Delta u(\mathbf{x},t)\right) v(\mathbf{x},t)\, d\mathbf{x}\, dt = v(\mathbf{y}_1,s_1) = G(\mathbf{y}_1,s_2;\mathbf{y}_2,s_1).$$

Therefore, (11.125) implies that $G(\mathbf{y}_2,s_2;\mathbf{y}_1,s_1) = G(\mathbf{y}_1,s_2;\mathbf{y}_2,s_1)$, or, changing the names of the variables,

$$G(\mathbf{x},t;\mathbf{y},s) = G(\mathbf{y},t;\mathbf{x},s) \text{ for all } \mathbf{x},\mathbf{y} \in \Omega,\ t,s \in (0,T).$$

This is what we wanted to prove.

Exercises

1. Let $\psi : \mathbf{R}^3 \to \mathbf{R}$. Use the chain rule to verify (11.113).

2. Complete the proof that (11.112) satisfies the wave equation in $\mathbf{R}^3$. Take advantage of the formulas (11.116) and the hint following those formulas.

3. Express the solution of (11.124) in two ways:
    (a) using an integral over $B_1(0)$ (the unit disk in $\mathbf{R}^2$);
    (b) using an integral over $B_{ct}(\mathbf{x})$ (the disk of radius ct, centered at $\mathbf{x}$, in $\mathbf{R}^2$).

4. Let G be the causal Green's function for the wave equation on a bounded domain Ω (in $\mathbf{R}^2$ or $\mathbf{R}^3$). Show that the translation property
$$G(\mathbf{x},t;\mathbf{y},s) = G(\mathbf{x},t-s;\mathbf{y},0)$$
holds.

5. Suppose $\psi : \mathbf{R}^3 \to \mathbf{R}$ is defined by
$$\psi(\mathbf{x}) = \begin{cases} 1, & \|\mathbf{x}\| < R, \\ 0 & \text{otherwise,} \end{cases}$$
and let u be the solution of (11.110). Prove that, for each $\mathbf{x} \in \mathbf{R}^3$, $u(\mathbf{x},t) = 0$ if $t < (\|\mathbf{x}\| - R)/c$ or $t > (\|\mathbf{x}\| + R)/c$.

6. Does the result of the previous exercise hold in $\mathbf{R}^2$? Explain.

7. Suppose that the eigenpairs of $-\Delta$, under Dirichlet conditions, on Ω are λ_n, ψ_n, $n=1,2,3,\ldots$. Express the Green's function for the wave equation on Ω (again, under Dirichlet conditions) in terms of the eigenvalues and eigenfunctions. Assume that $(\psi_n, \psi_n) = 1$ for all n.

11.8 Green's functions for the heat equation

11.8.1 The free-space Green's function

As in the case of the one-dimensional heat equation, the Green's function is best derived using the Fourier transform, which we do not discuss. Therefore, we will just state and verify the necessary solutions. The solutions in $\mathbf{R}^2$ and $\mathbf{R}^3$ are similar, so we will discuss them together. As usual, we will use Duhamel's principle, and begin by solving the IVP

$$\frac{\partial u}{\partial t} - k\Delta u = 0, \ \mathbf{x} \in \mathbf{R}^n, \ t > 0, \tag{11.129}$$
$$u(\mathbf{x}, 0) = f(\mathbf{x}, s), \ \mathbf{x} \in \mathbf{R}^n.$$

The solution can be written in terms of the Gaussian kernel, which is

$$S(\mathbf{x}, t) = \frac{1}{(4k\pi t)^{n/2}} e^{-\|\mathbf{x}\|^2/(4kt)}. \tag{11.130}$$

The solution of (11.129) is

$$v(x, t; s) = \int_{\mathbf{R}^n} S(\mathbf{x} - \mathbf{y}, t) f(\mathbf{y}, s) \, d\mathbf{y}. \tag{11.131}$$

The proof that (11.131) defines a solution of (11.129) proceeds just as in one spatial dimension (see Section 9.5). First, it can be shown directly that S satisfies the heat equation for $t > 0$ (see Exercise 11.8.1). Since we can differentiate under the integral, it follows that v, as defined by (11.131), also satisfies the heat equation (the fact that the heat equation is linear is critical to this reasoning). Clearly, $S(x, 0)$, and hence $v(x, 0; s)$, is formally undefined. However, as $t \to 0^+$, $S(\cdot, t)$ resembles a delta function more and more closely (for an example, see Figure 11.24), and hence $v(x, t; s) \to f(x, s)$ as $t \to 0^+$.

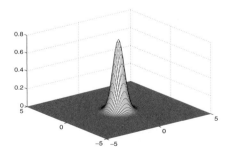

Figure 11.24. *The Gaussian kernel in $\mathbf{R}^2$ ($k = 1$, $t = 0.1$).*

We can now apply Duhamel's principle to construct the solution of

$$\frac{\partial u}{\partial t} - k\Delta u = f(\mathbf{x},t), \ \mathbf{x} \in \mathbf{R}^n, \ t > 0, \tag{11.132}$$
$$u(\mathbf{x},0) = 0, \ \mathbf{x} \in \mathbf{R}^n,$$

which is

$$u(x,t) = \int_0^t v(x,t-s;s)\,ds = \int_0^t \int_{\mathbf{R}^n} S(\mathbf{x}-\mathbf{y},t-s)f(\mathbf{y},s)\,d\mathbf{y}\,ds. \tag{11.133}$$

The Green's function for the heat equation in $\mathbf{R}^n$ is

$$G(\mathbf{x},t;\mathbf{y},s) = \begin{cases} S(\mathbf{x}-\mathbf{y},t-s), & s < t, \\ 0, & s > t. \end{cases} \tag{11.134}$$

The Green's function can also be written as $G(\mathbf{x},t;\mathbf{y},s) = H(t-s)S(\mathbf{x}-\mathbf{y},t-s)$, where H is the Heaviside function.

11.8.2 The Green's function on a bounded domain

We can derive the Green's function on a bounded domain Ω using an eigenfunction expansion. Let us assume that the eigenvalues and eigenfunctions of $-\Delta$ on Ω, under homogeneous Dirichlet conditions, are

$$\lambda_n, \ \psi_n(\mathbf{x}), \ n = 1,2,3,\ldots.$$

To find the Green's function, we wish to solve

$$\frac{\partial u}{\partial t} - k\Delta u = f(\mathbf{x},t), \ \mathbf{x} \in \Omega, \ t > 0,$$
$$u(\mathbf{x},0) = 0, \ \mathbf{x} \in \Omega, \tag{11.135}$$
$$u(\mathbf{x},t) = 0, \ \mathbf{x} \in \partial\Omega, \ t > 0.$$

We write the desired solution as

$$u(\mathbf{x},t) = \sum_{n=1}^{\infty} a_n(t)\psi_n(\mathbf{x}),$$

and we assume that

$$f(\mathbf{x},t) = \sum_{n=1}^{\infty} c_n(t)\psi_n(\mathbf{x}),$$

where the coefficients $c_n(t)$ are given by

$$c_n(t) = \int_\Omega \psi_n(\mathbf{y})f(\mathbf{y},t)\,d\mathbf{y}, \ n = 1,2,3,\ldots$$

11.8. Green's functions for the heat equation

(assuming that the eigenfunctions have been normalized). By the usual reasoning, the PDE is equivalent to

$$\sum_{n=1}^{\infty}\left\{\frac{da_n}{dt}(t)+k\lambda_n a_n(t)\right\}\psi_n(\mathbf{x})=\sum_{n=1}^{\infty}c_n(t)\psi_n(\mathbf{x}),$$

and the initial condition is

$$\sum_{n=1}^{\infty}a_n(0)\psi_n(\mathbf{x})=0.$$

We can therefore find a_n by solving

$$\frac{da_n}{dt}(t)+k\lambda_n a_n(t)=c_n(t),\ t>0,$$

$$a_n(0)=0.$$

The solution is

$$a_n(t)=\int_0^t e^{-k\lambda_n(t-s)}c_n(s)\,ds$$
$$=\int_0^t\int_\Omega e^{-k\lambda_n(t-s)}f(\mathbf{y},s)\,d\mathbf{y}\,s.$$

Therefore,

$$u(\mathbf{x},t)=\sum_{n=1}^{\infty}\left(\int_0^t\int_\Omega e^{-k\lambda_n(t-s)}f(\mathbf{y},s)\,d\mathbf{y}\,s\right)\psi_n(\mathbf{x})$$
$$=\int_0^t\int_\Omega\left(\sum_{n=1}^{\infty}e^{-k\lambda_n(t-s)}\psi_n(\mathbf{y})\psi_n(\mathbf{x})\right)f(\mathbf{y},s)\,d\mathbf{y}\,ds.$$

From this we see that the causal Green's function is

$$G(\mathbf{x},t;\mathbf{y},s)=\begin{cases}\sum_{n=1}^{\infty}e^{-k\lambda_n(t-s)}\psi_n(\mathbf{y})\psi_n(\mathbf{x}),&s<t,\\0,&s>t.\end{cases}\qquad(11.136)$$

Since the eigenvalues satisfy $\lambda_n\to\infty$ as $n\to\infty$ (see Section 12.7), it follows that this series converges for all $t>s$, although the convergence is slow if $t-s$ is sufficiently small.

Example 11.12. If Ω is the unit square in $\mathbf{R}^2$, then the causal Green's function for the heat equation is

$$G(\mathbf{x},t;\mathbf{y},s)=\begin{cases}\sum_{m=1}^{\infty}\sum_{n=1}^{\infty}e^{-k\lambda_{mn}(t-s)}\psi_{mn}(\mathbf{x})\psi_{mn}(\mathbf{y}),&s<t,\\0,&s>t,\end{cases}$$

where

$$\lambda_{mn}=(m^2+n^2)\pi^2,$$
$$\psi_{mn}(\mathbf{x})=\sin(m\pi x_1)\sin(n\pi x_2).$$

In Figure 11.25, we display $G(\mathbf{x},t;\mathbf{y},0)$ for $\mathbf{y}=(1/2,7/8)$, $t=0.002$, and $k=1$. ∎

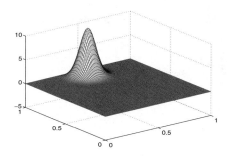

Figure 11.25. *The Green's function $G(\mathbf{x},t;\mathbf{y},0)$ in Example 11.12 with $t = 0.002$, $k = 1$, and $\mathbf{y} = (1/2, 7/8)$. A total of 400 terms of the (double) Fourier series was used to create this graph.*

Exercises

1. Verify that the Gaussian kernel (11.130) satisfies the homogeneous heat equation in $\mathbf{R}^n$.

2. Verify that the Gaussian kernel (11.130), for $n = 2$, satisfies
$$\int_{\mathbf{R}^2} S(x,t)\,dx = 1$$
for all $t > 0$.

3. In Section 7.6, we showed that it is not possible to solve the heat equation backward in time. That discussion introduced the backward heat equation, which in multiple spatial dimensions takes the form
$$-\frac{\partial u}{\partial t} - k\Delta u = f(\mathbf{x},t).$$
Show that we can solve the backward heat equation backward in time by explicitly deriving the solution to the final value problem
$$-\frac{\partial u}{\partial t} - k\Delta u = f(\mathbf{x},t), \ \mathbf{x} \in \Omega, \ 0 < t < T,$$
$$u(\mathbf{x},T) = \phi(\mathbf{x}), \ \mathbf{x} \in \Omega,$$
$$u(\mathbf{x},t) = 0, \ \mathbf{x} \in \partial\Omega, \ t > 0.$$
(Hint: Define $v(\mathbf{x},t) = u(-\mathbf{x},t)$ and find the IBVP that v satisfies.)

4. Find the Green's function for the backward heat equation (see the previous problem). That is, find $g(\mathbf{x},t;\mathbf{y},s)$ such that the solution to
$$-\frac{\partial u}{\partial t} - k\Delta u = f(\mathbf{x},t), \ \mathbf{x} \in \Omega, \ 0 < t < T,$$
$$u(\mathbf{x},T) = 0, \ \mathbf{x} \in \Omega,$$
$$u(\mathbf{x},t) = 0, \ \mathbf{x} \in \partial\Omega, \ t > 0,$$

is
$$u(\mathbf{x},t) = \int_0^T \int_\Omega g(\mathbf{x},t;\mathbf{y},s) f(\mathbf{y},s) d\mathbf{y}.$$

Assume that the Green's function G for the forward heat equation (under Dirichlet conditions) on Ω is known.

5. Let $u(\mathbf{x},t)$, $v(\mathbf{x},t)$ be smooth functions defined for $\mathbf{x} \in \overline{\Omega}$, $0 \leq t \leq T$. Prove that

$$\int_0^T \int_\Omega \left\{ u\left(\frac{\partial v}{\partial t} - k\Delta v\right) - v\left(-\frac{\partial u}{\partial t} - k\Delta u\right) \right\} d\mathbf{x}\, dt$$
$$= \int_\Omega (u(\mathbf{x},T) v(\mathbf{x},T) - u(\mathbf{x},0) v(\mathbf{x},0))\, d\mathbf{x} + k \int_0^T \int_\Omega \left(v\frac{\partial u}{\partial \mathbf{n}} - u\frac{\partial v}{\partial \mathbf{n}} \right) d\sigma_{\mathbf{x}}\, dt.$$

6. Using the results of the previous two exercises, find a formula for the solution of

$$\begin{aligned}
\frac{\partial v}{\partial t} - k\Delta v &= 0,\ \mathbf{x} \in \Omega,\ 0 < t < T, \\
u(\mathbf{x},0) &= 0,\ \mathbf{x} \in \Omega, \\
u(\mathbf{x},t) &= h(\mathbf{x},t),\ \mathbf{x} \in \partial\Omega,\ 0 < t < T.
\end{aligned} \qquad (11.137)$$

(Hint: Use the identity from the previous exercise, taking v to be the solution of the IBVP (11.137) and $u(\mathbf{x},t) = g(\mathbf{x},t;\mathbf{y},s)$, where g is the Green's function for the backward heat equation, derived in Exercise 4.)

11.9 Suggestions for further reading

The foundation of PDEs in two or more spatial dimensions is advanced calculus. Kaplan [37] gives a straightforward introduction; an alternative at the same level is Greenberg [27]. A more advanced treatment can be found in Marsden and Tromba [47].

All of the references cited in Sections 5.7 and 6.6 deal with PDEs in multiple spatial dimensions. More information about Bessel functions can be found in Folland [20] and Al-Gwaiz [1].

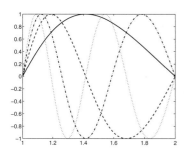

Chapter 12
More about Fourier Series

In Chapters 5–7, we introduced several kinds of Fourier series: the Fourier sine series, cosine series, quarter-wave sine series, quarter-wave cosine series, and the full Fourier series. These series were primarily used to represent the solution to differential equations, and their usefulness was based on two facts.

1. Each is based on an orthogonal sequence with the property that every continuous function can be represented in terms of this sequence.

2. The terms in the series represent eigenfunctions of certain simple differential operators (under various boundary conditions). This accounts for the fact that it is computationally tractable to determine a series representation of the solution to the corresponding differential equation.

In this chapter we will go deeper into the study of Fourier series. Specifically, we will consider the following questions.

1. What is the relationship among the various kinds of Fourier series?

2. How can a partial Fourier series be found and evaluated efficiently?

3. Under what conditions and in what sense can a function be represented by its Fourier series?

4. Can the Fourier series method be generalized to more complicated differential equations (including nonconstant coefficients and/or irregular geometry)?

Our discussion will justify many of the statements we made earlier in the book concerning the convergence of Fourier series. It also introduces the *fast Fourier transform* (FFT), which is an exciting and relatively recent development (from the last half of the twentieth century[67]) in the long history of Fourier series. The calculation of N Fourier coefficients would appear to require $O(N^2)$ operations (for reasons that we explain in Section 12.2). The FFT reduces the operation count to $O(N \log N)$, a considerable savings when N is large.

---

[67]The FFT was popularized in the 1960s by Cooley and Tukey [15]. However, the method was known to Gauss long before; see [33].

Some of our discussion about the convergence of Fourier series was anticipated in Chapter 10 in the context of Sturm–Liouville eigenvalue problems. However, the present chapter uses a more direct analysis to approach questions of convergence, and it can be read independently of Chapter 10.

We begin by introducing yet another type of Fourier series, the complex Fourier series, that is convenient both for analysis and for expressing the FFT.

12.1 The complex Fourier series

We saw in Chapter 4 that problems apparently involving only real numbers can sometimes be best addressed with techniques that use complex numbers. For example, if the characteristic roots r_1 and r_2 of the ODE

$$a\frac{d^2u}{dt^2} + b\frac{du}{dt} + cu = 0$$

are distinct, then the general solution is

$$u(t) = c_1 e^{r_1 t} + c_2 e^{r_2 t}.$$

This formula holds even if r_1 and r_2 are complex, in which case u is complex valued for most choices of c_1 and c_2. (We also saw, in Section 4.2.1, how to recover the general real-valued solution if desired.)

In a similar way, there are some advantages to using complex-valued eigenfunctions, even when solving differential equations involving only real-valued functions. Using Euler's formula,

$$e^{i\theta} = \cos\theta + i\sin\theta,$$

we see that

$$\frac{d}{dx}\left[e^{i\omega x}\right] = \frac{d}{dx}[\cos(\omega x) + i\sin(\omega x)] = -\omega\sin(\omega x) + i\omega\cos(\omega x)$$
$$= i\omega(\cos(\omega x) + i\sin(\omega x))$$
$$= i\omega e^{i\omega x},$$

and therefore

$$-\frac{d^2}{dx^2}\left[e^{i\omega x}\right] = -(i\omega)^2 e^{i\omega x} = \omega^2 e^{i\omega x}.$$

This suggests that the negative second derivative operator has complex exponential eigenfunctions.

We leave it to the exercises (see Exercise 12.1.4) to show that nonzero solutions to the BVP

$$-\frac{d^2u}{dx^2} = \lambda u, \ -\ell < x < \ell,$$
$$u(-\ell) = u(\ell), \qquad\qquad (12.1)$$
$$\frac{du}{dx}(-\ell) = \frac{du}{dx}(\ell)$$

12.1. The complex Fourier series

exist only for $\lambda = 0, \pi^2/\ell^2, \ldots, n^2\pi^2/\ell^2, \ldots$. The eigenvalue $\lambda = 0$ has eigenfunction 1 (the constant function), while each eigenvalue $\lambda = n^2\pi^2/\ell^2$ has two linearly independent eigenfunctions,
$$e^{i\pi nx/\ell}, e^{-i\pi nx/\ell}.$$

When $n = 0$, $e^{i\pi nx/\ell}$ reduces to the constant function 1 and $\lambda = n^2\pi^2/\ell^2$ reduces to $\lambda = 0$. Therefore, to simplify the notation, we will write the complete list of eigenvalue-eigenfunction pairs as
$$\frac{n^2\pi^2}{\ell^2}, e^{i\pi nx/\ell}, \; n = 0, \pm 1, \pm 2, \ldots. \tag{12.2}$$

In this section we use the eigenfunctions given in (12.2) to form Fourier series representing functions on the interval $(-\ell, \ell)$. Since Fourier series calculations are based on orthogonality, we must take a slight detour to discuss complex vector space and inner products. Some of the following results have been used earlier in the text, but only briefly in the course of demonstrating that a symmetric operator has only real eigenvalues.

12.1.1 Complex inner products

As we discussed in Section 3.1, a vector space is a set of objects (vectors), along with two operations, (vector) addition and scalar multiplication. To this point, we have used real numbers as the scalars; however, the complex numbers can also be used as scalars. For emphasis, when we use complex numbers as scalars, we refer to the vector space as a *complex vector space*.

The most common complex vector space is $\mathbf{C}^n$, complex n-space:
$$\mathbf{C}^n = \left\{ \begin{bmatrix} u_1 \\ u_2 \\ \vdots \\ u_n \end{bmatrix} : u_j \in \mathbf{C}, \; j = 1, 2, \ldots, n \right\}.$$

Just as for $\mathbf{R}^n$, addition and scalar multiplication are defined componentwise.

The only adjustment that must be made in working with complex vector spaces is in the definition of inner product. If u is a vector from a complex vector space, then it is permissible to multiply u by the imaginary unit i to get iu. But if there is an inner product $(\cdot, \cdot)$ on the space, and if the familiar properties of inner products hold, then
$$(iu, iu) = i^2(u, u) = -(u, u).$$

This suggests that, if u is nonzero, then either (u, u) or (iu, iu) is negative, contradicting one of the rules of inner products (and making it impossible to define a norm based on the inner product).

For this reason, the definition of inner product is modified for complex vector spaces.

Definition 12.1. *Let V be a complex vector space. An inner product on V is a function taking two vectors from V and producing a complex number, in such a way that the following three properties are satisfied:*

1. $(u, v) = \overline{(v, u)}$ *for all vectors u and v;*

2. $(\alpha u + \beta v, w) = \alpha(u,w) + \beta(v,w)$ and $(w, \alpha u + \beta v) = \overline{\alpha}(w,u) + \overline{\beta}(w,v)$ for all vectors u, v, and w and all complex numbers α and β;

3. $(u,u) \geq 0$ for all vectors u, and $(u,u) = 0$ if and only if u is the zero vector.

In the above definition, $\overline{z}$ denotes the complex conjugate of $z \in \mathbf{C}$. (If $z = x + iy$, with $x, y \in \mathbf{R}$, then $\overline{z} = x - iy$.)

For $\mathbf{C}^n$, the inner product is suggested by the fact that $|z| = \sqrt{z\overline{z}}$ for $z \in \mathbf{C}$. We define

$$(u,v)_{\mathbf{C}^n} = u \cdot \overline{v} = \sum_{j=1}^{n} u_i \overline{v}_i.$$

It is straightforward to verify that the properties of an inner product are satisfied. In particular,

$$(u,u)_{\mathbf{C}^n} = \sum_{j=1}^{n} u_i \overline{u}_i = \sum_{j=1}^{n} |u_i|^2 \geq 0.$$

If we now consider the space of complex-valued functions defined on an interval $[a,b]$, the same reasoning as in Section 3.4 leads to the complex L^2 inner product,

$$(f,g) = \int_a^b f(x)\overline{g(x)}\,dx.$$

(We will use the same notation as for the real L^2 inner product, since, when f and g are real valued, the presence of the complex conjugate in the formula has no effect, and the ordinary L^2 inner product is obtained.)

12.1.2 Orthogonality of the complex exponentials

A direct calculation now shows that the eigenfunctions given in (12.2) are orthogonal with respect to the (complex) L^2 inner product:

$$\begin{aligned}\left(e^{i\pi nx/\ell}, e^{m\pi ix/\ell}\right) &= \int_{-\ell}^{\ell} e^{i\pi nx/\ell} e^{-m\pi ix/\ell}\,dx = \int_{-\ell}^{\ell} e^{(n-m)\pi ix/\ell}\,dx \\ &= \left[\frac{e^{(n-m)\pi ix/\ell}}{(n-m)\pi i/\ell}\right]_{-\ell}^{\ell} \\ &= \frac{\ell}{(n-m)\pi i}\left(e^{(n-m)\pi i} - e^{-(n-m)\pi i}\right) \\ &= 0,\ m \neq n.\end{aligned}$$

The last step follows because $e^{(n-m)\theta i}$ is 2π-periodic. We also have

$$\left(e^{i\pi nx/\ell}, e^{i\pi nx/\ell}\right) = \int_{-\ell}^{\ell} e^{i\pi nx/\ell} e^{-i\pi nx/\ell}\,dx = \int_{-\ell}^{\ell} dx = 2\ell.$$

This shows that the L^2 norm of the eigenfunction is $\sqrt{2\ell}$.

12.1.3 Representing functions with complex Fourier series

If f is a continuous, complex-valued function defined on $[-\ell, \ell]$, then its *complex Fourier series* is

$$\sum_{n=-\infty}^{\infty} c_n e^{i\pi nx/\ell},$$

where

$$c_n = \frac{\left(f, e^{i\pi nx/\ell}\right)}{\left(e^{i\pi nx/\ell}, e^{i\pi nx/\ell}\right)} = \frac{1}{2\ell}\int_{-\ell}^{\ell} f(x)e^{-i\pi nx/\ell}\,dx.$$

These *complex Fourier coefficients* are computed according to the projection theorem; it follows that

$$\sum_{n=-N}^{N} c_n e^{i\pi nx/\ell}$$

is the element of the subspace

$$\text{span}\left\{e^{-i\pi Nx/\ell}, e^{-i\pi(N-1)x/\ell}, \ldots, e^{-i\pi x/\ell}, 1, e^{i\pi x/\ell}, \ldots, e^{i\pi Nx/\ell}\right\}$$

closest to f in the L^2 norm. Also, we will show in Section 12.6 that the complex Fourier series converges to f in the L^2 norm under mild conditions on f (in particular, if f is continuous).

12.1.4 The complex Fourier series of a real-valued function

If $f \in C[-\ell, \ell]$ is real valued, then, according to our assertions above, it can be represented by its complex Fourier series. We now show that, since f is real valued, each partial sum ($n = -N, -N+1, \ldots, N$) of the complex Fourier series is real, and moreover that the complex Fourier series is equivalent to the full Fourier series in this case.

So we suppose f is real valued and that c_n, $n = 0, \pm 1, \pm 2, \ldots$, are its complex Fourier coefficients. For reference, we will write the full Fourier series of f (see Section 6.3) as

$$a_0 + \sum_{n=1}^{\infty}\left(a_n \cos\left(\frac{n\pi x}{\ell}\right) + b_n \sin\left(\frac{n\pi x}{\ell}\right)\right),$$

where

$$a_0 = \frac{1}{2\ell}\int_{-\ell}^{\ell} f(x)\,dx,$$

$$a_n = \frac{1}{\ell}\int_{-\ell}^{\ell} f(x)\cos\left(\frac{n\pi x}{\ell}\right)dx, \, n = 1, 2, 3, \ldots,$$

$$b_n = \frac{1}{\ell}\int_{-\ell}^{\ell} f(x)\sin\left(\frac{n\pi x}{\ell}\right)dx, \, n = 1, 2, 3, \ldots.$$

Now,
$$c_0 = \frac{1}{2\ell}\int_{-\ell}^{\ell} f(x)\,dx = a_0.$$

For $n > 0$,
$$\begin{aligned}
c_n &= \frac{1}{2\ell}\int_{-\ell}^{\ell} f(x)e^{-i\pi nx/\ell}\,dx \\
&= \frac{1}{2\ell}\int_{-\ell}^{\ell} f(x)\left(\cos\left(\frac{n\pi x}{\ell}\right) - i\sin\left(\frac{n\pi x}{\ell}\right)\right)dx \\
&= \frac{1}{2\ell}\int_{-\ell}^{\ell} f(x)\cos\left(\frac{n\pi x}{\ell}\right)dx - \frac{i}{2\ell}\int_{-\ell}^{\ell} f(x)\sin\left(\frac{n\pi x}{\ell}\right)dx \\
&= \frac{1}{2}a_n - \frac{i}{2}b_n.
\end{aligned}$$

Similarly, with $n > 0$, we have
$$\begin{aligned}
c_{-n} &= \frac{1}{2\ell}\int_{-\ell}^{\ell} f(x)e^{i\pi nx/\ell}\,dx \\
&= \frac{1}{2\ell}\int_{-\ell}^{\ell} f(x)\left(\cos\left(\frac{n\pi x}{\ell}\right) + i\sin\left(\frac{n\pi x}{\ell}\right)\right)dx \\
&= \frac{1}{2\ell}\int_{-\ell}^{\ell} f(x)\cos\left(\frac{n\pi x}{\ell}\right)dx + \frac{i}{2\ell}\int_{-\ell}^{\ell} f(x)\sin\left(\frac{n\pi x}{\ell}\right)dx \\
&= \frac{1}{2}a_n + \frac{i}{2}b_n.
\end{aligned}$$

Therefore,
$$\begin{aligned}
c_n e^{i\pi nx/\ell} + c_{-n}e^{-i\pi nx/\ell} &= \left(\frac{1}{2}a_n - \frac{i}{2}b_n\right)\left(\cos\left(\frac{n\pi x}{\ell}\right) + i\sin\left(\frac{n\pi x}{\ell}\right)\right) \\
&\quad + \left(\frac{1}{2}a_n + \frac{i}{2}b_n\right)\left(\cos\left(\frac{n\pi x}{\ell}\right) - i\sin\left(\frac{n\pi x}{\ell}\right)\right) \\
&= a_n \cos\left(\frac{n\pi x}{\ell}\right) + b_n \sin\left(\frac{n\pi x}{\ell}\right).
\end{aligned}$$

(The reader should notice how all imaginary quantities sum to zero in this calculation.)

It follows that, for any $N \geq 0$,
$$\sum_{n=-N}^{N} c_n e^{i\pi nx/\ell} = a_0 + \sum_{n=1}^{N}\left(a_n \cos\left(\frac{n\pi x}{\ell}\right) + b_n \sin\left(\frac{n\pi x}{\ell}\right)\right).$$

This shows that every (symmetric) partial sum of the complex Fourier series of a real-valued function is real and also that the complex Fourier series is equivalent to the full Fourier series.

12.1. The complex Fourier series

Exercises

1. Let $f : [-1,1] \to \mathbf{R}$ be defined by $f(x) = 1 - x^2$. Compute the complex Fourier series
$$\sum_{n=-\infty}^{\infty} c_n e^{i\pi nx}$$
of f, and graph the errors
$$f(x) - \sum_{n=-N}^{N} c_n e^{i\pi nx}$$
for $N = 10, 20, 40$.

2. Let $f : [-\pi, \pi] \to \mathbf{R}$ be defined by $f(x) = x$. Compute the complex Fourier coefficients c_n, $n = 0, \pm 1, \pm 2, \ldots$, of f, and graph the errors
$$f(x) - \sum_{n=-N}^{N} c_n e^{inx}$$
for $N = 10, 20, 40$.

3. Let $f : [-1,1] \to \mathbf{C}$ be defined by $f(x) = e^{ix}$. Compute the complex Fourier coefficients c_n, $n = 0, \pm 1, \pm 2, \ldots$, of f, and graph the errors
$$f(x) - \sum_{n=-N}^{N} c_n e^{i\pi nx}$$
for $N = 10, 20, 40$. (Note: You will have to either graph the real and imaginary parts of the error separately or just graph the modulus of the error.)

4. Consider the negative second derivative operator under periodic boundary conditions on $[-\ell, \ell]$.

    (a) Show that each pair listed in (12.2) is an eigenpair.

    (b) Show that (12.2) includes *every* eigenvalue.

    (c) Show that *every* eigenfunction is expressible in terms of those given in (12.2).

    (d) Compare these results to those derived in Section 6.3, and explain why they are consistent.

5. Suppose that $f : [-\ell, \ell] \to \mathbf{C}$ is defined by $f(x) = g(x) + ih(x)$, where g and h are real-valued functions defined on $[-\ell, \ell]$. Show that the complex Fourier coefficients can be expressed in terms of the full Fourier coefficients of g and h.

6. Prove that, for any real numbers θ and λ,
$$e^{i(\theta + \lambda)} = e^{i\theta} e^{i\lambda}.$$
(Hint: Use Euler's formula and trigonometric identities.)

7. Let N be a positive integer, j a positive integer between 1 and $N-1$, and define vectors $\mathbf{u} \in \mathbf{R}^N$ and $\mathbf{v} \in \mathbf{R}^N$ by

$$u_n = \cos\left(\frac{jn\pi}{N}\right), \quad v_n = \sin\left(\frac{jn\pi}{N}\right), \quad n = 0, 1, \ldots, N-1.$$

(For this exercise, it is convenient to index the components of $\mathbf{x} \in \mathbf{R}^N$ as $0, 1, \ldots, N-1$, instead of $1, 2, \ldots, N$ as is usual.) The purpose of this exercise is to prove that

(a) $\mathbf{u} \cdot \mathbf{v} = 0$, that is, $\mathbf{u}$ and $\mathbf{v}$ are orthogonal.
(b) $\|\mathbf{u}\| = \|\mathbf{v}\| = \sqrt{N/2}$.

These results are based on the following trick: The sum

$$\sum_{n=0}^{N-1} \left(e^{\pi ijn/N}\right)^2 = \sum_{n=0}^{N-1} \left(e^{2\pi ij/N}\right)^n$$

is a finite geometric series, for which there is a simple formula. On the other hand,

$$\sum_{n=0}^{N-1} \left(e^{\pi ijn/N}\right)^2 \tag{12.3}$$

can be rewritten by replacing $e^{\pi ijn/N}$ by $\cos(jn\pi/N) + i\sin(jn\pi/N)$, expanding the square, and summing. Find the sum (12.3) both ways, and deduce the results given above.

12.2 Fourier series and the FFT

We now show that there is a very efficient way to estimate Fourier coefficients and evaluate (partial) Fourier series. We will use the BVP

$$-\kappa \frac{d^2 u}{dx^2} = f(x), \quad -\ell \leq x < \ell,$$
$$u(-\ell) = u(\ell), \tag{12.4}$$
$$\frac{du}{dx}(-\ell) = \frac{du}{dx}(\ell)$$

as our first example. To solve this by the method of Fourier series, we express u and f in complex Fourier series, say,

$$u(x) = \sum_{n=-\infty}^{\infty} c_n e^{i\pi nx/\ell}$$

($c_n, n = 0, \pm 1, \pm 2, \ldots$, unknown) and

$$f(x) = \sum_{n=-\infty}^{\infty} d_n e^{i\pi nx/\ell}$$

12.2. Fourier series and the FFT

$(d_n, n = 0, \pm 1, \pm 2, \ldots,$ known). The differential equation can then be written as

$$\sum_{n=-\infty}^{\infty} \frac{\kappa n^2 \pi^2}{\ell^2} c_n e^{i\pi n x/\ell} = \sum_{n=-\infty}^{\infty} d_n e^{i\pi n x/\ell}, \tag{12.5}$$

and we obtain

$$c_n = \frac{\ell^2 d_n}{\kappa n^2 \pi^2}, \; n = \pm 1, \pm 2, \ldots. \tag{12.6}$$

The reader will notice the similarity to how the problem was solved in Section 6.3.2. (The calculations, however, are simpler when using the complex Fourier series instead of the full Fourier series.) As in Section 6.3.2, the coefficient d_0 must be zero, that is,

$$\int_{-\ell}^{\ell} f(x)\,dx = 0$$

must hold, in order for a solution to exist. When this compatibility condition holds, the value of c_0 is not determined by the equation, and infinitely many solutions exist, differing only in the choice of c_0.

For analytic purposes, formula (12.6) may be all we need. For example, as we discuss in Section 12.5.1, (12.6) shows that the solution u is considerably smoother than the forcing function f (since the Fourier coefficients of u decay to zero faster than those of f). However, in many cases, we want to produce a numerical estimate of the solution u. We may wish, for example, to estimate the values of u on a grid so that we can graph the solution accurately. This requires three steps:

1. Compute the Fourier coefficients d_n, $n = 0, \pm 1, \pm 2, \ldots, N$, of f by evaluating the appropriate integrals.

2. Compute the Fourier coefficients c_n, $n = 0, \pm 1, \pm 2, \ldots, N$, of u from formula (12.6).

3. Evaluate the partial sum

$$u_n(x) = \sum_{n=-N}^{N} c_n e^{i\pi n x/\ell}$$

on a grid covering the interval $[-\ell, \ell]$, say $x_j = jh$, $j = 0, \pm 1, \pm 2, \ldots, \pm N$, $h = \ell/N$.

If N is chosen large enough, this will produce accurate estimates of $u(x_j)$, where $j = 0, \pm 1, \pm 2, \ldots, \pm N$.

Until this point, we have implicitly assumed that all of the calculations necessary to compute u would be done analytically (using various integration rules to compute the necessary Fourier coefficients, for example). This is not always possible (some integrals cannot be computed using elementary functions) and not always desirable when it is possible (because there may be more efficient methods). We will therefore consider how to estimate u.

12.2.1 Using the trapezoidal rule to estimate Fourier coefficients

We begin by estimating the integrals defining the Fourier coefficients of f:

$$d_n = \frac{1}{2\ell} \int_{-\ell}^{\ell} f(x) e^{-i\pi n x/\ell}\,dx.$$

A simple formula for estimating this integral is the so-called (composite) trapezoidal rule, which replaces the integral by a discrete sum that can be interpreted as the sum of areas of trapezoids:

$$\int_a^b g(x)\,dx = \frac{h}{2}\left(g(x_0) + 2\sum_{j=1}^{N-1} g(x_j) + g(x_N)\right) + O(h^2), \qquad (12.7)$$

where $h = (b-a)/N$ and $x_j = a + jh$, $j = 0, 1, 2, \ldots, N$. Figure 12.1 shows the geometric interpretation of the trapezoidal rule. Although the trapezoidal rule is generally only second order (that is, the error is $O(h^2)$), it is highly accurate for periodic functions.[68]

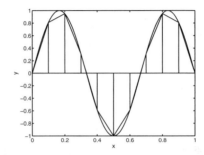

Figure 12.1. *Illustration of the trapezoidal rule for numerical integration; the (signed) area under the curve is estimated by the areas of the trapezoids.*

To apply the trapezoidal rule to the computation of d_n, we define the grid $x_j = jh$, $j = 0, \pm 1, \pm 2, \ldots, \pm N$, $h = \ell/N$. Then

$$d_n \doteq \frac{1}{2\ell}\frac{h}{2}\left(f(x_{-N})e^{-i\pi n x_{-N}/\ell} + 2\sum_{j=-N+1}^{N-1} f(x_j)e^{-i\pi n x_j/\ell} + f(x_N)e^{-i\pi n x_N/\ell}\right).$$

Since

$$e^{-i\pi n x_{-N}/\ell} = e^{i\pi n} = (-1)^n,\ e^{-i\pi n x_N/\ell} = e^{-i\pi n} = (-1)^n,$$

and

$$\frac{h}{2\ell} = \frac{1}{2N},\ -\frac{i\pi n x_j}{\ell} = -\frac{i\pi n j}{N},$$

we can simplify this to

$$d_n \doteq F_n = \frac{1}{2N}\sum_{j=-N}^{N-1} f_j e^{-i\pi nj/N},\ n = -N, -N+1, \ldots, 0, \ldots, N-1, \qquad (12.8)$$

---

[68] For details, see any book on numerical analysis, for example, [3].

12.2. Fourier series and the FFT

where

$$f_j = f(x_j), \quad j = -N+1, -N+2, \ldots, 0, \ldots, N-1,$$

and

$$f_{-N} = \frac{1}{2}(f(-\ell) + f(\ell)).$$

The reader should take note of the special definition of f_{-N}, which reduces to simply $f_{-N} = f(x_{-N}) = f(-\ell)$ when f is 2ℓ-periodic (so that $f(-\ell) = f(\ell)$).

The sequence of estimates $F_{-N}, F_{-N+1}, \ldots, F_{N-1}$ is essentially produced by applying the *discrete Fourier transform* to the sequence $f_{-N}, f_{-N+1}, \ldots, f_{N-1}$, as we now show.

12.2.2 The discrete Fourier transform

Just as a function defined on $[-\ell, -\ell]$ can be synthesized from functions, each having a distinct frequency, so a finite sequence can be synthesized from sequences with distinct frequencies.

Definition 12.2. *Let $a_0, a_1, \ldots, a_{M-1}$ be a sequence of real or complex numbers. Then the discrete Fourier transform (DFT) maps $a_0, a_1, \ldots, a_{M-1}$ to the sequence $A_0, A_1, \ldots, A_{M-1}$, where*

$$A_n = \frac{1}{M} \sum_{j=0}^{M-1} a_j e^{-2\pi i n j / M}, \quad n = 0, 1, \ldots, M-1. \tag{12.9}$$

The original sequence can be recovered by applying the *inverse DFT*:

$$a_j = \sum_{n=0}^{M-1} A_n e^{2\pi i n j / M}, \quad j = 0, 1, 2, \ldots, M-1. \tag{12.10}$$

The proof of this relationship is a direct calculation (see Exercise 12.2.5). We will often refer to the sequence $A_0, A_1, \ldots, A_{M-1}$ as the DFT of $a_0, a_1, \ldots, a_{M-1}$, although the correct way to express the relationship is that $A_0, A_1, \ldots, A_{M-1}$ is the image under the DFT of $a_0, a_1, \ldots, a_{M-1}$ (the DFT is the mapping, not the result of the mapping). This abuse of terminology is just a convenience and is quite common. As another convenience, we will write the sequence $a_0, a_1, \ldots, a_{M-1}$ as $\{a_j\}_{j=0}^{M-1}$, or even just $\{a_j\}$ (if the limits of j are understood).

We will now show that the relationship between the sequences $\{f_j\}_{j=-N}^{N-1}$ and $\{F_n\}_{n=-N}^{N-1}$ can be expressed in terms of the DFT. (The reader will recall that $F_n \doteq d_n$, the nth Fourier coefficient of the function f on $[-\ell, \ell]$.) Since $e^{i\theta}$ is 2π-periodic, we have

$$e^{-\pi i(j+2N)n/N} = e^{-\pi i j n/N - 2\pi i n} = e^{-\pi i j n/N} e^{-2\pi i n} = e^{-\pi i j n/N}.$$

We can therefore write, for $n = -N, -N+1, \ldots, N-1$,

$$F_n = \frac{1}{2N} \sum_{j=-N}^{N-1} f_j e^{-\pi i j n/N} = \frac{1}{2N} \left\{ \sum_{j=0}^{N-1} f_j e^{-\pi i j n/N} + \sum_{j=-N}^{-1} f_j e^{-\pi i j n/N} \right\}$$

$$= \frac{1}{2N} \left\{ \sum_{j=0}^{N-1} f_j e^{-\pi i j n/N} + \sum_{j=-N}^{-1} f_j e^{-\pi i (j+2N)n/N} \right\}$$

$$= \frac{1}{2N} \left\{ \sum_{j=0}^{N-1} f_j e^{-\pi i j n/N} + \sum_{j=N}^{2N-1} f_{j-2N} e^{-\pi i j n/N} \right\}.$$

If we define a sequence $\{\tilde{f}_j\}_{j=0}^{2N-1}$ by

$$\tilde{f}_j = \begin{cases} f_j, & j = 0, 1, \ldots, N-1, \\ f_{j-2N}, & j = N, N+1, \ldots, 2N-1, \end{cases}$$

then we have

$$F_n = \frac{1}{2N} \sum_{j=0}^{2N-1} \tilde{f}_j e^{-\pi i j n/N}, \; n = -N, -N+1, \ldots, N-1.$$

Moreover, by the periodicity of $e^{i\theta}$, we can write

$$F_n = \frac{1}{2N} \sum_{j=0}^{2N-1} \tilde{f}_j e^{-\pi i j (n+2N)/N}, \; n = -N, -N+1, \ldots, -1.$$

Defining the sequence $\{\tilde{F}_n\}_{n=0}^{2N-1}$ by

$$\tilde{F}_n = \begin{cases} F_n, & n = 0, 1, \ldots, N-1, \\ F_{n-2N}, & n = N, N+1, \ldots, 2N-1, \end{cases}$$

we have

$$\tilde{F}_n = \frac{1}{2N} \sum_{j=0}^{2N-1} \tilde{f}_j e^{-\pi i j n/N}, \; n = 0, 1, \ldots, 2N-1.$$

Thus $\{\tilde{F}_n\}$ is the DFT of $\{\tilde{f}_j\}$. With this rearrangement of terms understood, we can say that $\{F_n\}$ is the DFT of $\{f_j\}$.

To actually compute $\{F_n\}_{n=-N}^{N-1}$ from $\{f_j\}_{j=-N}^{N-1}$, we perform the following three steps.

1. Replace the sequence

$$f_{-N}, \ldots, f_{-1}, f_0, f_1, \ldots, f_{N-1}$$

with

$$f_0, f_1, \ldots, f_{N-1}, f_{-N}, \ldots, f_{-1},$$

and label the latter sequence as $\{\tilde{f}_j\}_{j=0}^{2N-1}$.

12.2. Fourier series and the FFT

2. Compute the DFT of $\{\tilde{f}_j\}_{j=0}^{2N-1}$ to get $\{\tilde{F}_n\}_{n=0}^{2N-1}$.

3. The desired sequence $\{F_n\}_{n=-N}^{N-1}$ is

$$\tilde{F}_N, \ldots, \tilde{F}_{2N-1}, \tilde{F}_0, \ldots, \tilde{F}_{N-1}.$$

The representation of the original sequence in terms of its DFT and the complex exponentials can be viewed as *trigonometric interpolation*, since it provides a combination of complex exponentials (and hence sines and cosines) that interpolates the sequence $f_{-N}, f_{-N+1}, \ldots, f_{N-1}$ and therefore the function f. To be precise, define a function $I : \mathbf{R} \to \mathbf{C}$ by

$$I(x) = \sum_{n=-N}^{N-1} F_n e^{i\pi nx/\ell}, \qquad (12.11)$$

where F_n, $n = 0, \pm 1, \pm 2, \ldots$, is defined by (12.8). Then I satisfies

$$I(x_j) = f(x_j), \quad j = -N+1, -N+2, \ldots, N-1, \qquad (12.12)$$

and

$$I(x_{-N}) = I(x_N) = \frac{1}{2}(f(x_{-N}) + f(x_N)) \qquad (12.13)$$

(see Exercise 12.2.4).

Example 12.3. We define $f : [-1,1] \to \mathbf{R}$ by $f(x) = x^3$. We will compute the interpolating function I of the previous paragraph with $N = 3$. The sequence $\{f_j\}$ is (approximately)

$$0, -0.29630, -0.037037, 0, 0.037037, 0.29630$$

(recall that $f_{-3} = (f(x_{-3}) + f(x_3))/2 = (f(-1) + f(1))/2 = (-1+1)/2 = 0$). The sequence $\{F_n\}$ is then given by[69]

$$F_{-3} \doteq 1.9466 \cdot 10^{-17} + 1.5519 \cdot 10^{-17}i,$$
$$F_{-2} \doteq 3.7066 \cdot 10^{-17} - 7.4842 \cdot 10^{-2}i,$$
$$F_{-1} \doteq -1.0606 \cdot 10^{-16} + 9.6225 \cdot 10^{-2}i,$$
$$F_0 \doteq 3.7007 \cdot 10^{-17},$$
$$F_1 \doteq 1.4501 \cdot 10^{-17} - 9.6225 \cdot 10^{-2}i,$$
$$F_2 \doteq -4.6788 \cdot 10^{-17} + 7.4842 \cdot 10^{-2}i.$$

In Figure 12.2, we display the function f, the sequence $\{f_j\}$, and the interpolating function

$$I(x) = \sum_{n=-3}^{2} F_n e^{i\pi nx/\ell}. \quad \blacksquare$$

---

[69] The calculation was performed in MATLAB using a built-in function for computing the DFT.

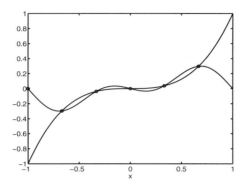

Figure 12.2. *The function $f(x) = x^3$, the sequence $\{f_j\}$, and the trigonometric interpolating function $I(x) = \sum_{n=-3}^{2} F_n e^{i\pi n x}$ (see Example 12.3).*

We now have two important facts.

1. By computing a DFT, we can estimate $2N$ of the complex Fourier coefficients of a known function.

2. Conversely, given the first $2N$ complex Fourier coefficients of a periodic function, we can estimate the function (on a regular grid with $2N + 1$ nodes) by computing an inverse DFT.

These facts are significant because there is a fast algorithm for computing the DFT; this algorithm is called, appropriately enough, the *fast Fourier transform* (FFT). There is a similar fast algorithm for the inverse DFT, which is simply called the inverse FFT.

A direct implementation of formula (12.9) requires $O((M)^2)$ arithmetic operations to compute a DFT. The FFT algorithm, on the other hand, can require as little as $O(M \log_2(M))$ operations to obtain the same result. The efficiency of the FFT depends on the prime factorization of M; if M is the product of small primes, then the FFT is very efficient. The ideal situation is that M be a power of 2; in this case, the $O(M \log_2(M))$ operation count applies. There are many books that explain how the FFT algorithm works, such as Kammler [36, Chapter 6]. We will not concern ourselves with the implementation of the algorithm in this text, but just regard the FFT as a black box for computing the DFT.[70]

Using the above results, we can devise an efficient algorithm for solving (approximately) the BVP (12.4).

1. Using the FFT, estimate the Fourier coefficients $d_{-N}, d_{-N+1}, \ldots, d_{N-1}$ of f.

2. Use (12.6) to estimate the corresponding Fourier series of the solution u.

3. Use the inverse FFT to estimate the values of u on the grid $-\ell, -\ell + h, \ldots, \ell - h$, $h = \ell/N$.

---

[70]The FFT is one of the most important computer algorithms for computational science (indeed, the original paper [15] announcing the algorithm is reputed to be the most widely cited mathematical paper of all time—see [36, page 295]). For this reason, implementations of the FFT exist on every major computer platform and it is also a feature of computer packages such as MATLAB, *Mathematica*, and *Maple*.

12.2. Fourier series and the FFT

Example 12.4. We use the above algorithm to estimate the solution of

$$-\frac{d^2u}{dx^2} = x^3, \quad -1 \le x < 1,$$
$$u(-1) = u(1), \qquad (12.14)$$
$$\frac{du}{dx}(-1) = \frac{du}{dx}(1).$$

The forcing function $f(x) = x^3$ satisfies the compatibility condition, as can easily be verified. An exact solution is $u(x) = (x - x^5)/20$, which corresponds to choosing $c_0 = 0$ in the Fourier series

$$u(x) = \sum_{n=-\infty}^{\infty} c_n e^{i\pi nx}.$$

We will use $N = 128 = 2^7$, which makes the FFT particularly efficient. Using (12.8), implemented by the FFT, we produce the estimates

$$F_{-128}, F_{-127}, \ldots, F_{127}.$$

We next compute the estimates of $c_{-128}, c_{-127}, \ldots, c_{127}$:

$$\tilde{c}_j = \frac{F_n}{n^2 \pi^2}, \quad n = \pm 1, \pm 2, \ldots,$$

and we take $\tilde{c}_0 = 0$. Finally, we use the inverse FFT to produce the estimates

$$\tilde{u}_j = \sum_{n=-128}^{127} \tilde{c}_n e^{i\pi jn/N}, \quad j = -128, -127, \ldots, 127.$$

We then have

$$\tilde{u}_j \doteq u\left(\frac{j}{128}\right), \quad j = -128, -127, \ldots, 127.$$

We graph the error in the computed solution in Figure 12.3. ∎

12.2.3 A note about using packaged FFT routines

There is more than one way to define the DFT, and therefore various implementations of the FFT may implement slightly different formulas. For example, the formulas for the DFT and the inverse DFT are asymmetric in that the factor of $1/M$ appears in the DFT (12.9) but not in the inverse DFT (12.10). However, some software packages put the factor of $1/M$ in the inverse DFT instead.[71] It is also possible to make symmetric formulas by putting a factor of $1/\sqrt{M}$ in each of the DFT and the inverse DFT.[72] One can also define the DFT while indexing from $-N$ to $N-1$, as in (12.8).

Given the diversity in definitions of the DFT and therefore of the FFT (which, as the reader should bear in mind, is just a fast algorithm for computing the DFT), it is essential that one knows which definition is being used before trying to apply a packaged FFT routine.

---
[71] MATLAB is one such package.
[72] *Mathematica* does this by default.

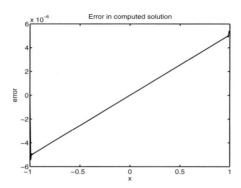

Figure 12.3. *Error in computed solution in Example* 12.4.

12.2.4 Fast transforms and other boundary conditions; the discrete sine transform

Fast transform methods are not restricted to problems with periodic boundary conditions. Discrete transforms can be defined that are useful for working with sine, cosine, quarter-wave sine, and quarter-wave cosine series, as well as with the full Fourier series. A comprehensive source of software for the corresponding fast transforms is [64].

As an example, we discuss the discrete sine transform (DST) and its use in solving problems with Dirichlet conditions. The DST of a sequence $f_1, f_2, \ldots, f_{N-1}$ is defined by

$$F_n = 2 \sum_{j=1}^{N-1} f_j \sin\left(\frac{\pi n j}{N}\right), \quad n = 0, 1, \ldots, N-1. \tag{12.15}$$

It can be shown that the DST is its own inverse, up to a multiplicative constant: applying the DST to $F_0, F_1, \ldots, F_{N-1}$ produces the original sequence multiplied by $2N$ (see Exercise 12.2.9).

In solving a BVP or an IBVP with Dirichlet conditions, the main calculations are

1. computing the Fourier sine coefficients of a given function;

2. computing the solution from its Fourier coefficients.

As with the DFT and complex Fourier series, we can solve these problems approximately using the DST. If $f \in C[0, \ell]$, then the Fourier sine coefficients of f are given by

$$c_n = \frac{2}{\ell} \int_0^\ell f(x) \sin\left(\frac{n\pi x}{\ell}\right) dx, \quad n = 1, 2, 3, \ldots.$$

Using the regular grid $x_j = jh$, $j = 0, 1, 2, \ldots, N$, $h = \ell/N$, and the trapezoidal rule, we obtain

$$c_n \doteq \frac{2}{\ell} \sum_{j=1}^{N-1} f(x_j) \sin\left(\frac{n\pi x_j}{\ell}\right) h = \frac{2}{N} \sum_{j=1}^{N-1} f_j \sin\left(\frac{\pi n j}{N}\right),$$

12.2. Fourier series and the FFT

where $f_j = f(x_j)$. The trapezoidal rule simplifies due to the fact that $\sin(0)$ and $\sin(n\pi)$ are both zero. We therefore see that the approximate Fourier sine coefficients are just a multiple of the DST of the sequence $f(x_1), f(x_2), \ldots, f(x_{N-1})$.

On the other hand, suppose the Fourier sine series of u is

$$u(x) = \sum_{n=1}^{\infty} c_n \sin\left(\frac{n\pi x}{\ell}\right)$$

and we know $c_1, c_2, \ldots, c_{N-1}$ (or approximations to them). We then have, for $j = 1, 2, \ldots, N-1$,

$$u(x_j) \doteq \sum_{n=1}^{N-1} c_n \sin\left(\frac{n\pi x_j}{\ell}\right) = \sum_{n=1}^{N-1} c_n \sin\left(\frac{\pi n j}{N}\right),$$

and we see that u can be estimated on a regular grid by another application of the DST.

12.2.5 Computing the DST using the FFT

As mentioned above, there are special programs available for computing transforms, such as the DST, using FFT-like algorithms. However, these programs are more specialized and therefore less widely available than the FFT.[73] Fortunately, the DST can be computed using the DFT (and hence the FFT) and a few additional manipulations. We will now explain how to do this.

We assume that $\{f_j\}_{j=1}^{N-1}$ is a sequence of real numbers and that we wish to compute the DST of $\{f_j\}$. We define a new sequence $\{\tilde{f}_j\}_{j=0}^{2N-1}$ by

$$\tilde{f}_j = \begin{cases} f_j, & j = 1, 2, \ldots, N-1, \\ -f_{2N-j}, & j = N+1, N+2, \ldots, 2N-1, \\ 0, & j = 0, N. \end{cases}$$

We then define $\{\tilde{F}_n\}_{n=0}^{2N-1}$ to be the DFT of $\{\tilde{f}_j\}$:

$$\tilde{F}_n = \frac{1}{2N}\sum_{j=0}^{2N-1} \tilde{f}_j e^{-\pi i j n/N} = \frac{1}{2N}\sum_{j=0}^{2N-1} \tilde{f}_j \cos\left(\frac{\pi j n}{N}\right) - \frac{i}{2N}\sum_{j=0}^{2N-1} \tilde{f}_j \sin\left(\frac{\pi j n}{N}\right).$$

Since we are interested in the DST, we will look at the imaginary part of $\tilde{F}_n$, which is (using the fact that $\tilde{f}_0 = \tilde{f}_N = 0$)

$$\operatorname{Im}(\tilde{F}_n) = -\frac{1}{2N}\sum_{j=0}^{2N-1} \tilde{f}_j \sin\left(\frac{\pi j n}{N}\right)$$

$$= -\frac{1}{2N}\left\{\sum_{j=0}^{N-1} \tilde{f}_j \sin\left(\frac{\pi j n}{N}\right) + \sum_{j=N+1}^{2N-1} \tilde{f}_j \sin\left(\frac{\pi j n}{N}\right)\right\}$$

$$= -\frac{1}{2N}\left\{\sum_{j=1}^{N-1} f_j \sin\left(\frac{\pi j n}{N}\right) - \sum_{j=1}^{N-1} f_{N-j} \sin\left(\frac{\pi (j+N) n}{N}\right)\right\}$$

---

[73] For example, MATLAB has an FFT command, but no fast DST command.

(using the fact that $\tilde{f}_{j+N} = f_{2N-N-j} = f_{N-j}$ for $j = 1, 2, \ldots, N-1$). We have

$$-\sum_{j=1}^{N-1} f_{N-j} \sin\left(\frac{\pi(j+N)n}{N}\right) = -\sum_{j=1}^{N-1} f_j \sin\left(\frac{\pi(2N-j)n}{N}\right)$$

$$= -\sum_{j=1}^{N-1} f_j \sin\left(2n\pi - \frac{\pi jn}{N}\right)$$

$$= \sum_{j=1}^{N-1} f_j \sin\left(\frac{\pi jn}{N}\right),$$

and so

$$\mathrm{Im}(\tilde{F}_n) = -\frac{1}{2N}\left\{ 2 \sum_{j=1}^{N-1} f_j \sin\left(\frac{\pi jn}{N}\right) \right\}.$$

It follows that the sequence

$$\{-2N\mathrm{Im}(\tilde{F}_n)\}_{n=1}^{N-1}$$

is the DST of $\{f_j\}$.

The reader should note that, when using the FFT to compute the DST, N should be a product of small primes for efficiency.

A similar technique allows one to compute the discrete cosine transform (DCT) using the FFT. The DCT is explored in Exercise 12.2.10.

Exercises

1. Define a sequence $\{f_j\}_{j=0}^{19}$ by

$$f_j = \sin\left(\frac{3\pi j}{10}\right).$$

Compute the DFT $\{F_n\}$ of $\{f_j\}$ and graph its magnitude.

2. Let $f : [-\pi, \pi] \to \mathbf{R}$ be defined by $f(x) = x^2$. Use the DFT to estimate the complex Fourier coefficients c_n of f, $n = -8, -7, \ldots, 7$. Compare with the exact values: $c_0 = \pi^2/3$, $c_n = 2(-1)^n/n^2$, $n = \pm 1, \pm 2, \ldots$.

3. Let $f : [-1, 1] \to \mathbf{R}$ be defined by $f(x) = x(1-x^2)$. The complex Fourier coefficients of f are

$$c_0 = 0, \quad c_n = \frac{6i(-1)^n}{n^3 \pi^3}, \quad n = \pm 1, \pm 2, \ldots.$$

Use the inverse DFT to estimate f on a grid, and graph both f and the computed estimate on $[-1, 1]$. Use $c_{-16}, c_{-15}, \ldots, c_{15}$.

4. Show that, with I defined by (12.11), the interpolation equations (12.12) and (12.13) hold.

12.2. Fourier series and the FFT

5. Let $a_0, a_1, \ldots, a_{N-1}$ be a sequence of real or complex numbers, and define

$$A_n = \frac{1}{N} \sum_{j=0}^{N-1} a_j e^{-2\pi i n j / N}, \quad n = 0, 1, \ldots, N-1.$$

Prove that

$$a_j = \sum_{n=0}^{N-1} A_n e^{2\pi i n j / N}, \quad j = 0, 1, \ldots, N-1.$$

Hint: Substitute the formula for A_n into

$$\sum_{n=0}^{N-1} A_n e^{2\pi i n j / N}$$

and interchange the order of summation. (A dummy index, say m, must be used in place of j in the formula for A_n.) Look for a geometric series and use

$$\sum_{n=0}^{K} r^n = \frac{r^{K+1} - 1}{r - 1}.$$

6. Suppose u satisfies periodic boundary conditions on the interval $[-\ell, \ell]$, and

$$\sum_{n=-\infty}^{\infty} c_n e^{i \pi n x / \ell}$$

is the complex Fourier series of u. Prove that

$$\sum_{n=-\infty}^{\infty} \frac{n^2 \pi^2}{\ell^2} c_n e^{i \pi n x / \ell}$$

is the complex Fourier series of $-d^2 u / dx^2$.

7. Use the DST to estimate the first 15 Fourier coefficients of the function $f(x) = x(1-x)$ on the interval $[0, 1]$. Compare to the exact values

$$a_n = -\frac{4(-1 + (-1)^n)}{n^3 \pi^3}, \quad n = 1, 2, 3, \ldots.$$

8. The Fourier sine coefficients of $f(x) = x(1-x)$ on $[0, 1]$ are

$$a_n = -\frac{4(-1 + (-1)^n)}{n^3 \pi^3}, \quad n = 1, 2, 3, \ldots.$$

Use the DST and $a_1, a_2, \ldots, a_{63}$ to estimate f on a grid with 63 evenly spaced points.

9. Let $f_0, f_1, \ldots, f_{N-1}$ be a sequence of real numbers, and define $F_0, F_1, \ldots, F_{N-1}$ by (12.15). Then define $g_0, g_1, \ldots, g_{N-1}$ by

$$g_j = 2\sum_{n=1}^{N-1} F_n \sin\left(\frac{\pi n j}{N}\right), \quad j = 0, 1, \ldots, N-1.$$

Show that $g_j = 2N f_j$, $j = 0, 1, \ldots, N-1$. Hint: Proceed according to the hint in Exercise 5. You can use the fact that the vectors

$$\begin{bmatrix} \sin\left(\frac{m\pi}{N}\right) \\ \sin\left(\frac{2m\pi}{N}\right) \\ \vdots \\ \sin\left(\frac{(N-1)m\pi}{N}\right) \end{bmatrix}, \begin{bmatrix} \sin\left(\frac{j\pi}{N}\right) \\ \sin\left(\frac{2j\pi}{N}\right) \\ \vdots \\ \sin\left(\frac{(N-1)j\pi}{N}\right) \end{bmatrix}$$

are orthogonal in $\mathbf{R}^{N-1}$ for $j, m = 1, 2, \ldots, N-1$, $j \neq m$ (see Exercise 3.5.7). Also, each of these vectors has norm $\sqrt{N/2}$ (see Exercise 12.1.7).

10. The DCT maps a sequence $\{f_j\}_{j=0}^N$ of real numbers to the sequence $\{F_n\}_{n=0}^N$, where

$$F_n = f_0 + 2\sum_{j=1}^{N-1} f_j \cos\left(\frac{\pi n j}{N}\right) + (-1)^n f_N, \quad n = 0, 1, \ldots, N.$$

(a) Reasoning as in Section 12.2.4, show how the DCT can be used to estimate the Fourier cosine coefficients of a function in $C[0, \ell]$.

(b) Reasoning as in Section 12.2.4, show how the DCT can be used to estimate a function from its Fourier cosine coefficients.

(c) Modifying the technique presented in Section 12.2.5, show how to compute the DCT using the DFT (and hence the FFT). (Hint: Given $\{f_j\}_{j=0}^N$, define $\{\tilde{f}_j\}_{j=-N}^{N-1}$ by

$$\tilde{f}_j = f_{|j|}, \quad j = -N, -N+1, \ldots, N-1,$$

and treat $\{\tilde{f}_j\}_{j=-N}^{N-1}$ by the three-step process on pages 544–545.)

(d) Show that the DCT is its own inverse, up to a constant multiple. To be precise, show that if the DCT is applied to a given sequence and then the DCT is applied to the result, one obtains $2N$ times the original sequence.

12.3 Relationship of sine and cosine series to the full Fourier series

In Section 12.1, we showed that the complex and full Fourier series are equivalent for a real-valued function. We will now show that both the Fourier cosine and the Fourier sine series can be recognized as special cases of the full Fourier series and hence of the

12.3. Relationship of sine and cosine series to the full Fourier series

complex Fourier series. This will show that the complex Fourier series is the most general concept.

To understand the relationships between the various Fourier series for real-valued functions, we must understand the following terms.

Definition 12.5. *Let $f : \mathbf{R} \to \mathbf{R}$. Then f is*

1. *odd if $f(-x) = -f(x)$ for all $x \in \mathbf{R}$;*

2. *even if $f(-x) = f(x)$ for all $x \in \mathbf{R}$;*

3. *periodic with period T if $T > 0$, $f(x+T) = f(x)$ for all $x \in \mathbf{R}$, and this condition does not hold for any smaller positive value of T.*

Examples of odd functions include polynomials with only odd powers and $\sin(x)$. Polynomials with only even powers and $\cos(x)$ are examples of even functions, while sine and cosine are the prototypical periodic functions (both have period 2π). The algebraic properties defining odd and even functions imply that the graph of an odd function is symmetric through the origin, while the graph of an even function is symmetric across the y-axis (see Figure 12.4).

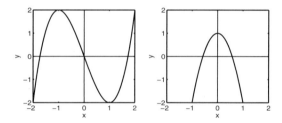

Figure 12.4. *Examples of odd (left) and even (right) functions.*

We will show that the full Fourier series of an odd function reduces to a sine series, and that the full Fourier series of an even function reduces to a cosine series. We need the following preliminary result.

Lemma 12.6.

1. *Suppose $f : \mathbf{R} \to \mathbf{R}$ is odd. Then*

$$\int_{-\ell}^{\ell} f(x)\,dx = 0.$$

2. *Suppose $f : \mathbf{R} \to \mathbf{R}$ is even. Then*

$$\int_{-\ell}^{\ell} f(x)\,dx = 2\int_{0}^{\ell} f(x)\,dx.$$

Proof. Suppose f is odd. Then

$$\int_{-\ell}^{\ell} f(x)\,dx = \int_{\ell}^{0} f(x)\,dx + \int_{0}^{\ell} f(x)\,dx.$$

Making the change of variables $x = -s$ in the first integral on the right, we obtain

$$\int_{-\ell}^{\ell} f(x)\,dx = -\int_{\ell}^{0} f(-s)\,ds + \int_{0}^{\ell} f(x)\,dx$$
$$= -\int_{0}^{\ell} f(s)\,ds + \int_{0}^{\ell} f(x)\,dx \text{ (since } f(-s) = -f(s)) = 0.$$

The result for even f is proved by making the same change of variables. □

We can now derive the main result.

Theorem 12.7. *Let $f \in L^2(-\ell, \ell)$, and suppose that*

$$a_0 + \sum_{n=1}^{\infty} \left\{ a_n \cos\left(\frac{n\pi x}{\ell}\right) + b_n \sin\left(\frac{n\pi x}{\ell}\right) \right\}$$

is the full Fourier series of f.

1. *If f is odd, then $a_n = 0$, $n = 0, 1, 2, \ldots$, and*

$$b_n = \frac{2}{\ell} \int_0^{\ell} f(x) \sin\left(\frac{n\pi x}{\ell}\right) dx.$$

That is, the full Fourier series (on $[-\ell, \ell]$) of an odd function is the same as its Fourier sine series (on $[0, \ell]$).

2. *If f is even, then $b_n = 0$, $n = 1, 2, 3, \ldots$, and*

$$a_0 = \frac{1}{\ell} \int_0^{\ell} f(x)\,dx,$$
$$a_n = \frac{2}{\ell} \int_0^{\ell} f(x) \cos\left(\frac{n\pi x}{\ell}\right) dx.$$

That is, the full Fourier series (on $[-\ell, \ell]$) of an even function is the same as its Fourier cosine series (on $[0, \ell]$).

Proof. If f is odd, then $f(x)\cos(n\pi x/\ell)$ is odd and $f(x)\sin(n\pi x/\ell)$ is even. This, together with the previous lemma, yields the first result.

If f is even, then $f(x)\cos(n\pi x/\ell)$ is even and $f(x)\sin(n\pi x/\ell)$ is odd. From this we obtain the second conclusion. □

12.3. Relationship of sine and cosine series to the full Fourier series

We can use the preceding result in the following fashion. Suppose we wish to understand the convergence of the Fourier sine series of $f : (0, \ell) \to \mathbf{R}$. Define $f_{odd} : (-\ell, \ell) \to \mathbf{R}$, the *odd extension* of f, by

$$f_{odd}(x) = \begin{cases} f(x), & 0 < x < \ell, \\ -f(-x), & -\ell < x < 0. \end{cases}$$

Then, by the previous theorem, the (full) Fourier series of f_{odd} is the sine series of f, so the convergence of the sine series can be understood in terms of the convergence of a related (full) Fourier series.

Similarly, given $f : (0, \ell) \to \mathbf{R}$, we define $f_{even} : (-\ell, \ell) \to \mathbf{R}$, the *even extension* of f, by

$$f_{even}(x) = \begin{cases} f(x), & 0 < x < \ell, \\ f(-x), & -\ell < x < 0. \end{cases}$$

The Fourier series of f_{even} is the cosine series of f, so, again, the convergence of the cosine series can be examined in terms of the convergence of a related Fourier series.

Figure 12.5 shows the odd and even extensions of $f : [-1, 1] \to \mathbf{R}$ defined by $f(x) = 1 + x$.

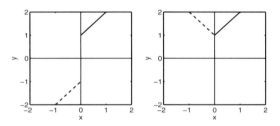

Figure 12.5. *The function $f(x) = 1 + x$ and its odd (left) and even (right) extensions.*

Exercises

1. Compute the Fourier sine series of $f(x) = x$ on the interval $[-1, 1]$ and graph the sum of the first 50 terms on the interval $[-3, 3]$. To what function does the series appear to converge? What is the period of this function?

2. Compute the Fourier cosine series of $f(x) = x$ on the interval $[-1, 1]$ and graph it on the interval $[-3, 3]$. To what function does the series appear to converge? What is the period of this function?

3. Suppose $f : [0, \ell] \to \mathbf{R}$ is continuous.
   (a) Under what conditions is f_{odd} continuous?
   (b) Under what conditions is f_{even} continuous?

4. Consider a function $f : [-\ell, \ell] \to \mathbf{R}$, and let c_n, $n = 0, \pm 1, \pm 2, \ldots$, be its complex Fourier coefficients. What special property do these coefficients have if
   (a) f is odd?
   (b) f is even?

5. Show how to relate the quarter-wave sine series of $f : [0,\ell] \to \mathbf{R}$ to the full Fourier series of a related function. (Hint: This other function will be defined on the interval $[-2\ell, 2\ell]$.)

6. Show how to relate the quarter-wave cosine series of $f : [0,\ell] \to \mathbf{R}$ to the full Fourier series of a related function. (Hint: This other function will be defined on the interval $[-2\ell, 2\ell]$.)

12.4 Pointwise convergence of Fourier series

Given the relationships that exist among the various Fourier series, it suffices to discuss the convergence of complex Fourier series. The convergence of other series, such as the Fourier sine series, will then follow directly.

If f is a real- or complex-valued function defined on $[-\ell, \ell]$, then the partial sums

$$f_N(x) = \sum_{n=-N}^{N} c_n e^{i\pi nx/\ell}$$

form a sequence of functions. Before discussing the convergence of these partial sums specifically, we describe different ways in which a sequence of functions can converge.

12.4.1 Modes of convergence for sequences of functions

Given any sequence $\{f_N\}_{N=1}^{\infty}$ of functions defined on $[a,b]$ and any (target) function f, also defined on $[a,b]$, we define three types of convergence of the sequence to f. That is, we assign three different meanings to "$f_N \to f$ as $N \to \infty$."

1. We say that $\{f_N\}$ converges *pointwise* to f on $[a,b]$ if, for each $x \in [a,b]$, we have $f_N(x) \to f(x)$ as $N \to \infty$ (i.e., $|f(x) - f_N(x)| \to 0$ as $N \to \infty$).

2. We say that $\{f_N\}$ converges to f in L^2 (or in the *mean-square sense*) if $\|f - f_N\| \to 0$ as $N \to \infty$. The reader will recall that the (L^2) norm of a real- or complex-valued function g on $[a,b]$ is

$$\|g\| = \sqrt{\int_a^b |g(x)|^2\, dx}.$$

   Therefore, $f_N \to f$ in the mean-square sense means that

$$\sqrt{\int_a^b |f(x) - f_N(x)|^2\, dx} \to 0 \text{ as } n \to \infty.$$

3. We say that $\{f_N\}$ converges to f *uniformly* on $[a,b]$ if

$$\max\{|f(x) - f_N(x)| : x \in [a,b]\} \to 0 \text{ as } N \to \infty.$$

12.4. Pointwise convergence of Fourier series

Actually, this definition is stated correctly only if all of the functions involved are continuous, so that the maximum is guaranteed to exist. The general definition is: $\{f_N\}$ converges to f *uniformly* on $[a,b]$ if, given any $\epsilon > 0$, there exists a positive integer N_ϵ such that, if $N \geq N_\epsilon$ and $x \in [a,b]$, then $|f(x) - f_N(x)| < \epsilon$. The intuitive meaning of this definition is that, given any small tolerance ϵ, by going far enough out in the sequence, f_N will approximate f to within this tolerance *uniformly*, that is, on the entire interval.

The following theorem shows that the uniform convergence of a sequence implies its convergence in the other two senses. This theorem is followed by examples that show that no other conclusions can be drawn in general.

Theorem 12.8. *Let $\{f_N\}$ be a sequence of complex-valued functions defined on $[a,b]$. If this sequence converges uniformly on $[a,b]$ to $f : [a,b] \to \mathbf{C}$, then it also converges to f pointwise and in the mean-square sense.*

Proof. See Exercise 12.4.5. □

Example 12.9. We define

$$g_N(x) = \begin{cases} Nx, & 0 \leq x < \frac{1}{N}, \\ 1 - N\left(x - \frac{1}{N}\right), & \frac{1}{N} \leq x < \frac{2}{N}, \\ 0, & \frac{2}{N} \leq x \leq 1. \end{cases}$$

Figure 12.6 shows the graphs of g_5, g_{10}, and g_{20}, which suggest that $\{g_N\}$ converges pointwise to the zero function. Indeed, $g_N(0) = 0$ for all N, so clearly $g_N(0) \to 0$, and if $0 < x \leq 1$, then for N sufficiently large, $2/N < x$. Therefore $g_N(x) = 0$ for all N sufficiently large, and so $g_N(x) \to 0$.

A direct calculation shows that $g_N \to 0$ in L^2:

$$\sqrt{\int_0^1 |g_N(x)|^2 \, dx} = \sqrt{\frac{2}{3N}} \to 0.$$

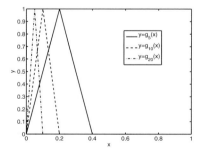

Figure 12.6. *The functions g_5, g_{10}, g_{20} (see Example 12.9).*

The convergence is not uniform, since

$$\max\{|g_N(x)-0| : x \in [0,1]\} = g\left(\frac{1}{N}\right) = 1 \text{ for all } N. \quad \blacksquare$$

Example 12.10. This example is almost the same as the previous one. We define

$$h_N(x) = \begin{cases} N^2 x, & 0 \le x < \frac{1}{N}, \\ N - N^2\left(x - \frac{1}{N}\right), & \frac{1}{N} \le x < \frac{2}{N}, \\ 0, & \frac{2}{N} \le x \le 1 \end{cases}$$

(see Figure 12.7). We have $h_N \to 0$ pointwise on $[0,1]$, and $\{h_N\}$ does not converge uniformly to 0 on $[0,1]$, by essentially the same arguments as in the previous example. In this example, however, the sequence also fails to converge in the L^2 norm. We have

$$\|h_N - 0\|_{L^2} = \sqrt{\int_0^1 |h_N(x)|^2 \, dx}$$

$$= \sqrt{\frac{2N}{3}} \to \infty. \quad \blacksquare$$

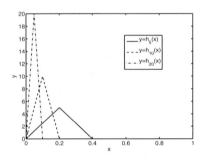

Figure 12.7. *The functions h_5, h_{10}, h_{20} (see Example 12.10).*

We saw in Theorem 12.8 that a uniformly convergent sequence also converges pointwise and in the mean-square sense. Example 12.9 shows that neither pointwise nor mean-square convergence implies uniform convergence, while Example 12.10 shows that pointwise convergence does not imply mean-square convergence. We will see, in the context of Fourier series, that mean-square convergence does not imply pointwise convergence either.

12.4.2 Pointwise convergence of the complex Fourier series

We now begin to develop conditions under which the Fourier series of a function converges pointwise, uniformly, and in the mean-square sense. We begin with pointwise convergence.

12.4. Pointwise convergence of Fourier series

The partial Fourier series as integration against a kernel

Our starting point is a direct calculation showing that a partial Fourier series of a function f can be written as the integral of f times a function approximating a delta function. The difficult part of the proof will be verifying the sifting property. (Delta functions and the sifting property were discussed in Section 9.1, but the essence of that discussion will be repeated here, so it is not necessary to have studied Section 9.1. The concepts from that section must be modified slightly here anyway, to deal with periodicity.)

We assume that the complex Fourier series of coefficients of $f : [-\ell, \ell] \to \mathbf{C}$ are

$$c_n = \frac{1}{2\ell} \int_{-\ell}^{\ell} f(s) e^{-i\pi ns/\ell}\, ds, \quad n = 0, \pm 1, \pm 2, \ldots.$$

We write

$$f_N(x) = \sum_{n=-N}^{N} c_n e^{i\pi nx/\ell},$$

and then substitute the formula for c_n and simplify:

$$f_N(x) = \sum_{n=-N}^{N} \left\{ \frac{1}{2\ell} \int_{-\ell}^{\ell} f(s) e^{-i\pi ns/\ell}\, ds \right\} e^{i\pi nx/\ell}$$

$$= \int_{-\ell}^{\ell} \left\{ \frac{1}{2\ell} \sum_{n=-N}^{N} e^{i\pi n(x-s)/\ell} \right\} f(s)\, ds.$$

We already see that the f_N can be written as

$$f_N(x) = \int_{-\ell}^{\ell} K_N(x-s) f(s)\, ds, \qquad (12.16)$$

where the *Dirichlet kernel* K_N is defined by

$$K_N(\theta) = \frac{1}{2\ell} \sum_{n=-N}^{N} e^{i\pi n\theta/\ell}.$$

With the formula for a finite geometric sum,

$$\sum_{n=k_1}^{k_2} r^n = \frac{r^{k_2+1} - r^{k_1}}{r - 1},$$

and using some clever manipulations, we can simplify the formula for K_N considerably:

$$\begin{aligned} K_N(\theta) &= \frac{1}{2\ell} \sum_{n=-N}^{N} e^{i\pi n\theta/\ell} = \frac{1}{2\ell} \sum_{n=-N}^{N} \left(e^{i\pi\theta/\ell}\right)^n \\ &= \frac{1}{2\ell} \frac{\left(e^{i\pi\theta/\ell}\right)^{N+1} - \left(e^{i\pi\theta/\ell}\right)^{-N}}{e^{i\pi\theta/\ell} - 1} \\ &= \frac{1}{2\ell} \frac{\left(e^{i\pi\theta/\ell}\right)^{N+1/2} - \left(e^{i\pi\theta/\ell}\right)^{-(N+1/2)}}{\left(e^{i\pi\theta/\ell}\right)^{1/2} - \left(e^{i\pi\theta/\ell}\right)^{-1/2}} \\ &= \frac{1}{2\ell} \frac{e^{i\pi(N+1/2)\theta/\ell} - e^{-i\pi(N+1/2)\theta/\ell}}{e^{i\pi(1/2)\theta/\ell} - e^{-i\pi(1/2)\theta/\ell}} \\ &= \frac{1}{2\ell} \frac{2i \sin\left(\frac{(N+\frac{1}{2})\pi\theta}{\ell}\right)}{2i \sin\left(\frac{\pi\theta}{2\ell}\right)} \\ &= \frac{\sin\left(\frac{(2N+1)\pi\theta}{2\ell}\right)}{2\ell \sin\left(\frac{\pi\theta}{2\ell}\right)}. \end{aligned}$$

The kernel K_N, for $N = 20$, is graphed in Figure 12.8.

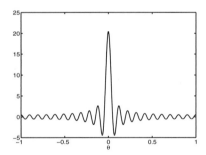

Figure 12.8. *The kernel K_{20}.*

Periodic convolution

The *convolution* of two functions $g : \mathbf{R} \to \mathbf{C}$ and $h : \mathbf{R} \to \mathbf{C}$ is defined by

$$(g*h)(x) = \int g(x-s)h(s)\,ds.$$

The integral sign without limits means integration over the entire real line (we do not discuss here conditions on g and h that guarantee that these integrals exists). The formula for f_N is similar to a convolution, except that the integration is over the finite interval $[-\ell, \ell]$. A fundamental property of an ordinary convolution is its symmetry; the change of variables

12.4. Pointwise convergence of Fourier series

that replaces $x-s$ by s yields

$$\int g(x-s)h(s)\,ds = \int g(s)h(x-s)\,ds.$$

A similar formula holds for integrals such as that in (12.16), provided the functions involved are periodic.

A direct calculation shows that if g, h are 2ℓ-periodic functions defined on **R**, then

$$\int_{-\ell}^{\ell} g(x-s)h(s)\,ds = \int_{x-\ell}^{x+\ell} g(s)h(x-s)\,ds = \int_{-\ell}^{\ell} g(s)h(x-s)\,ds.$$

The last step follows from the fact that $s \mapsto g(s)h(x-s)$ is 2ℓ-periodic, and so its integral over any interval of length 2ℓ is the same. We refer to

$$\int_{-\ell}^{\ell} g(x-s)f(s)\,ds$$

as the *periodic convolution* of g and h.

We can extend the idea of a periodic convolution to nonperiodic functions by using the *periodic extension* of a function. If g is defined on $[-\ell,\ell]$, then its *periodic extension* is the function g_{per} defined for all $x \in \mathbf{R}$ by

$$g_{per}(x) = \begin{cases} g(x), & -\ell < x < \ell, \\ g(x-2k\ell), & (2k-1)\ell < x < 2k\ell. \end{cases}$$

For example, in Figure 12.9 we graph the function g_{per}, where $g(x) = x$ on $[-1,1]$.

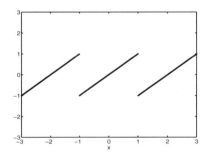

Figure 12.9. *The periodic extension of* $g : [-1,1] \to \mathbf{R}$, $g(x) = x$.

For any g and h defined on $[-\ell,\ell]$, periodic or not, we interpret the periodic convolution of g and h to be

$$\int_{-\ell}^{\ell} g_{per}(x-s)h_{per}(s)\,ds.$$

We will often write this convolution as simply

$$\int_{-\ell}^{\ell} g(x-s)h(s)\,ds,$$

but g and h are assumed to be replaced by their periodic extensions if necessary (that is, if they are not periodic).

The kernel K_N is 2ℓ-periodic. Provided the complex Fourier series of f converges, the limit is obviously a 2ℓ-periodic function (because each function $e^{i\pi nx/\ell}$ is 2ℓ-periodic). Therefore, it is natural to consider f_{per}, the periodic extension of f, and to interpret (12.16) as a periodic convolution. We then have

$$\int_{-\ell}^{\ell} K_N(x-s)f(s)\,ds = \int_{-\ell}^{\ell} K_N(s)f_{per}(x-s)\,ds, \qquad (12.17)$$

a fact which we will use below.

K_N as an approximate delta function

An examination of the graph of the kernel K_N (see Figure 12.8) shows that most of its "weight" is concentrated on a small interval around $\theta = 0$. The effect of this on the periodic convolution

$$\int_{-\ell}^{\ell} K_N(s)f_{per}(x-s)\,ds$$

is that the integral produces a weighted average of the values of f, with most of the weight on the values of $f(s)$ near $s = x$. Moreover, this effect is accentuated as $N \to \infty$, so that if f is regular enough, we obtain pointwise convergence. Thus K_N acts like an approximate delta function.

How regular must f be in order for convergence to occur? It turns out that mere continuity is not sufficient, as there are continuous functions on $[-\ell, \ell]$ whose Fourier series fail to converge at an infinite number of points. Some differentiability of f is necessary to guarantee convergence. On the other hand, we do not really want to require f to be continuous, since we often encounter discontinuities in practical problems, if not of f, then of f_{per}.

A suitable notion of regularity for our purposes is that of piecewise smoothness. We make the following definitions.

Definition 12.11. *Suppose f is a real- or complex-valued function defined on (a,b), except possibly at a finite number of points. We say that f has a jump discontinuity at $x_0 \in (a,b)$ if*

$$\lim_{x \to x_0^-} f(x) \text{ and } \lim_{x \to x_0^+} f(x)$$

both exist but are not equal. (Recall that if both one-sided limits exist and are equal to $f(x_0)$, then f is continuous at $x = x_0$.)

We say that f is piecewise continuous *on (a,b) if*

1. *f is continuous at all but a finite number of points in (a,b);*

2. *every discontinuity of f in (a,b) is a jump discontinuity;*

3. *$\lim_{x \to a^+} f(x)$ and $\lim_{x \to b^-} f(x)$ exist.*

12.4. Pointwise convergence of Fourier series

Finally, we say that f is piecewise smooth *on (a,b) if both f and df/dx are piecewise continuous on (a,b).*

An example of a piecewise smooth function is given in Figure 12.10.

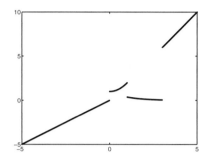

Figure 12.10. *A piecewise smooth function.*

Before we give the main result, we give two lemmas that will be useful.

Lemma 12.12.
$$\int_{-\ell}^{0} K_N(\theta)\,d\theta = \int_{0}^{\ell} K_N(\theta)\,d\theta = \frac{1}{2}. \tag{12.18}$$

Proof. This is proved by direct calculation (see Exercise 12.4.4). □

The next lemma is called Bessel's inequality and follows directly from the projection theorem.

Lemma 12.13. *Suppose $\{\phi_j : j = 1, 2, \ldots\}$ is an orthogonal sequence in an (infinite-dimensional) inner product space V, and $f \in V$. Then*
$$\sum_{j=1}^{\infty} \frac{|(f, \phi_j)|^2}{(\phi_j, \phi_j)} \leq \|f\|_V^2.$$

Proof. By the projection theorem, for each N,
$$f_N = \sum_{j=1}^{N} \frac{(f, \phi_j)}{(\phi_j, \phi_j)} \phi_j$$

is the element of the subspace $V_N = \text{span}\{\phi_1, \phi_2, \ldots, \phi_N\}$ closest to f, and f_N is orthogonal to $f - f_N$. Therefore, by the Pythagorean theorem,
$$\|f_N\|_V^2 + \|f - f_N\|_V^2 = \|f\|_V^2.$$

This implies that
$$\|f_N\|_V^2 \leq \|f\|_V^2 \text{ for all } N = 1, 2, 3, \ldots.$$

Moreover, since $\phi_1, \phi_2, \phi_3, \ldots$ are mutually orthogonal, we have (also by the Pythagorean theorem)

$$\|f_N\|_V^2 = \sum_{j=1}^N \frac{|(f,\phi_j)|^2}{(\phi_j,\phi_j)}.$$

Therefore,

$$\sum_{j=1}^N \frac{|(f,\phi_j)|^2}{(\phi_j,\phi_j)} \le \|f\|_V^2$$

holds for each N, and the result follows. $\square$

Since a Fourier series is based on an orthogonal series, Bessel's inequality can be applied to the sequence of Fourier coefficients. For instance, if the complex Fourier coefficients of $f \in L^2(-\ell,\ell)$ are $c_n, n = 0, \pm 1, \pm 2, \ldots$, then

$$c_n = \frac{(f, e^{i\pi nx/\ell})}{(e^{i\pi nx/\ell}, e^{i\pi nx/\ell})}$$

and

$$(e^{i\pi nx/\ell}, e^{i\pi nx/\ell}) = 2\ell.$$

Therefore,

$$\frac{|(f, e^{i\pi nx/\ell})|^2}{(e^{i\pi nx/\ell}, e^{i\pi nx/\ell})} = 2\ell |c_n|^2.$$

Bessel's inequality yields

$$2\ell \sum_{n=-\infty}^\infty |c_n|^2 \le \|f\|_{L^2}^2.$$

Similar results hold for Fourier sine or cosine coefficients.

We can now state and prove the main result concerning the pointwise convergence of Fourier series. The reader should note that the convergence of the Fourier series is naturally expressed in terms of the properties of f_{per} rather than those of f.

Theorem 12.14. *Suppose $f : (-\ell,\ell) \to \mathbf{C}$ is piecewise smooth. Then the complex Fourier series of f,*

$$\sum_{n=-\infty}^\infty c_n e^{i\pi nx/\ell},$$

converges to $f_{per}(x)$ if x is a point of continuity of f_{per}, and to

$$\frac{1}{2}\left[\lim_{s \to x^-} f_{per}(s) + \lim_{s \to x^+} f_{per}(s)\right]$$

if f_{per} has a jump discontinuity at x.

12.4. Pointwise convergence of Fourier series

Proof. As before, we write

$$f_N(x) = \sum_{n=-N}^{N} c_n e^{i\pi nx/\ell} = \int_{-\ell}^{\ell} K_N(s) f(x-s) \, ds.$$

If f_{per} is continuous at x, then

$$f_{per}(x) = \frac{1}{2}\left[\lim_{s \to x^-} f_{per}(s) + \lim_{s \to x^+} f_{per}(x)\right].$$

From now on, for conciseness, we will write f rather than f_{per} and remember to interpret $f(s)$ in terms of the periodic extension of f whenever s is outside of the interval $(-\ell, \ell)$. Also, we write

$$f(x-) = \lim_{s \to x^-} f(s), \quad f(x+) = \lim_{s \to x^+} f(s).$$

With this understanding, our task is to show that, for each $x \in [-\ell, \ell]$,

$$\lim_{N \to \infty} f_N(x) = \frac{1}{2}[f(x-) + f(x+)].$$

To do this, it suffices to show that

$$\lim_{N \to \infty} \int_{-\ell}^{0} K_N(s) f(x-s) \, ds = \frac{1}{2} f(x+) \tag{12.19}$$

and

$$\lim_{N \to \infty} \int_{0}^{\ell} K_N(s) f(x-s) \, ds = \frac{1}{2} f(x-). \tag{12.20}$$

We will prove that (12.20) holds; the proof of (12.19) is similar. Equation (12.20) is equivalent to

$$\lim_{N \to \infty} \left\{ \int_{0}^{\ell} K_N(s) f(x-s) \, ds - \frac{1}{2} f(x-) \right\} = 0.$$

By (12.18),

$$\frac{1}{2} f(x-) = f(x-) \int_{0}^{\ell} K_N(s) \, ds = \int_{0}^{\ell} K_N(s) f(x-) \, ds,$$

so

$$\int_{0}^{\ell} K_N(s) f(x-s) \, ds - \frac{1}{2} f(x-) = \int_{0}^{\ell} K_N(s)(f(x-s) - f(x-)) \, ds$$

$$= \frac{1}{2\ell} \int_{0}^{\ell} \frac{f(x-s) - f(x-)}{\sin\left(\frac{\pi s}{2\ell}\right)} \sin\left(\frac{(2N+1)\pi s}{2\ell}\right) ds.$$

We can recognize this integral as $1/(2\ell)$ times the L^2 inner product (on the interval $(0, \ell)$) of the functions
$$F_{(x)}(s) = \frac{f(x-s) - f(x-)}{\sin\left(\frac{\pi s}{2\ell}\right)}$$
and
$$\sin\left(\frac{(2N+1)\pi x}{2\ell}\right).$$

The sequence
$$\left\{\sin\left(\frac{(2N+1)\pi x}{2\ell}\right) : N = 0, 1, 2, \ldots\right\}$$
is an orthogonal sequence with respect to the L^2 inner product (see, for example, Exercise 5.2.2). Bessel's inequality then implies that
$$\sum_{N=0}^{\infty} \left|\left(F_{(x)}, \sin\left(\frac{(2N+1)\pi s}{2\ell}\right)\right)\right|^2 < \infty$$
and therefore
$$\frac{1}{2\ell}\int_0^\ell \frac{f(x-s) - f(x-)}{\sin\left(\frac{\pi s}{2\ell}\right)} \sin\left(\frac{(2N+1)\pi s}{2\ell}\right) ds = \left(F_{(x)}, \sin\left(\frac{(2N+1)\pi s}{2\ell}\right)\right) \to 0.$$

This is the desired result.

However, there is one problem with this argument. Bessel's inequality (Lemma 12.13) requires that $F_{(x)}$ and the orthogonal sequence belong to a common inner product space. We will take this space to be the space V of all piecewise continuous functions defined on $[0, \ell]$. Certainly the functions
$$\sin\left(\frac{(2N+1)\pi x}{2\ell}\right), \quad N = 0, 1, 2, \ldots,$$
belong to V (continuous functions are piecewise continuous). The proof, therefore, reduces to showing that $F_{(x)} \in V$.

Since f_{per} has only jump discontinuities, $F_{(x)}$ also has only jump discontinuities, except possibly at $s = 0$, where there is a zero in the denominator. If we show that
$$\lim_{s \to 0^+} F_{(x)}(s)$$
exists as a finite number, this will show that $F_{(x)}$ is piecewise continuous, and hence in V. Since f and df/dx each has at most a finite number of jump discontinuities, there is an interval $(0, \epsilon)$ such that the function $s \mapsto f(x-s)$ is continuous and differentiable on $(0, \epsilon)$. It follows from the mean value theorem that, for each $s \in (0, \epsilon)$, there exists $\gamma_s \in (0, 1)$ such that
$$f(x-s) - f(x-) = -\frac{df}{dx}(x - \gamma_s s)s.$$
Also, by the mean value theorem, there exists $\lambda_s \in (0, 1)$ such that
$$\sin\left(\frac{\pi s}{2\ell}\right) = \cos\left(\frac{\lambda_s \pi s}{2\ell}\right) \frac{\pi s}{2\ell}.$$

12.4. Pointwise convergence of Fourier series

Thus, for all $s > 0$ near zero, we have

$$F_{(x)}(s) = \frac{f(x-s) - f(x-)}{\sin\left(\frac{\pi s}{2\ell}\right)} = \frac{-\frac{df}{dx}(x - \gamma_s s)s}{\cos\left(\frac{\lambda_s \pi s}{2\ell}\right)\frac{\pi s}{2\ell}}$$

$$= -\frac{\frac{df}{dx}(x - \gamma_s s)}{\cos\left(\frac{\lambda_s \pi s}{2\ell}\right)\frac{\pi}{2\ell}}$$

$$\to -\frac{\frac{df}{dx}(x-)}{\frac{\pi}{2\ell}} \text{ as } s \to 0^+.$$

This shows that $\lim_{s \to 0^+} F_{(x)}(s)$ exists, which completes the proof. □

Example 12.15. We will compute the complex Fourier series of $g : [-1,1] \to \mathbf{R}$ defined by $g(x) = x$. Figure 12.9 shows that g_{per} is piecewise smooth, and smooth continuous except at integral values of x. We therefore expect that the Fourier series of g will converge to $g_{per}(x)$ except when x is an integer. At the points of discontinuity, the average of the left- and right-hand limits is 0, which will be the limit of the series at those points.

The Fourier coefficients of g are

$$c_0 = \frac{1}{2}\int_{-1}^{1} x\,dx = 0,$$

$$c_n = \frac{1}{2}\int_{-1}^{1} xe^{-\pi inx}\,dx = \frac{i(-1)^n}{n\pi}, \quad n = \pm 1, \pm 2, \ldots.$$

The partial Fourier series

$$\sum_{n=-N}^{N} c_n e^{\pi inx},$$

for $N = 10, 20, 40$, are shown in Figure 12.11. ∎

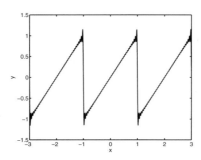

Figure 12.11. *The partial Fourier series* f_{40} *for Example* 12.15.

The following results can be derived immediately from the relationship between the complex Fourier series and various other forms of Fourier series.

Corollary 12.16.

1. *Suppose $f : (-\ell, \ell) \to \mathbf{R}$ is piecewise smooth. Then the full Fourier series of f,*

$$a_0 + \sum_{n=1}^{\infty} \left\{ a_n \cos\left(\frac{n\pi x}{\ell}\right) + b_n \sin\left(\frac{n\pi x}{\ell}\right) \right\},$$

   converges to $f_{per}(x)$ if x is a point of continuity of f_{per}, and to

$$\frac{1}{2}\left[\lim_{s \to x^-} f_{per}(s) + \lim_{s \to x^+} f_{per}(s) \right]$$

   if f_{per} has a jump discontinuity at x.

2. *Suppose $f : (0, \ell) \to \mathbf{R}$ is piecewise smooth, and let f_{odd} be the periodic, odd extension of f to $\mathbf{R}$ (that is, the periodic extension to $\mathbf{R}$ of the odd extension of f to $(-\ell, \ell)$). Then the Fourier sine series of f,*

$$\sum_{n=1}^{\infty} b_n \sin\left(\frac{n\pi x}{\ell}\right),$$

   converges to $f_{odd}(x)$ if x is a point of continuity of f_{odd}, and to

$$\frac{1}{2}\left[\lim_{s \to x^-} f_{odd}(s) + \lim_{s \to x^+} f_{odd}(s) \right]$$

   if f_{odd} has a jump discontinuity at x.

3. *Suppose $f : (0, \ell) \to \mathbf{R}$ is piecewise smooth, and let f_{even} be the periodic, even extension of f to $\mathbf{R}$ (that is, the periodic extension to $\mathbf{R}$ of the even extension of f to $(-\ell, \ell)$). Then the Fourier cosine series of f,*

$$a_0 + \sum_{n=1}^{\infty} a_n \cos\left(\frac{n\pi x}{\ell}\right),$$

   converges to $f_{even}(x)$ if x is a point of continuity of f_{even}, and to

$$\frac{1}{2}\left[\lim_{s \to x^-} f_{even}(s) + \lim_{s \to x^+} f_{even}(s) \right]$$

   if f_{even} has a jump discontinuity at x.

Proof. See Exercise 12.4.6. □

12.4. Pointwise convergence of Fourier series

Example 12.17. Let $f : (0,1) \to \mathbf{R}$ be defined by $f(x) = 1$. Then the periodic, odd extension of f, f_{odd}, is defined by

$$f_{odd}(x) = \begin{cases} 1 & \text{if } x \in (2k, 2k+1) \ (k \text{ an integer}), \\ -1 & \text{if } x \in (2k-1, 2k) \ (k \text{ an integer}) \end{cases}$$

(the so-called square wave). Thus f_{odd} is discontinuous at every integer value of x; at those points of discontinuity, the average of the left and right endpoints is zero. Figure 12.12 shows f_{odd} together with its partial sine series having 5, 10, 20, and 40 terms. As the above corollary guarantees, the sine series converges to f at every point of continuity (every nonintegral point, in this case) and to zero at every point of discontinuity. ∎

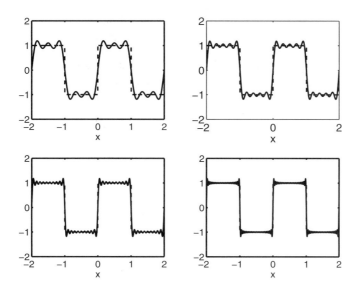

Figure 12.12. *The square-wave function from Example 12.17, together with its Fourier sine series with 5, 10, 20, and 40 terms.*

We can draw some further conclusions from Theorem 12.14 and Corollary 12.16.

Corollary 12.18.

1. *If $f : (-\ell, \ell) \to \mathbf{C}$ is continuous and piecewise smooth, then f_{per} is continuous everywhere except (possibly) at the points $\pm \ell, \pm 3\ell, \ldots$. It follows that the Fourier series of f converges to $f(x)$ for all $x \in (-\ell, \ell)$.*

2. *If $f : [-\ell, \ell] \to \mathbf{C}$ is continuous and piecewise smooth, and $f(-\ell) = f(\ell)$, then f_{per} is continuous everywhere, and so the Fourier series of f converges to $f(x)$ for all $x \in [-\ell, \ell]$ (including the endpoints).*

3. *If $f : [0, \ell] \to \mathbf{R}$ is continuous and piecewise smooth, then f_{odd} (the periodic extension of the odd extension of f) is continuous everywhere except (possibly) at the points*

$0, \pm \ell, \pm 2\ell, \ldots$. *It follows that the Fourier sine series of f converges to $f(x)$ for all $x \in (0,\ell)$.*

4. *If $f : [0,\ell] \to \mathbf{R}$ is continuous and piecewise smooth, and $f(0) = f(\ell) = 0$, then f_{odd} is continuous everywhere. It follows that the Fourier sine series of f converges to $f(x)$ for all $x \in [0,\ell]$ (including the endpoints).*

5. *If $f : [0,\ell] \to \mathbf{R}$ is continuous and piecewise smooth, then f_{even} (the periodic extension of the even extension of f) is continuous everywhere. It follows that the Fourier cosine series of f converges to $f(x)$ for all $x \in [0,\ell]$ (including the endpoints).*

Proof. The conclusions all follow directly from Corollary 12.16, except for the following: If $f : [0,\ell] \to \mathbf{R}$ is continuous, then f_{even} is continuous, and f_{odd} is continuous provided $f(0) = f(\ell) = 0$. The reader is asked to verify these conclusions in Exercise 12.4.7. □

When f_{per} (or f_{odd}, or f_{even}) is continuous everywhere, we can draw a stronger conclusion, as we show in the next section. The reader should note that, from the point of view of approximating a given function, the cosine series is more powerful than the sine series, since f_{even} is continuous everywhere for any continuous function $f : [0,\ell] \to \mathbf{R}$. Therefore, for example, Gibbs's phenomena cannot occur with the cosine series approximating such a function.

Exercises

1. Let $f : [-\pi,\pi] \to \mathbf{R}$ be defined by $f(x) = x^3$. Does the full Fourier series of f converge (pointwise) on $\mathbf{R}$? If so, what is the limit of the series?

2. Define $f : [-2,2] \to \mathbf{R}$ by $f(x) = x^3 + 1$. Write down the limits of

    (a) the Fourier sine series of f (regarded as a function defined on $[0,2]$),

    (b) the Fourier cosine series of f (regarded as a function defined on $[0,2]$), and

    (c) the full Fourier series of f.

3. Find an example of a function $f : [0,1] \to \mathbf{R}$ such that it is *not* the case that the Fourier cosine series of f converges to f at every $x \in [0,1]$.

4. Using the original formula for K_N,
$$K_N(\theta) = \frac{1}{2\ell} \sum_{n=-N}^{N} e^{i\pi n\theta/\ell},$$
prove that (12.18) holds.

5. Prove Theorem 12.8. For simplicity, assume all of the functions are continuous.

6. Prove Corollary 12.16.

7. Suppose $f : [0,\ell] \to \mathbf{R}$ is continuous. Prove that f_{even} is continuous everywhere, and that if $f(0) = f(\ell) = 0$, then f_{odd} is continuous everywhere.

8. (a) Extend Corollaries 12.16 and 12.18 to the case of the quarter-wave cosine series. (Hint: See Exercise 12.3.6.)

   (b) Define $f : [0,1] \to \mathbf{R}$ by $f(x) = 1 + x$. To what function does the quarter-wave cosine series of f converge?

9. (a) Extend Corollaries 12.16 and 12.18 to the case of the quarter-wave sine series. (Hint: See Exercise 12.3.5.)

   (b) Define $f : [0,1] \to \mathbf{R}$ by $f(x) = 1 + x$. To what function does the quarter-wave sine series of f converge?

12.5 Uniform convergence of Fourier series

In the previous section, we proved the pointwise convergence of the Fourier series of a piecewise smooth function. We will now consider conditions under which the convergence is actually uniform. The property of uniform convergence is very desirable, since it implies that a finite number of terms from the Fourier series can approximate the function accurately on the entire interval; in particular, uniform convergence rules out the possibility of Gibbs's phenomenon, which was introduced in Section 5.2.2 and will be discussed further below.

The rapidity with which a Fourier series converges, and, in particular, whether it converges uniformly or not, is intimately related to the rate at which its coefficients converge to zero. We address this question first.

12.5.1 Rate of decay of Fourier coefficients

In the following theorems, we will show that the Fourier coefficients c_n, $n = 0, \pm 1, \pm 2, \ldots$, of f satisfy

$$c_n = O\left(\frac{1}{|n|^k}\right), \tag{12.21}$$

where the exponent k is determined by the degree of smoothness of f. Recall that (12.21) means that there is a constant $M > 0$ such that

$$|c_n| \leq \frac{M}{|n|^k} \text{ for all } n.$$

The same condition that guarantees pointwise convergence of Fourier series also ensures that the Fourier coefficients decay—converge to zero—at least as fast as $1/n$, $n = 1, 2, 3, \ldots$.

Theorem 12.19. *Suppose that $f : (-\ell, \ell) \to \mathbf{C}$ is piecewise smooth, and let c_n, $n = 0, \pm 1, \pm 2, \ldots$, be the complex Fourier coefficients of f. Then*

$$c_n = O\left(\frac{1}{|n|}\right).$$

Proof. Since f is piecewise smooth, there is a finite number of discontinuities of f and/or df/dx. We label these points as

$$-\ell < x_1 < x_2 < \cdots < x_{m-1} < \ell$$

and write $x_0 = -\ell$, $x_m = \ell$. We then have

$$c_n = \frac{1}{2\ell}\int_{-\ell}^{\ell} f(x)e^{-i\pi nx/\ell}\,dx$$

$$= \frac{1}{2\ell}\sum_{j=1}^{m}\int_{x_{j-1}}^{x_j} f(x)e^{-i\pi nx/\ell}\,dx.$$

It then suffices to show that, for each $j = 1, 2, \ldots, m$,

$$\int_{x_{j-1}}^{x_j} f(x)e^{-i\pi nx/\ell}\,dx = O\left(\frac{1}{|n|}\right).$$

Since f is continuously differentiable on (x_{j-1}, x_j), and the limits of both f and df/dx exist at the endpoints of this interval, we can apply integration by parts to obtain

$$\int_{x_{j-1}}^{x_j} f(x)e^{-i\pi nx/\ell}\,dx = \left[f(x)\frac{e^{-i\pi nx/\ell}}{-i\pi n/\ell}\right]_{x_{j-1}}^{x_j} - \frac{i\ell}{\pi n}\int_{x_{j-1}}^{x_j}\frac{df}{dx}(x)e^{-i\pi nx/\ell}\,dx$$

$$= \frac{i\ell}{n\pi}\left(f(x_j-)e^{-i\pi nx_j/\ell} - f(x_{j-1}+)e^{-i\pi nx_{j-1}/\ell}\right)$$

$$- \frac{i\ell}{\pi n}\int_{x_{j-1}}^{x_j}\frac{df}{dx}(x)e^{-i\pi nx/\ell}\,dx.$$

Since each of these three terms is bounded by a constant times $1/|n|$, the result follows. We have used the fact that f and df/dx, being piecewise continuous, are both bounded on $(-\ell, \ell)$. $\square$

When f has some additional smoothness, its Fourier coefficients can be shown to decay more rapidly.

Theorem 12.20. *Suppose $f : (-\ell, \ell) \to \mathbf{C}$ has the property that its periodic extension f_{per} and the first $k-2$ derivatives of f_{per} are continuous, where $k \geq 2$, and the $(k-1)$st derivative of f is piecewise smooth. If c_n, $n = 0, \pm 1, \pm 2, \ldots$, are the complex Fourier coefficients of f, then*

$$c_n = O\left(\frac{1}{|n|^k}\right).$$

In particular, if f_{per} is continuous and its derivative is piecewise smooth, then

$$c_n = O\left(\frac{1}{|n|^2}\right).$$

Proof. Suppose f_{per} is continuous and df/dx is piecewise smooth. Then, by integration by parts, we have

$$\int_{-\ell}^{\ell} f(x)e^{-i\pi nx/\ell}\,dx = \left[f(x)\frac{e^{-i\pi nx/\ell}}{-i\pi n/\ell}\right]_{-\ell}^{\ell} - \frac{i\ell}{\pi n}\int_{-\ell}^{\ell}\frac{df}{dx}(x)e^{-i\pi nx/\ell}\,dx$$

$$= \frac{i\ell}{n\pi}\left(f(\ell-)e^{-i\pi n} - f(-\ell+)e^{i\pi n}\right) - \frac{i\ell}{\pi n}\int_{-\ell}^{\ell}\frac{df}{dx}(x)e^{-i\pi nx/\ell}\,dx.$$

12.5. Uniform convergence of Fourier series

Since f_{per} is continuous, $f(-\ell+) = f(\ell-)$, and $e^{i\pi n} = e^{-i\pi n}$ since $e^{i\theta}$ is 2π-periodic. Therefore, we obtain

$$c_n = -\frac{i}{2\pi n}\int_{-\ell}^{\ell}\frac{df}{dx}(x)e^{-i\pi nx/\ell}\,dx$$
$$= -\frac{i\ell}{\pi n}d_n,$$

where d_n, $n = 0, \pm 1, \pm 2, \ldots$, are the Fourier coefficients of df/dx. By Theorem 12.19, we have $d_n = O(1/|n|)$, and so $c_n = O(1/|n|^2)$, as desired.

The proof of the general case is similar; one shows that the Fourier coefficients of $d^{k-1}f/dx^{k-1}$ are of order $1/|n|$, the Fourier coefficients of $d^{k-2}f/dx^{k-2}$ are of order $1/|n|^2$, and so forth to obtain the desired result. □

Example 12.21.

1. Define $f : (-1,1) \to \mathbf{R}$ by $f(x) = x$. Then f_{per} has jump discontinuities (at $x = (2k-1)$, $k = 0, \pm 1, \pm 2, \ldots$), and, by Theorem 12.20, the Fourier coefficients of f must be of order $1/|n|$. In fact, the Fourier coefficients are

$$c_n = \frac{i(-1)^n}{\pi n},\ n = \pm 1, \pm 2, \ldots,$$

with $c_0 = 0$.

2. Define $f : (-1,1) \to \mathbf{R}$ by $f(x) = x^2$. Then f_{per} is continuous, but its derivative has jump discontinuities (at $x = (2k-1)$, $k = 0, \pm 1, \pm 2, \ldots$). By Theorem 12.20, the Fourier coefficients of f must be of order $1/|n|^2$. In fact, the Fourier coefficients are

$$c_n = \frac{2(-1)^n}{\pi^2 n^2},\ n = \pm 1, \pm 2, \ldots,$$

with $c_0 = 1/3$.

3. Define $f : (-1,1) \to \mathbf{R}$ by $f(x) = x^5 - 2x^3 + x$. Then f_{per} and its derivative are continuous, but its second derivative has jump discontinuities (at $x = (2k-1)$, $k = 0, \pm 1, \pm 2, \ldots$). By Theorem 12.20, the Fourier coefficients of f must be of order $1/|n|^3$. In fact, the Fourier coefficients are

$$c_n = \frac{-8i(n^2\pi^2 - 15)(-1)^n}{\pi^5 n^5},\ n = \pm 1, \pm 2, \ldots,$$

with $c_0 = 0$; we have $|c_n| \doteq 8/(\pi^3|n|^3)$ if $|n|$ is large.

4. Define $f : (-1,1) \to \mathbf{R}$ by $f(x) = x^4 - 2x^2 + 1$. Then f_{per} and its first two derivatives are continuous, but its third derivative has jump discontinuities (at $x = (2k-1)$, $k = 0, \pm 1, \pm 2, \ldots$). By Theorem 12.20, the Fourier coefficients of f must be of order

$1/|n|^4$. In fact, the Fourier coefficients are

$$c_n = \frac{-24(-1)^n}{\pi^4 n^4}, \, n = \pm 1, \pm 2, \ldots,$$

with $c_0 = 8/15$.

In Figure 12.13 we graph the error in approximating each of the above four functions by a truncated Fourier series with 41 terms (the constant term and the 20 lowest frequencies). The results are as expected; as f_{per} gets smoother, the convergence of the Fourier series to f becomes more rapid. ∎

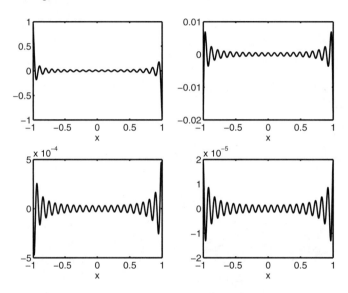

Figure 12.13. *The errors in approximating each function from Example* 12.21 *by its partial Fourier series with* 41 *terms:* $f(x) = x$ *(upper left),* $f(x) = x^2$ *(upper right),* $f(x) = x^5 - 2x^3 + x$ *(lower left),* $f(x) = x^4 - 2x^2 + 1$ *(lower right).*

12.5.2 Uniform convergence

It now follows immediately that, if f_{per} is sufficiently smooth, then its Fourier series converges uniformly to f.

Theorem 12.22. *Suppose* $f : (-\ell, \ell) \to \mathbf{C}$ *is continuous,* df/dx *is piecewise smooth, and* $f(-\ell) = f(\ell)$. *Then the partial sums of the complex Fourier series of* f *converge uniformly to* f *on* $(-\ell, \ell)$ *(and therefore to* f_{per} *on* $\mathbf{R}$).

Proof. The boundary conditions imply that f_{per} is continuous, so by Theorem 12.14, we know that the Fourier series of f converges pointwise to f_{per} on $\mathbf{R}$. We must show that the convergence is uniform. Since

$$f_{per}(x) = \sum_{n=-\infty}^{\infty} c_n e^{i\pi nx/\ell},$$

12.5. Uniform convergence of Fourier series

we have

$$f_{per}(x) - \sum_{n=-N}^{N} c_n e^{i\pi nx/\ell} = \sum_{n=-\infty}^{-N-1} c_n e^{i\pi nx/\ell} + \sum_{n=N+1}^{\infty} c_n e^{i\pi nx/\ell}.$$

Therefore, by Theorem 12.20, there is a constant M such that

$$\left| f_{per}(x) - \sum_{n=-N}^{N} c_n e^{i\pi nx/\ell} \right| = \left| \sum_{n=-\infty}^{-N-1} c_n e^{i\pi nx/\ell} + \sum_{n=N+1}^{\infty} c_n e^{i\pi nx/\ell} \right|$$

$$\leq \sum_{n=-\infty}^{-N-1} \left| c_n e^{i\pi nx/\ell} \right| + \sum_{n=N+1}^{\infty} \left| c_n e^{i\pi nx/\ell} \right|$$

$$\leq 2M \sum_{n=N+1}^{\infty} \frac{1}{n^2}$$

(using the fact that $\left| e^{i\pi nx/\ell} \right| = 1$). This holds for all $x \in \mathbf{R}$, and so

$$\left| f_{per}(x) - \sum_{n=-N}^{N} c_n e^{i\pi nx/\ell} \right| \leq 2M \sum_{n=N+1}^{\infty} \frac{1}{n^2} \text{ for all } x \in \mathbf{R}.$$

Since the infinite series

$$\sum_{n=1}^{\infty} \frac{1}{n^2}$$

is convergent, the tail of the series tends to zero, which shows that

$$\left| f_{per}(x) - \sum_{n=-N}^{N} c_n e^{i\pi nx/\ell} \right|$$

tends to zero at a rate that is independent of $x \in \mathbf{R}$. This proves the desired result. $\square$

As with pointwise convergence, the uniform convergence of other types of Fourier series can be deduced from their relationship to the complex Fourier series.

Corollary 12.23.

1. *Suppose $f : (-\ell, \ell) \to \mathbf{R}$ is continuous, df/dx is piecewise smooth, and $f(-\ell) = f(\ell)$. Then the full Fourier series of f converges uniformly to f on $(-\ell, \ell)$ (and therefore to f_{per} on $\mathbf{R}$).*

2. *Suppose $f : (0, \ell) \to \mathbf{R}$ is continuous, df/dx is piecewise smooth, $f(0) = 0$, and $f(\ell) = 0$. Then the Fourier sine series converges uniformly to f on $(-\ell, \ell)$ (and hence to f_{odd}, the periodic, odd extension of f, on all of $\mathbf{R}$).*

3. *Suppose $f : (0, \ell) \to \mathbf{R}$ is continuous and df/dx is piecewise smooth. Then the Fourier cosine series converges uniformly to f on $(-\ell, \ell)$ (and hence to f_{even}, the periodic, even extension of f, on all of $\mathbf{R}$).*

Proof.

1. This follows immediately from the equivalence of the full Fourier series and the complex Fourier series.

2. This follows from the fact that the sine series of f is the full Fourier series of f_{odd} and from the fact that the continuity of f and the boundary conditions $f(0) = f(\ell) = 0$ guarantee that f_{odd} is continuous.

3. This follows from the fact that the cosine series of f is the full Fourier series of f_{even} and from the fact that the continuity of f guarantees that f_{even} is continuous. □

The special property of the cosine series is important; no boundary conditions must be satisfied in order for the convergence to be uniform. Indeed, the relevant boundary conditions for a cosine series are Neumann conditions, which involve df/dx. The Fourier cosine coefficients of f are $O\left(1/|n|^3\right)$ or $O\left(1/|n|^2\right)$ according to whether f satisfies or does not satisfy homogeneous Neumann conditions (see Exercise 12.5.6), but in either case, the Fourier series converges uniformly to f.

Example 12.24. Let $f : [0,1] \to \mathbf{R}$ be defined by $f(x) = x^2$. Since f does not satisfy the Dirichlet conditions, the sine series does not converge uniformly on $[0,1]$. However, even though f does not satisfy the Neumann conditions either, the cosine series does converge uniformly on $[0,1]$. Figure 12.14 illustrates these results. ■

In Example 12.21, we showed several functions with progressively smoother periodic extensions. The smoother the periodic extension, the faster the Fourier coefficients decay

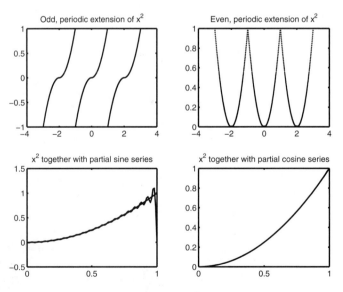

Figure 12.14. *Approximating $f(x) = x^2$ by sine and cosine series (40 terms in each series).*

12.5. Uniform convergence of Fourier series

to zero, and the better the partial Fourier series approximates the original function. We now give an extreme example of this.

Example 12.25. Define $f : (-1,1) \to \mathbf{R}$ by $f(x) = e^{\cos(2\pi x)}$. Then f_{per} and *all* of its derivatives are continuous. By Theorem 12.20, the Fourier coefficients of f converge to zero faster than any power of $1/|n|$. It is not possible to compute a formula for these coefficients (the integrals that must be computed are intractable), but we can estimate them using the techniques discussed in Section 12.2. Figure 12.15 shows the error in estimating f with 41 terms in its Fourier series. This graph should be compared with Figure 12.13. ∎

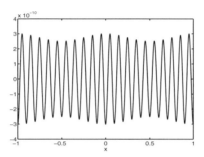

Figure 12.15. *The errors in approximating $f(x) = e^{\cos(2\pi x)}$ by its partial Fourier series with 41 terms (see Example 12.25).*

12.5.3 A note about Gibbs's phenomenon

A standard theorem in real analysis is the following: If a sequence of continuous functions converges uniformly, then the limit function is also continuous. Every partial Fourier series is a continuous function, so if f_{per} is discontinuous, it is impossible for its Fourier series to converge uniformly. Gibbs's phenomenon is the nonuniform convergence of the Fourier series near a jump discontinuity of f_{per}. Near the end of the 19th century, Gibbs showed that, near a jump discontinuity, the partial Fourier series overshoot the true value of f by about 9% of the jump[74] (cf. Exercise 12.5.5).

Exercises

1. Consider the following functions defined on $[-1,1]$:
$$f(x) = |x|,$$
$$g(x) = x - x^3,$$
$$h(x) = 1 + x.$$

   Rank f, g, and h in order of the speed of convergence of their Fourier series and illustrate with graphs.

---

[74]The history is briefly described by Kammler [36, page 45].

2. Consider the following functions defined on $[0,1]$:

$$f(x) = x^4 - x^3 + x^2 - x,$$
$$g(x) = e^x,$$
$$h(x) = x\cos(x).$$

For which function(s) does the Fourier sine series converge uniformly? For which does the Fourier cosine series?

3. Consider the following functions defined on $[0,1]$:

$$f(x) = x^3 + x,$$
$$g(x) = 1 - x^2.$$

For which function(s) does the Fourier quarter-wave sine series converge uniformly?

4. Find an example of a function $f : [-1,1] \to \mathbf{R}$ whose Fourier coefficients are

$$O\left(\frac{1}{|n|^4}\right)$$

but not

$$O\left(\frac{1}{|n|^5}\right).$$

(Hint: Choose f to be a fifth-degree polynomial.)

5. Consider the function $f : [-1,1] \to \mathbf{R}$ defined by

$$f(x) = \begin{cases} 0, & -1 \le x < 0, \\ 1, & 0 \le x \le 1. \end{cases}$$

Verify numerically that the overshoot in Gibbs's phenomenon is about 9% of the jump at $x = 0$.

6. Suppose $f : (0,\ell) \to \mathbf{R}$ is continuous and df/dx is piecewise smooth.

   (a) Prove that the periodic, even extension of f is continuous and therefore that the Fourier cosine coefficients of f are $O\left(1/|n|^2\right)$.

   (b) Prove that the Fourier cosine coefficients of f are $O\left(1/|n|^3\right)$ if and only if

$$\frac{df}{dx}(0) = \frac{df}{dx}(\ell) = 0.$$

12.6 Mean-square convergence of Fourier series

We now wish to show that, for a large class of functions $f : (-\ell, \ell) \to \mathbf{C}$, the complex Fourier series of f,

$$\sum_{n=-\infty}^{\infty} c_n e^{i\pi nx/\ell},$$

converges to f in the mean-square sense. Mean-square convergence means that if we write

$$f_N(x) = \sum_{n=-N}^{N} c_n e^{i\pi nx/\ell},$$

then

$$\|f - f_N\|_{L^2} \to 0 \text{ as } N \to \infty.$$

In contrast to pointwise or uniform convergence, it turns out that the mildest assumption about f guarantees mean-square convergence of the Fourier series.

In order to discuss mean-square convergence, we need f to have a finite L^2 norm:

$$\|f\| = \int_{-\ell}^{\ell} |f(x)|^2 \, dx < \infty.$$

It turns out that this essential assumption is the only condition required to guarantee the convergence of the Fourier series to f in the mean-square sense. To justify this statement, we must begin with some technical preliminaries. The technicalities are sufficiently subtle as to be beyond the scope of this book—the reader will have to accept certain assertions on faith.

12.6.1 The space $L^2(-\ell, \ell)$

We wish to identify the space of all functions $f : (-\ell, \ell) \to \mathbf{C}$ such that the complex Fourier series of f converges to f in the mean-square sense. Accepting the truth of the statement in the previous paragraph, it would be natural to define

$$L^2(-\ell, \ell) = \left\{ f : (-\ell, \ell) \to \mathbf{C} \ : \ \int_{-\ell}^{\ell} |f(x)|^2 \, dx < \infty \right\}. \tag{12.22}$$

There is more to this definition than meets the eye, however.

First of all, one of the essential properties of a norm is the following:

$$\|f\| = 0 \Rightarrow f = 0. \tag{12.23}$$

When f is a function, $f = 0$ means that f is the zero function: $f(x) = 0$ for all $x \in (-\ell, \ell)$. As long as we restrict ourselves to continuous functions, (12.23) holds.[75] However, a

---

[75] If f is continuous and not the zero function, then, by continuity, there must exist an interval $(a,b) \subset (-\ell, \ell)$ on which f is nonzero, and so $\int_{-\ell}^{\ell} |f(x)|^2 \, dx \geq \int_{a}^{b} |f(x)|^2 \, dx$ must be positive.

discontinuous function can be nonzero and yet have L^2 norm equal to zero. For example, consider the function $f : (-1, 1) \to \mathbf{R}$ defined by

$$f(x) = \begin{cases} 1, & x = 0, \\ 0, & x \neq 0. \end{cases}$$

This function is not the zero function, but

$$\int_{-1}^{1} |f(x)|^2 \, dx = 0$$

since the integrand is zero everywhere except at a single point. Therefore, in order for (12.23) to be satisfied, we have to agree that f and g are regarded as the same function provided

$$\int_{-\ell}^{\ell} |f(x) - g(x)|^2 \, dx = 0.$$

We will have more to say about this below.

There is another difficulty with (12.22). Since our interest is mean-square convergence of functions, the following property is desirable: If $\{f_N\}$ is a sequence of functions in $L^2(-\ell, \ell)$ and $f_N \to f$ in the mean-square sense, then $f \in L^2(-\ell, \ell)$. Is this property satisfied by $L^2(-\ell, \ell)$ as defined by (12.22)? The answer is, it depends on the definition of the integral used in (12.22). For example, any function with an infinite singularity fails to be Riemann integrable; as an example, consider the function $f : [-1, 1] \to \mathbf{R}$ defined by

$$f(x) = \frac{1}{|x|^{1/4}}.$$

This function is not Riemann integrable on $(-1, 1)$ because of the infinite discontinuity at $x = 0$. However, $|f(x)|^2$ has a finite area under its graph, as the following calculation shows:

$$\lim_{\epsilon \to 0^+} \left\{ \int_{-1}^{-\epsilon} |f(x)|^2 \, dx + \int_{\epsilon}^{1} |f(x)|^2 \, dx \right\} = \lim_{\epsilon \to 0^+} 4(1 - \sqrt{\epsilon}) = 4.$$

Therefore, we would like to include f in $L^2(-1, 1)$, which means that we must interpret the integral as an improper Riemann integral when f has a finite number of singularities.

However, allowing improper Riemann integrals is not enough, since it is possible to construct a sequence $\{f_N\}$ with the following properties:

1. f_N has N infinite discontinuities;

2. f_N is square integrable when the integral is interpreted as an improper Riemann integral;

3. $\{f_N\}$ converges in the mean-square sense to a function with an infinite number of discontinuities.

A function with an infinite number of discontinuities cannot be Riemann integrable, even in the improper sense.

12.6. Mean-square convergence of Fourier series

Faced with these difficulties, mathematicians eventually concluded that a better notion of integration was needed, which was created by Henri Lebesgue (the "L" in L^2). The definition of the Lebesgue integral is beyond the scope of this book, but we can describe its important features. The theory begins with a *measure* for subsets of $\mathbf{R}$ that agrees with our intuition for simple sets (for example, the measure of an interval $[a,b]$ is $b-a$). The Lebesgue measure is defined in a consistent fashion even for very complicated subsets of $\mathbf{R}$, but not for all subsets. Sets whose Lebesgue measure is defined are called (Lebesgue) *measurable*. Any set with measure zero is then negligible in a certain sense. Every set containing a finite number of points has measure zero.[76]

The Lebesgue integral is defined for functions that are *measurable*; the property of measurability is a regularity property like continuity or differentiability, but much weaker—functions with an infinite number of discontinuities can be measurable, for example. Giving the precise definitions of these concepts—measurable set, measurable function, Lebesgue integral—is beyond the scope of this book. The definitions, however, have the following consequences.

1. Two measurable functions f and g satisfy

$$\int_{-\ell}^{\ell} |f(x)-g(x)|^2 \, dx = 0$$

   if and only if f and g differ only on a set of measure zero. It is in this sense that a set of measure zero is negligible: functions differing only on a set of measure zero are regarded as being the same.

2. If f is Riemann integrable, then it is Lebesgue integrable, and the two integrals have the same value.

This last property assures us that whenever we are dealing with a continuous or piecewise continuous function, we can compute its integral using the familiar techniques of calculus. On the other hand, the Lebesgue integral is sufficiently powerful to handle pathological cases that arise inevitably when mean-square convergence is studied. We will discuss other properties of Lebesgue integration and $L^2(-\ell, \ell)$ below.

We can now define $L^2(-\ell, \ell)$ more precisely as the set of all measurable functions $f : (-\ell, \ell) \to \mathbf{C}$ satisfying

$$\int_{-\ell}^{\ell} |f(x)|^2 \, dx < \infty,$$

with the understanding that two functions that differ only on a set of measure zero represent the same element of the space.

In order to understand the L^2 theory of Fourier series, we need two more facts about $L^2(-\ell, \ell)$, facts that we cannot prove here, as their development would take us too far astray. Even stating the second fact takes a certain amount of work. The first fact, Theorem 12.26, allows us to prove that the Fourier series of every L^2 function converges to the function in the mean-square sense.

---

[76] Some sets having an infinite number of points also have measure zero. A countable set (a set that can be put in one-to-one correspondence with the integers $1, 2, 3, \ldots$) has measure zero. For example, the set of all rational numbers has measure zero. Some uncountable sets also have measure zero, although, obviously, such a set occupies a negligible part of the real line.

Theorem 12.26. *Let $f \in L^2(-\ell, \ell)$, and let ϵ be any positive real number (no matter how small). Then there is an infinitely differentiable function $g : [-\ell, \ell] \to \mathbf{C}$ such that*

$$\|f - g\|_{L^2} < \epsilon.$$

Moreover, g can be chosen to satisfy $g(-\ell) = g(\ell) = 0$.

This theorem simply says that any $L^2(-\ell, \ell)$ function can be approximated arbitrarily well, in the mean-square sense, using a smooth function satisfying Dirichlet conditions.

12.6.2 Mean-square convergence of Fourier series

Accepting the facts presented above about $L^2(-\ell, \ell)$, we can now prove the desired result.

Theorem 12.27. *If $f \in L^2(-\ell, \ell)$, then its Fourier series converges to f in L^2. That is, let*

$$\sum_{n=-\infty}^{\infty} c_n e^{i\pi n x/\ell}$$

be the complex Fourier series of f, and define

$$f_N(x) = \sum_{n=-N}^{N} c_n e^{i\pi n x/\ell}.$$

Then $\|f - f_N\|_{L^2} \to 0$ as $N \to \infty$.

Proof. We must show that for any $\epsilon > 0$, there exists a positive integer N_ϵ such that $\|f - f_N\|_{L^2} < \epsilon$ for every $N \geq N_\epsilon$. The proof is a typical $\epsilon/3$ argument—we show that f_N is close to f by using the triangle inequality with two intermediate functions. To be specific, let g be a smooth function satisfying $g(-\ell) = g(\ell)$ and

$$\|f - g\|_{L^2} < \frac{\epsilon}{3}.$$

We will show that f_N is close to f (for N sufficiently large) by showing that f_N is close to g_N (where g_N is the corresponding partial Fourier series for g) and g_N is close to g; we then use the fact that g is close to f by construction.

Let d_n, $n = 0, \pm 1, \pm 2, \ldots$, be the Fourier coefficients of g, and define

$$g_N(x) = \sum_{n=-N}^{N} d_n e^{i\pi n x/\ell}.$$

By Corollary 12.23, $g_N \to g$ uniformly on $[-\ell, \ell]$, and therefore there exists a positive integer N_ϵ such that

$$N \geq N_\epsilon, x \in [-\ell, \ell] \Rightarrow |g(x) - g_N(x)| < \frac{\epsilon}{3\sqrt{2\ell}}.$$

12.6. Mean-square convergence of Fourier series

It follows that

$$\|g - g_N\|_{L^2}^2 = \int_{-\ell}^{\ell} |g(x) - g_N(x)|^2 \, dx < \int_{-\ell}^{\ell} \frac{\epsilon^2}{3^2 2\ell} \, dx = \frac{\epsilon^2}{3^2},$$

and so $\|g - g_N\|_{L^2} < \frac{\epsilon}{3}$. Finally, we have that $f_N - g_N$ is the partial Fourier series of the function $f - g$, and hence that $f_N - g_N$ is the projection onto the space spanned by

$$e^{-i\pi N x/\ell}, e^{-i\pi (N-1) x/\ell}, \ldots, e^{i\pi N x/\ell}.$$

It follows that

$$\|f_N - g_N\|_{L^2} \leq \|f - g\|_{L^2} < \frac{\epsilon}{3}.$$

Thus we obtain

$$N \geq N_\epsilon \Rightarrow \|f - f_N\|_{L^2} \leq \|f - g\|_{L^2} + \|g - g_N\|_{L^2} + \|g_N - f_N\|_{L^2} < \frac{\epsilon}{3} + \frac{\epsilon}{3} + \frac{\epsilon}{3} = \epsilon.$$

This completes the proof. □

Now that we have obtained this fundamental result, we can draw some further conclusions. First, a question of terminology: the above theorem shows that the complex exponentials $e^{i\pi n x/\ell}$, $n = 0, \pm 1, \pm 2, \ldots$, are *complete* in $L^2(-\ell, \ell)$.

Definition 12.28. *Let V be an (infinite-dimensional) inner product space. We say that an orthogonal sequence $\{\phi_j\}_{j=1}^{\infty}$ is* complete *if every $\mathbf{v} \in V$ satisfies*

$$\mathbf{v} = \sum_{j=1}^{\infty} \frac{(\mathbf{v}, \phi_j)}{(\phi_j, \phi_j)} \phi_j,$$

where the convergence is in the sense of the norm on V.

The completeness of an orthogonal sequence in an infinite-dimensional inner product space is thus analogous to the property of spanning for a finite collection of vectors in a finite-dimensional vector space. Moreover, orthogonality implies linear independence, so a complete orthogonal sequence for an infinite-dimensional vector space is analogous to a basis for a finite-dimensional vector space.

A result that is sometimes useful is the following.

Theorem 12.29. *Let $\{\phi_j : j = 1, 2, \ldots\}$ be a complete orthogonal sequence in an inner product space V. Then*

$$\mathbf{v} \in V, \ (\mathbf{v}, \phi_j)_V = 0 \text{ for all } j = 1, 2, 3, \ldots \Rightarrow \mathbf{v} = 0. \tag{12.24}$$

Proof. By the completeness of $\{\phi_j : j = 1, 2, \ldots\}$, we have, for any $\mathbf{v} \in V$,

$$\mathbf{v} = \sum_{j=1}^{\infty} \frac{(\mathbf{v}, \phi_j)_V}{(\phi_j, \phi_j)_V} \phi_j.$$

It follows immediately that if $(\mathbf{v}, \phi_j)_V = 0$ for all $j = 1, 2, 3, \ldots$, then $\mathbf{v} = 0$. □

This result justifies a fact that we have used constantly in developing Fourier series methods for BVPs: Two functions are equal if and only if they have the same Fourier coefficients.

Corollary 12.30. *Let $\{\phi_j : j = 1,2,\ldots\}$ be a complete orthogonal sequence in an inner product space V. Then, for any $\mathbf{u}, \mathbf{v} \in V$,*

$$\mathbf{u} = \mathbf{v} \Leftrightarrow (\mathbf{u}, \phi_j)_V = (\mathbf{v}, \phi_j)_V \text{ for all } j = 1,2,3,\ldots.$$

Proof. Obviously, if $\mathbf{u} = \mathbf{v}$, then $(\mathbf{u}, \phi_j)_V = (\mathbf{v}, \phi_j)_V$ for all j. On the other hand, if $(\mathbf{u}, \phi_j)_V = (\mathbf{v}, \phi_j)_V$ for all j, then $\mathbf{w} = \mathbf{u} - \mathbf{v}$ satisfies

$$(\mathbf{w}, \phi_j)_V = 0 \text{ for all } j = 1,2,3,\ldots.$$

By the preceding theorem, this implies that $\mathbf{w} = 0$, that is, that $\mathbf{u} = \mathbf{v}$. □

Finally, because all other Fourier series (sine, cosine, full, quarter-wave sine, quarter-wave cosine) are special cases of the complex Fourier series, all of the above results extend to these other Fourier series as well. For example, we define $L^2(0, \ell)$ to be the space of all measurable functions $f : (0, \ell) \to \mathbf{R}$ such that

$$\int_0^\ell |f(x)|^2 \, dx < \infty.$$

Then, if $f \in L^2(0, \ell)$, the Fourier sine series of f converges to f in the mean-square sense (see Exercise 12.6.6).

We have now completed the basic theory used in the earlier chapters of this book. In the next section, we discuss one further point for the sake of tying up a loose end.

12.6.3 Cauchy sequences and completeness

We assume that $\{\phi_j\}_{j=1}^\infty$ is an orthogonal sequence in a complex inner product space V and also, without loss of generality, that each ϕ_j has norm one (so the sequence is orthonormal). If $\mathbf{v} \in V$ and

$$a_j = (\mathbf{v}, \phi_j)_V = \frac{(\mathbf{v}, \phi_j)_V}{(\phi_j, \phi_j)_V}, \; j = 1,2,3,\ldots,$$

then, by Bessel's inequality,

$$\sum_{j=1}^\infty |a_j|^2 < \infty.$$

We now consider the converse: If $a_1, a_2, a_3, \ldots$ is *any* sequence of complex numbers satisfying

$$\sum_{j=1}^\infty |a_j|^2 < \infty,$$

12.6. Mean-square convergence of Fourier series

does the series

$$\sum_{j=1}^{\infty} a_j \phi_j \tag{12.25}$$

converge to a vector in V? To consider this question, we write

$$\mathbf{u}_n = \sum_{j=1}^{n} a_j \phi_j$$

and note that, if $m > n$, then

$$\mathbf{u}_n - \mathbf{u}_m = \sum_{j=1}^{n} a_j \phi_j - \sum_{j=1}^{m} a_j \phi_j = \sum_{j=n+1}^{m} a_j \phi_j.$$

It follows that

$$\|\mathbf{u}_n - \mathbf{u}_m\|_V^2 = \left\| \sum_{j=n+1}^{m} a_j \phi_j \right\|_V^2 = \sum_{j=n+1}^{m} |a_j|^2 \leq \sum_{j=n+1}^{\infty} |a_j|^2$$

(using the orthogonality of $\{\phi_1, \phi_2, \phi_3, \ldots\}$). Since we are assuming that (12.25) holds, we conclude that

$$\sum_{j=n+1}^{\infty} |a_j|^2 \to 0 \text{ as } n \to \infty.$$

It follows that $\|\mathbf{u}_n - \mathbf{u}_m\|_V \to 0$ as $m, n \to \infty$. This is the mark of a sequence that "ought" to converge.

Definition 12.31. *Let V be a normed vector space, and suppose $\{\mathbf{u}_n\}$ is a sequence in V. We say that $\{\mathbf{u}_n\}$ is* Cauchy *if, for any $\epsilon > 0$, there exists a positive integer N such that*

$$m, n \geq N \Rightarrow \|\mathbf{u}_n - \mathbf{u}_m\|_V < \epsilon,$$

or, in brief, $\|\mathbf{u}_n - \mathbf{u}_m\|_V \to 0$ as $m, n \to \infty$.

The terms of a Cauchy sequence "bunch up," as do the terms of a convergent sequence. Indeed, it is easy to show that every convergent sequence is a Cauchy sequence. On the other hand, the converse (does every Cauchy sequence converge?) depends on the space as well as on the norm of the space.

Example 12.32. Define a sequence of rational numbers $\{x_n\}$ by the rule that x_n is the number obtained by truncating the decimal expansion of π after n digits (so $x_1 = 3, x_2 = 3.1, x_3 = 3.14$, and so on). Then $\{x_n\}$ is certainly Cauchy (if $m > n$, then $|x_m - x_n| < 10^{-n+1}$). However, the question of whether $\{x_n\}$ converges depends on the space under consideration. If the space is $\mathbf{Q}$, the set of rational numbers, then $\{x_n\}$ fails to converge. On the other hand, if the space is $\mathbf{R}$, then $\{x_n\}$ converges, and the limit is the irrational number π. ∎

This last example may seem rather trivial; it just points out the fact that the real number system contains numbers that are not rational, and that without the irrational numbers there are "holes" in the system. The next example is more relevant to our study.

Example 12.33. Define $f : [0,1] \to \mathbf{R}$ by

$$f(x) = \begin{cases} 1, & x \in [0, \tfrac{1}{2}), \\ 0, & x \in [\tfrac{1}{2}, 1]. \end{cases}$$

Let $f_N : [0,1] \to \mathbf{R}$ be the function defined by the partial Fourier sine series of f with N terms. Then $\|f - f_N\|_{L^2} \to 0$ as $n \to \infty$, so $\{f_N\}$ is convergent and hence Cauchy. However, if the space under consideration is $C[0,1]$, the space of all continuous functions defined on the interval $[0,1]$, and the norm is the L^2 norm, then $\{f_N\}$ is still Cauchy and yet is not convergent (since $f \notin C[0,1]$). ∎

We see that the second example is analogous to the first. The space $C[0,1]$, which is a subspace of $L^2(0,1)$, has some "holes" in it, at least when the norm is taken to be the L^2 norm.

We have a name for a space in which every Cauchy sequence converges.

Definition 12.34. *Let V be a normed vector space. If every Cauchy sequence in V converges to an element of V, then we say that V is* complete.

We now have two different (and unrelated) uses for the word "complete." An orthogonal sequence in an inner product space can be complete, and a normed vector space can be complete.

The above examples show that $\mathbf{Q}$ is not complete, and neither is $C[0,1]$ under the L^2 norm.[77] On the other hand, $\mathbf{R}$ is complete (as is $\mathbf{R}^n$). Also, the space $L^2(0,1)$ is complete, the proof of which is beyond the scope of this book.

Theorem 12.35. *Both real and complex $L^2(a,b)$ are complete spaces.*

We can now answer the question we posed at the beginning of this section: If $\{\phi_j\}$ is an orthonormal sequence in a complex inner product space V and $\{a_j\}$ is a sequence of complex numbers satisfying $\sum_{j=1}^{\infty} |a_j|^2 < \infty$, does $\sum_{j=1}^{\infty} a_j \phi_j$ converge to an element of V? The answer is that if V is complete, the convergence is guaranteed.

Theorem 12.36. *Let V be a complete (complex) inner product space, and let $\phi_1, \phi_2, \phi_3, \ldots$ be an orthonormal sequence in V. If $a_1, a_2, a_3, \ldots$ is a sequence of complex numbers satisfying $\sum_{j=1}^{\infty} |a_j|^2 < \infty$, then $\sum_{j=1}^{\infty} a_j \phi_j$ converges to an element of V.*

Proof. Since V is complete, it suffices to show that the sequence $\{\mathbf{v}_n\}$ of partial sums,

$$\mathbf{v}_n = \sum_{j=1}^{n} a_j \phi_j, \ n = 1, 2, 3, \ldots,$$

---

[77] A standard result from analysis is that $C[0,1]$ *is* complete under a different norm, namely, the norm of uniform convergence: $\|f\|_\infty = \max\{|f(x)| : x \in [0,1]\}$. This result is related to the fact that if a sequence of continuous functions converges uniformly to a function, the limit function must also be continuous.

12.6. Mean-square convergence of Fourier series

is Cauchy. If $m > n$, then

$$\mathbf{v}_m - \mathbf{v}_n = \sum_{j=n+1}^{m} a_j \phi_j,$$

and, since the sequence $\{\phi_j\}$ is orthonormal, we have

$$\|\mathbf{v}_m - \mathbf{v}_n\|_V^2 = \sum_{j=n+1}^{m} |a_j|^2 \leq \sum_{j=n+1}^{\infty} |a_j|^2.$$

Since the series $\sum_{j=1}^{\infty} |a_j|^2$ is convergent, its tail must converge to zero, which shows that $\|\mathbf{v}_m - \mathbf{v}_n\|_V \to 0$ as $n, m \to \infty$. This completes the proof. □

The conclusion of all this is that if a (complex) inner product space contains a complete orthonormal sequence $\{\phi_j\}$, then there is a one-to-one correspondence between V and the space

$$\ell^2 = \left\{ \{a_j\}_{j=1}^{\infty} : a_1, a_2, \ldots \in \mathbf{C}, \sum_{j=1}^{\infty} |a_j|^2 < \infty \right\}.$$

The vector $\mathbf{v} \in V$ corresponds to $\{(\mathbf{v}, \phi_j)_V\} \in \ell^2$. In certain regards, then, V and ℓ^2 are really the same space, with the elements given different names. We say that V and ℓ^2 are *isomorphic*.

In particular, $L^2(-\ell, \ell)$ and ℓ^2 are isomorphic, and $L^2(0, \ell)$ is isomorphic to real ℓ^2.

Exercises

1. Consider the sine series

$$\sum_{n=1}^{\infty} \frac{1}{n} \sin(n\pi x).$$

   (a) Explain why this series converges to a function $f \in L^2(0, 1)$.

   (b) Graph the partial sine series with N terms, for various values of N, and guess the function f. Then verify your guess by calculating the Fourier sine coefficients of f.

2. Consider the series

$$\sum_{n=-\infty}^{\infty} c_n e^{in\pi x},$$

   where $c_n = 1/n$ for $n \neq 0$, and $c_0 = 0$.

   (a) Explain why this series converges to a function f in (complex) $L^2(-1, 1)$.

   (b) Graph the partial series with $2N+1$ terms, for various values of N, and guess the function f. Then verify your guess by calculating the Fourier coefficients of f.

3. Does the sine series

$$\sum_{n=1}^{\infty} \frac{1}{\sqrt{n}} \sin(n\pi x)$$

converge in the mean-square sense to a function $f \in L^2(0,1)$? Why or why not?

4. Consider the function $f : [0,1] \to \mathbf{R}$ defined by $f(x) = x^{1/4}(1-x)$, and its Fourier sine series

$$\sum_{n=1}^{\infty} a_n \sin(n\pi x).$$

   (a) Which of the convergence theorems of this chapter apply to f? What kind of convergence is guaranteed: pointwise, uniform, mean-square?

   (b) Estimate $a_1, a_2, \ldots, a_{63}$ using some form of numerical integration. (One option is to use the DST, as described in Section 12.2.4. Another possibility is to use the formula

   $$a_n = 2 \int_0^1 x^{1/4}(1-x) \sin(n\pi x)\, dx$$

   and some form of numerical integration. In this case, because of the singularity at $x = 0$, it is a good idea to change variables so that the integrand is smooth. Use $x = s^4$.)

   (c) Graph f together with its partial Fourier sine series. Also graph the difference between the two functions.

5. (a) For what values of k does the function $f : [0,1] \to \mathbf{R}$ defined by $f(x) = x^k$ belong to $L^2(0,1)$? (Consider both positive and negative values of k.)

   (b) Estimate the first 63 Fourier sine coefficients of $f(x) = x^{-1/4}$.

   (c) Graph $f(x) = x^{-1/4}$, together with its partial Fourier sine series (with 63 terms). Also graph the difference between the two functions.

6. Prove that if $f \in L^2(0,\ell)$, then the Fourier sine series of f converges to f in the mean-square sense.

7. Let $\{\mathbf{v}_j\}$ be a sequence of vectors in a normed vector space V, and suppose $\mathbf{v}_j \to \mathbf{v} \in V$. Prove that $\{\mathbf{v}_j\}$ is Cauchy.

12.7 A note about general eigenvalue problems

We will now briefly discuss the general eigenvalue problem

$$\begin{aligned} -\nabla \cdot (k(\mathbf{x})\nabla u) &= \lambda u \text{ in } \Omega, \\ u &= 0 \text{ on } \partial\Omega. \end{aligned} \qquad (12.26)$$

12.7. A note about general eigenvalue problems

We assume that Ω is a bounded domain in $\mathbf{R}^2$ with a smooth boundary, and that the coefficient k is a smooth function of $\mathbf{x}$ satisfying $k(\mathbf{x}) \geq k_0 > 0$ for all $\mathbf{x} \in \Omega$.

We have already seen, for a constant coefficient and the case in which Ω is a rectangular domain or a disk, that (12.26) has an infinite sequence of solutions. Moreover, the eigenpairs had certain properties that we could verify directly: the eigenvalues were positive, the eigenfunctions were orthogonal, and so forth.

The properties we observed in Sections 11.2 and 11.3 extend to the general eigenvalue problem (12.26), although they cannot be verified by directly computing the eigenpairs. Instead, the results can be deduced from some fairly sophisticated mathematical arguments that are largely beyond the scope of this book.[78] Therefore, we will not attempt to justify all of our statements in this section.

We define $K_D : C_D^2(\overline{\Omega}) \to C(\overline{\Omega})$ by $K_D u = -\nabla \cdot (k(\mathbf{x}) \nabla u)$, so that (12.26) can be written as simply $K_D u = \lambda u$. The following results hold.

1. All eigenvalues of K_D are real and positive. To show this, we assume that $K_D u = \lambda u$ and $(u, u) = 1$. Then

$$\begin{aligned}
\lambda = \lambda(u, u) = (\lambda u, u) = (K_D u, u) &= \int_\Omega (K_D u) \overline{u} \\
&= -\int_\Omega \nabla \cdot (k(\mathbf{x}) \nabla u) \overline{u} \\
&= \int_\Omega k(\mathbf{x}) \nabla u \cdot \nabla \overline{u} - \int_{\partial \Omega} k(\mathbf{x}) \overline{u} \frac{\partial u}{\partial \mathbf{n}} \\
&= \int_\Omega k(\mathbf{x}) \nabla u \cdot \nabla \overline{u} \\
&= \int_\Omega k(\mathbf{x}) \|\nabla u\|^2.
\end{aligned}$$

(The reader should notice that, since we cannot assume a priori that λ and u are real, we use the complex L^2 inner product.) This shows that λ is a nonnegative real number. Moreover,

$$\begin{aligned}
\lambda = 0 &\Rightarrow \int_\Omega k(\mathbf{x}) \|\nabla u\|^2 = 0 \\
&\Rightarrow \nabla u = 0 \text{ (since } k(\mathbf{x}) > 0) \\
&\Rightarrow u(\mathbf{x}) = \text{constant} \\
&\Rightarrow u(\mathbf{x}) = 0,
\end{aligned}$$

where the last step follows from the Dirichlet boundary conditions. Therefore, K_D has only positive eigenvalues.

2. Eigenfunctions of K_D corresponding to distinct eigenvalues are orthogonal. As we have seen before, the orthogonality of eigenfunctions depends only on the symmetry of the operator, which is easy to verify using integration by parts (see Exercise 1).

---

[78] Some of these arguments were sketched, for Sturm–Liouville eigenvalue problems in one spatial dimension, in Chapter 10.

3. The operator K_D has an infinite sequence $\{\lambda_n\}_{n=1}^\infty$ of eigenvalues satisfying $0 < \lambda_1 \leq \lambda_2 \leq \lambda_3 \leq \cdots$ and $\lambda_n \to \infty$ as $n \to \infty$. In the sequence $\{\lambda_n\}$, each eigenvalue is repeated according to the dimension of the associated eigenspace, that is, according to the number of linearly independent eigenfunctions associated with that eigenvalue. In particular, each eigenvalue corresponds to only finitely many linearly independent eigenfunctions.

    It is always possible to replace a basis for a finite-dimensional subspace by an orthogonal basis. Therefore, all of the eigenfunctions of K_D can be taken to be orthogonal. We will assume that $\{\psi_n\}_{n=1}^\infty$ is an orthogonal sequence satisfying
    $$K_D \psi_n = \lambda_n \psi_n, \ n = 1, 2, 3, \ldots.$$

4. The set of eigenfunctions $\{\psi_n\}$ is a complete orthogonal sequence in $L^2(\Omega)$: For each $f \in L^2(\Omega)$, the series
    $$\sum_{n=1}^\infty \frac{(f, \psi_n)}{(\psi_n, \psi_n)} \psi_n \tag{12.27}$$
    converges in the mean-square sense to f. The space $L^2(\Omega)$ is defined as was $L^2(a,b)$ (informally, it is the space of square-integrable functions defined on Ω, with the understanding that if two functions differ only on a set of measure zero, then they are regarded as the same). The series (12.27) is called a *generalized Fourier series* for f.

For specific domains, it may be more convenient to enumerate the eigenvalues and eigenfunctions in a doubly indexed list rather than a singly indexed list as suggested above. For example, the eigenvalue/eigenfunction pairs of the negative Laplacian on the unit square are
$$\lambda_{mn} = (m^2 + n^2)\pi^2, \ \psi_{mn}(\mathbf{x}) = \sin(m\pi x_1)\sin(n\pi x_2), \ m, n = 1, 2, 3, \ldots.$$

It is possible (although not necessarily useful) to order the λ_{mn} and ψ_{mn} in (singly indexed) sequences.

The usefulness of the above facts for many computational tasks is limited, since, for most domains Ω and coefficients $k(\mathbf{x})$, it is not possible to obtain the eigenvalues and eigenfunctions analytically (that is, in "closed form"). However, as we have seen before, the eigenpairs give some information that can be useful in its own right. It may be useful to expend some effort in computing a few eigenpairs numerically. We illustrate this with an example.

Example 12.37. Consider a membrane that at rest occupies the domain Ω, and suppose that the (unforced) small transverse vibrations of the membrane satisfy the IBVP
$$\begin{aligned} \frac{\partial^2 u}{\partial t^2} - c^2 \Delta u &= 0, \ \mathbf{x} \in \Omega, \ t > 0, \\ u(\mathbf{x}, 0) &= \phi(\mathbf{x}), \ \mathbf{x} \in \Omega, \\ \frac{\partial u}{\partial t}(\mathbf{x}, 0) &= \gamma(\mathbf{x}), \ \mathbf{x} \in \Omega, \\ u(\mathbf{x}, t) &= 0, \ \mathbf{x} \in \partial\Omega, \ t > 0. \end{aligned} \tag{12.28}$$

12.7. A note about general eigenvalue problems

The question we wish to answer is, What is the lowest frequency at which the membrane will resonate? This could be an important consideration in designing a device that contained such a membrane.

The key point is that, if we knew the eigenvalues and eigenvectors of the negative Laplacian on Ω (subject to Dirichlet conditions), then we could solve the IBVP just as in the Fourier series method. Suppose the eigenpairs are

$$\lambda_n, \psi_n(\mathbf{x}), \, n = 1, 2, 3, \ldots.$$

Then we can write the solution $u(\mathbf{x}, t)$ as

$$u(\mathbf{x}, t) = \sum_{n=1}^{\infty} a_n(t) \psi_n(\mathbf{x}).$$

Substituting into the PDE yields

$$\sum_{n=1}^{\infty} \left\{ \frac{d^2 a_n}{dt^2}(t) + \lambda_n c^2 a_n(t) \right\} \psi_n(\mathbf{x}) = 0,$$

and so $a_n(t)$ satisfies

$$\frac{d^2 a_n}{dt^2} + \lambda_n c^2 a_n = 0, \, n = 1, 2, 3, \ldots.$$

Therefore,

$$a_n(t) = b_n \cos\left(c\sqrt{\lambda_n} t\right) + c_n \sin\left(c\sqrt{\lambda_n} t\right),$$

with the coefficients b_n and c_n determined by the initial conditions. Therefore, the fundamental frequency—the smallest natural frequency—of the membrane is $c\sqrt{\lambda_1}/(2\pi)$. ∎

This example shows that we would gain some useful information by computing the smallest eigenvalue of the operator on Ω. We will revisit this example in Section 13.4, where we use finite element methods to estimate the smallest eigenvalue of various domains.

Exercises

1. Prove that the operator K defined above is symmetric.

2. Let the eigenvalues and eigenvectors of K_D be

$$\lambda_n, \psi_n(\mathbf{x}), \, n = 1, 2, 3, \ldots,$$

as in the text. Suppose that $u \in C_D^2(\Omega)$ is represented in a generalized Fourier series as follows:

$$u(\mathbf{x}) = \sum_{n=1}^{\infty} a_n \psi_n(\mathbf{x}).$$

Explain why the generalized Fourier series of $K_D u$ is equal to

$$\sum_{n=1}^{\infty} \lambda_n a_n \psi_n(\mathbf{x}).$$

3. Consider a metal plate occupying a domain Ω in $\mathbf{R}^2$. Suppose that the plate is heated to a constant 5 degrees Celsius, the top and bottom of the plate are perfectly insulated, and the edges of the plate are fixed at 0 degrees. Let $u(\mathbf{x},t)$ be the temperature at $\mathbf{x} \in \Omega$ after t seconds. Using the first eigenvalue and eigenfunction of the negative Laplacian on Ω, give a simple estimate of $u(\mathbf{x},t)$ that is valid for large t (your answer will also depend on the material constants ρ, c, and κ describing the plate). What must be true in order that a single eigenpair suffices to define a good estimate?

12.8 Suggestions for further reading

Kammler [36] and Folland [20] contain a wealth of information about Fourier series and other aspects of Fourier analysis. Folland's book is classical in nature, while Kammler delves into a number of applications of modern interest.

Briggs and Henson [9] give a comprehensive introduction to the discrete Fourier transform and cover some applications, while Van Loan [66] provides an in-depth treatment of the mathematics of the FFT. Other texts that explain the FFT algorithm include Walker [67] and Brigham [10].

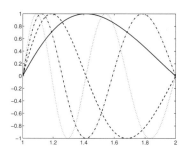

Chapter 13
More about Finite Element Methods

In this chapter, we will look more deeply into finite element methods for solving steady-state boundary value problems (BVPs). Finite element methods form a vast area, and even by the end of this chapter, we will have only scratched the surface. Our goal is modest: We wish to give the reader a better idea of how finite element methods are implemented in practice, and also to give an overview of the convergence theory.

The main tasks involved in applying the finite element method are

- defining a mesh on the computational domain;
- computing the stiffness matrix $\mathbf{K}$ and the load vector $\mathbf{f}$;
- solving the finite element equation $\mathbf{Ku} = \mathbf{f}$.

We will mostly ignore the first task, except to provide examples for simple domains. Mesh generation is an area of study in its own right, and delving into this subject is beyond the scope of this book. We begin by addressing issues involved in computing the stiffness matrix and load vector, including data structures for representing and manipulating the mesh. We will concentrate on two-dimensional problems, triangular meshes, and piecewise linear finite elements, as these are sufficient to illustrate the main ideas. Next, in Section 13.2, we discuss methods for solving the finite element equation $\mathbf{Ku} = \mathbf{f}$, specifically, on algorithms for taking advantage of the sparsity of this system of equations. In Section 13.3, we provide a brief outline of the convergence theory for Galerkin finite element methods. Finally, we close this chapter with a discussion of finite element methods for solving eigenvalue problems, such as those suggested in Section 12.7.

13.1 Implementation of finite element methods

There are several issues that must be resolved in order to efficiently implement the finite element method in a computer program, including the following:

- Data structures for representation of the mesh.
- Efficient computation of the various integrals that define the entries in the stiffness matrix and load vector.

- Algorithms for assembling the stiffness matrix and the load vector.

We will now discuss these issues.

13.1.1 Describing a triangulation

Before we can describe data structures and algorithms, we must develop notation for the computational mesh. First of all, we will assume that the domain Ω on which the BVP is posed is a bounded polygonal domain in $\mathbf{R}^2$, so that it can be triangulated.[79] Let

$$\mathcal{T}_h = \{T_i \,:\, i = 1, 2, \ldots, L\}$$

be the collection of triangles in the mesh under consideration, and let

$$\mathcal{N}_h = \{\mathbf{n}_j \,:\, j = 1, 2, \ldots, M\}$$

be the set of nodes of the triangles in $\mathcal{T}_h$. (Standard notation in finite element methods labels each mesh with h, the length of the longest side of any triangle in the mesh. Similarly, the space of piecewise linear functions relative to that mesh is denoted V_h, and an arbitrary element of V_h as v_h. We will now adopt this standard notation.)

To make this discussion as concrete as possible, we will use as an example a regular triangular mesh, defined on the unit square, consisting of 32 triangles and 25 nodes. This mesh is shown in Figure 13.1.

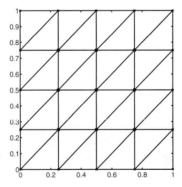

Figure 13.1. *A triangulation of the unit square.*

In order to perform the necessary calculations, we must know which nodes are associated with which triangles. Of course, each triangle has three nodes; we need to know the indices of these three nodes in the list $\mathbf{n}_1, \mathbf{n}_2, \ldots, \mathbf{n}_M$. We define the mapping e by the

---

[79]If the domain Ω has curved boundaries, then, in the context of piecewise linear finite elements, the boundary must be approximated by a polygonal curve made up of edges of triangles. This introduces an additional error into the approximation. When using higher-order finite elements, for example, piecewise quadratic or piecewise cubic functions, it is straightforward to approximate the boundary by a piecewise polynomial curve of the same degree as is used for the finite element functions, thus reducing the error associated with approximating the boundary. This technique is called the *isoparametric* method. We will ignore all such questions here by restricting ourselves to polygonal domains.

13.1. Implementation of finite element methods

property that the nodes of triangle T_i are $\mathbf{n}_{e(i,1)}, \mathbf{n}_{e(i,2)}, \mathbf{n}_{e(i,3)}$. (So e is a function of two variables; the first must take an integral value from 1 to L, and the second must take one of the integers $1, 2, 3$.)

In our sample mesh, there are 32 triangles ($L = 32$), and they are enumerated as in Figure 13.2. The $M = 25$ nodes are enumerated as shown in Figure 13.3, which explicitly shows the indices of the nodes belonging to each triangle. For example,

$$e(10,1) = 6, \; e(10,2) = 7, \; e(10,3) = 12,$$

expressing the fact that the vertices of triangle 10 are nodes 6, 7, and 12.

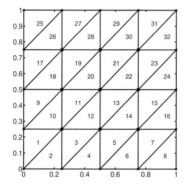

Figure 13.2. *Enumerating the triangles in the sample mesh.*

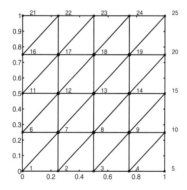

Figure 13.3. *Enumerating the nodes in the sample mesh.*

Each of the standard basis functions for the space of continuous piecewise linear functions corresponds to one node in the triangulation. The basis functions correspond to the degrees of freedom in the finite element approximation. However, not every node in the triangulation need give rise to a degree of freedom; some may be associated with Dirichlet boundary conditions which determine the weight given to the corresponding basis function. For this reason, it is not sufficient to identify the set of all nodes in the triangulation; we must also determine the *free nodes*—those that do not correspond to a Dirichlet condition. We will label the free nodes as $\mathbf{n}_{f_1}, \mathbf{n}_{f_2}, \ldots, \mathbf{n}_{f_N}$.

Referring again to the sample mesh of Figure 13.1, if we assume that this mesh will be used in a Dirichlet problem, only the interior nodes correspond to degrees of freedom. Thus $N = 9$, and the free nodes can be enumerated in the same order as the set of all nodes (left-to-right and bottom-to-top). This yields

$$f_1 = 7, \ f_2 = 8, \ f_3 = 9, \ f_4 = 12, \ f_5 = 13, \ f_6 = 14, \ f_7 = 17, \ f_8 = 18, \ f_9 = 19.$$

We now define the standard basis function ϕ_i by the conditions that ϕ_i is a continuous piecewise linear function relative to the mesh $\mathcal{T}_h$, and that

$$\phi_i(\mathbf{n}_j) = \delta_{ij} = \begin{cases} 1, & i = j, \\ 0, & i \neq j. \end{cases}$$

The finite element space—the space of continuous piecewise linear functions (relative to the mesh $\mathcal{T}_h$) that satisfy any given homogeneous Dirichlet conditions—is then

$$V_h = \mathrm{span}\{\phi_{f_1}, \phi_{f_2}, \ldots, \phi_{f_N}\}.$$

13.1.2 Computing the stiffness matrix

We assume that the weak form of the BVP in question has been written in the form

$$u \in V, \ a(u,v) = (f,v) \text{ for all } v \in V$$

and that the Galerkin problem takes the form

$$u_h \in V_h, \ a(u_h,v) = (f,v) \text{ for all } v \in V_h.$$

We need to compute the stiffness matrix $\mathbf{K} \in \mathbf{R}^{N \times N}$, where

$$K_{ij} = a\left(\phi_{f_j}, \phi_{f_i}\right).$$

By design, most of these values will be zero; let us focus on a fixed i between 1 and N, and ask the question, For which values of j will K_{ij} be nonzero?

As an example, consider $i = 5$ for our sample mesh. The support of $\phi_{13} = \phi_{f_5}$ is shown in Figure 13.4. The question we must answer is, What other basis functions ϕ_{f_j} have a support that intersects the support of ϕ_{13}? (The intersection must have a positive area; if the intersection is just an edge of a triangle, this will not lead to a nonzero contribution to $\mathbf{K}$.) It is easy to see that the desired basis functions correspond to nodes of triangles of which $\mathbf{n}_{13}$ is itself a node, that is, to the nodes of the triangles shaded in Figure 13.4.

It is possible to store all of the necessary connectivity information (that is, to store, along with each node, a list of nodes belonging to a common triangle). With this information, one can compute the matrix $\mathbf{K}$ by rows, according to the following algorithm:

for $i = 1, 2, \ldots, N$
    for each $j = 1, 2, \ldots, N$ such that the supports of ϕ_{f_j} and ϕ_{f_i} overlap,
        compute $K_{ij} = a\left(\phi_{f_j}, \phi_{f_i}\right)$.

However, this is probably not the most common approach, because it is more efficient to perform all the calculations pertaining to a given triangle at one time. Therefore, many

13.1. Implementation of finite element methods

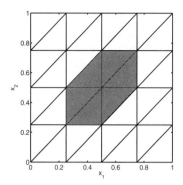

Figure 13.4. *The support of ϕ_{13}.*

codes use a triangle-oriented approach rather than the node-oriented algorithm described above. We will now explain how to implement a triangle-oriented algorithm.

The integral represented by $a(\phi_{f_j}, \phi_{f_i})$ is naturally decomposed into the sum of several integrals, one for each triangular element making up the domain of integration. It is convenient to make this decomposition, as ϕ_{f_i} and ϕ_{f_j} are given by different formulas on the different elements. It is then natural to compute, for each element T_k, the contributions to various entries of the stiffness matrix **K**.

For example, consider $i = 5$, $j = 1$ in our sample mesh. Then $a(\phi_{f_j}, \phi_{f_i})$ is the sum of two integrals, one over triangle T_{11} and the other over T_{12}:

$$a(\phi_{f_1}, \phi_{f_5}) = \int_{T_{11}} (\ldots) dx + \int_{T_{12}} (\ldots) dx.$$

There is no reason that these two integrals must be computed together, as long as both contributions to K_{51} are eventually accumulated.

Following the above reasoning, a standard algorithm for computing **K** is to loop over the *elements* (i.e., the triangles) in the mesh (rather than over the nodes) and compute the contributions to **K** from each element in turn. Since each triangle has three nodes, a given element contributes to up to 9 entries in **K**. There is one piece of information, however, that is not available unless we make a point to record it in the data structure: the "inverse" of the mapping $i \mapsto f_i$. In the course of assembling the stiffness matrix, we need to be able to decide if a given node $\mathbf{n}_j$ is a free node (that is, if $j = f_i$ for some i) and, if so, to determine i. So let us define a mapping $j \mapsto g_j$ by

$$g_j = \begin{cases} i & \text{if } j = f_i, \\ 0 & \text{if } j \neq f_i \text{ for all } i = 1, 2, \ldots, N. \end{cases}$$

The algorithm then takes the following form:

for $k = 1, 2, \ldots, L$
    for $i = 1, 2, 3$
        for $j = 1, 2, 3$
            if $p = g_{e(k,i)} \neq 0$ and $q = g_{e(k,j)} \neq 0$
                add the contribution to K_{pq} obtained by integrating over T_k.

The condition that $g_{e(k,i)} \neq 0$ indicates that the ith vertex ($i = 1, 2, 3$) of triangle T_k is a free node, and similarly for $g_{e(k,j)}$. The integral mentioned in the last line of this algorithm is the contribution to $K_{pq} = a\left(\phi_{e(k,j)}, \phi_{e(k,i)}\right)$ from the element T_k. Also, it should be pointed out that, in the case that **K** is symmetric (as it is in the examples considered in this book), part of the work can be eliminated: we just compute the entries K_{pq} with $q \geq p$, and then assign the value of K_{pq} to K_{qp}.

We have described above the information needed to carry out this algorithm. It is easy to store this information in arrays for use in a computer program. A simple data structure would consist of four arrays:[80]

1. **NodeList**: An $M \times 2$ array; the ith row consists of the x and y coordinates of node $\mathbf{n}_i$.

2. **ElementList**: An $L \times 3$ array; the kth row consists of $e(k,1), e(k,2), e(k,3)$.

3. **FNodePtrs**: An $N \times 1$ array; the ith entry is f_i.

4. **NodePtrs**: An $M \times 1$ array; the ith entry is g_i.

By way of example, Table 13.1 shows these arrays for the sample mesh of Figure 13.1. The information recorded in these arrays will suffice for straightforward finite element problems with homogeneous Dirichlet or Neumann boundary conditions. For problems with inhomogeneous Dirichlet conditions, it would be convenient to add an array **CNodePtrs**, analogous to **FNodePtrs** but specifying the constrained nodes. We would also need to modify **NodePtrs** in this case to indicate not only that a node is constrained, but also to indicate its index in the list of constrained nodes. For problems with inhomogeneous Neumann conditions, it may also be necessary to record whether a given edge belongs to the boundary. The **ElementList** array can be augmented by columns containing flags indicating whether the edges lie on the boundary. (Or, of course, these flags can be stored in independent arrays, if desired.) This extension is left to the reader.

13.1.3 Computing the load vector

Computing the load vector **f** is similar to computing the stiffness matrix, only easier. We loop over the elements and, for each node $\mathbf{n}_{f_i}$ of a given triangle, compute the contribution to f_i. The value f_i will be the sum of integrals over one or more triangles, and this sum is accumulated as we loop over the triangular elements:

for $k = 1, 2, \ldots, L$
   for $i = 1, 2, 3$
      if $p = g_{e(k,i)} \neq 0$
         add the contribution to f_p obtained by integrating over T_k.

13.1.4 Quadrature

Up to this point, we have computed all integrals exactly, perhaps with the help of a computer package such as *Mathematica* or *Maple*. However, this would be difficult to incorporate

---

[80]Using modern techniques, such as modular programming or object-oriented programming, one might define a structure or a class to represent a mesh.

13.1. Implementation of finite element methods

Table 13.1. *The data structure for the mesh of Figure* 13.1.

ElementList

1	1	6	7
2	1	2	7
3	2	7	8
4	2	3	8
5	3	8	9
6	3	4	9
7	4	9	10
8	4	5	10
9	6	11	12
10	6	7	12
11	7	12	13
12	7	8	13
13	8	13	14
14	8	9	14
15	9	14	15
16	9	10	15
17	11	16	17
18	11	12	17
19	12	17	18
20	12	13	18
21	13	18	19
22	13	14	19
23	14	19	20
24	14	15	20
25	16	21	22
26	16	17	22
27	17	22	23
28	17	18	23
29	18	23	24
30	18	19	24
31	19	24	25
32	19	20	25

NodeList

1	0	0
2	0.2500	0
3	0.5000	0
4	0.7500	0
5	1.0000	0
6	0	0.2500
7	0.2500	0.2500
8	0.5000	0.2500
9	0.7500	0.2500
10	1.0000	0.2500
11	0	0.5000
12	0.2500	0.5000
13	0.5000	0.5000
14	0.7500	0.5000
15	1.0000	0.5000
16	0	0.7500
17	0.2500	0.7500
18	0.5000	0.7500
19	0.7500	0.7500
20	1.0000	0.7500
21	0	1.0000
22	0.2500	1.0000
23	0.5000	1.0000
24	0.7500	1.0000
25	1.0000	1.0000

NodePtrs

1	0
2	0
3	0
4	0
5	0
6	0
7	1
8	2
9	3
10	0
11	0
12	4
13	5
14	6
15	0
16	0
17	7
18	8
19	9
20	0
21	0
22	0
23	0
24	0
25	0

FNodePtrs

1	7
2	8
3	9
4	12
5	13
6	14
7	17
8	18
9	19

into a general-purpose computer program for finite elements. Moreover, it is not necessary to compute the various integrals exactly, and it may be more efficient to use numerical integration (quadrature). It is only necessary to choose a quadrature rule that introduces an error small enough that it does not affect the rate of convergence of the error in the approximate solution to zero as the mesh is refined.

A quadrature rule is an approximation of the form

$$\int_R f \doteq \sum_{i=1}^{k} w_i f(s_i, t_i), \tag{13.1}$$

where $w_1, w_2, \ldots, w_k$, the *weights*, are real numbers and $(s_1,t_1), (s_2,t_2), \ldots, (s_k,t_k)$, the *nodes*, are points in the domain of integration R. Equation (13.1) defines a k-point quadrature rule.

Choosing a quadrature rule

There are two issues that must be resolved: the choice of a quadrature rule and its implementation for an arbitrary triangle. We begin by discussing the quadrature rule. A useful way to classify quadrature rules is by their *degree of precision*. While it may not be completely obvious at first, it is possible to define rules of the form (13.1) that give the exact value when applied to certain polynomials. A rule has degree of precision d if the rule is exact for all polynomials of degree d or less. Since both integration and a quadrature rule of the form (13.1) are linear in f, it suffices to consider only monomials.

As a simple example, consider the rule

$$\int_{-1}^{1} f(x)\,dx \doteq 2f(0). \tag{13.2}$$

This rule has degree of precision 1, since

$$p(x) = 1 \Rightarrow \int_{-1}^{1} p(x)\,dx = 2 = 2p(0),$$

$$p(x) = x \Rightarrow \int_{-1}^{1} p(x)\,dx = 0 = 2p(0),$$

$$p(x) = x^2 \Rightarrow \int_{-1}^{1} p(x)\,dx = \frac{2}{3} \neq 2p(0).$$

Equation (13.2) is the one-point Gaussian quadrature rule. For one-dimensional integrals, the n-point Gaussian quadrature rule is exact for polynomials of degree $2n - 1$ or less. For example, the two-point Gaussian quadrature rule, which has degree of precision 3, is

$$\int_{-1}^{1} f(x)\,dx \doteq f\left(-\frac{1}{\sqrt{3}}\right) + f\left(\frac{1}{\sqrt{3}}\right).$$

The Gaussian quadrature rules are defined on the *reference interval* $[-1, 1]$; to apply the rules on a different interval requires a simple change of variables.

In multiple dimensions, it is also possible to define quadrature rules with a given degree of precision. Of course, things are more complicated because of the variety of geometries that are possible. We will exhibit rules for triangles with degrees of precision 1 and 2. These rules will be defined for the *reference triangle* T^R with vertices $(0,0)$, $(1,0)$, and $(0,1)$ (see Figure 13.5). The following one-point rule has degree of precision 1:

$$\int_{T^R} f \doteq \frac{1}{2} f\left(\frac{1}{3}, \frac{1}{3}\right) \tag{13.3}$$

(see Exercise 13.1.1), while the following two-point rule has degree of precision 2:

$$\int_{T^R} f \doteq \frac{1}{6}\left(f\left(\frac{1}{6}, \frac{1}{6}\right) + f\left(\frac{2}{3}, \frac{1}{6}\right) + f\left(\frac{1}{6}, \frac{2}{3}\right)\right) \tag{13.4}$$

(see Exercise 13.1.2).

13.1. Implementation of finite element methods

Figure 13.5. *The reference triangle T^R.*

What degree of precision is necessary in applying the finite element method? Here is a rule of thumb: The quadrature rule should compute the exact entries in the stiffness matrix in the case of a PDE with constant coefficients. Such a quadrature rule will then be accurate enough, even in a problem with (smooth) nonconstant coefficients, that the accuracy of the finite element method will not be significantly degraded.

When using piecewise linear functions, the stiffness matrix is assembled from integrals of the form

$$\int_T k \nabla \phi_j \cdot \nabla \phi_i,$$

and the integrand is constant when k is constant (a linear function has a constant gradient). Therefore, the one-point rule will give exact results for a constant-coefficient problem, and, by our rule of thumb, the same rule is adequate for nonconstant-coefficient problems.

If we were to use piecewise quadratic functions, then $\nabla \phi_j \cdot \nabla \phi_i$ would be piecewise quadratic as well, and so we would need a quadrature rule with degree of precision 2. The three-point rule (13.4) would be appropriate in that case.

Integrating over an arbitrary triangle

There is still the following technical detail to discuss: How can we apply a quadrature rule defined on the reference triangle T^R to an integral over an arbitrary triangle T? We assume that T has vertices (p_1, p_2), (q_1, q_2), and (r_1, r_2). We can then map $(z_1, z_2) \in T^R$ to $(x_1, x_2) \in T$ by the linear mapping

$$\begin{aligned} x_1 &= p_1 + (q_1 - p_1)z_1 + (r_1 - p_1)z_2, \\ x_2 &= p_2 + (q_2 - p_2)z_1 + (r_2 - p_2)z_2. \end{aligned} \quad (13.5)$$

The mapping (13.5) sends $(0,0)$, $(1,0)$, and $(0,1)$ to (p_1, p_2), (q_1, q_2), and (r_1, r_2), respectively. We will denote this mapping by $\mathbf{x} = \mathbf{F}(\mathbf{y})$, and we note that the Jacobian matrix of $\mathbf{F}$ is

$$\mathbf{J} = \begin{bmatrix} q_1 - p_1 & r_1 - p_1 \\ q_2 - p_2 & r_2 - p_2 \end{bmatrix}.$$

We can now apply the rule for a change of variables in a multiple integral:[81]

$$\int_T f(\mathbf{x}) \, d\mathbf{x} = \int_{T^R} f(\mathbf{F}(\mathbf{y})) \, |\det(\mathbf{J})| \, d\mathbf{y}.$$

---

[81] This rule is explained in calculus texts such as Gillett [22], or more fully in advanced calculus texts such as Kaplan [37].

As the determinant of $\mathbf{J}$ is the constant

$$j = \det(\mathbf{J}) = (q_1 - p_1)(r_2 - p_2) - (q_2 - p_2)(r_1 - p_1),$$

it is easy to apply the change of variables and then any desired quadrature rule. For example, the three-point rule (13.4) would be applied as follows:

$$\int_T f \doteq \frac{|j|}{6}\left(f\left(x_1^{(1)}, x_2^{(1)}\right) + f\left(x_1^{(2)}, x_2^{(2)}\right) + f\left(x_1^{(3)}, x_2^{(3)}\right)\right),$$

$$\left(x_1^{(1)}, x_2^{(1)}\right) = (p_1 + (q_1 - p_1)/6 + (r_1 - p_1)/6, p_2 + (q_2 - p_2)/6 + (r_2 - p_2)/6),$$

$$\left(x_1^{(2)}, x_2^{(2)}\right) = (p_1 + 2(q_1 - p_1)/3 + (r_1 - p_1)/6, p_2 + 2(q_2 - p_2)/3 + (r_2 - p_2)/6),$$

$$\left(x_1^{(3)}, x_2^{(3)}\right) = (p_1 + (q_1 - p_1)/6 + 2(r_1 - p_1)/3, p_2 + (q_2 - p_2)/6 + 2(r_2 - p_2)/3).$$

Exercises

1. Verify that (13.3) produces the exact integral for the monomials $1, x, y$.

2. Verify that (13.4) produces the exact integral for the monomials $1, x, y, x^2, xy, y^2$.

3. Let T be the triangular region with vertices $(1,0), (2,0)$, and $(3/2, 1)$, and let $f : T \to \mathbf{R}$ be defined by $f(\mathbf{x}) = x_1 x_2^2$. Compute

$$\int_T f$$

directly, and also by transforming the integral to an integral over T^R, as suggested in the text. Verify that you obtain the same result.

4. A common way to create a mesh on a domain Ω is to establish a coarse mesh, and then to refine it until it is suitable. A simple way to refine a triangular mesh is to replace each triangle with four triangles by connecting the midpoints of the three edges of the original triangle. An example is shown in Figure 13.6, in which a mesh with two triangles is refined to a mesh with eight triangles by this method. This technique of replacing a triangle with four triangles can be repeated several times until the mesh is fine enough.

   (a) Consider the coarsest mesh in Figure 13.6. Describe the mesh using the notation in the text, and also compute the arrays **NodeList**, **ElementList**, **FNodePtrs**, **NodePtrs**. Assume that the mesh will be used for a Dirichlet problem.

   (b) Repeat for the finer mesh in Figure 13.6.

   (c) Refine the (right-hand) mesh from Figure 13.6 one more time, and compute the arrays describing the refined mesh.

   Note: There is not a unique right answer to this exercise, since the nodes and elements can be ordered in various ways.

13.2. Solving sparse linear systems

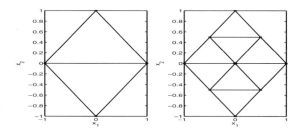

Figure 13.6. *A coarse mesh (left), refined in the standard fashion (right). (Note: The mesh is a triangulation of the rhombus in both figures; the circumscribed box just indicates the axes.)*

5. Explain how to modify the information stored in the data structure suggested in the text (specifically, by adding a **CNodePtrs** array and modifying **NodePtrs**) to allow for the solution of inhomogeneous Dirichlet problems.

6. Consider the data structure suggested in the text. What information is lacking that is needed to solve an inhomogeneous Neumann problem? Explain.

13.2 Solving sparse linear systems

As we have mentioned several times, the finite element method is a practical numerical method, even for large two- and three-dimensional problems, because the matrices that it produces have mostly zero entries. A matrix is called *sparse* if a large percentage of its entries are known to be zero. In this section, we wish to briefly survey the subject of solving sparse linear systems.

Methods for solving linear systems can be divided into two categories. A method that will produce the exact solution in a finite number of steps is called a *direct* method. (Actually, when implemented on a computer, even a direct method will not compute the exact solution because of unavoidable roundoff error. To be precise, a direct method will compute the exact solution in a finite number of steps, provided it is implemented in exact arithmetic.) The most common direct methods are variants of Gaussian elimination. Below, we discuss modifications of Gaussian elimination for sparse matrices.

An algorithm that computes the exact solution to a linear system only as the limit of an infinite sequence of approximations is called an *iterative* method. There are many iterative methods in use; we will discuss the most popular: the *conjugate gradient* (CG) algorithm. We also touch on the topic of preconditioning, the art of transforming a linear system so as to obtain faster convergence with an iterative method.

13.2.1 Gaussian elimination for dense systems

In order to have some standard of comparison, we first briefly discuss the standard Gaussian elimination algorithm for the solution of dense systems. Our interest is in the operation count—the number of arithmetic operations necessary to solve an $n \times n$ dense system. The basic algorithm is simple to write down. In the following description, we assume that no row interchanges are required, as these do not use any arithmetic operations anyway.[82] The following pseudocode solves $\mathbf{Ax} = \mathbf{b}$, where $\mathbf{A} \in \mathbf{R}^{n \times n}$ and $\mathbf{b} \in \mathbf{R}^n$, overwriting the values of $\mathbf{A}$ and $\mathbf{b}$:

for $i = 1, 2, \ldots, n-1$
    for $j = i+1, i+2, \ldots, n$
        $A_{ji} \leftarrow \frac{A_{ji}}{A_{ii}}$
        $b_j \leftarrow b_j - A_{ji} b_i$
        for $k = i+1, i+2, \ldots, n$
            $A_{jk} \leftarrow A_{jk} - A_{ji} A_{ik}$

for $i = n, n-1, \ldots, 1$
    $b_i \leftarrow \left(b_i - \sum_{j=i+1}^{n} A_{ij} b_j \right) / A_{ii}$.

The first (nested) loop is properly called Gaussian elimination; it systematically eliminates variables to produce an upper triangular system that is equivalent to the original system. The second loop is called *back substitution*; it solves the last equation, which now has only one variable, for x_n, substitutes this value in the preceding equation and solves for x_{n-1}, and so forth. The pseudocode above overwrites the right-hand-side vector $\mathbf{b}$ with the solution $\mathbf{x}$.

The Gaussian elimination part of the algorithm is equivalent (when, as we have assumed, no row interchanges are required) to factoring $\mathbf{A}$ into the product of an upper triangular matrix $\mathbf{U}$ and a *unit* lower triangular matrix (that is, a lower triangular matrix with all ones on the diagonal) $\mathbf{L}$: $\mathbf{A} = \mathbf{LU}$. As the algorithm is written above, the matrix $\mathbf{A}$ is overwritten with the values of $\mathbf{U}$ (on and above the diagonal) and $\mathbf{L}$ (below the diagonal—there is no need to store the diagonal of $\mathbf{L}$, since it is known to consist of all ones). At the same time, the first loop above effectively replaces $\mathbf{b}$ by $\mathbf{L}^{-1}\mathbf{b}$. The back substitution phase is then equivalent to replacing $\mathbf{L}^{-1}\mathbf{b}$ by $\mathbf{U}^{-1}\mathbf{L}^{-1}\mathbf{b} = \mathbf{A}^{-1}\mathbf{b}$. The factorization $\mathbf{A} = \mathbf{LU}$ is simply called the LU factorization.

It is not difficult to count the number of arithmetic operations required by Gaussian elimination and back substitution. It turns out to be

$$O\left(\frac{2}{3}n^3\right),$$

with most of the operations used to factor $\mathbf{A}$ into $\mathbf{LU}$ (the computation of $\mathbf{L}^{-1}\mathbf{b}$ and then $\mathbf{U}^{-1}\mathbf{L}^{-1}\mathbf{b}$ requires only $O(2n^2)$ operations). Exercise 13.2.1 asks the reader to verify these results.

When the matrix $\mathbf{A}$ is symmetric and positive definite[83] (SPD), then one can take advantage of symmetry to factor the matrix as $\mathbf{A} = \mathbf{LL}^T$, where $\mathbf{L}$ is lower triangular. This factorization is called the *Cholesky factorization*. (The Cholesky factor $\mathbf{L}$ is not the same

---

[82] In general, Gaussian elimination is numerically unstable unless *partial pivoting* is used. Partial pivoting is the technique of interchanging rows to get the largest possible pivot entry. This ensures that all of the multipliers appearing in the algorithm are bounded above by one, and in virtually every case, that roundoff error does not increase unduly. There are special classes of matrices, most notably the class of symmetric positive definite matrices, for which Gaussian elimination is provably stable with no row interchanges.

[83] The reader should recall that a symmetric matrix $\mathbf{A} \in \mathbf{R}^{n \times n}$ is called positive definite if $\mathbf{x} \in \mathbf{R}^n$, $\mathbf{x} \neq \mathbf{0} \Rightarrow \mathbf{x} \cdot \mathbf{Ax} > 0$. The matrix $\mathbf{A}$ is positive definite if and only if all of its eigenvalues are positive.

13.2. Solving sparse linear systems

Table 13.2. *Time required to solve an $n \times n$ dense linear system on a personal computer.*

n	Time (s)
500	0.0219
1000	0.157
2000	0.880
4000	5.64
8000	38.0

matrix **L** appearing in the LU factorization—in particular, it does not have all ones on the diagonal—but it is closely related.) The symmetry of **A** makes the Cholesky factorization less expensive than the LU factorization; the operation count is $O(n^3/3)$. For simplicity (because the Cholesky factorization is less familiar and harder to describe than the LU factorization), we will discuss the direct solution of linear systems in terms of Gaussian elimination and the LU factorization. However, the reader should bear in mind that the stiffness matrix **K** is SPD, and so the Cholesky factorization is preferred in practice.

The operation count shows how the computation time increases as the size of the system increases—doubling the size of the system causes an 8-fold increase in computation time. As a frame of reference, Table 13.2 gives the time required to solve an $n \times n$ linear system on a reasonably good personal computer at the time of this writing.[84]

With the rapid improvements in computer hardware of the last two decades, it might seem that concerns about algorithmic efficiency are less significant than they used to be. Indeed, many problems that required mainframe computers 25 years ago can now easily be solved on inexpensive personal computers. However, with an operation count of $O(2n^3/3)$, it is not difficult to encounter problems that exceed available computing power. For example, consider the discretization of a three-dimensional cube. If there are N divisions in each dimension, then the size of the system will be on the order of $N^3 \times N^3$, and, if a dense linear system of this order is to be solved (or even if a sparse linear system is solved by a dense method, that is, a method that does not take advantage of sparsity), then the operation count will be $O(N^9)$. Merely to refine the mesh by a factor of 2 in each dimension will increase the computation time by a factor of 512. (We are not even considering the memory requirements.) Thus, given the times in Table 13.2, to solve a problem on a $20 \times 20 \times 20$ grid would take less than a minute; to solve the same problem on a $40 \times 40 \times 40$ grid would take more than 5 hours!

This discussion should convince the reader that algorithmic efficiency is critical for solving some realistic problems. We now discuss the solution of sparse systems; a comparison of operation counts will show the advantages of a method, such as finite elements, that leads to sparse linear systems.

13.2.2 Direct solution of banded systems

When the mesh employed in the finite element method has a regular structure, the resulting stiffness matrix tends to be *banded*. The entries of a banded matrix are zero except for those in a band close to the main diagonal. The precise definition is the following.

---

[84] Of course, the rate of improvement of CPUs being what it is, this will no longer be considered a reasonably good CPU by the time this book is published!

Definition 13.1. *Let* $\mathbf{A} \in \mathbf{R}^{n \times n}$. *We say that* $\mathbf{A}$ *is* banded *with* half-bandwidth p *if*
$$|i - j| > p \Rightarrow A_{ij} = 0.$$

As an example, consider the stiffness matrix arising from solving Poisson's equation:
$$\begin{aligned} -\Delta u &= f(\mathbf{x}) \text{ in } \Omega, \\ u &= 0 \text{ on } \partial\Omega. \end{aligned} \tag{13.6}$$

Suppose Ω is the unit square ($\Omega = \{\mathbf{x} \in \mathbf{R}^2 : 0 < x_1 < 1, 0 < x_2 < 1\}$), and we apply the finite element method with a regular triangulation of Ω and piecewise linear finite element functions. Dividing the x and y intervals, $[0,1]$, into n subintervals each, we obtain $2n^2$ elements and $(n+1)^2$ nodes. Only the interior nodes are free, so the stiffness matrix is $(n-1)^2 \times (n-1)^2$. As we showed in Example 11.10, a typical free node n_{f_i} interacts with nodes
$$n_{f_{i-n+1}}, n_{f_{i-1}}, n_{f_i}, n_{f_{i+1}}, n_{f_{i+n-1}}$$
(also with nodes $n_{f_{i-n}}$ and $n_{f_{i+n}}$; however, with constant coefficients, the corresponding entries of $\mathbf{K}$ turn out to be zero due to cancellation), so that the subdiagonals of $\mathbf{A}$ indexed by $-n+1, -1$ and the superdiagonals indexed by $1, n-1$, along with the main diagonal, contain nonzero entries. Thus $\mathbf{A}$ is banded with half-bandwidth $n-1$. See Figure 13.7 for the sparsity pattern of $\mathbf{A}$.

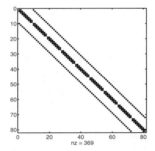

Figure 13.7. *The sparsity pattern of the discrete Laplacian (200 triangular elements).*

It is completely straightforward to apply Gaussian elimination with back substitution to a banded system. Indeed, the typical step of the algorithm is just as in the dense case, except that the inner loops run over a limited range of indices—keeping the computation within the band. Here is the algorithm:

for $i = 1, 2, \ldots, n-1$
    for $j = i+1, i+2, \ldots, \min\{i+p, n\}$
        $A_{ji} \leftarrow \frac{A_{ji}}{A_{ii}}$
        $b_j \leftarrow b_j - A_{ji} b_i$
        for $k = i+1, i+2, \ldots, \min\{i+p, n\}$
            $A_{jk} \leftarrow A_{jk} - A_{ji} A_{ik}$

for $i = n, n-1, \ldots, 1$
    $b_i \leftarrow \left(b_i - \sum_{j=i+1}^{n} A_{ij} b_j \right) / A_{ii}.$

(Here, again, we assume that no row interchanges are required.)

13.2. Solving sparse linear systems

It is easy to show that the above algorithm requires $O(2p^2n)$ operations (see Exercise 13.2.2)—a considerable savings if $p \ll n$. For example, consider solving Poisson's equation in two dimensions, as discussed above. The discrete Laplacian (under Dirichlet conditions) is an $(n-1)^2 \times (n-1)^2$ banded matrix with half-bandwidth $n-1$. The cost of solving such a banded system is $O(2(n-1)^4)$, as opposed to a cost of

$$O\left(\frac{2}{3}(n-1)^6\right)$$

if the standard Gaussian elimination algorithm is used. (For SPD matrices, these operation counts can be divided by two.)

The cost of a direct method performed on a sparse matrix is controlled, not merely by the number of nonzeros in the matrix, but by the *fill-in* that occurs during the course of the algorithm. Fill-in is said to have occurred whenever an entry that is originally zero becomes nonzero during the course of the algorithm. In factoring a banded matrix, fill-in occurs within the bands, as can be seen by inspecting the **L** and **U** factors (see Figure 13.8). Entries that become nonzero during the course of the algorithm must be included in subsequent calculations, increasing the cost.

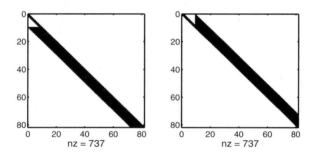

Figure 13.8. *The sparsity pattern of the* **L** *(left) and* **U** *(right) factors of the discrete Laplacian (200 triangular elements).*

13.2.3 Direct solution of general sparse systems

When **A** is sparse but not banded, it is more difficult to predict the cost of a direct method. The cost is governed by the degree of fill-in, and when the matrix is not banded, the fill-in is unpredictable. Figure 13.9 shows a symmetric sparse matrix **A** with the nonzeros placed randomly, and also shows the resulting fill-in. The original matrix has a density of 9.46%, while the lower triangular factor of **A** has a density of over 61%.

There are a number of methods for reordering the rows and columns of **A** in order to reduce fill-in. We will not explain such methods here; the interested reader can consult the book by Davis [16] for details.

Sparse matrices whose sparsity pattern is unstructured are particularly suited to iterative methods. Iterative methods only require the ability to compute matrix-vector products, such as **Ap** (and possibly, for nonsymmetric matrices, the product $\mathbf{A}^T\mathbf{p}$). Therefore, the particular sparsity pattern is relatively unimportant; the cost of the matrix-vector product is mostly governed by the density of the matrix (the percentage of nonzero entries).

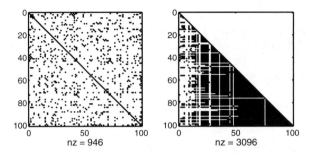

Figure 13.9. *A random sparse matrix (left) and its lower triangular factor (right).*

13.2.4 Iterative solution of sparse linear systems

Even when $\mathbf{A}$ is sparse, it may be too costly, in terms of arithmetical operations or memory (or both), to solve $\mathbf{Ax} = \mathbf{b}$ directly. An alternative is to use an *iterative* algorithm, which produces a sequence of increasingly accurate approximations to the true solution. Although the exact solution generally cannot be obtained except in the limit (that is, an infinite number of steps is required to get the exact solution), in many cases relatively few steps are needed to produce an approximate solution that is sufficiently accurate. Indeed, we should keep in mind that, in the context of solving differential equations, the "exact" solution to $\mathbf{Ax} = \mathbf{b}$ is not really the exact solution—it is only an approximation to the true solution of the differential equation. Therefore, as long as the iterative algorithm introduces errors no larger than the discretization errors, it is perfectly satisfactory.

Many iterative algorithms have been developed, and it is beyond the scope of this book to survey them. We will content ourselves with outlining the CG method, the most popular algorithm for solving SPD systems. (The stiffness matrix $\mathbf{K}$ from a finite element problem, being a Gram matrix, is SPD—see Exercise 13.2.6.) We will also briefly discuss *preconditioning*, a method for accelerating convergence.

The CG method is actually an algorithm for minimizing a quadratic form. If $\mathbf{A} \in \mathbf{R}^{n \times n}$ is SPD and $\phi : \mathbf{R}^n \to \mathbf{R}$ is defined by

$$\phi(\mathbf{x}) = \frac{1}{2}\mathbf{x} \cdot \mathbf{A}\mathbf{x} - \mathbf{b} \cdot \mathbf{x}, \tag{13.7}$$

then a direct calculation (see Exercise 13.2.3) shows that

$$\nabla \phi(\mathbf{x}) = \mathbf{A}\mathbf{x} - \mathbf{b}. \tag{13.8}$$

Therefore, the unique stationary point of ϕ is $\mathbf{x} = \mathbf{A}^{-1}\mathbf{b}$. Moreover, a consideration of the second derivative matrix shows that this stationary point is the global minimizer of ϕ (a quadratic form defined by an SPD matrix is analogous to a scalar quadratic $ax^2 + bx + c$ with $a > 0$—see Figure 13.10). Therefore, solving $\mathbf{Ax} = \mathbf{b}$ and minimizing ϕ are equivalent.

We can thus apply any iterative minimization algorithm to ϕ, and, assuming it works, it will converge to the desired value of $\mathbf{x}$. A large class of minimization algorithms are *descent* methods based on a *line search*. Such algorithms are based on the idea of a *descent direction*: Given an estimate $\mathbf{x}^{(i)}$ of the solution, a descent direction $\mathbf{p}$ is a direction such

13.2. Solving sparse linear systems

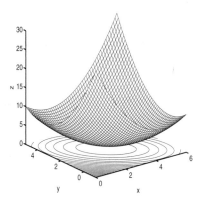

Figure 13.10. *The graph of a quadratic form defined by a positive definite matrix. The contours of the function are also shown.*

that, starting from $\mathbf{x}^{(i)}$, ϕ decreases in the direction of $\mathbf{p}$. This means that, for all $\alpha > 0$ sufficiently small,

$$\phi\left(\mathbf{x}^{(i)} + \alpha \mathbf{p}\right) < \phi\left(\mathbf{x}^{(i)}\right)$$

holds. Equivalently, this means that the directional derivative of ϕ at $\mathbf{x}^{(i)}$ in the direction of $\mathbf{p}$ is negative; that is,

$$\nabla \phi(\mathbf{x}^{(i)}) \cdot \mathbf{p} < 0.$$

Given a descent direction, a line search algorithm will seek to minimize ϕ along the ray $\{\mathbf{x}^{(i)} + \alpha \mathbf{p} : \alpha \geq 0\}$ (that is, it will search along this "line," which is really a ray).

Since ϕ is quadratic, it is particularly easy to perform the line search—along a one-dimensional subset, ϕ reduces to a scalar quadratic. Indeed,

$$\begin{aligned}
\phi(\mathbf{x}^{(i)} + \alpha \mathbf{p}) &= \frac{1}{2}(\mathbf{x}^{(i)} + \alpha \mathbf{p}) \cdot \mathbf{A}(\mathbf{x}^{(i)} + \alpha \mathbf{p}) - \mathbf{b} \cdot (\mathbf{x}^{(i)} + \alpha \mathbf{p}) \\
&= \frac{1}{2}\mathbf{x}^{(i)} \cdot \mathbf{A}\mathbf{x}^{(i)} + \alpha \mathbf{p} \cdot \mathbf{A}\mathbf{x}^{(i)} + \frac{\alpha^2}{2}\mathbf{p} \cdot \mathbf{A}\mathbf{p} - \mathbf{b} \cdot \mathbf{x}^{(i)} - \alpha \mathbf{b} \cdot \mathbf{p} \\
&= \frac{\alpha^2}{2}\mathbf{p} \cdot \mathbf{A}\mathbf{p} + \alpha \left(\mathbf{p} \cdot \mathbf{A}\mathbf{x}^{(i)} - \mathbf{p} \cdot \mathbf{b}\right) + \phi(\mathbf{x}^{(i)}) \\
&= \frac{\alpha^2}{2}\mathbf{p} \cdot \mathbf{A}\mathbf{p} - \alpha \mathbf{p} \cdot \left(\mathbf{b} - \mathbf{A}\mathbf{x}^{(i)}\right)
\end{aligned}$$

(the symmetry of $\mathbf{A}$ was used to combine the terms $\mathbf{x}^{(i)} \cdot \mathbf{A}\mathbf{p}/2$ and $\mathbf{p} \cdot \mathbf{A}\mathbf{x}^{(i)}/2$). The minimum is easily seen to occur at

$$\alpha = \frac{\mathbf{p} \cdot \left(\mathbf{b} - \mathbf{A}\mathbf{x}^{(i)}\right)}{\mathbf{p} \cdot \mathbf{A}\mathbf{p}}. \tag{13.9}$$

How should the descent direction be chosen? The obvious choice is the *steepest descent* direction

$$p = -\nabla \phi(\mathbf{x}^{(i)}),$$

since the directional derivative of ϕ at $\mathbf{x}^{(i)}$ is as negative as possible in this direction. The resulting algorithm (choose a starting point, move to the minimum in the steepest descent direction, calculate the steepest descent direction at that new point, and repeat) is called the *steepest descent* algorithm. It is guaranteed to converge to the minimizer $\mathbf{x}$ of ϕ, that is, to the solution of $\mathbf{Ax} = \mathbf{b}$. However, it can be shown that the steepest descent method converges slowly, especially when the eigenvalues of $\mathbf{A}$ differ greatly in magnitude (when this is true, the matrix $\mathbf{A}$ is said to be *ill-conditioned*).

For an example of a line search in the steepest descent direction, see Figure 13.11 and Exercise 13.2.7.

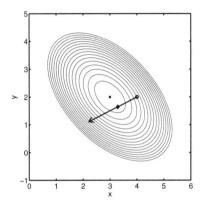

Figure 13.11. *The contours of the quadratic form from Figure* 13.10. *The steepest descent direction from* $\mathbf{x} = (4,2)$ *(marked by the "o") is indicated, along with the minimizer in the steepest descent direction (marked by "♦"). The desired (global) minimizer is marked by "x."*

Example 13.2. To test the algorithms described in this section, we will use the BVP

$$-\Delta u = f(\mathbf{x}) \text{ in } \Omega,$$
$$u = 0 \text{ on } \partial\Omega, \qquad (13.10)$$

where Ω is the unit square and

$$f(\mathbf{x}) = -2\left(-x_1^2 + x_1^3 - (-1+x_2)x_2 + 3x_1(-1+x_2)x_2\right).$$

The exact solution is

$$u(\mathbf{x}) = x_1^2 x_2 (1-x_1)(1-x_2).$$

We establish a regular mesh on Ω by dividing the x_1 and x_2 intervals into 64 subintervals each. This results in $2 \cdot 64^2$ triangles and $63^2 = 3969$ free nodes. The finite element equation, $\mathbf{Ku} = \mathbf{f}$, is therefore 3969×3969 (a fairly large system, but very sparse).

We apply the steepest descent algorithm to solve $\mathbf{Ku} = \mathbf{f}$. One hundred steps, starting with the zero vector as the initial estimate, produce an estimate of $\mathbf{u}$ that differs from the

13.2. Solving sparse linear systems

exact value **u** by about 76% in the Euclidean norm. The error in the corresponding piecewise linear function, compared to the exact solution u, is about 72% in the energy norm. By comparison, the exact solution **u** of $\mathbf{Ku} = \mathbf{f}$ corresponds to a piecewise linear function that has an error of about 3% in the energy norm.

Clearly these results are not very good. If we take 1000 steps instead of 100, the errors become about 25% in the Euclidean norm and 23% in the energy norm. It appears that we can obtain an accurate answer using the steepest descent algorithm, but only by taking a large number of steps. ∎

13.2.5 The conjugate gradient algorithm

The CG algorithm is another descent algorithm that is usually a great improvement over the steepest descent method. The problem with the steepest descent method is that, while the steepest descent direction is *locally* optimal, from a global point of view, the search directions are poorly chosen. Indeed, it can be shown that successive search directions are orthogonal (see Exercise 13.2.4). It is not efficient to approach the desired solution via orthogonal steps (after all, the shortest path between two points follows a straight line).

The CG algorithm defines the successive search directions to satisfy a pleasing global property—basically that each step preserves the optimality property of previous steps. To be precise, after k steps of CG, the estimated solution is the minimizer of ϕ over the k-dimensional subset spanned by the first k search directions.

It is rather difficult to derive the CG algorithm—the final form results from several nonobvious simplifications. We will content ourselves with showing the critical step: the computation of the search direction. We will assume that the initial estimate of the solution is $\mathbf{x}^{(0)} = \mathbf{0}$, that the first k search directions are $\mathbf{p}^{(1)}, \mathbf{p}^{(2)}, \ldots, \mathbf{p}^{(k)}$, and that after k steps we have determined $\alpha_1, \alpha_2, \ldots, \alpha_k$ so that

$$\mathbf{x}^{(k)} = \sum_{i=1}^{k} \alpha_i \mathbf{p}^{(i)}$$

solves

$$\min\left\{\phi(\mathbf{x}) \ : \ \mathbf{x} \in \text{span}\{\mathbf{p}^{(1)}, \ldots, \mathbf{p}^{(k)}\}\right\}.$$

We now wish to find a new search direction $\mathbf{p}^{(k+1)}$ with the following property: If

$$\mathbf{x}^{(k+1)} = \mathbf{x}^{(k)} + \alpha_{k+1} \mathbf{p}^{(k+1)},$$

where α_{k+1} solves

$$\min_{\alpha} \phi\left(\mathbf{x}^{(k)} + \alpha \mathbf{p}^{(k+1)}\right),$$

then $\mathbf{x}^{(k+1)}$ solves

$$\min\left\{\phi(\mathbf{x}) \ : \ \mathbf{x} \in \text{span}\{\mathbf{p}^{(1)}, \ldots, \mathbf{p}^{(k+1)}\}\right\}. \tag{13.11}$$

It is not clear that such a $\mathbf{p}^{(k+1)}$ can be found; however, it can be, as we now show.

To solve (13.11), we must find $\beta_1, \beta_2, \ldots, \beta_{k+1}$ such that

$$\phi\left(\sum_{i=1}^{k} \beta_i \mathbf{p}^{(i)} + \beta_{k+1} \mathbf{p}^{(k+1)}\right)$$

is as small as possible. (We separate the last term, $\beta_{k+1}\mathbf{p}^{(k+1)}$, from the sum because we already know how to make

$$\phi\left(\sum_{i=1}^{k} \beta_i \mathbf{p}^{(i)}\right)$$

as small as possible.) Some straightforward algebra shows that

$$\phi\left(\sum_{i=1}^{k} \beta_i \mathbf{p}^{(i)} + \beta_{k+1}\mathbf{p}^{(k+1)}\right) = \phi\left(\sum_{i=1}^{k} \beta_i \mathbf{p}^{(i)}\right) \\ + \beta_{k+1}\mathbf{p}^{(k+1)} \cdot \mathbf{A}\left(\sum_{i=1}^{k} \beta_i \mathbf{p}^{(i)}\right) + \phi\left(\beta_{k+1}\mathbf{p}^{(k+1)}\right).$$

Here is the crucial observation: If we can choose $\mathbf{p}^{(k+1)}$ so that

$$\mathbf{p}^{(k+1)} \cdot \mathbf{A}\left(\sum_{i=1}^{k} \beta_i \mathbf{p}^{(i)}\right) = \sum_{i=1}^{k} \left(\beta_i \mathbf{p}^{(k+1)} \cdot \mathbf{A}\mathbf{p}^{(i)}\right)$$

is zero, then

$$\phi\left(\sum_{i=1}^{k} \beta_i \mathbf{p}^{(i)} + \beta_{k+1}\mathbf{p}^{(k+1)}\right) = \phi\left(\sum_{i=1}^{k} \beta_i \mathbf{p}^{(i)}\right) + \phi\left(\beta_{k+1}\mathbf{p}^{(k+1)}\right).$$

The minimization problem is then "decoupled." That is, we can independently choose $\beta_1, \beta_2, \ldots, \beta_k$ to minimize

$$\phi\left(\sum_{i=1}^{k} \beta_i \mathbf{p}^{(i)}\right) \tag{13.12}$$

and β_{k+1} to minimize

$$\phi\left(\beta_{k+1}\mathbf{p}^{(k+1)}\right),$$

and the resulting $\beta_1, \beta_2, \ldots, \beta_{k+1}$ will be the solution of (13.11). This is what we want, since we already have computed the minimizer of (13.12).

Our problem then reduces to finding $\mathbf{p}^{(k+1)}$ to satisfy

$$\sum_{i=1}^{k} \left(\beta_i \mathbf{p}^{(k+1)} \cdot \mathbf{A}\mathbf{p}^{(i)}\right) = 0.$$

It is certainly sufficient to satisfy

$$\mathbf{p}^{(k+1)} \cdot \mathbf{A}\mathbf{p}^{(i)} = 0, \ i = 1, 2, \ldots, k.$$

13.2. Solving sparse linear systems

We can assume by induction that

$$\mathbf{p}^{(j)} \cdot \mathbf{A}\mathbf{p}^{(i)} = 0, \; i, j = 1, 2, \ldots, k, \; i \neq j. \tag{13.13}$$

We can recognize condition (13.13) as stating that the vectors $\mathbf{p}^{(1)}, \mathbf{p}^{(2)}, \ldots, \mathbf{p}^{(k)}$ are orthogonal with respect to the inner product[85]

$$(\mathbf{x}, \mathbf{y})_{\mathbf{A}} = \mathbf{x} \cdot \mathbf{A}\mathbf{y}.$$

To compute the search direction $\mathbf{p}^{(k+1)}$, then, we just take a descent direction and subtract off its component lying in the subspace

$$S_k = \mathrm{span}\left\{\mathbf{p}^{(1)}, \mathbf{p}^{(2)}, \ldots, \mathbf{p}^{(k)}\right\};$$

the result will be orthogonal to each of the vectors $\mathbf{p}^{(1)}, \mathbf{p}^{(2)}, \ldots, \mathbf{p}^{(k)}$. We will use the steepest descent direction $\mathbf{r} = -\nabla\phi(\mathbf{x}^{(k)})$ to generate the new search direction. To achieve the desired orthogonality, we must compute the component of $\mathbf{r}$ in S_k by projecting $\mathbf{r}$ onto S_k in the inner product defined by $\mathbf{A}$. We therefore take

$$\mathbf{p}^{(k+1)} = \mathbf{r} - \sum_{i=1}^{k} \frac{\mathbf{r} \cdot \mathbf{A}\mathbf{p}^{(i)}}{\mathbf{p}^{(i)} \cdot \mathbf{A}\mathbf{p}^{(i)}} \mathbf{p}^{(i)}.$$

For reasons that we cannot explain here, most of the inner products $\mathbf{r} \cdot \mathbf{A}\mathbf{p}^{(i)}$ are zero, and the result is

$$\mathbf{p}^{(k+1)} = \mathbf{r} - \frac{\mathbf{r} \cdot \mathbf{A}\mathbf{p}^{(k)}}{\mathbf{p}^{(k)} \cdot \mathbf{A}\mathbf{p}^{(k)}} \mathbf{p}^{(k)}. \tag{13.14}$$

We can then use the formula from (13.9) to find the minimizer α_{k+1} in the direction of $\mathbf{p}^{(k+1)}$, and we will have $\mathbf{x}^{(k+1)}$.

By taking advantage of the common features of formulas (13.9) and (13.14) (and using some other simplifications), we can express the CG algorithm in the following efficient form. (The vector $\mathbf{b} - \mathbf{A}\mathbf{x}^{(k)}$ is called the *residual* in the equation $\mathbf{A}\mathbf{x} = \mathbf{b}$—it is the amount by which the equation fails to be satisfied.)

$\mathbf{r} = \mathbf{b} - \mathbf{A}\mathbf{x}$ (* Compute the initial residual *)
$\mathbf{p} \leftarrow \mathbf{r}$ (* Compute the initial search direction *)
$c_1 \leftarrow \mathbf{r} \cdot \mathbf{r}$
for $k = 1, 2, \ldots$
  $\mathbf{v} \leftarrow \mathbf{A}\mathbf{p}$
  $c_2 \leftarrow \mathbf{p} \cdot \mathbf{v}$
  $\alpha \leftarrow \frac{c_1}{c_2}$ (* Solve the one-dimensional minimization problem *)
  $\mathbf{x} \leftarrow \mathbf{x} + \alpha\mathbf{p}$ (* Update the estimate of the solution *)
  $\mathbf{r} \leftarrow \mathbf{r} - \alpha\mathbf{v}$ (* Compute the new residual *)
  $c_3 \leftarrow \mathbf{r} \cdot \mathbf{r}$
  $\beta \leftarrow \frac{c_3}{c_1}$
  $\mathbf{p} \leftarrow \beta\mathbf{p} + \mathbf{r}$ (* Compute the new search direction *)
  $c_1 \leftarrow c_3$

---

[85] It can be shown that, since $\mathbf{A}$ is positive definite, $\mathbf{x} \cdot \mathbf{A}\mathbf{y}$ defines an inner product on $\mathbf{R}^n$; see Exercise 13.2.5.

The reader should note that only a single matrix-vector product is required at each step of the algorithm, making it very efficient. We emphasize that we have not derived all of the steps in the CG algorithm.[86]

The name "conjugate gradient" is derived from the fact that many authors refer to the orthogonality of the search directions, in the inner product defined by $\mathbf{A}$, as $\mathbf{A}$-*conjugacy*. Therefore, the key step is to make the (negative) *gradient* direction *conjugate* to the previous search directions.

Example 13.3. We apply 100 steps of the CG method to the system $\mathbf{Ku} = \mathbf{f}$ from Example 13.2. The result differs from the exact solution $\mathbf{u}$ by about 0.001% in the Euclidean norm, and the corresponding piecewise linear function is just as accurate (with error of about 3% in the energy norm) as that obtained from solving $\mathbf{Ku} = \mathbf{f}$ exactly. ∎

13.2.6 Convergence of the CG algorithm

The CG algorithm is constructed so that the kth estimate, $\mathbf{x}^{(k)}$, of the solution $\mathbf{x}$ minimizes ϕ over the k-dimensional subspace $S_k = \text{span}\{\mathbf{p}^{(1)}, \mathbf{p}^{(2)}, \ldots, \mathbf{p}^{(k)}\}$. Therefore, $\mathbf{x}^{(n)}$ minimizes ϕ over an n-dimensional subspace of $\mathbf{R}^n$, that is, over all of $\mathbf{R}^n$. It follows that $\mathbf{x}^{(n)}$ must be the desired solution: $\mathbf{x}^{(n)} = \mathbf{x}$.

Because of this observation, the CG algorithm can be regarded as a direct method—it computes the exact solution after a finite number of steps (at least when performed in floating point arithmetic). There are two reasons, though, why this property is irrelevant.

1. In floating point arithmetic, the computed search directions will not actually be $\mathbf{A}$-conjugate (due to the accumulation of roundoff errors), and so, in fact, $\mathbf{x}^{(n)}$ may differ significantly from $\mathbf{x}$.

2. Even apart from the issue of roundoff errors, CG is not used as a direct method for the simple reason that n steps is too many! We look to iterative methods when n is very large, making Gaussian elimination too expensive. In such a case, an iterative method must give a reasonable approximation in much less than n iterations, or it also is too expensive. The CG algorithm is useful precisely because it can give very good results in a relatively small number of iterations.

The rate of convergence of CG is related to the *condition number* of $\mathbf{A}$, which is defined as the ratio of the largest eigenvalue of $\mathbf{A}$ to the smallest. When the condition number is relatively small (that is, when $\mathbf{A}$ is *well-conditioned*), CG will converge rapidly. The algorithm also works well when the eigenvalues of $\mathbf{A}$ are clustered into a few groups. In this case, even if the largest eigenvalue is much larger than the smallest, CG will perform well. The worst case for CG is a matrix $\mathbf{A}$ whose eigenvalues are spread out over a wide range.

13.2.7 Preconditioned CG

It is often possible to replace a matrix $\mathbf{A}$ with a related matrix whose eigenvalues are clustered, and for which CG will converge quickly. This technique is called *preconditioning*,

---

[86]For a complete derivation and discussion of the CG algorithm, see [26], for example.

13.2. Solving sparse linear systems

and it requires that one find a matrix $\mathbf{M}$ (the *preconditioner*) that is somehow similar to $\mathbf{A}$ (in terms of its eigenvalues) but is much simpler to invert. At each step of the preconditioned conjugate gradient (PCG) algorithm, it is necessary to solve an equation of the form $\mathbf{Mq} = \mathbf{r}$.

Preconditioners can be found in many different ways, but most require an intimate knowledge of the matrix $\mathbf{A}$. For this reason, there are few general-purpose methods. One method that is often used is to define a preconditioner from an *incomplete factorization* of $\mathbf{A}$. An incomplete factorization is a factorization (like Cholesky) in which fill-in is limited by fiat. Another method for constructing preconditioners is to replace $\mathbf{A}$ by a simpler matrix (perhaps arising from a simpler PDE) that can be inverted by FFT methods.

Exercises

1. Suppose $\mathbf{A} \in \mathbf{R}^{n \times n}$. Determine the exact number of arithmetic operations required for the computation of $\mathbf{A} = \mathbf{LU}$ via Gaussian elimination. Further count the number of operations required to compute $\mathbf{L}^{-1}\mathbf{b}$ and $\mathbf{U}^{-1}\mathbf{L}^{-1}\mathbf{b}$. Verify the results given in the text. The following formulas will be useful:

$$\sum_{i=1}^{n} i = \frac{n(n+1)}{2},$$

$$\sum_{i=1}^{n} i^2 = \frac{n(n+1)(2n+1)}{6}.$$

2. Suppose $\mathbf{A} \in \mathbf{R}^{n \times n}$ is banded with half-bandwidth p. Determine the exact number of arithmetic operations required to factor $\mathbf{A}$ into $\mathbf{LU}$.

3. Let $\mathbf{A} \in \mathbf{R}^{n \times n}$ be symmetric, and suppose $\mathbf{b} \in \mathbf{R}^n$. Define ϕ as in (13.7). Show that (13.8) holds. (Hint: One method is to write out

$$\phi(\mathbf{x}) = \frac{1}{2} \sum_{i=1}^{n} \sum_{j=1}^{n} A_{ij} x_j x_i - \sum_{i=1}^{n} b_i x_i$$

and then show that

$$\frac{\partial \phi}{\partial x_i}(\mathbf{x}) = \sum_{j=1}^{n} A_{ij} x_j - b_i.$$

Another method is to show that

$$\phi(\mathbf{x}+\mathbf{y}) = \phi(\mathbf{x}) + (\mathbf{Ax} - \mathbf{b}) \cdot \mathbf{y} + O\left(\|\mathbf{y}\|^2\right).$$

In either case, the symmetry of $\mathbf{A}$ is essential.)

4. Let $\mathbf{A} \in \mathbf{R}^{n \times n}$ be SPD, let $\mathbf{b}, \mathbf{y} \in \mathbf{R}^n$ be given, and suppose α^* solves

$$\min_{\alpha} \phi(\mathbf{y} - \alpha \nabla \phi(\mathbf{y})),$$

where

$$\phi(\mathbf{x}) = \frac{1}{2}\mathbf{x} \cdot \mathbf{A}\mathbf{x} - \mathbf{b} \cdot \mathbf{x}.$$

Show that

$$\nabla\phi(\mathbf{y} - \alpha^*\nabla\phi(\mathbf{y}))$$

is orthogonal to $\nabla\phi(\mathbf{y})$.

5. Suppose $\mathbf{A} \in \mathbf{R}^{n \times n}$ is SPD. Show that

$$(\mathbf{x},\mathbf{y})_\mathbf{A} = \mathbf{x} \cdot \mathbf{A}\mathbf{y}$$

defines an inner product on $\mathbf{R}^n$.

6. Let $\{\mathbf{v}_1, \mathbf{v}_2, \ldots, \mathbf{v}_n\}$ be a linearly independent set of vectors in an inner product space, and let $\mathbf{G} \in \mathbf{R}^{n \times n}$ be the corresponding Gram matrix:

$$G_{ij} = (\mathbf{v}_j, \mathbf{v}_i), \ i, j = 1, 2, \ldots, n.$$

Prove that $\mathbf{G}$ is SPD. (See the hint for Exercise 3.4.6.)

7. The quadratic form shown in Figures 13.10 and 13.11 is

$$\phi(\mathbf{x}) = \frac{1}{2}\mathbf{x} \cdot \mathbf{A}\mathbf{x} - \mathbf{b} \cdot \mathbf{x} + 20,$$

where

$$\mathbf{A} = \begin{bmatrix} 2 & 1 \\ 1 & 2 \end{bmatrix}, \mathbf{b} = \begin{bmatrix} 8 \\ 7 \end{bmatrix}.$$

(a) Perform one step of the steepest descent algorithm, beginning at the point $\mathbf{x}^{(0)} = (4,2)$. Compute the steepest descent direction at $\mathbf{x}^{(0)}$ and the minimizer along the line defined by the steepest descent direction. Compare your results to Figure 13.11.

(b) Now compute the second step of the steepest descent algorithm. Do you arrive at the exact solution of $\mathbf{A}\mathbf{x} = \mathbf{b}$?

(c) Perform two steps of CG, beginning with $\mathbf{x}^{(0)} = \mathbf{0}$. Do you obtain the exact solution?

8. When applying an iterative method, a *stopping criterion* is essential: One must decide, on the basis of some computable quantity, when the computed solution is accurate enough that one can stop the iteration. A common stopping criterion is to stop when the *relative residual*,

$$\frac{\|\mathbf{A}\mathbf{x}^{(k)} - \mathbf{b}\|}{\|\mathbf{b}\|},$$

falls below some predetermined tolerance ϵ.

(a) Write a program implementing the CG algorithm with the above stopping criterion.

(b) Use your program to solve the finite element equation $\mathbf{Ku} = \mathbf{f}$ arising from the BVP (13.10) for meshes of various sizes. For a fixed value of ϵ, how does the number of iterations required depend on the number of unknowns (free nodes)?

9. Explain how the CG algorithm must be modified if the initial estimate $\mathbf{x}^{(0)}$ is not the zero vector. (Hint: Think of using the CG algorithm to compute $\mathbf{y} = \mathbf{x} - \mathbf{x}^{(0)}$. What linear system does $\mathbf{y}$ satisfy?)

13.3 An outline of the convergence theory for finite element methods

We will now present an overview of the convergence theory for Galerkin finite element methods for BVPs. Our discussion is little more than an outline, describing the kinds of analysis that are required to prove that the finite element method converges to the true solution as the mesh is refined. We will use the Dirichlet problem

$$-\nabla \cdot (k(\mathbf{x})\nabla u) = f(\mathbf{x}) \text{ in } \Omega,$$
$$u = 0 \text{ on } \partial\Omega \quad (13.15)$$

as our model problem.

13.3.1 The Sobolev space $H_0^1(\Omega)$

We begin by addressing the question of the proper setting for the weak form of the BVP. In Section 11.4, we derived the weak form as

$$\text{find } u \in C_D^2(\overline{\Omega}) \text{ such that } a(u,v) = (f,v) \text{ for all } v \in C_D^2(\overline{\Omega}). \quad (13.16)$$

However, as we pointed out in Section 5.6 (for the one-dimensional case), the choice of $C_D^2(\overline{\Omega})$ for the space of test functions is not really appropriate for our purposes, since the finite element method uses functions that are not smooth enough to belong to $C_D^2(\overline{\Omega})$. Moreover, the statement of the weak form does not require that the test functions be so smooth.

We can define the appropriate space of test functions by posing the following question: What properties must the test functions have in order that the variational equation

$$a(u,v) = (f,v)$$

be well defined? The energy inner product is essentially the sum of the L^2 inner products

$$\left(\frac{\partial u}{\partial x_1}, \frac{\partial v}{\partial x_1}\right), \left(\frac{\partial u}{\partial x_2}, \frac{\partial v}{\partial x_2}\right).$$

(The coefficient of the PDE, $k(\mathbf{x})$, is also involved. However, if we assume that k is smooth and bounded, then, for the purposes of this discussion, it may as well be constant, and therefore can be ignored.) It suffices, therefore, that the partial derivatives of u and v belong to $L^2(\Omega)$, and it is natural to define the space

$$H^1(\Omega) = \left\{ u \in L^2(\Omega) : \frac{\partial u}{\partial x_1}, \frac{\partial u}{\partial x_2} \in L^2(\Omega) \right\}. \tag{13.17}$$

An issue that immediately arises is the definition of the partial derivatives of u. Need $\partial u/\partial x_1$, $\partial u/\partial x_2$ exist at every point of Ω? If so, then the use of piecewise linear functions is still ruled out.

The classical definition of partial derivatives (as presented in calculus books and courses), in terms of limits of difference quotients, is purely local. There is another way to define derivatives that is more global in nature, and therefore more tolerant of certain kinds of singularities. We begin by defining the space $C_0^\infty(\Omega)$ to be the set of all infinitely differentiable functions whose support[87] does not intersect the boundary of Ω. (That is, the support of a $C_0^\infty(\Omega)$ function is strictly in the interior of Ω.) If f is any smooth function defined on Ω and $\phi \in C_0^\infty(\Omega)$, then, by integration by parts,

$$\int_\Omega \frac{\partial f}{\partial x_1} \phi = -\int_\Omega f \frac{\partial \phi}{\partial x_1} \tag{13.18}$$

(the boundary integral must vanish since ϕ and all of its derivatives are zero on the boundary). We can *define* a derivative of $f \in L^2(\Omega)$ by this equation: If there is a function $g \in L^2(\Omega)$ such that

$$\int_\Omega g\phi = -\int_\Omega f \frac{\partial \phi}{\partial x_1} \text{ for all } \phi \in C_0^\infty(\Omega),$$

then g is called the (weak) partial derivative of f with respect to x_1, and denoted $\partial f/\partial x_1$. The definition of $\partial f/\partial x_2$ in the weak sense is entirely analogous.

We use the same notation for the weak partial derivatives as we do for the classical partial derivatives; it can be proved that if the classical partial derivatives of f exist, then so do the weak partial derivatives, and the two are the same. The definition (13.17) is to be interpreted in terms of weak derivatives. The space $H^1(\Omega)$ is an example of a *Sobolev space*.

It can be shown that continuous piecewise linear functions belong to $H^1(\Omega)$, while, for example, discontinuous piecewise linear functions do not. Exercise 13.3.2 explores this question in one dimension; a more complete justification is beyond the scope of this book.

The standard inner product for $H^1(\Omega)$ is

$$(f,g)_{H^1} = \int_\Omega \left(fg + \frac{\partial f}{\partial x_1}\frac{\partial g}{\partial x_1} + \frac{\partial f}{\partial x_2}\frac{\partial g}{\partial x_2} \right),$$

and the induced norm is

$$\|f\|_{H^1} = \sqrt{\int_\Omega \left(f^2 + \left(\frac{\partial f}{\partial x_1}\right)^2 + \left(\frac{\partial f}{\partial x_2}\right)^2 \right)}.$$

---

[87]The reader will recall that the support of a function is the closure of the set on which the function is nonzero.

13.3. An outline of the convergence theory for finite element methods

A function g is a good approximation to f in the H^1 sense if and only if g is a good approximation to f and ∇g is a good approximation to ∇f. If we were to measure closeness in the L^2 norm, on the other hand, ∇g need not be close to ∇f. For an example of the distinction (in one dimension, which is easier to visualize), see Figure 13.12.

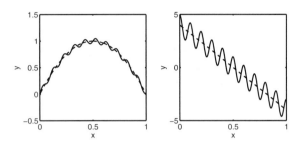

Figure 13.12. *Two functions belonging to $H^1(0,1)$ (left) and their derivatives (right). The difference in the two functions is less than 5% in the L^2 norm, but it is more than 91% in the H^1 norm.*

Just as the expression $a(u,v)$ makes sense as long as u and v belong to $H^1(\Omega)$, the right-hand side (f,v) of the variational equation is well defined as long as f belongs to $L^2(\Omega)$. It is in this sense that the variational equation $a(u,v) = (f,v)$ is called the weak form of the BVP: the equation makes sense under relatively weak assumptions on the functions involved.

We now define the subspace $H_0^1(\Omega)$ of $H^1(\Omega)$ by

$$H_0^1(\Omega) = \left\{ u \in H^1(\Omega) : u = 0 \text{ on } \partial\Omega \right\}.$$

Then, by the so-called *trace theorem*, $H_0^1(\Omega)$ is a well-defined and closed subspace of $H^1(\Omega)$. Moreover, it can be shown that $H^1(\Omega)$ (and hence $H_0^1(\Omega)$) is complete under the H^1 norm. (The notion of completeness was defined in Section 12.6.) Moreover, the energy inner product is equivalent to the H^1 norm on the subspace $H_0^1(\Omega)$, which implies that $H_0^1(\Omega)$ is complete under the energy norm as well. It then follows from the *Riesz representation theorem* that the weak form of the BVP, which we now pose as

$$\text{find } u \in H_0^1(\Omega) \text{ such that } a(u,v) = (f,v) \text{ for all } v \in H_0^1(\Omega), \qquad (13.19)$$

has a unique solution for each $f \in L^2(\Omega)$. When f is continuous and there is a classical solution u to the BVP, then u is also the unique solution to the weak form of the BVP.

The various analytical results mentioned in this section, such as the trace theorem, completeness of the Sobolev space $H^1(\Omega)$, and the Riesz representation theorem, are discussed in advanced books on finite elements, such as [8].

13.3.2 Best approximation in the energy norm

As we have seen, the weak form of the BVP (13.15) has a unique solution $u \in H_0^1(\Omega)$ for each $f \in L^2(\Omega)$. Moreover, given a triangulation $\mathcal{T}$ of Ω, the space V_h of continuous piecewise linear functions relative to $\mathcal{T}$, satisfying the Dirichlet boundary conditions, is a finite-dimensional subspace of $H_0^1(\Omega)$. We can therefore apply the Galerkin method to compute the best approximation from V_h, in the energy norm, to the true solution u.

13.3.3 Approximation by piecewise polynomials

We have seen (at least in outline) that the weak form (13.19) has a unique solution u, and that we can compute the best approximation to u from the finite element space V_h. The next question is, How good is the best approximation from V_h? This is a question of piecewise polynomial approximation, and the simplest way to answer it is to consider the piecewise linear interpolant of the true solution, which is (using the usual notation)

$$u_I(\mathbf{x}) = \sum_{i=1}^{N} u(\mathbf{x}_{f_i}) \phi_i(\mathbf{x}),$$

where $\mathbf{x}_{f_1}, \mathbf{x}_{f_2}, \ldots, \mathbf{x}_{f_N}$ are the free nodes of the mesh. If we can determine how well u_I approximates u in the energy norm, then, since $u_I \in V_h$ and the finite element approximation u_h is the best approximation to u from V_h, we know that the error in u_h is no greater.

It can be shown that, if u has some extra smoothness (in particular, if u is a solution of the strong form of the BVP, so that it is twice continuously differentiable), then

$$\|u - u_I\|_E \leq Ch,$$

where h is the length of the largest side of any triangle in the mesh $\mathcal{T}_h$ and C is a constant that depends on the particular solution u but is independent of the mesh.[88] It follows that

$$\|u - u_h\|_E \leq Ch \qquad (13.20)$$

also holds, since $\|u - u_h\|_E \leq \|u - u_I\|_E$.

Since the energy norm is equivalent to the H^1 norm, we also have

$$\|u - u_h\|_{H^1} \leq Ch$$

(although for a possibly different value of C).

This completes the outline of the basic convergence theory for piecewise linear Galerkin finite elements.

13.3.4 Elliptic regularity and L^2 estimates

It would be desirable to have an estimate on the error in the L^2 norm. This can be obtained by a standard "duality" argument, provided "elliptic regularity" holds. Assuming that the coefficient k is smooth, when the domain Ω has a smooth boundary, or when Ω is convex[89] with a piecewise smooth boundary, the solution of (13.15) is smoother than the right-hand

---

[88] There are some restrictions on the nature of the meshes $\mathcal{T}_h$ as $h \to 0$, basically that the triangles do not become arbitrarily "skinny."

[89] A set Ω is called convex if, whenever the endpoints of a line segment lie in Ω, then the entire line segment lies in Ω.

13.3. An outline of the convergence theory for finite element methods

side f by two degrees of differentiability. Thus, if f lies in L^2, then u and its partial derivatives up to order two lie in L^2. This property (that the solution is smoother than the right-hand side) is referred to as elliptic regularity. As long as elliptic regularity holds, it can be shown (by the duality argument mentioned above) that

$$\|u - u_h\|_{L^2} \leq Ch\|u - u_h\|_E,$$

so that

$$\|u - u_h\|_{L^2} \leq Ch^2.$$

In each of the three previous inequalities, the constant C represents a generic constant (not necessarily the same in each inequality) that can depend on the particular solution u and the domain Ω but is independent of h.

We will now illustrate the convergence theory with an example.

Example 13.4. Suppose Ω is the unit square, and consider the BVP

$$-\nabla \cdot (k(\mathbf{x})\nabla u) = f(\mathbf{x}) \text{ in } \Omega,$$
$$u = 0 \text{ on } \partial\Omega,$$

where $k(\mathbf{x}) = 1 + x_1 x_2^2$ and f is chosen so that the exact solution is

$$u(\mathbf{x}) = x_1(1 - x_1)\sin(\pi x_2).$$

Using piecewise linear finite element functions on a sequence of regular meshes, the following errors are obtained:

$\frac{h}{\sqrt{2}}$	$\|u - u_h\|_{H^1}$	$\|u - u_h\|_{L^2}$
$5.0000 \cdot 10^{-1}$	$4.0757 \cdot 10^{-1}$	$7.0613 \cdot 10^{-2}$
$2.5000 \cdot 10^{-1}$	$2.2369 \cdot 10^{-1}$	$2.2713 \cdot 10^{-2}$
$1.2500 \cdot 10^{-1}$	$1.1441 \cdot 10^{-1}$	$6.0681 \cdot 10^{-3}$
$6.2500 \cdot 10^{-2}$	$5.7524 \cdot 10^{-2}$	$1.5429 \cdot 10^{-3}$

From these results, we see that, when h is reduced by a factor of 2, the error in the H^1 norm is reduced by a factor of approximately 2, while the error in the L^2 norm is reduced by a factor of approximately 4. This is consistent with the asymptotic error results given above. ∎

Exercises

Exercises 5 and 6 require code implementing piecewise linear Galerkin finite elements for two-dimensional problems, and also a routine for solving the system $\mathbf{Ku} = \mathbf{f}$.[90]

---

[90] A simple collection of MATLAB codes is available on the book's web site, http://www.siam.org/books/ot122.

1. Suppose $\phi \in C_0^\infty(\Omega)$ and $f : \overline{\Omega} \to \mathbf{R}$ is smooth. Prove that
$$\int_\Omega \frac{\partial f}{\partial x_i}\phi = -\int_\Omega f \frac{\partial \phi}{\partial x_i}$$
for $i = 1, 2$. What would the boundary integral be in this integration-by-parts formula if it were not assumed that ϕ is zero on $\partial \Omega$?

2. The purpose of this exercise is to illustrate why *continuous* piecewise linear functions belong to $H^1(0, 1)$, but discontinuous piecewise linear functions do not.

    (a) Define $f : [0, 1] \to \mathbf{R}$ by
    $$f(x) = \begin{cases} x, & 0 \le x \le \frac{1}{2}, \\ 1 - x, & \frac{1}{2} < x \le 1. \end{cases}$$
    Then f is differentiable (in the classical sense) everywhere except at $x = 1/2$, and
    $$\frac{df}{dx}(x) = \begin{cases} 1, & 0 < x < \frac{1}{2}, \\ -1, & \frac{1}{2} < x < 1. \end{cases}$$
    Show that f belongs to $H^1(0, 1)$ and that its weak derivative is as defined above by verifying that
    $$\int_0^1 \frac{df}{dx}(x)\phi(x)\,dx = -\int_0^1 f(x)\frac{d\phi}{dx}(x)\,dx$$
    for every $\phi \in C_0^\infty(0, 1)$. (Hint: Start with the integral on the right, rewrite it as the sum of two integrals, and apply integration by parts to each.)

    (b) Define $g : [0, 1] \to \mathbf{R}$ by
    $$g(x) = \begin{cases} x, & 0 \le x \le \frac{1}{2}, \\ 2 - x, & \frac{1}{2} < x \le 1. \end{cases}$$
    Then g, just like f, is differentiable (in the classical sense) except at $x = 1/2$, and
    $$\frac{dg}{dx}(x) = \begin{cases} 1, & 0 < x < \frac{1}{2}, \\ -1, & \frac{1}{2} < x < 1. \end{cases}$$
    However, $g \notin H^1(0, 1)$. Show that it is *not* the case that
    $$\int_0^1 \frac{dg}{dx}(x)\phi(x)\,dx = -\int_0^1 g(x)\frac{d\phi}{dx}(x)\,dx$$
    for every $\phi \in C_0^\infty(0, 1)$.

3. Let Ω be the unit square and define $f, g \in H^1(\Omega)$ by
$$f(\mathbf{x}) = 1 + x_1 + x_2,$$
$$g(\mathbf{x}) = f(\mathbf{x}) + \sin(m\pi x_1)\sin(n\pi x_2).$$
Compute the relative difference in f and g in the L^2 and H^1 norms.

4. The exact solution of the BVP

$$-\frac{d^2u}{dx^2} = 12x^2 - 6x, \ 0 < x < 1,$$
$$u(0) = 0,$$
$$u(1) = 0$$

is $u(x) = x^3(1-x)$. Consider the regular mesh with three elements, the subintervals $[0, 1/3]$, $[1/3, 2/3]$, and $[2/3, 1]$.

   (a) Compute the piecewise linear finite element approximation to u, and call it v.

   (b) Compute the piecewise linear interpolant of u, and call it w.

   (c) Compute the error in v and w (as approximations to u) in the energy norm. Which do you expect to be smaller? Which is actually smaller?

5. Let Ω be the unit square, define $k(\mathbf{x}) = 1 + x_1$, and define

$$f(\mathbf{x}) = 4x_1x_2 - 4x_1x_2^2 - 2x_1^3 + 2x_1 - x_2^2 + x_2.$$

Using the Galerkin finite element method with continuous piecewise linear functions, estimate the solution of

$$-\nabla \cdot (k(\mathbf{x})\nabla u) = f(\mathbf{x}), \ \mathbf{x} \in \Omega,$$
$$u(\mathbf{x}) = 0, \ \mathbf{x} \in \partial\Omega,$$

on a sequence of increasingly finer meshes. Verify the rates of convergence of the errors, in both the L^2 and energy norms, to zero. The exact solution of the BVP is $u(\mathbf{x}) = x_1(1-x_1)x_2(1-x_2)$.

6. Repeat Exercise 5 with Ω equal to the unit disk and

$$k(\mathbf{x}) = 2 + x_1x_2, \ f(\mathbf{x}) = 8x_1x_2 + 8.$$

The solution to the BVP is $u(\mathbf{x}) = 1 - x_1^2 - x_2^2$. (Note: Since Ω is not a polygonal domain, the computational mesh will only approximate the true domain.)

13.4 Finite element methods for eigenvalue problems

In Section 12.7, we described how eigenvalues and eigenfunctions exist for nonconstant-coefficient problems on (more or less) arbitrary domains, but we gave no techniques for finding them. We will now show how finite element methods can be used to estimate eigenpairs. This will be a fitting topic to end this book, as it ties together our two main themes: (generalized) Fourier series methods and finite element methods.

We will consider the following model problem:

$$-\nabla \cdot (k(\mathbf{x})\nabla u) = \lambda u \text{ in } \Omega,$$
$$u = 0 \text{ on } \partial\Omega. \tag{13.21}$$

We follow the usual procedure to derive the weak form: multiply by a test function, and integrate by parts:

$$-\nabla \cdot (k(\mathbf{x})\nabla u) = \lambda u, \ \mathbf{x} \in \Omega$$
$$\Rightarrow -\nabla \cdot (k(\mathbf{x})\nabla u)v = \lambda uv, \ \mathbf{x} \in \Omega, \ v \in H_0^1(\Omega)$$
$$\Rightarrow -\int_\Omega \nabla \cdot (k(\mathbf{x})\nabla u)v = \lambda \int_\Omega uv, \ v \in H_0^1(\Omega)$$
$$\Rightarrow \int_\Omega k(\mathbf{x})\nabla u \cdot \nabla v = \lambda \int_\Omega uv, \ v \in H_0^1(\Omega).$$

In the last step, the boundary term vanishes because of the boundary conditions satisfied by the test function v. The weak form of the eigenvalue problem is therefore

$$u \in H_0^1(\Omega), \ a(u,v) = \lambda(u,v) \text{ for all } v \in H_0^1(\Omega). \tag{13.22}$$

We now apply Galerkin's method with piecewise linear finite elements. We write V_h for the space of continuous piecewise linear functions, satisfying Dirichlet conditions, relative to a given mesh $\mathcal{T}_h$. As usual, $\{\phi_1, \phi_2, \ldots, \phi_n\}$ will represent the standard basis for V_h. Galerkin's method now reduces to

$$u_h \in V_h, \ a(u_h, v) = \lambda(u_h, v) \text{ for all } v \in V_h$$

or to

$$u_h = \sum_{j=1}^n \alpha_j \phi_j, \ a(u_h, \phi_i) = \lambda(u_h, \phi_i), \ i = 1, 2, \ldots, n.$$

Substituting the formula for u_h, we obtain

$$a\left(\sum_{j=1}^n \alpha_j \phi_j, \phi_i\right) = \lambda \left(\sum_{j=1}^n \alpha_j \phi_j, \phi_i\right), \ i = 1, 2, \ldots, n$$
$$\Rightarrow \sum_{j=1}^n a(\phi_j, \phi_i)\alpha_j = \lambda \sum_{j=1}^n (\phi_j, \phi_i)\alpha_j, \ i = 1, 2, \ldots, n$$
$$\Rightarrow \mathbf{Ku} = \lambda \mathbf{Mu},$$

where $\mathbf{K} \in \mathbf{R}^{n \times n}$ and $\mathbf{M} \in \mathbf{R}^{n \times n}$ are the stiffness and mass matrices, respectively, encountered before. If we find λ and $\mathbf{u}$ satisfying $\mathbf{Ku} = \lambda \mathbf{Mu}$, then λ and the corresponding piecewise linear function u_h will approximate an eigenpair of the differential operator.

We do not have the space here to discuss the accuracy of eigenvalues and eigenfunctions approximated by this method (however, Section 10.3 discusses the analogous theory for the Sturm–Liouville eigenvalue problem in one dimension). Only the smaller eigenvalues computed by the above method will be reliable estimates of the true eigenvalues. To keep our discussion on an elementary level, we will restrict our attention to the computation of the smallest eigenvalue. (The reader will recall that this smallest eigenvalue can be quite important, for example, in mechanical vibrations. It defines the fundamental frequency of a system modeled by the wave equation.)

13.4. Finite element methods for eigenvalue problems

Given two matrices $\mathbf{K} \in \mathbf{R}^{n \times n}$ and $\mathbf{M} \in \mathbf{R}^{n \times n}$, the problem

$$\mathbf{u} \in \mathbf{R}^n,\ \mathbf{u} \neq 0,\ \lambda \in \mathbf{C},\ \mathbf{Ku} = \lambda \mathbf{Mu}$$

is called a *generalized eigenvalue problem*. It can be converted to an ordinary eigenvalue problem in a couple of ways. For example, when $\mathbf{M}$ is invertible, multiplying both sides of the equation by $\mathbf{M}^{-1}$ yields

$$\mathbf{M}^{-1}\mathbf{Ku} = \lambda \mathbf{u},$$

which is the ordinary eigenvalue problem for the matrix $\mathbf{M}^{-1}\mathbf{K}$. However, this transformation has several drawbacks, not the least of which is that $\mathbf{M}^{-1}\mathbf{K}$ will usually not be symmetric even if both $\mathbf{K}$ and $\mathbf{M}$ are symmetric. (The reader should recall that symmetric matrices have many special properties with respect to the eigenvalue problem: real eigenvalues, orthogonal eigenvectors, etc.)

The preferred method for converting $\mathbf{Ku} = \lambda \mathbf{Mu}$ into an ordinary eigenvalue problem uses the Cholesky factorization. Every SPD matrix can be written as the product of a nonsingular lower triangular matrix times its transpose. The mass matrix, being a Gram matrix, is SPD (see Exercise 13.2.6), so there exists a nonsingular lower triangular matrix $\mathbf{L}$ such that $\mathbf{M} = \mathbf{LL}^T$. We can then rewrite the problem as follows:

$$\mathbf{Ku} = \lambda \mathbf{Mu}$$
$$\Rightarrow \mathbf{Ku} = \lambda \mathbf{LL}^T \mathbf{u}$$
$$\Rightarrow \mathbf{L}^{-1}\mathbf{Ku} = \lambda \mathbf{L}^T \mathbf{u}$$
$$\Rightarrow \mathbf{L}^{-1}\mathbf{KL}^{-T}\left(\mathbf{L}^T \mathbf{u}\right) = \lambda \left(\mathbf{L}^T \mathbf{u}\right)$$
$$\Rightarrow \mathbf{Ax} = \lambda \mathbf{x},\ \mathbf{A} = \mathbf{L}^{-1}\mathbf{KL}^{-T},\ \mathbf{x} = \mathbf{L}^T \mathbf{u}.$$

This is an ordinary eigenvalue problem for $\mathbf{A}$, and $\mathbf{A}$ is symmetric because $\mathbf{K}$ is. This shows that the generalized eigenvalue problem $\mathbf{Ku} = \lambda \mathbf{Mu}$ has real eigenvalues. The eigenvectors corresponding to distinct eigenvalues are orthogonal when expressed in the transformed variable $\mathbf{x}$, but not necessarily in the original variable $\mathbf{u}$. However, given distinct eigenvalues λ and μ and corresponding eigenvectors $\mathbf{u}$ and $\mathbf{v}$, the piecewise linear functions u_h and v_h with nodal values given by $\mathbf{u}$ and $\mathbf{v}$ are orthogonal in the L^2 norm (see Exercise 13.4.1).

There is a drawback to using the above method for converting the generalized eigenvalue problem to an ordinary one: There is no reason why $\mathbf{A}$ should be sparse, even though $\mathbf{K}$ and $\mathbf{M}$ are sparse. For this reason, the most efficient algorithms for the generalized eigenvalue problem treat the problem directly instead of using the above transformation. However, the transformation is useful if one has access only to software for the ordinary eigenvalue problem.

Example 13.5. We will estimate the smallest eigenvalue and the corresponding eigenfunction of the negative Laplacian (subject to Dirichlet conditions) on the unit square Ω. The reader will recall that the exact eigenpair is

$$\lambda_{11} = 2\pi^2,\ \psi_{11}(\mathbf{x}) = \sin(\pi x_1)\sin(\pi x_2).$$

We establish a regular mesh on Ω, with $2n^2$ triangles, $(n+1)^2$ nodes, and $(n-1)^2$ free nodes. We compute the stiffness matrix $\mathbf{K}$ and the mass matrix $\mathbf{M}$ and solve the generalized

Table 13.3. *Errors in the finite element estimates of the smallest eigenvalue and corresponding eigenfunction of the negative Laplacian on the unit square. The errors in the eigenfunction are computed in the energy norm.*

n	Error in λ_{11}	Error in ψ_{11}
2	12.261	1.5528
4	3.1266	0.84754
8	0.76634	0.43315
16	0.19058	0.21772
32	0.047583	0.10900

eigenvalue problem using the function **eig** from MATLAB. The results, for various values of n, are shown in Table 13.3. The reader should notice that the eigenvalue estimate is converging faster than the eigenfunction estimate. This is typical. As the results suggest, the error in the eigenvalue is $O(h^2)$, while the error in the eigenfunction is $O(h)$ (measured in the energy norm). ∎

The reader will recall from Section 12.7 that the fundamental frequency of a membrane occupying a domain Ω is $c\sqrt{\lambda_1}/(2\pi)$, where λ_1 is the smallest eigenvalue of the negative Laplacian on Ω (under Dirichlet conditions) and c is the wave speed. Therefore, the fundamental frequency of a square membrane of area 1 is

$$\frac{c\sqrt{2\pi^2}}{2\pi} = \frac{c}{\sqrt{2}} \doteq 0.7071c.$$

Using finite elements, we can compute the fundamental frequency for membranes of other shapes. In the next example, we consider an equilateral triangle with area 1.

Example 13.6. Consider the triangle Ω whose vertices are

$$\left(-\frac{1}{3^{1/4}}, 0\right), \left(\frac{1}{3^{1/4}}, 0\right), \left(0, 3^{1/4}\right).$$

This triangle is equilateral and has area 1. We use the finite element method to estimate the smallest eigenvalue of the negative Laplacian on this domain, using 5 successive regular grids. The coarsest has 16 triangles and is shown in Figure 13.13. The remaining grids are obtained by refining the original grid in the standard fashion (see Exercise 13.1.4). The finest mesh has 4096 triangles. The estimates of the smallest eigenvalue, as obtained on the successively finer meshes, are

$$27.7128, 23.9869, 23.0873, 22.8662, 22.8112.$$

We conclude that the smallest eigenvalue is $\lambda_1 \doteq 22.8$, and the fundamental frequency is

$$\frac{c\sqrt{\lambda_1}}{2\pi} \doteq 0.7601c. \quad \blacksquare$$

In Exercise 4 below, the reader is asked to study the fundamental frequency of a membrane in the shape of a regular n-gon having area 1.

13.4. Finite element methods for eigenvalue problems

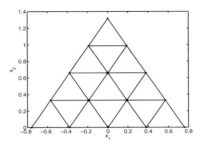

Figure 13.13. *The coarsest mesh from Example* 13.6.

Exercises

Exercises 2–4 require code implementing piecewise linear Galerkin finite elements for two-dimensional problems, and also a routine for solving the generalized eigenvalue problem $\mathbf{Ku} = \lambda \mathbf{Mu}$.

1. (a) Suppose $\mathbf{u}, \mathbf{v} \in \mathbf{R}^n$ contain the nodal values of piecewise linear functions $u_h, v_h \in V_h$, respectively. Explain how to compute the L^2 inner product of u_h and v_h from $\mathbf{u}$ and $\mathbf{v}$.

    (b) Show that if $\mathbf{u}$ and $\mathbf{v}$ satisfy
    $$\mathbf{Ku} = \lambda \mathbf{Mu}, \quad \mathbf{Kv} = \mu \mathbf{Mv},$$
    where $\lambda \neq \mu$ and $\mathbf{K}$ and $\mathbf{M}$ are the stiffness and mass matrices, respectively, then the corresponding piecewise linear functions $u_h, v_h \in V_h$ are orthogonal in the L^2 inner product.

2. Let $k(\mathbf{x}) = 1 + x_1 x_2$. Use finite elements to estimate the smallest eigenvalue of the operator $-\nabla \cdot (k(\mathbf{x}) \nabla u)$ on the unit square.

3. Repeat Exercise 2 using $k(\mathbf{x}) = 2 + x_1$ on the square with vertices $(0,1), (1,0), (-1,0), (0,-1)$.

4. This exercise requires that you write a program to assemble the stiffness and mass matrix for $-\Delta$ on an arbitrary triangulation of a polygonal domain Ω. You will also need access to a routine to solve the generalized eigenvalue problem.

    (a) Repeat Example 13.6, replacing the triangular domain by a domain bounded by a regular pentagon with area 1.

    (b) Let Ω be the polygonal domain with vertices $(1,-1), (1,1), (-1,1), (-1,0), (0,0), (0,-1)$ (listed in counterclockwise order). Use finite elements to estimate the smallest eigenvalue of $-\Delta$, subject to Dirichlet conditions, on Ω.

    (c) Let $\lambda_1^{(n)}$ be the smallest eigenvalue of $-\Delta$ on a region having area 1 and bounded by a regular n-gon. Form a conjecture about
    $$\lim_{n \to \infty} \lambda_1^{(n)}.$$

(d) Test your hypothesis by repeating Example 13.6 for a regular n-gon, choosing n to be the largest integer that is practical. (Whatever value of n you choose, you have to create one or more meshes on the corresponding n-gon.)

13.5 Suggestions for further reading

An excellent introduction to the theory of finite element methods is the book by Brenner and Scott [8] mentioned earlier. Ciarlet [13] is another outstanding reference. Strang and Fix [61] also discuss the convergence theory for finite elements applied to time-dependent PDEs and eigenvalue problems.

Most finite element references discuss the computer implementation of finite elements in only general terms. Readers wishing to learn more about this issue can consult the *Texas Finite Element Series* [4] and the author's text [23].

As mentioned early in this chapter, mesh generation is an important area of study in its own right; it is treated by Knupp and Steinberg [39].

Appendix A
Proof of Theorem 3.47

Theorem 3.47. *Let $\mathbf{A} \in \mathbf{R}^{n \times n}$ be symmetric, and suppose $\mathbf{A}$ has an eigenvalue λ of (algebraic) multiplicity k (meaning that λ is a root of multiplicity k of the characteristic polynomial of $\mathbf{A}$). Then $\mathbf{A}$ has k linearly independent eigenvectors corresponding to λ.*

Proof. We argue by induction on n. The result holds trivially for $n = 1$. We assume it holds for symmetric $(n-1) \times (n-1)$ matrices, and suppose $\mathbf{A} \in \mathbf{R}^{n \times n}$ is symmetric and has an eigenvalue λ of (algebraic) multiplicity k. There exists $\mathbf{x}_1 \neq 0$ such that $\mathbf{A}\mathbf{x}_1 = \lambda \mathbf{x}_1$. We can assume that $\|\mathbf{x}_1\| = 1$ (since $\mathbf{x}_1/\|\mathbf{x}_1\|$ is also an eigenvector of $\mathbf{A}$), and we extend $\mathbf{x}_1$ to an orthonormal basis $\{\mathbf{x}_1, \mathbf{x}_2, \ldots, \mathbf{x}_n\}$. Then

$$\mathbf{X}^T \mathbf{A} \mathbf{X} = \begin{bmatrix} \mathbf{x}_1^T \mathbf{A} \mathbf{x}_1 & \mathbf{x}_1^T \mathbf{A} \mathbf{x}_2 & \cdots & \mathbf{x}_1^T \mathbf{A} \mathbf{x}_n \\ \mathbf{x}_2^T \mathbf{A} \mathbf{x}_1 & \mathbf{x}_2^T \mathbf{A} \mathbf{x}_2 & \cdots & \mathbf{x}_2^T \mathbf{A} \mathbf{x}_n \\ \vdots & \vdots & \ddots & \vdots \\ \mathbf{x}_n^T \mathbf{A} \mathbf{x}_1 & \mathbf{x}_n^T \mathbf{A} \mathbf{x}_2 & \cdots & \mathbf{x}_n^T \mathbf{A} \mathbf{x}_n \end{bmatrix},$$

where $\mathbf{X} \in \mathbf{R}^{n \times n}$ is defined by $\mathbf{X} = [\mathbf{x}_1 | \mathbf{x}_2 | \cdots | \mathbf{x}_n]$. We have

$$\mathbf{x}_i^T \mathbf{A} \mathbf{x}_1 = \mathbf{x}_i^T (\lambda \mathbf{x}_1) = \lambda \mathbf{x}_i^T \mathbf{x}_1 = \lambda \delta_{i1}$$

and

$$\begin{aligned} \mathbf{x}_1^T \mathbf{A} \mathbf{x}_i &= (\mathbf{A}\mathbf{x}_1)^T \mathbf{x}_i \text{ (since } \mathbf{A} \text{ is symmetric)} \\ &= \lambda \mathbf{x}_1^T \mathbf{x}_i \\ &= \lambda \delta_{i1}. \end{aligned}$$

(The symbol δ_{ij} is 1 if $i = j$ and 0 if $i \neq j$.) Therefore,

$$\mathbf{X}^T \mathbf{A} \mathbf{X} = \begin{bmatrix} \lambda & 0 & \cdots & 0 \\ 0 & \mathbf{x}_2^T \mathbf{A} \mathbf{x}_2 & \cdots & \mathbf{x}_2^T \mathbf{A} \mathbf{x}_n \\ \vdots & \vdots & \ddots & \vdots \\ 0 & \mathbf{x}_n^T \mathbf{A} \mathbf{x}_2 & \cdots & \mathbf{x}_n^T \mathbf{A} \mathbf{x}_n \end{bmatrix}$$

$$= \begin{bmatrix} \lambda & 0 \\ 0 & \mathbf{B} \end{bmatrix},$$

where $\mathbf{B} \in \mathbf{R}^{(n-1) \times (n-1)}$ is defined by

$$B_{ij} = \mathbf{x}_{i+1}^T \mathbf{A} \mathbf{x}_{j+1}.$$

The matrix $\mathbf{B}$ is symmetric. Also,

$$\begin{aligned} \det(\lambda \mathbf{I} - \mathbf{X}^T \mathbf{A} \mathbf{X}) &= \det(\mathbf{X}^T (\lambda \mathbf{I} - \mathbf{A}) \mathbf{X}) \\ &= \det(\mathbf{X}^T) \det(\lambda \mathbf{I} - \mathbf{A}) \det(\mathbf{X}) \\ &= \det(\lambda \mathbf{I} - \mathbf{A}), \end{aligned}$$

since $\mathbf{X}^T = \mathbf{X}^{-1}$ and therefore $\det(\mathbf{X}^T) = \det(\mathbf{X})^{-1}$. It follows that

$$\begin{aligned} \det(\lambda \mathbf{I} - \mathbf{A}) &= \begin{vmatrix} \lambda - \lambda & 0 \\ 0 & \lambda \mathbf{I} - \mathbf{B} \end{vmatrix} \\ &= (\lambda - \lambda) \det(\lambda \mathbf{I} - \mathbf{B}). \end{aligned}$$

Since λ is an eigenvalue of $\mathbf{A}$ of multiplicity k, it must be an eigenvalue of $\mathbf{B}$ of multiplicity $k - 1$, and therefore, by the induction hypothesis, $\mathbf{B}$ has $k - 1$ linearly independent eigenvectors $\mathbf{y}_2, \mathbf{y}_3, \ldots, \mathbf{y}_k$ corresponding to λ. We define

$$\tilde{\mathbf{y}}_i = \begin{bmatrix} 0 \\ \mathbf{y}_i \end{bmatrix} \in \mathbf{R}^n$$

for $i = 2, 3, \ldots, k$ and define $\mathbf{u}_i = \mathbf{X} \tilde{\mathbf{y}}_i$. Then

$$\begin{aligned} \mathbf{A} \mathbf{u}_i &= \mathbf{X} \begin{bmatrix} \lambda & 0 \\ 0 & \mathbf{B} \end{bmatrix} \mathbf{X}^T \mathbf{u}_i \\ &= \mathbf{X} \begin{bmatrix} \lambda & 0 \\ 0 & \mathbf{B} \end{bmatrix} \mathbf{X}^T \mathbf{X} \begin{bmatrix} 0 \\ \mathbf{y}_i \end{bmatrix} \\ &= \mathbf{X} \begin{bmatrix} \lambda & 0 \\ 0 & \mathbf{B} \end{bmatrix} \begin{bmatrix} 0 \\ \mathbf{y}_i \end{bmatrix} \\ &= \mathbf{X} \begin{bmatrix} 0 \\ \lambda \mathbf{y}_i \end{bmatrix} \\ &= \lambda \mathbf{X} \begin{bmatrix} 0 \\ \mathbf{y}_i \end{bmatrix} \\ &= \lambda \mathbf{u}_i \end{aligned}$$

Appendix A. Proof of Theorem 3.47

(the columns of $\mathbf{X}$ are orthonormal, so $\mathbf{X}^T\mathbf{X} = \mathbf{I}$; this is why the factor of $\mathbf{X}^T\mathbf{X}$ disappeared in the above calculation). This shows that $\mathbf{u}_2, \mathbf{u}_3, \ldots, \mathbf{u}_k$ are eigenvectors of $\mathbf{A}$ corresponding to λ. Moreover,

$$\mathbf{u}_1 = \mathbf{X}\begin{bmatrix} 1 \\ \mathbf{0} \end{bmatrix} = \mathbf{x}_1$$

($\mathbf{0}$ is the zero vector in $\mathbf{R}^{n-1}$), from which we can deduce that $\{\mathbf{u}_1, \mathbf{u}_2, \ldots, \mathbf{u}_k\}$ is a linearly independent set of k eigenvectors of $\mathbf{A}$ corresponding to λ. This completes the proof. □

Exercise

Under the hypotheses of Theorem 3.47 prove that $\mathbf{A}$ cannot have more than k linearly independent eigenvectors corresponding to λ by proving that (in the notation of the above proof) if $\mathbf{u}$ is any eigenvector of $\mathbf{A}$ corresponding to λ, then $\mathbf{u} \in \{\mathbf{u}_1, \mathbf{u}_2, \ldots, \mathbf{u}_k\}$. (Hint: Let $\mathbf{u} = \mathbf{X}\mathbf{z}$, and define $\mathbf{y} = (z_2, z_2, \ldots, z_n)$. Prove that $\mathbf{y}$ is an eigenvector of $\mathbf{B}$ and hence a linear combination of $\mathbf{y}_2, \mathbf{y}_3, \ldots, \mathbf{y}_k$. Use this to derive the desired result.)

Appendix B
Shifting the Data in Two Dimensions

B.1 Inhomogeneous Dirichlet conditions on a rectangle

The method of shifting the data can be used to transform an inhomogeneous Dirichlet problem to a homogeneous Dirichlet problem. In two dimensions, this technique works just as it did for a one-dimensional problem (see Section 5.3.4), although in two dimensions it is more difficult to find a function satisfying the boundary conditions. We consider the BVP

$$-\Delta u = f(\mathbf{x}), \; \mathbf{x} \in \Omega, \tag{B.1}$$

$$u(x) = \begin{cases} g_1(x_1), & \mathbf{x} \in \Gamma_1, \\ g_2(x_2), & \mathbf{x} \in \Gamma_2, \\ g_3(x_1), & \mathbf{x} \in \Gamma_3, \\ g_4(x_2), & \mathbf{x} \in \Gamma_4, \end{cases} \tag{B.2}$$

where Ω is the rectangle defined in (11.16) and $\partial\Omega = \Gamma_1 \cup \Gamma_2 \cup \Gamma_3 \cup \Gamma_4$, as in (11.19). We will assume that the boundary data are continuous, so

$$g_1(\ell_1) = g_2(0), \; g_2(\ell_2) = g_3(\ell_1), \; g_3(0) = g_4(\ell_2), \; g_4(0) = g_1(0).$$

Suppose we find a function p defined on $\overline{\Omega}$ and satisfying $p(\mathbf{x}) = g(\mathbf{x})$ for all $\mathbf{x} \in \partial\Omega$. We then define $v = u - p$ and note that $-\Delta v = -\Delta u + \Delta p = f(\mathbf{x}) + \Delta p(\mathbf{x})$ and $v(\mathbf{x}) = u(\mathbf{x}) - p(\mathbf{x}) = 0$ for all $\mathbf{x} \in \partial\Omega$ (since p satisfies the same Dirichlet conditions that u is to satisfy). We can then solve

$$-\Delta v = \tilde{f}(\mathbf{x}), \; \mathbf{x} \in \Omega,$$
$$v(x) = 0, \; \mathbf{x} \in \partial\Omega,$$

where $\tilde{f}(\mathbf{x}) = f(\mathbf{x}) + \Delta p(\mathbf{x})$, to find v, and then u will be given by $u = v + p$.

We now describe a method (admittedly rather tedious) for computing a function p that satisfies the given Dirichlet conditions. We first note that there is a polynomial of the form $q(x_1, x_2) = a + bx_1 + cx_2 + dx_1x_2$ which assumes the desired boundary values at the corners:

$$q(0,0) = g_1(0) = g_4(0),$$
$$q(\ell_1,0) = g_1(\ell_1) = g_2(0),$$
$$q(\ell_1,\ell_2) = g_2(\ell_2) = g_3(\ell_1),$$
$$q(0,\ell_2) = g_3(0) = g_4(\ell_2).$$

A direct calculation shows that

$$a = g_1(0),$$
$$b = \frac{g_1(\ell_1) - g_1(0)}{\ell_1},$$
$$c = \frac{g_4(\ell_2) - g_4(0)}{\ell_2},$$
$$d = \frac{g_2(\ell_2) + g_1(0) - g_1(\ell_1) - g_4(\ell_2)}{\ell_1 \ell_2}.$$

We then define

$$h_1(x_1) = g_1(x_1) - \left(g_1(0) + \frac{g_1(\ell_1) - g_1(0)}{\ell_1} x_1\right),$$
$$h_2(x_2) = g_2(x_2) - \left(g_2(0) + \frac{g_2(\ell_2) - g_2(0)}{\ell_2} x_2\right),$$
$$h_3(x_1) = g_3(x_1) - \left(g_3(0) + \frac{g_3(\ell_1) - g_3(0)}{\ell_1} x_1\right),$$
$$h_4(x_2) = g_4(x_2) - \left(g_4(0) + \frac{g_4(\ell_2) - g_4(0)}{\ell_2} x_2\right).$$

We have thus replaced each g_i by a function h_i which differs from g_i by a linear function, and which has value zero at the two endpoints:

$$h_1(0) = h_1(\ell_1) = h_2(0) = h_2(\ell_2) = h_3(\ell_1) = h_3(0) = h_4(\ell_2) = h_4(0) = 0.$$

Finally, we define

$$p(x_1,x_2) = (a + bx_1 + cx_2 + dx_1x_2) + \left(h_1(x_1) + \frac{h_3(x_1) - h_1(x_1)}{\ell_2} x_2\right)$$
$$+ \left(h_4(x_2) + \frac{h_2(x_2) - h_4(x_2)}{\ell_1} x_1\right).$$

The reader should notice how the second term interpolates between the boundary values on Γ_1 and Γ_3, while the third term interpolates between the boundary values on Γ_2 and Γ_4. In order for these two terms not to interfere with each other, it is necessary that the boundary data be zero at the corners. It was for this reason that we transformed the g_i's into the h_i's. The first term in the formula for p undoes this transformation.

B.1. Inhomogeneous Dirichlet conditions on a rectangle

It is straightforward to verify that p satisfies the desired boundary conditions. For example,

$$p(x_1, 0) = a + bx_1 + c0 + dx_1 0 + h_1(x_1) + \frac{h_3(x_1) - h_1(x_1)}{\ell_2} 0$$
$$+ h_4(0) + \frac{h_2(0) - h_4(0)}{\ell_1} x_1$$
$$= a + bx_1 + h_1(x_1) \text{ (since } h_4(0) = h_2(0) = 0)$$
$$= g_1(0) + bx_1 + g_1(x_1) - (g_1(0) + bx_1)$$
$$= g_1(x_1).$$

Thus the boundary condition on Γ_1 is satisfied. Similar calculations show that p satisfies the desired boundary conditions on the other parts of the boundary.

Example 3.47. We will solve the BVP

$$-\Delta u = 0, \; \mathbf{x} \in \Omega, \tag{B.3}$$

$$u(x) = \begin{cases} g_1(x_1) = x_1^2, & \mathbf{x} \in \Gamma_1, \\ g_2(x_2) = 1 - x_2^2, & \mathbf{x} \in \Gamma_2, \\ g_3(x_1) = (1 - x_1)^2, & \mathbf{x} \in \Gamma_3, \\ g_4(x_2) = x_2, & \mathbf{x} \in \Gamma_4 \end{cases} \tag{B.4}$$

by the above method, where Ω is the unit square:

$$\Omega = \left\{ (x_1, x_2) \in \mathbf{R}^2 \; : \; 0 < x_1 < 1, \; 0 < x_2 < 1 \right\}.$$

To compute p, we define

$$a = g_1(0) = 0,$$
$$b = \frac{g_1(1) - g_1(0)}{1} = 1,$$
$$c = \frac{g_4(1) - g_4(0)}{1} = 1,$$
$$d = \frac{g_2(1) + g_1(0) - g_1(1) - g_4(1)}{1 \cdot 1} = -2,$$
$$h_1(x_1) = x_1^2 - (0 + x_1) = x_1^2 - x_1,$$
$$h_2(x_2) = 1 - x_2^2 - (1 + (-1)x_2) = x_2 - x_2^2,$$
$$h_3(x_1) = (1 - x_1)^2 - (1 + (-1)x_1) = x_1^2 - x_1,$$
$$h_4(x_2) = x_2 - (0 + x_2) = 0.$$

We can then define

$$p(x_1, x_2) = x_1 + x_2 - 2x_1 x_2 + x_1^2 - x_1 + 0 \cdot x_2 + 0 + (x_2 - x_2^2) x_1$$
$$= x_2 - x_1 x_2 + x_1^2 - x_1 x_2^2.$$

Having computed p, we define $v = u - p$, so that

$$-\Delta v = -\Delta u + \Delta p = 0 + \Delta p = 2 - 2x_1, \quad \mathbf{x} \in \Omega,$$

and $v(\mathbf{x}) = 0$, $\mathbf{x} \in \partial\Omega$. We then see that

$$v(x_1, x_2) = \sum_{m=1}^{\infty} \sum_{n=1}^{\infty} \frac{c_{mn}}{(m^2 + n^2)\pi^2} \sin(m\pi x_1) \sin(n\pi x_2),$$

where

$$c_{mn} = 4 \int_0^1 \int_0^1 (2 - 2x_1) \sin(m\pi x_1) \sin(n\pi x_2) \, dx_2 \, dx_1$$
$$= \frac{8(1 - (-1)^n)}{mn\pi^2}, \quad m, n = 1, 2, 3, \ldots.$$

We then obtain the following formula for u:

$$u(x_1, x_2) = x_2 - x_1 x_2 + x_1^2 - x_1 x_2^2 + \sum_{m=1}^{\infty} \sum_{n=1}^{\infty} \frac{8(1 - (-1)^n)}{mn(m^2 + n^2)\pi^4} \sin(m\pi x_1) \sin(n\pi x_2).$$

The solution is shown in Figure B.1. ∎

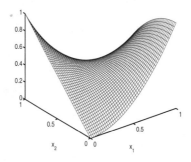

Figure B.1. *The solution to the BVP (B.3), approximated by a Fourier series with 400 terms.*

B.2 Inhomogeneous Neumann conditions on a rectangle

We can also apply the technique of shifting the data to transform a BVP with inhomogeneous Neumann conditions into a related BVP with homogeneous Neumann conditions. However, the details are somewhat more involved than in the Dirichlet case.

Suppose the Neumann conditions are

$$\frac{\partial u}{\partial n}(x) = \begin{cases} g_1(x_1), & \mathbf{x} \in \Gamma_1, \\ g_2(x_2), & \mathbf{x} \in \Gamma_2, \\ g_3(x_1), & \mathbf{x} \in \Gamma_3, \\ g_4(x_2), & \mathbf{x} \in \Gamma_4. \end{cases}$$

B.2. Inhomogeneous Neumann conditions on a rectangle

We first note that this is equivalent to

$$-\frac{\partial u}{\partial x_2}(x) = g_1(x_1), \ \mathbf{x} \in \Gamma_1,$$

$$\frac{\partial u}{\partial x_1}(x) = g_2(x_2), \ \mathbf{x} \in \Gamma_2,$$

$$\frac{\partial u}{\partial x_2}(x) = g_3(x_1), \ \mathbf{x} \in \Gamma_3,$$

$$-\frac{\partial u}{\partial x_1}(x) = g_4(x_2), \ \mathbf{x} \in \Gamma_4.$$

We make the following observation: If there is a twice-continuously differentiable function u satisfying the given Neumann conditions, then, since

$$\frac{\partial^2 u}{\partial x_1 \partial x_2} = \frac{\partial^2 u}{\partial x_2 \partial x_1},$$

we have

$$-\frac{\partial^2 u}{\partial x_1 \partial x_2}(x_1, 0) = \frac{dg_1}{dx_1}(x_1)$$

and

$$-\frac{\partial^2 u}{\partial x_2 \partial x_1}(0, x_2) = \frac{dg_4}{dx_2}(x_2),$$

which together imply that

$$\frac{dg_1}{dx_1}(0) = \frac{dg_4}{dx_2}(0).$$

By similar reasoning, we have all of the following conditions:

$$\frac{dg_1}{dx_1}(0) = \frac{dg_4}{dx_2}(0), \tag{B.5}$$

$$-\frac{dg_1}{dx_1}(\ell_1) = \frac{dg_2}{dx_2}(0), \tag{B.6}$$

$$\frac{dg_2}{dx_2}(\ell_2) = \frac{dg_3}{dx_1}(\ell_1), \tag{B.7}$$

$$-\frac{dg_4}{dx_2}(\ell_2) = \frac{dg_3}{dx_1}(0). \tag{B.8}$$

We will assume that (B.5) holds.

We now explain how to compute a function p that satisfies the desired Neumann conditions. The method is similar to that used to shift the data in a Dirichlet problem: we will "interpolate" between the Neumann conditions in each dimension and arrange things so that the two interpolations do not interfere with each other. We use the fact that

$$\psi(s) = -as + \frac{a+b}{2\ell}s^2 \tag{B.9}$$

satisfies
$$\frac{\psi}{ds}(0) = -a, \quad \frac{\psi}{ds}(\ell) = b. \tag{B.10}$$

The first step is to transform the boundary data g_1 to a function h_1 satisfying
$$\frac{dh_1}{dx_1}(0) = \frac{dh_1}{dx_1}(\ell_1) = 0,$$

and similarly for g_2, g_3, g_4 and h_2, h_3, h_4. Since these derivatives of the boundary data at the corners are (plus or minus) the mixed partial derivatives of the desired function at the corners, it suffices to find a function $q(x_1, x_2)$ satisfying the conditions

$$\frac{\partial^2 q}{\partial x_1 \partial x_2}(0,0) = -\frac{dg_1}{dx_1}(0),$$
$$\frac{\partial^2 q}{\partial x_1 \partial x_2}(\ell_1, 0) = -\frac{dg_1}{dx_1}(\ell_1),$$
$$\frac{\partial^2 q}{\partial x_1 \partial x_2}(0, \ell_2) = \frac{dg_3}{dx_1}(0),$$
$$\frac{\partial^2 q}{\partial x_1 \partial x_2}(\ell_1, \ell_2) = \frac{dg_2}{dx_2}(\ell_2).$$

We can satisfy these conditions with a function of the form
$$q(x_1, x_2) = ax_1 x_2 + bx_1^2 x_2 + cx_1 x_2^2 + dx_1^2 x_2^2.$$

The reader can verify that the necessary coefficients are
$$a = -\frac{dg_1}{dx_1}(0),$$
$$b = \frac{1}{2\ell_1}\left(\frac{dg_1}{dx_1}(0) - \frac{dg_1}{dx_1}(\ell_1)\right),$$
$$c = \frac{1}{2\ell_2}\left(\frac{dg_3}{dx_1}(0) + \frac{dg_1}{dx_1}(0)\right),$$
$$d = \frac{1}{4\ell_1 \ell_2}\left(\frac{dg_2}{dx_2}(\ell_2) + \frac{dg_1}{dx_1}(\ell_1) - \frac{dg_3}{dx_1}(0) - \frac{dg_1}{dx_1}(0)\right).$$

If p is to satisfy the desired Neumann conditions, then $p - q = h_i$ on Γ_i, where
$$h_1(x_1) = g_1(x_1) + ax_1 + bx_1^2,$$
$$h_2(x_2) = g_2(x_2) - (a + 2\ell_1 b)x_2 - (c + 2\ell_1 d)x_2^2,$$
$$h_3(x_1) = g_3(x_1) - (a + 2\ell_2 c)x_1 - (b + 2\ell_2 d)x_1^2,$$
$$h_4(x_2) = g_4(x_2) + ax_2 + cx_2^2.$$

We can now define $p - q$ by the interpolation described by (B.9), (B.10):
$$p(x_1, x_2) = q(x_1, x_2) - h_1(x_1)x_2 + \frac{h_3(x_1) + h_1(x_1)}{2\ell_2}x_2^2 - h_4(x_2)x_1 + \frac{h_2(x_2) + h_4(x_2)}{2\ell_1}x_1^2.$$

B.2. Inhomogeneous Neumann conditions on a rectangle

Then p satisfies the original Neumann conditions, as the interested reader can verify directly. We will illustrate this computation with an example.

Example 3.47. We will solve the BVP

$$-\Delta u = 0, \ \mathbf{x} \in \Omega, \tag{B.11}$$

$$\frac{\partial u}{\partial \mathbf{n}}(x) = \begin{cases} g_1(x_1) = -1, & \mathbf{x} \in \Gamma_1, \\ g_2(x_2) = x_2^2, & \mathbf{x} \in \Gamma_2, \\ g_3(x_1) = 1 + x_1^2, & \mathbf{x} \in \Gamma_3, \\ g_4(x_2) = -\frac{2}{3}, & \mathbf{x} \in \Gamma_4, \end{cases} \tag{B.12}$$

by the above method, where Ω is the unit square:

$$\Omega = \{(x_1, x_2) \in \mathbf{R}^2 \ : \ 0 < x_1 < 1, \ 0 < x_2 < 1\}.$$

To compute p, we define

$$a = -\frac{dg_1}{dx_1}(0) = 0,$$

$$b = \frac{1}{2}\left(\frac{dg_1}{dx_1}(0) - \frac{dg_1}{dx_1}(1)\right) = 0,$$

$$c = \frac{1}{2}\left(\frac{dg_3}{dx_1}(0) + \frac{dg_1}{dx_1}(0)\right) = 0,$$

$$d = \frac{1}{4}\left(\frac{dg_2}{dx_2}(1) + \frac{dg_1}{dx_1}(1) - \frac{dg_3}{dx_1}(0) - \frac{dg_1}{dx_1}(0)\right) = \frac{1}{2},$$

$$h_1(x_1) = g_1(x_1) = -1,$$

$$h_2(x_2) = g_2(x_2) - x_2^2 = 0,$$

$$h_3(x_1) = g_3(x_1) - x_1^2 = 1,$$

$$h_4(x_2) = g_4(x_2) = -\frac{2}{3}.$$

We can then define

$$p(x_1, x_2) = q(x_1, x_2) - h_1(x_1)x_2 + \frac{h_3(x_1) + h_1(x_1)}{2\ell_2}x_2^2 - h_4(x_2)x_1 + \frac{h_2(x_2) + h_4(x_2)}{2\ell_1}x_1^2$$

$$= \frac{1}{2}x_1^2 x_2^2 + x_2 + \frac{2}{3}x_1 - \frac{1}{3}x_1^2.$$

Having computed p, we define $v = u - p$, so that

$$-\Delta v = -\Delta u + \Delta p = 0 + \Delta p = x_1^2 + x_2^2 - \frac{2}{3}, \ \mathbf{x} \in \Omega,$$

and

$$\frac{\partial v}{\partial \mathbf{n}}(\mathbf{x}) = 0, \ \mathbf{x} \in \partial\Omega.$$

We can write the solution $v(x_1, x_2)$ in the form

$$u(x_1, x_2) = a_{00} + \sum_{m=1}^{\infty} a_{m0} \cos(m\pi x_1) + \sum_{n=1}^{\infty} a_{0n} \cos(n\pi x_2)$$
$$+ \sum_{m=1}^{\infty} \sum_{n=1}^{\infty} a_{mn} \cos(m\pi x_1) \cos(n\pi x_2).$$

With $f(x_1, x_2) = x_1^2 + x_2^2 - 2/3$, we have

$$f(x_1, x_2) = c_{00} + \sum_{m=1}^{\infty} c_{m0} \cos(m\pi x_1) + \sum_{n=1}^{\infty} c_{0n} \cos(n\pi x_2)$$
$$+ \sum_{m=1}^{\infty} \sum_{n=1}^{\infty} c_{mn} \cos(m\pi x_1) \cos(n\pi x_2),$$

where

$$c_{00} = \int_0^1 \int_0^1 f(x_1, x_2) dx_2 dx_1 = 0,$$
$$c_{m0} = 2 \int_0^1 \int_0^1 f(x_1, x_2) \cos(m\pi x_1) dx_2 dx_1 = \frac{4(-1)^m}{m^2 \pi^2}, \ m = 1, 2, 3, \ldots,$$
$$c_{0n} = 2 \int_0^1 \int_0^1 f(x_1, x_2) \cos(n\pi x_2) dx_2 dx_1 = \frac{4(-1)^n}{n^2 \pi^2}, \ n = 1, 2, 3, \ldots,$$
$$c_{mn} = 4 \int_0^1 \int_0^1 f(x_1, x_2) \cos(m\pi x_1) \cos(n\pi x_2) dx_2 dx_1 = 0, \ m, n = 1, 2, 3, \ldots.$$

It then follows that

$$v(x_1, x_2) = a_{00} + \sum_{m=1}^{\infty} \frac{4(-1)^m}{m^4 \pi^4} \cos(m\pi x_1) + \sum_{n=1}^{\infty} \frac{4(-1)^n}{n^4 \pi^4} \cos(n\pi x_2),$$

and $u(x_1, x_2) = v(x_1, x_2) + p(x_1, x_2)$. ∎

Bibliography

[1] M. A. Al-Gwaiz. *Sturm-Liouville Theory and Its Applications*. Springer-Verlag, London, 2008.

[2] Stuart S. Antman. The equations for large vibrations of strings. *American Mathematical Monthly*, 87:359–370, 1980.

[3] Kendall E. Atkinson. *An Introduction to Numerical Analysis*, 2nd edition. John Wiley & Sons, New York, 1989.

[4] Eric B. Becker, Graham F. Carey, and J. Tinsley Oden. *Finite Elements*. The Texas Finite Series I. Prentice–Hall, Englewood Cliffs, NJ, 1981. Six volumes; volumes 2–5 by Graham F. Carey and J. Tinsley Oden.

[5] T. A. Bick. *Elementary Boundary Value Problems*. Marcel Dekker, New York, 1993.

[6] P. Bochev and R. B. Lehoucq. On the finite element solution of the pure Neumann problem. *SIAM Review*, 47:50–66, 2005.

[7] William E. Boyce and Richard C. DiPrima. *Elementary Differential Equations and Boundary Value Problems*, 9th edition. John Wiley & Sons, New York, 2008.

[8] Susanne C. Brenner and L. Ridgway Scott. *The Mathematical Theory of Finite Element Methods*, 3rd edition. Springer-Verlag, New York, 2007.

[9] W. L. Briggs and V. E. Henson. *The DFT: An Owner's Manual for the Discrete Fourier Transform*. SIAM, Philadelphia, 1995.

[10] E. O. Brigham. *The Fast Fourier Transform and Its Applications*. Prentice–Hall, Englewood Cliffs, NJ, 1988.

[11] Michael A. Celia and William G. Gray. *Numerical Methods for Differential Equations: Fundamental Concepts for Scientific and Engineering Applications*. Prentice–Hall, Englewood Cliffs, NJ, 1992.

[12] Ruel V. Churchill and James Ward Brown. *Fourier Series and Boundary Value Problems*, 7th edition. McGraw–Hill, New York, 2006.

[13] Philippe G. Ciarlet. *The Finite Element Method for Elliptic Problems*. North–Holland, Amsterdam, 1978. Reprinted by SIAM, Philadelphia, 2002.

[14] Earl A. Coddington and Norman Levinson. *Theory of Ordinary Differential Equations.* McGraw–Hill, New York, 1955.

[15] J. W. Cooley and J. W. Tukey. An algorithm for the machine computation of complex Fourier series. *Mathematics of Computation*, 19:297–301, 1965.

[16] Timothy A. Davis. *Direct Methods for Sparse Linear Systems.* SIAM, Philadelphia, 2006.

[17] James W. Demmel. *Applied Numerical Linear Algebra.* SIAM, Philadelphia, 1997.

[18] Dean G. Duffy. *Green's Functions with Applications.* CRC Press, Boca Raton, FL 2001.

[19] G. B. Folland. *Introduction to Partial Differential Equations*, 2nd edition. Princeton University Press, Princeton, NJ, 1995.

[20] Gerald B. Folland. *Fourier Analysis and Its Applications.* Brooks/Cole, Pacific Grove, CA, 1992.

[21] George E. Forsythe, Michael A. Malcolm, and Cleve B. Moler. *Computer Methods for Mathematical Computations.* Prentice–Hall, Englewood Cliffs, NJ, 1977.

[22] Philip Gillett. *Calculus and Analytic Geometry.* D. C. Heath, Lexington, MA, 1984.

[23] Mark S. Gockenbach. *Understanding and Implementing the Finite Element Method.* SIAM, Philadelphia, 2006.

[24] Mark S. Gockenbach. *Finite-Dimensional Linear Algebra.* CRC Press, Boca Raton, FL, 2010.

[25] Jack L. Goldberg and Merle Potter. *Differential Equations: A Systems Approach.* Prentice–Hall, Upper Saddle River, NJ, 1997.

[26] Gene H. Golub and Charles F. Van Loan. *Matrix Computations*, 3rd edition. Johns Hopkins University Press, Baltimore, 1996.

[27] Michael Greenberg. *Advanced Engineering Mathematics*, 2nd edition. Prentice–Hall, Upper Saddle River, NJ, 1998.

[28] M. E. Gurtin. *An Introduction to Continuum Mechanics.* Academic Press, New York, 1981.

[29] Richard Haberman. *Mathematical Models: Mechanical Vibrations, Population Dynamics, and Traffic Flow.* Prentice–Hall, Englewood Cliffs, NJ, 1977. Reprinted by SIAM, Philadelphia, 1998.

[30] Richard Haberman. *Elementary Applied Partial Differential Equations: With Fourier Series and Boundary Value Problems*, 4th edition. Prentice–Hall, Upper Saddle River, NJ, 2003.

[31] William W. Hager. *Applied Numerical Linear Algebra.* Prentice–Hall, Englewood Cliffs, NJ, 1988.

[32] J. Ray Hanna and John H. Rowland. *Fourier Series, Transforms, and Boundary Value Problems*, 2nd edition. John Wiley & Sons, New York, 1990.

[33] M. T. Heideman, D. H. Johnson, and C. S. Burrus. Gauss and the history of the fast Fourier transform. *Archive for History of Exact Science*, 34:265–277, 1985.

[34] Morris Hirsch and Stephen Smale. *Differential Equations, Dynamical Systems, and Linear Algebra*. Academic Press, New York, 1974.

[35] Claes Johnson. *Numerical Solution of Partial Differential Equations by the Finite Element Method*. Cambridge University Press, Cambridge, UK, 1987.

[36] David W. Kammler. *A First Course in Fourier Analysis*. Prentice–Hall, Upper Saddle River, NJ, 2000.

[37] Wilfred Kaplan. *Advanced Calculus*, 5th edition. Addison–Wesley, Reading, MA, 2002.

[38] David Kincaid and Ward Cheney. *Numerical Analysis*, 2nd edition. Brooks/Cole, Pacific Grove, CA, 1996.

[39] Patrick M. Knupp and Stanly Steinberg. *The Fundamentals of Grid Generation*. CRC Press, Boca Raton, FL, 1993.

[40] J. D. Lambert. *Numerical Methods for Ordinary Differential Systems*. John Wiley & Sons, Chichester, UK, 1991.

[41] David C. Lay. *Linear Algebra and Its Applications*, updated 3rd edition. Addison Wesley Longman, Reading, MA, 2005.

[42] Randall J. LeVeque. *Numerical Methods for Conservation Laws*. Birkhäuser, Basel, 1992.

[43] Randall J. LeVeque. *Finite Difference Methods for Ordinary and Partial Differential Equations*. SIAM, Philadelphia, 2007.

[44] David R. Lide, editor. *CRC Handbook of Chemistry and Physics*. CRC Press, Boca Raton, FL, 1998.

[45] C. C. Lin and L. A. Segel. *Mathematics Applied to Deterministic Problems in the Natural Sciences*. Macmillan, New York, 1974. Reprinted by SIAM, Philadelphia, 1988.

[46] J. David Logan. *An Introduction to Nonlinear Partial Differential Equations*. John Wiley & Sons, New York, 1994.

[47] Jerrold E. Marsden and Anthony J. Tromba. *Vector Calculus*, 5th edition. W. H. Freeman, New York, 2003.

[48] R. Mattheij and J. Molenaar. *Ordinary Differential Equations in Theory and Practice*. John Wiley & Sons Ltd., Chichester, UK, 1996. Reprinted by SIAM, Philadelphia, 2002.

[49] R. M. M. Mattheij, S. W. Rienstra, and J. H. M. ten Thije Boonkkamp. *Partial Differential Equations: Modeling, Analysis, Computation.* SIAM, Philadelphia, 2005.

[50] Robert C. McOwen. *Partial Differential Equations: Methods and Applications*, 2nd edition. Prentice–Hall, Upper Saddle River, NJ, 2002.

[51] Carl D. Meyer. *Matrix Analysis and Applied Linear Algebra.* SIAM, Philadelphia, 2000.

[52] K. W. Morton and D. F. Mayers. *Numerical Solution of Partial Differential Equations*, 2nd edition. Cambridge University Press, Cambridge, UK, 2005.

[53] W. T. Reid. *Sturmian Theory for Ordinary Differential Equations.* Springer-Verlag, New York, 1980.

[54] Michael Rennardy and Robert C. Rogers. *An Introduction to Partial Differential Equations*, 2nd edition. Springer-Verlag, New York, 2009.

[55] G. F. Roach. *Green's Functions*, 2nd edition. Cambridge University Press, Cambridge, UK, 1982.

[56] L. F. Shampine and M. W. Reichelt. The MATLAB ODE suite. *SIAM Journal on Scientific Computing*, 18:1–22, 1997.

[57] Lawrence E. Shampine. *Numerical Solutions of Ordinary Differential Equations.* Chapman & Hall, New York, 1994.

[58] Ivar Stakgold. *Boundary Value Problems of Mathematical Physics.* MacMillan, New York, 1967. Reprinted by SIAM, Philadelphia, 2000.

[59] Ivar Stakgold. *Green's Functions and Boundary Value Problems*, 2nd edition. Wiley-Interscience, New York, 1997.

[60] Gilbert Strang. *Linear Algebra and Its Applications*, 4th edition. Thomson Brooks/Cole, Belmont, CA, 2005.

[61] Gilbert Strang and George J. Fix. *An Analysis of the Finite Element Method.* Wellesley-Cambridge Press, Wellesley, MA, 1988.

[62] Walter A. Strauss. *Partial Differential Equations: An Introduction*, 2nd edition. John Wiley & Sons, New York, 2007.

[63] John C. Strikwerda. *Finite Difference Schemes and Partial Differential Equations*, 2nd edition. SIAM, Philadelphia, 2004.

[64] Paul N. Swarztrauber. FFTPACK, 1985. Available at http://www.netlib.org.

[65] Lloyd N. Trefethen and David Bau, III. *Numerical Linear Algebra.* SIAM, Philadelphia, 1997.

[66] Charles Van Loan. *Computational Frameworks for the Fast Fourier Transform.* SIAM, Philadelphia, 1992.

[67] J. S. Walker. *Fast Fourier Transform*, 2nd edition. CRC Press, Boca Raton, FL, 1996.

[68] H. F. Weinberger. *A First Course in Partial Differential Equations with Complex Variables and Transform Methods*. Blaisdell, New York, 1965.

[69] E. C. Zachmanoglou and Dale W. Thoe. *Introduction to Partial Differential Equations with Applications*. Dover, New York, 1986.

[70] Anton Zettl. *Sturm-Liouville Theory*. American Mathematical Society, Providence, RI, 2005.

[71] Dennis G. Zill. *A First Course in Differential Equations*, 9th edition. Thomson Brooks/Cole, Belmont, CA, 2008.

[72] Antoni Zygmund. *Trigonometric Series*, 2nd edition. Cambridge University Press, Cambridge, UK, 1959.

Index

advection equation, 29
    in three dimensions, 446
advection-diffusion equation, 31
aluminum, 200
angle between two vectors, 58
area integral, 439
automatic step control, 119

back substitution, 604
backward Euler method, 126, 242
backward heat equation, 306
banded matrix, 495, 606
bar
    circular, 225
    elastic, 20
    heat flow in, 9
basis, 53
    orthogonal, 59
    orthonormal, 59
Bessel function
    of order n, 476
    properties, 476
Bessel's equation
    of order n, 474
Bessel's inequality, 563
    best approximation, 62, 142, 151, 175, 454
    in the energy norm, 620
bilinear form, 174
    symmetric, 174
body force, 21
boundary conditions
    Dirichlet, 13
    essential, 245, 286, 496
    for the advection equation, 30
    for the diffusion equation, 19
    for the heat equation, 13
    homogeneous, 13
    inhomogeneous, 13, 157, 189, 252
    mixed, 13
    natural, 245, 286, 496
    Neumann, 13
    periodic, 226
    Robin, 139, 388
        nonphysical case, 415
    time dependent, 206
boundary value problem (BVP), 14
    on a rectangular domain, 455
Burgers's equation, 33

$C[a,b]$, 37
$C^2(\overline{\Omega})$, 447
$C^k[a,b]$, 37
$C_0^\infty(\Omega)$, 618
$C_D^2(\overline{\Omega})$, 447
$C_N^2(\overline{\Omega})$, 449
carbon monoxide, 203
Cauchy sequence, 585
causality, 375, 383
caustic, 331
central difference, 298
CFL condition, 302
chain rule, 470
change of variables
    in a multiple integral, 601
characteristic polynomial
    of a second-order ODE, 85
    of a square matrix, 69
characteristics
    method of, 313
    of a PDE, 310
Cholesky factorization, 604
chromium, 257
closure, 447
codomain, 36

coefficients
    constant versus nonconstant, 4
    discontinuous, 243
compact operator, 428
    eigenvectors of, 431
compact set, 428
compatibility condition, 220, 223, 230, 460
    for a singular linear system, 48
complete
    normed vector space, 586
    space, 431
complete orthogonal sequence, 583
completeness
    of $H^1(\Omega)$, 619
    of L^2, 586
complex numbers, 534
concentration gradient, 17
conditioning
    of a matrix, 610
conjugate directions, 614
conjugate gradients
    algorithm, 613
    convergence, 614
conservation law, 32
conservation of energy, 166
continuously differentiable function, 37
convection equation, 29
convection-diffusion equation, 31
convergence
    mean-square, 144, 556
    pointwise, 556
    uniform, 556
convolution, 560
    periodic, 561
copper, 210

d'Alembert's solution to the wave equation, 261
data structure
    for describing a triangulation, 598
delta function, 338
dense matrix, 179
density, 10
descent direction, 608

determinant, 69
differential equation, 1
    autonomous, 118
    constant-coefficient, 4, 97
    converting to a first-order system, 79
    first-order linear, 91
    homogeneous, 2
    inhomogeneous, 3
    linear, 2
    nonconstant-coefficient, 4
    nonlinear, 3
    order, 2
    ordinary, 1
    partial, 1
    right-hand side, 4
    scalar, 4
    system, 4, 97
differential operator
    symmetric, 134
differentiating an integral, 11
diffusion, 16
diffusion coefficient, 17
diffusion equation, 17
    inhomogeneous, 203
dimension, 53
Dirac delta function, 338
directional derivative, 310
Dirichlet condition, 13
Dirichlet kernel, 559
discrete cosine transform (DCT), 552
discrete Fourier transform (DFT), 543
discrete sine transform (DST), 548
    and the FFT, 549
displacement function, 20
divergence
    operator, 439
    theorem, 439
domain
    of a function, 35
domain of dependence, 263
domain of influence, 264
dot product, 48, 57
    complex, 71, 536
Duhamel's principle, 104, 363, 372, 520
dynamic wave, 31

Index

eigenfunction
    complex-valued, 534
    of a differential operator, 135
eigenfunction expansion, 385
eigenpairs
    on a disk, 479
    under Dirichlet conditions, 141
    under mixed boundary conditions, 145, 147
    under Neumann conditions, 215
    under periodic boundary conditions, 227
    under Robin conditions, 410
eigenspace, 70
eigenvalue, 69, 123
    of a differential operator, 135
eigenvalue problem
    generalized, 398, 625
    Sturm–Liouville, 385
eigenvector, 69
elliptic regularity, 620
energy inner product, 175, 488
energy norm, 175
Euclidean n-space, 36
    complex, 535
Euler equation, 93
Euler's formula, 57, 86, 534
Euler's method, 112, 241
    backward, 126, 242
    improved, 114
even extension, 555
even function, 553
    and the full Fourier series, 554
existence, 219
    of solutions to a linear system, 42
explicit method, 126

fast Fourier transform (FFT), 533, 546
Fick's law, 17
fill-in, 607
finite difference, 298
finite difference method
    2–2 scheme for the wave equation, 301
finite element method, 166
    for eigenvalue problems, 624
    for the heat equation, 237

    for the Poisson equation, 487
    for the wave equation, 281
    under Robin conditions, 418
forward difference, 298
Fourier coefficients, 144
    complex, 537
    cosine, 215
    rate of decay, 571
    sine, 144
Fourier series, 144, 149
    complex, 537
    cosine, 215
    double sine series, 453
    for the heat equation, 195
    for the wave equation, 266
    full, 229, 232, 537
    in three dimensions, 465
    method of, 159
    on a disk, 469
    on a rectangular domain, 450
    pointwise convergence, 564
    sine, 144
Fourier's law, 440
Fourier's law of heat conduction, 12
fourth-order Runge–Kutta method (RK4), 116
Fredholm alternative, 48, 222
free nodes, 488
function, 35
function space, 37
fundamental frequency, 626
    of a circular drum, 484
    of a vibrating string, 268
fundamental theorem of calculus, 11

Galerkin method, 166, 174, 238, 281
Gaussian elimination, 46
    for a banded matrix, 606
    operation count, 56, 604, 615
    pseudocode, 604
Gaussian kernel, 370, 527
general solution
    of the wave equation, 260
generalized eigenvalue problem, 398
generalized function, 338
Gibbs's phenomenon, 143, 454, 570, 577
gold, 231

gradient, 440
Gram matrix, 243
Gram–Schmidt procedure, 73
gravitational constant, 25
gravity, 25
Green's first identity, 448
Green's function, 335
    causal, 362
        interpretation, 365
    for a BVP, 337
    for the heat equation, 368
        free-space, 528
    for the Laplacian
        free-space (2D), 500
        free-space (3D), 505
        on a bounded domain, 508
        on a disk, 511
    for the wave equation, 376
        free-space (3D), 520, 522
Green's second identity, 514
grid, 111, 181, 299
    irregular, 119
guitar, 270

$H^1(\Omega)$, 618
$H_0^1(\Omega)$, 619
harmonic function, 517
harmonics, 401
heat capacity, 10
heat equation, 9, 12
    and finite elements, 237
    backward, 306
    in three dimensions, 441
    in two dimensions, 443
    inhomogeneous, 13
heat flow, 9
    steady-state, 14
heat flux, 11
Heaviside function, 356
homogeneous
    versus heterogeneous, 13
Hooke's law, 21
Hookean, 21
Hopf–Cole transformation, 34
Huygen's principle, 523

implicit method, 126
incomplete factorization, 615

initial condition
    for ODEs, 80
    for the heat equation, 13
    for the wave equation, 23
initial value problem (IVP), 81
    for an ODE, 81
    for the heat equation, 369
    for the wave equation, 259
initial-boundary value problem
    (IBVP), 13
    for the heat equation, 13
    for the wave equation, 266
inner product, 58, 174
    complex, 535
    complex L^2, 536
    energy, 175
inner product space, 59
integral
    area, 439
    iterated, 438
    line, 439
    surface, 439
    volume, 438
integral operator, 352, 426
integrating an ODE, 111
integrating factor, 91
integration by parts, 134, 171, 281,
    447, 618
interpolant
    piecewise linear, 193, 620
inverse
    of a differential operator, 341
inverse discrete Fourier transform, 543
inviscid Burgers's equation, 33, 327
iron, 198
iterated integral, 438
iterative method
    for solving a linear system, 608

Jacobian matrix, 601
jump discontinuity, 562

kinematic wave, 31
kinetic energy, 168

L^2 inner product, 61
$L^2(\Omega)$, 579
Laplace's equation, 445

Laplacian, 441
    in polar coordinates, 472
    in spherical coordinates, 505
lead, 235
Lebesgue
    integral, 581
    measure, 581
line integral, 439
line search, 609
linear combination, 52
linear operator, 38
linear operator equation, 39
linear system
    algebraic, 35
linearly independent, 55
load vector, 176
    computation of, 598
Lotka–Volterra predator-prey model, 117
LU factorization, 604

mapping, 35
mass matrix, 238, 281
matrix, 35
    banded, 495, 606
    dense, 179
    Gram, 63, 243
    identity, 49
    inverse, 49
    invertible, 49
    mass, 281
    nonsingular, 49
    positive definite, 75
    singular, 249
    sparse, 179, 182, 495
    stiffness, 281
    symmetric, 70
    tridiagonal, 184
maximum principle, 518
mean value theorem, 566
mean-square, 61
mean-square convergence, 144, 556
    of complex Fourier series, 582
measurable
    function, 581
    set, 581
mesh, 181
method of characteristics, 313

method of images, 512
    for the wave equation, 379
method of lines
    for the heat equation, 240
middle C, 269
modulus
    of elasticity, 21
    Young's, 21
multiplicity
    of eigenvalues, 69

natural frequency
    of a string, 268
Neumann condition, 13
Newton, 24
Newton's law of cooling
    and boundary conditions, 19
Newton's second law of motion, 21, 27
nodal values, 182
nodes, 181
nondimensionalization, 201
norm, 61
    energy, 175
    of an operator, 429
normal derivative, 442
normal equations, 63
normal modes
    of a vibrating string, 268
null space
    of a matrix, 45
    trivial, 46

odd extension, 380, 555
odd function, 553
    and the full Fourier series, 554
    regular, 344
one-way wave equation, 31
operator, 35
    compact, 428
    derivative, 39
    differential, 40, 132
    linear, 38
    matrix, 39
    wave, 259
order
    of a differential equation, 2
    of a numerical method, 113
ordinary differential equation (ODE), 1

orthogonality
    of functions, 61
    of vectors, 59

partial derivative
    weak, 618
partial differential equation (PDE), 1
partial pivoting, 604
particular solution, 88
Pascal, 24
    first-order linear, 319
    first-order quasi-linear, 319
    first-order semilinear, 319
periodic convolution, 561
periodic extension, 561
periodic function, 553
piano, 272
piecewise continuous, 562
piecewise linear
    approximation, 620
    basis function, 182
    function, 181
    interpolant, 193, 239
piecewise polynomials, 166
piecewise smooth, 563
pointwise convergence, 556
    of a complex Fourier series, 564
    of a Fourier cosine series, 568
    of a Fourier sine series, 568
    of a full Fourier series, 568
Poisson integral formula, 516
Poisson's equation, 444
    on a disk, 482
polar coordinates, 470
positive definite matrix, 75
potential energy, 166
    elastic, 166
    external, 168
    gravitational, 168
    minimal, 170
power series solution
    of Bessel's equation, 474
precompact set, 428
preconditioned conjugate gradients, 615
pressure, 24
prime factorization
    and the FFT, 546

principle of virtual work, 170
product rule, 447
programming finite elements, 495
projection theorem, 175, 537
Pythagorean theorem, 58, 563

quadrature, 111
    choosing a rule, 601
    general formula, 599
    one-point Gauss rule, 600
    one-point rule on a triangle, 600
    over an arbitrary triangle, 601
    three-point rule for a triangle, 600
quarter-wave cosine series, 147
quarter-wave sine series, 146

radial function, 501
range
    of a function, 35
    of a linear operator, 42
reciprocity, 341, 352, 375, 383, 510
resonance, 269, 295
Riemann integral, 580
Riesz representation theorem, 619
Runge–Kutta method, 115
Runge–Kutta–Fehlberg method, 119

scalar, 36, 535
semidiscretization in space, 240
separation of variables, 207, 451
    in polar coordinates, 472
shifting the data, 158, 189, 204, 273, 463
    on a rectangle, 633
shock, 330
sifting property, 338
silver, 235
Simpson's rule, 115
sink
    heat, 12
snapshot
    temperature, 198
Sobolev space, 618
solution
    particular, 88
source
    heat, 12
span, 55

Index

sparse matrix, 179, 495
sparsity pattern, 495, 606
specific heat, 10
    and temperature, 18
spectral method, 150
    for a system of ODEs, 102
spectral theorem for symmetric matrices, 73, 629
speed
    of a wave, 262
square wave, 569
stability, 125
    of a time-stepping scheme, 283
standing wave, 268
steady-state
    and Neumann conditions, 218
    temperature, 207
steepest descent
    algorithm, 610
    direction, 609
stiff differential equation, 124, 240
stiff system of ODEs, 124
stiffness
    of a bar, 21
stiffness matrix, 176, 281
    algorithm for computing, 597
    singular, 249
strain, 21
string
    elastic, 26
    sagging, 28
strong form of a BVP, 170
Sturm comparison theorem, 434
Sturm–Liouville eigenvalue problem, 385
    theory, 425
subspace, 37
    of $\mathbf{R}^n$, 43
superposition, 260
    principle of, 50, 89
support of a function, 184
supremum, 429
surface integral, 439
symmetry
    and eigenvalues, 72
    of a differential operator, 134
    of a matrix, 70
    of the Laplacian, 448

temperature gradient, 12
test function, 171
thermal conductivity, 12
time step, 111
time-stepping, 111
trace theorem, 619
transient behavior, 124
translation, 375, 383
transpose
    and the dot product, 70
    of a matrix, 48
trapezoidal rule, 542
traveling wave, 31
triangulation, 488
    data structure for, 598
    description of, 594
trigonometric interpolation, 545

uniform convergence, 556
    of a complex Fourier series, 574
    of a Fourier cosine series, 575
    of a Fourier sine series, 575
    of a full Fourier series, 575
    of continuous functions, 577
    of cosine series, 576
uniqueness, 219
    of solutions to a linear system, 45
unit normal vector, 438

variation of parameters, 88, 103
variational form of a BVP, 166
vector, 35, 36
    function as, 132
vector field, 439
vector space
    complex, 535
    definition, 36
vibrating membrane, 444
virtual work, 170
volume integral, 438

wave
    dynamic, 31
    kinematic, 31
    left-moving, 262
    right-moving, 262
    traveling, 31

wave equation, 22
    acoustic, 443
    and numerical methods, 279
    backward in time, 306
    d'Alembert's solution, 261
    general solution, 260
    in an infinite domain, 259
    in three dimensions, 443
    in two dimensions, 444
    on a disk, 483
    on a square, 457
    with a localized source, 293
wave operator, 259
wave speed, 262
weak derivative, 618
weak form
    of a BVP, 166, 170
    of a BVP in two or three dimensions, 487
    of a Sturm–Liouville problem, 396
    of the heat equation, 237
    of the wave equation, 281
Wronskian, 83

Young's modulus, 21